# Physics of Protein Folding, Misfolding, and Intrinsic Disorder: A Themed Issue in Honour of Professor Vladimir Uversky on the Occasion of His 60th Birthday

# Physics of Protein Folding, Misfolding, and Intrinsic Disorder: A Themed Issue in Honour of Professor Vladimir Uversky on the Occasion of His 60th Birthday

Editors

**Prakash Kulkarni**
**Stefania Brocca**
**Keith Dunker**
**Sonia Longhi**

MDPI

Basel • Beijing • Wuhan • Barcelona • Belgrade • Novi Sad • Cluj • Manchester

*Editors*
Prakash Kulkarni
City of Hope National
Medical Center
Duarte
CA
USA

Stefania Brocca
University of Milan-Bicocca
Milan
Italy

Keith Dunker
Indiana University School of
Medicine
Indianapolis
IN
USA

Sonia Longhi
CNRS & Aix-Marseille
University
Marseille
France

*Editorial Office*
MDPI
St. Alban-Anlage 66
4052 Basel, Switzerland

This is a reprint of articles from the Special Issue published online in the open access journal *Biomolecules* (ISSN 2218-273X) (available at: https://www.mdpi.com/journal/biomolecules/special_issues/60th_birthday).

For citation purposes, cite each article independently as indicated on the article page online and as indicated below:

| Lastname, A.A.; Lastname, B.B. Article Title. *Journal Name* **Year**, *Volume Number*, Page Range. |
|---|

**ISBN 978-3-0365-9112-4 (Hbk)**
**ISBN 978-3-0365-9113-1 (PDF)**
**doi.org/10.3390/books978-3-0365-9113-1**

# Contents

# About the Editors

**Prakash Kulkarni**

Prof. Prakash Kulkarni is a Research Professor and Director of Translational Research, Department of Medical Oncology, and holds a secondary appointment as Research Professor in the Department of Systems Biology at the City of Hope National Medical Centre in Duarte, California. After receiving his PhD in biochemistry from India, he completed his postdoctoral training in biochemistry at the Indian Institute of Science and, subsequently, in cell biology at New York University School of Medicine. He began his independent academic career as an Assistant Professor of urology and oncology at Johns Hopkins University School of Medicine, where he was named the Irene and Bernard L. Schwartz Scholar of the Patrick C Walsh Prostate Research Fund. He then worked as a Research Associate Professor to the W. M. Keck Laboratory for Structural Biology, University of Maryland Institute for Bioscience and Biotechnology Research. Prior to Johns Hopkins, Prof. Kulkarni held Staff Scientist positions in the Division of Chemistry & Chemical Engineering and Division of Biology & Biological Engineering at the California Institute of Technology and in the Department of Genetics at the Yale University School of Medicine. He is Associate Editor-in-Chief of *Biomolecules*. His research interests are interdisciplinary and are focused on understanding how conformational dynamics of intrinsically disordered proteins contribute to phenotypic switching, especially in the evolution of multicellularity, in cancer, and in non-genetic heterogeneity. He is a Fellow of the Royal Society of Biology, UK.

**Stefania Brocca**

Stefania Brocca is an Associate Professor of Biochemistry in the Department of Biotechnology and Biosciences, at the University of Milano-Bicocca, Milano, Italy. She is a biologist and obtained a PhD in Biotechnology from the University of Milano (Italy) in 1994. During post-doctoral training at the Institut fuer Technische Biochemie at the University of Stuttgart (Germany), she acquired expertise in recombinant protein production using non-conventional yeasts. In 2005, she was hired as a researcher at the University of Milan-Bicocca and affiliated to the Department of Biotechnology and Biosciences, where she lectures for the Biology and Biotechnology courses. She has served as Guest Editor for Special Issues of *Biomolecules* and the *International Journal of Molecular Sciences* and as a reviewer for several journals in the field of biochemistry, microbiology and biophysics. Her scientific interests include various functional and structural aspects of different classes of proteins, from hydrolytic enzymes to disordered proteins. These studies require the integration of biochemical and biophysical techniques and the design and production of recombinant proteins, starting from natural or synthetic gene sequences.

**Keith Dunker**

A. Keith Dunker received a B.S in Chemistry from the University of California Berkeley in 1965, then an M.S. in Physics in 1967 and a PhD in Biophysics in 1969 from the University of Wisconsin Madison. He was a post-doctoral fellow at Yale University from 1969 to 1973, then a Research Associate in Virology at the Sloan Kettering Center for Cancer Research in New York from 1973 to 1975. Subsequently, he was Assistant Professor, Associate Professor and Professor in Biochemistry at Washington State University from 1975 to 2003. In 2003, he took a position as the founding Director of the Center for Computational Biology and Bioinformatics and as a Professor of Biochemistry and Molecular Biology at the Indiana University School of Medicine, where he spent the following years and where he remains as an Emeritus Professor. He has authored over 300 papers and co-edited more than 30 Special Issues and books. He is a highly cited researcher with an h-index of 109 and over 58,000 citations. In 2021 Keith Dunker was elected Fellow of the International Society for Computational Biology as "a pioneer of the development and application of statistical and computational methods to understand the prevalence, the patterns of evolution, and the functional repertoire of intrinsically disordered proteins across all domains of life".

**Sonia Longhi**

Sonia Longhi is Director of Research at the Center for National Scientific Research (CNRS). Since 2005, she has been the head of the "Structural Disorder and Molecular Recognition" group within the AFMB laboratory, Marseille. She obtained a PhD in molecular biology from the University of Milan (Italy) in 1993. During post-doctoral training at the AFMB lab, she acquired expertise in structural biology. In 1999, she was recruited as a tenured senior scientist within the CNRS and affiliated to the AFMB lab. She received an HDR in Structural Virology from the University of Aix-Marseille I in 2003. She is a member of the editorial board of *Biomolecules* and has served as Guest Editor for several Special Issues of *Biomolecules and the International Journal of Molecular Sciences*. Her scientific focus is on health relevant intrinsically disordered proteins (IDPs) and the mechanistic and functional aspects of the interactions they establish with partners. To date, she has authored > 150 scientific publications and edited a book on measles virus nucleoprotein. She also co-edited (with Prof. Vladimir Uversky) a book on the experimental approaches to characterize IDPs and a book on structural disorder within viral proteins (Wiley). In 2013, she was awarded the CNRS prize for scientific excellence and, in 2019, the CNRS prize for PhD student supervision.

# Preface

The discovery of intrinsically disordered proteins (IDPs) has changed our perception of proteins as existing either in their native, folded and, hence, active state or in a denatured, unfolded, and non-functional state. We now know that proteins may exist as dynamic conformational ensembles with little or no secondary and tertiary structure and yet be functional. Although apparently posing a challenge to Anfinsen's postulate, IDPs in fact represent the limits of the structure–function paradigm.

IDPs also pose an additional challenge. Structural disorder is not easily tractable/detectable experimentally using techniques and approaches traditionally used by structuralists, and these techniques need to be enriched and complemented by those typically used by polymer physicists and chemists. Indeed, for the same reason, IDPs, which are prevalent in all three kingdoms of life and comprise almost half the human proteome, are often referred to as constituents of the 'dark' matter of biology. Over the past almost thirty years, studies on IDPs have fueled crosstalk between biochemists, structuralists, and molecular biologists on the one hand and physicists and chemists on the other. These different communities have therefore faced the need to understand each other while speaking different languages.

Prof. Vladimir Uversky, one of the pioneers in the IDP field, realised the necessity of this dialogue and published an eloquent and thought-provoking paper: 'Natively unfolded proteins: a point where biology waits for physics' (Protein Sci. 2002). Today, twenty years later, to stimulate reflections on this issue and ponder the role of physical sciences in unravelling the structure–function paradigm, Biomolecules will be publishing this Special Issue on the occasion of Prof. Uversky's 60th birthday, celebrating his many contributions to the IDP field.

**Prakash Kulkarni, Stefania Brocca, Keith Dunker, and Sonia Longhi**
*Editors*

*biomolecules*

MDPI

*Editorial*

# Per Aspera ad Chaos: Vladimir Uversky's Odyssey through the Strange World of Intrinsically Disordered Proteins

**Prakash Kulkarni [1,2,*], Stefania Brocca [3,*], A. Keith Dunker [4,*] and Sonia Longhi [5,*]**

1 Department of Medical Oncology, City of Hope National Medical Center, Duarte, CA 91010, USA
2 Department of Systems Biology, City of Hope National Medical Center, Duarte, CA 91010, USA
3 Department of Biotechnology and Biosciences, University of Milan-Bicocca, 20126 Milan, Italy
4 Biochemistry & Molecular Biology, Indiana University School of Medicine, Indianapolis, IN 46202, USA
5 Architecture and Function of Biological Macromolecules (AFMB), UMR 7257, Aix Marseille University and CNRS, 13288 Marseille, France
* Correspondence: pkulkarni@coh.org (P.K.); stefania.brocca@unimib.it (S.B.); kedunker@iu.edu (A.K.D.); sonia.longhi@univ-amu.fr (S.L.)

## 1. Introduction

Until the late 1990s, we believed that protein function required a unique, well-defined 3D structure encrypted in the amino acid sequence. However, over the past two decades, we have witnessed a protein 'renaissance'. We learned that proteins with unique 3D structures can switch folds (referred to as 'metamorphic' or shapeshifting proteins), and a new class of proteins, called intrinsically disordered proteins (IDPs), that lack structure but can either transition to order, or in many cases, be functional in the absence of any structure ('fuzzy' logic), was discovered. Thus, contrary to the hitherto prevailing dogma, it became apparent that protein structure cannot be envisaged merely as binary states; rather it is a continuum of conformations including the propensity to form amyloid fibrils and encode information for transgenerational inheritance. Furthermore, in the last decade we realized that many IDPs and prion-like proteins have the potential to undergo liquid-liquid phase separation (LLPS), a process underlying the formation of so-called proteinaceous membrane-less organelles (PMLOs). Because PMLOs serve as master regulators of the cell, the propensity of IDPs to undergo LLPS means that a role for IDPs in evolution of life on earth can be argued, especially prebiotic evolution.

Thus, it follows that a new era in the protein field has dawned in which IDPs have attracted a lot of interest. Nonetheless, IDPs also pose a major challenge as they are not easily tractable experimentally using techniques and approaches traditionally used by protein scientists. Because of this, IDPs, which are prevalent in all three domains of life and comprise almost half the human proteome, are often referred to as constituents of 'dark matter' in biology.

## 2. A Special Issue in Honour of Vladimir Uversky

This Special Issue of Biomolecules, "Physics of Protein Folding, Misfolding, and Intrinsic Disorder: A Themed Issue in Honour of Professor Vladimir Uversky on the Occasion of His 60th Birthday", is a dedication to one of the discoverers of IDPs [1]. This collection is a small token of the respect, admiration, and affection the contributors have for Prof. Vladimir (Volodya) Uversky. It is also a celebration of his illustrious career, and his accomplishments, and contributions to the IDP field that have inspired so many minds worldwide.

This Special Issue presents the state of the art as it emerges from the contribution of the community of IDP researchers (IDPers), who have responded to the invitation to give credit to the pioneering work of Prof. Vladimir Uversky aimed at defining the class of disordered proteins and at promoting the attention of scientists toward the existence of "non-globular proteins".

**Citation:** Kulkarni, P.; Brocca, S.; Dunker, A.K.; Longhi, S. Per Aspera ad Chaos: Vladimir Uversky's Odyssey through the Strange World of Intrinsically Disordered Proteins. *Biomolecules* **2023**, *13*, 1015. https://doi.org/10.3390/biom13061015

Received: 12 June 2023
Accepted: 15 June 2023
Published: 19 June 2023

The papers in this collection show the advancement of our knowledge through the application of an integrative structural approach and witnesses at the same time the interest of the IDPer community toward new concepts (i.e., liquid-liquid phase separation) and new methodological frontiers (i.e., the application of machine learning and artificial intelligence to disorder prediction and modelling). This Special Issue collects 20 original research articles and 5 reviews, divided into 4 main areas (see below), which we defined for the purpose of this editorial by drawing on cloud analysis applied to the whole list of keywords (Figure 1).

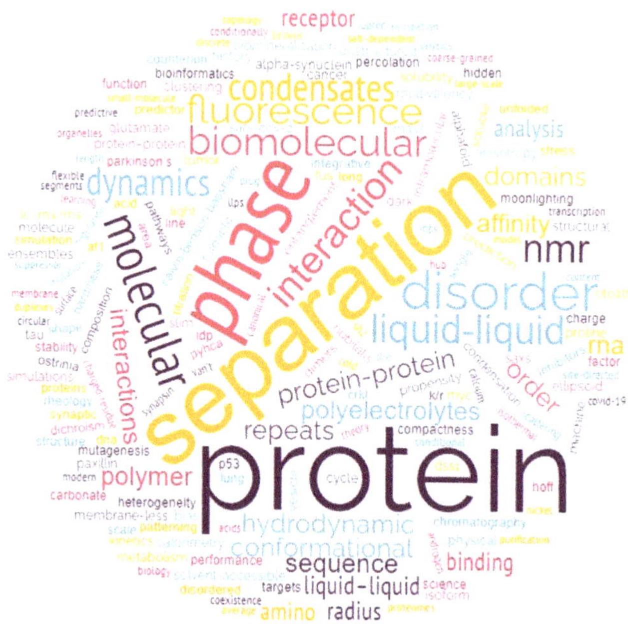


**Figure 1.** Cloud analysis of key words of papers of the Special Issue in honour of Prof. Vladimir Uversky.

## 3. The Different Areas and the Distribution of Articles within Them

*Intrinsic disorder characterization and methodological development*—Integrative structural biology, which is the application of multiple experimental and computational methods, has emerged as an essential approach to understanding IDP phenomena. The collection of papers and reviews in this section embodies this philosophy to study IDPs. This appears as one of the most intense fields of study, with 6 scientific articles and one review [2–8] (Table 1).

*Phase separation*—Liquid-liquid phase separation (LLPS) represents one of the major functional areas in which IDPs are involved and has been attracting a lot of interest in the scientific community in recent years. This is also evident from the numbers of contributions on this topic received in our SI, with five scientific articles and three reviews [9–16] regarding both computational prediction and biochemical analysis of condensation propensity.

*Binding mode and properties of IDPs/IDRs*—IDPs are remarkably well suited to interact with other proteins and are often found at the center of interaction hubs. Understanding the subtle mechanisms by which such interactions are triggered and regulated is of paramount importance to decipher their pathophysiological role, but also for exploring the potential for pharmacological approaches targeting IDPs. Under this theme, readers will find five scientific articles and one review [17–22].

**Table 1.** Content of the Special Issue in honour of Prof. Vladimir Uversky. Four main topics can be identified into which the 25 articles are distributed.

| | |
|---|---|
| **Intrinsic Disorder Characterization & Methodological Development** | |
| A Novel Tandem-Tag Purification Strategy for Challenging Disordered Proteins | Mészáros et al., 2022 [2] |
| Illuminating Intrinsically Disordered Proteins with Integrative Structural Biology | Evans et al., 2023 [3] |
| Distribution of Charged Residues Affects the Average Size and Shape of Intrinsically Disordered Proteins | Bianchi et al., 2022 [4] |
| Identification of Intrinsically Disordered Proteins and Regions in a Non-Model Insect Species *Ostrinia nubilalis* (Hbn.) | Avramov et al., 2022 [5] |
| NMR Reveals Specific Tracts within the Intrinsically Disordered Regions of the SARS-CoV-2 Nucleocapsid Protein Involved in RNA Encountering | Pontoriero et al., 2022 [6] |
| The Ni(II)-Binding Activity of the Intrinsically Disordered Region of Human NDRG1, a Protein Involved in Cancer Development | Beniamino et al., 2022 [7] |
| Deciphering the Alphabet of Disorder—Glu and Asp Act Differently on Local but Not Global Properties | Roesgaard et al., 2022 [8] |
| **Phase Separation** | |
| An Interpretable Machine-Learning Algorithm to Predict Disordered Protein Phase Separation Based on Biophysical Interactions | Cai et al., 2022 [9] |
| Quantifying Coexistence Concentrations in Multi-Component Phase-Separating Systems Using Analytical HPLC | Bremer et al., 2022 [10] |
| In-Silico Analysis of pH-Dependent Liquid-Liquid Phase Separation in Intrinsically Disordered Proteins | Pintado-Grima et al., 2022 [11] |
| Different Forms of Disorder in NMDA-Sensitive Glutamate Receptor Cytoplasmic Domains Are Associated with Differences in Condensate Formation | Basak et al., 2022 [12] |
| Effects of Mass Change on Liquid–Liquid Phase Separation of the RNA-Binding Protein Fused in Sarcoma | Dong et al., 2023 [13] |
| Topological Considerations in Biomolecular Condensation | Das and Deniz, 2023 [14] |
| Reorganization of Cell Compartmentalization Induced by Stress | Fefilova et al., 2022 [15] |
| The Role of Intrinsically Disordered Proteins in Liquid–Liquid Phase Separation during Calcium Carbonate Biomineralization | Tarczewska et al., 2022 [16] |
| **Binding Mode and Properties of IDPs/IDRs** | |
| Portability of a Small-Molecule Binding Site between Disordered Proteins | Jaiprashad et al., 2022 [17] |
| The Role of Membrane Affinity and Binding Modes in Alpha-Synuclein Regulation of Vesicle Release and Trafficking | Das et al., 2022 [18] |
| Sequence Properties of an Intramolecular Interaction that Inhibits p53 DNA Binding | Gregory & Daughdrill, 2022 [19] |

**Table 1.** *Cont.*

| | |
|---|---|
| Folding and Binding Mechanisms of the SH2 Domain from Crkl | Nardella et al., 2022 [20] |
| Linker Length Drives Heterogeneity of Multivalent Complexes of Hub Protein LC8 and Transcription Factor ASCIZ | Walker et al., 2023 [21] |
| A Trajectory of Discovery: Metabolic Regulation by the Conditionally Disordered Chloroplast Protein, CP12 | Gérard et al., 2022 [22] |
| **Modeling of IDPs by Conventional and Advanced Bioinformatic Tools** | |
| Digging into the 3D Structure Predictions of AlphaFold2 with Low Confidence: Disorder and Beyond | Bruley et al., 2022 [23] |
| The Difference in Structural States between Canonical Proteins and Their Isoforms Established by Proteome-Wide Bioinformatics Analysis | Osmanli et al., 2022 [24] |
| Conformational Analysis of Charged Homo-Polypeptides | Bigman and Levy, 2023 [25] |
| Compositional Bias of Intrinsically Disordered Proteins and Regions and Their Predictions | Zhao and Kurgan, 2022 [26] |

*Predicting and modeling of IDPs by conventional and advanced bioinformatic tools*—Significant progress has been made over the past decade in the development of bioinformatics tools for predicting and modeling structural disorder. In addition to conventional predictors, which often make use of compositional bias analysis, one increasingly finds artificial intelligence-based programs that enable accurate and rapid analysis of entire proteomes. Four articles can be found in this section of SI [23–26].

In that respect, note that half of the papers focused on phase separation address the phenomenon with computational and bioinformatic tools.

## 4. The Impact of Vladimir Uversky on the Scientific Community and Our Careers

The full range of Volodya's contributions is so vast to describe that it would be a disservice to him if we even tried to do so in this editorial. Nonetheless, despite the reputation he has earned, Volodya is one of the most humble, respectful, and caring scientists you will come across. A few statistics concerning Volodya and the profound influence he has had on the field is worth mentioning to justify how flabbergasted one may feel when meeting him for the first time; Volodya has published ~1150 research articles and reviews, ~110 book chapters, edited or co-edited >25 books, edited 5 book series that include 45 volumes, guest edited countless Special Issues for various Journals, and mentored/advised ~185 undergraduate, masters, graduate students, postdoctoral fellows and visiting faculty over his career spanning over 3 decades. As of 2021, Volodya had co-authored papers with >15,000 researchers from >2750 institutions in 90 countries around the globe!

As Guest Editors of this Special Issue, we are indeed delighted to honor an esteemed colleague and friend, but to illustrate his humility, dedication, and gregarious personality, we have taken the liberty to share a few personal anecdotal notes with the readers.

Prakash Kulkarni: I have known Volodya for almost a decade and we have published several papers together. However, if I were to highlight 1 or 2 papers that I would consider as the most significant and thought provoking among them, the one elaborating the concept of IDPs as complex systems, would be one. Beginning 2010, together with Govindan Rangarajan, my mathematics collaborator, we were working on the IDP conformational noise hypothesis. At the time, I was toying with the idea that IDPs that are critical in events such as phenotypic switching and cell fate determination, may be viewed as dynamical systems. Our conformational noise paper [27] was published on 19 December 2012, and thus, we figured we may resume working in January 2013 right after the Christmas holidays.

However, we were surprised when, on 23 December 2012, we saw Volodya's paper also alluding to IDPs as edge-of-chaos systems [28]! In retrospect, we should have anticipated this knowing very well that Volodya is a physicist!

At any rate, I decided that we should reach out to him and perhaps, eventually join forces to explore these initial ideas further. And so, we did. And true to his impeccable reputation, he not only agreed to hear me out, but also shared his insight and some ideas on how we may want to approach the problem. The paper we published in Chem Rev together with Rangarajan and several of our other colleagues was the culmination of this enduring spirit of cooperation [29]. A couple of other papers that we published together which I cherish immensely are the role of IDPs in evolution using the beak of the finch and the origin of multicellularity in the green algae, as paradigms [30,31]. The thrill and the excitement we shared when we conceived all these ideas prior to putting them down as formal manuscripts is hard to describe in words. Nonetheless, I have not had an opportunity to meet Volodya in person thus far and I look forward to that day in earnest.

Stefania Brocca: I first came into contact with Volodya in 2008, when the existence of IDPs already seemed to be fairly accepted by the international scientific community, but still none of the researchers in my department had had the opportunity to come across them. The opportunity to collaborate with Volodya jumped out through a study on Sic1, a yeast cell cycle regulator, as part of a project of which Prof. Lilia Alberghina was PI. So, we started a biochemical characterization project on a protein that was "strange" compared to those hitherto manipulated by myself and colleagues in my department. The first question to be answered concerned the prediction of disorder. I first turned to Dr. Sonia Longhi, a friend and former lab colleague in Prof. Marina Lotti's group, who did not hesitate to suggest contacting Volodya. I had little confidence that a "super-star" scientist of his stature could devote time to an obscure project of an unknown Italian researcher. Encouraged by Sonia, I wrote to him, and Volodya, to my surprise, not only responded immediately and effectively, but treated me with unparalleled helpfulness and friendliness. Volodya was a co-author of that paper [32], which was the first of a series of papers [33–37]. These collaborative efforts also involved other colleagues in my Department, namely Prof. Rita Grandori, Prof. Silvia Maria Doglia, Dr. Antonino Natalello, and Dr. Carlo Santambrogio, experts in biophysical techniques such as native mass spectrometry and Fourier transform infrared spectroscopy. The collaboration with Volodya was instrumental in the creation and consolidation of our multidisciplinary team for the study of IDPs. Alongside IDPs, Volodya has also been involved in studies on protein folding and aggregation, starting to be recognized as an issue of biological and biotechnological interest at that time [38,39]. I have met Volodya personally only a couple of times. Nevertheless, his energy, enthusiasm and availability make him an ever and effectively present member of my Dept's team of IDPers.

Alan Keith Dunker: In early 1999 the *Proceedings of the National Academy of Sciences*, USA (PNAS) asked me to review a manuscript submitted by Volodya and co-workers. This paper showed that structured proteins and IDPs could be separated by a straight line on a plot of a protein's net charge ($Y$-axis) versus its overall hydropathy ($X$-axis) using the Kyte and Doolittle hydropathy scale. On this "Charge-Hydropathy Plot", IDPs are distinguished from structured proteins by their higher net charge and reduced hydrophobicity.

By early 1999 we had already published our early predictors of protein disorder and we were working on improving them. Our published and unpublished data supported the overall findings in Volodya's PNAS submittal, but Volodya had missed our papers. After pointing out these missing references, we gave Volodya's submittal a very strong positive review. To my great surprise, this paper was not accepted by PNAS. Several years later one of the world's leading biochemists told me that he rejected Volodya's PNAS submittal and that he regarded this as one of his biggest mistakes. Indeed, that preeminent biochemist has published multiple papers on IDPs!

Volodya's PNAS submittal was eventually published in late 2000 in the journal Proteins [40]. In this version of his paper, Volodya not only included the missing references to our work, but he also showed that, for a particular set of IDPs, his Charge-Hydropathy Plot and our predictions agreed with each other.

Sometime after "meeting" Volodya via his PNAS paper submission, I contacted him and arranged to meet him in person at the Annual Meeting of the Biophysical Society in San Francisco on 23 February 2002. We expressed our interest in working together.

In July 2003, I moved from Washington State University to Indiana University School of Medicine to become the founding Director of the Center for Computational Biology and Bioinformatics. Recruiting Volodya became a very high priority. He joined me in Indiana in 2004 and remained with me there until 2010. During this period and continuing to the present, we have published more than 100 papers together. I have never met anyone who works even half as hard as Volodya does.

At Volodya's suggestion, we revisited his Charge-Hydropathy Plot to test whether alternative hydropathy scales might improve the predictions. We compared the Kyte-Doolittle scale to 18 other published scales, the best of which proved to be a scale developed by Robert Guy. In addition, we developed a new scale called IDP-Hydropathy. The balanced accuracies of the Charge-Hydropathy Plot using each of these scales are 79 $\pm$ 9% (Kyte-Doolittle), 84 $\pm$ 9% (Guy), and 90 $\pm$ 7% (IDP-Hydropathy), where balanced accuracy is given by (%Correct Structure + %Correct Disorder)/2 [41].

Even though Volodya left Indiana in 2010, we continue to collaborate [42,43]. Words cannot express how much Volodya has helped me and our students over our many years together.

Sonia Longhi: My first contacts with Volodya date back to 23 years ago, when my PhD student at that time contacted Volodya to clarify a doubt, we had on how to compute the actual hydrodynamic radius versus the one expected for an IDP from the gel filtration profile. We were very impressed by the rapidity of his answer and by his accessibility and patience. Thanks to his clarification and explanation, we could bring the last finishing touches to the study we were carrying out [44]. Shortly thereafter, we benefited from his deep knowledge of IDPs and co-authored a review on the assessment of protein disorder and induced folding that is still nowadays highly cited [45]. This was the first of a total of 13 co-publications over years, including a comprehensive review on intrinsic disorder [46] and three publications that also involved another guest editor of this Special Issue (i.e., Prof. Stefania Brocca) [34,36,37]. On top of that, we co-edited a book on experimental methods for IDPs [47] and a book on structural disorder within viruses [48]. In the context of all these collaborations over the years I have been dumbfounded by his working abilities and reactivity. His responses were so prompt that I had the impression he was working in the neighboring office, and we were actually chatting. This has now become a sort of joke between us, whereby in the last exchange, no more than a few days ago, I noticed that he was "getting slower" (his answer took three minutes instead of the usual 30 s!), a "slowness" that Volodya ascribed to the fact that he had just turned 60!

Our first encounter dates back to 2007, on the occasion of the first meeting of the IDPs subgroup of the Biophysical Society in Baltimore, a subgroup which was created at the initiative of Volodya and Prof. Keith Dunker. Since then, I have had multiple occasions to meet him, including when he accepted my invitation to visit my lab and give a highly appreciated seminar in 2011. Over the years, he proved to be always extremely supportive and eager to help. His work has been very inspiring to me. Having started myself working in the field of IDPs in the early 2000s, I can perfectly appreciate the difficulties that he must have encountered to make the concept of disorder accepted in the scientific community. I am very admirative of his perseverance and of all the efforts he did to make this concept be adopted by protein scientists. Without his efforts and energies, I doubt I would have dared to make the decision of becoming an "IDPer". Thanks, Volodya, for having been such a pioneer and for your guiding role!

We wish Volodya a very happy 60th birthday, and a long healthy life in the years to come, and keenly look forward to seeing more discoveries by him.

**Author Contributions:** Conceptualization, P.K., S.B., A.K.D. and S.L.; writing—original draft preparation, P.K., S.B., A.K.D. and S.L.; writing—review and editing P.K., S.B., A.K.D. and S.L. All authors have read and agreed to the published version of the manuscript.

**Acknowledgments:** We sincerely thank all the authors for their valuable contributions to the Special Issue, to the Journal's Editorial Office for kindly agreeing to our idea to organize this Special Issue.

**Conflicts of Interest:** The authors declare no conflict of interest.

## References

1. Uversky, V.N.; Kulkarni, P. Intrinsically disordered proteins: Chronology of a discovery. *Biophys. Chem.* **2021**, *279*, 106694. [CrossRef] [PubMed]
2. Mészáros, A.; Muwonge, K.; Janvier, S.; Ahmed, J.; Tompa, P. A Novel Tandem-Tag Purification Strategy for Challenging Disordered Proteins. *Biomolecules* **2022**, *12*, 1566. [CrossRef]
3. Evans, R.; Ramisetty, S.; Kulkarni, P.; Weninger, K. Illuminating Intrinsically Disordered Proteins with Integrative Structural Biology. *Biomolecules* **2023**, *13*, 124. [CrossRef]
4. Bianchi, G.; Mangiagalli, M.; Barbiroli, A.; Longhi, S.; Grandori, R.; Santambrogio, C.; Brocca, S. Distribution of Charged Residues Affects the Average Size and Shape of Intrinsically Disordered Proteins. *Biomolecules* **2022**, *12*, 561. [CrossRef]
5. Avramov, M.; Schád, É.; Révész, Á.; Turiák, L.; Uzelac, I.; Tantos, Á.; Drahos, L.; Popović, Ž.D. Identification of Intrinsically Disordered Proteins and Regions in a Non-Model Insect Species Ostrinia nubilalis (Hbn.). *Biomolecules* **2022**, *12*, 592. [CrossRef]
6. Pontoriero, L.; Schiavina, M.; Korn, S.M.; Schlundt, A.; Pierattelli, R.; Felli, I.C. NMR Reveals Specific Tracts within the Intrinsically Disordered Regions of the SARS-CoV-2 Nucleocapsid Protein Involved in RNA Encountering. *Biomolecules* **2022**, *12*, 929. [CrossRef]
7. Beniamino, Y.; Cenni, V.; Piccioli, M.; Ciurli, S.; Zambelli, B. The Ni(II)-Binding Activity of the Intrinsically Disordered Region of Human NDRG1, a Protein Involved in Cancer Development. *Biomolecules* **2022**, *12*, 1272. [CrossRef]
8. Roesgaard, M.A.; Lundsgaard, J.E.; Newcombe, E.A.; Jacobsen, N.L.; Pesce, F.; Tranchant, E.E.; Lindemose, S.; Prestel, A.; Hartmann-Petersen, R.; Lindorff-Larsen, K.; et al. Deciphering the Alphabet of Disorder—Glu and Asp Act Differently on Local but Not Global Properties. *Biomolecules* **2022**, *12*, 1426.
9. Cai, H.; Vernon, R.M.; Forman-Kay, J.D. An Interpretable Machine-Learning Algorithm to Predict Disordered Protein Phase Separation Based on Biophysical Interactions. *Biomolecules* **2022**, *12*, 1131. [CrossRef] [PubMed]
10. Bremer, A.; Posey, A.E.; Borgia, M.B.; Borcherds, W.M.; Farag, M.; Pappu, R.V.; Mittag, T. Quantifying Coexistence Concentrations in Multi-Component Phase-Separating Systems Using Analytical HPLC. *Biomolecules* **2022**, *12*, 1480. [CrossRef] [PubMed]
11. Pintado-Grima, C.; Bárcenas, O.; Ventura, S. In-Silico Analysis of pH-Dependent Liquid-Liquid Phase Separation in Intrinsically Disordered Proteins. *Biomolecules* **2022**, *12*, 974. [CrossRef] [PubMed]
12. Basak, S.; Saikia, N.; Kwun, D.; Choi, U.B.; Ding, F.; Bowen, M.E. Different Forms of Disorder in NMDA-Sensitive Glutamate Receptor Cytoplasmic Domains Are Associated with Differences in Condensate Formation. *Biomolecules* **2023**, *13*, 4. [CrossRef] [PubMed]
13. Dong, W.; Tang, C.; Chu, W.-T.; Wang, E.; Wang, J. Effects of Mass Change on Liquid–Liquid Phase Separation of the RNA-Binding Protein Fused in Sarcoma. *Biomolecules* **2023**, *13*, 625. [PubMed]
14. Das, D.; Deniz, A.A. Topological Considerations in Biomolecular Condensation. *Biomolecules* **2023**, *13*, 151. [CrossRef]
15. Fefilova, A.S.; Antifeeva, I.A.; Gavrilova, A.A.; Turoverov, K.K.; Kuznetsova, I.M.; Fonin, A.V. Reorganization of Cell Compartmentalization Induced by Stress. *Biomolecules* **2022**, *12*, 1441. [CrossRef]
16. Tarczewska, A.; Bielak, K.; Zoglowek, A.; Sołtys, K.; Dobryszycki, P.; Ożyhar, A.; Różycka, M. The Role of Intrinsically Disordered Proteins in Liquid–Liquid Phase Separation during Calcium Carbonate Biomineralization. *Biomolecules* **2022**, *12*, 1266.
17. Jaiprashad, R.; De Silva, S.R.; Fred Lucena, L.M.; Meyer, E.; Metallo, S.J. Portability of a Small-Molecule Binding Site between Disordered Proteins. *Biomolecules* **2022**, *12*, 1887. [CrossRef]
18. Das, T.; Ramezani, M.; Snead, D.; Follmer, C.; Chung, P.; Lee, K.Y.; Holowka, D.A.; Baird, B.A.; Eliezer, D. The Role of Membrane Affinity and Binding Modes in Alpha-Synuclein Regulation of Vesicle Release and Trafficking. *Biomolecules* **2022**, *12*, 1816. [CrossRef]
19. Gregory, E.; Daughdrill, G.W. Sequence Properties of an Intramolecular Interaction that Inhibits p53 DNA Binding. *Biomolecules* **2022**, *12*, 1558. [CrossRef]
20. Nardella, C.; Toto, A.; Santorelli, D.; Pagano, L.; Diop, A.; Pennacchietti, V.; Pietrangeli, P.; Marcocci, L.; Malagrinò, F.; Gianni, S. Folding and Binding Mechanisms of the SH2 Domain from Crkl. *Biomolecules* **2022**, *12*, 1014. [CrossRef]
21. Walker, D.R.; Jara, K.A.; Rolland, A.D.; Brooks, C.; Hare, W.; Swansiger, A.K.; Reardon, P.N.; Prell, J.S.; Barbar, E.J. Linker Length Drives Heterogeneity of Multivalent Complexes of Hub Protein LC8 and Transcription Factor ASCIZ. *Biomolecules* **2023**, *13*, 404. [CrossRef]
22. Gérard, C.; Carrière, F.; Receveur-Bréchot, V.; Launay, H.; Gontero, B. A Trajectory of Discovery: Metabolic Regulation by the Conditionally Disordered Chloroplast Protein, CP12. *Biomolecules* **2022**, *12*, 1047. [CrossRef]
23. Bruley, A.; Mornon, J.-P.; Duprat, E.; Callebaut, I. Digging into the 3D Structure Predictions of AlphaFold2 with Low Confidence: Disorder and Beyond. *Biomolecules* **2022**, *12*, 1467. [CrossRef] [PubMed]

24. Osmanli, Z.; Falgarone, T.; Samadova, T.; Aldrian, G.; Leclercq, J.; Shahmuradov, I.; Kajava, A.V. The Difference in Structural States between Canonical Proteins and Their Isoforms Established by Proteome-Wide Bioinformatics Analysis. *Biomolecules* **2022**, *12*, 1610. [CrossRef]
25. Bigman, L.S.; Levy, Y. Conformational Analysis of Charged Homo-Polypeptides. *Biomolecules* **2023**, *13*, 363. [CrossRef] [PubMed]
26. Zhao, B.; Kurgan, L. Compositional Bias of Intrinsically Disordered Proteins and Regions and Their Predictions. *Biomolecules* **2022**, *12*, 888. [CrossRef] [PubMed]
27. Mahmoudabadi, G.; Rajagopalan, K.; Getzenberg, R.H.; Hannenhalli, S.; Rangarajan, G.; Kulkarni, P. Intrinsically disordered proteins and conformational noise: Implications in cancer. *Cell Cycle* **2013**, *12*, 26–31. [CrossRef]
28. Uversky, V.N. Unusual biophysics of intrinsically disordered proteins. *Biochim. Biophys. Acta* **2013**, *1834*, 932–951. [CrossRef]
29. Kulkarni, P.; Bhattacharya, S.; Achuthan, S.; Behal, A.; Jolly, M.K.; Kotnala, S.; Mohanty, A.; Rangarajan, G.; Salgia, R.; Uversky, V. Intrinsically Disordered Proteins: Critical Components of the Wetware. *Chem. Rev.* **2022**, *122*, 6614–6633. [CrossRef]
30. Kulkarni, P.; Mohanty, A.; Salgia, R.; Uversky, V.N. Intrinsically disordered BMP4 morphogen and the beak of the finch: Co-option of an ancient axial patterning system. *Int. J. Biol. Macromol.* **2022**, *219*, 366–373. [CrossRef]
31. Kulkarni, P.; Behal, A.; Mohanty, A.; Salgia, R.; Nedelcu, A.M.; Uversky, V.N. Co-opting disorder into order: Intrinsically disordered proteins and the early evolution of complex multicellularity. *Int. J. Biol. Macromol.* **2022**, *201*, 29–36. [CrossRef]
32. Brocca, S.; Samalíková, M.; Uversky, V.N.; Lotti, M.; Vanoni, M.; Alberghina, L.; Grandori, R. Order propensity of an intrinsically disordered protein, the cyclin-dependent-kinase inhibitor Sic1. *Proteins* **2009**, *76*, 731–746. [CrossRef]
33. Uversky, V.N.; Santambrogio, C.; Brocca, S.; Grandori, R. Length-dependent compaction of intrinsically disordered proteins. *FEBS Lett.* **2012**, *586*, 70–73. [CrossRef]
34. Testa, L.; Brocca, S.; Santambrogio, C.; D'Urzo, A.; Habchi, J.; Longhi, S.; Uversky, V.N.; Grandori, R. Extracting structural information from charge-state distributions of intrinsically disordered proteins by non-denaturing electrospray-ionization mass spectrometry. *Intrinsically Disord Proteins* **2013**, *1*, e25068. [CrossRef]
35. Daniels, M.J.; Nourse, J.B., Jr.; Kim, H.; Sainati, V.; Schiavina, M.; Murrali, M.G.; Pan, B.; Ferrie, J.J.; Haney, C.M.; Moons, R.; et al. Cyclized NDGA modifies dynamic α-synuclein monomers preventing aggregation and toxicity. *Sci. Rep.* **2019**, *9*, 2937. [CrossRef] [PubMed]
36. Brocca, S.; Grandori, R.; Longhi, S.; Uversky, V. Liquid-Liquid Phase Separation by Intrinsically Disordered Protein Regions of Viruses: Roles in Viral Life Cycle and Control of Virus-Host Interactions. *Int. J. Mol. Sci.* **2020**, *21*, 9045. [CrossRef] [PubMed]
37. Bianchi, G.; Brocca, S.; Longhi, S.; Uversky, V.N. Liaisons dangereuses: Intrinsic Disorder in Cellular Proteins Recruited to Viral Infection-Related Biocondensates. *Int. J. Mol. Sci.* **2023**, *24*, 2151. [CrossRef]
38. Uversky, V. Fundamentals of Protein Folding. In *Protein Aggregation in Bacteria: Functional and Structural Properties of Inclusion Bodies in Bacterial Cells*; Doglia, S., Lotti, M., Eds.; John Wiley & Sons, Inc.: Hoboken, NJ, USA, 2014.
39. Natalello, A.; Ami, D.; Doglia, S. Protein aggregation studied in intact cells by Fourier transform infrared spectroscopy. In *Methods in Protein Structure and Stability Analysis*; Uversky, V., Permyakov, E.A., Eds.; Nova Science Publishers: New York, NY, USA, 2007.
40. Uversky, V.N.; Gillespie, J.R.; Fink, A.L. Why are "natively unfolded" proteins unstructured under physiologic conditions? *Proteins* **2000**, *41*, 415–427. [CrossRef] [PubMed]
41. Huang, F.; Oldfield, C.J.; Xue, B.; Hsu, W.L.; Meng, J.; Liu, X.; Shen, L.; Romero, P.; Uversky, V.N.; Dunker, A. Improving protein order-disorder classification using charge-hydropathy plots. *BMC Bioinform.* **2014**, *15* (Suppl. 17), S4. [CrossRef]
42. Bondos, S.E.; Dunker, A.K.; Uversky, V.N. Intrinsically disordered proteins play diverse roles in cell signaling. *Cell Commun. Signal* **2022**, *20*, 20. [CrossRef]
43. Piersimoni, L.; Abd El Malek, M.; Bhatia, T.; Bender, J.; Brankatschk, C.; Calvo Sánchez, J.; Dayhoff, G.W.; Di Ianni, A.; Figueroa Parra, J.O.; Garcia-Martinez, D.; et al. Lighting up Nobel Prize-winning studies with protein intrinsic disorder. *Cell Mol. Life Sci.* **2022**, *79*, 449. [CrossRef] [PubMed]
44. Karlin, D.; Longhi, S.; Receveur, V.; Canard, B. The N-terminal domain of the phosphoprotein of morbilliviruses belongs to the natively unfolded class of proteins. *Virology* **2002**, *296*, 251–262. [CrossRef] [PubMed]
45. Receveur-Bréchot, V.; Bourhis, J.M.; Uversky, V.N.; Canard, B.; Longhi, S. Assessing protein disorder and induced folding. *Proteins Struct. Funct. Bioinform.* **2006**, *62*, 24–45. [CrossRef] [PubMed]
46. Habchi, J.; Tompa, P.; Longhi, S.; Uversky, V.N. Introducing Protein Intrinsic Disorder. *Chem. Rev.* **2014**, *114*, 6561–6588. [CrossRef] [PubMed]
47. Uversky, V.N.; Longhi, S. *Instrumental Analysis of Intrinsically Disordered Proteins: Assessing Structure and Conformation*; John Wiley and Sons: Hoboken, NJ, USA, 2010.
48. Uversky, V.N.; Longhi, S. *Flexible Viruses: Structural Disorder in Viral Proteins*; Joh Wiley & Sons: Hoboken, NJ, USA, 2012.

biomolecules

MDPI

Article

# Effects of Mass Change on Liquid–Liquid Phase Separation of the RNA-Binding Protein Fused in Sarcoma

Weiqian Dong [1,2], Chun Tang [3,4,*], Wen-Ting Chu [1,*], Erkang Wang [1,2] and Jin Wang [5,*]

1 State Key Laboratory of Electroanalytical Chemistry, Changchun Institute of Applied Chemistry, Chinese Academy of Sciences, Changchun 130022, China
2 School of Applied Chemistry and Engineering, University of Science and Technology of China, Hefei 230029, China
3 Beijing National Laboratory for Molecular Sciences, College of Chemistry and Molecular Engineering, Beijing 100871, China
4 Peking-Tsinghua Center for Life Sciences, Academy for Advanced Interdisciplinary Studies, Peking University, Beijing 100871, China
5 Department of Chemistry and Physics, Stony Brook University, Stony Brook, NY 11794-3400, USA
* Correspondence: tang_chun@pku.edu.cn (C.T.); wtchu@ciac.ac.cn (W.-T.C.); jin.wang.1@stonybrook.edu (J.W.)

**Abstract:** In recent years, many experimental and theoretical studies of protein liquid–liquid phase separation (LLPS) have shown its important role in the processes of physiology and pathology. However, there is a lack of definite information on the regulation mechanism of LLPS in vital activities. Recently, we found that the intrinsically disordered proteins with the insertion/deletion of a non-interacting peptide segment or upon isotope replacement could form droplets, and the LLPS states are different from the proteins without those. We believed that there is an opportunity to decipher the LLPS mechanism with the mass change perspective. To investigate the effect of molecular mass on LLPS, we developed a coarse-grained model with different bead masses, including mass 1.0, mass 1.1, mass 1.2, mass 1.3, and mass 1.5 in atomic units or with the insertion of a non-interacting peptide (10 aa) and performed molecular dynamic simulations. Consequently, we found that the mass increase promotes the LLPS stability, which is based on decreasing the z motion rate and increasing the density and the inter-chain interaction of droplets. This insight into LLPS by mass change paves the way for the regulation and relevant diseases on LLPS.

**Keywords:** liquid–liquid phase separation; coarse-grained simulation; molecular mass; LLPS stability; FUS

**Citation:** Dong, W.; Tang, C.; Chu, W.-T.; Wang, E.; Wang, J. Effects of Mass Change on Liquid–Liquid Phase Separation of the RNA-Binding Protein Fused in Sarcoma. *Biomolecules* **2023**, *13*, 625. https://doi.org/10.3390/biom13040625

Academic Editors: Ugo Bastolla and Jörg Gsponer

Received: 21 November 2022
Revised: 13 March 2023
Accepted: 24 March 2023
Published: 30 March 2023



## 1. Introduction

Liquid–liquid phase separation (LLPS) plays an essential role in cell survival, which is a physicochemical phenomenon in which a solution of proteins and/or nucleic acids concentrates into a distinct, dense phase in equilibrium with a dilute phase depleted in macromolecules [1,2]. Although the concept of membraneless compartments inside cells such as the nucleolus were described as early as the 1830s [3], recently mounting evidence on the wide-ranging roles that biomolecular condensates, including the nucleolus, nuclear speckles, stress granules, Cajal bodies, and P bodies [4–7], are viewed as critical in regulating diverse cellular function have reignited interest in the behaviors of biological LLPS [8]. The functions of biomolecular condensates referred to as membraneless organelles (MLOs) include cell signaling, nuclear transcription, RNA splicing and processing, and DNA sensing and damage repair [3,5–7,9–13]. Importantly, dysregulation of LLPS has been associated with the pathogenesis of neurodegenerative diseases, including amyotrophic lateral sclerosis (ALS), frontotemporal dementia (FTD), and Alzheimer's, Parkinson's, and Huntington's diseases [6,14–18]. Although there is no study which can decipher conclusively the cellular and pathologic basis of the diseases, the unifying observation of abnormal protein inclusions in postmortem tissue may suggest that one well-characterized cellular feature of neurodegenerative disease is the deposition of protein aggregates in affected brain regions [19].

Proteins that undergo LLPS tend to be the intrinsically disordered protein (IDP) or contain the intrinsically disordered region (IDR) which lacks a defined secondary structure [20]. Uversky, Dunker et al. opened the door to the investigation of IDPs [21,22], and Uversky et al. firstly proposed that IDPs serve as important drivers of intracellular LLPS based on the comprehensive assessment of protein intrinsic disorder predisposition by in silico predictors [23]. Recently, Uversky et al. developed a novel web platform named BIAPSS, which can uncover the sequence-encoded signals of proteins capable of undergoing LLPS [24]. IDRs are typically enriched in charged, polar, and/or aromatic amino acids and contain amino acids such as glycine and proline that may convey some structural information [6]. Based on the specific composition and the abundance of amino acids, IDRs can be further classified into arginine/glycine-rich (RG/RGG) domains, phenylalanine/glycine (FG) domains, and prion-like domains (PrLDs), which respectively engage in weak multivalent interactions responsible for driving phase transitions [6].

RNA-binding protein fused in sarcoma (FUS) is a canonical IDP for neurodegenerative diseases, which is mis-localized to cytoplasmic inclusions in degenerating neurons with the onset of ALS and FTD [25]. Furthermore, the FUS is an important model for investigating the LLPS behavior of IDPs/IDRs, and there are abundant studies for LLPS using the FUS model. Some functional MLOs containing FUS are modulated by the recognition of FUS to special RNA fragments [6,16,26]. Kang et al. found that the LLPS of FUS whose aggregation leads to ALS/FTD is enhanced at low concentrations for ATP but is dissolved at high concentrations [27]. In addition, the MD simulations results of Aida et al. have revealed that ATP affects LLPS of FUS by promoting both hydration and solubilization of FUS [28]. On the other hand, Levone et al. found that FUS-dependent LLPS is the requirement of the activation of the cellular DNA damage response (DDR) [11]. The studies of Lao et al. have shown in atomistic detail how phosphorylation inhibits FUS LLPS and reverses the FUS gel/solid phase toward the liquid phase [29]. Bock et al. found that N-terminal acetylation of FUS LC promotes phase separation and reduces aggregation in *E. coli* [30]. Yoshizawa et al. found that the importin karyopherin-beta 2/transportin-1 inhibits LLPS of FUS [31]. In addition, some studies found that environmental factors including pH, molecular crowder [32], temperature [33,34], salt concentration [35], and osmotic pressure [36,37] also affect FUS LLPS and aggregation.

At a given temperature T, higher mass leads to slower thermal motions for the beads, which shows the average effect of mass at the macroscopic level. However, it is unclear how the mass of IDP affects LLPS at the molecular level. Three common techniques to study IDPS that form condensates are solution NMR spectroscopy, small-angle X-ray scattering (SAXS), and Förster resonance energy transfer (FRET), but all of these are relatively low-resolution methods [38,39]. Due to the lack of persistent secondary structures, multiple fuzzy conformations, difficulty in aligning low-complexity regions (LCRs), of obtaining structural properties of droplets, and of choosing appropriate mutations for IDPs, our current molecular understanding of LLPS through experimental approaches is still restrictive [39]. In comparison, molecular dynamics (MD) simulations provide an insightful route to characterize the dynamics of LLPS on atomic and microsecond scales and to generate detailed information on conformational ensembles of IDPs and the contacts formed within a condensate composed of IDP molecules [40]. Best et al. developed a coarse-grained simulation method to determine thermodynamic phase diagrams of IDPs [41] and characterized phase boundaries and material properties for 20 diverse IDP sequences [42]. Additionally, Best et al. used the coarse-grained models to determine the hydrophobicity scale, which can predict LLPS of a given protein and confirms the importance of pi–pi interactions in LLPS [43]. Uversky et al. demonstrated that conformational dynamics of IDPs can rewire the regulatory networks by combining experimental measurements with coarse-grained simulations [44]. There are two ways that the molecular mass of an IDP would change: isotope replacement or the insertion/deletion of a non-interacting peptide segment. In this study, both methods are applied. So, to elucidate the accurate mass effect on LLPS at molecular level, we develop different models based on the two segments of FUS, including a prion-like domain of 50-residue length and an RGG domain of 50-residue length and perform coarse-grained MD simulations. Our results provide the detailed mechanism how IDPs mass change affects the LLPS behaviors of FUS segments.

## 2. Materials and Methods

### 2.1. Simulation System

To our knowledge, there is an effect of chain length on phase diagram [41], and Best et al. found that the results of the slab method and Monte Carlo method of sampling phase coexistence are in good agreement, especially for the proteins whose chain length is equal to 20 or 50 [42]. The major splicing isoform of FUS consists of 526 residues, as reported, and the intrinsically disordered domain of proteins is crucial for the formation of droplets for FUS proteins, which are prion-like domain, RGG1 domain, RGG2 domain, and RGG3 domain [45]. Considering RGG3 domain (FUS 453-501) is about 50 aa, in this study, we selected two amino acid sequences for comparison, which were truncated as 1–50 residue and 453–502 residue in the FUS amino acid sequence, denoted as PLD and RGG, respectively. Next, molecular dynamics simulations with coarse-grained and slab models [41] are used to capture the behavior of the IDPs with or without LLPS. We used the tool of SMOG website to simplify the process of transforming the PDB structure to the coarse-grained model provided for GROMACS [46]. In our coarse-grained model, each amino acid residue is represented by a single bead, using its $C_\alpha$ position, and all beads of a protein sequence have the same mass (shown in Figure 1) [47]. To investigate the effort of isotope labels for the behavior of IDPs in LLPS, the mass of each bead of normal protein is set as 1.00. In contrast, the bead mass of isotope-labeled protein is set as 1.20. As shown in Figure 1, normal PLD chain model (PLD 1.0), isotope-labeled PLD chain model (PLD 1.2), normal RGG chain model (RGG 1.0), and isotope-labeled RGG chain model (RGG 1.2) were treated as four simulation systems. In addition, we supplemented mass 1.1, mass 1.3, and mass 1.5 systems to verify conclusions from the comparison of the normal FUS (mass 1.0) systems and the isotope-labeled FUS (mass 1.2) systems. In each system, 200 identical FUS chains ($n = 200$) were added in the simulation box. Hence, the total number of beads in each system is 10,000.

In order to investigate the effect of IDPs with the insertion/deletion of a non-interacting peptide segment to LLPS from mass change perspective, we constructed four models, as shown in Figure 1, referred to as PLD-tG1, PLD-tG2, RGG-tG1, RGG-tG2. PLD-tG1, and RGG-tG1, adding 10 glycine amino acids to the end of the PLD sequence and RGG sequence. PLD-tG2 and RGG-tG2 add 5 glycine amino acids to the head and the end of both the PLD sequence and RGG sequence. In each system, 200 identical FUS chains with glycines insertion ($n = 200$) were added in the simulation box. Hence the total number of beads in each system is 12,000. The masses of these four models are the same as the mass 1.2 system models.

In this study, our model incorporates a potential energy function including bonded potential, 12-6 Lennard-Jones (LJ) potential, and Debye-Hückel potential to represent bonding, backbone rigidity, Van der Waals interactions as well as electrostatic interactions, where the bonded potential is classical harmonic model and is given by

$$U_b(r_{ij}) = K_b(r_{ij} - r_0)^2 \tag{1}$$

where the bond constant $K_b$ is taken to be 20,000 kJ·nm$^{-2}$·mol$^{-1}$ and the equilibrium bond length $r_0$ is equal to 3.8 Å. The standard Lennard − Jones potential is given by

$$U_{LJ} = \varepsilon\left[\left(\frac{\sigma}{r}\right)^{12} - \left(\frac{\sigma}{r}\right)^{6}\right] \tag{2}$$

where the parameter $\sigma$ is the "finite distance", $\sigma_{ij}$ is the optimal distance between beads $i$ and $j$ that are in contact with each other. We consider the parameter σ equals a constant; that is, the $\sigma$ is 10 Å (about 2.6 a, a = 3.8 Å is the mean bond length). In addition, we performed a series of Langevin dynamics simulations on isolated normal RGG and found that when the parameter ε is equal to 0.001, 0.01, 0.1 kJ/mol ($\sigma$ = 1.0 nm), respectively, the head-to-tail distance (D) results are similar and show that the isolated protein chain is in the disorder state (see Supplementary Materials Figure S1). Considering our simulations are not completely quantitative, only qualitative, the energy parameter ε is set as 0.001 kJ/mol to make sure to capture the behavior of FUS chains in LLPS and LLPS disappears eventually [48]. For the glycine-inserted FUS chains, there are two ε parameters, one is the scale of LJ interaction between the residues of the FUS chains, which is set as 0.001 kJ/mol. The other is the scale of LJ interaction between an inserted residue and another residue (referred as ε-insert) and is set as 0.00001 kJ/mol, which indicates that the inserted peptide is non-interacting. We performed a series of test simulations with three ε-insert parameters for glycine-inserted FUS chains and found the ε-insert parameters as 0.0001, 0.00001 or 0.000001. The results are all similar (shown in Figure 2, Figures S2 and S3). This validated that the effect of the interaction changed by glycine-inserted peptide to LLPS is negligible.

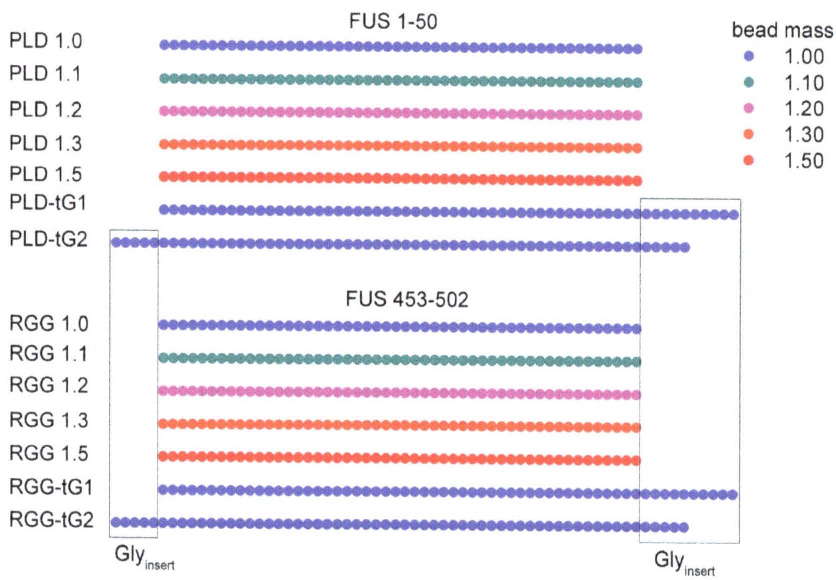


**Figure 1.** The schematic diagram of different FUS chain models. The different colors represent different bead masses. The beads within rectangle area represent inserted glycine residues.

In addition, the Debye – Hückel potential is given by

$$V_{\text{Debye-Hückel}} = K_{coulomb}B(\kappa)\sum_{i,j}\frac{q_i q_j \exp(-\kappa r_{ij})}{\epsilon r_{ij}} \tag{3}$$

where $K_{coulomb}$ = 138.94 kJ·mol$^{-1}$·nm·e$^{-2}$ is the electric conversion factor; $B(\kappa)$ is the salt-dependent coefficient; $\kappa^{-1}$ is the Debye screening length, which is directly dependent on the solvent ionic strength (IS)/salt concentration $C_{salt}$ ($\kappa \approx 3.2\sqrt{C_{salt}}$); $\epsilon$ is the dielectric constant, which was set to 80 during the simulations to mimic the solvent medium (water); $q_i$ and $q_j$ are the charges of beads i and j. In our model, aspartic acid and glutamic acid have a negative charge, $q = -1$, and lysine and arginine have a positive charge, $q = +1$. Other residues were set to $q = 0$. Thus, the PLD chain possesses 2 negative charges and the RGG chain possesses 6 negative charges and 9 positive charges. In order to investigate the role of electrostatic interaction for the phase separation with changed beads mass, we consider two extreme conditions with one where salt concentration is 10 mM ($C_{salt}$ = 0.01 M), which represents there being almost no effect of electrostatic screening, and the other where there is no charge interaction.

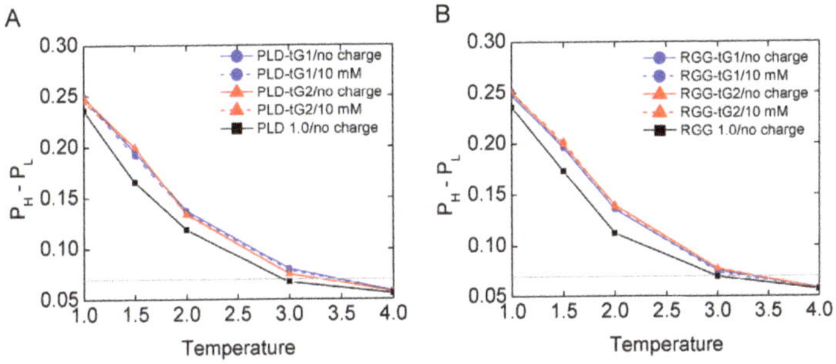


**Figure 2.** $P_H - P_L$ of PLD-tG (**A**) and RGG-tG (**B**) in different solvents changes with temperature when ε-insert=0.00001 kJ/mol. The $P_H$ and $P_L$ values are calculated with the last 1000 τ simulation data as $P_H - P_L$ of all the simulations reaches equilibrium after 3000 τ (see Figures S4–S7). The gray line is $P_H - P_L$ = 0.07.

After the initial equilibrium (10 ns NVT and 10 ns NPT simulations), we changed the simulation box by elongating the z dimension to 300 nm (z = 300 nm) and for 10,000 bead systems shortening both the x and y to 31 nm (x = 31 nm, y = 31 nm) or for 12,000 bead systems shortening both the x and y to 34 nm (x = 34 nm, y = 34 nm) [41]. Compared with the cubic box approach [49], using slab method reduces the simulation cost and does not affect results [41,50]. Then, 5 μs long-time simulations are conducted to all FUS chains systems at multiple temperatures with two strength-of-charge interactions using constant temperature and volume with a Langevin thermostat with 2.0 fs time step and 1.0 $ps^{-1}$ friction coefficient. In order to cover the overall process from LLPS to phase-separation disappearance, we used a series temperature from 100 K to 400 K (100, 150, 200, 300, and 400 K) in Gromacs. For the reduced unit in the coarse-grained model, we set the unit temperature ($T_0$) and unit time (τ) to 100 K and 1 ns in Gromacs. As a result, the simulation temperatures correspond to 1.0, 1.5, 2.0, 3.0, and 4.0 $T_0$, and the simulation length of each trajectory corresponds to 5000 τ. In the simulation, all the scales including the length scale, time scale, mass scale, and energy scale are based on theory and used as reduced unit, so the simulation temperature/time cannot be equal to the real temperature directly [51,52]. In order to avoid misunderstanding, we did not mention K in following figures involving the temperature.

*2.2. Data Analysis*

We introduced the maximum difference of local density of beads in the box to describe the extent of phase-separation. The local density of beads is determined by the proportion ($P_\gamma = m_\gamma/N$, γ = 1 ... 30) of the bead number (denote as $m_\gamma$) of each window in the amount (N) of beads of the box, where the window is the order coordinate set by cutting the z axis into 30 windows (γ, the length of γ is 10 nm) and then clustering each bead of the specified window according to the z coordinate of the bead. For the bead density distribution function of z, we calculate the difference between the highest ($P_H$) and the lowest ($P_L$) values. When LLPS occurs, the protein solution emerges, and demixing and a phenome of the condensed-phase and dilute-phase coexisting in solution is observed, which can be characterized by the difference ($P_H$-$P_L$) of the densities of the two coexisting phases and the value greater than LLPS more obviously.

We introduced the z motion rate of the chains in each model to investigate the role of variant bead mass in the velocity perspective as the same reason to calculate the flux for the droplet boundary. In order to distinguish condensed-phase vs. dilute-phase simply based on a boundary line, we set the center of the system (condensed-phase with LLPS) at the zero point of x, y, and z axes and took the distance/displacement of each chain (center of mass (CM)) to the zero point on the z axis ($|z|_n = |z_n - z_0|$, $n = 1 \ldots 200$) as the coordinate of each chain. Hence, the condensed-phase is below and the dilute-phase is above for a boundary line when LLPS occurs. As the same as above, we cut the $|z|$ into 30 windows as reaction coordinate. Each window length is 5 nm. Subsequently, the z motion rate is calculated by averaging the change rates of z coordinate of each chain in the box. The z motion rate equation is given by

$$v_m = \frac{\sum_{t_0}^{t_m} \sum_{n=1}^{200} (|z_n(t_{m+1}) - z_n(t_m)|)}{200 \times 2000\Delta t} \tag{4}$$

where $t_0$ is 4000 $\tau$, $t_m$ limit is 5000 $\tau$, $z_n(t_m)$ is the z coordinate of the $n$ chain at time $t_m$, $\Delta t = (t_{m+1} - t_m)$ is 0.5 $\tau$.

The flux equation is given by

$$N_t = \frac{\sum_{n=1}^{200} f(z_n(t))}{0.5\tau} \tag{5}$$

$$f(z_n(t)) = \begin{cases} 1, \ \left(z_n(t) - \left(2750 - z_{bundary-line}\right)\right) \times \left(z_n(t+1) - \left(2750 - z_{bundary-line}\right)\right) < 0 \\ or \ \left(z_n(t) - \left(2750 + z_{bundary-line}\right)\right) \times \left(z_n(t+1) - \left(2750 + z_{bundary-line}\right)\right) < 0 \\ or \ z_n(t) \times z_n(t+1) < 0 \\ 0, \ others \end{cases} \tag{6}$$

where $z_n(t)$ is the z coordinate of the $n$ chain at time $t$, and $z_{bundary-line}$ is 250 angstroms, which is the distance between the center of condensed-phase and the boundary line.

In addition, we analyzed the electrostatic interactions by calculating the amount of intra-chain electrostatic contact ($E_{intra}$) and the amount of inter-chain electrostatic contact ($E_{inter}$). $E_{intra}$ is defined by the number of intra-chain contacts when the pairwise ($C_\alpha$-$C_\alpha$) distance between residues of having opposite charge within same chain was less than 12 Å. Similarly, $E_{inter}$ is defined by the number of inter-chain contacts when the pairwise ($C_\alpha$-$C_\alpha$) distance between residues having opposite charges in different chain was less than 15 Å. Considering the charge amount of the chain for all systems and the charge environment in two salt concentration solvents, only chains containing RGG sequence and in 10 mM salt concentration solvent take possession of $E_{intra}$ and $E_{inter}$.

## 3. Results and Discussion

### *3.1. The Mass Effect on LLPS Stability*

In order to investigate the system phase property, we firstly calculated the difference between $P_H$ and $P_L$ ($P_H$-$P_L$) as a function of time $\tau$ with an overall 5000 $\tau$ simulation time for all the simulations trajectories. As shown in Figures S4–S17, all the simulation models have reached equilibrium after 3000 $\tau$. To further confirm that the systems are in equilibrium, as shown in Figure S18, we calculated the $P_H - P_L$ average value of every 100 $\tau$ simulation time in last 1000 $\tau$ data and found that the $P_H - P_L$ values remain stable over the last 1000 $\tau$ simulation time. So, it is safe to say that the simulation systems are in equilibrium in last 1000 $\tau$ simulation time. We analyzed the last 1000 $\tau$ trajectories, representing the ensemble average values of the equilibrated simulations. As shown in Figure 3 and Figure S19, both normal FUS chains and isotope-labeled FUS chains have a decreasing trend of $P_H - P_L$ values as the temperature increases. The heavier mass FUS chains systems have a greater $P_H - P_L$ value at the low temperature (LLPS occurs), which indicates that the LLPS of the heavier mass FUS chains systems are more stable. For example, Figure 3A,B suggest that the temperature of an obvious LLPS ($P_H - P_L > 0.15$) for PLD 1.0 chains without charge interaction, PLD 1.0 chains with charge interaction, PLD

1.2 chains without charge interaction, PLD 1.2 chains with charge interaction, RGG 1.0 chains without charge interaction, RGG 1.0 chains with charge interaction, RGG 1.2 chains without charge interaction, and RGG 1.2 chains with charge interaction is below 1.7 $T_0$, 1.7 $T_0$, 2.0 $T_0$, 2.0 $T_0$, 1.7 $T_0$, 1.7 $T_0$, 2.0 $T_0$, and 2.0 $T_0$, respectively. The FUS chain system's mass increase from 1.0 to 1.2 enlarges the temperature range about 0.3 $T_0$ for the emergence of the obvious LLPS ($P_H - P_L > 0.15$), while charge effects on LLPS are not as significant as the mass effect, as shown in Figure 3. At the same time, when the temperature increases, the $P_H$ value decreases and the $P_L$ value increases (see Figure S20). Furthermore, we found that the LLPS disappears at high temperature (critical temperature, $T_{Cr}$) when the density difference is negligible in the protein solution. The critical temperature $T_{Cr}$ can be obtained by the Flory–Huggins theory or fitting by

$$\rho_H - \rho_L = A(T_{Cr} - T)^{\beta} \quad (7)$$

where $\beta$ is the critical exponent, and $A$ is a protein-specific fitting parameter [41]. When $P_H - P_L = 0$, the temperature of the phase diagram equals the critical temperature ($T_{Cr}$). However, the absolute zero point of $P_H - P_L$ cannot be obtained from the simulations. Thus, we set the threshold to be 0.07. LLPS disappears when $P_H - P_L < 0.07$. In detail, the $T_{Cr}$ value is only relevant to residue mass, regardless of the salt concentration in the models. The $T_{Cr}$ values of PLD 1.0 chains and RGG 1.0 chains are 2.9 $T_0$, and those of PLD 1.2 chains and RGG 1.2 chains are about 3.6 $T_0$. Hence, it is easy to confirm that the FUS chain system's mass increase from 1.0 to 1.2 increases the $T_{Cr}$ by 0.7 $T_0$. In addition, as shown in Figure 4, the critical temperature increases as the mass of the systems increases.

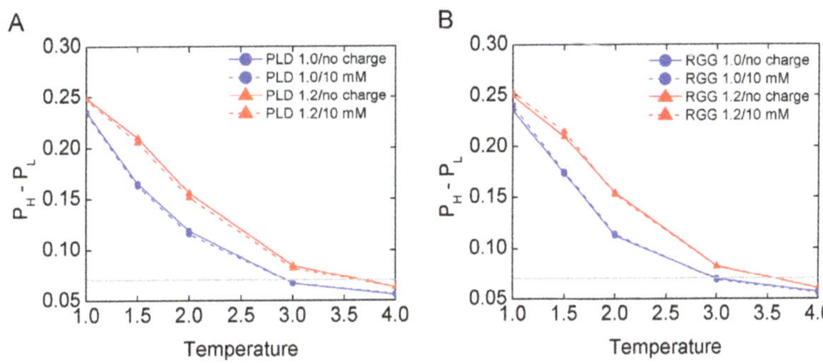


**Figure 3.** $P_H - P_L$ of PLD (**A**) and RGG (**B**) with bead mass 1.0 and 1.2 in different solvents changes with temperature. The $P_H$ and $P_L$ values are calculated with the last 1000 τ simulation data as $P_H - P_L$ of all the simulations reaches equilibrium after 3000 τ (see Supplementary Materials). The gray line is $P_H - P_L$ =0.07.

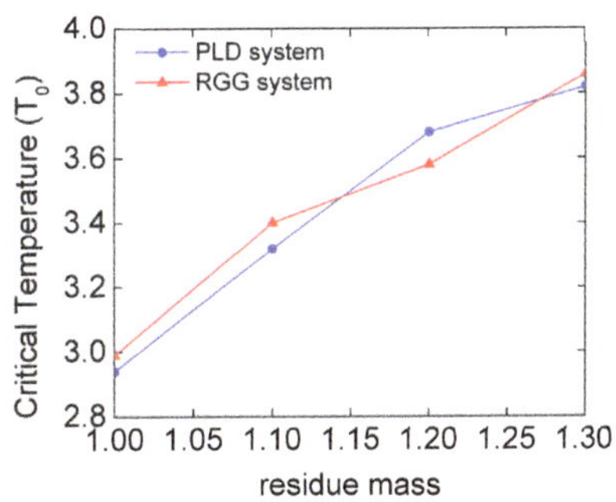


**Figure 4.** Critical temperature ($T_{Cr}$) of LLPS changes with PLD system mass or RGG system mass.

As shown in Figure 2, there is no obvious difference of $P_H - P_L$ value changes with temperature between the FUS segments with different modes of glycine peptide insertion, while the same is true for that in different solvents. Compared with normal FUS chains, the glycine-inserted FUS chains have a greater $P_H - P_L$ value at the low temperatures (LLPS occurs). This indicates that the LLPS of the glycine-inserted FUS chains are more stable. For example, Figure 2A,B suggest that the temperature of an obvious LLPS ($P_H - P_L > 0.15$) for PLD 1.0, PLD-tG1, RGG 1.0, and RGG-tG1 is below 1.7 $T_0$, 1.9 $T_0$, 1.7 $T_0$, and 1.9 $T_0$, respectively. The FUS chain with glycine insertion enlarges the temperature range about 0.2 $T_0$ for the emergence of the obvious LLPS ($P_H - P_L > 0.15$). Additionally, The $T_{cr}$ values of PLD 1.0 chains and RGG 1.0 chains are 2.9 $T_0$, and those of PLD-tG1 and RGG-tG1 are about 3.5 $T_0$ and 3.3 $T_0$, respectively. In summary, both the isotope labeling and the peptide insertion lead to the mass increase and promote the LLPS stability. Thus, the isotope labeling promotes the greater LLPS stability.

### *3.2. The Mechanism of Mass Effect on LLPS*

Considering that each chain undergoes the stochastic dynamics in the simulation, motion and diffusion of beads may slow down when the mass increases from 1.00 to 1.50. In this simulation, to quantify the motion on the z axis of the slab model, we calculated the average z motion rate of all the 200 chains. As shown in Figure 5 and Figure S21, the results suggest that the z motion rate increases as the temperature increases. In addition, the z motion rate of the mass of heavier FUS chains is always smaller than that of the normal FUS chains at the same temperature. For example, at 2.0 $T_0$, the z motion rate of PLD 1.2 chains without charge is 79.6 Å/τ, lower than that of PLD 1.0 chains (88.8 Å/τ). As shown in Figure 5C,D, the $P_H - P_L$ values decrease as the z motion rates increase for normal and isotope-labeled models. The results suggest that the z motion rate is strongly correlated with the stability extent of the LLPS, and lower rate of z motion favors the formation and stability of LLPS.

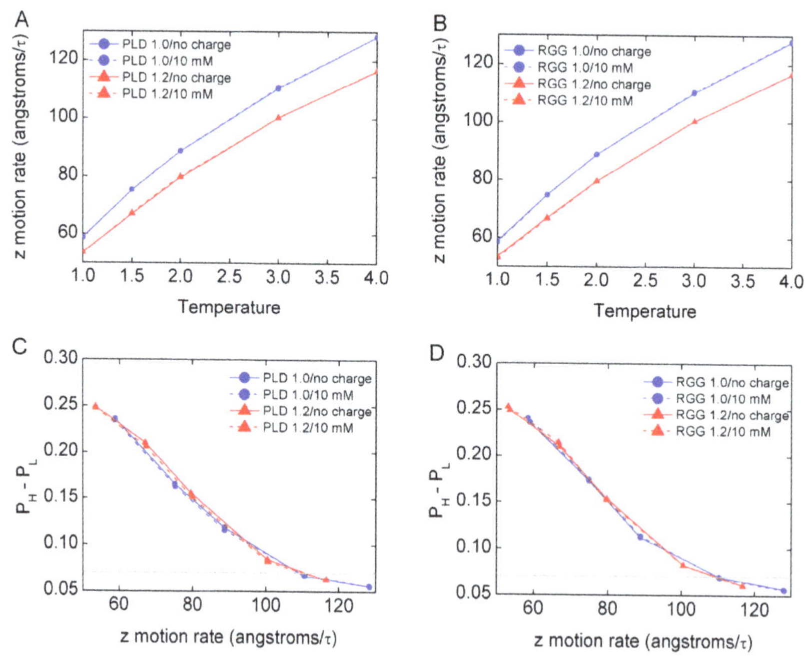


**Figure 5.** The z motion rate of PLD (**A**) and RGG (**B**) with bead mass 1.0 and 1.2 in different solvents changes with temperature; $P_H - P_L$ of PLD (**C**) and RGG (**D**) with bead mass 1.0 and 1.2 in different solvents change with z motion rate. The z motion rates are calculated by averaging 200 chains in the box with the last 1000 τ simulation data. The gray line is $P_H - P_L = 0.07$.

In order to quantify the diffusion of molecules between condensed-phase and dilute-phase, we calculated the average flux of the chains across the boundary line of the condensed-phase during last 1000 τ time. As shown in Figure 6A,B, the flux increases as temperature rises for all models. The flux values of isotope-labeled FUS chains are smaller than that of normal FUS chains at the same temperature. For example, at 2.0 $T_0$, the flux of PLD 1.2 chains without charge is 8.6 chains/τ, lower than that of PLD 1.0 chains (11.2 chains/τ). As shown in Figure 6C,D, the $P_H - P_L$ values decrease as the flux values increase for normal and isotope-labeled models. As shown in Figure S22, we calculated the average flux during the last 100 τ simulation time to confirm that our systems are in equilibrium and the results are reliable. As a result, the flux is strongly correlated with the stability extent of LLPS and lower flux values favor LLPS formation and stability, whose results are equivalent to the z motion rate's.

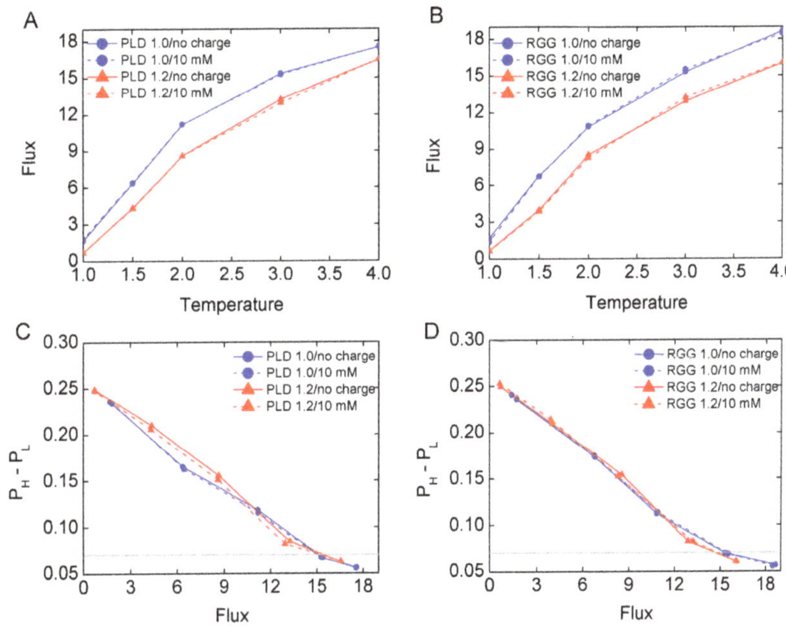


**Figure 6.** The flux on the boundary line. (**A**) The flux of PLD 1.0 and PLD 1.2 in different solvents changes with temperature. (**B**) The flux of RGG 1.0 and RGG 1.2 in different solvents changes with temperature. (**C**) $P_H - P_L$ of PLD 1.0 and PLD 1.2 in different solvents change with flux. (**D**) $P_H - P_L$ of RGG 1.0 and RGG 1.2 in different solvents change with flux. The fluxes are calculated by averaging the last 1000 τ simulation data, and the boundary line is $|z| = 25$ nm. The gray line is $P_H - P_L = 0.07$.

As shown in Figure S23A,B, the results suggest that the z motion rate increases as the temperature increases, and there is no obvious difference for different inserted modes and different solvents. In addition, the z motion rate of glycine-inserted FUS chains is always smaller than that of normal FUS chains at the same temperature. For example, at 2.0 $T_0$, the z motion rate of PLD-tG1 chains is 80.5 Å/τ, lower than that of PLD 1.0 chains (88.8 Å/τ). As shown in Figure S23C,D, the $P_H - P_L$ values decrease as the z motion rates increase for glycine-inserted models, which are the same as the isotope-labeled FUS chains.

As shown in Figure S24A,B, the flux increases as the temperature rises for the glycine-inserted models. The flux values of glycine-inserted FUS chains are smaller than that of the normal FUS chains at the same temperature. For example, at 2.0 $T_0$, the flux of PLD-tG1 chains is 9.5 chains/τ, lower than that of PLD 1.0 chains (11.2 chains/τ). As shown in Figure S24C,D, the $P_H - P_L$ values decrease as the flux values increase for the glycine-inserted models.

In addition, we use the distribution of probability of FUS chains as a function of displacement |z| to describe the degree of chain aggregation. As shown in Figures 7 and S25, in the droplet, the probability of the mass heavier of FUS chains is greater than that of normal FUS chains, which indicates that the heavier FUS chains have more concentrated distribution than normal FUS chains at the low temperature (LLPS occurs). For example, at 1.0 $T_0$, when |z| = 0 nm, the probability of PLD 1.2 chains is 0.23 greater than that of PLD 1.0 chains (0.21). As shown in Figure S26, the distribution of the probability of glycine-inserted FUS chains is similar to that of the isotope-labeled FUS chains and compared with normal FUS chains. We found that at the lower temperature (LLPS occurs), the probability of glycine-inserted FUS chains distributed in droplets (|z| < 25 nm) is greater than that of normal FUS chains. For example, at 1.0 $T_0$, when |z| = 0 nm, the probability of PLD-tG1 chains is 0.23 greater than that of PLD 1.0 chains (0.21). In summary, the mass increase leads to FUS chains being more concentrated in the droplets.

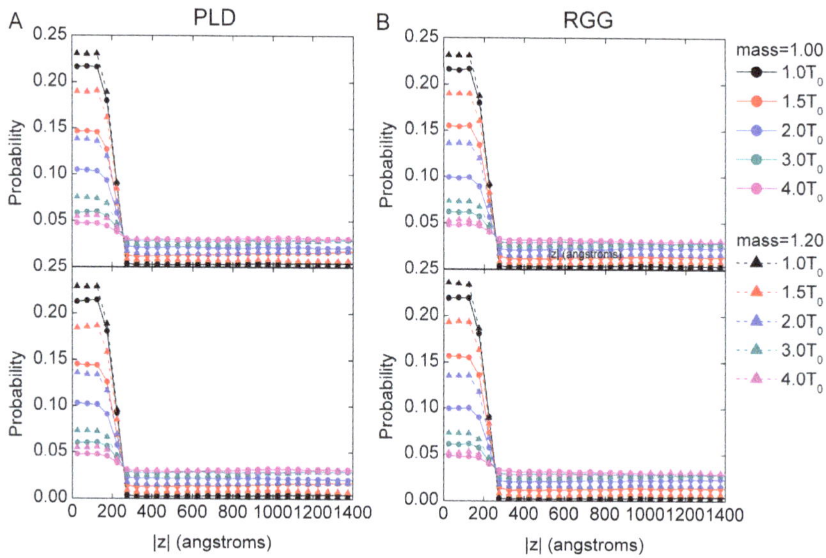


**Figure 7.** The distribution of probability of PLD (**A**) and RGG (**B**) with bead mass 1.0 and 1.2 as a function of displacement |z|. Data without charges and at 10 mM salt concentration solvent are shown in the upper panels and the bottom panels, respectively. The probability is the average value of window along |z| during the last 1000 τ simulation data.

*3.3. The Effect of Mass Increase on the Conformation of Chains and the Electrostatic Contact*

We calculated the mean end-to-end distance (D) of 200 chains to show the conformation changes at different conditions. As shown in Figures 8 and S27, there is a trend that the mean D decreases as the temperature increases in normal and isotope-labeled models. The effect of the mass increases on the mean D can be negligible. This indicates that the mass of chains does not have a significant effect on the conformation of an individual molecule. Intriguingly, the mean D shows difference with different charge patterns. The PLD chains with 10 mM salt concentration solvent have a slightly greater mean D than that without charge interactions. In contrast, the RGG chains with 10 mM salt concentration solvent have a smaller mean D than that without charge interaction. The results suggest that the electrostatic interactions help RGG chains to fold a bit.

In order to distinguish the chain conformations in the condensed-phase and the dilute-phase, we calculated the distributions of D along the displacement |z|. As shown in Figure 9, there is no significant difference between normal and isotope-labeled FUS chains.

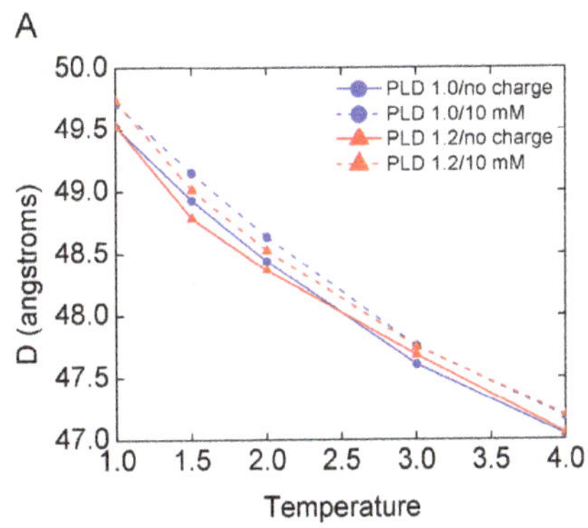


**Figure 8.** Head-to-tail distance (D) of PLD (**A**) and RGG (**B**) chains with bead mass 1.0 and 1.2 as a function of temperature. The D value is calculated by the mean value of the 200 chains in the system during the last 1000 $\tau$ simulation data.

For the glycine-inserted FUS chains, we calculated the head-to-tail distance of PLD sequence or RGG sequence excluding the glycine-inserted peptide. As shown in Figure S28, the mean D decreases as the temperature increases in glycine-inserted models. In addition, PLD-tG2 and RGG-tG2 have greater mean D values than that of PLD-tG1 and RGG-tG1, respectively. This indicates that different modes of insertion influence the head-to-tail distance of the FUS chains. Comparing with normal FUS chains and isotope-labeled FUS chains, the mean D values of glycine-inserted FUS chains are greater than that of normal and isotope-labeled FUS chains at the same temperature. We believe that the difference is not caused by the mass increase. In summary, the mass increase hardly affects the mean D of FUS chains and the conformation of FUS chains.

As shown in Figures 10 and S29, there is no significant difference for the intra-chain electrostatic contacts ($E_{intra}$) between the normal RGG chains and the isotope-labeled RGG chains. If the conformation of the chain is curved, the $E_{intra}$ value will be high. Therefore, the $E_{intra}$ value correlates to the D values negatively. In the condensed phase, the $E_{inter}$ value correlates to the local probability of chains positively (as shown in Figures 7, 10, S25 and S29 and $P_H$ in Figure S20). The mass increase leads to the increase of the local probability of chains in the condensed-phase (as shown in Figures 7 and S25 and $P_H$ in Figure S20). As a result, in the condensed-phase, the $E_{inter}$ values of the isotope-labeled RGG chains are higher than that of the normal RGG chains. For example, $E_{inter}$ of RGG 1.2 chains at $|z| = 0$ is 1.62 ($T = 1.0\ T_0$), and by contrast, that of RGG 1.0 chains is 1.49.

As shown in Figure S30, there is no significant difference for the intra-chain electrostatic contacts ($E_{intra}$) and inter-chain ($E_{inter}$) electrostatic contacts between RGG-tG1 chains and RGG-tG2 chains. However, compared with the normal RGG chains and the isotope-labeled RGG chains, both the intra-chain ($E_{intra}$) and inter-chain ($E_{inter}$) electrostatic contacts of RGG-tG1 and RGG-tG2 are smaller. For example, at 1.0 $T_0$, when $|z| = 0$ nm, $E_{intra}$ of RGG-tG1 is 4.32 smaller than that of RGG 1.2 chains (5.42), $E_{inter}$ of RGG-tG1 is 1.21 smaller than that of RGG 1.2 chains (1.62). We believe that the difference is not caused by the mass increase. Considering that the critical temperature of LLPS for RGG-tG chains is greater than that of RGG 1.0 chains and smaller than that of RGG 1.2 chains, we speculated that the electrostatic contact increase is not the main cause of the increased LLPS stability.

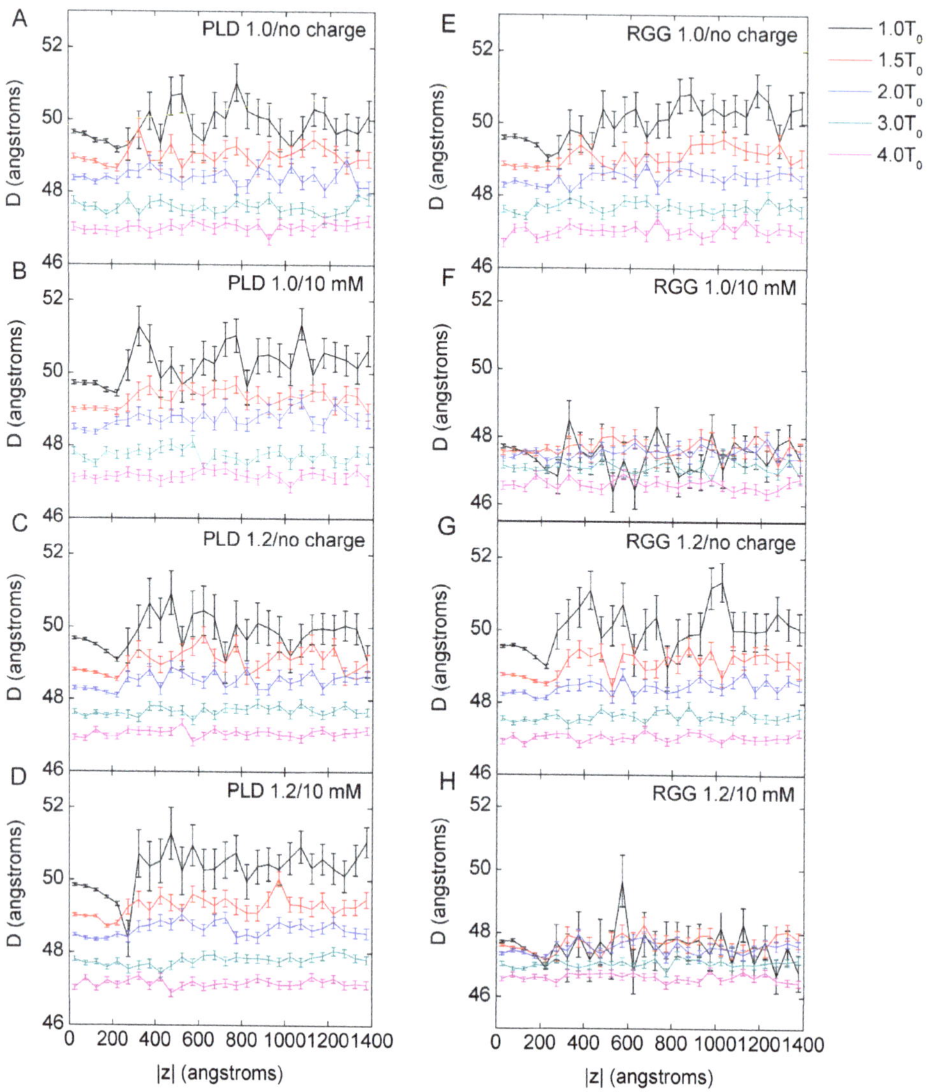


**Figure 9.** Head-to-tail distance (D) of PLD and RGG as a function of displacement |z|. (**A**) The head-to-tail distance of PLD 1.0 with no charge interaction. (**B**) The head-to-tail distance of PLD 1.0 with 10 mM salt solvent. (**C**) The head-to-tail distance of PLD 1.2 with no charge interaction. (**D**) The head-to-tail distance of PLD 1.2 with 10 mM salt solvent. (**E**) The head-to-tail distance of RGG 1.0 with no charge interaction. (**F**) The head-to-tail distance of RGG 1.0 with 10 mM salt solvent. (**G**) The head-to-tail distance of RGG 1.2 with no charge interaction. (**H**) The head-to-tail distance of RGG 1.2 with 10 mM salt solvent. Mean D of the chains in each window along |z| and standard error ($\sigma$) values during the last 1000 $\tau$ simulation data are illustrated in this figure. Considering the effect of boundary, the data with displacement z higher than 1400 are not calculated.

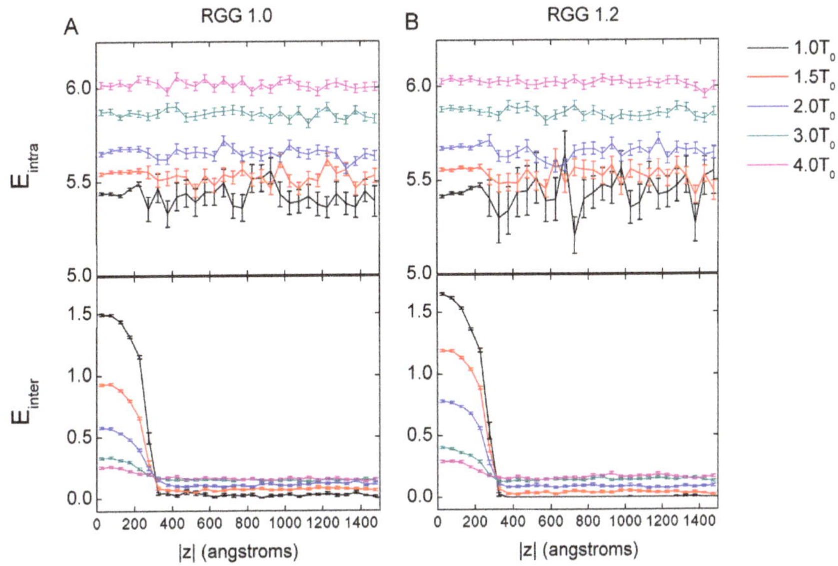


**Figure 10.** Mean number intra-chain ($E_{intra}$) of and inter-chain ($E_{inter}$) electrostatic contacts of RGG 1.0 (**A**) and RGG 1.2 (**B**) as a function of ∣z∣. Average and standard error values during the last 1000 τ simulation data are illustrated in this figure.

## 4. Conclusions

In this study we performed Langevin dynamics simulations to gain insight into the effects and the mechanism of FUS chain mass increase in LLPS. The study was inspired by one of our observations during NMR sample preparation, in which $^{15}N$, $^{13}C$, $^{2}H$-isotope-labeled FUS RGG and another aggregation-prone protein exhibited greater tendency to coacervate than unlabeled protein of the same concentration under the same buffer conditions and temperature. Our simulation results suggest that the mass increase of FUS chain promotes the level of LLPS stability, but different mass increase methods have different devotion to LLPS stability. For the critical transition temperature ($T_{Cr}$) where the LLPS start to emerge, the value of RGG 1.2 chains is 0.7 $T_0$ higher than that of RGG 1.0 chains, while the value of RGG-tG1 is 0.4 $T_0$ higher than that of RGG 1.0 chains. Based on our simulations, the details of how the FUS chain mass change affects the behavior of LLPS at various temperatures and ionic strength are vividly revealed at the molecular level. We found that, in the same environment, the z motion rate of chains of the mass 1.2 system and glycine-inserted system is lower than that of the mass 1.0 system, and the flux of the mass 1.2 system and glycine-inserted system is lower than that of the mass 1.0 system. Therefore, lower z motion rate and lower flux are beneficial to LLPS stability. Furthermore, using the distribution of probability of FUS chains as a function of displacement ∣z∣, the results reveal that the mass increase will increase the degree of chain aggregation at the same temperature, and the chains of the mass 1.2 system and glycine-inserted system both are more concentrated than that of the mass 1.0 system. The mass increase hardly affects the head-to-tail distance (D) of FUS chains. In addition, we have noted that the mass increase by the isotope replacement is favorable to strengthen the inter-chain electrostatic contacts in the condensed-phase, hardly affect the intra-chain electrostatic contacts and the head-to-tail distance of the chains. The effect of the mass increase by glycine insertion on the intra-chain electrostatic contact and the inter-chain electrostatic contact is fuzzy and will be studied in the future. We believe that FUS chain mass increase leads to the increase of the inter-chain electrostatic contact from more concentrated distribution of FUS chains, while the electrostatic contact increase is not the main cause for increased LLPS stability.

Consequently, we believe that the mass increase promotes the LLPS stability, which is based on decreasing the z motion rate, increasing the density and the inter-chain interaction of droplets.

Our findings highlight the importance of residue mass change of IDPs on LLPS. Such residue mass change often emerges in the NMR experiments used to explore the information on structures of IDPs. In our study, these changed mass FUS models may provide enlightenment towards understanding the roles of isotope-labeling effects in modulating LLPS. In addition, it is helpful to test more systems with simulation and to elaborate results from the IDP chain mass perspective for investigating the mechanism of LLPS. This may pave the way for ameliorating phase-separation-related pathologies, which will be our future work direction.

**Supplementary Materials:** The following supporting information can be downloaded at: https://www.mdpi.com/article/10.3390/biom13040625/s1, Figure S1: Head-to-tail distance (D) of RGG with different parameters $\varepsilon$ ($\sigma$ = 1.0 nm, $\varepsilon$ = 0.001, 0.01, 0.1, 0.8 kJ/mol); Figure S2: $P_H - P_L$ of PLD-tG (A) and RGG-tG (B) in different solvents changes with temperature when $\varepsilon$-insert = 0.0001 kJ/mol. The $P_H$ and $P_L$ values are calculated with the last 1000 $\tau$ simulation data as $P_H - P_L$ of all the simulations reaches equilibrium after 3000 $\tau$. The gray line is $P_H - P_L$ = 0.07; Figure S3: $P_H - P_L$ of PLD-tG (A) and RGG-tG (B) in different solvents changes with temperature when $\varepsilon$-insert = 0.000001 kJ/mol. The $P_H$ and $P_L$ values are calculated with the last 1000 $\tau$ simulation data as $P_H - P_L$ of all the simulations reaches equilibrium after 3000 $\tau$. The gray line is $P_H - P_L$ = 0.07; Figure S4: $P_H - P_L$ of PLD-tG1 as a function of simulation time. The upper panel shows the simulations without charge interactions (no charge); the bottom panel shows the simulations at 10 mM salt concentration; Figure S5: $P_H - P_L$ of PLD-tG2 as a function of simulation time. The upper panel shows the simulations without charge interactions (no charge); the bottom panel shows the simulations at 10 mM salt concentration; Figure S6: $P_H - P_L$ of RGG-tG1 as a function of simulation time. The upper panel shows the simulations without charge interactions (no charge); the bottom panel shows the simulations at 10 mM salt concentration; Figure S7: $P_H - P_L$ of RGG-tG2 as a function of simulation time. The upper panel shows the simulations without charge interactions (no charge); the bottom panel shows the simulations at 10 mM salt concentration; Figure S8: $P_H - P_L$ of PLD 1.0 as a function of simulation time. The upper panel shows the simulations without charge interactions (no charge); the bottom panel shows the simulations at 10mM salt concentration; Figure S9: $P_H - P_L$ of PLD 1.1 as a function of simulation time. The upper panel shows the simulations without charge interactions (no charge); the bottom panel shows the simulations at 10 mM salt concentration; Figure S10: $P_H - P_L$ of PLD 1.2 as a function of simulation time. The upper panel shows the simulations without charge interactions (no charge); the bottom panel shows the simulations at 10mM salt concentration; Figure S11: $P_H - P_L$ of PLD 1.3 as a function of simulation time. The upper panel shows the simulations without charge interactions (no charge); the bottom panel shows the simulations at 10 mM salt concentration; Figure S12: $P_H - P_L$ of PLD 1.5 as a function of simulation time. The upper panel shows the simulations without charge interactions (no charge); the bottom panel shows the simulations at 10 mM salt concentration; Figure S13: $P_H - P_L$ of RGG 1.0 as a function of simulation time. The upper panel shows the simulations without charge interactions (no charge); the bottom panel shows the simulations at 10mM salt concentration; Figure S14: $P_H - P_L$ of RGG 1.1 as a function of simulation time. The upper panel shows the simulations without charge interactions (no charge); the bottom panel shows the simulations at 10 mM salt concentration; Figure S15: $P_H - P_L$ of RGG 1.2 as a function of simulation time. The upper panel shows the simulations without charge interactions (no charge); the bottom panel shows the simulations at 10mM salt concentration; Figure S16: $P_H - P_L$ of RGG 1.3 as a function of simulation time. The upper panel shows the simulations without charge interactions (no charge); the bottom panel shows the simulations at 10 mM salt concentration; Figure S17: $P_H - P_L$ of RGG 1.5 as a function of simulation time. The upper panel shows the simulations without charge interactions (no charge); the bottom panel shows the simulations at 10 mM salt concentration; Figure S18: $P_H - P_L$ average as a function of simulation time. Every point represents the $P_H - P_L$ average value of 100 $\tau$ simulation time; Figure S19: $P_H - P_L$ of PLD (A) and RGG (B) with bead mass 1.1, 1.3, and 1.5 in different solvents changes with temperature. The $P_H$ and $P_L$ values are calculated with the last 1000 $\tau$ simulation data as $P_H - P_L$ of all the simulations reaches equilibrium after 3000 $\tau$ (see Supplementary Materials). The gray line is $P_H - P_L$ = 0.07; Figure S20:

Phase diagrams of PLD (A) and RGG (B) with $P_H$ and $P_L$ as a function of temperature. (A) Phase diagrams of PLD 1.0 and PLD 1.2 at different solvents. (B) Phase diagrams of RGG 1.0 and RGG 1.2 at different solvents. Here, $P_H$ and $P_L$ are the highest and the lowest points of FUS segment residue distribution along the $|z|$ axis; Figure S21: The z motion rate of PLD (A) and RGG (B) with bead mass 1.1, 1.3, and 1.5 in different solvents changes with temperature. The z motion rates are calculated by averaging 200 chains in the box with the last 1000 $\tau$ simulation data; Figure S22: The flux of last 100 $\tau$ on the boundary line. (A) The flux of PLD 1.0 and PLD 1.2 in different solvents changes with temperature. (B) The flux of RGG 1.0 and RGG 1.2 in different solvents changes with temperature. (C) $P_H - P_L$ of PLD 1.0 and PLD 1.2 in different solvents changes with flux. (D) $P_H - P_L$ of RGG 1.0 and RGG 1.2 in different solvents changes with flux. The fluxes are calculated by averaging the last 100 $\tau$ simulation data, and the boundary line is $|z| = 25$ nm. The gray line is $P_H - P_L = 0.07$; Figure S23: The z motion rate of PLD-tG (A) and RGG-tG (B) in different solvents changes with temperature; $P_H - P_L$ of PLD-tG (C) and RGG-tG (D) in different solvents changes with z motion rate. The z motion rates are calculated by averaging 200 chains in the box with the last 1000 $\tau$ simulation data. The gray line is $P_H - P_L = 0.07$; Figure S24: The flux on the boundary line. (A) The flux of PLD-tG in different solvents changes with temperature. (B) The flux of RGG-tG in different solvents changes with temperature. (C) $P_H - P_L$ of PLD-tG in different solvents changes with flux. (D) $P_H - P_L$ of RGG-tG in different solvents changes with flux. The fluxes are calculated by averaging the last 1000 $\tau$ simulation data, and the boundary line is $|z| = 25$ nm. The gray line is $P_H - P_L = 0.07$; Figure S25: The distribution of probability of PLD (A) and RGG (B) with bead mass 1.1 and 1.3 as a function of displacement $|z|$. Data without charges and at 10mM salt concentration solvent are shown in the up panels and the bottom panels, respectively. The probability is the average value of window along $|z|$ during the last 1000 $\tau$ simulation data; Figure S26: The distribution of probability of PLD-tG (A) and RGG-tG (B) as a function of displacement $|z|$. Data without charges and at 10 mM salt concentration solvent are shown in the up panels and the bottom panels, respectively. The probability is the average value of window along $|z|$ during the last 1000 $\tau$ simulation data; Figure S27: Head-to-tail distance (D) of PLD (A) and RGG (B) chains with bead mass 1.1, 1.3 and 1.5 as a function of temperature. The D value is calculated by the mean value of the 200 chains in the system during the last 1000 $\tau$ simulation data; Figure S28: Head-to-tail distance (D) of PLD-tG (A) and RGG-tG (B) chains as a function of temperature. The D value is calculated by the mean value of the 200 chains in the system during the last 1000 $\tau$ simulation data; Figure S29: Mean number intra-chain ($E_{intra}$) of and inter-chain ($E_{inter}$) electrostatic contacts of RGG 1.1 (A) and RGG 1.3 (B) as a function of $|z|$. Average and standard error values during the last 1000 $\tau$ simulation data are illustrated in this figure; Figure S30: Mean number intra-chain ($E_{intra}$) and inter-chain ($E_{inter}$) electrostatic contacts of RGG-tG1 (A) and RGG-tG2 (B) as a function of $|z|$. Average and standard error values during the last 1000 $\tau$ simulation data are illustrated in this figure.

**Author Contributions:** Conceptualization, C.T., J.W. and W.-T.C.; methodology, W.D., J.W. and W.-T.C.; analysis, W.D., J.W. and W.-T.C.; writing—original draft preparation, W.D., J.W. and W.-T.C.; writing—review and editing, W.D., J.W., W.-T.C., C.T. and E.W. All authors have read and agreed to the published version of the manuscript.

**Funding:** W.D., E.W., C.T. and W.-T.C. are funded by National Natural Science Foundation of China (Grant Nos. 32171244, 12234019, 21721003), Youth Innovation Promotion Association CAS (Grant No. 2020231), and Jilin Province Science and Technology Development Plan (Grant No. 20200301009RQ).

**Institutional Review Board Statement:** Not applicable.

**Informed Consent Statement:** Not applicable.

**Data Availability Statement:** Not applicable.

**Acknowledgments:** W.D. and W.-T.C. thank the Network and Computing Center, Changchun Institute of Applied Chemistry, Chinese Academy of Science and the Computing Center of Jilin Province for computational support.

**Conflicts of Interest:** The authors declare no conflict of interest.

## References


1. An, H.; de Meritens, C.R.; Shelkovnikova, T.A. Connecting the "dots": RNP granule network in health and disease. *Biochim. Biophys. Acta-Mol. Cell Res.* **2021**, *1868*, 119058. [CrossRef] [PubMed]
2. Murthy, A.C.; Fawzi, N.L. The (un)structural biology of biomolecular liquid-liquid phase separation using NMR spectroscopy. *J. Biol. Chem.* **2020**, *295*, 2375–2384. [CrossRef] [PubMed]
3. Alberti, S.; Gladfelter, A.; Mittag, T. Considerations and Challenges in Studying Liquid-Liquid Phase Separation and Biomolecular Condensates. *Cell* **2019**, *176*, 419–434. [CrossRef]
4. Sawner, A.S.; Ray, S.; Yadav, P.; Mukherjee, S.; Panigrahi, R.; Poudyal, M.; Patel, K.; Ghosh, D.; Kummerant, E.; Kumar, A.; et al. Modulating alpha-Synuclein Liquid-Liquid Phase Separation. *Biochemistry* **2021**, *60*, 3676–3696. [CrossRef]
5. Spegg, V.; Altmeyer, M. Biomolecular condensates at sites of DNA damage: More than just a phase. *DNA Repair* **2021**, *106*, 103179. [CrossRef] [PubMed]
6. Mann, J.R.; Donnelly, C.J. RNA modulates physiological and neuropathological protein phase transitions. *Neuron* **2021**, *109*, 2663–2681. [CrossRef]
7. Sehgal, P.B.; Westley, J.; Lerea, K.M.; DiSenso-Browne, S.; Etlinger, J.D. Biomolecular condensates in cell biology and virology: Phase-separated membraneless organelles (MLOs). *Anal. Biochem.* **2020**, *597*, 113691. [CrossRef]
8. Garaizar, A.; Sanchez-Burgos, I.; Collepardo-Guevara, R.; Espinosa, J.R. Expansion of Intrinsically Disordered Proteins Increases the Range of Stability of Liquid-Liquid Phase Separation. *Molecules* **2020**, *25*, 4705. [CrossRef]
9. Vodnala, M.; Choi, E.B.; Fong, Y.W. Low complexity domains, condensates, and stem cell pluripotency. *World J. Stem Cells* **2021**, *13*, 416–438. [CrossRef]
10. Reber, S.; Jutzi, D.; Lindsay, H.; Devoy, A.; Mechtersheimer, J.; Levone, B.R.; Domanski, M.; Bentmann, E.; Dormann, D.; Muhlemann, O.; et al. The phase separation-dependent FUS interactome reveals nuclear and cytoplasmic function of liquid-liquid phase separation. *Nucleic Acids Res.* **2021**, *49*, 7713–7731. [CrossRef]
11. Levone, B.R.; Lenzken, S.C.; Antonaci, M.; Maiser, A.; Rapp, A.; Conte, F.; Reber, S.; Mechtersheimer, J.; Ronchi, A.E.; Muhlemann, O.; et al. FUS-dependent liquid-liquid phase separation is important for DNA repair initiation. *J. Cell Biol.* **2021**, *220*, e202008030. [CrossRef] [PubMed]
12. Su, J.M.; Wilson, M.Z.; Samuel, C.E.; Ma, D. Formation and Function of Liquid-Like Viral Factories in Negative-Sense Single-Stranded RNA Virus Infections. *Viruses* **2021**, *13*, 126. [CrossRef] [PubMed]
13. Noda, N.N.; Wang, Z.; Zhang, H. Liquid-Liquid phase separation in autophagy. *J. Cell Biol.* **2020**, *219*, 1–13. [CrossRef] [PubMed]
14. Peng, Q.; Tan, S.M.; Xia, L.Z.; Wu, N.Y.; Oyang, L.; Tang, Y.Y.; Su, M.; Luo, X.; Wang, Y.; Sheng, X.W.; et al. Phase separation in Cancer: From the Impacts and Mechanisms to Treatment potentials. *Int. J. Biol. Sci.* **2022**, *18*, 5103–5122. [CrossRef] [PubMed]
15. Milicevic, K.; Rankovic, B.; Andjus, P.R.; Bataveljic, D.; Milovanovic, D. Emerging Roles for Phase Separation of RNA-Binding Proteins in Cellular Pathology of ALS. *Front. Cell Dev. Biol.* **2022**, *10*, 840256. [CrossRef]
16. Yoneda, R.; Ueda, N.; Kurokawa, R. m(6)A Modified Short RNA Fragments Inhibit Cytoplasmic TLS/FUS Aggregation Induced by Hyperosmotic Stress. *Int. J. Mol. Sci.* **2021**, *22*, 11014. [CrossRef] [PubMed]
17. Darling, A.L.; Shorter, J. Combating deleterious phase transitions in neurodegenerative disease. *Biochim. Biophys. Acta-Mol. Cell Res.* **2021**, *1868*, 118984. [CrossRef]
18. Farina, S.; Esposito, F.; Battistoni, M.; Biamonti, G.; Francia, S. Post-Translational Modifications Modulate Proteinopathies of TDP-43, FUS and hnRNP-A/B in Amyotrophic Lateral Sclerosis. *Front. Mol. Biosci.* **2021**, *8*, 693325. [CrossRef]
19. Hayashi, Y.; Ford, L.K.; Fioriti, L.; McGurk, L.; Zhang, M. Liquid-Liquid Phase Separation in Physiology and Pathophysiology of the Nervous System. *J. Neurosci.* **2021**, *41*, 834–844. [CrossRef]
20. Lin, Y.-H.; Song, J.; Forman-Kay, J.D.; Chan, H.S. Random-phase-approximation theory for sequence-dependent, biologically functional liquid-liquid phase separation of intrinsically disordered proteins. *J. Mol. Liq.* **2017**, *228*, 176–193. [CrossRef]
21. Uversky, V.N. Per aspera ad chaos: A personal journey to the wonderland of intrinsic disorder. *Biochem. J.* **2021**, *478*, 3015–3024. [CrossRef] [PubMed]
22. Dunker, A.K.; Lawson, J.D.; Brown, C.J.; Williams, R.M.; Romero, P.; Oh, J.S.; Oldfield, C.J.; Campen, A.M.; Ratliff, C.M.; Hipps, K.W.; et al. Intrinsically disordered protein. *J. Mol. Graph. Model.* **2001**, *19*, 26–59. [CrossRef] [PubMed]
23. Uversky, V.N.; Kuznetsova, I.M.; Turoverov, K.K.; Zaslavsky, B. Intrinsically disordered proteins as crucial constituents of cellular aqueous two phase systems and coacervates. *FEBS Lett.* **2015**, *589*, 15–22. [CrossRef] [PubMed]
24. Badaczewska-Dawid, A.E.; Uversky, V.N.; Potoyan, D.A. BIAPSS: A Comprehensive Physicochemical Analyzer of Proteins Undergoing Liquid-Liquid Phase Separation. *Int. J. Mol. Sci.* **2022**, *23*, 6204. [CrossRef] [PubMed]
25. Portz, B.; Lee, B.L.; Shorter, J. FUS and TDP-43 Phases in Health and Disease. *Trends Biochem. Sci.* **2021**, *46*, 550–563. [CrossRef] [PubMed]
26. Ishiguro, A.; Ishihama, A. Essential Roles and Risks of G-Quadruplex Regulation: Recognition Targets of ALS-Linked TDP-43 and FUS. *Front. Mol. Biosci.* **2022**, *9*, 957502. [CrossRef] [PubMed]
27. Kang, J.; Lim, L.Z.; Song, J.X. ATP enhances at low concentrations but dissolves at high concentrations liquid-liquid phase separation (LLPS) of ALS/FTD-causing FUS. *Biochem. Biophys. Res. Commun.* **2018**, *504*, 545–551. [CrossRef]
28. Aida, H.; Shigeta, Y.; Harada, R. The role of ATP in solubilizing RNA-binding protein fused in sarcoma. *Proteins* **2022**, *90*, 1606–1612. [CrossRef]

29. Lao, Z.; Dong, X.; Liu, X.; Li, F.; Chen, Y.; Tang, Y.; Wei, G. Insights into the Atomistic Mechanisms of Phosphorylation in Disrupting Liquid-Liquid Phase Separation and Aggregation of the FUS Low-Complexity Domain. *J. Chem. Inf. Model* **2022**, *62*, 3227–3238. [CrossRef]
30. Bock, A.S.; Murthy, A.C.; Tang, W.S.; Jovic, N.; Shewmaker, F.; Mittal, J.; Fawzi, N.L. N-terminal acetylation modestly enhances phase separation and reduces aggregation of the low-complexity domain of RNA-binding protein fused in sarcoma. *Protein Sci.* **2021**, *30*, 1337–1349. [CrossRef]
31. Yoshizawa, T.; Ali, R.; Jiou, J.; Fung, H.Y.J.; Burke, K.A.; Kim, S.J.; Lin, Y.; Peeples, W.B.; Saltzberg, D.; Soniat, M.; et al. Nuclear Import Receptor Inhibits Phase Separation of FUS through Binding to Multiple Sites. *Cell* **2018**, *173*, 693–705.e22. [CrossRef]
32. Kaur, T.; Alshareedah, I.; Wang, W.; Ngo, J.; Moosa, M.M.; Banerjee, P.R. Molecular Crowding Tunes Material States of Ribonucleoprotein Condensates. *Biomolecules* **2019**, *9*, 71. [CrossRef] [PubMed]
33. Dong, X.; Bera, S.; Qiao, Q.; Tang, Y.; Lao, Z.; Luo, Y.; Gazit, E.; Wei, G. Liquid-Liquid Phase Separation of Tau Protein Is Encoded at the Monomeric Level. *J. Phys. Chem. Lett.* **2021**, *12*, 2576–2586. [CrossRef]
34. Dignon, G.L.; Zheng, W.; Kim, Y.C.; Mittal, J. Temperature-Controlled Liquid-Liquid Phase Separation of Disordered Proteins. *ACS Cent. Sci.* **2019**, *5*, 821–830. [CrossRef] [PubMed]
35. Maity, H.; Baidya, L.; Reddy, G. Salt-Induced Transitions in the Conformational Ensembles of Intrinsically Disordered Proteins. *J. Phys. Chem. B* **2022**, *126*, 5959–5971. [CrossRef] [PubMed]
36. Li, S.J.; Yoshizawa, T.; Yamazaki, R.; Fujiwara, A.; Kameda, T.; Kitahara, R. Pressure and Temperature Phase Diagram for Liquid-Liquid Phase Separation of the RNA-Binding Protein Fused in Sarcoma. *J. Phys. Chem. B* **2021**, *125*, 6821–6829. [CrossRef]
37. Kitahara, R.; Yamazaki, R.; Ide, F.; Li, S.J.; Shiramasa, Y.; Sasahara, N.; Yoshizawa, T. Pressure-Jump Kinetics of Liquid-Liquid Phase Separation: Comparison of Two Different Condensed Phases of the RNA-Binding Protein, Fused in Sarcoma. *J. Am. Chem. Soc.* **2021**, *143*, 19697–19702. [CrossRef]
38. Burke, K.A.; Janke, A.M.; Rhine, C.L.; Fawzi, N.L. Residue-by-Residue View of In Vitro FUS Granules that Bind the C-Terminal Domain of RNA Polymerase II. *Mol. Cell* **2015**, *60*, 231–241. [CrossRef]
39. Bari, K.J.; Prakashchand, D.D. Fundamental Challenges and Outlook in Simulating Liquid-Liquid Phase Separation of Intrinsically Disordered Proteins. *J. Phys. Chem. Lett.* **2021**, *12*, 1644–1656. [CrossRef]
40. Zheng, W.; Dignon, G.L.; Jovic, N.; Xu, X.; Regy, R.M.; Fawzi, N.L.; Kim, Y.C.; Best, R.B.; Mittal, J. Molecular Details of Protein Condensates Probed by Microsecond Long Atomistic Simulations. *J. Phys. Chem. B* **2020**, *124*, 11671–11679. [CrossRef]
41. Dignon, G.L.; Zheng, W.; Kim, Y.C.; Best, R.B.; Mittal, J. Sequence determinants of protein phase behavior from a coarse-grained model. *PLoS Comput. Biol.* **2018**, *14*, e1005941. [CrossRef] [PubMed]
42. Dignon, G.L.; Zheng, W.; Best, R.B.; Kim, Y.C.; Mittal, J. Relation between single-molecule properties and phase behavior of intrinsically disordered proteins. *Proc. Natl. Acad Sci. USA* **2018**, *115*, 9929–9934. [CrossRef] [PubMed]
43. Dannenhoffer-Lafage, T.; Best, R.B. A Data-Driven Hydrophobicity Scale for Predicting Liquid-Liquid Phase Separation of Proteins. *J. Phys. Chem. B* **2021**, *125*, 4046–4056. [CrossRef]
44. Lin, X.; Kulkarni, P.; Bocci, F.; Schafer, N.P.; Roy, S.; Tsai, M.Y.; He, Y.; Chen, Y.; Rajagopalan, K.; Mooney, S.M.; et al. Structural and Dynamical Order of a Disordered Protein: Molecular Insights into Conformational Switching of PAGE4 at the Systems Level. *Biomolecules* **2019**, *9*, 77. [CrossRef] [PubMed]
45. Lenard, A.J.; Zhou, Q.; Madreiter-Sokolowski, C.; Bourgeois, B.; Habacher, H.; Khanna, Y.; Madl, T. EGCG Promotes FUS Condensate Formation in a Methylation-Dependent Manner. *Cells* **2022**, *11*, 592. [CrossRef]
46. Noel, J.K.; Whitford, P.C.; Sanbonmatsu, K.Y.; Onuchic, J.N. SMOG@ctbp: Simplified deployment of structure-based models in GROMACS. *Nucleic Acids Res.* **2010**, *38*, W657–W661. [CrossRef] [PubMed]
47. Das, S.; Muthukumar, M. Microstructural Organization in α-Synuclein Solutions. *Macromolecules* **2022**, *55*, 4228–4236. [CrossRef]
48. Chu, W.T.; Wang, J. Thermodynamic and sequential characteristics of phase separation and droplet formation for an intrinsically disordered region/protein ensemble. *PLoS Comput. Biol.* **2021**, *17*, e1008672. [CrossRef]
49. Hazra, M.K.; Levy, Y. Charge pattern affects the structure and dynamics of polyampholyte condensates. *Phys. Chem. Chem. Phys.* **2020**, *22*, 19368–19375. [CrossRef]
50. Silmore, K.S.; Howard, M.P.; Panagiotopoulos, A.Z. Vapour-liquid phase equilibrium and surface tension of fully flexible Lennard-Jones chains. *Mol. Phys.* **2017**, *115*, 320–327. [CrossRef]
51. Kouza, M.; Li, M.S.; O'Brien, E.P.J.; Hu, C.K.; Thirumalai, D. Effect of finite size on cooperativity and rates of protein folding. *J. Phys. Chem. A* **2006**, *110*, 671–676. [CrossRef] [PubMed]
52. Jackson, J.; Nguyen, K.; Whitford, P.C. Exploring the balance between folding and functional dynamics in proteins and RNA. *Int. J. Mol. Sci.* **2015**, *16*, 6868–6889. [CrossRef] [PubMed]

*biomolecules*

MDPI

*Article*

# Linker Length Drives Heterogeneity of Multivalent Complexes of Hub Protein LC8 and Transcription Factor ASCIZ

**Douglas R. Walker [1], Kayla A. Jara [1], Amber D. Rolland [2,3], Coban Brooks [1], Wendy Hare [1], Andrew K. Swansiger [2], Patrick N. Reardon [1,4], James S. Prell [2,5] and Elisar J. Barbar [1,*]**

1 Department of Biochemistry and Biophysics, Oregon State University, Corvallis, OR 97331, USA
2 Department of Chemistry and Biochemistry, University of Oregon, Eugene, OR 97403, USA
3 Institute of Molecular Biology, University of Oregon, Eugene, OR 97403, USA
4 NMR Facility, Oregon State University, Corvallis, OR 97331, USA
5 Materials Science Institute, University of Oregon, Eugene, OR 97403, USA
* Correspondence: barbare@oregonstate.edu

**Abstract:** LC8, a ubiquitous and highly conserved hub protein, binds over 100 proteins involved in numerous cellular functions, including cell death, signaling, tumor suppression, and viral infection. LC8 binds intrinsically disordered proteins (IDPs), and although several of these contain multiple LC8 binding motifs, the effects of multivalency on complex formation are unclear. Drosophila ASCIZ has seven motifs that vary in sequence and inter-motif linker lengths, especially within subdomain QT2–4 containing the second, third, and fourth LC8 motifs. Using isothermal-titration calorimetry, analytical-ultracentrifugation, and native mass-spectrometry of QT2–4 variants, with methodically deactivated motifs, we show that inter-motif spacing and specific motif sequences combine to control binding affinity and compositional heterogeneity of multivalent duplexes. A short linker separating strong and weak motifs results in stable duplexes but forms off-register structures at high LC8 concentrations. Contrastingly, long linkers engender lower cooperativity and heterogeneous complexation at low LC8 concentrations. Accordingly, two-mers, rather than the expected three-mers, dominate negative-stain electron-microscopy images of QT2–4. Comparing variants containing weak-strong and strong-strong motif combinations demonstrates sequence also regulates IDP/LC8 assembly. The observed trends persist for trivalent ASCIZ subdomains: QT2–4, with long and short linkers, forms heterogeneous complexes, whereas QT4–6, with similar mid-length linkers, forms homogeneous complexes. Implications of linker length variations for function are discussed.

**Keywords:** hub proteins; intrinsic disorder; multivalency; transcription factor; linker length; heterogeneity; dimers; duplexes

**Citation:** Walker, D.R.; Jara, K.A.; Rolland, A.D.; Brooks, C.; Hare, W.; Swansiger, A.K.; Reardon, P.N.; Prell, J.S.; Barbar, E.J. Linker Length Drives Heterogeneity of Multivalent Complexes of Hub Protein LC8 and Transcription Factor ASCIZ. *Biomolecules* **2023**, *13*, 404. https://doi.org/10.3390/biom13030404

Academic Editors: Prakash Kulkarni, Stefania Brocca, Keith Dunker and Sonia Longhi

Received: 18 January 2023
Revised: 8 February 2023
Accepted: 18 February 2023
Published: 21 February 2023



## 1. Introduction

Hub proteins, which interact with many different proteins in an organism, gained recognition around the turn of the century as highly important and often essential parts of an organism's proteome [1]. Jeong et al. showed in Saccharomyces cerevisiae that 0.7% of proteins interact with 15 or more other proteins, but a single deletion in 62% of these proved to be lethal, three times more than for proteins with a small number of protein partners. Hub proteins can be subdivided according to their structure and their clients. Intrinsic disorder plays a major role in enabling flexibility and promiscuity in hub binding [2]. Thus, hub proteins can be organized into three broad classes: (1) completely disordered interacting with ordered proteins, (2) partially disordered interacting with ordered proteins, and (3) fully ordered interacting with intrinsically disordered proteins. The third class of hub proteins often induces folding of a short linear binding motif in their partner proteins upon binding. Proteins that fit into this class of hub protein include calmodulin, actin, Cdk2, 14-3-3 [2], RCD1-RST [3], Keap1 [4], and LC8 [5]. Due to the structural flexibility of the binding groove of class three hubs and the variability allowed in the partner sequence, they tend to interact with a greater number and wider variety of proteins than those in other classes.

Recognition of the prevalence of intrinsically disordered proteins and protein regions (IDPs/IDRs) and their roles as biologically active proteins has rapidly grown [6]. IDPs and IDRs are characterized by low sequence diversity, lack of hydrophobic residues, abundance of charged residues, and areas of sequence repeats. Due in part to their high number of charged residues, as well as their abundance of short linear binding motifs, disordered regions are promiscuous in their binding interactions and facilitate the formation of many complex large protein assemblies [7]. IDPs/IDRs are also extremely functionally diverse, and in addition to their structural plasticity and dynamic conformational flexibility, they often interact with their binding partners multivalently.

Because they interact with IDPs, it is not uncommon that class three hubs will interact with their partners multivalently. An actin filament in red cell membranes will bind to between five and seven 4.1R proteins [8]. Keap1 binds at two locations on NRF2 to facilitate ubiquitination [4]. The C-terminal domain of calmodulin binds melittin in the absence of $Ca^{2+}$, but upon the addition of $Ca^{2+}$, the N-terminal domain also binds [9]. LC8 binds many of its protein partners multivalently (Figure 1A); however, it is unique in its large number of both multivalent partners and binding motifs on a single protein; for instance, ASCIZ contains 11–16 LC8 recognition sites [10] (Figures 1C and 2A). Compared to monovalent interactions, in which ligands bind a partner at a single site, multivalent interactions involve linked associations of ligands binding to multiple sites [11–13]. Multivalent IDP assemblies are considered to belong to one of the following groups: binary complexes, IDP single-chain scaffolds, IDP duplex scaffolds, higher-order IDP associations, and IDP multi-site collective binding ligands [13]. LC8, the focus of this work, folds as a homodimer and assembles IDPs into duplex scaffolds which are composed of two IDP chains connected by one or more bivalent partners with two symmetrical binding sites and/or by self-association interactions within the chain [12,13]. Cases in which the same dimeric ligand binds multiple sites across disordered chains are common for partners of LC8 (Figure 1) [13–15].

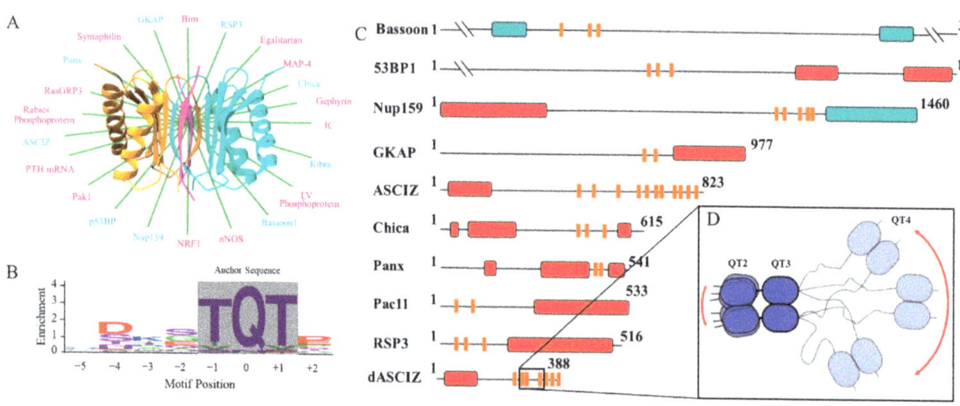


**Figure 1.** LC8 hub, binding motif, and multivalent partners. (**A**) Ribbon diagram of the LC8 dimer showing each monomer (orange and cyan) bound to disordered peptides (magenta) that adopt β-strand structure upon binding in LC8's binding groove (Protein Data Bank code 2P2T, from *D. melanogaster*). Ribbon diagram is overlayed on a star display of a selection of LC8 partner proteins. Red font denotes monovalent partners, while blue denotes multivalent partners. (**B**) Amino acid enrichment for each position in the LC8 binding motif. The TQT anchor is boxed in gray. (**C**) Multivalent LC8 binding partners. Sequence-based predictions of order (red boxes), disorder (black lines), coiled-coil (blue boxes), and LC8 binding motifs (orange bars) are shown. PSIPRED [16] was used to predict order and disorder. Paracoil2 [17] was used to predict coiled-coils. LC8 binding motif locations are shown for Bassoon [18], 53BP1 [19], NUP159 [20], GKAP [21], ASCIZ [22],

Chica [23], Panx [24], Pac11 [25], RSP3 [26], and dASCIZ [27]. Panel adapted from Clark et al. [28]. (**D**) Zoom of QT2–4 from dASCIZ showing the effect of a long linker on flexibility. Panel adapted from Reardon et al. [29].

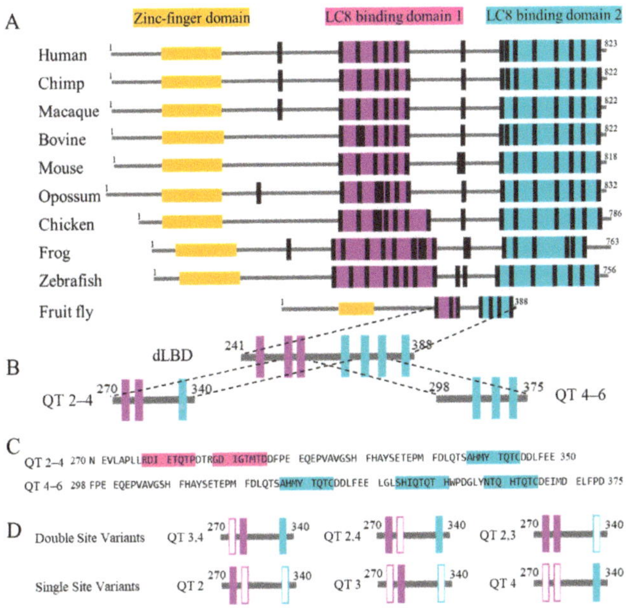


**Figure 2.** Alignment of ASCIZ homologs' domain architecture and dASCIZ constructs. (**A**) Comparison of 10 ASCIZ homologs, aligned to show the similarity of the LBDs and the linker connecting LBD1 to LBD2. (**B**) Drosophila ASCIZ LC8 binding region and the constructs utilized in this study, including QT2–4 and QT4–6. (**C**) Sequences of QT2–4 and QT4–6. (**D**) Variants that systematically abolish either one (**top**) or two (**bottom**) LC8 recognition motifs from QT2–4 through mutation of the anchor triplet to AAA. Hollow boxes indicate sites that have been mutated. Construct nomenclature denotes the binding sites left intact in each construct.

Within the LC8 homodimer, two symmetrical binding grooves allow LC8 to duplex its intrinsically disordered binding partners (Figure 1A) [11,13,30]. LC8 is an essential protein in animal proteomes [31–33] and is confirmed to partner with more than 100 different client IDPs; among these are IDPs performing functions such as intracellular transport [34,35], nuclear pore formation [20], viral interactions [36–38], tumor suppression [39], and transcription [10,27,29,40]. LC8 partner proteins share a short (eight amino acid) linear recognition motif that mediates binding to LC8 [5,15,37]. The binding motif allows some variation; however, it is typically anchored by a threonine-glutamine-threonine (TQT) sequence (Figure 1B) [28]. Although it is common for LC8 partners to contain multiple LC8 binding motifs, one unique example is its own transcription factor, ASCIZ (ATMIN-Substrate Chk-Interacting $Zn^{2+}$ finger) [40,41], which contains an astonishing eleven experimentally verified LC8 recognition motifs within its human homolog [10].

Importantly, in vivo and biophysical studies have characterized ASCIZ as a transcription factor and concentration regulator of LC8 [27,42,43]. ASCIZ is thought to act as a concentration sensor that fine-tunes LC8 transcription by interacting with LC8 via a dynamic ensemble of unsaturated bound complexes [10]. Unlike Nup159 (containing five LC8 binding sites) from yeast, which forms rigid stacked complexes, ASCIZ instead forms heterogeneous complexes as visualized by negative-stain EM analysis [10]. Drosophila (with seven LC8 sites) and human ASCIZ studies show that ASCIZ/LC8 interactions display both positive and negative cooperativity to enable this heterogeneous complexation. Such heterogeneity may be due to the disordered linkers between LC8 binding sites in ASCIZ that vary considerably in length (Figure 2). LC8Pred [15] predicts sixteen LC8 binding sites within human ASCIZ (five more than have been experimentally verified [10]). As shown in Figure 2, these binding sites can be roughly grouped into two LC8 binding domains (LBDs), with a few additional sites flanked by extensive linker regions. This trend holds true across all investigated homologs, even that from Drosophila, which contains the shortest linker between adjacent LBDs at thirty amino acids in length. This conservation of mixed long and short intra-motif spacing suggests a functional purpose to promote dynamic complexation and enable LC8 concentration sensing.

A multivalent subdomain of Drosophila ASCIZ, QT2–4, which contains the second, third, and fourth sequential LC8 binding sites, serves as a model system to probe both the highest variety in intra-site linker length and highest variability in LC8 motif strength within dASCIZ (Figure 2). Unique among experimentally verified LC8 motifs, Drosophila ASCIZ possesses an LC8 binding motif containing a TMT (QT3) rather than the canonical TQT anchor (Figure 2). Our recent studies utilizing QT2–4 provided the first evidence of in-register binding during LC8/IDP complex assembly and suggested that linker length contributes to modulating the flexibility and LC8 occupancy in multivalent LC8/IDP complexes in general [29]. The work presented here utilizes single- and double-site variants of QT2–4 to investigate the interplay of linker length and motif specificity in the regulation of dynamic, multivalent LC8 complexes. Experiments comparing the biophysical analysis of QT2–4 to QT4–6, which contains mid-length linkers, confirmed the conclusions from the variants' study.

## 2. Materials and Methods

### 2.1. Cloning, Protein Expression, and Purification

Cloning of Drosophila ASCIZ QT2–4 (residues 271–341) with various mutations of recognition motifs was performed using either the QuikChange Lightening mutagenesis kit (Agilent) or the Q5 site-directed mutagenesis kit (New England Biolabs). The resulting constructs verified by sequencing are QT2, QT3, QT4, QT2,3, QT2,4, and QT3,4, where the number(s) following 'QT' indicate which LC8 recognition motif(s) remain and have not been mutated to AAA. Drosophila LC8 and ASCIZ proteins were expressed and purified according to previously published protocols [10,29]. Briefly, constructs were expressed in frame with a hexahistidine tag, Protein A solubility tag (for ASCIZ constructs), and a cleavage site for the tobacco etch virus (TEV) enzyme. All constructs were transformed into *Escherichia coli* Rosetta (DE3) cells (Merck KGaA, Darmstadt, Germany) and expressed at 37 °C in LB or ZYM-5052 autoinduction media. Recombinant protein expression was induced with 0.4 mM IPTG (for LB cultures) and growth continued at 26 °C for 16 h. Cells were harvested and purified under either regular (LC8) or denaturing (ASCIZ constructs) conditions using the TALON His-Tag purification protocol (Clontech). The solubility tag and/or hexahistadine tag were cleaved by TEV protease and the proteins were further purified using strong anion exchange chromatography (Bio-Rad, Hercules, California) followed by gel filtration on a SuperdexTM 75 gel filtration column (GE Health). Purity was assessed by SDS-polyacrylamide gels.

### 2.2. Isothermal Titration Calorimetry

Binding thermodynamics of the QT/LC8 interactions were obtained with a MicroCal VP-ITC microcalorimeter (Malvern Panalytical, Malvern, UK). All experiments were obtained at 25 °C and with protein samples in a buffer composed of 50 mM sodium phosphate, 50 mM sodium chloride, 1 mM sodium azide, 5 mM β-mercaptoethanol, pH 7.5. Each experiment was started with a 2 µL injection, followed by 27 to 33 injections of 10 µL. Experiments were conducted with QT variants in the sample cell at concentrations ranging from 20–50 µM and LC8 in the syringe at concentrations ranging from 400–500 µM. Experiments for the single site constructs resulted in calculated Brandt parameter values (c values) of 5.4, 1.4, and 3.2 for QT2, QT3, and QT4, respectively, indicating that the thermodynamic parameters for each interaction are almost out of an acceptable range for reliability. The data were processed using Origin 7.0 and fit to a simple, single set of sites binding model; however, these systems are more complicated because LC8 is a dimer binding two IDP chains. Data for the double site variants were also fit using the sequential binding sites and two sets of sites models to address the failure of the single set of sites model in satisfactorily representing the stoichiometry of binding. The reported data are from two independent experiments. In all cases, the data were reproducible. The reported concentrations are expected to have a 5–10% uncertainty in protein concentrations that were determined by absorbance measurement at 280 nm.

### 2.3. Size Exclusion Chromatography Multiangle Light Scattering

SEC-MALS was performed using a GE Healthcare AKTA FPLC with a Wyatt Technology DAWN multiple-angle light scattering and Optilab refractive index system. Experiments were performed on a GE life sciences Superdex200 10/300 GL column at room temperature equilibrated with 50 mM sodium phosphate, 0.4 M NaCl, 1 mM $NaN_3$, 5 mM β-mercaptoethanol, pH 7.5 buffer at a flow rate of 0.75 mL/min. QT2–4 and QT4–6 were both prepared at 90 μM and mixed with LC8 at 300 μM resulting in a 1:3.3 ratio. Data were analyzed with Wyatt Technology ASTRA software package, version 8.

### 2.4. Analytical Ultracentrifugation

SV-AUC was performed using a Beckman Coulter Optima XL-A analytical ultracentrifuge equipped with absorbance optics. LC8 was mixed with each double site variant at ratios of 1:1, 1:2, 1:3, 1:4, 1:5, and 1:6 (molar ratio of QT:LC8). Solutions were prepared with 60 μM QT construct protein concentration. Buffer conditions for SV-AUC analysis were 20 mM Tris-HCl, 50 mM NaCl, 5 mM Tris(2-carboxyethyl)phosphine, 1 mM sodium azide, pH 7.5. The complexes were loaded into standard, 12-mm pathlength, two-channel sectored centerpieces and centrifuged at 42,000 rpm and 20 °C. A total of 300 scans were acquired with no interscan delay. The wavelength used to measure each set of experiments was chosen such that the absorbance of the sample at the given wavelength, between 280 and 302 nm, was in the 0.6–1.1 range. The data were fit to a c(S) distribution using the software SEDFIT [44]. Buffer density was calculated to be 1.0009 g/mL using Sednterp [45].

### 2.5. Native Electrospray Ionization Mass Spectrometry (Native ESI-MS)

All native mass spectra were acquired as previously described using a Waters Synapt G2-Si mass spectrometer equipped with a nanoelectrospray ionization source [29]. The instrumental settings used were as follows: source at ambient temperature, sample cone collision energy of 25 V, trap collision energy of 25 V, transfer collision energy of 5 V, and trap gas flow rate of 7–7.5 mL/min. Spectra shown represent the summation of data scans acquired over a period of 5 min. A native mass spectrum was acquired for each individual protein sample at 25 μM and used to determine accurate monomer masses. Complexes were formed by mixing LC8 with each QT2–4 mutant at a 2:1 LC8:QT molar ratio to achieve a final total protein concentration of 25 μM. After allowing complex formation to occur overnight at 4 °C, native mass spectra were acquired for each 25 μM sample, as well as for a dilution series of each at total protein concentrations of 15 μM, 10 μM, 5 μM, 1 μM, and 500 nM in 200 mM ammonium acetate at pH 7.4. After peaks in the native mass spectra were assigned, the areas of each peak were integrated with IGOR Pro 9. The summed area of each species' various charge state peaks was used to determine relative abundances, which were then normalized to the LC8 dimer abundance for each spectrum.

## 3. Results

### 3.1. Interactions of QT2–4 Single Site Variants with LC8

We created three variants (QT2, QT3, and QT4) in which two out of three native LC8 recognition motifs in the QT2–4 construct were abolished by replacing the three TQT anchor residues with AAA so that each binding site could be studied individually while maintaining the context of the longer, disordered chain (Figure 2D). ITC at 25 °C was used to characterize the thermodynamics of these variants' interaction with LC8. All single-site variant isotherms were fit using Origin's "single set of sites" (SSS) model. As shown in Table S1, ITC experiments of QT2 (Figure 3A) and QT4 (Figure 3C) with LC8 yield modest dissociation constants ($K_d$) of 9.3 μM and 15 μM, respectively. The similarity of these dissociation constants is expected as the QT2 and QT4 LC8 binding sites share the canonical TQT motif anchor, and the slight affinity preference for the QT2 site supports previous results that indicate this site as the first to bind in the context of QT2–4 [29]. As expected, the interaction of QT3 with LC8 (Figure 3B) yields a much weaker binding affinity ($K_d$ of 36 μM), supporting previous data on short peptides [10]. Interestingly, the ΔH and TΔS values for QT4 (−16.1 and −9.5 kcal/mol) are significantly different from those of QT2 (−10.7 and −3.9 kcal/mol) despite both containing the TQT anchoring motif. This suggests that the composition of the motif outside of the TQT anchor and/or the distance the binding site lies from the closest terminus, 9 versus 15 residues for QT4 and QT2, respectively, impact the thermodynamics of LC8 binding. For all three variants, the ΔG values are between −6 and −7 kcal/mol (Table S1). It is worth noting that these fits ignore the context of LC8-driven duplex formation and only represent a model in which the variant is already duplexed.

### 3.2. *Interactions of QT2–4 Double Site Variants with LC8*

ITC experiments measuring interactions of the double site QT2–4 variants (QT2,3, QT3,4, and QT2,4) (Figure 2D) with LC8 illustrate how pairs of LC8 binding motifs interact to stabilize the duplex formed (Figure 3D–F). Since all three isotherms display a single binding step, we first modeled the binding events using SSS (Table S1). These fits produce results that indicate that the overall binding of each protein is improved compared to the single-site variants, with lower dissociation constants and more negative free energy values. However, the N values associated with these fits are a poor representation of the reality of the complex formed.

The failure of SSS to accurately fit the N expected from these isotherms, particularly the QT2,4 and QT3,4 proteins (even the intact construct, QT2–4, although to a lesser degree) indicates that the shape of the isotherm contains multiple, convoluted sigmoidal curves. This raises the possibility that the sites in each of the double-site variants might be interacting with LC8 completely independently from one another and that the apparent increase in binding strength of each protein could simply be due to the higher concentration of LC8 binding sites in the double-site variants than exist in the single-site variants. To investigate this, we used the thermodynamic values obtained from the isotherms of the single-site variants to simulate the expected isotherms if the two sites involved in each double-site variant interact with LC8 completely independently of one another (Figure 3G–I). None of the simulated isotherms match the experimental isotherms of the double-site variants; rather, each simulated isotherm indicates weaker binding than is seen by experiment, indicating that for each of the variants, the two intact sites bind cooperatively. Due to the low c-values of the motifs in the simulated isotherms at the conditions used for the experimental isotherms at ~2.0, 1.3, and 0.56, a high degree of expected uncertainties precludes more detailed analysis.

After confirming that the isotherms of the double-site variants are not representative of completely independent LC8 binding sites, we then fit the isotherms using Origin's "sequential binding sites" (SBS) and "two sets of sites" (TSS) models (Table S1). SBS represents the system using a given number of binding sites that always bind to the partner in the same order. For our system, the QT2,3 protein is fit as if site 2 always binds before site 3. This assumption is reasonable because the $K_d$ values of each site differ by a factor of at least 5 [46]. SBS is an imperfect fit for this system, for the same reasons previously mentioned for the fit of SSS to the single-site variants, but additionally, because concentrations may vary from measurement by up to 10% and true $K_d$ values may not differ enough that the binding can be realistically expected to follow a strict progression. However, due to the presence of two disparate sites on these proteins, SBS is more appropriate than SSS and produces values that can be roughly compared to one another. TSS does not assume a binding order, and it is able to slightly adjust for concentration errors by varying the N value associated with each binding site. However, the model still must assume the QT2–4 variants are already duplexed. Additionally, because this model utilizes and reports so many parameters, the fits inherently possess higher error in each value than for the other fitting methods, as all of the terms will co-vary.

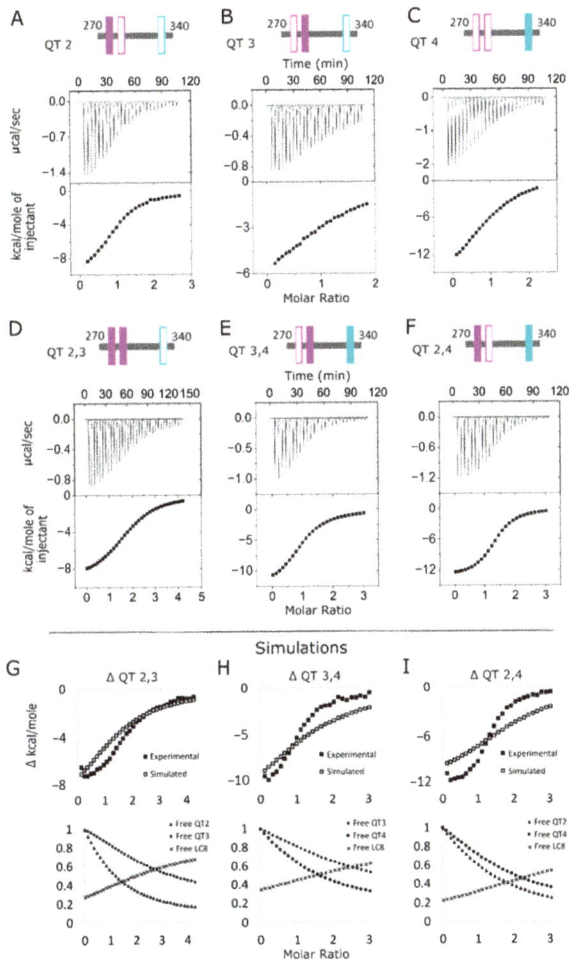


**Figure 3.** LC8-ASCIZ interactions monitored by ITC. (**A–F**) Representative thermograms of the titration of LC8 into QT2–4 variants corresponding to the single-site constructs QT2 (**A**), QT3 (**B**), and QT4 (**C**), and the double site constructs QT2,3 (**D**), QT3,4 (**E**), and QT2,4 (**F**). (**G–I**) Simulated isotherms overlaid on experimental isotherms for each of the double-site variants: QT2,3 (**G**), QT3,4 (**H**), and QT2,4 (**I**). Isotherms were simulated using ΔH and $K_d$ values obtained from single-site isotherms. Fractions of free sites at each point in the simulated isotherms are shown below each, respectively.

While it is true that the imperfection of the suitability of the models to our system (as illustrated in Figure S1) means that we cannot compare the precise values of the fits, we can compare the relative magnitude of the values in question. The resultant thermodynamic values map nicely to the expectation that binding affinity is increased for each double-site variant above what would be expected from independent sites. In both models, in 5 of the

6 sites in the double-site variants, the $K_d$ is reduced compared to the same binding site in its respective single-site variant, ranging from a factor of 2 up to a factor of ~100 times improvement. The one site that breaks this pattern is QT3 in QT3,4, which exhibits no change in affinity. Subtle in the SBS fits, but made much more obvious in the fits to TSS, is the fact that these isotherms contain relatively little information about the second binding site in each double-site variant. While this is especially true for QT3,4 (with SBS $K_d$ = 40 ± 9 μM and TSS $K_d$ = 25.7 ± 13.5 μM), the large error values for N and ΔH in each weaker binding site in the TSS fits are indicative thereof. The N value of 0.2 ± 3.5 for QT3 in QT3,4 indicates incomplete binding, the N value of 0.6 ± 0.6 for QT4 in QT2,4 may indicate incomplete binding or a concentration adjustment, and the other four N values likely are only different from 1 to adjust for concentration errors. Strikingly, the $K_d$ values are mostly consistent between SBS and TSS, with the one obvious deviation from this being QT4 in QT2,4. SBS indicates that the context of being coupled with QT2 increases the affinity of QT4 by 5-fold, whereas TSS indicates that it is barely stabilized at all by this context. According to SBS, $K_d$ stabilization trends with the identity of the anchoring motif, in which stronger TQT motifs are strongly stabilized in the context of double-site variants, whereas the weaker TMT anchored motif is stabilized weakly or not at all in this context. TSS trends moreso with the linker length, in which a long linker stabilizes the stronger binding site greatly (10–20 fold) and the weaker binding site modestly (1.3–1.5 fold), while a short linker results in stabilizing the stronger site slightly less (~7 fold) and the weaker site slightly more so (~2 fold), for a more balanced interaction. While neither SBS nor TSS adequately models the double-site variants, analyzing the results of both provides the most complete picture of the binding events in this system as attainable by ITC.

As a reminder, SBS is a reasonable approach if the $K_d$ values involved are different by at least 5-fold. The $K_d$ values for LC8 binding to QT2–4 at QT2 and QT4 are not different enough for the model assumptions to be reasonable; however, because QT2–4 contains three binding sites, the other models cannot assess this system at all. Thus, the values derived must be considered cautiously. With this in mind, the values fit to the QT2–4 isotherm (reanalyzed for the purpose of this discussion [10]) may reveal interesting insights into this complex. For instance, the addition of a third binding site reduces the affinity of each binding site in relation to the double-site variants (barring QT3 in QT3,4 and possibly QT4 in QT2,4). Because this effect is seen from each of the double-site variants in comparison to the intact three-site protein indicates that multiple factors play into this property. Comparison of both variants with long linkers between intact binding sites (QT2,4 and QT3,4) to the QT2–4 construct suggests that the inclusion of a weak motif between two relatively strong motifs results in steric pressure on the duplex when the third LC8 attempts to intercalate between the other two sites that are already bound. Contrastingly, when QT2,3 is compared to QT2–4, the addition of site 4 introduces a long linker into the context of binding, and this linker results in a reduction of the affinity of both QT2 and QT3. This suggests that the long linker also contributes to the negative cooperativity and additional heterogeneity of LC8 binding to ASCIZ. The evidence of multiple sources of heterogeneity is particularly interesting when considering the role ASCIZ plays in sensing and regulating the cellular concentration of LC8. It follows that the various contributors to allostery in ASCIZ binding LC8 allow ASCIZ to experience a wider variety of bound states in response to a broad range of LC8 concentrations, an important feature for a quality cellular concentration sensor.

### 3.3. *Complex Formation Monitored by Sedimentation Velocity Analytical Ultracentrifugation (SV-AUC)*

To further investigate the heterogeneity of complexes formed between the QT2–4 double-site variants and LC8, we used SV-AUC to track QT/LC8 complex assembly. For these experiments, peaks indicate the presence of LC8, whether alone or in complex, because the extinction coefficient of QT2–4 at 280 nm is too small to be measured. SV-AUC analysis of the double site constructs in complex with LC8 show that the proteins are in dynamic equilibrium at molar ratios of QT:LC8 up to 1:5 for QT3,4 (Figure 4A), 1:3 for QT2,4 (Figure 4B), and above 1:6 for QT2,3 (Figure 4C) and that the complexes formed by each variant at each LC8 ratio vary from one another in their sedimentation coefficients. Complexes formed with QT3,4 have sedimentation coefficients of ~2.5, ~3.8, ~3.9, ~4.1, and ~4.25 S at QT:LC8 ratios of 1:1, 1:2, 1:3, 1:4, and 1:5, respectively (Figure 4A). Complexes formed with QT2,4 have sedimentation coefficients of ~2.8, ~3.9, and ~4.1 S at ratios of 1:1, 1:2, and 1:3 (Figure 4B). Finally, complexes formed with QT2,3 have sedimentation coefficients that increase approximately linearly from ~3.3 to ~4.3 S along the measured ratios (Figure 4C) and, based on the trend, may continue to grow at higher ratios.

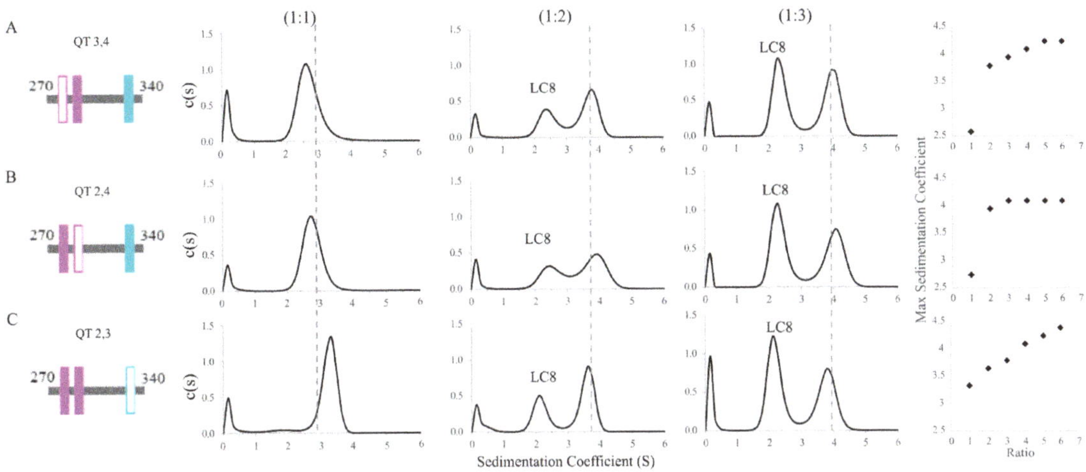


**Figure 4.** Sedimentation velocity analytical ultracentrifugation of double site ASCIZ constructs bound to LC8. SV-AUC titrations of QT3,4 (**A**), QT2,4 (**B**), and QT2,3 (**C**) with LC8 at three separate molar ratios of QT:LC8 (1:1, 1:2, 1:3) and a plot of LC8:QT ratio vs. complex sedimentation coefficient up to a ratio of 1:6. When applicable, populations corresponding to free LC8 are labeled. The dashed lines correspond to the location of the peak seen in QT2–4 at the same ratio.

The sizeable shift in sedimentation coefficient for QT3,4 and QT2,4 complexes between the ratios of 1:1 and 1:2 is indicative of the convolution of complex with free LC8. Knowing that free LC8 has a peak at $S \approx 2$, the complex peaks in QT3,4 and QT2,4 are likely close to $S = 3.5$. QT2,3 at 1:1, however, has almost no free LC8 peak, which indicates nearly complete binding of the available LC8. These values are consistent with the shifts seen for each of the subsequent titration points, which indicate the equilibrium of the mixture moving toward a fully bound 2:1 (LC8:QT) complex. While QT2,3 is the most efficient at binding LC8 early in the titration, QT2,4 plateaus at the earliest titration point (1:3) and QT3,4 plateaus at a 1:5 ratio at a slightly higher sedimentation coefficient. The sedimentation coefficient of the complex peak for QT2,3/LC8 at the 1:2 ratio is lower than those seen for QT3,4 and QT2,4; this can be explained by the close proximity of QT2 and QT3 to the N-terminus which leaves a long, unbound tail which increases the frictional ratio of the complex (Figure 4C). The continuing increasing value of S in the QT2,3 AUC titration at high concentrations of LC8 may suggest an alternative binding mode that begins to occur at high concentrations of LC8, such as an offset structure that allows three LC8 dimers to bind a pair of QT2,3 chains. This structure, while perhaps not intuitive, is favored as per Le Châtelier's principle, in which a greater number of partially bound LC8 dimers becomes more favorable than a fewer number of fully bound LC8 dimers. However, SV-AUC cannot directly inform on the stoichiometry of complexes formed.

### 3.4. Complex Formation Monitored by Native ESI-MS and EM

Using native electrospray ionization (ESI)-MS, measurements of individual protein mixtures with LC8 allow for the characterization of complex stoichiometries. Similar experiments were used previously to study the complex formation of QT2–4 with LC8 [29]. Accurate mass determination for each protein matches closely with the expected masses of each sequence (Figure S2, Table S2). Upon conducting dilution series, QT3,4, QT2,4, and QT2,3 each remain as monomeric chains while LC8 is overwhelmingly dimeric in solution.

Further native mass spectra acquired for 2:1 mixtures of LC8 with double-site variants indicated the same complex stoichiometries exist for each variant (Figure 5, Table S1). The four detected complexes correspond to species with variant:LC8 ratios of 1:2, 1:4, 2:4, and 2:6. These results mimic those determined for wildtype QT2–4/LC8 complexes, as both the expected "in-register" complex (2:4) and an "off-register" complex (2:6) are present. Note that QT2,3 exhibits the greatest population of 2:6 complex consistent with the hypothesis that it forms off-register complexes more readily than the other two variants. Of note, in-register complexes are always detected in greater abundance than off-register complexes for all double-site constructs, but the persistence of in- and off-register complexes at low concentrations indicates they are each naturally occurring rather than spurious [29] (Figure 5).

While QT3,4, QT2,4, and QT2,3 all form the same set of complexes with LC8, the detected abundances vary between the systems (Figure 5B). Of the four complexes detected, the 2:4 in-register complex (one QT duplex, two LC8 dimers) is the most abundant species formed by QT2,3 at nearly all concentrations studied, indicating high cooperativity between sites QT2 and QT3. In contrast, both QT2,4 and QT3,4 form the intermediate 1:2 complex (one QT chain, one LC8 dimer) as the most abundant species detected across all concentrations. These results indicate that LC8 binding to QT2,3 is more cooperative than QT2,4 or QT3,4. This is consistent with ITC and SV-AUC results presented above.

The complex species identified with native ESI-MS also provide evidence for a potential mechanism of complex formation. First, an LC8 dimer binds to a single chain, forming the 1:2 species. This is followed either by binding a second LC8 dimer, resulting in a 1:4 complex, or by binding to another 1:2 species and rearrangement to a symmetric 2:4 species. If the former path occurs, a second protein strand is subsequently recruited to form the expected in-register 2:4 complex. Misalignment of the second strand by either pathway would allow a third LC8 dimer to bind, resulting in an off-register 2:6 complex (one QT duplex, three LC8 dimers). Figure 5C depicts the proposed mechanism of assembly and ensembles of complexes formed by each variant.

Electron microscopy (EM) images (Figure 5D) collected of mixtures of LC8 with QT2–4 match the conclusions made from MS data. Relative proportions of strings of two, three, and four LC8 dimers observed by EM are plotted and indicate a large excess of species with two LC8 dimers attached by QT2–4 proteins and small amounts of species with three or four LC8 dimers. From the AUC and MS results, it seems reasonable to conclude that the majority of the species with two LC8 dimers are bound at QT2 and QT3.

### 3.5. Comparison of the Complex Heterogeneity of LC8 Bound to QT2–4 versus QT4–6

To further test the conclusions gleaned from the variants of the QT2–4 construct, we compared the intact QT2–4 construct to a different ASCIZ subdomain, QT4–6 (Figure 2C). Previous ITC [10] has shown QT4–6 to bind LC8 more tightly with an N of 3 and $K_d$ of 1.0 μM, compared to an N of 2.7 and $K_d$ of 4.1 μM for QT2–4 (Table S1). To further this comparison, we characterized the complexation of each construct with LC8 by AUC titration and Size Exclusion Chromatography MultiAngle Light Scattering (SEC-MALS). AUC titrations illustrate that the QT2–4 complex (Figure 6A) forms later in the titration than the QT4–6 complex (Figure 6B). As described previously for QT2,4 and QT3,4 AUC, the peak seen in QT2–4 at 1:1, with sedimentation coefficient 3.0, is evidence of a convolution of lower occupancy complex with free LC8, whereas QT4–6 traces do not exhibit free LC8 until the 1:3 ratio. Furthermore, the LC8 peaks in QT2–4 do not line up with the LC8 alone trace at any titration point, while the QT4–6 LC8 free peak lines up consistently, indicating that QT2–4 contains a small population of low occupied complex even at high titrations while the same is not true for QT4–6. Lastly, the QT4–6 titration shows saturation by the 1:4 ratio, while that is not observed for QT2–4, evident by the position of the LC8 free peak. These together indicate that QT4–6 binds LC8 highly cooperatively and uniformly but that QT2–4 binds LC8 much more heterogeneously.

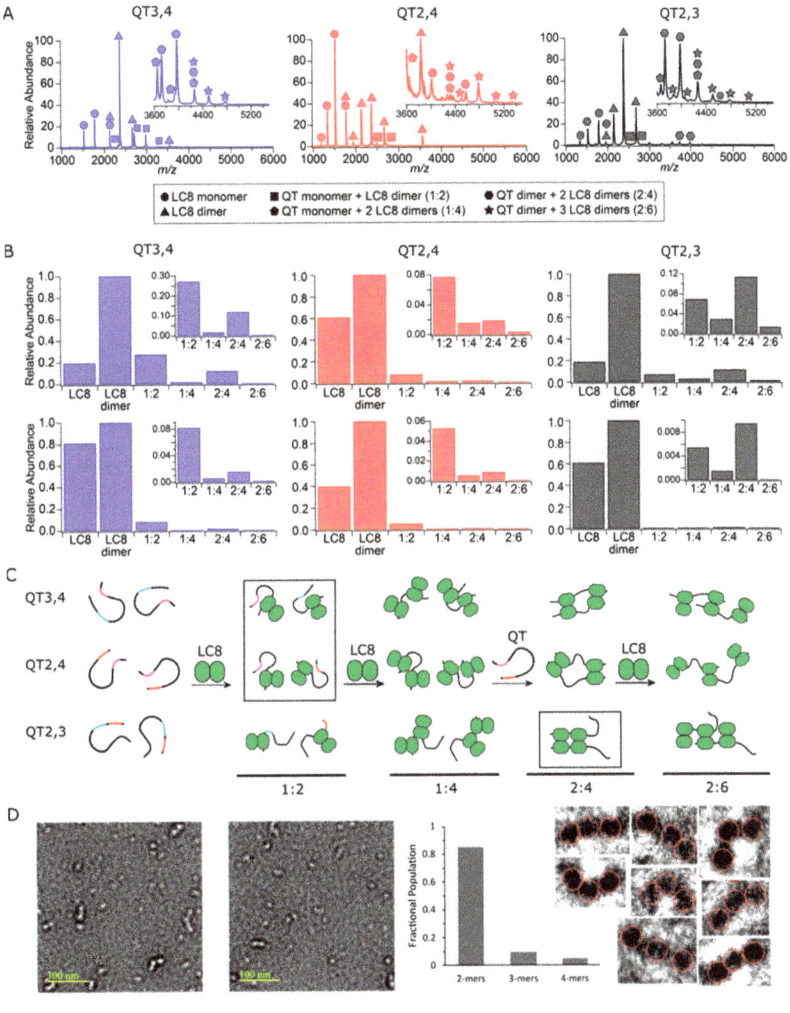


**Figure 5.** QT/LC8 complex species and abundance distributions determined by ESI-MS and EM. (**A**) Native mass spectra of double-site variants at 25 μM are shown with individual and complex species labeled. (**B**) Abundance distributions of species detected at 25 μM (**top**) and 5 μM (**bottom**) for each double-site variant are shown. (**C**) Monomeric chains of each double site variant with QT2, QT3, and QT4 sites color coded. Upon addition of LC8, 1:2, 1:4, 2:4, and 2:6 complexes form. The most abundant complex species for each QT construct is boxed. (**D**) Two representative EM images out of 50 captured of QT2–4/LC8 mixture in which bright dots in the images indicate LC8 dimers. Plotted relative populations of complexes seen in EM, and zoomed negatives of all eight 3-mers observed in the 50 images.

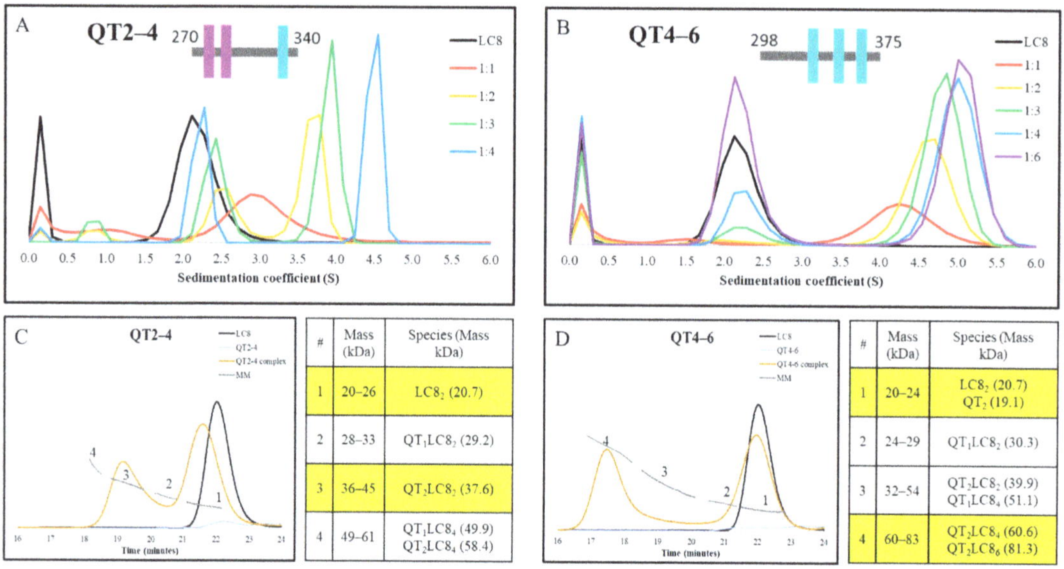


| # | Mass (kDa) | Species (Mass kDa) |
|---|---|---|
| 1 | 20–26 | $LC8_2$ (20.7) |
| 2 | 28–33 | $QT_1LC8_2$ (29.2) |
| 3 | 36–45 | $QT_2LC8_2$ (37.6) |
| 4 | 49–61 | $QT_1LC8_4$ (49.9) $QT_2LC8_4$ (58.4) |

| # | Mass (kDa) | Species (Mass kDa) |
|---|---|---|
| 1 | 20–24 | $LC8_2$ (20.7) $QT_2$ (19.1) |
| 2 | 24–29 | $QT_1LC8_2$ (30.3) |
| 3 | 32–54 | $QT_2LC8_2$ (39.9) $QT_1LC8_4$ (51.1) |
| 4 | 60–83 | $QT_2LC8_4$ (60.6) $QT_2LC8_6$ (81.3) |

E

| QT2 | Peptide $K_d$ | QT3 | Peptide $K_d$ | QT4 | Peptide $K_d$ | QT5 | Peptide $K_d$ | QT6 | Peptide $K_d$ |
|---|---|---|---|---|---|---|---|---|---|
| RDIETQTP | 7 ± 0.4 | GDIGTMTD | Weak | AHMYTQTC | 15 ± 0.8 | SHIQTQTH | 11 ± 0.6 | NTQHTQTC | 20 ± 1.0 |

**Figure 6.** AUC, SEC-MALS, and peptide $K_d$ comparisons for the constructs QT2–4 and QT4–6. AUC titration of (**A**) QT2–4 (replotted data from Reardon et al. [29]) and (**B**) QT4–6. SEC-MALS chromatogram and mass key for ranges numbered as shown on the chromatogram for (**C**) QT2–4 and (**D**) QT4–6 as single proteins and in complex with LC8. LC8 alone trace is shown plotted with both proteins for reference. Highlighted regions of the mass key emphasize the major peaks seen in the SEC-MALS traces of the mixture. (**E**) Measured $K_d$ values for peptides representing the five binding sites [10] represented across the two analyzed constructs.

SEC-MALS analysis of the QT2–4 construct complexed with LC8 (Figure 6C) indicates that the major species involves duplexed QT2–4 linked by one LC8 dimer ($QT_2LC8_2$). However, peak 1, which contains free LC8, does not align with LC8 when run alone, indicating that a significant amount of complex disassociated on the column and that the complex upon injection may have been the $QT_2LC8_4$ complex. Contrastingly, the major species in the complex of QT4–6 with LC8 (Figure 6D) is a mixture of a QT4–6 duplex bridged by two or three LC8 dimers. The width of the LC8 free peak in QT4–6 indicates a minor population of complex dissociated on the column. These results are again consistent with the conclusion that QT2–4 forms a more heterogeneous complex than QT4–6.

Analysis of the binding motifs present in each construct by ITC of peptides has been conducted previously [10] (Figure 6E). $K_d$ values indicate that the difference observed between these two constructs cannot be attributed simply to a better set of binding motifs in QT4–6 than is present in QT2–4; in fact, the opposite might be claimed wherein the QT2 motif is much more favorable for LC8 binding than any of the other motifs involved in either construct. Thus, if the incomplete binding and heterogeneous behavior of the QT2–4 construct cannot be attributed to motif stability and specificity, then it must be attributed to the varying linker lengths that are found in that construct. This also indicates this region as an origin of the dynamic ensemble that is observed in dASCIZ and its homologs in general.

## 4. Discussion

LC8 commonly forms duplex scaffold assemblies with its many multivalent IDP partners [13–15], and cases in which the IDP ligand contains multiple binding sites for

LC8 continue to emerge. However, the contribution of multivalency to complex stability and heterogeneity is not fully understood. Variability in both motif specificity and linker lengths between motifs are well represented in Drosophila ASCIZ, especially within the QT2–4 subdomain. Serving as a model system, this construct contains both the shortest and longest linkers between LC8 binding sites as well as an uncommon TMT LC8 anchor motif. A recent study utilizing QT2–4 provided the first confirmation of in-register binding during LC8/IDP complex assembly and showcased the role that linker length plays in modulating the flexibility of such complexes [29]. The work presented here expands on these results by investigating how the interplay of linker length and motif specificity regulate the compositional heterogeneity of dynamic, multivalent LC8 duplexes. Additionally, a comparison of QT2–4 to QT4–6, another construct from Drosophila ASCIZ, further illustrates the regulatory role played by short and long linkers.

### 4.1. Two LC8 Binding Sites Are Cooperative, but a Third Site Is Negatively Cooperative

ITC experiments of single site variants provide motif-specific binding affinities in the context of the QT2–4 disordered chain for QT2, QT3, and QT4 (Figure 3A–C). QT2 and QT4 show similar binding affinities (9.3 and 15 μM, respectively), while QT3 is considerably weaker (36 μM). This is expected because QT3 contains a TMT anchor that is weaker than the TQT anchors found in QT2 and QT4. The slight favorability for QT2 is consistent with prior evidence which indicates QT2 as the initial site of LC8 binding within QT2–4 [29]. With double site variants, ITC indicates variability in LC8 binding, but always with positive cooperativity (Figure 3D–F). When a third binding site is introduced, affinity decreases for all binding sites. In the QT2–4 system, we are unable to distinguish if location or motif specificity plays a larger role in imparting negative cooperativity in a triple-site, multivalent IDP compared to a double-site. However, because of the sizeable decrease in affinity shown here, we conclude that both properties are likely at play. In fact, motif specificity seems to be tuned by multivalency: stronger motifs benefit more from a second binding site but are also hindered more so by a third binding site. The discussion of linker length effects, however, is more intricate.

### 4.2. QT2,3 Forms Stable Complexes with LC8 More Readily than Do QT3,4 and QT2,4

Though the double-site variants have the same number of binding sites for LC8 dimers, it is clear they differ in their complex assembly. Although $K_d$ values calculated by all three fitting methods (SSS, SBS, and TSS) indicate that QT2,4 forms the most cooperative complex in comparison to the single site variants, the N values calculated by SSS and TSS both indicate that QT2,3 binds LC8 more completely than the other two constructs. This is further substantiated by the AUC and native MS results in which no excess LC8 is present at low titration points of QT2,3 AUC and where MS shows a greater proportion of QT2,3 is complexed in a 2:4 stoichiometry (QT:LC8) than is seen for the other variants. We attribute this degree of cooperativity to the very short linker in QT2,3 which is 3 residues long, compared to the longer linkers, 30 and 41 residues in length. This indicates homogeneous binding to QT2,3 at low LC8:QT ratios compared to QT2,4 and QT3,4, which both bind heterogeneously at these ratios.

However, at higher ratios of LC8:QT, the aforementioned trends persist and imply heterogeneous binding of LC8 to QT2,3 in these concentration regimes. The AUC results for QT2,3 at titration points of 1:4, 1:5, and 1:6 exhibit a continued increase in sedimentation coefficient further than expected. Combined with the results from MS that show that QT2,3 forms more of the 2:6 (QT:LC8) complex than the other two variants, this suggests that this complex is becoming more populated in the higher titration points of the AUC which leads to an increased sedimentation coefficient. We hypothesize that at high LC8 concentrations, the QT2,3 complex assembles as shown in Figure 5C, in which the chains slide into an offset registration and two of the LC8 are only half bound. The extremely short linker may enable this complex to be stabilized by lateral contacts between adjacent LC8 dimers, perhaps via van der Waals interactions. This mode of complexation can be further explained with the application of Le Châtelier's Principle to this system in which higher ratios of LC8 put pressure on the equilibrium to favor a higher population of partially bound LC8 over a small population of fully bound LC8 accompanied by a large population of unbound LC8. It is unsurprising that no evidence of daisy-chaining is seen in our experiments because steric hindrance would likely preclude any chain from binding to the free side of the half-bound LC8 dimers, let alone the entropic penalty of binding a stiff chain of LC8 dimers. This effect is not seen in the variants with longer linkers because they lose the lateral contacts between LC8 dimers and result in long, extended structures (Figure 5C).

### *4.3. Linker Length Is More Important than Motif Specificity for Determining Heterogeneity of LC8 Binding*

While we have discussed that the very short linker in QT2,3 induces heterogeneous binding at higher LC8 concentrations, it is also true that the long linker present in the other variants induces heterogeneous binding, especially at lower concentrations of LC8. ITC (Figure 3) and AUC (Figure 4) both indicate incomplete binding of LC8 through the low N values fit by SSS and TSS to ITC and the presence of free LC8 at low titration points by ITC. Moreover, although the double-site variants each form the same four complex species in solution as determined by native MS, at stoichiometries of 1:2, 1:4, 2:4, and 2:6 (QT:LC8), they vary significantly in their detected abundance (Table S2, Figure 5). Single chain complexes (1:2 and 1:4) are more abundant for the variants with long linkers (QT3,4 and QT2,4) than for QT2,3. For QT2,4 and QT3,4, these single-chain complexes are also more abundant than the duplex species, even at the highest concentration tested (Figure 5). This indicates that increasing the linker length between LC8 binding sites disrupts duplex formation of IDP multivalent complexes. Similarly, EM results indicate an overwhelming proportion of species with only two LC8 dimers bound to QT2–4 protein strands, presumably bound to the QT2 and QT3 binding sites. Additionally, a comparison of SEC-MALS (Figure 6) of QT2–4, which contains the lengthy linker, and QT4–6, which contains moderate-length linkers, shows that QT4–6 assembles as a dimer with either two or three LC8 dimers bound at the same conditions in which QT2–4 dimers only bind to one LC8 dimer. Of note, the QT2–4 complex peak is broad and the LC8 peak is shifted, indicating a more heterogeneous mixture of complexes and a more dynamic assembly than is seen with the QT4–6 construct. Interestingly, these differences cannot be explained by differences between motif specificities involved in each construct because a comparison of the motifs between QT2–4 and QT4–6 indicates similar binding strengths (Figure 6E). It is worth noting, however, that motif specificity remains important to complex formation. The weak-binding TMT motif in QT3,4 causes a lower overall LC8 binding affinity by ITC compared to QT2,4, which contains a similar length linker (Figure 3E, Table S1) and requires a greater ratio of LC8 to reach saturation by AUC (Figure 4). Additionally, the weak TMT motif is likely part of what enables the offset structure proposed for QT2,3 through dynamic binding to LC8. Comparison of the double-site variants and the intact constructs QT2–4 and QT4–6 suggests a "Goldilocks" zone for linker length in regard to non-heterogeneous binding wherein the short 3 residue linker and the long 30 and 41 residue linkers result in heterogeneous binding, but the mid-length 6 and 9 residue linkers are not associated with heterogeneous binding. Together, these results highlight the importance of both linker length and motif specificity and determine their interplay as a regulation mechanism for IDP/LC8 multivalent complex assembly.

We have shown that short linkers and long linkers both contribute to heterogeneity, while 6 and 9 amino acid linkers result in homogeneous complexation. However, we have not established the barrier between a "mid-length" linker and a long linker. In dASCIZ, between QT6 and QT7, there is a 12 amino acid linker and between QT1 and QT2, there is a 16 amino acid linker. Previous research shows that the QT4–7 and QT1–3 constructs have binding affinities that fall between those of QT4–6 and 2–4 [10]. Of note, this means that QT1–3 has a lower affinity than QT2,3 and that QT4–7 has a lower affinity than QT4–6. While these constructs are not identical in their contextual residues, the differences do suggest that the addition of QT1 and QT7 both result in poorer binding systems. This may indicate that 12 and 16 amino acid linkers are already long enough that they begin to induce heterogeneity in LC8/ASCIZ complexation. Further study of these constructs and of double site variants of these and QT4–6 (QT4,6 contains a 23 amino acid linker) will help to elucidate the barrier between mid-length linkers, which lead to homogeneous binding, and long linkers, which induce heterogeneity at low LC8 concentrations.

Interestingly, human ASCIZ contains a run of 4 LC8 binding sites (Figure 2) with linkers of 1, 6, and 24 residues, respectively, and quite intriguingly, the second of these binding sites has a TMT anchor sequence. While we by no means believe this to be the only source of heterogeneity in LC8 binding to hASCIZ, we hypothesize that it is a strong contributor to the formation of the dynamic complex that has been described for this system [10].

## 5. Conclusions

Herein we show that binding of LC8 to multivalent QT2–4 variants is complex and governed more strongly by the length of disordered linkers between LC8 binding sites than by LC8 motif specificity. Cooperativity between multivalent sites is positive for double-site variants but negative for the three-site construct QT2–4. Additionally, the multivalent constructs with short linkers between sites resulted in stable saturated LC8/IDP assemblies that are readily formed in solution compared to constructs with longer linkers that showed a greater propensity for the formation of unsaturated complexes. Comparison of constructs with similar linker lengths, but variability in motif specificity, emphasize that both properties are involved to varying degrees in regulating IDP/LC8 complex assembly. While our initial hypothesis that long linkers contribute to heterogeneous binding was validated by our findings, it is also evident from our experiments that very short linkers similarly contribute to heterogeneous LC8 binding at high concentrations, matching observations that ASCIZ/LC8 complexes are heterogeneous at all concentrations. These findings are important for understanding the behavior of the hASCIZ/LC8 complex and suggest regions that should be studied further, which may contribute to heterogeneity. In particular, the long linker between LBD1 and LBD2, as emphasized in Figure 2, but also the region between F641 and N750 containing a 1 residue linker, a mid-length linker, a long linker, and a TMT anchored LC8 binding site. Our work is also applicable to the study of other ordered hubs binding their partners and to IDPs with multiple partner binding sites, whether for one or multiple distinct partners. Partner binding will be regulated by the lengths of the disordered linkers between each site and the strength of the binding sites involved.

**Supplementary Materials:** The following supporting information can be downloaded at: https://www.mdpi.com/article/10.3390/biom13030404/s1, Figure S1: Algorithm models; Figure S2: Mass spectra of each of the individual species analyzed; Table S1: Thermodynamic parameter fits of the QT-LC8 interactions; Table S2: Observed versus expected masses for individual proteins and complexes.

**Author Contributions:** Conceptualization, D.R.W., K.A.J. and E.J.B.; formal analysis, D.R.W., K.A.J., A.D.R. and P.N.R.; investigation, D.R.W., K.A.J., A.D.R., C.B., W.H., A.K.S. and P.N.R.; resources, J.S.P. and E.J.B.; writing—original draft preparation, D.R.W. and K.A.J.; writing—review and editing, D.R.W., K.A.J., A.D.R., P.N.R., J.S.P. and E.J.B.; visualization, D.R.W., K.A.J. and A.D.R.; supervision, E.J.B. and J.S.P.; project administration, D.R.W., K.A.J. and E.J.B.; funding acquisition, E.J.B. All authors have read and agreed to the published version of the manuscript.

**Funding:** This work is funded by the National Institutes of Health, grant number R01GM141733.

**Data Availability Statement:** The data presented in this study are available upon request from the corresponding author.

**Acknowledgments:** The authors acknowledge Seth Pickney for work in cloning the mutant constructs, Russell McFarland for his assistance in preparing EM images for analysis, and Aidan Estelle for helpful discussion with the ITC analysis.

**Conflicts of Interest:** The authors declare no conflict of interest. The funders had no role in the design of the study; in the collection, analyses, or interpretation of data; in the writing of the manuscript; or in the decision to publish the results.

## References

1. Jeong, H.; Mason, S.P.; Barabási, A.-L.; Oltvai, Z.N. Lethality and Centrality in Protein Networks. *Nature* **2001**, *411*, 41–42. [CrossRef]
2. Dunker, A.K.; Cortese, M.S.; Romero, P.; Iakoucheva, L.M.; Uversky, V.N. Flexible Nets. The Roles of Intrinsic Disorder in Protein Interaction Networks. *FEBS J.* **2005**, *272*, 5129–5148. [CrossRef]
3. Jaspers, P.; Blomster, T.; Brosche, M.; Salojarvi, J.; Ahlfors, R.; Vainonen, J.P.; Reddy, R.A.; Immink, R.; Angenent, G.; Turck, F.; et al. Unequally Redundant RCD1 and SRO1 Mediate Stress and Developmental Responses and Interact with Transcription Factors. *Plant J.* **2009**, *60*, 268–279. [CrossRef]
4. Cino, E.A.; Killoran, R.C.; Karttunen, M.; Choy, W.-Y. Binding of Disordered Proteins to a Protein Hub. *Sci. Rep.* **2013**, *3*, 2305. [CrossRef]
5. Barbar, E. Dynein Light Chain LC8 Is a Dimerization Hub Essential in Diverse Protein Networks. *Biochemistry* **2008**, *47*, 503–508. [CrossRef]
6. Uversky, V.N. Intrinsically Disordered Proteins and Their "Mysterious" (Meta)Physics. *Front. Phys.* **2019**, *7*, 10. [CrossRef]
7. Van der Lee, R.; Buljan, M.; Lang, B.; Weatheritt, R.J.; Daughdrill, G.W.; Dunker, A.K.; Fuxreiter, M.; Gough, J.; Gsponer, J.; Jones, D.T.; et al. Classification of Intrinsically Disordered Regions and Proteins. *Chem. Rev.* **2014**, *114*, 6589–6631. [CrossRef]
8. Baines, A.J.; Lu, H.-C.; Bennett, P.M. The Protein 4.1 Family: Hub Proteins in Animals for Organizing Membrane Proteins. *Biochimica et Biophysica Acta (BBA)—Biomembranes* **2014**, *1838*, 605–619. [CrossRef]
9. Newman, R.A.; Van Scyoc, W.S.; Sorensen, B.R.; Jaren, O.R.; Shea, M.A. Interdomain Cooperativity of Calmodulin Bound to Melittin Preferentially Increases Calcium Affinity of Sites I and II. *Proteins* **2008**, *71*, 1792–1812. [CrossRef]
10. Clark, S.; Myers, J.B.; King, A.; Fiala, R.; Novacek, J.; Pearce, G.; Heierhorst, J.; Reichow, S.L.; Barbar, E.J. Multivalency Regulates Activity in an Intrinsically Disordered Transcription Factor. *eLife* **2018**, *7*, e36258. [CrossRef]
11. Forsythe, H.M.; Barbar, E. Chapter Seven—The Role of Dancing Duplexes in Biology and Disease. In *Dancing Protein Clouds: Intrinsically Disordered Proteins in the Norm and Pathology, Part C*; Uversky, V.N., Ed.; Progress in Molecular Biology and Translational Science; Academic Press: Cambridge, MA, USA, 2021; Volume 183, pp. 249–270.
12. Barbar, E.; Nyarko, A. Polybivalency and Disordered Proteins in Ordering Macromolecular Assemblies. *Semin. Cell Dev. Biol.* **2015**, *37*, 20–25. [CrossRef] [PubMed]
13. Clark, S.A.; Jespersen, N.; Woodward, C.; Barbar, E. Multivalent IDP Assemblies: Unique Properties of LC8-Associated, IDP Duplex Scaffolds. *FEBS Lett.* **2015**, *589*, 2543–2551. [CrossRef] [PubMed]
14. Jespersen, N.; Barbar, E. Emerging Features of Linear Motif-Binding Hub Proteins. *Trends Biochem. Sci.* **2020**, *45*, 375–384. [CrossRef] [PubMed]
15. Jespersen, N.; Estelle, A.; Waugh, N.; Davey, N.E.; Blikstad, C.; Ammon, Y.-C.; Akhmanova, A.; Ivarsson, Y.; Hendrix, D.A.; Barbar, E. Systematic Identification of Recognition Motifs for the Hub Protein LC8. *Life Sci. Alliance* **2019**, *2*, e201900366. [CrossRef] [PubMed]
16. McGuffin, L.J.; Bryson, K.; Jones, D.T. The PSIPRED Protein Structure Prediction Server. *Bioinformatics* **2000**, *16*, 404–405. [CrossRef] [PubMed]
17. McDonnell, A.V.; Jiang, T.; Keating, A.E.; Berger, B. Paircoil2: Improved Prediction of Coiled Coils from Sequence. *Bioinformatics* **2006**, *22*, 356–358. [CrossRef]
18. Fejtova, A.; Davydova, D.; Bischof, F.; Lazarevic, V.; Altrock, W.D.; Romorini, S.; Schöne, C.; Zuschratter, W.; Kreutz, M.R.; Garner, C.C.; et al. Dynein Light Chain Regulates Axonal Trafficking and Synaptic Levels of Bassoon. *J. Cell Biol.* **2009**, *185*, 341–355. [CrossRef]
19. Howe, J.; Weeks, A.; Reardon, P.; Barbar, E. Multivalent Binding of the Hub Protein LC8 at a Newly Discovered Site in 53BP1. *Biophys. J.* **2022**, *121*, 4433–4442. [CrossRef]
20. Nyarko, A.; Song, Y.; Nováček, J.; Žídek, L.; Barbar, E. Multiple Recognition Motifs in Nucleoporin Nup159 Provide a Stable and Rigid Nup159-Dyn2 Assembly. *J. Biol. Chem.* **2013**, *288*, 2614–2622. [CrossRef]
21. Rodríguez-Crespo, I.; Yélamos, B.; Roncal, F.; Albar, J.P.; Ortiz de Montellano, P.R.; Gavilanes, F. Identification of Novel Cellular Proteins That Bind to the LC8 Dynein Light Chain Using a Pepscan Technique. *FEBS Lett.* **2001**, *503*, 135–141. [CrossRef]
22. Rapali, P.; García-Mayoral, M.F.; Martínez-Moreno, M.; Tárnok, K.; Schlett, K.; Albar, J.P.; Bruix, M.; Nyitray, L.; Rodriguez-Crespo, I. LC8 Dynein Light Chain (DYNLL1) Binds to the C-Terminal Domain of ATM-Interacting Protein (ATMIN/ASCIZ) and Regulates Its Subcellular Localization. *Biochem. Biophys. Res. Commun.* **2011**, *414*, 493–498. [CrossRef] [PubMed]
23. Dunsch, A.K.; Hammond, D.; Lloyd, J.; Schermelleh, L.; Gruneberg, U.; Barr, F.A. Dynein Light Chain 1 and a Spindle-Associated Adaptor Promote Dynein Asymmetry and Spindle Orientation. *J. Cell Biol.* **2012**, *198*, 1039–1054. [CrossRef]

24. Eastwood, E.L.; Jara, K.A.; Bornelöv, S.; Munafò, M.; Frantzis, V.; Kneuss, E.; Barbar, E.J.; Czech, B.; Hannon, G.J. Dimerisation of the PICTS Complex via LC8/Cut-up Drives Co-Transcriptional Transposon Silencing in Drosophila. *eLife* **2021**, *10*, e65557. [CrossRef] [PubMed]
25. Hall, J.; Karplus, P.A.; Barbar, E. Multivalency in the Assembly of Intrinsically Disordered Dynein Intermediate Chain. *J. Biol. Chem.* **2009**, *284*, 33115–33121. [CrossRef] [PubMed]
26. Gupta, A.; Diener, D.R.; Sivadas, P.; Rosenbaum, J.L.; Yang, P. The Versatile Molecular Complex Component LC8 Promotes Several Distinct Steps of Flagellar Assembly. *J. Cell Biol.* **2012**, *198*, 115–126. [CrossRef]
27. Zaytseva, O.; Tenis, N.; Mitchell, N.; Kanno, S.; Yasui, A.; Heierhorst, J.; Quinn, L.M. The Novel Zinc Finger Protein DASCIZ Regulates Mitosis in *Drosophila* via an Essential Role in Dynein Light-Chain Expression. *Genetics* **2014**, *196*, 443–453. [CrossRef]
28. Clark, S.; Nyarko, A.; Löhr, F.; Karplus, P.A.; Barbar, E. The Anchored Flexibility Model in LC8 Motif Recognition: Insights from the Chica Complex. *Biochemistry* **2016**, *55*, 199–209. [CrossRef]
29. Reardon, P.N.; Jara, K.A.; Rolland, A.D.; Smith, D.A.; Hoang, H.T.M.; Prell, J.S.; Barbar, E.J. The Dynein Light Chain 8 (LC8) Binds Predominantly "in-Register" to a Multivalent Intrinsically Disordered Partner. *J. Biol. Chem.* **2020**, *295*, 4912–4922. [CrossRef]
30. Nyarko, A.; Hare, M.; Hays, T.S.; Barbar, E. The Intermediate Chain of Cytoplasmic Dynein Is Partially Disordered and Gains Structure upon Binding to Light-Chain LC8. *Biochemistry* **2004**, *43*, 15595–15603. [CrossRef]
31. Petryszak, R.; Keays, M.; Tang, Y.A.; Fonseca, N.A.; Barrera, E.; Burdett, T.; Füllgrabe, A.; Fuentes, A.M.-P.; Jupp, S.; Koskinen, S.; et al. Expression Atlas Update—an Integrated Database of Gene and Protein Expression in Humans, Animals and Plants. *Nucleic Acids Res.* **2016**, *44*, D746–D752. [CrossRef]
32. Chen, Y.-M.; Gerwin, C.; Sheng, Z.-H. Dynein Light Chain LC8 Regulates Syntaphilin-Mediated Mitochondrial Docking in Axons. *J. Neurosci.* **2009**, *29*, 9429–9438. [CrossRef]
33. Wang, X.; Olson, J.R.; Rasoloson, D.; Ellenbecker, M.; Bailey, J.; Voronina, E. LC8 Dynein Light Chain Promotes Localization and Function of PUF Protein FBF-2 in Germline Progenitor Cells. *Development* **2016**, dev.140921. [CrossRef]
34. Makokha, M.; Hare, M.; Li, M.; Hays, T.; Barbar, E. Interactions of Cytoplasmic Dynein Light Chains Tctex-1 and LC8 with the Intermediate Chain IC74. *Biochemistry* **2002**, *41*, 4302–4311. [CrossRef]
35. Nyarko, A.; Barbar, E. Light Chain-Dependent Self-Association of Dynein Intermediate Chain. *J. Biol. Chem.* **2011**, *286*, 1556–1566. [CrossRef] [PubMed]
36. Raux, H.; Flamand, A.; Blondel, D. Interaction of the Rabies Virus P Protein with the LC8 Dynein Light Chain. *J. Virol.* **2000**, *74*, 10212–10216. [CrossRef] [PubMed]
37. Jespersen, N.E.; Leyrat, C.; Gérard, F.C.; Bourhis, J.-M.; Blondel, D.; Jamin, M.; Barbar, E. The LC8-RavP Ensemble Structure Evinces A Role for LC8 in Regulating Lyssavirus Polymerase Functionality. *J. Mol. Biol.* **2019**, *431*, 4959–4977. [CrossRef] [PubMed]
38. Rodriguez Galvan, J.; Donner, B.; Veseley, C.H.; Reardon, P.; Forsythe, H.M.; Howe, J.; Fujimura, G.; Barbar, E. Human Parainfluenza Virus 3 Phosphoprotein Is a Tetramer and Shares Structural and Interaction Features with Ebola Phosphoprotein VP35. *Biomolecules* **2021**, *11*, 1603. [CrossRef] [PubMed]
39. Jurado, S.; Gleeson, K.; O'Donnell, K.; Izon, D.J.; Walkley, C.R.; Strasser, A.; Tarlinton, D.M.; Heierhorst, J. The Zinc-Finger Protein ASCIZ Regulates B Cell Development via DYNLL1 and Bim. *J. Exp. Med.* **2012**, *209*, 1629–1639. [CrossRef] [PubMed]
40. Jurado, S.; Conlan, L.A.; Baker, E.K.; Ng, J.-L.; Tenis, N.; Hoch, N.C.; Gleeson, K.; Smeets, M.; Izon, D.; Heierhorst, J. ATM Substrate Chk2-Interacting Zn2+ Finger (ASCIZ) Is a Bi-Functional Transcriptional Activator and Feedback Sensor in the Regulation of Dynein Light Chain (DYNLL1) Expression. *J. Biol. Chem.* **2012**, *287*, 3156–3164. [CrossRef]
41. Rapali, P.; Radnai, L.; Süveges, D.; Harmat, V.; Tölgyesi, F.; Wahlgren, W.Y.; Katona, G.; Nyitray, L.; Pál, G. Directed Evolution Reveals the Binding Motif Preference of the LC8/DYNLL Hub Protein and Predicts Large Numbers of Novel Binders in the Human Proteome. *PLoS ONE* **2011**, *6*, e18818. [CrossRef]
42. King, A.; Hoch, N.C.; McGregor, N.E.; Sims, N.A.; Smyth, I.M.; Heierhorst, J. Dynll1 Is Essential for Development and Promotes Endochondral Bone Formation by Regulating Intraflagellar Dynein Function in Primary Cilia. *Hum. Mol. Genet.* **2019**, *28*, 2573–2588. [CrossRef] [PubMed]
43. Jurado, S.; Smyth, I.; van Denderen, B.; Tenis, N.; Hammet, A.; Hewitt, K.; Ng, J.-L.; McNees, C.J.; Kozlov, S.V.; Oka, H.; et al. Dual Functions of ASCIZ in the DNA Base Damage Response and Pulmonary Organogenesis. *PLoS Genet* **2010**, *6*, e1001170. [CrossRef]
44. Schuck, P. Size-Distribution Analysis of Macromolecules by Sedimentation Velocity Ultracentrifugation and Lamm Equation Modeling. *Biophys. J.* **2000**, *78*, 1606–1619. [CrossRef] [PubMed]
45. Harding, S.E.; Rowe, A.J.; Horton, J.C. Analytical Ultracentrifugation in Biochemistry and Polymer Science. 1992. Available online: https://www.sciencedirect.com/science/article/abs/pii/0003267095904018?via%3Dihub (accessed on 15 January 2023).
46. MicroCal. *ITC Data Analysis in Origin: Tutorial Guide*; MicroCal, LLC: Northampton, MS, USA, 2004.

*biomolecules*

MDPI

*Article*

# Conformational Analysis of Charged Homo-Polypeptides

**Lavi S. Bigman and Yaakov Levy** *

Department of Chemical and Structural Biology, Weizmann Institute of Science, Rehovot 7610001, Israel
* Correspondence: koby.levy@weizmann.ac.il

**Abstract:** Many proteins have intrinsically disordered regions (IDRs), which are often characterized by a high fraction of charged residues with polyampholytic (i.e., mixed charge) or polyelectrolytic (i.e., uniform charge) characteristics. Polyelectrolytic IDRs include consecutive positively charged Lys or Arg residues (K/R repeats) or consecutive negatively charged Asp or Glu residues (D/E repeats). In previous research, D/E repeats were found to be about five times longer than K/R repeats and to be much more common in eukaryotes. Within these repeats, a preference is often observed for E over D and for K over R. To understand the greater prevalence of D/E over K/R repeats and the higher abundance of E and K, we simulated the conformational ensemble of charged homo-polypeptides (polyK, polyR, polyD, and polyE) using molecular dynamics simulations. The conformational preferences and dynamics of these polyelectrolytic polypeptides change with changes in salt concentration. In particular, polyD and polyE are more sensitive to salt than polyK and polyR, as polyD and polyE tend to adsorb more divalent cations, which leads to their having more compact conformations. We conclude with a discussion of biophysical explanations for the relative abundance of charged amino acids and particularly for the greater abundance of D/E repeats over K/R repeats.

**Keywords:** intrinsically disordered proteins; polyelectrolytes; D/E repeats; K/R repeats; molecular dynamics simulations

## 1. Introduction

The intrinsically disordered regions (IDRs) of proteins are linked to various biological functions [1–4] and are often characterized by highly charged amino acid content. The more highly charged content of IDRs compared with foldable sequences favors interactions with the solvent and may disfavor their folding into a unique three-dimensional structure [5–8]. The structural and dynamic properties of IDRs depend on their charge composition. IDRs differ from each other with respect to the fraction of positively and negatively charged residues they contain, their overall net charges, and the organization or pattern of charges along the IDRs. Charge composition and organization are expected to determine the biophysical characteristics and function of IDRs. For example, it was shown that changing charge organization in the IDRs of DNA-binding proteins can tune binding affinity to DNA and the diffusion coefficient for linear diffusion along DNA [9–12]. In another example, the charge pattern was shown to have a pronounced effect on the ability of IDRs to form condensates via liquid–liquid phase separation [13,14] and on the stability of a complex formed between two highly but oppositely charged intrinsically disordered proteins (IDPs) [15].

IDRs are found to span a wide range of net charges, with the net charge per residue ranging between −1 and +1. For many IDRs, the net charge per residue is close to zero, reflecting the presence of a similar number of negatively and positively charged residues (i.e., polyampholytic IDRs). Other IDRs are highly charged, and their net charge per residue deviates from zero. It was reported that highly negatively charged IDRs are longer and more highly charged than positively charged IDRs [16]. A particularly interesting group of IDRs are those with net charge close to −1 or +1. In these cases, the fraction of negatively or positively charged residues is close to unity. These IDRs, which are quite

**Citation:** Bigman, L.S.; Levy, Y. Conformational Analysis of Charged Homo-Polypeptides. *Biomolecules* **2023**, *13*, 363. https://doi.org/10.3390/biom13020363

Academic Editors: Prakash Kulkarni, Stefania Brocca, Keith Dunker and Sonia Longhi

Received: 13 December 2022
Revised: 6 February 2023
Accepted: 8 February 2023
Published: 15 February 2023

homogenously charged and are thus classified as polyelectrolytes, sometimes include residues of opposite charges or neutral residues. Some polyelectrolytic IDRs have charge density of unity. Additional polyelectrolytes that are essential to function may include non-protein biopolymers. For example, inorganic polyphosphate [17] or matriglycans [18] are long negatively charged polyelectrolytic polymers composed of phosphates and saccharide building blocks and are involved in various distinctive functions.

The polyelectrolytic IDRs of proteins are positively charged when comprising repeating Lys (K) or Arg (R) residues (K/R repeats), whereas they are negatively charged when comprising repeating Asp (D) or Glu (E) residues (D/E repeats). A recent study [16] showed that many proteins include such repeats and that D/E repeats are more common than K/R. In eukaryotic genomes, ~10% of proteins have D/E repeats containing at least five residues; however, only ~5% of K/R repeats are at least five-residue long. D/E repeats are even more favored in longer polyelectrolytic IDRs. In various eukaryotes, about 1–2% of proteins include D/E repeats longer than 10 residues, but the population of K/R repeats containing 10 or more residues is zero (see Figure 1) [16]. Several proteins include 40–50-residue D/E repeats, but K/R repeats longer than 10 residues are not found in any organism. Several possible explanations have been proposed for why negatively charged D/E repeats are longer and more common than positively charged K/R repeats, including suggestions that K/R repeats are more prone to proteolysis [19] and that they may slow down translation kinetics in the ribosome because its exit tunnel is negatively charged [20,21].

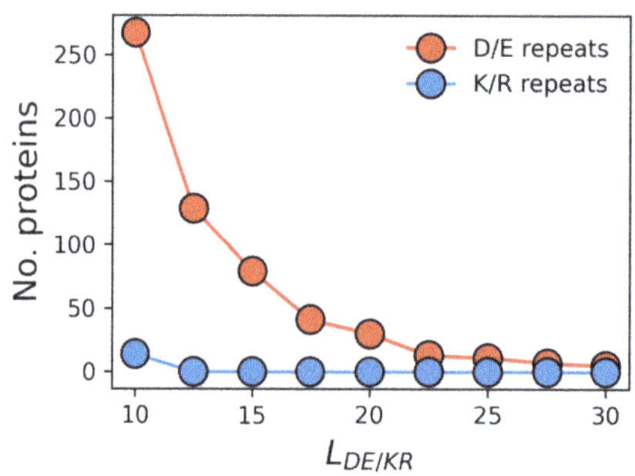


**Figure 1.** Occurrence of proteins with negatively or positively charged polyelectrolytic intrinsically disordered regions (IDRs) in the human proteome. Protein abundance is shown for proteins with D/E or K/R repeats of various lengths, as represented by $L_{DE/KR}$ (i.e., the number of charged residues in the negatively or positively charged homo-polypeptides). The indicated number of proteins (out of the 20,600 proteins in the human proteome) is a cumulative value for all D/E or K/R repeat lengths up to the value of the corresponding $L_{DE/KR}$. The shortest repeat length in this analysis is a repeat of 10 residues.

The strong preference for D/E repeats over the K/R repeats is accompanied by a preference for E over D. In the human proteome, the frequency ratio of n(E)/n(D) is 3.1 in D/E repeats longer than 10 residues, whereas the overall ratio for human proteins of any length is 1.5. Similar values were found for the mouse proteome [16]. In K/R repeats, the n(K)/n(R) ratio is 1.7, and it is ~1 in all human proteins. The strong preference for D/E repeats over K/R and for E over D is supported but cannot be fully explained by the total concentrations of these amino acids as free solvated molecules in the cell (the concentrations of E, D, R, and K are 96 nM, 4.2 nM, 0.57 nM, and 0.4 nM, respectively) [22].

To elucidate the observed differences in the abundance and length of D/E and K/R repeats as well as the greater abundance of E in these repeats, here, we examined conformational ensembles of polyelectrolytic homo-polypeptides comprising D, E, K, or R. Using atomistic molecular dynamics (MD) simulations, we investigated the molecular biophysics of these homo-polypeptides to address whether the observed differences in D/E and K/R repeats may have a biophysical origin.

## 2. Materials and Methods

### *All-Atom Molecular Dynamics Simulations*

To quantify the biophysical properties of polyelectrolytes, we constructed polypeptides of length of 30 amino acids that were homo-repeats of aspartate (polyD), glutamate (polyE), arginine (polyR), or lysine (polyK). As a control, we constructed a polypeptide with consecutive repeats of glycine and serine, termed polyGS. The polypeptides were initially modeled as linear chains in PyMol, with more realistic conformations achieved during the MD simulations.

The conformational dynamics of the polypeptides were studied using all-atom MD simulations. The simulations were performed using GROMACS [23] version 2022. The molecular system was solvated in a box with periodic boundary conditions containing pre-equilibrated TIP3P water molecules, as implemented in the Charmm36m force field. Three salt concentrations were investigated. The salt concentration referred to as 0 M represents a neutral system, which was obtained by modeling the polyelectrolytes in an environment that included sufficient $Na^+$ or $Cl^-$ counterions to neutralize the charges on the homo-polypeptide amino acid residues. The other two salt concentrations involved modeling the polyelectrolytes in a low-salt (125 mM NaCl or $MgCl_2$) or high-salt (250 mM NaCl or $MgCl_2$) environment. We used the Charmm36m [24] force field. The LINCS algorithm [25] was used to control bonds during the simulation. The leapfrog algorithm was employed with steps of 2 fs.

The temperature was controlled at 300 K using a modified scheme of the theBerendsen thermostat [26]. The system was minimized using the steepest descent algorithm. Next, the system was equilibrated under an NVT ensemble and an NPT ensemble (100 ps each phase). Production runs were executed at a constant pressure (1 atm) for 200 ns. We ran each system to obtain five repeats at three NaCl concentrations (0 mM, low, and high) and five further repeats at two $MgCl_2$ concentrations (low and high) for an accumulated simulation time of 25 µs.

Data analysis was performed using in-house python scripts. Principal Component Analysis (PCA) was performed as implemented in MDAnalysis [27].

## 3. Results

### *3.1. D/E Repeats Are More Common Than K/R Repeats*

The bioinformatic analysis of the human proteome revealed that there are more proteins with negatively charged IDRs (D/E repeats) than with positively charged IDRs (K/R repeats) [16]. Figure 1 shows the number of proteins containing D/E or K/R repeats of various lengths ($L_{DE/KR}$). Each data point in Figure 1 corresponds to all repeats with length $\leq L_{DE/KR}$. The shortest repeat considered in this analysis is of length 10. Figure 1 shows that for a length threshold of 10 consecutive residues, there are >250 proteins with D/E repeats but only ~10 proteins with K/R repeats. For all repeat lengths, a greater number of IDRs contain D/E repeats compared with K/R repeats. Similar results were reported for 22 different proteomes [16].

### *3.2. Dimensions of Polyelectrolytic Polypeptides*

Guided by the observation that D/E repeats are often longer than K/R repeats, we explored the possibility that the preference for negatively charged polyelectrolytes over positively charged polyelectrolytes has a biophysical origin. For that purpose, we constructed 30-residue models of homo-polypeptides of polyelectrolytes containing either negatively (i.e., polyD and polyE) or positively (i.e., polyK and polyR) charged residues. The conformational ensemble of each of the homo-polypeptides was sampled using atomistic all-atom MD simulations that were analyzed to quantify their biophysical characteristics. As a control, we also simulated a polypeptide with 15 consecutive pairs of glycine and serine to produce a 30-residue polyGS.

Importantly, whereas the radius of gyration (Rg) of charged polypeptides was found to be in the range of 20–25 Å in the absence of salt and at both salt concentrations, the Rg of the polyGS control was found to be only ~10 Å. Thus, it appears that the more-extended dimensions of polyD/E and polyK/R are due to their polyelectrolytic nature. Moreover, the simulated ensembles of the polyelectrolytic polypeptides reveal differences between them. With respect to the negatively charged polypeptides, the Rg of polyE is larger than that of polyD ($Rg_{polyE} > Rg_{polyD}$). For the positively charged polypeptides, polyK is more expanded than polyR ($Rg_{polyK} > Rg_{polyR}$) (Figure 2A). Electrostatic repulsions between the charged amino acid residues of the polyelectrolytic polypeptides provide a possible physical explanation for the greater expansion of the polyelectrolytic polypeptides compared with the uncharged control (Figure 2A), whereas the screening of these repulsions by salt may explain the decrease in the value of Rg with the increase in the concentrations of NaCl from 0 to 0.25 M (Figure 2B).

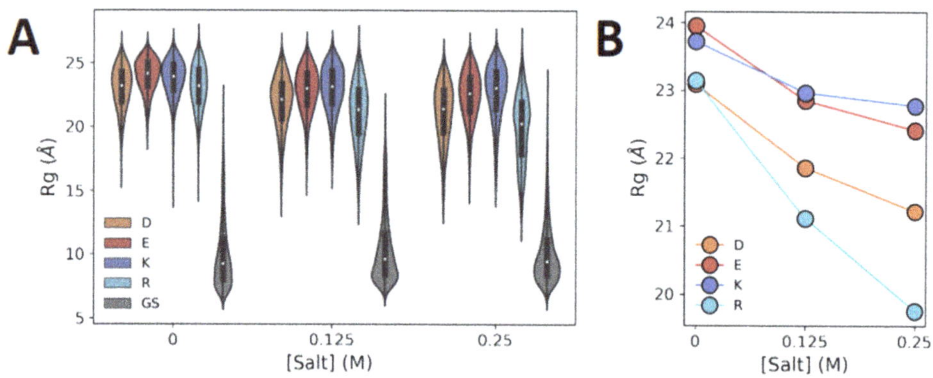


**Figure 2.** Dimensions of charged homo-polypeptides. (**A**) Violin plots of the Rg values of polyD, polyE, polyK, and polyR polyelectrolytic polypeptides, each constituting 30 residues, at three NaCl concentrations: 0 M, 0.125 M, and 0.25 M. The simulations at 0 M salt concentration included counterions to neutralize the charges of the homo-polypeptides. A polypeptide with 15 GS repeats was also simulated, as a control. The violin plots are colored according to amino acid identity, as indicated by the key. (**B**) Mean Rg of each charged homo-polypeptide as a function of NaCl concentration.

Although the Rg analysis illustrates a clear difference between negatively charged polyD and polyE and an even greater difference between positively charged polyR and polyK, there is no clear difference between negatively and positively charged polyelectrolytes.

### *3.3. Conformational Ensemble of Polyelectrolytic Polypeptides*

To further quantify the differences between different polyelectrolytic polypeptides, we performed PCA to elucidate the conformational ensemble of each system. Figure 3 shows the projection on the first two PCs of polyD (top row, orange circles) and polyR (bottom row, cyan circles) with no salt (left panels) and at a high salt concentration (right panels). As a control, we show in the background of each panel (gray circles) the projection on the first two PCs for the corresponding polyGS system. The PCA shows that the conformational ensembles of polyD and polyR are more restricted in low salt concentrations than high salt concentrations because of the greater screening of electrostatic repulsions in the presence of salt that allows a larger conformational space to be sampled with both more compact conformations than those sampled at low salt concentrations. The polyGS control samples a larger conformational space, which can be understood based on the absence of inter-residue electrostatic repulsions, thus a more flexible conformational ensemble. For the polyelectrolytes, the conformational ensemble is more restricted, likely due to electrostatic repulsions. The compaction observed upon the increase in salt concentration is illustrated to the right of each PCA by the presentation of a selected conformation for each system.

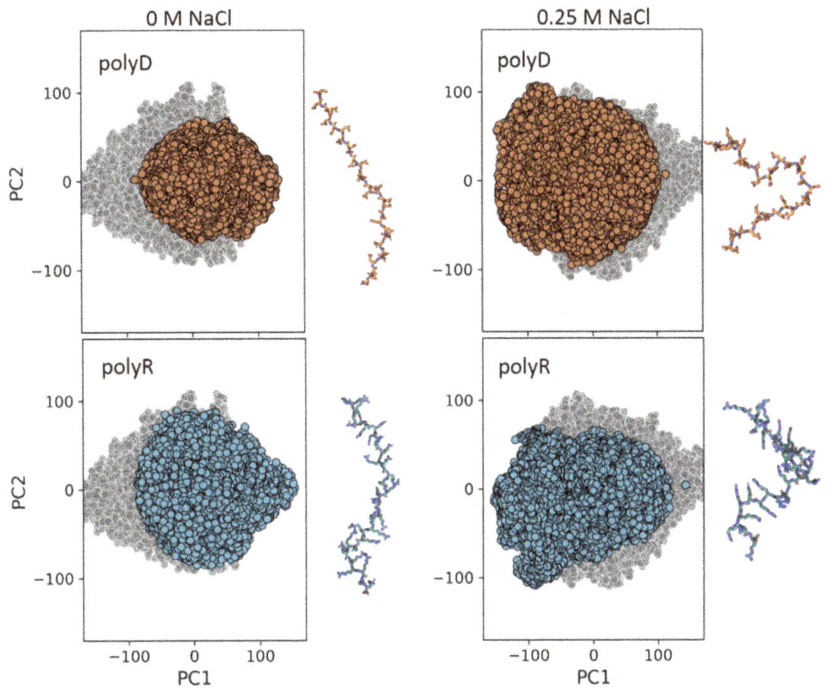


**Figure 3.** Conformational ensemble of charged homo-polypeptides. Projection of the first two principal components (PCs) from principal component analysis (PCA) of polyD (orange) and polyR (cyan) at NaCl concentrations of 0 M (**left**) and of 0.25 M (**right**). The projection for polyGS (gray) at the corresponding salt concentration is shown in the background of each panel for reference. Adjacent to each panel, a representative conformation is shown for each polyelectrolyte.

### 3.4. Flory Exponents and Relaxation Times

In addition to the conformational properties of the polyelectrolytic homo-polypeptides, their polymeric properties may also depend on their chemical nature. According to Flory [28], the Rg of a polymer scales with the number of bonds in the polymer (N) and an exponent ν, $Rg \propto N^{\nu}$. Due to the fractal nature of proteins in a good solvent, a similar relation can be obtained by calculating Rg as a function of the inter-residue distance in a single chain [29]. Hence, we use $Rg \propto |i-j|^{\nu}$, where $|i-j|$ is the sequence separation between two residues in the substituent chain. Hence, by plotting Rg against $|i-j|$ on a log–log scale, the Flory exponent can be derived from the slope (Figure 4A, right panel). Polymer theory predicts a scaling of ν = 1/3 for a compact polymer, ν = 2/3 for a random coil polymer, and ν = 1 for an extended conformation.

We used this relation to derive the Flory exponent for the simulated polyelectrolytes at three different salt concentrations (Figure 4A). With no salt and at both salt concentrations, the value of ν for the charged polypeptides lies in the range of 0.8–0.9, which is very similar to the value expected for a polyelectrolyte in an extended conformation because of extensive inter-residue charge repulsions. By contrast, the value of ν for polyGS is ~0.5, which is similar to the value expected for a random coil polymer. The Flory exponents are smaller for polyD and polyR than for polyE and polyK, in agreement with their Rg behavior (Figure 2). The Flory exponent decreases at higher salt concentration for all polyelectrolytes, but the effect is the largest for polyR.

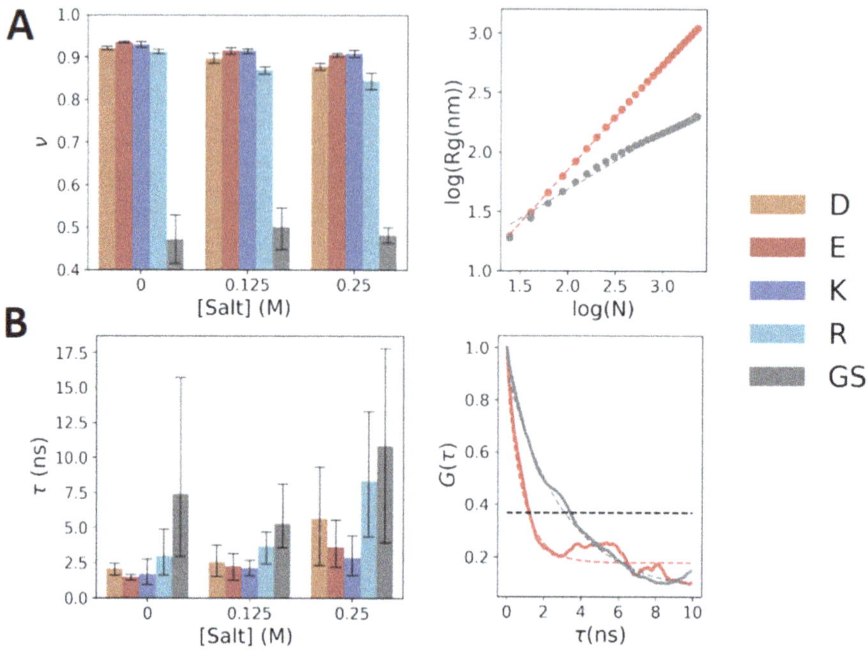


**Figure 4.** Polymeric properties of charged homo-polypeptides. (**A**) (Left) Flory exponent, $\upsilon$, of the five simulated polypeptides at three salt concentrations, extracted from the relation Rg ~ $|i\text{-}j|^{\upsilon}$ (see main text for details). Error bars are the standard deviation of $\upsilon$ obtained from five independent simulations for each polypeptide. (Right) Representative example of the extraction of $\upsilon$ from the slope when plotting Rg versus $|i\text{-}j|$ on a log–log plot. Data are shown for polyE (red circles) and polyGS (gray circles), and the dashed line is the best linear fit. (**B**) Relaxation times for Rg at three different salt concentrations. Values of $\tau$ were extracted by fitting the auto-correlation function, G(t), of Rg to a single exponential function (example on right panel for polyE and polyGS).

The differences among polyD, polyE, polyR, and polyK were also demonstrated when quantifying polypeptide dynamics by analyzing the relaxation times, $\tau$, of Rg (Figure 4B), calculated by fitting the auto-correlation function of Rg to a single exponential function (Figure 4B, right). Higher relaxation times are indicative of slower conformation sampling. The relaxation times increase with salt concentration, which can be rationalized by reducing the electrostatic repulsion among the charged homo-polypeptides. Figure 4B shows that the relaxation times are higher for polyD and polyR than for polyE and polyK, with polyGS having the largest $\tau$ value irrespective of salt concentration.

### *3.5. Sensitivity to Cation Valency Is Greater for D/E Repeats Than for K/R Repeats*

An important question remains as to whether there is a direct connection between the salt concentration and the biophysics of the polyelectrolytes. To address this question, we plotted the mean Rg of each polyelectrolyte as a function of the number of ions adsorbed on the polypeptide (Figure 5A). Each point in Figure 5A was obtained using simulations at different salt concentrations, increasing from left to right. For polyD and polyE, the x-axis shows the number of $Na^+$ (filled circles) or $Mg^{2+}$ (empty circles) ions, and for polyK and polyR, the x-axis shows the number of $Cl^-$ ions. Rg decreases as the number of adsorbed ions on the polypeptides increases, that is, the dimensions of the polypeptides decrease because the salt ions screen the electrostatic repulsions between neighboring amino acids. The positively charged polyelectrolytes (i.e., polyK and polyR) adsorb, on average, twice as many ions as their negatively charged counterparts (i.e., polyD and polyE), even though polyK is as compact as polyE. The greater compactness of polyD compared with polyE can be explained by the higher number of $Na^+$ adsorbed on the former. However, the greater compaction of polyR compared with polyK cannot simply be explained by different extents of ion adsorption.

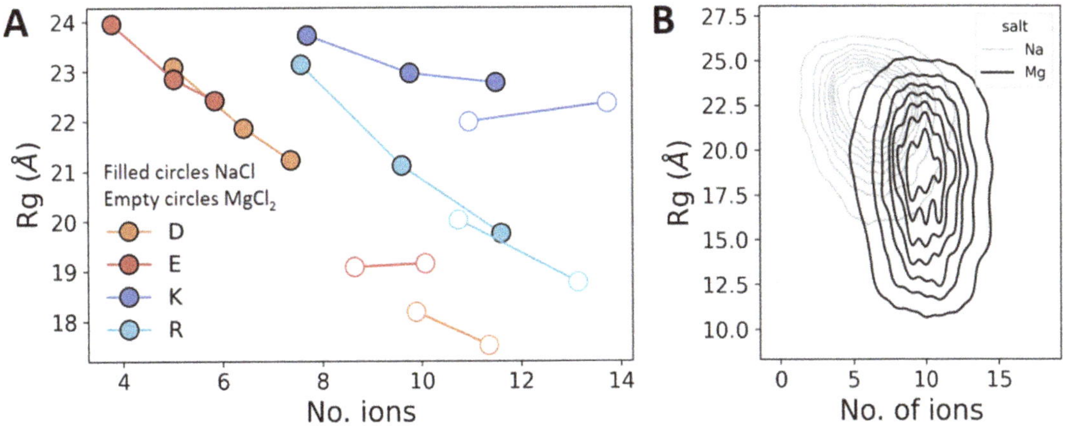


**Figure 5.** Ion adsorption on charged homo-polypeptides. (**A**) Mean Rg for each system as a function of the mean number of ions adsorbed on each charged homo-polypeptide. The three data points for each charged homo-polypeptide were obtained using simulations at three different concentrations of NaCl (0 M, 0.125 M, and 0.25 M) and two salt concentrations for $MgCl_2$ (0.125 M and 0.25 M). The highest number of adsorbed ions for each system corresponds to simulations at a salt concentration of 0.25 M, with the lowest number of adsorbed ions being found at a salt concentration of 0 M. Filled and empty circles correspond to NaCl and $MgCl_2$, respectively. (**B**) Two-dimensional distribution of Rg versus number of adsorbed sodium (blue) or magnesium (orange) ions for polyE when simulated in the presence of 0.125 M NaCl or $MgCl_2$. Ion adsorption is defined based on a cutoff distance of 4 Å of the ions from any peptide atom, and the number of adsorbed ions is quantified by averaging the ions that satisfy the cutoff throughout the analyzed trajectory.

Thus far, we did not observe any significant difference between negatively and positively charged polyelectrolytes. However, a plot of Rg against the number of ions for polyelectrolytes in the presence of NaCl compared with $MgCl_2$ shows that the Rg values of polyD and polyE decrease from ~22 Å in NaCl to 18 Å in $MgCl_2$, whereas for polyK and polyR, the Rg values are less affected by changing the cations from monovalent $Na^+$ to divalent $Mg^{2+}$ (Figure 5A, filled vs. empty circles). We note that the adsorption of ions on uncharged peptides (i.e., polyGS) is negligible. The number of adsorbed mono- or divalent ions on polyGS ranges between 0 and 1 ions, regardless of the ionic strength.

Figure 5B shows a representative 2D distribution of polyE at 0.125 M NaCl and 0.125 M $MgCl_2$, again showing the strong effect of cation valency on Rg for a negatively charged polyelectrolytic IDR.

## 4. Discussion and Conclusions

In this study, we investigated the conformational and polymeric properties of two positively charged homo-polypeptides (polyK and polyR) and two negatively charged homo-polypeptides (polyD and polyE). These charged homo-polypeptides are similar to polyelectrolytic sequences found in natural proteins, which often comprise repeats of K or R and of D or E. Some natural polyelectrolytic IDRs have high charge density per residue, but it is lower than unity, as they comprise neutral residues or residues with opposite charge. Here, we only focused on polyanionic and polycationic sequences, which are widespread in natural proteins. These stretches are often attached to folded domains and thus affect their function [30,31]. Quantifying the molecular biophysics of isolated polyelectrolytic peptides is essential towards understanding their role in biomolecular function, for example, via intra- or inter-molecular binding to other domains, either folded or disordered.

The current computational characterization of polyK, polyR, polyD, and polyE was motivated by a recent bioinformatic study that showed substantial differences between D/E and K/R repeats. K/R repeats were found to be much shorter and less common than D/E repeats. Although several potential biological explanations have been suggested to address these differences, here, we quantify their conformational properties to examine the possibility that the bias towards D/E repeats has a molecular biophysical basis.

Atomistic MD simulations show that the conformations adopted by the four charged homo-polypeptides are extended compared with typical neutral IDP conformations. This is illustrated by their respective mean Rg values, which are at least two times greater for polyelectrolytic peptides than for polyGS. The extended conformations are also reflected in the Flory exponent values of ~0.9 for polyelectrolytes compared with ~0.5 for polyGS. The difference between polyelectrolytic peptides and uncharged IDPs originates, as expected, from intra-molecular electrostatic repulsions, which also lead to a smaller conformational space. This electrostatic repulsion can be modulated by increasing the salt concentration. Increasing the concentration of NaCl results in the polyelectrolytic peptides adopting more compact conformations, with a lower Flory exponent, as well as in greater conformational heterogeneity.

Our study reveals some differences between the two positively charged homo-polypeptides and between the two negatively charged homo-polypeptides. Within the positively charged pair, polyR is more compact than polyK, whereas within the negatively charged pair, polyD is more compact than polyE. In addition, polyR is more sensitive to salt concentrations than polyK. This greater response to salt is also found for polyD compared with polyE, but to a lesser extent. The effect of salt on polyR and polyD correlates with the higher tendency of these polyelectrolytes to adsorb ions ($Na^+$ and $Cl^-$ by polyD and polyR, respectively).

Furthermore, a clear difference between the positively (polyK and polyR) and negatively (polyD and polyE) charged homo-polypeptides is observed when the simulation involves a divalent cation ($Mg^{+2}$). Although all homo-polypeptides adsorb a similar number of ions when simulated in the presence of $MgCl_2$, the negatively charged homo-polypeptides become much more compact compared with the effect observed when simulated with NaCl. Recently, a computational study of the solvation of isolated D, E, K, and R reported a more favorable hydration free energy for D and E than for K and R [32]. Furthermore, the heat capacities of the hydration of D and E have an opposite sign to those of K and R. The negative heat capacities of D and E have been attributed to differences in the hydration structure and the propagation of these effects beyond the first hydration shell. Our study also shows a higher tendency of D to adsorb both monovalent and divalent cations than E. This is in accordance with a recent study showing a greater number of calcium ions next to D than next to E, which was argued to explain their different roles in biomineralization processes [33].

In summary, alongside biological explanations for the abundance of D/E repeats over K/R repeats as possibly arising from their providing greater resistance to proteolysis or enabling more efficient translation by the ribosome [16], the current study also identifies biophysical differences between them. D/E repeats may have a more favorable solvation energy but are also more sensitive to cation valency and its effects on their degree of compaction. The abundance of polyelectrolytic peptides in various proteins may suggest that the understanding of their functional role is incomplete. The function and biophysical characteristics of polyelectrolytic peptides should be further addressed in the future both for polyelectrolytic homo- and hetero-peptides. The effect of the composition and pattern of Asp and Glu in polyelectrolytic hetero-peptides (or of Arg and Lys in polyelectrolytic hetero-peptides) on the biological function of polyelectrolytic peptides is unclear and may correspond to their specificity.

**Author Contributions:** Conceptualization, Y.L. and L.S.B.; methodology, Y.L. and L.S.B.; formal analysis, L.S.B.; investigation, L.S.B.; writing—review and editing, Y.L. and L.S.B.; visualization, L.S.B.; supervision, Y.L.; project administration, Y.L.; funding acquisition, Y.L. All authors have read and agreed to the published version of the manuscript.

**Funding:** This work was supported by Israel–United States Binational Science Foundation (2020624), by Israel Science Foundation (2072/22), and by a research grant from Estate of Betty Weneser.

**Institutional Review Board Statement:** Not applicable.

**Informed Consent Statement:** Not applicable.

**Data Availability Statement:** Not applicable.

**Conflicts of Interest:** The authors declare no conflict of interest.

## References

1. Van Der Lee, R.; Buljan, M.; Lang, B.; Weatheritt, R.J.; Daughdrill, G.W.; Dunker, A.K.; Fuxreiter, M.; Gough, J.; Gsponer, J.; Jones, D.T.; et al. Classification of Intrinsically Disordered Regions and Proteins. *Chem. Rev.* **2014**, *114*, 6589–6631. [CrossRef] [PubMed]
2. Uversky, V.N. The most important thing is the tail: Multitudinous functionalities of intrinsically disordered protein termini. *FEBS Lett.* **2013**, *587*, 1891–1901. [CrossRef] [PubMed]
3. Oldfield, C.J.; Dunker, A.K. Intrinsically Disordered Proteins and Intrinsically Disordered Protein Regions. *Annu. Rev. Biochem.* **2014**, *83*, 553–584. [CrossRef]
4. Das, R.K.; Ruff, K.M.; Pappu, R.V. Relating sequence encoded information to form and function of intrinsically disordered proteins. *Curr. Opin. Struct. Biol.* **2015**, *32*, 102–112. [CrossRef]
5. Uversky, V.N.; Gillespie, J.R.; Fink, A.L. Why are "natively unfolded" proteins unstructured under physiologic conditions? *Proteins* **2000**, *41*, 415–427. [CrossRef] [PubMed]
6. Muller-Spath, S.; Soranno, A.; Hirschfeld, V.; Hofmann, H.; Ruegger, S.; Reymond, L.; Nettels, D.; Schuler, B. From the Cover: Charge interactions can dominate the dimensions of intrinsically disordered proteins. *Proc. Natl. Acad. Sci. USA* **2010**, *107*, 14609–14614. [CrossRef]
7. Hofmann, H.; Soranno, A.; Borgia, A.; Gast, K.; Nettels, D.; Schuler, B. Polymer scaling laws of unfolded and intrinsically disordered proteins quantified with single-molecule spectroscopy. *Proc. Natl. Acad. Sci. USA* **2012**, *109*, 16155–16160. [CrossRef]
8. Bianchi, G.; Longhi, S.; Grandori, R.; Brocca, S. Relevance of Electrostatic Charges in Compactness, Aggregation, and Phase Separation of Intrinsically Disordered Proteins. *Int. J. Mol. Sci.* **2020**, *21*, 6208. [CrossRef]
9. Vuzman, D.; Azia, A.; Levy, Y. Searching DNA via a "Monkey Bar" Mechanism: The Significance of Disordered Tails. *J. Mol. Biol.* **2010**, *396*, 674–684. [CrossRef]
10. Vuzman, D.; Levy, Y. DNA search efficiency is modulated by charge composition and distribution in the intrinsically disordered tail. *Proc. Natl. Acad. Sci. USA* **2010**, *107*, 21004–21009. [CrossRef]
11. Vuzman, D.; Levy, Y. Intrinsically disordered regions as affinity tuners in protein–DNA interactions. *Mol. Biosyst.* **2011**, *8*, 47–57. [CrossRef] [PubMed]
12. Bigman, L.S.; Greenblatt, H.M.; Levy, Y. What Are the Molecular Requirements for Protein Sliding along DNA? *J. Phys. Chem. B* **2021**, *125*, 3119–3131. [CrossRef] [PubMed]
13. Hazra, M.K.; Levy, Y. Charge pattern affects the structure and dynamics of polyampholyte condensates. *Phys. Chem. Chem. Phys.* **2020**, *22*, 19368–19375. [CrossRef] [PubMed]
14. Hazra, M.K.; Levy, Y. Biophysics of Phase Separation of Disordered Proteins Is Governed by Balance between Short- And Long-Range Interactions. *J. Phys. Chem. B* **2021**, *125*, 2202–2211. [CrossRef] [PubMed]
15. Hazra, M.K.; Levy, Y. Affinity of disordered protein complexes is modulated by entropy–energy reinforcement. *Proc. Natl. Acad. Sci. USA* **2022**, *119*, e2120456119. [CrossRef] [PubMed]
16. Bigman, L.S.; Iwahara, J.; Levy, Y. Negatively Charged Disordered Regions are Prevalent and Functionally Important Across Proteomes. *J. Mol. Biol.* **2022**, *434*, 167660. [CrossRef]
17. Xie, L.; Jakob, U. Inorganic polyphosphate, a multifunctional polyanionic protein scaffold. *J. Biol. Chem.* **2019**, *294*, 2180–2190. [CrossRef]
18. Yoshida-Moriguchi, T.; Campbell, K.P. Matriglycan: A novel polysaccharide that links dystroglycan to the basement membrane. *Glycobiology* **2015**, *25*, 702–713. [CrossRef]
19. Hosaka, M.; Nagahama, M.; Kim, W.; Watanabe, T.; Hatsuzawa, K.; Ikemizu, J.; Murakami, K.; Nakayama, K. Arg-X-Lys/Arg-Arg motif as a signal for precursor cleavage catalyzed by furin within the constitutive secretory pathway. *J. Biol. Chem.* **1991**, *266*, 12127–12130. [CrossRef]
20. Leininger, S.E.; Rodriguez, J.; Vu, Q.V.; Jiang, Y.; Li, M.S.; Deutsch, C.; O'Brien, E.P. Ribosome Elongation Kinetics of Consecutively Charged Residues Are Coupled to Electrostatic Force. *Biochemistry* **2021**, *60*, 3223–3235. [CrossRef]

21. Lu, J.; Deutsch, C. Electrostatics in the Ribosomal Tunnel Modulate Chain Elongation Rates. *J. Mol. Biol.* **2008**, *384*, 73–86. [CrossRef] [PubMed]
22. Milo, R.; Phillips, R. *Cell Biology by the Numbers*; Garland Science: New York City, NY, USA, 2015. [CrossRef]
23. Pronk, S.; Páll, S.; Schulz, R.; Larsson, P.; Bjelkmar, P.; Apostolov, R.; Shirts, M.R.; Smith, J.C.; Kasson, P.M.; Van Der Spoel, D.; et al. GROMACS 4.5: A high-throughput and highly parallel open source molecular simulation toolkit. *Bioinformatics* **2013**, *29*, 845–854. [CrossRef] [PubMed]
24. Huang, J.; Rauscher, S.; Nawrocki, G.; Ran, T.; Feig, M.; de Groot, B.L.; Grubmüller, H.; MacKerell, A.D., Jr. CHARMM36m: An improved force field for folded and intrinsically disordered proteins. *Nat. Methods* **2017**, *14*, 71–73. [CrossRef]
25. Hess, B.; Bekker, H.; Berendsen, H.J.C.; Fraaije, J.G.E.M. LINCS: A linear constraint solver for molecular simulations. *J. Comput. Chem.* **1997**, *18*, 1463–1472. [CrossRef]
26. Bussi, G.; Donadio, D.; Parrinello, M. Canonical sampling through velocity rescaling. *J. Chem. Phys.* **2007**, *126*, 014101. [CrossRef] [PubMed]
27. Michaud-Agrawal, N.; Denning, E.J.; Woolf, T.B.; Beckstein, O. MDAnalysis: A toolkit for the analysis of molecular dynamics simulations. *J. Comput. Chem.* **2011**, *32*, 2319–2327. [CrossRef]
28. Flory, P.J. *Principles of Polymer Chemistry*; Cornell University Press: Ithaca, NY, USA, 1953.
29. Vitalis, A.; Wang, X.; Pappu, R.V. Quantitative Characterization of Intrinsic Disorder in Polyglutamine: Insights from Analysis Based on Polymer Theories. *Biophys. J.* **2007**, *93*, 1923–1937. [CrossRef]
30. Wang, X.; Greenblatt, H.M.; Bigman, L.S.; Yu, B.; Pletka, C.C.; Levy, Y.; Iwahara, J. Dynamic Autoinhibition of the HMGB1 Protein via Electrostatic Fuzzy Interactions of Intrinsically Disordered Regions. *J. Mol. Biol.* **2021**, *433*, 167122. [CrossRef]
31. Wang, X.; Bigman, L.; Greenblatt, H.; Yu, B.; Levy, Y.; Iwahara, J. Negatively charged, intrinsically disordered regions can accelerate target search by DNA-binding proteins. *Nucleic Acid Res.* **2023**, gkad045. [CrossRef]
32. Fossat, M.J.; Zeng, X.; Pappu, R.V. Uncovering Differences in Hydration Free Energies and Structures for Model Compound Mimics of Charged Side Chains of Amino Acids. *J. Phys. Chem. B* **2021**, *125*, 4148–4161. [CrossRef]
33. Lemke, T.; Edte, M.; Gebauer, D.; Peter, C. Three Reasons Why Aspartic Acid and Glutamic Acid Sequences Have a Surprisingly Different Influence on Mineralization. *J. Phys. Chem. B* **2021**, *125*, 10335–10343. [CrossRef] [PubMed]

*biomolecules*

MDPI

*Article*

# Different Forms of Disorder in NMDA-Sensitive Glutamate Receptor Cytoplasmic Domains Are Associated with Differences in Condensate Formation

**Sujit Basak [1], Nabanita Saikia [2], David Kwun [1], Ucheor B. Choi [3], Feng Ding [4] and Mark E. Bowen [1,*]**

1 Department of Physiology and Biophysics, Stony Brook University, Stony Brook, NY 11794, USA
2 Department of Chemistry, Navajo Technical University, Crownpoint, NM 87313, USA
3 Quantum-Si, Inc., Guilford, CT 06437, USA
4 Department of Physics and Astronomy, Clemson University, Clemson, SC 29634-0978, USA
* Correspondence: mark.bowen@stonybrook.edu

**Abstract:** The N-methyl-D-aspartate (NMDA)-sensitive glutamate receptor (NMDAR) helps assemble downstream signaling pathways through protein interactions within the postsynaptic density (PSD), which are mediated by its intracellular C-terminal domain (CTD). The most abundant NMDAR subunits in the brain are GluN2A and GluN2B, which are associated with a developmental switch in NMDAR composition. Previously, we used single molecule fluorescence resonance energy transfer (smFRET) to show that the GluN2B CTD contained an intrinsically disordered region with slow, hop-like conformational dynamics. The CTD from GluN2B also undergoes liquid–liquid phase separation (LLPS) with synaptic proteins. Here, we extend these observations to the GluN2A CTD. Sequence analysis showed that both subunits contain a form of intrinsic disorder classified as weak polyampholytes. However, only GluN2B contained matched patterning of arginine and aromatic residues, which are linked to LLPS. To examine the conformational distribution, we used discrete molecular dynamics (DMD), which revealed that GluN2A favors extended disordered states containing secondary structures while GluN2B favors disordered globular states. In contrast to GluN2B, smFRET measurements found that GluN2A lacked slow conformational dynamics. Thus, simulation and experiments found differences in the form of disorder. To understand how this affects protein interactions, we compared the ability of these two NMDAR isoforms to undergo LLPS. We found that GluN2B readily formed condensates with PSD-95 and SynGAP, while GluN2A failed to support LLPS and instead showed a propensity for colloidal aggregation. That GluN2A fails to support this same condensate formation suggests a developmental switch in LLPS propensity.

**Keywords:** glutamate receptor; intrinsically disordered protein; discrete molecular dynamics; single molecule fluorescence; liquid-liquid phase separation

**Citation:** Basak, S.; Saikia, N.; Kwun, D.; Choi, U.B.; Ding, F.; Bowen, M.E. Different Forms of Disorder in NMDA-Sensitive Glutamate Receptor Cytoplasmic Domains Are Associated with Differences in Condensate Formation. *Biomolecules* **2023**, *13*, 4. https://doi.org/10.3390/biom13010004

Academic Editors: Prakash Kulkarni, Stefania Brocca, Keith Dunker and Sonia Longhi

Received: 21 October 2022
Revised: 5 December 2022
Accepted: 10 December 2022
Published: 20 December 2022



## 1. Introduction

The N-methyl–D-Aspartate (NMDA)-sensitive glutamate receptor (NMDAR) plays a pivotal role in excitatory synaptic transmission and synaptic plasticity, which impacts learning, memory, and cognition [1,2]. NMDARs are heterotetrametric formed from two GluN1 and two GluN2 subunits, which can be GluN2A, GluN2B, or mixtures of different isoforms [3,4]. NMDARs have four structurally-separable domains: the extracellular amino terminal domain (ATD), the ligand binding domain (LBD), the transmembrane domain (TMD), and the intracellular C-terminal domain (CTD). These domains work together, enabling NMDARs to function as a ligand-gated ion channel. The binding of glutamate and glycine to the extracellular LBDs propagates a conformational change leading to the opening (i.e., gating) of the ion conduction pore in the transmembrane domain [3,5]. The gating propensity is further modulated by both the extracellular ATD [6] as well as the intracellular CTD [7,8].

Whole-exome sequencing revealed that mutations in NMDARs are associated with neuropsychiatric disorders [9]. While the majority of mutations are found within the LBD and TMD, several disease-associated mutations fall within the CTDs of GluN2A and GluN2B [10,11]. Knowledge of NMDAR structure is necessary to understand the molecular basis of these disorders. The structure of NMDARs is almost entirely known, from structural studies of the extracellular and transmembrane domains [12,13]. Thus, our understanding of the mechanism for ligand-induced gating is nearly complete [4]. However, structural information about the intracellular CTD has proved elusive due the presence of intrinsic disorder [14–16]. In GluN2A and GluN2B, the CTD is the largest single domain in the protein and appears to be split into two separate subdomains (CTD1 and CTD2) by a central palmitoylation motif [17] (Figure 1A). The full CTDs have never been characterized due to their limited solubility. Previously, we confirmed experimentally the presence of intrinsic disorder in CTD2 from GluN2B and identified slow timescale conformational dynamics [18]. However, no information is available for GluN2A.

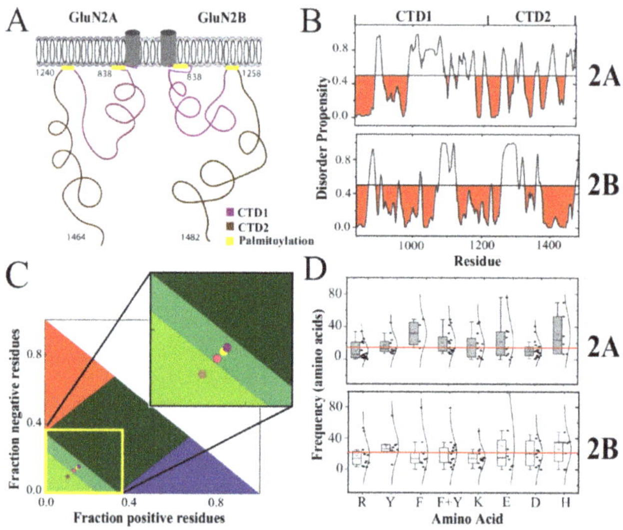


**Figure 1. Prediction and classification of intrinsic disorder in the cytoplasmic domains of the GluN2A and GluN2B.** (**A**) Cartoon schematic of domain organization in the intracellular C-terminal domain (CTD) of the GluN2A and GluN2B subunits of the NMDA receptor. The CTD is connected to the M4 helix within the transmembrane domain [3]. The essential palmitoylation sites (**yellow**) mediate attachment to the membrane [17]. The subdomains demarked by palmitoylation are termed CTD1 (**purple**) and CTD2 (**brown**). (**B**) The disorder propensity from PONDR is plotted for the CTDs of GluN2A and GluN2B with regions predicted to be order-prone (PONDR scores < 0.5) highlighted in **red**. (**C**) Classification of Intrinsically Disordered Ensemble Regions (CIDER) analysis [19] of the CTD subdomains. Colored regions indicate conformational classes of IDPs showing the boundaries for positive polyelectrolytes (**red**), negative polyelectrolytes (**blue**), strong polyampholytes (**dark green**) intermediate polyampholytes (**mint green**) and weak polyampholytes (**pea green**, lower left) [20]. Circles representing the CTD subdomains are placed based on their classification by CIDER analysis. The CTD1 of GluN2A (**yellow circle**) and CTD2 of GluN2A (**magenta circle**) are classified as intermediate polyampholytes. The CTD1 of GluN2B (**purple circle**) is also classified as an intermediate polyampholyte. However, CTD2 of GluN2B (**pink circle**) is classified as a weak polyampholyte. (**D**) The separation between residues within CTD2A and CTD2B are represented by a boxplot with the Gaussian distribution of its recurrence. Shown are the distributions for arginine (R), tyrosine (Y), phenylalanine (F), total aromatics (F + Y), lysine (K), aspartate (D), glutamate (E) and histidine (H). The mean frequency, of all the above-mentioned residues within each isoform, is highlighted with a red line. The standard deviation for the boxplot indicated by black bars.

In addition to allosteric modulation of gating, the CTD plays a major role in the formation of postsynaptic signaling complexes through interactions with the scaffolding protein post synaptic density protein of 95 kDa (PSD-95) [21] along with numerous other signaling proteins [22]. Thus, the CTD plays a role in the initiation of signaling cascades, which is separate from its role in ion channel gating [23,24]. Recent reports have shown that PSD-95 and the GluN2B CTD are capable of liquid–liquid phase separation (LLPS) in vitro with a recombinant synGAP [25–28]. Proteins containing intrinsic disorder are key players in LLPS because exposed aromatic "sticker" residues enable multivalent interactions [29–31]. The postsynapse has long been known to contain condensates, which have been termed the postsynaptic density (PSD) [32–34]. The formation of condensates in both the presynapse and postsynapse have been linked to LLPS [35,36].

Here, we compared the CTD2 domains from GluN2A and GluN2B using sequence analysis, discrete molecular dynamics simulations (DMD), and single molecule FRET (smFRET). Analysis of the amino acid sequences suggested differences between the subunits in the form of disorder [20]. DMD revealed differences in polypeptide compaction, with GluN2A favoring extended states while GluN2B remained globular. We did not observe any slow timescale dynamics in single molecule fluorescence measurements GluN2A, which we previously observed in GluN2B [37]. To understand how these differences in disorder affected protein interactions, we compared GluN2A and 2B for the ability to undergo LLPS using sedimentation and differential interference contrast (DIC) microscopy. This revealed that GluN2A was not capable of supporting LLPS while GluN2B lowers the concentration regime for phase separation with PSD-95 and synGAP [25]. Given the developmental switch in these receptor isoforms [38,39], this would imply an associated switch in LLPS propensity at the synaptic membrane with a higher propensity for LLPS during early development and then decreasing LLPS propensity as GluN2A comes to predominate.

## 2. Materials and Methods

### 2.1. Protein Purification

The C-terminal domain 2 (CTD2) of GluN2A (residues 1239–1464, CTD2A) and of GluN2B (residues 1259–1482, CTD2B) from *Rattus norvegicus* were expressed in the Rosetta strain of *Escherichia coli* (MilliporeSigma, Burlington, MA, USA) from the expression vector pPROEX HTB (ThermoFisher Scientific, Waltham, MA, USA), which imparts an N-teminal 6-His tag [15,40]. The CTD2 cell pellets were lysed under denaturing and reducing conditions, which were maintained during affinity purification. For CTD2B, the protein was eluted in denaturant free buffer [40], but for GluN2A, the protein was maintained in a nondenaturing concentration of urea (2 M) to prevent aggregation. The 6-His tags were removed using tobacco etch virus protease (TEV), which is unaffected by 2M urea. Subsequent rounds of cation exchange and size exclusion chromatography on Superdex S-200 (Cytiva, Marlborough, MA, USA) were used to obtain protein purity of 95% or greater as verified using SDS-PAGE. Full-length PSD-95 from *Rattus norvegicus* was expressed in the Rosetta 2 strain of *E. coli* and purified by a combination of Ni-affinity, anion exchange, and size exclusion chromatography as previously described [41]. The recombinant construct containing the N-terminal coil-coiled (CC) fused to the PSD-95 binding motif (PBM) of synGAP was a kind gift from Mingjie Zhang and was expressed and purified as described [25].

To enable fluorescent labeling, we used two native cysteines in CTD2A (C1239 and C1412) with the three remaining native cysteines (C1241, C1387, and C1448) changed to serine through classic site-directed mutagenesis as confirmed by DNA sequencing. For CTD2B, there was not a suitable native cysteine pair, so we introduced a cysteine at S1273 and paired this with a native cysteine at 1445. The two remaining native cysteines in CTD2B (C1394 and C1455) were mutated to serine.

### 2.2. Single Molecule Total Internal Reflection Fluorescence (smTIRF) Microscopy

The purified CTD2s were randomly labeled with an equimolar ratio of Alexa Fluor 555 C5 maleimide and Alexa Fluor 647 C2 maleimide (Thermo Fisher Scientific) overnight at 4 °C in 25 mM HEPES, pH 7.4, 300 mM NaCl, and 0.5 mM tris (2-carboxyethyl) phosphine (TCEP). All buffers included 2M urea for CTD2A. Unconjugated dye was removed by desalting with Sephadex G50 (Cytiva, Marlborough, MA, USA) followed by dialysis. Fluorescently labeled CTD2s were N-terminally biotinylated by adding a 5-fold molar excess NHS-LC-Biotin with a ~2 nm spacer (ThermoFisher Scientific, Waltham, MA, USA) in 50 mM potassium phosphate buffer at pH 6.5 to direct the reaction to the N-terminus. The reaction mixture was incubated at 4 °C overnight, followed by desalting to remove free biotin.

Biotinylated proteins were attached via streptavidin to a quartz slide passivated with biotinylated BSA and a mixture of Biolipidure 203 and 206 (NOF AMERICA Corporation, White Plains, NY, USA). Alternating illumination using diode lasers at 532 nm (Laser Quantum) and 640 nm (Coherent) allowed for the identification of optically resolved single molecules containing one donor and one acceptor. Samples were excited using prism-based TIRF. Images were acquired on an Olympus IX-71 microscope with a 60X-1.2 NA water-immersion objective. Fluorescence emission collected from donor and acceptor were spectrally separated using an optosplit emission image splitter (Cairn Research, Faversham, UK) and relayed onto a single Andor iXon EMCCD camera (Andor Technology, Ltd., Belfast, UK). Data were collected at 10 frames/second. All smFRET measurements were performed in 50 mm tris, 200 mM NaCl, pH 7.4, and supplemented with 1 mM cyclooctatetraene, 0.8% *w/v* glucose, 7.5 units/mL glucose oxidase, and 1000 units/mL catalase. Microscopy data were analyzed using MATLAB to correlate donor and acceptor images, extract single molecule intensity time traces, and calculate FRET efficiency [42].

### 2.3. Discrete Molecular Dynamics (DMD) Simulations

DMD is a molecular dynamics algorithm that has been shown to have high predictive power and sampling efficiency in studying conformational dynamics of IDPs [43–45]. Details of DMD methods can be found in [46,47]. To sample the conformational free energy landscape efficiently, we performed replica exchange DMD simulations with 18 neighboring replicas in the temperature range of 275–360 K. Both proteins started from an extended conformation and reached equilibrium quickly in DMD simulations as indicated by the distributions of the radius of gyration and secondary structure contents (Figure 2). We used the conformations sampled in rxDMD within the temperature range of 300–310 K to compare the conformational difference between CTD2A and CTD2B.

The secondary structure was calculated using the DSSP program. The hydrogen bond was considered to be formed when the N···O distance was within 3.5 Å, and the N–H···O angle was more than 120°. A pairwise residue contact was defined as the distance between the heavy atoms from the main chain or side chain of two nonsequential residues within 0.65 nm.

### 2.4. Measurement of Turbidity

Full-length PSD-95, CC-PBM from SynGAP, and CTD2A or CTD2B were mixed at 1:1:1 ratio in 20 mM tris, 150 mM NaCl, at pH 7.4 buffer with each protein at a final concentration of 20 μM. Samples were equilibrated in polypropylene microfuge tubes at room temperature (25 °C) for 5 min before measurements. Transmittance of aliquots removed from the incubation was measured at 550 nm in quartz cuvettes with a 1 cm path length using an Agilent model 8453 UV-Vis spectrophotometer. Turbidity was calculated as percentage of transmittance. The time-dependent turbidity formation seen with CTD2A was slower than condensate formation induced by the ternary mixture. As such, we can remove urea from CTD2A for condensate formation experiments without the appearance of CTD2A precipitation.

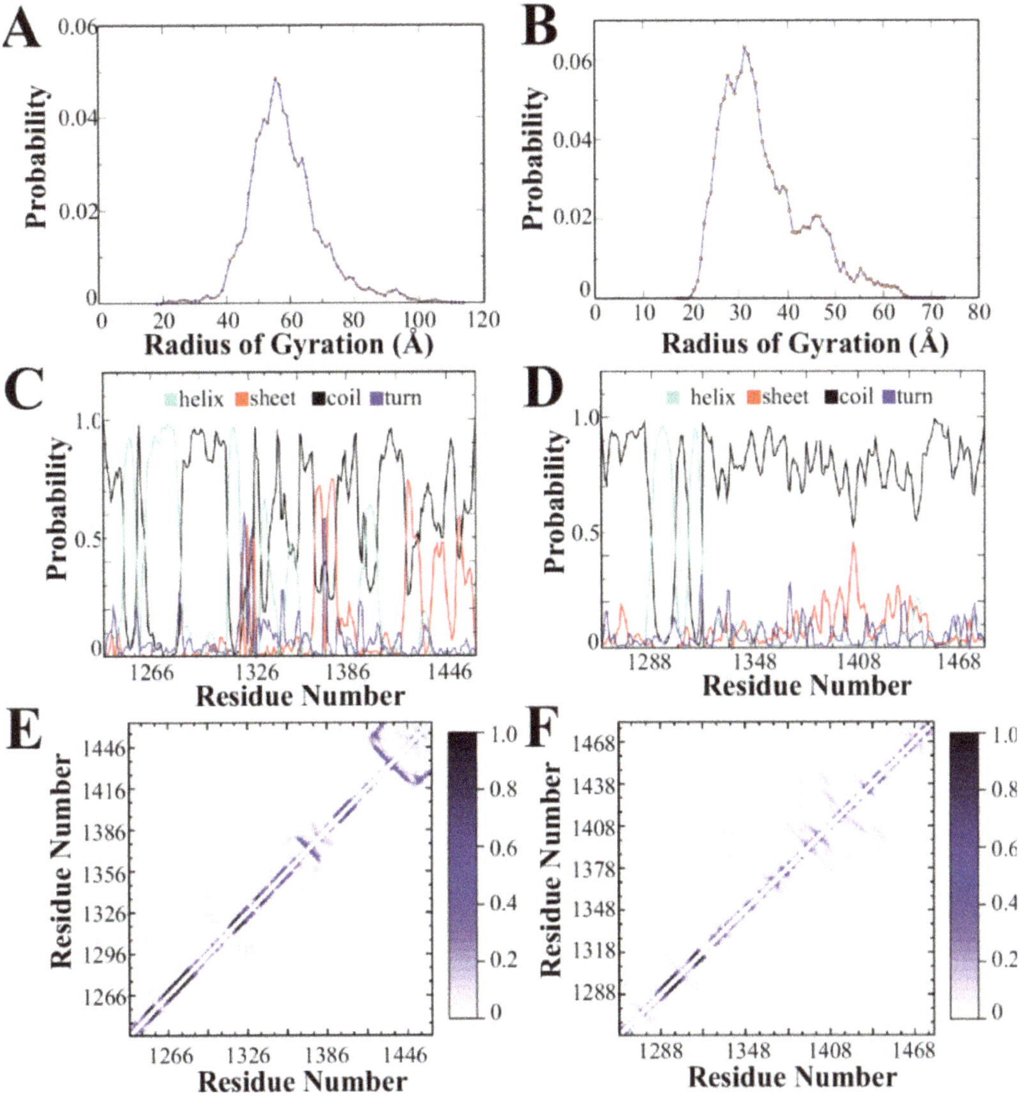


**Figure 2. Discrete molecular dynamics simulation of the CTD2 subdomains from GluN2A and GluN2B.** (**A**) Distribution of the radius of gyration derived from simulations for GluN2A and (**B**) GluN2B. GluN2A favors more extended states. (**C**) The calculated per residue probabilities of different secondary structures based on occupancy observed during simulations for GluN2A and (**D**) GluN2B. Shown are the probability of an individual residue adopting α-helical (light blue), β-sheet (red), random coil (black), and turn (dark blue) conformations. Random coil was the dominant secondary structure for both CTD2s, although GluN2A showed more stable secondary structural elements. (**E**) The pairwise residue-contact frequency maps show the intramolecular interactions observed in simulations of the CTD2 from GluN2A and (**F**) GluN2B. The associated color scale gives the probability of contact between two residues. GluN2A showed stable short-range interactions involved in stabilizing the local, ordered secondary structures. GluN2B showed more long-range contacts.

### 2.5. Sedimentation Analysis of Condensates

Full-length PSD-95, SynGAP CC-PBM, and CTD2A or CTD2B were mixed at 1:1:1 ratio in 20 mM tris-HCl, 150 mM NaCl, at pH 7.4 buffer with each protein at a final concentration of 20 µM. Samples were equilibrated in polypropylene microfuge tubes at room temperature (25 °C) for 5 min before sedimentation at 13,200× $g$ for 1 min. The isolated pellets were suspended in the original volume of buffer. Then, supernatant and pellet fractions were boiled in Laemmli buffer containing dithiothreitol and resolved by SDS-PAGE on 4–20% gradient gels, which were stained with Coomassie Brilliant Blue R-250 (Bio-Rad).

### 2.6. Differential Interference Contrast (DIC) and Fluorescence Microscopy

Full-length PSD-95, SynGAP CC-PBM, and CTD2A or CTD2B were mixed at 1:1:1 ratio in 20 mM tris-HCl, 150mM NaCl, at pH 7.4 with each protein at a final concentration of 20 µM. Samples were equilibrated in 8 chamber slides (Nunc, Lab-TEK II) at room temperature (25 °C) for 5 min before imaging. The chamber was passivated with BSA to avoid nonspecific interactions with the coverslip. The samples were imaged on a Nikon Eclipse Ti-E microscope with a 100X 1.4 NA oil-immersion objective equipped with prisms for DIC imaging and a Nikon total internal reflection fluorescence (TIRF) excitation module connected to a fiber-coupled laser launch. Images were recorded with an iXon electron multiplying charge-coupled device camera (Andor Technology, Ltd., Belfast, UK) and analyzed using Nikon Elements for background subtraction.

Proteins containing two unique cysteine residues (taken from our previous work [15,41]) were expressed and purified as described for wild type proteins. Full-length PSD-95, containing the mutations S398C and R492C, was labeled with Alexa 488 maleimide. CTD2B, containing the mutations S1273C and C1445, was labeled with Alexa 647 maleimide. The labeled proteins were isolated from the free dye by desalting with Sephadex G-50. The labeling efficiency was >98% for both proteins as determined with absorbance spectroscopy using the calculated extinction coefficients.

For imaging, 300 nM of labeled protein was used along with full-length PSD-95, CC-PBM from SynGAP, and CTD2B at 1:1:1 ratio in 20 mM Tris-HCl, 150 mM NaCl, at pH 7.4 at 20 µM concentration. Samples were mixed in 8 chamber slides (Nunc, Lab-TEK II) at room temperature (25 °C) for 5 min before imaging. The chamber was passivated with BSA to avoid nonspecific interactions with the surface. Laser excitation was introduced using highly inclined and laminated optical sheet (HILO) microscopy [48]. The samples were imaged with visible light using a DIC prism, HILO at 488 nm, and HILO at 642 nm. Fluorescence emission was separated from laser excitation using a 405/488/561/642 multi-band filter set (Chroma Technology Corp). Images were recorded with an iXon electron multiplying charge-coupled device camera (Andor Technology, Ltd.) and analyzed using Nikon Elements for background subtraction.

## 3. Results

### 3.1. Primary Sequence Analysis

The GluN2A and GluN2B subunits from *Rattus norvegicus* share a 71% sequence identity throughout their ordered extracellular and transmembrane domains. However, the sequence conservation drops to only 31% sequence identity within the CTDs. The sequence similarity is higher at 47% because the overall chemical composition is similar with a high proportion of serine and asparagine. Both CTDs contain two conserved cysteine clusters, which have been shown to be sites of palmitoylation that lead to membrane attachment once post-translationally modified [49,50]. Thus, the CTDs from both GluN2A and GluN2B share this organization of two subdomains demarcated by internal palmitoylation clusters, which we have termed CTD1 and CTD2 (Figure 1A). Excluding the palmitoylation motifs, the 26% sequence identity within CTD1 is slightly lower than within CTD2 at 36% identity (Table 1).

**Table 1. Sequence analysis of the GluN2A and GluN2B cytoplasmic domains.** The amino acid sequences for the CTDs from GluN2A and GluN2B were analyzed with Classification of Intrinsically Disordered Ensemble Regions (CIDER) [19] The CIDER analysis was performed on the full CTD or the individual subdomains as indicated. The **Sequence** indicates the residue numbers used as boundaries for the analyses of individual subdomains. The **Kappa** value measures the segregation of positive and negative charges within the polypeptide. A kappa value of one indicates a perfect segregation of charge while a value of zero is perfectly mixed. The Fraction of Charged Residues (**FCR**) indicates the ratio of residues containing positive or negative charge to the total number of residues. The Net Charge per Residue (**NCPR**) is the difference between the fraction of positively charged residues and the fraction of negatively charged residues [51]. The **Hydropathy** value reports the mean hydropathy across the indicated polypeptide sequence. The Kyte-Doolittle hydropathy was rescaled from 0 (hydrophilic) and 1 (hydropathy) and then calculated for groups of five residues using a scanning window.

| Protein | Sequence | Kappa | FCR | NCPR | Hydropathy |
|---|---|---|---|---|---|
| **GluN2A** | 838–1464 | 0.158 | 0.264 | 0.018 | 3.5 |
| **GluN2B** | 838–1482 | 0.183 | 0.245 | 0.025 | 3.6 |
| **CTD1A** | 873–1211 | 0.138 | 0.286 | 0.009 | 3.3 |
| **CTD1B** | 874–1212 | 0.176 | 0.292 | 0.009 | 3.4 |
| **CTD2A** | 1243–1462 | 0.204 | 0.261 | 0.009 | 3.6 |
| **CTD2B** | 1250–1482 | 0.221 | 0.206 | 0.026 | 3.8 |

Previously, we used PONDR to show that the CTD from GluN2B was predicted to contain intrinsically disordered regions (IDRs) [18,52,53]. Here, we used PONDR to compare the distribution of IDRs within GluN2A and GluN2B (Figure 1B). From this analysis, we observed that both subunits have similar predictions of an order-forming region after the transmembrane domain, which is broken up by an IDR. In GluN2B, CTD1 is predicted to be order-prone to around residue 1075 with only short disordered motifs. In contrast, the GluN2A CTD1 is predicted to contain a long IDR between residues 915 and 987. Both isoforms also have a prediction of an IDR at the beginning of CTD2, which is longer in GluN2B. However, the distal half of CTD2 in GluN2B is predicted to be order-prone until just before the C-terminus. In contrast, the distal half of CTD2 in GluN2A contains a mixture of short disorder and order-prone motifs. The C-terminus of both isoforms contains the PSD-95 binding motif. The two isoforms differ with GluN2A containing a predicted IDR preceding the C-terminal PSD-95 binding motif, while GluN2B is predicted to be order-prone.

To provide more detail on the differences in sequence features between these isoforms, we performed Classification of Intrinsically Disordered Ensemble Regions (CIDER) for the CTDs [19]. Interestingly, the CTDs from both GluN2A and GluN2B have a relatively low fraction of charged residues (FCR) and low net charge per residue (NCPR) for a protein containing intrinsic disorder (Table 1), which is often associated with a high FCR [51,54,55]. In CTD1, the FCR was comparable for both isoforms and the low NCPR values classify them as weak polyampholytes (Figure 1C). The segregation of positive and negative charges within the polypeptide (kappa [20]) was 28% higher in the GluN2B CTD1, although both isoforms were relatively well-mixed. The CTD2 from GluN2A showed a 27% higher FCR than GluN2B (0.206 and 0.261, respectively). Surprisingly, the GluN2B CTD2 showed an almost 3-fold higher NCPR than GluN2A (NCPR = 0.026 and 0.009, respectively). GluN2B also had a slightly higher kappa value in both subdomains, indicating a higher degree of charge segregation. Overall, both CTD2s have fewer charged residues compared to CTD1, but the charges are more segregated in CTD2, which can influence the form of disorder [20,31].

According to our CIDER analysis, both CTDs are classified as disordered globules rather than extended polymers [20]. Both CTD1 and CTD2 from GluN2A, along with CTD1 of GluN2B, lie on the border between strong and weak polyampholytes, which makes their conformational behavior hard to predict. In contrast, the CTD2 of GluN2B is classified as a weak polyampholyte (Figure 1C, inset). Protein sequences in this region of the CIDER plot have a high tendency to form collapsed globules [56]. Based on this analysis, the amino acid sequence of CTD2 from GluN2B appears to have evolved to adopt a different form of intrinsic disorder.

Recombinant constructs based on the CTD2 subdomain of GluN2B have been shown capable of participating in LLPS [25–28]. LLPS in IDPs has been linked to amino acid patterning, particularly of aromatic and arginine residues, which participate in cation–$\pi$ and $\pi$–$\pi$ interactions [29,31]. Similarly, the distribution of charged residues has been linked to the form of intrinsic disorder [20]. To analyze residue patterning within CTD2, we plotted the separation between repeated amino acids in boxplot format along with the Gaussian distribution of their frequency (Figure 1D). Both GluN2A and GluN2B have a similar number of aromatic residues within CTD2 with similar frequency. In GluN2A, these tend to be tyrosine, whereas in GluN2B, phenylalanine predominates. In GluN2A, the frequency of arginine residues is half that of the aromatic residues ($10 \pm 10$ compared to $21 \pm 14$, respectively; $p = 0.015$), while in GluN2B, the frequency of arginine and aromatic residues is the same ($19 \pm 17$ compared to $22 \pm 18$, respectively). Additionally, GluN2A has a higher density of negatively charged residues along with fewer lysines, resulting in only four unpaired arginine residues. In contrast, GluN2B has fewer negatively charged residues along with more lysines, which results in eight unpaired arginine residues. Thus, the GluN2B CTD2 has matched arginine and aromatic residue patterning that appears favorable for the cation–$\pi$ interactions, which support LLPS, while GluN2A appears to be dominated by electrostatic interactions resulting in the low NCPR.

### *3.2. Discrete Molecular Dynamics*

Based on amino acid sequence analysis, the CTD2 domains were predicted to adopt different forms of intrinsic disorder (Figure 1C). To understand how this difference manifests in the conformational free energy landscapes, we used replica-exchange discrete molecular dynamics (rxDMD) with 18 replicas, running at different temperatures, for a combined simulation time of 8.0 µs. The predictive power of DMD with the enhanced sampling of replica-exchange is well suited to describing the energy landscape of IDPs and folded proteins [43–45,57]. We performed rxDMD simulations for the CTD2 subdomain from both GluN2A and GluN2B as free polypeptides (Figure 2). Both proteins showed a highly dynamic and variable conformation. Examination of the radius of gyration ($R_g$) for the individual conformations sampled during the rxDMD trajectory revealed that GluN2A favored extended conformations starting at 40 Å but extending to 100 Å (Figure 2A). In contrast, GluN2B favored compact states with mode radii around 30 Å. However, the GluN2B $R_g$ distribution did contain a second peak with extended states out to 60 Å (Figure 2B). By analyzing all the snapshots from the DMD trajectory, we could calculate the secondary structural propensity along the polypeptide chain, which agreed well with PONDR predictions. Both CTD2 started with a pair of short $\alpha$-helices, followed by a disordered region, which is periodically interrupted by structured elements in GluN2A but continues uninterrupted in GluN2B (Figure 2C,D). The low propensity for secondary structure in GluN2B is in good agreement with our previous circular dichroism measurements [40].

Examination of the pairwise contact maps from rxDMD revealed few persistent long-range interactions in either CTD2s, as expected for IDPs (Figure 2E,F). However, comparison of the contact frequency maps revealed differences in medium- and short-range contacts (~20 to 60 residue separation), which were less pronounced in GluN2B. In contrast, GluN2A showed a central region with persistent contacts suggesting an order-prone domain. Additionally, the pairwise contact map shows the strongest short-range contacts in GluN2A at the C-terminus (Figure 2E). Thus, rxDMD found that both CTD2s share a helical region following the palmitoylation motif but diverge after this point. In GluN2A, there is a mixture of secondary structural elements along with two regions showing persistent contacts, which is in good general agreement with PODR predictions. In contrast, GluN2B was largely disordered throughout its length with minimal persistent contacts.

### 3.3. Single Molecule Fluorescence Resonance Energy Transfer (smFRET)

We previously used smFRET to show that recombinant CTD2 from GluN2B (CTD2B) displayed slow timescale conformational dynamics, which we termed hop-like intramolecular diffusion [15,18,37,40]. To probe conformational dynamics in GluN2A with smFRET, we created a recombinant CTD2 from GluN2A (CTD2A), which retained two native cysteines (C1239-C1412) for a separation of 173 residues. For CTD2B, there were no native cysteines with similar separation, so we paired the S1273C mutation with the native C1445 to achieve a separation of 172 residues. Thus, the contour length of the polypeptide between the points of measurement are similar for both protein constructs.

We immediately noticed differences between the CTD2 constructs during recombinant expression. While the CTD2B is highly soluble, CTD2A was prone to self-association displaying a slow accumulation of colloidal turbidity over time that eventually led to precipitation. We found that inclusion of urea was sufficient to forestall this process during protein handling and could be removed before any measurements. Proteins were randomly labeled to completion with an equimolar mixture of the Alexa 555 donor and the Alexa 647 acceptor dyes. The labeled protein was then selectively biotinylated at the N-terminus and attached to a passivated microscope slide that was functionalized with streptavidin. Once the proteins were surface attached, we removed all urea by rinsing and made measurements under urea-free conditions. The optical resolution between molecules allowed no possibility of intramolecular aggregation. Samples were excited using prism-based total internal reflection with alternating laser excitation to identify single molecules containing an active donor-acceptor pair.

Examination of the individual time traces for CTD2A revealed steady intensity until photobleaching but individual molecules persisted in high, medium, or low FRET states (Figure 3A). This is in stark contrast to the stochastic intensity transitions that we have repeatedly observed for CTD2B (Figure 3B) [18,40]. When we accumulated the molecules into population histograms, we observed that CTD2A showed three well-resolved peaks in the distribution: a low FRET peak encompassing 24% of the population along with a broader peak at intermediate FRET with 24% occupancy and a predominant high FRET peak with 52% occupancy (Figure 3C). In contrast, the population histogram for CTD2B showed a wide distribution that was fit by two broad peaks at low FRET and intermediate FRET (20% and 80% occupancy, respectively) without a distinct peak at high FRET (Figure 3D). The low and intermediate FRET peaks were of similar efficiency in both isoforms but much narrower in CTD2A than CTD2B, suggesting differences in the rate of conformational exchange [58,59]. The conformational dynamics of these IDPs are orders of magnitude faster than the time resolution of data collection (10 Hz). As such, the histograms represent the time-averaged distribution of states and do not provide information about the underlying rapid dynamics.

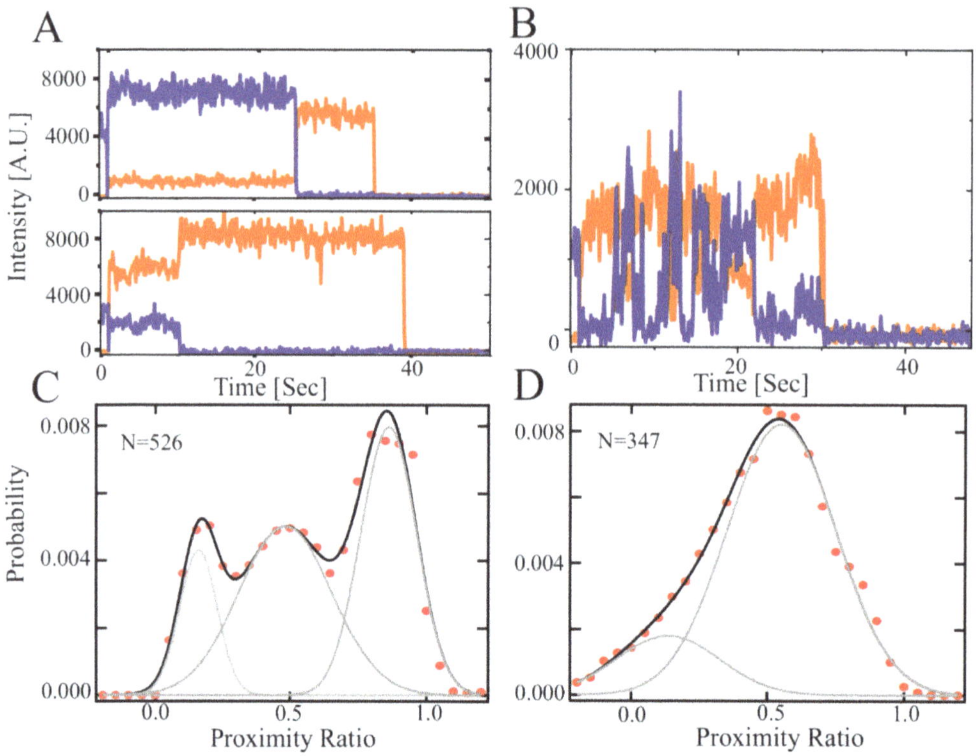


**Figure 3. Single molecule FRET measurements of the CTD2 subdomains from GluN2A and GluN2B.** Representative single molecule intensity time traces for the CTD2 subdomains. (**A**) Representative GluN2A molecules in low and high FRET states. Emission of donor (**orange**) and acceptor (**blue**) fluorophores show stable intensity in GluN2A but vary between molecules within the population. (**B**) Representative GluN2B molecules showing slow timescale, anticorrelated changes in intensity, which is the predominant state as observed previously [15,18,40]. (**C**) Population histogram of raw FRET efficiency (**proximity ratio**) accumulated from each frame captured before photobleaching for the CTD2 subdomain of GluN2A and (**D**) GluN2B. Shown are the experimental data (**red circles**) along with the global fit (**black line**). The number of individual states from global fitting (**grey lines**) differed. GluN2A adopted three states while GluN2B was well fit with a two state model containing wider peaks (Table 2). The number of molecules analyzed is indicated in each panel.

**Table 2. Analysis of the population histograms from smFRET.** The FRET efficiency was calculated from each recorded frame before photobleaching for all molecules containing a single, active donor acceptor pair. These FRET efficiency values were then accumulated into population histograms (Figure 3). These population histograms were fit to a multistate model with an increasing number of Gaussian functions to minimize the fitting statistics. CTD2A required 3 Gaussian states while CTD2B only required 2 Gaussian states. The **Mean** reports the maxima of the Gaussian peak while the **Width** reports the full-width at half height of the Gaussian peak. For these parameters, we report the SEM for three replicate measurements. We also include a simple calculation of the time-averaged distance between the donor and acceptor fluorophores (**$\langle R_{DA} \rangle$**) in nm for each FRET state based on a self-avoiding random walk (SAW) polymer model [60].

| | Mean | Width | $\langle R_{DA} \rangle$ | Mean | Width | $\langle R_{DA} \rangle$ | Mean | Width | $\langle R_{DA} \rangle$ |
|---|---|---|---|---|---|---|---|---|---|
| **CTD2A** | 0.21 ± 0.04 | 0.11 ± 0.01 | 8.4 ± 0.6 | 0.46 ± 0.08 | 0.17 ± 0.01 | 6.0 ± 0.6 | 0.85 ± 0.01 | 0.15 ± 0.02 | 3.5 ± 0.1 |
| **CTD2B** | 0.2 ± 0.01 | 0.33 ± 0.03 | 8.6 ± 0.2 | 0.55 ± 0.05 | 0.28 ± 0.03 | 5.3 ± 0.3 | NA | NA | NA |

Surprisingly, we observed a higher FRET state in CTD2A, suggesting a collapsed state, which was not observed in CTD2B. Such collapsed states were not observed in rxDMD simulations of CTD2A. However, CTD2A was directionally attached via the N-terminus to a passivated surface for measurements, which mimics the membrane attachment that occurs upon palmitoylation, while rxDMD simulations were of free protein. We previously showed that CTD2B favored more condensed states when directionally attached to a surface relative to the conformation in solution [15].

### 3.4. Condensate Formation

Previously, recombinant constructs based on CTD2 from GluN2B have been shown to undergo LLPS in vitro with the synaptic scaffold PSD-95 and a redesigned construct based on synGAP that fuses the Coiled-Coil domain to the PSD-95 Binding Motif (CC-PBM) [26,28]. However, condensate formation has not been examined with GluN2A. We examined condensate formation by monitoring transmittance at 550 nm to measure the turbidity of protein mixtures. The formation of condensates is highly sensitive to solvent conditions [61], so we performed all experiments at room temperature (25 °C) in tris-buffered saline (20 mM tris 150 mM NaCl pH 7.4). All the individual proteins showed 100% transmittance at 20 μM, which indicates a lack of condensate formation. Among all the binary protein combinations, only PSD-95 with GluN2B CTD2B showed any turbidity as a binary mixture with transmittance at 36%, which agrees well with previous binary LLPS experiments [28]. As expected, the ternary solution of PSD-95, CC-PBM, and CTD2B showed a very low transmittance of 5% indicating robust condensate formation (Figure 4A). In contrast, CTD2A showed no signs of turbidity under the exact same conditions.

To examine the protein composition of the condensates, the ternary mixtures containing PSD-95 and CC-PBM with CTD2A or CTD2B were centrifuged to separate the condensed phase from the dilute phase [62]. The sedimented pellets were dissolved in the same volume as the original supernatant and then resolved with SDS-PAGE to examine the partitioning of individual proteins into condensates. As expected from turbidity measurements, there was no protein pellet for CTD2A (Figure 4B), whereas the condensates isolated using CTD2B contained both PSD-95 and CC-PBM. To provide further evidence, we examined the ternary solution of a 1:1:1 ratio containing PSD-95, CC-PBM, and CTD2 using differential interference contrast (DIC) microscopy. We observed that 20 μM CTD2A remained clear (Figure 4C). In contrast, the ternary mixture with 20 μM CTD2B formed a dispersion of spherical droplets with a range of diameters (Figure 4D). To provide additional confirmation that CTD2B was located within the droplets, we used cysteine variants from our previous work [15,41] to label PSD-95 Alexa 488 and label CTD2B with Alexa 647. We performed two-color imaging by including 300 nM of each labeled protein to the ternary solution of a 1:1:1 ratio PSD-95, CC-PBM, and CTD2. Both labeled proteins localized to the same droplet. Thus, all droplets visible by DIC contained CTD2B and PSD-95 (Figure S1) in agreement with our SDS-PAGE analysis (Figure 4B).

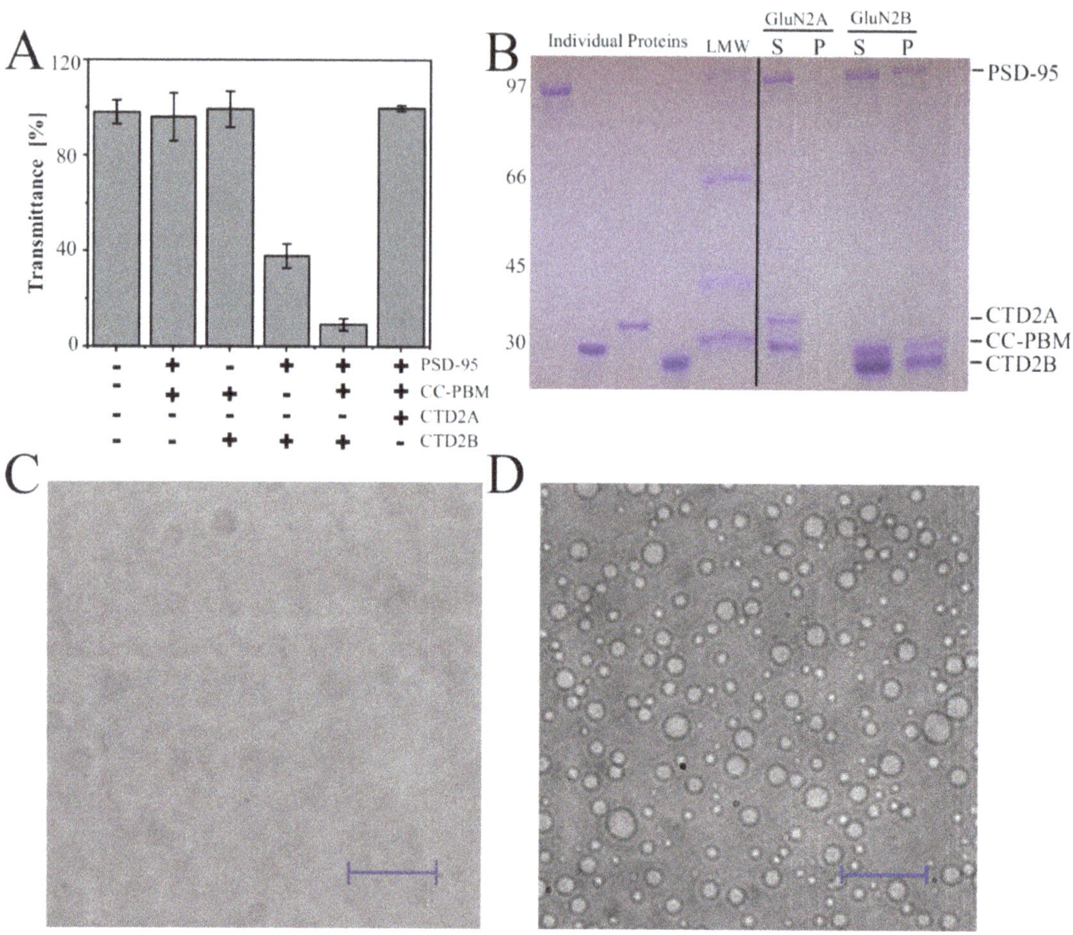


**Figure 4. Condensate formation by the CTD2 subdomains from GluN2A and GluN2B.** (**A**) Measurement of turbidity at 550 nm for the binary and ternary protein mixtures indicated beneath the panel. Samples contained 20 μM of each protein including full-length PSD-95, the CC-PBM fusion from synGAP, and the CTD2A domain from GluN2A or the CTD2B domain from GluN2B. CTD2B shows maximal turbidity while the same concentration of CTD2A remains clear. (**B**) Analysis of protein composition in the condensed phase isolated by sedimentation. Samples were resolved using SDS-PAGE. Left, the individual proteins were run separately followed by the low molecular weight markers (**LMW**). Right, soluble (**S**) and pellet (**P**) fractions from sedimentation of ternary mixtures containing a 1:1:1 ratio of PSD-95, CC-PBM, and CTD2 at 20 μM for CTD2A (**left**) and CTD2B (**right**). The molecular weights are indicated to the left of the gel (in kDa). The identity of each protein band is indicated to the right of the gel (**C**,**D**). Representative images from differential interference contrast (DIC) microscopy of the same ternary protein mixtures used for sedimentation analysis. (**C**) CTD2A does not form droplets, although some scattering is observed at high contrast. (**D**) CTD2B forms droplets with a range of different sizes. The scale bars are 100 μm.

## 4. Discussion

The NMDA receptor is an obligate heterotetramer containing two GluN1 subunits and two GluN2 subunits, which are predominantly GluN2A or GluN2B in the cortex and hippocampus [3,4]. The ordered extracellular and transmembrane domains in NMDARs form a ligand–gated ion channel. Despite high sequence conservation in these domains, receptors containing only GluN2A are functionally distinct from those containing only GluN2B, both in terms of their channel properties [63] and also in their downstream signal transduction [22]. The most variable domain in NMDAR subunits is the intracellular CTD, which has evolved to be the largest domain in GluN2A and GluN2B [16,64]. Despite the low sequence conservation in their CTDs, these two isoforms share a similar arrangement of two "domains" demarcated by palmitoylation sites [17]. Whether these are truly domains in the structural sense remains unclear.

The CTD subdomains from GluN2A and GluN2B share little sequence homology with each other or other known proteins. The CTDs are predicted to contain a mixture of order-forming and disordered motifs (Figure 1B). The entire GluN2A CTD and CTD1 from GluN2B share a similar amino acid composition with a low net charge on the boundary of strong and weak polyampholytes (Figure 1C), which makes their conformational behavior hard to predict. In contrast, CTD2 from GluN2B was classified as a weak polyampholyte, mostly due to differences in amino acid patterning, which favors collapsed states. Weak polyampholytes form globule or tadpole-like conformations while strong polyampholytes can form coil-like conformations or admixtures [20].

While simulations have been used to understand ligand binding and gating in NMDARs [65–68], we present the first simulations involving the CTD. In agreement with our CIDER classification of CTD2 from GluN2A and GluN2B into different conformational classes, our rxDMD simulations revealed large differences in polypeptide extension and secondary structural propensity. PONDR prediction of GluN2A showed a mixture of ordered and strongly disordered motifs, which agrees well with the interspersion of $\alpha$-helical and $\beta$-sheet conformation within a framework of random coil. These local structural elements give rise to strong short-range interactions in the contact frequency map, particularly around the PSD-95 binding motif (Figure 2E). The presence of local structured elements in GluN2A has the effect of increasing the net polypeptide expansion by preventing a globular collapse, which is largely what we observed in GluN2B. There were almost no persistent intramolecular contacts in GluN2B (Figure 2E). Thus, rxDMD observed a collapsed globule with almost no secondary structure that remained highly dynamic. This seems at odds with the PONDR prediction of an order-prone domain within the GluN2B CTD2 (Figure 1B). However, GluN2B had a much smaller $R_g$ suggesting that the PONDR prediction may be identifying the propensity to undergo globular collapse rather than becoming ordered through persistent contacts.

To date, only recombinant constructs based on CTD2 from GluN2B have been characterized experimentally. Here, we present experimental characterization of CTD2 from GluN2A. We found that CTD2A was poorly soluble compared to CTD2B. The slow accumulation of turbidity in CTD2A during protein handling was prevented with urea that was removed before any measurements. Using camera detection to measure smFRET, the fast conformational dynamics were time-averaged, which would result in a single time-averaged peak for random coil-like IDPs [58,59]. However, we saw three distinct, narrow peaks in the population histogram for CTD2A. Single molecules showed a stable energy transfer until photobleaching, suggesting a static heterogeneity across the population. This is in contrast to CTD2B, which showed two broad peaks with dynamic, anticorrelated intensity transitions at the single molecule level (Figure 3B,D). Thus, CTD2A lacks the slow timescale stochastic transitions seen in CTD2B (and other IDPs) using smFRET [37].

We are hesitant to interpret the changes in energy transfer in terms of distance given the dynamic environment of the fluorophores. However, simple calculations based on a self-avoiding walk (SAW) polymer model suggest similar polypeptide extension for the low and intermediate FRET states in both isoforms (Table 2). We also observed a high FRET peak in CTD2A, which suggests a compact state that was not present in CTD2B. The origins of this are not clear given the more extended conformations seen in rxDMD (Figure 2A). In contrast to DMD, where CTD2 was free at both ends, we attached CTD2A to the surface using N-terminal biotinylation, which in some ways mimics the directional membrane attachment from palmitoylation by restricting the conformational space. Previously, we showed that directional attachment of CTD2B to the surface favored polypeptide compaction [15], which may be the origin of this effect in CTD2A.

GluN2A and GluN2B generate different signaling outcomes due in part to differences in protein interactions with the CTDs [23,69]. For GluN2B, these interactions involve liquid-liquid phase separation with PSD-95 and synGAP [25,26,28]. We used the same synGAP construct, which contains only ~12% of the native protein including the coiled-coil (CC) domain, which drives synGAP multimerization, and the PSD-95 binding motif (PBM) [25]. We were able to reproduce the published results with CTD2B but did not see condensates with CTD2A using three different measurements for condensate formation: turbidity, sedimentation with SDS-PAGE, and DIC microscopy. Thus, CTD2A is more likely than CTD2B to self-associate into a colloidal suspension but less likely to participate in LLPS with PSD-95 and synGAP. This could be due to differences in the interaction with PSD-95, which we did not directly confirm. Both CTD2A and CTD2B contain the identical PSD-95 binding motif at their C-termini. However, DMD simulations found that CTD2A showed strong mid-range contacts in this region, which could affect PSD-95 binding. It is also possible that the difference in LLPS arises from the sequence patterning we identified in CTD2B, which would be more favorable for cation–π interactions (Figure 1D). This may help support condensate formation [29,31].

There is a developmental transition in NMDA receptor composition with GluN2A replacing GluN2B at mature synapses, which is driven by gene expression rather than the properties of the CTD [38]. Nonetheless, this transition in isoforms could lead to a difference in LLPS propensity similar to what we observed (Figure 4). Our observation that CTD2A favors the formation of colloidal condensates and eventual solid aggregation would support a liquid to solid phase transition in the postsynapse during development. Indeed, the postsynaptic density of mature synapses, which was one of the early condensates to be identified, had the appearance of a semi-solid in electron micrographs [70,71].

**Supplementary Materials:** The following supporting information can be downloaded at: https://www.mdpi.com/article/10.3390/biom13010004/s1, Figure S1: Confirmation of the protein components within LLPS droplets. To confirm the protein composition within the condensed droplets formed during LLPS, we fluorescently labeled individual components and performed multicolor imaging. PSD-95 was labeled with Alexa 488; the CTD2 domain from GluN2B (CTD2B) was labeled with Alexa 647 while CC-PBM from synGAP was unlabeled. No droplet formation occurs without CC-PBM. The ternary mixture contained a 1:1:1 ratio of PSD-95, CTD2B and CC-PBM at 20 μM, which included 300 nM of each labeled protein. Shown are representative images of the ternary mixture from Differential Interference Contrast (DIC) microscopy (left); from excitation at 488 nm to detect PSD-95 (center) and excitation at 642 nm excitation to detect CTD2B (right). This shows that all droplets detected by DIC contain both PSD-95 and CTD2, which agrees well with sedimentation analysis (Figure 4B).

**Author Contributions:** Conceptualization, M.E.B., U.B.C. and F.D.; methodology, U.B.C. and F.D; software, M.E.B., F.D. and N.S.; validation, M.E.B. and F.D.; formal analysis, S.B., N.S. and M.E.B.; investigation, S.B., N.S. and D.K.; resources, M.E.B. and F.D.; data curation, M.E.B. and F.D.; writing—original draft preparation, M.E.B.; writing—review and editing, M.E.B., S.B., U.B.C. and F.D.; visualization, S.B., N.S. and M.E.B.; supervision, M.E.B. and F.D.; project administration, M.E.B.; funding acquisition, M.E.B. and F.D. All authors have read and agreed to the published version of the manuscript.

**Funding:** The project has received funding from the National Institutes of Health (MH081923 to M.E.B. and R35GM119691 and R35GM145409 to F.D.).

**Institutional Review Board Statement:** Not applicable.

**Informed Consent Statement:** Not applicable.

**Data Availability Statement:** Data from DMD simulations is publically available at https://dlab.clemson.edu/research/NMDA-CTDs accessed on 21 October 2022.

**Acknowledgments:** The authors would like to thank Fang Wu, Lori Saxena, Daniel Sunko, and Leon Zheng for technical assistance. We benefited from helpful discussions with Keith Weninger, Hugo Sanabria, and Vladimir Uversky.

**Conflicts of Interest:** The authors declare no conflict of interest.

## References

1. Li, F.; Tsien, J.Z. Memory and the NMDA receptors. *N Engl. J. Med.* **2009**, *361*, 302–303. [CrossRef]
2. Newcomer, J.W.; Farber, N.B.; Olney, J.W. NMDA receptor function, memory, and brain aging. *Dialogues Clin. Neurosci.* **2000**, *2*, 219–232. [CrossRef] [PubMed]
3. Hansen, K.B.; Yi, F.; Perszyk, R.E.; Furukawa, H.; Wollmuth, L.P.; Gibb, A.J.; Traynelis, S.F. Structure, function, and allosteric modulation of NMDA receptors. *J. Gen. Physiol.* **2018**, *150*, 1081–1105. [CrossRef] [PubMed]
4. Hansen, K.B.; Wollmuth, L.P.; Bowie, D.; Furukawa, H.; Menniti, F.S.; Sobolevsky, A.I.; Swanson, G.T.; Swanger, S.A.; Greger, I.H.; Nakagawa, T.; et al. Structure, Function, and Pharmacology of Glutamate Receptor Ion Channels. *Pharm. Rev.* **2021**, *73*, 298–487. [CrossRef] [PubMed]
5. Zhu, S.; Gouaux, E. Structure and symmetry inform gating principles of ionotropic glutamate receptors. *Neuropharmacology* **2017**, *112(Pt A)*, 11–15. [CrossRef]
6. Yuan, H.; Hansen, K.B.; Vance, K.M.; Ogden, K.K.; Traynelis, S.F. Control of NMDA receptor function by the NR2 subunit amino-terminal domain. *J. Neurosci.* **2009**, *29*, 12045–12058. [CrossRef] [PubMed]
7. Punnakkal, P.; Jendritza, P.; Kohr, G. Influence of the intracellular GluN2 C-terminal domain on NMDA receptor function. *Neuropharmacology* **2012**, *62*, 1985–1992. [CrossRef] [PubMed]
8. Petit-Pedrol, M.; Groc, L. Regulation of membrane NMDA receptors by dynamics and protein interactions. *J. Cell Biol.* **2021**, *220*, e202006101. [CrossRef]
9. XiangWei, W.; Jiang, Y.; Yuan, H. De Novo Mutations and Rare Variants Occurring in NMDA Receptors. *Curr. Opin. Physiol.* **2018**, *2*, 27–35. [CrossRef]
10. Mota Vieira, M.; Nguyen, T.A.; Wu, K.; Badger, J.D., 2nd; Collins, B.M.; Anggono, V.; Lu, W.; Roche, K.W. An Epilepsy-Associated GRIN2A Rare Variant Disrupts CaMKIIalpha Phosphorylation of GluN2A and NMDA Receptor Trafficking. *Cell Rep.* **2020**, *32*, 108104. [CrossRef]
11. Liu, S.; Zhou, L.; Yuan, H.; Vieira, M.; Sanz-Clemente, A.; Badger, J.D., 2nd; Lu, W.; Traynelis, S.F.; Roche, K.W. A Rare Variant Identified Within the GluN2B C-Terminus in a Patient with Autism Affects NMDA Receptor Surface Expression and Spine Density. *J. Neurosci.* **2017**, *37*, 4093–4102. [CrossRef]
12. Karakas, E.; Furukawa, H. Crystal structure of a heterotetrameric NMDA receptor ion channel. *Science* **2014**, *344*, 992–997. [CrossRef]
13. Lee, C.H.; Lu, W.; Michel, J.C.; Goehring, A.; Du, J.; Song, X.; Gouaux, E. NMDA receptor structures reveal subunit arrangement and pore architecture. *Nature* **2014**, *511*, 191–197. [CrossRef]
14. Regan, M.C.; Romero-Hernandez, A.; Furukawa, H. A structural biology perspective on NMDA receptor pharmacology and function. *Curr. Opin. Struct Biol.* **2015**, *33*, 68–75. [CrossRef]
15. Choi, U.B.; Xiao, S.; Wollmuth, L.P.; Bowen, M.E. Effect of Src kinase phosphorylation on disordered C-terminal domain of N-methyl-D-aspartic acid (NMDA) receptor subunit GluN2B protein. *J. Biol. Chem.* **2011**, *286*, 29904–29912. [CrossRef]
16. Ryan, T.J.; Emes, R.D.; Grant, S.G.N.; Komiyama, N.H. Evolution of NMDA receptor cytoplasmic interaction domains: Implications for organisation of synaptic signalling complexes. *BMC Neurosci.* **2008**, *9*, 14. [CrossRef]
17. Hayashi, T.; Thomas, G.M.; Huganir, R.L. Dual palmitoylation of NR2 subunits regulates NMDA receptor trafficking. *Neuron* **2009**, *64*, 213–226. [CrossRef]
18. Choi, U.B.; McCann, J.J.; Weninger, K.R.; Bowen, M.E. Beyond the random coil: Stochastic conformational switching in intrinsically disordered proteins. *Structure* **2011**, *19*, 566–576. [CrossRef]
19. Holehouse, A.S.; Das, R.K.; Ahad, J.N.; Richardson, M.O.; Pappu, R.V. CIDER: Resources to Analyze Sequence-Ensemble Relationships of Intrinsically Disordered Proteins. *Biophys. J.* **2017**, *112*, 16–21. [CrossRef]
20. Das, R.K.; Pappu, R.V. Conformations of intrinsically disordered proteins are influenced by linear sequence distributions of oppositely charged residues. *Proc. Natl. Acad. Sci. USA* **2013**, *110*, 13392–13397. [CrossRef]
21. Lin, Y.; Skeberdis, V.A.; Francesconi, A.; Bennett, M.V.; Zukin, R.S. Postsynaptic density protein-95 regulates NMDA channel gating and surface expression. *J. Neurosci.* **2004**, *24*, 10138–10148. [CrossRef]
22. Sun, Y.; Xu, Y.; Cheng, X.; Chen, X.; Xie, Y.; Zhang, L.; Wang, L.; Hu, J.; Gao, Z. The differences between GluN2A and GluN2B signaling in the brain. *J. Neurosci. Res.* **2018**, *96*, 1430–1443. [CrossRef]
23. Ishchenko, Y.; Carrizales, M.G.; Koleske, A.J. Regulation of the NMDA receptor by its cytoplasmic domains: (How) is the tail wagging the dog? *Neuropharmacology* **2021**, *195*, 108634. [CrossRef]
24. Aow, J.; Dore, K.; Malinow, R. Conformational signaling required for synaptic plasticity by the NMDA receptor complex. *Proc. Natl. Acad. Sci. USA* **2015**, *112*, 14711–14716. [CrossRef]
25. Zeng, M.; Shang, Y.; Araki, Y.; Guo, T.; Huganir, R.L.; Zhang, M. Phase Transition in Postsynaptic Densities Underlies Formation of Synaptic Complexes and Synaptic Plasticity. *Cell* **2016**, *166*, 1163–1175.e12. [CrossRef]

26. Zeng, M.; Chen, X.; Guan, D.; Xu, J.; Wu, H.; Tong, P.; Zhang, M. Reconstituted Postsynaptic Density as a Molecular Platform for Understanding Synapse Formation and Plasticity. *Cell* **2018**, *174*, 1172–1187.e16. [CrossRef]
27. Hosokawa, T.; Liu, P.W.; Cai, Q.; Ferreira, J.S.; Levet, F.; Butler, C.; Sibarita, J.B.; Choquet, D.; Groc, L.; Hosy, E.; et al. CaMKII activation persistently segregates postsynaptic proteins via liquid phase separation. *Nat. Neurosci.* **2021**, *24*, 777–785. [CrossRef]
28. Vistrup-Parry, M.; Chen, X.; Johansen, T.L.; Bach, S.; Buch-Larsen, S.C.; Bartling, C.R.O.; Ma, C.; Clemmensen, L.S.; Nielsen, M.L.; Zhang, M.; et al. Site-specific phosphorylation of PSD-95 dynamically regulates the postsynaptic density as observed by phase separation. *iScience* **2021**, *24*, 103268. [CrossRef]
29. Wang, J.; Choi, J.M.; Holehouse, A.S.; Lee, H.O.; Zhang, X.; Jahnel, M.; Maharana, S.; Lemaitre, R.; Pozniakovsky, A.; Drechsel, D.; et al. A Molecular Grammar Governing the Driving Forces for Phase Separation of Prion-like RNA Binding Proteins. *Cell* **2018**, *174*, 688–699.e16. [CrossRef]
30. Boeynaems, S.; Alberti, S.; Fawzi, N.L.; Mittag, T.; Polymenidou, M.; Rousseau, F.; Schymkowitz, J.; Shorter, J.; Wolozin, B.; van den Bosch, L.; et al. Protein Phase Separation: A New Phase in Cell Biology. *Trends Cell Biol.* **2018**, *28*, 420–435. [CrossRef]
31. Martin, E.W.; Holehouse, A.S.; Peran, I.; Farag, M.; Incicco, J.J.; Bremer, A.; Grace, C.R.; Soranno, A.; Pappu, R.V.; Mittag, T. Valence and patterning of aromatic residues determine the phase behavior of prion-like domains. *Science* **2020**, *367*, 694–699. [CrossRef] [PubMed]
32. Palay, S.L. Synapses in the central nervous system. *J. Biophys. Biochem. Cytol.* **1956**, *2* (Suppl. 4), 193–202. [CrossRef] [PubMed]
33. Palade, G. Electron microscope observations of interneuronal and neuromuscular synapses. *Anat. Rec* **1954**, *118*, 335–336.
34. De Robertis, E.; Bennett, H.S. Submicroscopic vesicular component in the synapse. *Fed. Proc.* **1954**, *13*, 170.
35. Wu, X.; Ganzella, M.; Zhou, J.; Zhu, S.; Jahn, R.; Zhang, M. Vesicle Tethering on the Surface of Phase-Separated Active Zone Condensates. *Mol. Cell* **2021**, *81*, 13–24.e7. [CrossRef]
36. Chen, X.; Wu, X.; Wu, H.; Zhang, M. Phase separation at the synapse. *Nat. Neurosci.* **2020**, *23*, 301–310. [CrossRef]
37. Choi, U.B.; Sanabria, H.; Smirnova, T.; Bowen, M.E.; Weninger, K.R. Spontaneous Switching among Conformational Ensembles in Intrinsically Disordered Proteins. *Biomolecules* **2019**, *9*, 114. [CrossRef]
38. McKay, S.; Ryan, T.J.; McQueen, J.; Indersmitten, T.; Marwick, K.F.M.; Hasel, P.; Kopanitsa, M.V.; Baxter, P.S.; Martel, M.A.; Kind, P.C.; et al. The Developmental Shift of NMDA Receptor Composition Proceeds Independently of GluN2 Subunit-Specific GluN2 C-Terminal Sequences. *Cell Rep.* **2018**, *25*, 841–851.e4. [CrossRef]
39. Wyllie, D.J.; Livesey, M.R.; Hardingham, G.E. Influence of GluN2 subunit identity on NMDA receptor function. *Neuropharmacology* **2013**, *74*, 4–17. [CrossRef]
40. Choi, U.B.; Kazi, R.; Stenzoski, N.; Wollmuth, L.P.; Uversky, V.N.; Bowen, M.E. Modulating the intrinsic disorder in the cytoplasmic domain alters the biological activity of the N-methyl-D-aspartate-sensitive glutamate receptor. *J. Biol. Chem.* **2013**, *288*, 22506–22515. [CrossRef]
41. McCann, J.J.; Zheng, L.; Rohrbeck, D.; Felekyan, S.; Kühnemuth, R.; Sutton, R.B.; Seidel, C.A.; Bowen, M.E. Supertertiary structure of the synaptic MAGuK scaffold proteins is conserved. *Proc. Natl. Acad. Sci. USA* **2012**, *109*, 15775–15780. [CrossRef] [PubMed]
42. Choi, U.B.; Weninger, K.R.; Bowen, M.E. Immobilization of proteins for single-molecule fluorescence resonance energy transfer measurements of conformation and dynamics. *Methods Mol. Biol.* **2012**, *896*, 3–20. [PubMed]
43. Yanez Orozco, I.S.; Mindlin, F.A.; Ma, J.; Wang, B.; Levesque, B.; Spencer, M.; Adariani, S.R.; Hamilton, G.; Ding, F.; Bowen, M.E.; et al. Identifying weak interdomain interactions that stabilize the supertertiary structure of the N-terminal tandem PDZ domains of PSD-95. *Nat. Commun.* **2018**, *9*, 3724. [CrossRef] [PubMed]
44. Basak, S.; Sakia, N.; Dougherty, L.; Guo, Z.; Wu, F.; Mindlin, F.; Lary, J.W.; Cole, J.L.; Ding, F.; Bowen, M.E. Probing Interdomain Linkers and Protein Supertertiary Structure In Vitro and in Live Cells with Fluorescent Protein Resonance Energy Transfer. *J. Mol. Biol.* **2021**, *433*, 166793. [CrossRef] [PubMed]
45. Hamilton, G.L.; Saikia, N.; Basak, S.; Welcome, F.S.; Wu, F.; Kubiak, J.; Zhang, C.; Hao, Y.; Seidel, C.A.M.; Ding, F.; et al. Fuzzy supertertiary interactions within PSD-95 enable ligand binding. *Elife* **2022**, *11*, e77242. [CrossRef]
46. Ding, F.; Tsao, D.; Nie, H.; Dokholyan, N.V. Ab Initio Folding of Proteins with All-Atom Discrete Molecular Dynamics. *Structure* **2008**, *16*, 1010–1018. [CrossRef]
47. Shirvanyants, D.; Ding, F.; Tsao, D.; Ramachandran, S.; Dokholyan, N.V. Discrete Molecular Dynamics: An Efficient and Versatile Simulation Method for Fine Protein Characterization. *J. Phys. Chem. B* **2012**, *116*, 8375–8382. [CrossRef]
48. Tokunaga, M.; Imamoto, N.; Sakata-Sogawa, K. Highly inclined thin illumination enables clear single-molecule imaging in cells. *Nat. Methods* **2008**, *5*, 159–161. [CrossRef]
49. Craven, S.E.; El-Husseini, A.E.; Bredt, D.S. Synaptic targeting of the postsynaptic density protein PSD-95 mediated by lipid and protein motifs. *Neuron* **1999**, *22*, 497–509. [CrossRef]
50. El-Husseini, A.E.; Craven, S.E.; Chetkovich, D.M.; Firestein, B.L.; Schnell, E.; Aoki, C.; Bredt, D.S. Dual palmitoylation of PSD-95 mediates its vesiculotubular sorting, postsynaptic targeting, and ion channel clustering. *J. Cell Biol.* **2000**, *148*, 159–172. [CrossRef]
51. Mao, A.H.; Crick, S.L.; Vitalis, A.; Chicoine, C.L.; Pappu, R.V. Net charge per residue modulates conformational ensembles of intrinsically disordered proteins. *Proc. Natl. Acad. Sci. USA* **2010**, *107*, 8183–8188. [CrossRef] [PubMed]
52. Romero, P.; Obradovic, Z.; Li, X.; Garner, E.C.; Brown, C.J.; Dunker, A.K. Sequence complexity of disordered protein. *Proteins* **2001**, *42*, 38–48. [CrossRef] [PubMed]
53. Xue, B.; Dunbrack, R.L.; Williams, R.W.; Dunker, A.K.; Uversky, V.N. PONDR-FIT: A meta-predictor of intrinsically disordered amino acids. *Biochim. Biophys. Acta* **2010**, *1804*, 996–1010. [CrossRef] [PubMed]

54. Weathers, E.A.; Paulaitis, M.E.; Woolf, T.B.; Hoh, J.H. Reduced amino acid alphabet is sufficient to accurately recognize intrinsically disordered protein. *FEBS Lett.* **2004**, *576*, 348–352. [CrossRef]
55. Uversky, V.N.; Gillespie, J.R.; Fink, A.L. Why are "natively unfolded" proteins unstructured under physiologic conditions? *Proteins* **2000**, *41*, 415–427. [CrossRef]
56. Tomasso, M.E.; Tarver, M.J.; Devarajan, D.; Whitten, S.T. Hydrodynamic Radii of Intrinsically Disordered Proteins Determined from Experimental Polyproline II Propensities. *PLoS Comput Biol.* **2016**, *12*, e1004686. [CrossRef]
57. Saikia, N.; Yanez-Orozco, I.S.; Qiu, R.; Hao, P.; Milikisiyants, S.; Ou, E.; Hamilton, G.L.; Weninger, K.R.; Smirnova, T.I.; Sanabria, H.; et al. Integrative structural dynamics probing of the conformational heterogeneity in synaptosomal-associated protein 25. *Cell Rep. Phys. Sci.* **2021**, *2*, 100616. [CrossRef]
58. Gopich, I.V.; Szabo, A. FRET efficiency distributions of multistate single molecules. *J. Phys. Chem. B* **2010**, *114*, 15221–15226. [CrossRef]
59. Gopich, I.V.; Szabo, A. Single-molecule FRET with diffusion and conformational dynamics. *J. Phys. Chem. B* **2007**, *111*, 12925–12932. [CrossRef]
60. Zheng, W.; Zerze, G.H.; Borgia, A.; Mittal, J.; Schuler, B.; Best, R.B. Inferring properties of disordered chains from FRET transfer efficiencies. *J. Chem. Phys.* **2018**, *148*, 123329. [CrossRef]
61. Alberti, S.; Gladfelter, A.; Mittag, T. Considerations and Challenges in Studying Liquid-Liquid Phase Separation and Biomolecular Condensates. *Cell* **2019**, *176*, 419–434. [CrossRef] [PubMed]
62. Mitrea, D.M.; Chandra, B.; Ferrolino, M.C.; Gibbs, E.B.; Tolbert, M.; White, M.R.; Kriwacki, R.W. Methods for Physical Characterization of Phase-Separated Bodies and Membrane-less Organelles. *J. Mol. Biol.* **2018**, *430*, 4773–4805. [CrossRef] [PubMed]
63. Paoletti, P.; Bellone, C.; Zhou, Q. NMDA receptor subunit diversity: Impact on receptor properties, synaptic plasticity and disease. *Nat. Rev. Neurosci.* **2013**, *14*, 383–400. [CrossRef]
64. Ryan, T.J.; Kopanitsa, M.V.; Indersmitten, T.; Nithianantharajah, J.; Afinowi, N.O.; Pettit, C.; Stanford, L.E.; Sprengel, R.; Saksida, L.M.; Bussey, T.J.; et al. Evolution of GluN2A/B cytoplasmic domains diversified vertebrate synaptic plasticity and behavior. *Nat. Neurosci.* **2013**, *16*, 25–32. [CrossRef] [PubMed]
65. Iacobucci, G.J.; Wen, H.; Helou, M.; Liu, B.; Zheng, W.; Popescu, G.K. Cross-subunit interactions that stabilize open states mediate gating in NMDA receptors. *Proc. Natl. Acad. Sci. USA* **2021**, *118*, e2007511118. [CrossRef] [PubMed]
66. Zheng, W.; Wen, H.; Iacobucci, G.J.; Popescu, G.K. Probing the Structural Dynamics of the NMDA Receptor Activation by Coarse-Grained Modeling. *Biophys. J.* **2017**, *112*, 2589–2601. [CrossRef]
67. Pang, X.; Zhou, H.X. Structural modeling for the open state of an NMDA receptor. *J. Struct. Biol.* **2017**, *200*, 369–375. [CrossRef] [PubMed]
68. Sinitskiy, A.V.; Pande, V.S. Computer Simulations Predict High Structural Heterogeneity of Functional State of NMDA Receptors. *Biophys. J.* **2018**, *115*, 841–852. [CrossRef]
69. Hardingham, G. NMDA receptor C-terminal signaling in development, plasticity, and disease. *F1000Res* **2019**, *8*. [CrossRef]
70. Cohen, R.S.; Blomberg, F.; Berzins, K.; Siekevitz, P. The structure of postsynaptic densities isolated from dog cerebral cortex. I. Overall morphology and protein composition. *J. Cell Biol.* **1977**, *74*, 181–203. [CrossRef]
71. Petersen, J.D.; Chen, X.; Vinade, L.; Dosemeci, A.; Lisman, J.E.; Reese, T.S. Distribution of postsynaptic density (PSD)-95 and $Ca^{2+}$/calmodulin-dependent protein kinase II at the PSD. *J. Neurosci.* **2003**, *23*, 11270–11278. [CrossRef] [PubMed]

*biomolecules*

MDPI

*Article*

# Portability of a Small-Molecule Binding Site between Disordered Proteins

**Rajesh Jaiprashad [1,†], Sachith Roch De Silva [1,†], Lisette M. Fred Lucena [1], Ella Meyer [1] and Steven J. Metallo [1,2,*]**

1 Department of Chemistry, Georgetown University, Washington, DC 20057, USA
2 Institute for Soft Matter Synthesis and Metrology (ISMSM), Georgetown University, Washington, DC 20057, USA
* Correspondence: sjm24@georgetown.edu
† These authors contributed equally to this work.

**Abstract:** Intrinsically disordered proteins (IDPs) are important in both normal and disease states. Small molecules can be targeted to disordered regions, but we currently have only a limited understanding of the nature of small-molecule binding sites in IDPs. Here, we show that a minimal small-molecule binding sequence of eight contiguous residues derived from the Myc protein can be ported into a different disordered protein and recapitulate small-molecule binding activity in the new context. We also find that the residue immediately flanking the binding site can have opposing effects on small-molecule binding in the different disordered protein contexts. The results demonstrate that small-molecule binding sites can act modularly and are portable between disordered protein contexts but that residues outside of the minimal binding site can modulate binding affinity.

**Keywords:** intrinsically disordered proteins; Myc; protein-protein interaction; drug targets; SLiM; small-molecule inhibitors

**Citation:** Jaiprashad, R.; De Silva, S.R.; Fred Lucena, L.M.; Meyer, E.; Metallo, S.J. Portability of a Small-Molecule Binding Site between Disordered Proteins. *Biomolecules* **2022**, *12*, 1887. https://doi.org/10.3390/biom12121887

Academic Editors: Prakash Kulkarni, Stefania Brocca, Keith Dunker and Sonia Longhi

Received: 4 October 2022
Accepted: 9 December 2022
Published: 16 December 2022

**Publisher's Note:** MDPI stays neutral with regard to jurisdictional claims in published maps and institutional affiliations.



## 1. Introduction

Proteins exist along a conformational spectrum from fully folded and well-structured proteins to unstructured or intrinsically disordered proteins (IDPs) [1,2]. Many proteins lie between these two endpoints and contain both ordered regions as well as substantial (>40 amino acids) intrinsically disordered regions (IDRs) [3,4]. While structured regions fluctuate around a clear average conformation, IDPs and IDRs exist as a rapidly fluctuating series of conformations [5]. An IDR can be described as an ensemble of conformations with low energy barriers for interconversion [6]. Protein disorder is found throughout biological systems and is particularly prevalent in complex eukaryotes [7]. Within cells, IDPs and IDRs perform many crucial functions and are particularly prevalent in signal transduction and transcriptional control with greater than 80% of transcription factors predicted to be partially or completely disordered [8,9].

Proteins containing disordered regions are also overrepresented in pathological conditions such as cancer and neurodegenerative diseases [10,11]. A contributing reason for a central role of IDRs in both normal cellular functions and in pathologies is the ability of IDRs to act as sites of molecular recognition [12–14]. Within cells, the formation of many biomolecular condensates has been shown to be driven by molecular recognition functions of disordered proteins [15]. Through dynamic and multivalent interactions with other proteins or with nucleic acids, typically RNA, IDRs are able to mediate the formation and properties of many of the biomolecular condensates in cells [16,17]. These membraneless organelles function in crucial processes such as RNA splicing, modulation of reaction rates, and transcription control, among others [18].

While many IDRs participate in highly dynamic interactions, IDRs can participate in protein–protein interactions (PPIs) with a range of affinities and kinetic stabilities [19]. Interactions also occur with a range of disorder present in the complex. Certain IDRs undergo coupled folding and binding in the formation of a complex [20]. Some IDRs adopt different conformations when bound to different partners [21]. Other IDRs form complexes while remaining disordered [22]. Within larger disordered domains, portions of sequence that mediate protein–protein interactions via coupled folding and binding to structured partners are referred to as molecular recognition features (MoRFs) [13]. MoRFs were recognized as potentially useful starting points in developing inhibitors of PPIs [23] and can be predicted within disordered sequences [24,25]. Post-translational modifications (PTM) often involve recognition of a disordered modification site. Of the characterized phosphorylation sites, 84% percent are in disordered regions [26]. These PTM sites are an example of short (3–10 residues) recognition sequences that are found in disordered regions and that can mediate specific domain interactions. These short sequences, which overlap with MoRFs, are called short linear motifs, SLiMs [27]. In both MoRFs and SLiMs, the disordered nature of the target is important in allowing access to its binding partner. The sequences are not sequestered in a folded context and therefore are available for binding with access to the chemical moieties along the entire sequence [28].

In addition to mediating interactions between biomolecules, disordered regions were also found to support binding by small molecules. Early studies involved targeting of the disordered, monomeric bHLHZip region of the c-Myc oncoprotein (Myc) with the goal of interfering with the coupled folding and binding of Myc to its obligate heterodimerization partner Max [29,30]. Myc is dysregulated in a majority of human cancers [31] and even transient inhibition of Myc activity can cause cancer cells to differentiate [32]. Consequently, Myc activity has been targeted in a wide array of mechanisms [33–35]. The crucial biological function of Myc drove the direct targeting of Myc, in spite of its disordered character, and caused it to become an early test case for the direct targeting of disordered proteins with small molecules [36]. Subsequently, a range of disordered proteins with a variety of functions have been demonstrated as targets of small molecules with a concentration on transcription factors and neurological disease-related targets [37–40]. Despite progress, with an increasing scope of small-molecule IDP interactions reported, we still do not have a clear understanding of the major factors controlling what constitutes a disordered sequence that supports small-molecule affinity, nor do we know how binding site specificity is achieved in these interactions that appear to remain dynamic and exposed to solvent in the complex.

In order to better understand the binding of small molecules to disordered sequences, we sought to investigate potential parallels between small-molecule IDR interactions and SLiM interactions with partner proteins. Both SLiMs and disordered small-molecule binding sites consist of short linear sequences that mediate specific binding with an interaction partner, either a protein partner or a molecular partner [27,36]. We sought to determine if disordered small-molecule binding sites could recapitulate the ability of SLiM sequences to recognize their specific binding partner in a modular fashion, using the same (or similar) recognition sequence embedded in different protein contexts to bind to the same partner [41]. Here, we ported a specific small-molecule recognition sequence between two disordered proteins and demonstrated that the small-molecule binding function moved along with the sequence. Further, we found that residues flanking the binding site modulated binding affinity as in other IDR recognition motifs.

## 2. Materials and Methods

### 2.1. $Myc_{353–437}$, MaxRH, Max, and $Myc_{402–412}$ Purification

The coding sequences for $Myc_{353–437}$, MaxRH, P21 Max, and P22 Max were designed to include a hexahistidine (6xHis) tag, and a tobacco etch virus (TEV) recognition site immediately prior to the protein coding region (Figure S1). The $Myc_{353–437}$ coding sequence was inserted into a pET23d+ plasmid (Genscript) while MaxRH was inserted into a pET24d+ plasmid (Genscript). Max isoforms (P21 and P22) were expressed from previously described pET151D-TOPO plasmids [42]. The $Myc_{353–437}$ A401E, E410N, and MaxRH-N78E mutants were generated using QuickChange Lightning Mutagenesis (Agilent) following the manufacturer's protocol. MaxRH-Y115F/Y123F mutagenesis was conducted on the MaxRH plasmid by Genscript.

The 6xHis-tagged proteins were expressed in BL21(DE3) pLysS *E. coli* cells (Invitrogen) under autoinducing conditions following a protocol by Studier [43]. The cells were grown in a medium containing 1% *w*/*v* N-Z amine, 0.5% *w*/*v* yeast extract, 25 mM $Na_2HPO_4$, 25 mM $KH_2PO_4$, 50 mM $NH_4Cl$, 5 mM $Na_2SO_4$, 2 mM $MgSO_4$, 0.5 % *v*/*v* glycerol, 0.05 % *w*/*v* glucose, 0.2 % *w*/*v* lactose and a trace-metals mix of 10 μM $FeCl_3$, 4 μM $CaCl_2$, 2 μM $MnCl_2$, 2 μM $ZnSO_4$, 0.4 μM $CoCl_2$, 0.4 μM $CuCl_2$, and 0.4 μM $NiCl_2$. A single colony of the bacterial culture was grown for 18 h at 37 °C in a shaking incubator at 200 rpm. Cells were collected by centrifugation at 9000 rpm for 30 min using a Sorvall RC 6+ centrifuge (Thermo Scientific, Marietta, OH, USA). The supernatant was discarded, and cells were lysed by sonication in 50 mL of lysis buffer containing 8 M urea, 100 mM Tris-HCl, and 10 mM sodium phosphate at pH 8.0. Cell debris was removed by centrifugation at 18,000 rpm for 30 min. The lysate was loaded onto a nickel nitriloacetate (Ni-NTA) affinity resin (GoldBio) column to purify the proteins using a pH gradient, where the column was equilibrated with lysis buffer at pH 8, and non-specific proteins were removed using a wash buffer at pH 6.4 (8 M urea, 100 mM Tris-HCl, and 10 mM sodium phosphate). An elution buffer at pH 4.5 (8 M urea, 100 mM Tris-HCl, and 10 mM sodium phosphate) was then used to elute 6xHis-tagged proteins bound to the Ni-NTA column. The elutions were pooled and buffer exchanged into 50 mM MES at pH 5 using 3000 MW cutoff Amicon ultrafiltration units (Millipore-Sigma, Billerica, MA, USA). After buffer exchanging into MES, the protein was incubated overnight with a TEV protease to cleave the 6xHis-tag. His-tag cleavage was confirmed using SDS-PAGE. Once cleavage was confirmed, the TEV protease activity was quenched by adding urea, and the protein was then buffer exchanged into 8M urea lysis buffer at pH 8. A second Ni-NTA column was used to remove uncleaved protein, 6xHis-tag, and 6xHis-tagged TEV. The cleaved protein was dialyzed against HPLC grade water containing 0.1% trifluoroacetic acid (TFA) using a 3000 MW cutoff dialysis membrane. All proteins were further purified through reverse phase HPLC (Vydac-C18) with a water/acetonitrile gradient containing 0.1% TFA and purified to >95% purity. The proteins were lyophilized and stored at −20 °C. The $Myc_{402–412}$ peptide, Ac-YILSVQAEEQK-$NH_2$, was synthesized by Genscript using solid phase peptide synthesis. The peptide was reconstituted in HPLC grade water and further purified through reverse phase HPLC (Vydac-C18) with a water/acetonitrile gradient and 0.1% TFA (Sigma-Aldrich, Saint Louis, MI, USA). The peptide was lyophilized and stored at −20 °C.

### 2.2. Preparation, Characterization and pKa Determination of 34RH

The small molecule (Z)-4-((4-oxo-2-thioxothiazolidin-5-ylidene)methyl)benzoic acid (hereafter referred to as 34RH) was previously synthesized according to established procedures, and the structure was confirmed by $^1H$ and $^{13}C$ NMR (Figure S6) using a 400 MHz spectrometer [44]. The dry compound was stored at 4 °C. Stock solutions (1 mM) of 34RH were made using either dimethyl sulfoxide (DMSO) or ethanol and stored at −20 °C. Fluorescence and dynamic light scattering (DLS) experiments were performed using DMSO, while circular dichroism (CD) was conducted using ethanol to avoid the high absorbance of DMSO at short wavelengths. DLS measurements were performed on 34RH using an LS Spectrophotometer (LS Instruments) at 25 °C (Figure S3). The dispersant viscosity was set for water. Samples were analyzed in 1xPBS (pH 7.4) and 5% total DMSO. The compound was serially diluted two-fold from 100 μM to a final concentration of 6.25 μM. Samples were placed into 5 mm cylindrical glass cuvettes (LS Instruments) and measured using a 600 nm laser at a 90° angle for 20 s. The data were analyzed using LS Spectrophotometer software provided by the manufacturer.

The imide pKa of 34RH was determined using UV/Vis by measuring the absorbance of 10 μM of 34RH in 1xPBS at various pH values using an Agilent 8453 UV/Vis spectrophotometer (Figure S2). The absorbance at 327 nm for each pH value was fit to a curve using the

Henderson-Hasselbalch (shown in Equations (1) and (2)), where c is the pKa, x is the pH, a is the signal for the fully protonated acid, and b is the signal for the fully ionized base [45].

$$\text{pKa} - \text{pH} = \log \frac{[\text{acid}]}{[\text{base}]} \tag{1}$$

$$y = \frac{a + b*10^{(c-x)}}{1 + 10^{(c-x)}} \tag{2}$$

### 2.3. *Tyrosine Fluorescence Quenching Assay*

The lyophilized protein (or peptide) was reconstituted in ultrapure water and incubated to room temperature for at least 1 hr. The protein or peptide was then sterile filtered using a pre-wet 0.2 µm polyethylene sulfone (PES) filter (VWR), and the concentration was determined by the absorbance at 274 nm using the extinction coefficient per tyrosine of $\varepsilon_{274}$ = 1470 $M^{-1}cm^{-1}$. The final stock concentrations ranged from 50 to 100 µM.

For excess 34RH tyrosine fluorescence quenching experiments, samples were prepared with the following buffer components: sterile filtered water, 1xPBS (137 mM NaCl, 2.7 mM KCl, 4.3 mM $Na_2HPO_4$, 1.4 mM $KH_2PO_4$), and 5% total DMSO. The pH of the buffer was adjusted to 6 or 7.4 depending on the experiment performed. Three separate samples were prepared in 1xPBS and 5% total DMSO: one containing 50 µM 34RH alone, one containing 1 µM of protein and 50 µM 34RH, and one containing 1 µM protein alone. For the 34RH containing samples, the compound was delivered from a 1 mM DMSO stock. These samples were then serially diluted two-fold from 50 µM 34RH to 1.56 µM 34RH. All samples for fluorescence measurements maintained a final 5% DMSO concentration.

The samples were incubated for at least 15 min before fluorescence was measured. The samples were excited at 274 nm, and the emission spectra were obtained from 285 to 340 nm using 5 nm excitation and 5 nm emission slits using a Horiba Fluoromax 4 fluorometer. The fluorescence data were background corrected using buffer (for samples with only protein) or small molecule (for samples containing protein and 34RH) (Figure S4). The absorbance due to 34RH and protein at the excitation and emission wavelengths can suppress the observed fluorescence intensity to give rise to the inner filter effect [46]. To account for this suppression, we corrected the fluorescence signals using Equation (3). This correction accounted for the fluorescence suppression due to the absorbance of both 34RH and the protein [47].

$$F_{corr} = F_{obs} * 10^{\frac{(Aex + Aem)}{2}} \tag{3}$$

Here, the corrected fluorescence is $F_{corr}$, the background-subtracted observed fluorescence is $F_{obs}$, and $A_{ex}$ and $A_{em}$ are the total absorbance at the excitation and emission wavelengths, respectively. The amount of protein fluorescence quenched by 34RH at a particular concentration was calculated using $F_{corr}$ at λ = 304 nm (Equation (4)).

$$\text{Fraction Quenched} = 1 - \frac{\text{Fcorr}_{304\text{nm}} \text{ of (Protein } + \text{34RH)}}{\text{Fcorr}_{304\text{nm}} \text{ of Protein only}} \tag{4}$$

The quenching data were fit to a Langmuir binding isotherm using Equation (5), from which the dissociation constant ($K_D$) was obtained [48]. Here, Qmax describes the maximum fraction quenched, $[L]_T$ is the total 34RH concentration, and $[P]_T$ is the total protein concentration. The total concentration of the protein was 1 µM for all fluoresce experiments conducted at a constant protein concentration.

$$\text{Fraction Quenched} = \text{Qmax}* \left[ \frac{([L]_T + [P]_T + K_D) - \sqrt{(([L]_T + [P]_T + K_D)^2 - 4 * [L]_T * [P]_T)}}{2 * [L]_T} \right] \tag{5}$$

*2.4. Circular Dichroism (CD)*

Samples containing $Myc_{353–437}$, $Myc_{402–412}$ peptide, MaxRH, Max isoforms, or mutants in the absence and presence of excess concentrations of 34RH were prepared in 1xCD buffer (50 mM KF, 4.3 mM $NaH_2PO_4$, 1.4 mM $KH_2PO_4$, 5% ethanol). The pH of the buffer was adjusted to either 6 or 7.4 depending on the experiment conducted. The compound was delivered from a 1 mM ethanol stock solution. Samples were incubated for 1 hr before measurement. The far UV-spectra of the proteins and peptide were recorded in a quartz cuvette with a path length of 0.1 cm using a Jasco J720 spectropolarimeter. The samples were scanned from 270 to 195 nm with an increment of 1 nm, constant bandwidth of 10 nm, and a scanning speed of 1 nm per minute. After subtracting the buffer signal, the raw data in millidegrees was converted to mean residue ellipticity (MRE).

## 3. Results

*3.1. Binding of the Small Molecule 34RH to the Myc Target Site*

Previously, Yin and coworkers demonstrated that the small molecule 10058-F4 (1RH) disrupted Myc-Max dimerization [30]. Subsequently, we identified the specific interaction site of 1RH within the disordered, monomeric Myc bHLHZip domain [49]. In this study, we use the previously reported 1RH-derivative, 34RH—which maintains the core structure of 1RH, while replacing an ethyl group with a carboxylic acid moiety on the phenyl ring (Figure 1A) [44]. At neutral pH, 34RH is present primarily in the dianionic form as the pKa of the imide group of the rhodanine heterocycle is 5.3 ± 0.3 (Figure S2) and shows good solubility based on dynamic light scattering (Figure S3).

The binding site of 1RH in $Myc_{353–437}$ had been previously localized to within residues 402 to 412 [49]. In $Myc_{353–437}$, the only fluorescent residue (Tyr or Trp) is $Tyr_{402}$ located in the binding site. We and others have demonstrated that the interaction with 1RH causes quenching of $Tyr_{402}$ [49,50]. Here, we exploited this tyrosine fluorescence to evaluate binding of the 34RH molecule to $Myc_{353–437}$. Upon addition of 34RH to $Myc_{353–437}$, we observed that the $Myc_{353–437}$ fluorescence was quenched (Figure 1B).

The observed fluorescence quenching was titratable, and the 34RH binding affinity to $Myc_{353–437}$ was determined by monitoring tyrosine fluorescence as a function of 34RH concentration. The quenching data was fit to a Langmuir binding isotherm yielding a dissociation constant ($K_D$) of 3.9 ± 1.3 μM (Figure 1C). Notably, the dissociation constant obtained for 34RH and $Myc_{353–437}$ is comparable to the previously determined $K_D$ for 1RH and $Myc_{353–437}$ of 5.3 ± 0.7 μM [49]. In addition to titrations with 34RH in excess over $Myc_{353–437}$, we performed titrations with equimolar concentrations of $Myc_{353–437}$ and 34RH, where we observed that the $Myc_{353–437}$ fluorescence was quenched to a comparable extent, and we obtained a similar $K_D$ of 5.9 ± 0.8 μM (Figure S5).

We performed circular dichroism (CD) experiments with and without 34RH to determine if the addition of 34RH altered the average conformation of $Myc_{353–437}$. The CD spectrum of $Myc_{353–437}$ indicated that the domain was largely disordered with some α-helical character, as expected from NMR experiments on Myc [51–53]. Those NMR experiments indicated that the Myc sequence was predominantly random coil but with partial helical character, particularly in the region around residues 360–370 and with strong helical character from residues 416–422. Comparison of the CD spectra of $Myc_{353–437}$ with and without the addition of the small molecule indicated that 34RH did not substantially alter the average conformation of the protein (Figure 1D).

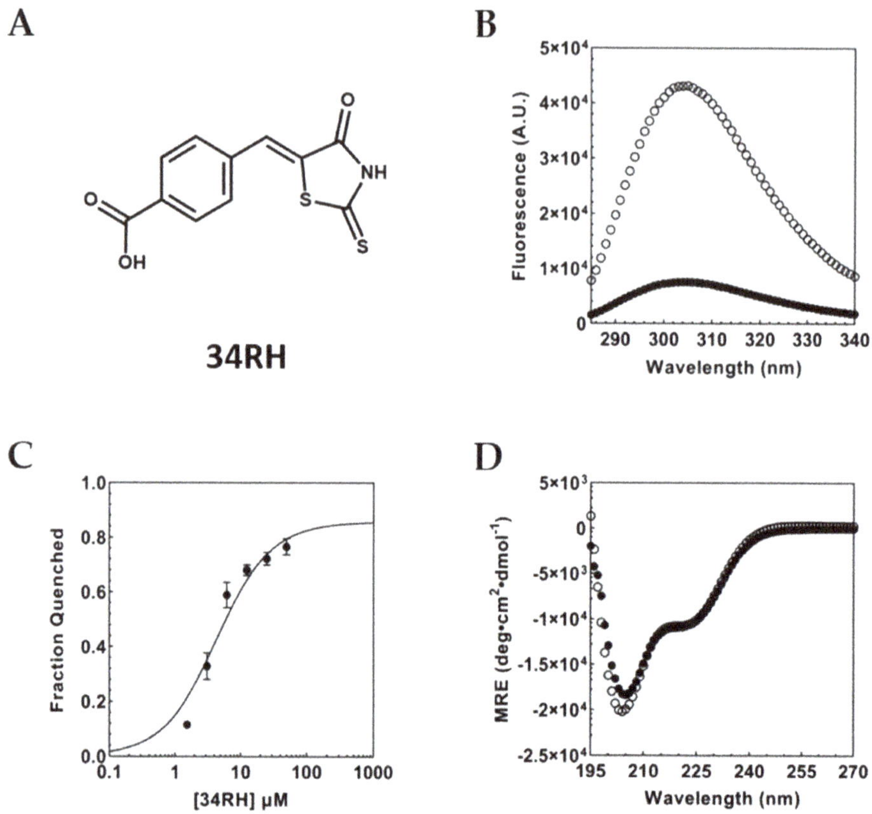


**Figure 1.** (**A**) Structure of 34RH. (**B**) Inner filter corrected fluorescence emission spectrum of 1 μM $Myc_{353–437}$ with (black circles) and without (white circles) 50 μM 34RH in 1xPBS at 25 °C, pH 7.4. (**C**) Equilibrium titration of 1 μM $Myc_{353–437}$ with excess 34RH fit to a Langmuir binding isotherm, $K_D = 3.9 \pm 1.3$ μM. Error bars represent the standard error of three independent trials. (**D**) Circular dichroism of 2.5 μM $Myc_{353–437}$ with (black circles) and without (white circles) 50 μM 34RH in 1xCD buffer.

### *3.2. Binding of 34RH to the $Myc_{402–412}$ Peptide*

Our previous studies showed that small molecules can bind to short contiguous segments in $Myc_{353–437}$ [42]. Guided by mutations and truncations, we demonstrated that the small molecule 1RH could bind to the short peptide sequence $Myc_{402–412}$ [49]. Here, we used this peptide, $Y_{402}ILSVQAEEQK_{412}$, to determine the affinity of 34RH for the isolated binding site. As with $Myc_{353–437}$, binding of 34RH to the peptide was monitored via Tyr fluorescence quenching (Figure 2A). In the context of the peptide, we again observed strong fluorescence quenching and titratable binding. From the data, we obtained a $K_D$ of $11.5 \pm 1.2$ μM, within three-fold of the affinity determined for $Myc_{353–437}$. The dissociation constant for the isolated peptide sequence is similar to the previously reported binding affinities of 1RH for the $Myc_{402–412}$ peptide of between 13 and 14 μM [49,50].

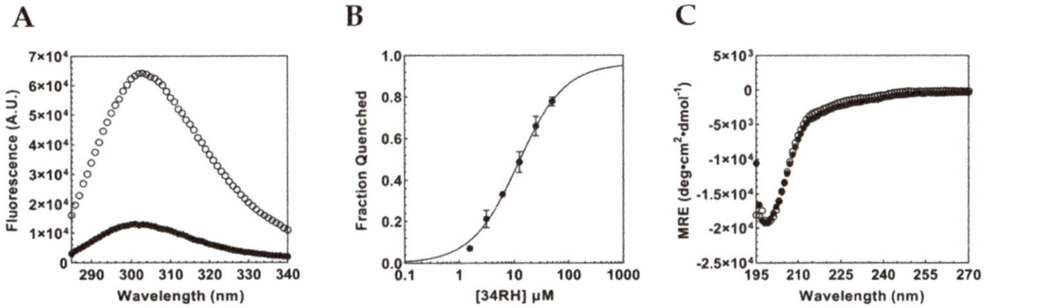


**Figure 2.** (**A**) Inner filter corrected fluorescence emission spectrum of 1 μM $Myc_{402–412}$ peptide with (black circles) and without (white circles) 50 μM 34RH. (**B**) Fraction quenched titration curve for 1 μM $Myc_{402–412}$ peptide and 34RH fitted to a Langmuir binding isotherm yielding a $K_D = 11.5 \pm 1.2$ μM. Error bars represent the standard error of three independent trials. (**C**) CD spectrum of 2.5 μM $Myc_{402–412}$ peptide with (black circles) and without (white circles) 50 μM 34RH.

To monitor the conformation of the peptide upon introducing 34RH, we performed CD. We observed that the peptide displayed a single negative MRE at 202 nm, indicating a predominantly random-coil conformation. Upon addition of 34RH, the peptide does not exhibit perturbations to the structural ensemble, as observed by the near identical CD spectra with and without the compound. The result with 34RH contrasts with that of the previous data with 1RH, where the addition of 1RH induced a substantial shift in the peptide's secondary structure [49]. The lower concentration of the peptide (2.5 μM versus 20 μM) and the charged nature of 34RH potentially account for the differences in the structural perturbation. Our results illustrate that the small molecule 34RH can bind to a short segment of $Myc_{353–437}$ independent of the entire protein domain and without imparting significant structural alterations. Furthermore, 34RH can bind to the random coil, indicating that a disordered eleven-residue peptide is sufficient for the binding of the small molecule.

### 3.3. Portability of the Small-Molecule IDP Binding Site

Short linear motifs (SLiMs) or eukaryotic linear motifs (ELMs) use the same or closely related sequences to bind partner proteins in different contexts [27]. The short linear binding site of 34RH is similar to a SliM since binding occurs independently of the larger context while maintaining affinity. If 34RH binding to the peptide sequence is truly independent of the overall context, we should be able to move the binding sequence into a different protein and recapitulate 34RH binding activity in that new context. Here, we chose Max, a heterodimerization partner of Myc [54] previously shown not to interact with 1RH [30], to receive the ported binding sequence. The canonical isoform of Max (P22 Max) is 160 amino acids in length and shares a 38% sequence identity with Myc in the bHLHZip region. Max has a short N-terminal disordered region and a longer disordered C-terminus [55]. We aligned $Myc_{353–437}$ and Max and compared the binding site region (Figure 3). The comparison indicated that the Max sequence, $Y_{70}$IQYMRR$K_{77}$, aligned with the binding site in Myc. Beyond the first two residues of this site, the Max sequence lacks identity with Myc in the binding region. We wanted to mutate a minimal set of amino acids in Max to form the small molecule binding sequence. Previously, we determined that the 370–409 sequence of Myc, but not 353–405, could bind to 1RH [49]. Together with the $Myc_{402–412}$ binding data, we used this information to demarcate the minimal binding site of $Y_{402}$ to $E_{409}$. Therefore, we mutated six residues ($Q_{73}$YMRR$K_{77}$) in Max, in order to match the 402–409 region of $Myc_{353–437}$ (Figure 3). This new construct, termed MaxRH, contained what we postulated to be a minimal, functional 34RH binding sequence (-YILSVQAE-) ported into Max.

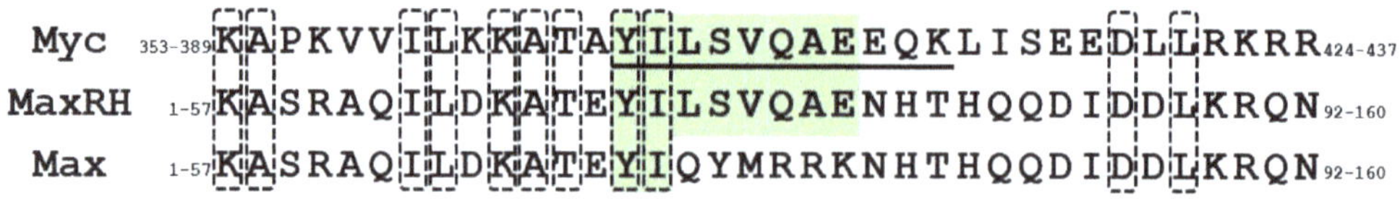


**Figure 3.** Alignment of the binding site region of $Myc_{353–437}$ with MaxRH and Max. Outlined residues are identical. Underlined residues denote the $Myc_{402–412}$ sequence. Highlighted green residues represent the overlap of the minimal binding site between the proteins.

As previously described for $Myc_{353–437}$, we monitored 34RH binding to MaxRH via tyrosine fluorescence quenching (Figure 4A). MaxRH contains three tyrosine residues in total, one in the binding sequence ($Tyr_{70}$) and two in the disordered C-terminus ($Tyr_{115}$ and $Tyr_{123}$). We expect $Tyr_{115}$ and $Tyr_{123}$ not to quench in the presence of 34RH while $Tyr_{70}$, in the generated binding site, should exhibit titratable quenching. If we successfully ported over the 34RH binding site, we would observe titratable quenching but with a lower maximum fraction quenched (in comparison to $Myc_{353–437}$) due to $Y_{115}$ and $Y_{123}$ retaining their fluorescence. In the presence of 50 μM 34RH, MaxRH tyrosine fluorescence is quenched (Figure 4A). Titration of a constant concentration of MaxRH with 34RH yielded a binding curve with a dissociation constant of 23.4 ± 1.1 μM (Figure 4B) and the expected lower maximum quenching. The $K_D$ for MaxRH:34RH indicates that 34RH can bind to the ported sequence in a new context, albeit with reduced affinity. We also tested MaxRH with and without 50 μM of 34RH using CD (Figure 4C). The CD of MaxRH in the absence of compound showed a spectrum similar to $Myc_{353–437}$, indicative of a random coil with partial helical character. The addition of 50 μM of 34RH did not change the conformation of MaxRH. The CD spectrum of MaxRH is consistent with it being a monomer at 1 μM, presumably due to the introduced Myc residues reducing the homodimer stability of the parental P22 Max sequence [54,56]. In order to isolate the fluorescence of the tyrosine in the binding site from the signal of the two C-terminal tyrosine residues in MaxRH, we mutated these residues to phenylalanine to generate MaxRH-Y115F/Y123F. We observed that MaxRH-Y115F/Y123F fluorescence was quenched with 50 μM 34RH (Figure 4D) on par with the quenching seen with $Myc_{353–437}$ and $Myc_{402–412}$ confirming that the Tyr in the binding site of MaxRH is the residue quenched upon binding and that the quenching is similar to that seen in the native Myc context. From the titration of MaxRH-Y115F/Y123F with 34RH, we obtained a $K_D$ of 14.9 ± 1.9 μM (Figure 4E). We also obtained the CD spectra of MaxRH-Y115F/Y123F in the presence and absence of 50 μM 34RH and confirmed that the protein remains disordered even in the presence of the small molecule (Figure 4F).

We next tested P22 Max to verify that Max does not bind to 34RH. We observed that the tyrosine fluorescence of P22 Max does not exhibit titratable quenching with 34RH (Figure 4G,H). The CD of P22 Max with and without 34RH indicated that 34RH does not alter the CD of P22 Max. The spectra do, however, exhibit a substantially greater helical character of P22 Max, indicative of homodimer formation (Figure 4I) [57]. In a homodimer, $Tyr_{115}$ and $Tyr_{123}$ would still be expected to be accessible to 34RH; however, $Tyr_{70}$ and adjacent residues would likely be occluded by the dimer structure.

To control for binding interactions of 34RH with the P22 Max sequence in a monomeric state, titrations were conducted at pH 6. The lower pH disfavors dimer formation leading to monomeric P22 Max [58]. At pH 6, CD results with P22 Max indicated a substantial loss in helical character, with a spectrum similar to $Myc_{353–437}$ and MaxRH, and consistent with the monomeric form of P22 Max (Figure 5A). Fluorescence experiments with 34RH and P22 Max were performed at pH 6 (Figure 5D) and again showed no titratable quenching of P22 Max fluorescence. To confirm binding still occurs under these conditions, MaxRH fluorescence quenching and CD were measured at pH 6 (Figure 5B,E). At pH 6, MaxRH still bound to 34RH and actually improved in affinity with a dissociation constant of 9.1 ± 3.9 μM while remaining disordered as observed via CD. As a further control, we also tested for binding to the 151 residue P21 isoform of Max. The nine-residue difference at the N-terminus (prior to the bHLHZip) between the two Max isoforms is associated with a weaker homodimerization constant for P21 Max [59]. The CD spectrum of P21 Max at pH 7.4 was consistent with a monomeric state with no indication of perturbation in the presence of 34RH (Figure 5C). The tyrosine fluorescence of P21 Max versus 34RH concentration was similar to results with P22 Max showing no titratable quenching (Figure 5F). These results indicated that the native Max sequence does not interact with 34RH in regions around its tyrosine residues and demonstrated that the 34RH binding function was ported into the Max context by introduction of a minimal binding sequence.

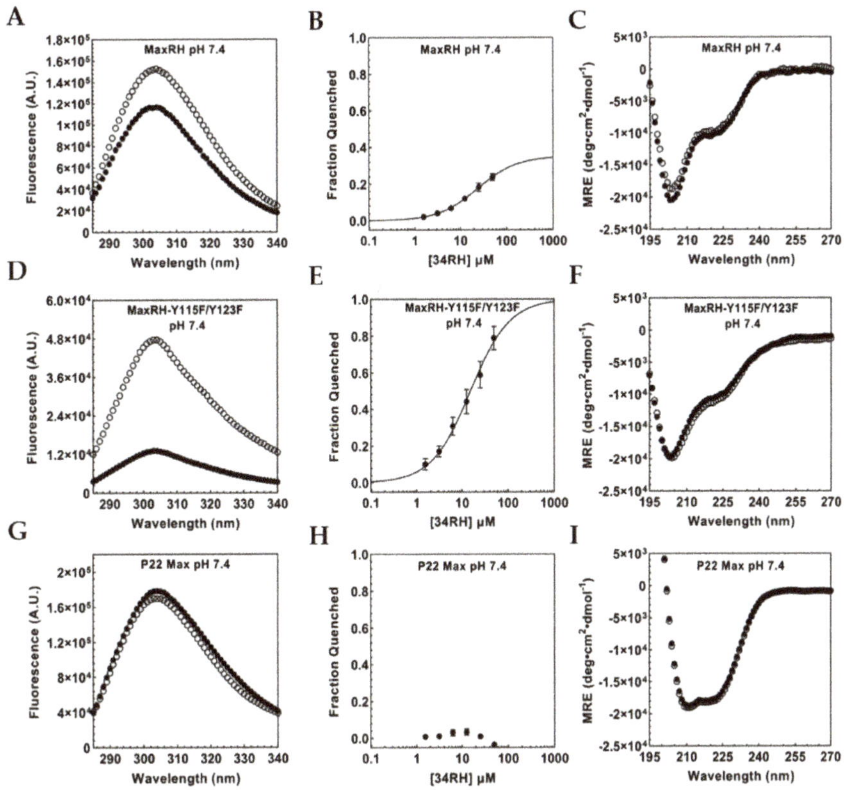


**Figure 4.** (**A**) Inner filter corrected fluorescence emission spectrum of 1 μM MaxRH with (black circles) and without (white circles) 50 μM 34RH. (**B**) Fraction quenched titration curve for 1 μM MaxRH with 34RH fit to a Langmuir binding isotherm, $K_D$ = 23.4 ± 1.1 μM (**C**) CD spectrum of 1 μM MaxRH with (black circles) with without (white circles) 50 μM 34RH. (**D**) Inner filter corrected fluorescence emission spectrum of 1 μM MaxRH-Y115F/Y123F with (black circles) and without (white circles) 50 μM 34RH. (**E**) Fraction quenched titration curve for 1 μM MaxRH-Y115F/Y123F with 34RH fit to a Langmuir binding isotherm, $K_D$ = 14.9 ± 1.9 μM (**F**) CD spectrum of 1 μM MaxRH-Y115F/Y123F with (black circles) with without (white circles) 50 μM 34RH. (**G**) Inner filter corrected fluorescence emission spectrum of 1 μM P22 Max with (black circles) and without (white circles) 50 μM 34RH. (**H**) Fraction quenched titration curve for 1 μM P22 Max with 34RH (**I**) CD spectrum of 1 μM P22 Max with (black circles) with without (white circles) 50 μM 34RH at pH 7.4.

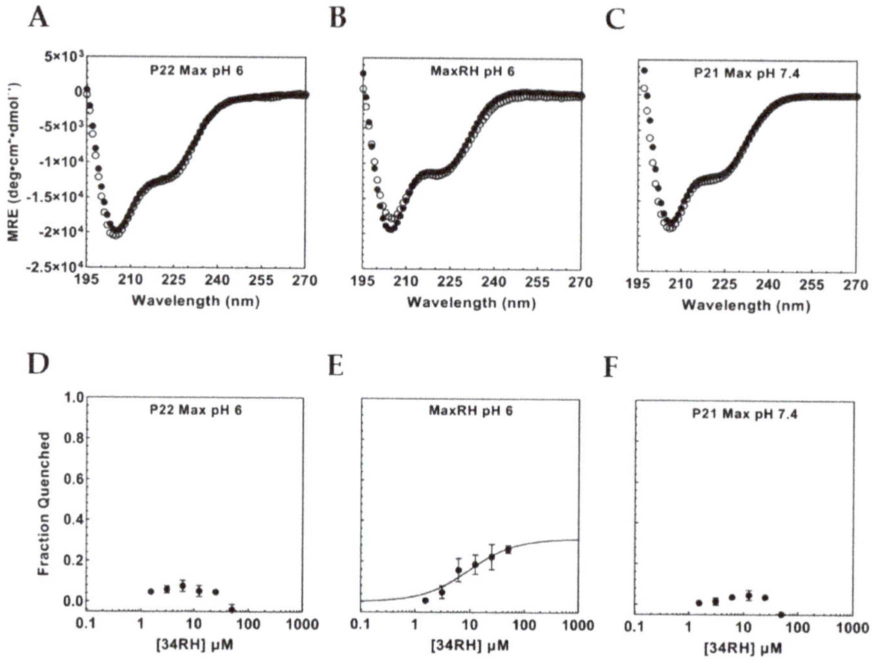


**Figure 5.** (**A**) CD of 1 μM P22 Max at pH 6 with (black circles) and without (white circles) 50 μM 34RH. (**B**) CD of 1 μM MaxRH at pH 6 with (black circles) and without (white circles) 50 μM 34RH. (**C**) CD of 4 μM P21 Max at pH 7.4 with (black circles) and without (white circles) 50 μM 34RH. (**D**) Fraction quenched titration curve for 1 μM P22 Max with 34RH at pH 6. (**E**) Fraction quenched titration curve for 1 μM MaxRH with 34RH at pH 6 fitted to a Langmuir binding isotherm, $K_D = 9.1 \pm 3.9$ μM (**F**) Fraction quenched titration curve for 1 μM P21 Max with 34RH at pH 7.4. Error bars represent the standard error of three independent trials.

### *3.4. Flanking Residues Modulate 34RH Binding*

At pH 7.4, the $K_D$ obtained for MaxRH:34RH binding was notably higher than the value observed with $Myc_{353–437}$. The $Myc_{353–437}$ and MaxRH sequences differ in the flanking residues directly adjacent to the binding site. Using point mutations, we wanted to examine the impact of flanking residues on the minimal binding site in $Myc_{353–437}$ and MaxRH. At the C-terminus, MaxRH has an asparagine directly adjacent to the binding site while $Myc_{353–437}$ has a glutamic acid. We wanted to test if mutating $N_{78}$ in MaxRH to a glutamic acid would improve binding. The MaxRH-N78E mutant extended the MaxRH:$Myc_{353–437}$ identity by one residue flanking the binding site (-$Y_{70}$ILSVQAE$E_{78}$-). Surprisingly, the mutation caused a complete loss of observable binding with no titratable tyrosine quenching (Figure 6A). Since a flanking Asn permitted 34RH binding in MaxRH while Glu eliminated it, the reciprocal mutation was tested in $Myc_{353–437}$. The construct $Myc_{353–437}$ E410N was tested for binding to 34RH. Here again, a relatively conservative change in the residue flanking the binding site eliminated binding to 34RH (Figure 6B). The identity of the C-terminal flanking residue had opposing effects in the Myc and MaxRH contexts. These sequences diverge on the C-terminal side of the binding site showing little sequence identity (Figure 3). At the N-terminal side, however, five out of seven residues adjacent to the binding site are identical between Myc and MaxRH. Directly flanking the Tyr of the binding site, MaxRH has a glutamic acid while Myc has an alanine. We constructed $Myc_{353–437}$ A401E to determine if the same flanking residue would be permissive of binding in both protein contexts at the N-terminal side. The titration of $Myc_{353–437}$ A401E with

34RH caused no detectable binding (Figure 6C). Here again, we observed a flanking residue that was permissive of binding in one context but eliminated binding in the other.

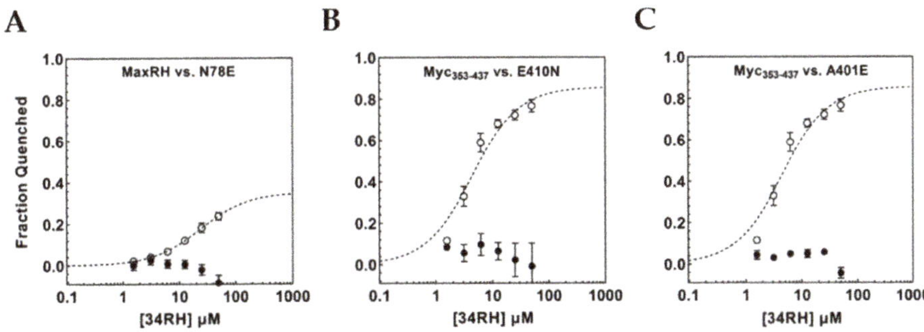


**Figure 6.** (**A**) MaxRH-N78E fluorescence quenching titration curve (black circles) overlaid with MaxRH curve (white circles). (**B**) $Myc_{353–437}$ E410N fluorescence quenching titration curve (black circles) overlaid with $Myc_{353–437}$ curve (white circles). (**C**) $Myc_{353–437}$ A401E fluorescence quenching titration curve (black circles) overlaid with $Myc_{353–437}$ curve (white circles). Error bars represent the standard error of three independent trials.

## 4. Discussion

Short stretches of disordered regions have been shown to bind to small molecules with at least micromolar affinity [40]. SLiMs also engage in molecular recognition via short, localized sequences, are typically present in disordered regions, and typically bind to their partner proteins with micromolar affinity. An inherent characteristic of SLiMs is their modularity and resulting portability [27]. Based on analogous aspects between SLiMs and small-molecule binding sites in disordered proteins, we believed that small-molecule binding sites could also show portability and allow their binding function to move between protein contexts as the short binding sequence is moved.

Using the small molecule 34RH, we demonstrated that the binding observed in the context of $Myc_{353–437}$ is maintained with only a moderate (3-fold) change in affinity for the binding site in the isolated peptide sequence $Myc_{402–412}$, similar to what was previously observed for the 1RH compound [49]. NMR data from Panova and coworkers have indicated that $Myc_{353–437}$ is expected to be compact with paramagnetic relaxation enhancement (PRE) data showing contacts between residue 400–412 and 360–380, along with some predicted helical character (~20%) in the 400–412 region [51]. In contrast, the $Myc_{402–412}$ peptide is a random coil that lacks a surrounding protein context and so is devoid of additional contacts with the protein sequence. Despite these differences, the affinity for 34RH in the two contexts differs by less than 0.7 kcal $mol^{-1}$, indicating substantial modularity to the small molecule binding sequence.

By mutating six residues in Max to produce MaxRH, we transferred a small molecule binding site into a new protein context and could observe binding. The affinity of 34RH was about 6-fold weaker than in the Myc context (2-fold weaker relative to the peptide). At pH 6, the binding of 34RH to MaxRH improved 2.5-fold to a $K_D$ of 9.1 μM. Kizilsavas and coworkers had studied monomeric Max via NMR under similar conditions (pH 5.5) and found the sequence to be disordered but highly compact [60]. These results demonstrate that small molecule binding sites can exhibit portability between disordered protein contexts. Furthermore, the binding can be robust to variations in the conformational propensity and surrounding protein environment with only several-fold variation in affinity when the binding site is in a very compact disordered domain (Max), a partially ordered domain with tertiary contacts (Myc), or in a short peptide sequence. The protein context can tune the binding, but in the absence of a disorder to order transition [28], it does not appear to be a major factor or even a necessary component for binding [40].

The eight-residue sequence from Myc (YILSVQAE) was found to be sufficient to transfer binding function when placed in the context of the Max sequence; however, in both Myc and MaxRH, binding was very sensitive to the identity of the immediately flanking residue at both ends of the sequence. In MaxRH, mutating the C-terminal flanking Asn to Glu eliminated detectable binding while in the Myc context we observed a reciprocal effect. Mutating the flanking Glu to Asn eliminated binding to Myc. A residue that was permissive of binding in one context was prohibitive in the other. At the N-terminal end of the binding sequence in $Myc_{353–437}$, we observed a similar effect; mutating the native Ala to a Glu, which is present in the equivalent position in MaxRH, eliminated binding. Truncations can define a minimal necessary sequence for small-molecule binding to a peptide but that may not be sufficient for binding in a given protein domain context. Flanking residues have been shown previously to influence the binding of adjacent disordered sequences [61,62]. Despite remaining disordered in the complex, small-molecule binding affinity can also be strongly influenced by flanking residues.

Here, we show that a disordered small-molecule binding site can be ported between disordered protein contexts and retain its binding function. This finding supports the idea that if we are able to identify minimal sequences that can bind small molecules, then these sequences are likely to retain their binding function when in the context of various disordered domains. However, we also find that residues flanking the set of necessary binding residues can influence binding, with the same flanking residue being either permissive or prohibitive of binding depending on the broader protein context. While the influence of flanking residues increases the complexity of identifying the small-molecule binding sites, it also increases the specificity of the binding site by increasing the sequence requirements needed to achieve binding

**Supplementary Materials:** The following supporting information can be downloaded at: https://www.mdpi.com/article/10.3390/biom12121887/s1, Figure S1: Myc, MaxRH, P22 Max, and P21 Max constructs aligned using CLUSTAL O (1.2.4) multiple sequence alignment [63]; Figure S2: UV/Vis characterization of 34RH [45]; Figure S3: Dynamic Light Scattering of 34RH in 1xPBS (buffer) at pH 7.4; Figure S4: $Myc_{353–437}$ and 34RH interaction monitored via tyrosine fluorescence quenching; Figure S5: Equimolar $Myc_{353–437}$ and 34RH interaction monitored via tyrosine fluorescence quenching [46,49]; Figure S6: NMR characterization of 34RH.

**Author Contributions:** L.M.F.L., R.J., S.R.D.S. and S.J.M. designed the study; E.M., L.M.F.L., R.J. and S.R.D.S. conducted experiments; all authors analyzed the results; R.J., S.R.D.S. and S.J.M. wrote the manuscript. All authors reviewed the manuscript. All authors have read and agreed to the published version of the manuscript.

**Funding:** This work was supported by the National Institutes of Health (1R15GM120698) and by the Institute of Soft Matter Synthesis and Metrology (ISMSM), Georgetown University.

**Institutional Review Board Statement:** Not applicable.

**Informed Consent Statement:** Not applicable.

**Data Availability Statement:** Not applicable.

**Acknowledgments:** The authors gratefully acknowledge Edward V. Prochownik for P21 Max and P22 Max expression plasmids.

**Conflicts of Interest:** The authors declare no conflict of interest.

## References


1. Uversky, V.N. Natively unfolded proteins: A point where biology waits for physics. *Protein Sci.* **2002**, *11*, 739–756. [CrossRef]
2. Fisher, C.K.; Stultz, C.M. Protein Structure along the Order-Disorder Continuum. *J. Am. Chem. Soc.* **2011**, *133*, 10022–10025. [CrossRef]
3. Uversky, V.N. Intrinsically Disordered Proteins and Their "Mysterious" (Meta) Physics. *Front. Phys.* **2019**, *7*, 10. [CrossRef]
4. Dunker, A.K.; Brown, C.J.; Lawson, J.D.; Iakoucheva, L.M.; Obradovic, Z. Intrinsic disorder and protein function. *Biochemistry* **2002**, *41*, 6573–6582. [CrossRef] [PubMed]
5. Fenwick, R.B.; Esteban-Martin, S.; Salvatella, X. Understanding biomolecular motion, recognition, and allostery by use of conformational ensembles. *Eur. Biophys. J. Biophy.* **2011**, *40*, 1339–1355. [CrossRef]
6. Turoverov, K.K.; Kuznetsova, I.M.; Uversky, V.N. The protein kingdom extended: Ordered and intrinsically disordered proteins, their folding, supramolecular complex formation, and aggregation. *Prog. Biophys. Mol. Bio.* **2010**, *102*, 73–84. [CrossRef] [PubMed]
7. Galea, C.A.; High, A.A.; Obenauer, J.C.; Mishra, A.; Park, C.G.; Punta, M.; Schllessinger, A.; Ma, J.; Rost, B.; Slaughter, C.A.; et al. Large-Scale Analysis of Thermostable, Mammalian Proteins Provides Insights into the Intrinsically Disordered Proteome. *J. Proteome Res.* **2009**, *8*, 211–226. [CrossRef] [PubMed]
8. Liu, J.G.; Perumal, N.B.; Oldfield, C.J.; Su, E.W.; Uversky, V.N.; Dunker, A.K. Intrinsic disorder in transcription factors. *Biochemistry* **2006**, *45*, 6873–6888. [CrossRef]
9. Wright, P.E.; Dyson, H.J. Intrinsically disordered proteins in cellular signalling and regulation. *Nat. Rev. Mol. Cell Biol.* **2015**, *16*, 18–29. [CrossRef]
10. Babu, M.M.; van der Lee, R.; de Groot, N.S.; Gsponer, J. Intrinsically disordered proteins: Regulation and disease. *Curr. Opin. Struc. Biol.* **2011**, *21*, 432–440. [CrossRef]
11. Uversky, V.N.; Oldfield, C.J.; Dunker, A.K. Intrinsically disordered proteins in human diseases: Introducing the D2 concept. *Annu. Rev. Biophys.* **2008**, *37*, 215–246. [CrossRef] [PubMed]
12. Fuxreiter, M.; Tompa, P.; Simon, I.; Uversky, V.N.; Hansen, J.C.; Asturias, F.J. Malleable machines take shape in eukaryotic transcriptional regulation. *Nat. Chem. Biol.* **2008**, *4*, 728–737. [CrossRef] [PubMed]
13. Mohan, A.; Oldfield, C.J.; Radivojac, P.; Vacic, V.; Cortese, M.S.; Dunker, A.K.; Uversky, V.N. Analysis of molecular recognition features (MoRFs). *J. Mol. Biol.* **2006**, *362*, 1043–1059. [CrossRef] [PubMed]
14. Neduva, V.; Russell, R.B. Linear motifs: Evolutionary interaction switches. *Febs. Lett.* **2005**, *579*, 3342–3345. [CrossRef]
15. Banani, S.F.; Lee, H.O.; Hyman, A.A.; Rosen, M.K. Biomolecular condensates: Organizers of cellular biochemistry. *Nat. Rev. Mol. Cell Bio.* **2017**, *18*, 285–298. [CrossRef]
16. Roden, C.; Gladfelter, A.S. RNA contributions to the form and function of biomolecular condensates. *Nat. Rev. Mol. Cell Biol.* **2021**, *22*, 183–195. [CrossRef]
17. Weber, S.C.; Brangwynne, C.P. Getting RNA and Protein in Phase. *Cell* **2012**, *149*, 1188–1191. [CrossRef]
18. Lyon, A.S.; Peeples, W.B.; Rosen, M.K. A framework for understanding the functions of biomolecular condensates across scales. *Nat. Rev. Mol. Cell Biol.* **2021**, *22*, 215–235. [CrossRef]
19. Shammas, S.L.; Crabtree, M.D.; Dahal, L.; Wicky, B.I.M.; Clarke, J. Insights into Coupled Folding and Binding Mechanisms from Kinetic Studies. *J. Biol. Chem.* **2016**, *291*, 6689–6695. [CrossRef]
20. Wright, P.E.; Dyson, H.J. Linking folding and binding. *Curr. Opin. Struc. Biol.* **2009**, *19*, 31–38. [CrossRef]
21. van der Lee, R.; Buljan, M.; Lang, B.; Weatheritt, R.J.; Daughdrill, G.W.; Dunker, A.K.; Fuxreiter, M.; Gough, J.; Gsponer, J.; Jones, D.T.; et al. Classification of Intrinsically Disordered Regions and Proteins. *Chem. Rev.* **2014**, *114*, 6589–6631. [CrossRef] [PubMed]
22. Tompa, P.; Fuxreiter, M. Fuzzy complexes: Polymorphism and structural disorder in protein-protein interactions. *Trends Biochem. Sci.* **2008**, *33*, 2–8. [CrossRef] [PubMed]
23. Cheng, Y.; LeGall, T.; Oldfield, C.J.; Mueller, J.P.; Van, Y.Y.; Romero, P.; Cortese, M.S.; Uversky, V.N.; Dunker, A.K. Rational drug design via intrinsically disordered protein. *Trends Biotechnol.* **2006**, *24*, 435–442. [CrossRef] [PubMed]
24. Malhis, N.; Gsponer, J. Computational identification of MoRFs in protein sequences. *Bioinformatics* **2015**, *31*, 1738–1744. [CrossRef]
25. Disfani, F.M.; Hsu, W.L.; Mizianty, M.J.; Oldfield, C.J.; Xue, B.; Dunker, A.K.; Uversky, V.N.; Kurgan, L. MoRFpred, a computational tool for sequence-based prediction and characterization of short disorder-to-order transitioning binding regions in proteins. *Bioinformatics* **2012**, *28*, I75–I83. [CrossRef]
26. Burgi, J.; Xue, B.; Uversky, V.N.; van der Goot, F.G. Intrinsic Disorder in Transmembrane Proteins: Roles in Signaling and Topology Prediction. *PLoS ONE* **2016**, *11*, e0158594. [CrossRef]
27. Davey, N.E.; Van Roey, K.; Weatheritt, R.J.; Toedt, G.; Uyar, B.; Altenberg, B.; Budd, A.; Diella, F.; Dinkel, H.; Gibson, T.J. Attributes of short linear motifs. *Mol. Biosyst.* **2012**, *8*, 268–281. [CrossRef]
28. Davey, N.E. The functional importance of structure in unstructured protein regions. *Curr. Opin. Struc. Biol.* **2019**, *56*, 155–163. [CrossRef]
29. Berg, T.; Cohen, S.B.; Desharnais, J.; Sonderegger, C.; Maslyar, D.J.; Goldberg, J.; Boger, D.L.; Vogt, P.K. Small-molecule antagonists of Myc/Max dimerization inhibit Myc-induced transformation of chicken embryo fibroblasts. *Proc. Nat. Acad. Sci. USA* **2002**, *99*, 3830–3835. [CrossRef]
30. Yin, X.; Giap, C.; Lazo, J.S.; Prochownik, E.V. Low molecular weight inhibitors of Myc-Max interaction and function. *Oncogene* **2003**, *22*, 6151–6159. [CrossRef]
31. Dang, C.V. MYC on the Path to Cancer. *Cell* **2012**, *149*, 22–35. [CrossRef] [PubMed]

32. Soucek, L.; Whitfield, J.; Martins, C.P.; Finch, A.J.; Murphy, D.J.; Sodir, N.M.; Karnezis, A.N.; Swigart, L.B.; Nasi, S.; Evan, G.I. Modelling Myc inhibition as a cancer therapy. *Nature* **2008**, *455*, 679–683. [CrossRef] [PubMed]
33. Whitfield, J.R.; Soucek, L. The long journey to bring a Myc inhibitor to the clinic. *J. Cell Biol.* **2021**, *220*, e202103090. [CrossRef] [PubMed]
34. Llombart, V.; Mansour, M.R. Therapeutic targeting of "undruggable" MYC. *Ebiomedicine* **2022**, *75*, 103756. [CrossRef] [PubMed]
35. Madden, S.K.; de Araujo, A.D.; Gerhardt, M.; Fairlie, D.P.; Mason, J.M. Taking the Myc out of cancer: Toward therapeutic strategies to directly inhibit c-Myc. *Mol. Cancer* **2021**, *20*, 3. [CrossRef]
36. Metallo, S.J. Intrinsically disordered proteins are potential drug targets. *Curr. Opin. Chem. Biol.* **2010**, *14*, 481–488. [CrossRef]
37. Ruan, H.; Sun, Q.; Zhang, W.L.; Liu, Y.; Lai, L.H. Targeting intrinsically disordered proteins at the edge of chaos. *Drug Discov. Today* **2019**, *24*, 217–227. [CrossRef]
38. Santofimia-Castano, P.; Rizzuti, B.; Xia, Y.; Abian, O.; Peng, L.; Velazquez-Campoy, A.; Neira, J.L.; Iovanna, J. Targeting intrinsically disordered proteins involved in cancer. *Cell Mol. Life Sci.* **2020**, *77*, 1695–1707. [CrossRef]
39. Chen, J.L.; Liu, X.R.; Chen, J.H. Targeting Intrinsically Disordered Proteins through Dynamic Interactions. *Biomolecules* **2020**, *10*, 743. [CrossRef]
40. Biesaga, M.; Frigole-Vivas, M.; Salvatella, X. Intrinsically disordered proteins and biomolecular condensates as drug targets. *Curr. Opin. Chem. Biol.* **2021**, *62*, 90–100. [CrossRef]
41. Benz, C.; Ali, M.; Krystkowiak, I.; Simonetti, L.; Sayadi, A.; Mihalic, F.; Kliche, J.; Andersson, E.; Jemth, P.; Davey, N.E.; et al. Proteome-scale mapping of binding sites in the unstructured regions of the human proteome. *Mol. Syst. Biol.* **2022**, *18*, e10584. [CrossRef] [PubMed]
42. Hammoudeh, D.I.; Follis, A.V.; Prochownik, E.V.; Metallo, S.J. Multiple independent binding sites for small-molecule inhibitors on the oncoprotein c-Myc. *J. Am. Chem. Soc.* **2009**, *131*, 7390–7401. [CrossRef]
43. Studier, F.W. Protein production by auto-induction in high-density shaking cultures. *Protein Expres. Purif.* **2005**, *41*, 207–234. [CrossRef] [PubMed]
44. Wang, H.; Hammoudeh, D.I.; Follis, A.V.; Reese, B.E.; Lazo, J.S.; Metallo, S.J.; Prochownik, E.V. Improved low molecular weight Myc-Max inhibitors. *Mol. Cancer Ther.* **2007**, *6*, 2399–2408. [CrossRef] [PubMed]
45. Luiz, F.C.L.; Louro, S.R.W. Acid-base equilibrium of drugs in time-resolved fluorescence measurements: Theoretical aspects and expressions for apparent pK(a) shifts. *J. Photoch. Photobio. A* **2011**, *222*, 10–15. [CrossRef]
46. Lakowicz, J.R. *Principles of Fluorescence Spectroscopy*; Springer: New York, NY, USA, 2006.
47. Dobrev, V.S.; Fred, L.M.; Gerhart, K.P.; Metallo, S.J. Characterization of the Binding of Small Molecules to Intrinsically Disordered Proteins. *Method Enzym.* **2018**, *611*, 677–702. [CrossRef]
48. Jarmoskaite, I.; AlSadhan, I.; Vaidyanathan, P.P.; Herschlag, D. How to measure and evaluate binding affinities. *eLife* **2020**, *9*, e57264. [CrossRef]
49. Follis, A.V.; Hammoudeh, D.I.; Wang, H.B.; Prochownik, E.V.; Metallo, S.J. Structural Rationale for the Coupled Binding and Unfolding of the c-Myc Oncoprotein by Small Molecules. *Chem. Biol.* **2008**, *15*, 1149–1155. [CrossRef]
50. Heller, G.T.; Aprile, F.A.; Bonomi, M.; Camilloni, C.; De Simone, A.; Vendruscolo, M. Sequence Specificity in the Entropy-Driven Binding of a Small Molecule and a Disordered Peptide. *J. Mol. Biol.* **2017**, *429*, 2772–2779. [CrossRef]
51. Panova, S.; Cliff, M.J.; Macek, P.; Blackledge, M.; Jensen, M.R.; Nissink, J.W.M.; Embrey, K.J.; Davies, R.; Waltho, J.P. Mapping Hidden Residual Structure within the Myc bHLH-LZ Domain Using Chemical Denaturant Titration. *Structure* **2019**, *27*, 1537–1546. [CrossRef]
52. Macek, P.; Cliff, M.J.; Embrey, K.J.; Holdgate, G.A.; Nissink, J.W.M.; Panova, S.; Waltho, J.P.; Davies, R.A. Myc phosphorylation in its basic helix-loop-helix region destabilizes transient -helical structures, disrupting Max and DNA binding. *J. Biol. Chem.* **2018**, *293*, 9301–9310. [CrossRef]
53. Sammak, S.; Hamdani, N.; Gorrec, F.; Allen, M.D.; Freund, S.M.V.; Bycroft, M.; Zinzalla, G. Crystal Structures and Nuclear Magnetic Resonance Studies of the Apo Form of the c-MYC:MAX bHLHZip Complex Reveal a Helical Basic Region in the Absence of DNA. *Biochemistry* **2019**, *58*, 3144–3154. [CrossRef] [PubMed]
54. Nair, S.K.; Burley, S.K. X-ray structures of Myc-Max and Mad-Max recognizing DNA. Molecular bases of regulation by proto-oncogenic transcription factors. *Cell* **2003**, *112*, 193–205. [CrossRef] [PubMed]
55. Pursglove, S.E.; Fladvad, M.; Bellanda, M.; Moshref, A.; Henriksson, M.; Carey, J.; Sunnerhagen, M. Biophysical properties of regions flanking the bHLH-Zip motif in the p22 Max protein. *Biochem. Bioph. Res. Com.* **2004**, *323*, 750–759. [CrossRef]
56. Brownlie, P.; Ceska, T.A.; Lamers, M.; Romier, C.; Stier, G.; Teo, H.; Suck, D. The crystal structure of an intact human Max-DNA complex: New insights into mechanisms of transcriptional control. *Structure* **1997**, *5*, 509–520. [CrossRef]
57. Naud, J.F.; McDuff, F.O.; Sauve, S.; Montagne, M.; Webb, B.A.; Smith, S.P.; Chabot, B.; Lavigne, P. Structural and thermodynamical characterization of the complete p21 gene product of Max. *Biochemistry* **2005**, *44*, 12746–12758. [CrossRef] [PubMed]
58. Tchan, M.C.; Weiss, A.S. Asn(78) and His(81) form a destabilizing locus within the Max HLH-LZ homodimer. *Febs. Lett.* **2001**, *509*, 177–180. [CrossRef] [PubMed]
59. Zhang, H.; Fan, S.J.; Prochownik, E.V. Distinct roles for MAX protein isoforms in proliferation and apoptosis. *J. Biol. Chem.* **1997**, *272*, 17416–17424. [CrossRef]
60. Kizilsavas, G.; Ledolter, K.; Kurzbach, D. Hydrophobic Collapse of the Intrinsically Disordered Transcription Factor Myc Associated Factor X. *Biochemistry* **2017**, *56*, 5365–5372. [CrossRef]

61. Crabtree, M.D.; Borcherds, W.; Poosapati, A.; Shammas, S.L.; Daughdrill, G.W.; Clarke, J. Conserved Helix-Flanking Prolines Modulate Intrinsically Disordered Protein:Target Affinity by Altering the Lifetime of the Bound Complex. *Biochemistry* **2017**, *56*, 2379–2384. [CrossRef]
62. Das, R.K.; Crick, S.L.; Pappu, R.V. N-Terminal Segments Modulate the alpha-Helical Propensities of the Intrinsically Disordered Basic Regions of bZIP Proteins. *J. Mol. Biol.* **2012**, *416*, 287–299. [CrossRef] [PubMed]
63. Sievers, F.; Wilm, A.; Dineen, D.; Gibson, T.J.; Karplus, K.; Li, W.; Lopez, R.; McWilliam, H.; Remmert, M.; Söding, J.; et al. Fast, Scalable Generation of High-Quality Protein Multiple Sequence Alignments Using Clustal Omega. *Mol. Syst. Biol.* **2011**, *7*, 539. [CrossRef] [PubMed]

biomolecules

Article

# The Role of Membrane Affinity and Binding Modes in Alpha-Synuclein Regulation of Vesicle Release and Trafficking

Tapojyoti Das [1,2], Meraj Ramezani [3,4], David Snead [1,5], Cristian Follmer [6], Peter Chung [7,8], Ka Yee Lee [7], David A. Holowka [3], Barbara A. Baird [3] and David Eliezer [1,*]

1 Department of Biochemistry, Weill Cornell Medical College, New York, NY 10065, USA
2 Department of Structural Biology, St. Jude Children's Research Hospital, Memphis, TN 38105, USA
3 Department of Chemistry and Chemical Biology, Cornell University, Ithaca, NY 14853, USA
4 Metabolism Program, The Broad Institute of MIT & Harvard, Cambridge, MA 02142, USA
5 Department of Biochemistry and Molecular Biology, Johns Hopkins University, Baltimore, PA 21205, USA
6 Laboratory of Biological Chemistry of Neurodegenerative Disorders, Department of Physical-Chemistry, Institute of Chemistry, Federal University of Rio de Janeiro, Rio de Janeiro 22290-240, Brazil
7 Department of Chemistry, James Franck Institute, Institute of Biophysical Dynamics, The University of Chicago, Chicago, IL 60637, USA
8 Department of Physics and Astronomy and Department of Chemistry, University of Southern California, Los Angeles, CA 90089, USA
* Correspondence: dae2005@med.cornell.edu

**Abstract:** Alpha-synuclein is a presynaptic protein linked to Parkinson's disease with a poorly characterized physiological role in regulating the synaptic vesicle cycle. Using RBL-2H3 cells as a model system, we earlier reported that wild-type alpha-synuclein can act as both an inhibitor and a potentiator of stimulated exocytosis in a concentration-dependent manner. The inhibitory function is constitutive and depends on membrane binding by the helix-2 region of the lipid-binding domain, while potentiation becomes apparent only at high concentrations. Using structural and functional characterization of conformationally selective mutants via a combination of spectroscopic and cellular assays, we show here that binding affinity for isolated vesicles similar in size to synaptic vesicles is a primary determinant of alpha-synuclein-mediated potentiation of vesicle release. Inhibition of release is sensitive to changes in the region linking the helix-1 and helix-2 regions of the N-terminal lipid-binding domain and may require some degree of coupling between these regions. Potentiation of release likely occurs as a result of alpha-synuclein interactions with undocked vesicles isolated away from the active zone in internal pools. Consistent with this, we observe that alpha-synuclein can disperse vesicles from in vitro clusters organized by condensates of the presynaptic protein synapsin-1.

**Keywords:** alpha-synuclein; membrane; synaptic vesicle; synapsin; Parkinson's

**Citation:** Das, T.; Ramezani, M.; Snead, D.; Follmer, C.; Chung, P.; Lee, K.Y.; Holowka, D.A.; Baird, B.A.; Eliezer, D. The Role of Membrane Affinity and Binding Modes in Alpha-Synuclein Regulation of Vesicle Release and Trafficking. *Biomolecules* **2022**, *12*, 1816. https://doi.org/10.3390/biom12121816

Academic Editors: Prakash Kulkarni, Stefania Brocca, Keith Dunker, Sonia Longhi and Jean-Marie Ruysschaert

Received: 21 September 2022
Accepted: 27 November 2022
Published: 5 December 2022



## 1. Introduction

Alpha-synuclein is a 140-amino-acid protein genetically and pathologically linked to Parkinson's disease (PD). First described by James Parkinson in 1817, PD is a progressive neurodegenerative disorder that affects more than 6 million people globally [1], with an increasing prevalence in the aging population. The disease is characterized by a triad of clinical symptoms—resting tremor, bradykinesia and rigidity—resulting from a progressive loss of dopaminergic neurons in the substantia nigra in the basal ganglia of the midbrain. Although the disease was clinically described long ago, a causal link between alpha-synuclein and PD was only established 180 years later [2].

In PD, alpha-synuclein is a major constituent of characteristic intraneuronal deposits known as Lewy bodies and Lewy neurites in the form of highly ordered amyloid fibril aggregates [3]. Despite great efforts to clarify the relation of these aggregates and their formation to PD, their role in the etiology of PD remains poorly understood. Relatively less effort has been expended to understand the normal physiological functions of alpha-synuclein because the relevance of such functions to disease are unclear. Nevertheless, based on a variety of observations in different contexts, a consensus has emerged that alpha-synuclein is a modulator of synaptic function. Before being linked to PD, alpha-synuclein was described as a novel protein enriched in synaptic vesicle (SV) preparations in the giant neurons of the electric ray *Torpedo californica* [4]. Early studies in songbirds suggested a role in song learning [5], while studies of knockout (KO) models have reported diverse effects on postsynaptic potentiation [6], paired pulse and frequency facilitation [7], SV pool size [8], 4-aminopyridine responses [9], increased neurotransmitter release [10–12] and other more subtle impairments [13,14]. The relatively few studies that examined KO of all three synuclein family members (alpha-, beta- and gamma) consistently showed an increase in synaptic transmission [12,15], suggesting an inhibitory role for the synucleins. Overexpression of alpha-synuclein in various contexts has also generally been reported to lead to a decrease in neurotransmitter release [7,16–20], although a few studies have instead reported increased release [6,13].

SV release is a highly regulated process culminating in the fusion of docked vesicles with the synaptic plasma membrane, a process that is itself regulated by a number of proteins and protein complexes [21]. SVs are organized in physiological pools defined by shared functional properties [22,23] and differing in their ability to release in response to a stimulus. The classical three-pool model comprises a readily releasable pool that is immediately released on stimulation, a recycling pool that is released upon moderate stimulation and a reserve pool that is only mobilized and released upon intense or repetitive stimulation of the neuron [22]. Synucleins have been reported to alter SV pool sizes in functional studies [16] as well as to affect vesicle clustering as observed in ultrastructural studies [24].

Structurally, alpha-synuclein has been shown to be an intrinsically disordered protein, lacking secondary and tertiary structure when isolated in solution. In the presence of membranes, its lipid-binding domain (residues 1–94) acquires an alpha-helical structure, which can take the form of a single extended helix spanning the entire lipid-binding domain or of two shorter helices (helix-1 spanning residues 1–37 and helix-2, spanning residues 45–94) linked by a non-helical and somewhat flexible linker. While membrane binding of alpha-synuclein has long been assumed to be important for its regulation of SV trafficking and fusion, a mechanistic understanding of the underlying structure–function relationships has remained elusive. Recently, we developed a non-neuronal model system for alpha-synuclein function using assays of calcium-triggered exocytosis of recycling vesicles in RBL-2H3 cells, which are used as a model for immune system secretory cells. We found that alpha-synuclein functions as both an inhibitor and a potentiator of vesicle release in a concentration-dependent manner, within the range of reported physiological concentrations. Using structure–function studies of specific mutants, we suggested that different lipid-binding modes are associated with inhibition and potentiation, with the broken-helix conformation being critical for inhibitory function and the extended-helix state mediating potentiation of vesicle release. We also found that potentiation of release at high alpha-synuclein concentrations is associated with dispersal of vesicles from the endocytic recycling compartment, a membranous organelle typically localized in the perinuclear region, to the cell periphery.

Despite these novel insights, critical questions remain regarding the link between specific membrane-bound conformations and alpha-synuclein function. Given the proposed roles of the broken- and extended-helix states, we noted that the linker region between helix-1 and helix-2 (residues 38–44) is ideally situated to regulate alpha-synuclein conformation and function. Indeed, we previously reported that phosphorylation of Y39 within the linker region by c-Abl kinase alters the lipid-binding conformation of alpha-synuclein [25]. Interestingly, dysregulation of c-Abl activity in phosphorylation alpha-synuclein at residue Y39 has been linked to PD [26–28], providing a potential link between the physiological functions and pathological roles of the protein. To probe the role of the linker region and of the broken- and extended-helix states of alpha-synuclein in the regulation of exocytosis, we report here studies of two conformationally selective mutants of alpha-synuclein, correlating their effects on alpha-synuclein structure when bound to membranes and membrane mimetics with their functional effects in assays of vesicle release in RBL-2H3 cells. We find that stabilizing the extended-helix state of alpha-synuclein enhances potentiation of release. More generally, this activity correlates with binding affinity for isolated vesicles. Surprisingly, promoting the broken-helix state via helix-perturbing mutations in the linker region reduced inhibition of vesicle release, indicating that additional factors beyond two flexibly coupled helices contribute to this activity of the protein.

## 2. Methods

### 2.1. Expression and Purification of Alpha-Synuclein Variants

The wild-type human alpha-synuclein sequence, cloned into a pT7.7 vector, a kind gift from Peter Lansbury [29,30], was used as the starting point for all alpha-synuclein expression constructs. Site-directed mutagenesis was performed using QuikChange site-directed mutagenesis kit (Agilent), and all resulting sequences were confirmed by DNA sequencing. To express N-terminal acetylated proteins in *E. coli* BL21(DE3) cells, the alpha-synuclein plasmid was co-transformed with a pNatB plasmid, containing the yeast *N*-acetyltransferase complex NatB, a kind gift from Prof. Daniel Mulvihill (University of Kent, Canterbury, UK) via Prof. Elizabeth Rhoades (University of Pennsylvania, Philadelphia, PA, USA) [31,32]. Transformed BL21(DE3) cells were grown in either LB media (for unlabeled proteins) or M9 minimal media supplemented with $^{15}$N-NH$_4$Cl and either unlabeled or $^{13}$C-labeled glucose as the sole nitrogen and carbon source to produce $^{15}$N- or $^{15}$N$^{13}$C-labeled recombinant proteins for NMR experiments. Protein expression was induced by addition of IPTG to 0.84 mM at 0.6 OD, and optical density was monitored until a peak was reached (1.2–1.5 OD) about 2–3 h later. Cultures were harvested by centrifugation at 10,000× *g* for 15 min at 4 °C, and the pellet was stored frozen at −20 °C or below until purification.

Thawed bacterial pellets were resuspended in 50 mL of lysis buffer (20 mM Tris-Cl at pH 8.0, 1 mM EDTA, 1 mM PMSF) by vortexing, sonicated on wet ice for a total of 12 min (2 times 6 min with stirring in between) and ultracentrifuged for 45 min at 200,000× *g* (40,000 rpm on a Beckman 50.2 Ti rotor), and the supernatant was collected. The pH of the ultracentrifuge supernatant was lowered to 3.5 using 1M HCl and centrifuged at 40,000× *g* for 15 min at 4 °C to remove contaminating proteins, as previously described [33,34]. The pH was readjusted to 7.5 using 1M NaOH, and alpha-synuclein was precipitated by salting-out method using ammonium sulfate at 50% saturation (0.291 g/mL). The precipitate was collected by centrifugation at 40,000× *g* for 15 min at 4 °C, dissolved in 25 mL Tris-Cl buffer (pH 8) and dialyzed through a 3.5 kDa cutoff membrane against water, changing the dialysate twice. After dialysis, the protein was flash-frozen in liquid nitrogen and lyophilized to store indefinitely. A truncated version of the 3AE mutant, used for obtaining resonance assignments in the SDS-bound state, was produced by inserting a stop codon at position 102. Purification proceeded for the full-length protein until cell lysis and ultracentrifugation. Subsequently, 1% *w*/*v* streptomycin sulfate was added to the ultracentrifuge supernatant (0.5 g for 50 mL supernatant), which was stirred at 4 °C for 30 min to precipitate nucleic acids, and then centrifuged at 40,000× *g* at 4 °C for 15 min. The supernatant was subjected to two successive ammonium sulfate cuts, starting with addition of 0.116 g/mL, followed by centrifugation, collecting the supernatant and adding an additional 0.129 g/mL to precipitate the alpha-synuclein-containing fraction. The pellet was resuspended in 50 mL of lysis buffer (20 mM Tris-Cl at pH 8.0, 1 mM EDTA, 1 mM PMSF) and dialyzed overnight against 1L of dialysis buffer (25 mM Tris-Cl pH 8, 20 mM NaCl, 1 mM EDTA) at 4 °C. The protein solution was further subjected to cation exchange chromatography using a CM Sepharose column equilibrated with ion exchange Buffer A (25 mM Tris-Cl pH 8, 20 mM NaCl, 1 mM EDTA) before loading the sample. Sample fractions were eluted with a salt gradient from 0 to 40% ion exchange buffer B (25 mM

Tris-Cl pH 8, 1 M NaCl, 1 mM EDTA) over 4 column volumes. Fractions were run on 18% SDS-PAGE to determine those containing the purified protein, which were pooled, dialyzed against 5% acetic acid overnight and passed through a 0.22 μM membrane filter to remove particulate impurities. The sample was then injected onto a reversed phase C4 column using a Waters 2690 separation module and eluted with a gradient of 20% to 100% HPLC Buffer B (90% acetonitrile and 0.1% trifluoroacetic acid in water) in HPLC Buffer A (0.1% trifluoroacetic acid in water) while monitoring the eluate for protein absorbance at 229 nm. Major peaks from HPLC were collected separately and analyzed by SDS-PAGE. Protein-containing fractions were spun in a vacuum concentrator for 2 h to remove acetonitrile, dialyzed in water to remove residual TFA, flash frozen in liquid nitrogen and lyophilized for storage.

### 2.2. Lipid Vesicle Preparation and Characterization

Synthetic lipids were purchased as stock solutions in chloroform (Avanti Polar Lipids, Alabaster, AL, USA), and appropriate volumes were mixed at a molar ratio of DOPC/DOPE/DOPS = 60:25:15 to mimic the composition of native SVs [35–37]. The lipids were dried under nitrogen or argon flow for 20 min while rolling the tube to form a thin film on the side, followed by drying in a vacuum concentrator for a further 2 h. If not immediately needed, the tube was flushed with nitrogen or argon, sealed with paraffin film and stored at −20 °C for up to 3 days. The lipid film was hydrated in an appropriate volume of NMR buffer (10 mM $Na_2HPO_4$, 100 mM NaCl, 10% $D_2O$, pH 6.8) and vortexed vigorously to generate a cloudy suspension, which was sonicated using a bath sonicator (Elmasonic P30H) at room temperature, 37 kHz frequency and 100 W power for 20–30 min until visually clear. Clarity of the preparation ensures that the vesicles are sufficiently small (<100 nm diameter) to become non-scattering to visible light. Next, the vesicle suspension was ultracentrifuged at 150,000× *g* (60,000 rpm on a Sorvall S120-AT2 rotor) to pellet larger particles. The supernatant was carefully removed and collected in a fresh tube and used within 2 days to prepare NMR samples by adding to protein stock solution at specific lipid/protein ratios. In order to determine concentration of lipids in vesicles, we performed a phosphate assay based on a modified version of the Rouser assay for phospholipids [38,39]. The size distribution of vesicles was measured for select SUV preparations using dynamic light scattering, which showed that the vesicle preparation method consistently produces vesicles of 30–50 nm diameter.

### 2.3. NMR

To prepare samples for NMR, lyophilized proteins were weighed out and dissolved in NMR buffer (10 mM $Na_2HPO_4$, 100 mM NaCl, 10% $D_2O$, pH 6.8), pH readjusted to 6.8 and filtered using a 100 kDa cutoff centrifugal filter (MilliporeSigma, Burlington, MA, USA) to remove any higher molecular weight species/aggregates. Protein concentration was estimated from absorbance at 280 nm using a molar extinction coefficient of 5120 $M^{-1}cm^{-1}$ [30]. NMR experiments were conducted using Bruker 600 MHz, 700 MHz, 800 MHz and 900 MHz NMR instruments with triple resonance gradient-equipped cryoprobes. Pulse sequences, except for the DEST experiments, were derived from the standard Bruker library. NMR tubes were either 5 mM thin-walled precision tubes (Wilmad-LabGlass, Vineland, NJ, USA) or 3 mM Bruker SampleJet NMR tubes (Bruker, Billerica, MA, USA). Experiments were conducted at 10 °C for free-state and lipid vesicle-binding experiments and at 40 °C for samples containing SDS. For relaxation experiments, the temperature was set at 13.5 °C because the temperature control of one of the spectrometers used was better at this temperature. NMR raw data conversion and processing were conducted using NMRPipe [40], and data analysis and visualization were performed using CCPNmr Analysis [41].

$^{15}N$ $R_2$ relaxation rates were measured by conducting a series of HSQC-like experiments with variable time delays when the amide nitrogen magnetization is transverse, with a number of Carr–Purcell–Meiboom–Gill (CPMG) refocusing elements interspersed.

Each CPMG delay–pulse–delay element duration was 16.32 ms, and variable numbers of such pulses (0 to 18) corresponded to a maximum of 294 ms relaxation delay. The multiplier array for the number of CPMG pulses was scrambled, and the delay between each complete pulse sequence was set at 4 s to prevent sample overheating. A single data point in the mid-range relaxation delay was chosen and measured in triplicate to obtain an error estimate. The signal intensity for each amide peak was plotted as a function of the time delay and fit to a single exponential to extract the $R_2$ relaxation rate.

Resonances of vesicle-bound residues are broadened beyond direct detection, but this broad resonance can be selectively saturated using a narrow bandwidth saturation pulse far away from the corresponding free-state resonance frequency. When this saturated bound state exchanges with the free state, the signal from the free state decreases consequently. This technique is referred to as Dark-state Exchange Saturation Transfer (DEST) and, in combination with $R_2$ relaxation experiments, is a powerful method to probe the kinetics of exchange processes occurring at timescales ranging from approximately 10 ms to 1 s [42]. Using these experiments, it is possible to uncover which residues are interacting with the membrane, to evaluate kinetic models for the system and to extract the kinetic rates of membrane interactions.

DEST and $R_2$ relaxation experiments were performed on either a 700 MHz or a 900 MHz spectrometer based on instrument availability, with data from a single spectrometer used for each individual sample. DEST experiments were performed using a saturation pulse of 900 ms on the $^{15}$N channel at resonance offsets ranging from −30 kHz to +30 kHz, using two different saturation bandwidths (400 Hz and 175 Hz on the 700 MHz spectrometer and 500 Hz and 200 Hz on the 900 MHz spectrometer). For each bandwidth, the saturation pulse power was calculated from the $^{15}$N 90° pulse length assuming an ideal linear amplifier.

For both $R_2$ relaxation and DEST experiments, peak picking, annotation and height measurements and (for $R_2$ relaxation) exponential decay modeling were performed using the NMRPipe suite using scripts originally developed by Fawzi et al. [42] and subsequently modified by us to accept different input formats. Joint fitting of $R_2$ relaxation and DEST profiles to the different kinetic models was carried out using the DESTfit MATLAB script [42], which essentially fits the experimental values to a homogenous form of the McConnell equations, describing a single spin in two-site exchange at chemical equilibrium between a free state with low $R_2$ and an SUV-bound state with larger $R_2$ in the presence of a continuous-wave saturation field [43,44]. The models tested include a simple two-state model for membrane binding where the protein exchanges between a free state and a fully bound state and the overall exchange process are the same for each residue (a single global apparent on-rate, $k_{on}^{app}$, and a single global off-rate, $k_{off}$) and a pseudo-two-state model, as described by Fawzi et al. [42], where bound-state conformations are divided in two subsets, with the $i^{th}$ residue either in direct contact with the vesicle surface or tethered to the vesicle surface by other nearby residues that are in direct contact (Figure S10). The global apparent on-rate, $k_{on}^{app}(i)$, is then the sum of the apparent on-rates for binding in the tethered contact mode, $k_1^{app}(i)$, and in the direct contact mode, $k_2^{app}(i)$, and the global off-rate, $k_{off} = k_{-1} = k_{-2}$ is considered the same for these two states for all residues. The residue-specific equilibrium between the tethered and direct contact states, described by $K_3(i) = k_2^{app}(i)/k_1^{app}(i)$, then relates the populations of the tethered contact and direct contact states. The pseudo-two-state model can also incorporate direct interconversion between the tethered and the direct contact states with rate constants $k_3$ and $k_{-3}$, but this interconversion only becomes relevant if the rates involved are faster than the off-rate $k_{off}$. In our case, addition of $k_3$ and $k_{-3}$ to the fitting parameters did not improve the fits for any of the variants, and the resulting values were lower than $k_{off}$, indicating that interconversion is relatively slow and is not contributing meaningfully to our measurements.

### 2.4. Paramagnetic Relaxation Enhancement Experiments

To incorporate a nitroxide spin label at a single-cysteine mutant of alpha-synuclein, $^{15}$N-labeled alpha-synuclein was dissolved in 1mL of PRE buffer (10 mM $Na_2HPO_4$, 100 mM NaCl, pH 6.8) at a concentration of 200–400 μM. Then, the spin label reagent MTSL (S-(1-oxyl-2,2,5,5-tetramethyl-2,5-dihydro-1H-pyrrol-3-yl)methyl methanesulfonothioate) was added at a 30-fold molar excess (from a 300 mM stock solution in acetonitrile) and allowed to react for a couple of hours at room temperature. Next, the unbound spin label was removed by dialyzing against 1 L of PRE buffer through a 3.5 kDa cutoff membrane, with one change of dialysate. Finally, 10% *v*/*v* $D_2O$ was added to prepare the sample for NMR experiments. A matched control sample was also prepared by adding DTT to a final concentration of 5 mM to detach the spin label by reducing the disulfide bond.

### 2.5. Tryptophan Fluorescence

For each of the alpha-synuclein variants used in the study, single tryptophan mutants at position 4 (F4W) were created by site-directed mutagenesis. Samples were prepared using 0.1 μM of protein and a series of intermediate lipid concentrations of 10, 1, 0.1 and 0.01 mM, which were further diluted in twofold steps to fill in the intervening concentration ranges. A lipid-free protein sample was also prepared for each of the variants. A protein-free lipid concentration series of samples was also prepared to measure reference spectra at each lipid concentration, which were subtracted from corresponding measurements with protein to account for absorption and scattering due to lipids alone. To assess consistency of measurements performed over multiple days using multiple batches of lipid SUV preparations, wild-type alpha-synuclein (F4W) was measured as an internal control on all the days. Fluorescence spectra were recorded using a Spectramax M5 fluorimeter (Molecular Devices, Silicon Valley, CA, USA) using a transparent quartz cuvette (Starna cells # 9F-Q-10, excitation path length 10 mM, emission path length 4 mM) using excitation at 280 nm. Emission was recorded from 300–500 nm using 10 nm steps, with the PMT potential difference set at 900 V, 60 flashes averaged.

Collecting at 10 nm resolution enabled the collection of a large data set, but the low resolution precluded a direct estimate of peak position and intensity. To determine these, we fit each fluorescence spectrum using a biparametric log-normal model that was described for organic fluorophores, including tryptophan residues in proteins (Figure S3A) [45,46]. The model, originally described by Siano and Metzler [47], postulates that fluorescence emission spectrum of single organic fluorophores empirically correspond to a log-normal distribution on the frequency axis. It is expressed as follows:

$$I_\nu = \begin{cases} I_m \cdot e^{\left\{-\frac{\ln 2}{\ln^2 \rho} \cdot \ln^2\left(\frac{a-\nu}{a-\nu_m}\right)\right\}} \text{ at } \nu < a \\ 0 \text{ at } \nu \geq a \end{cases} \tag{1}$$

where $I_m$ is the maximum fluorescence intensity at $\nu_m$, the wavenumber of the fluorescence maximum; $\nu$ is the wavenumber, which is the reciprocal of the wavelength $\lambda$; $\rho$ is the band asymmetry parameter, which can be described as

$$\rho = \frac{\nu_m - \nu_-}{\nu_+ - \nu_m} \tag{2}$$

in which $\nu_+$ and $\nu_-$ are the wavenumber positions at the left and right half-maximal amplitudes, respectively; $a$ is the function limiting point, described as

$$a = \nu_m + \frac{H\rho}{\rho^2 - 1}, \text{ where the bandwidth } H = \nu_+ - \nu_- \tag{3}$$

Plots of emission maxima ($\lambda_{max}$) versus peak width can be used to assess the environmental heterogeneity of the emitting tryptophan species [48]. In a homogeneous environment, the plot should fall roughly on a straight line that has been experimentally determined using free zwitterionic tryptophan in various solvents with different degrees of hydrophobicity [48,49]. A plot of our data at different lipid concentrations reveals that at the highest lipid concentration (where tryptophan fluorescence is blue-shifted due to membrane interactions), where all alpha-synuclein variants except for the A30P/V70P double mutant reaches saturation binding (Figure 3B), the emission spectrum of tryptophan falls on the line indicating a homogeneous population in a fully hydrophobic environment (Figure S2B, leftmost data points). As expected, at intermediate lipid concentrations, our data lie above the empirical line, signifying a higher-than-expected width compared to a homogeneous species, consistent with a heterogeneous population made up of free-state and bound-state species. Interestingly, the free-state spectra also indicate heterogeneity (Figure S2B, rightmost data points), likely resulting from the fact that the N-terminal ~10 residues of N-terminally acetylated alpha-synuclein sample partly helical conformations [50–52], resulting in a mixture of conformational states with varying degrees of solvent accessibility at position 4.

In order to plot the fluorescence-monitored binding curves as the change in bound fraction of the protein at different lipid concentrations, the bound fraction was estimated for each lipid concentration. Since the time scales of protein–membrane interactions are much slower than the fluorescence measurement time scale, spectra at intermediate lipid concentrations were fit to a linear combination of two spectra: free-state protein spectrum (no lipids) and fully bound state protein spectrum (at the highest concentration of lipids) (Figure S3). In the case of the A30P/V70P/F4W mutant, which did not reach saturation binding even at the highest lipid concentration, the WT/F4W spectrum with 10 mM lipids was used instead. Fits of the resulting curves to a bimolecular binding model between protein molecules and binding sites on lipid vesicles [53], in which the binding site concentration is related to the total lipid concentration by a proportionality constant, $B_{max}$ (maximum binding sites per lipid molecule; $1/B_{max}$ is then the minimum number of lipids per biding site), were performed using an R implementation of quadratic programming to solve for the bound fraction. The minimum number of lipids per binding site, $1/B_{max}$, was determined for the WT protein as the lipid concentration divided by the bound protein concentration, the latter of which was determined from the NMR intensity plots (Figure 2C) as the fraction of bound protein, defined as the median value for the bound fraction of the N-terminal 9 residues, multiplied by the total protein concentration. The conditions under which the NMR intensity plots were obtained (2.5 mM lipid concentration) are in the saturating regime (Figure 3B), where every binding site on the vesicle surface is occupied by a protein molecule.

*2.6. Exocytosis*

RBL-2H3 cells were cultured as monolayers in minimal essential medium (Invitrogen Corp, Carlsbad, CA, USA) with 20% fetal bovine serum (Atlanta Biologicals, Atlanta, GA, USA) and 10 µg/mL of gentamicin sulfate (Invitrogen), as previously described [54]. Adherent cells were harvested by treatment with Trypsin-EDTA (0.05%) for 8–10 min 3–5 days after passage. RBL-2H3 cells are continuously cultured in the Baird–Holowka laboratory, routinely checked for normal function, and frozen stocks are thawed for fresh cultures as warranted.

A pcDNA 3.0 vector for cell expression of human WT alpha-synuclein was obtained as a gift from Dr. Chris Rochet (Purdue University). Plasmids for cell expression of alpha-synuclein mutants (A30P, V70P, A30PV70P, 3AE, 4G) were created from this vector by site-directed mutagenesis using Phusion High-Fidelity DNA Polymerase (New England Biolabs, Ipswich, MA, USA), and all mutations were confirmed by DNA sequencing. Plasmids for cell expression of VAMP8-pHluorin and mCherry-Rab11 were created as previously described [55,56].

RBL-2H3 cells were harvested 3–5 days after passage, and $5 \times 10^6$ cells were suspended in 0.5 mL of cold electroporation buffer (137 mM NaCl, 2.7 mM KCl, 1 mM $MgCl_2$, 1 mg/mL glucose, 20 mM HEPES, pH 7.4). Co-transfections used a reporter plasmid DNA (5 µg VAMP8-pHluorin), together with 5 µg (low expression) or 25 µg (high expression) of human alpha-synuclein (or empty vector) plasmid DNA. We found previously that cells transfected with the two constructs express both or none, such that the fluorescent VAMP8-pHluorin construct could be used as a reporter for cells co-transfected with the non-fluorescent alpha-synuclein construct [57].

For all conditions, cells were electroporated at 280 V and 950 µF using Gene Pulser X (Bio-Rad, Hercules, California, USA). Then, cells were immediately resuspended in 6 mL of medium and cultured for 24 h to recover; the medium was changed after live cells became adherent (1–3 h). For exocytosis experiments, the cell suspensions were added to three different MatTek dishes (2 mL/dish) (MatTek Corporation, Ashland, MA, USA) for recovery.

After the electroporation recovery period and prior to imaging, cells were washed once and then incubated for 5 min at 37 °C with buffered saline solution (BSS: 135 mM NaCl, 5 mM KCl, 1 mM $MgCl_2$, 1.8 mM $CaCl_2$, 5.6 mM glucose, 20 mM HEPES, pH 7.4). VAMP8-pHluorin fluorescence was monitored for 20 s prior to addition of 250 nM thapsigargin, and after 6–8 min of stimulation, 50 mM $NH_4Cl$ was added. Cells were monitored by confocal microscopy (Zeiss 710) using a heated, 40× water immersion objective. VAMP8-pHluorin was excited using the 488 nM line of a krypton/argon laser and viewed with a 502–551 nm band-pass filter. Representative movies for a control and for experiments with the WT, 4G and 3AE mutants are available upon request, and 3AE mutants are included as Supplementary Material.

Offline image analysis was conducted using Fiji ImageJ [58]. Regions of interest (ROIs) were manually drawn around individual cells from which time traces of VAMP8-pHluorin fluorescence were obtained. In case of stage movement, the ROIs were translated accordingly. For each cell, the average fluorescence of 5 frames each of before stimulation ($F_{basal}$), after peak of stimulated exocytosis ($F_{stimulated}$) and after $NH_4Cl$ administration ($F_{total}$) were used to measure fraction of vesicles exocytosed (*Exo*) using the following equation:

$$Exo = \frac{F_{stimulated} - F_{basal}}{F_{total} - F_{basal}} \quad (4)$$

For each condition (variant and expression level), outliers in the measured per-cell exocytosis values were excluded using the 1.5 times interquartile range method before comparison and statistical tests for significance. Analysis of the data without outlier removal did not impact the statistical comparisons between the conditions.

### 2.7. Distribution of Recycling Endosomes

RBL-2H3 cells were prepared and electroporated as described above with 5 µg of mCherry-Rab11 plasmid DNA to label recycling endosomes [56] and either 5 µg or 25 µg of alpha-synuclein (WT or mutant) or 3 µg of control plasmid DNA. Samples were fixed and immunostained with anti-alpha-synuclein antibody 3H2897 (Santa Cruz Biotechnology Cat# sc-69977) after the recovery period and imaged confocally using a 63× oil immersion objective, selecting a plane near the middle of the cell including both plasma membrane and perinuclear regions. Fluorescence from immunostained alpha-synuclein was used to select sufficiently bright cells, and mCherry-Rab11 fluorescence was quantified and used to determine the relative distributions of REs using Fiji ImageJ [58]. Fluorescence in a shell of ~800 nm thickness, drawn to include the plasma membrane and a small region extending inward, was divided by the total cellular mCherry-Rab11 fluorescence to yield the percentage of REs proximal to the plasma membrane.

### 2.8. Expression Levels of Alpha-Synuclein Using Immunocytochemistry

To directly assess protein expression levels, we quantified and compared alpha-synuclein immunostaining under conditions of high expression for the WT and linker mutants in RBL-2H3 cells (Figure S11). All three variants exhibited similar expression levels, confirming that the linker mutants are not aberrantly expressed or degraded. In addition, for the 4G mutant, we compared immunostaining at low and high expression levels, confirming an approximately 3- to 5-fold difference between the two conditions, as previously reported for the WT and other variants including A30P and V70P [57]. We previously demonstrated that VAMP8 fluorescence is a highly reliable reporter of alpha-synuclein expression levels in our system [57]. Having established by immunostaining that both the linker mutants express normally, we also visually examined VAMP8 fluorescence levels in each individual experiment to confirm they were consistent with the expected high or low expression levels.

### 2.9. Effects of Alpha-Synuclein on Enrichment of Lipid Vesicles in Synapsin-1 Condensates

The construct for expression of human synapsin-1-eGFP in mammalian cells (pEGFP-C1 vector) was a kind gift from Dragomir Milovanovic and Pietro De Camilli (Yale University). Full-length synapsin-1-eGFP was expressed in Expi293 cells and purified as described [59]. Briefly, after transfection of cells, proteins were expressed for 3 days and harvested/lysed in buffer A (25 mM Tris-HCl at pH 7.4, 300 mM NaCl, 0.5 mM TCEP) with requisite protease inhibitors. To clarify lysates, samples were spun at 17,000× $g$ and filtered. Samples were then run through a nickel column (His60 Ni Superflow Resin, Takara Biosciences, San Jose, CA, USA) and eluted with Buffer A + 400 mM imidazole. Samples were then run through a size exclusion column (HiLoad 16/600 Superdex 200 pg, Cytiva Life Sciences, Marlborough, MA, USA) in Buffer B (25 mM Tris-HCl at pH 7.4, 150 mM NaCl, 0.5 mM TCEP) and subsequently concentrated. All purification steps took place at or near 4 °C.

Rhodamine-labeled SUVs were prepared as described above but using a molar ratio of DOPC/DOPE/DOPS/lissamine-rhodamine PE = 60:24:15:1 and rehydrated using the reaction buffer (25 mM Tris-HCl, 150 mM NaCl, pH 7.4). Unlabeled N-terminally acetylated WT alpha-synuclein, prepared as described above, was reconstituted from lyophilized stock in reaction buffer (25 mM Tris-HCl, 150 mM NaCl, pH 7.4). PEG-8000 was added from a stock concentration of 40% ($w/v$) in $H_2O$. The glass slides and cover glasses were chemically cleaned, aminosilanized and passivated by reacting with NHS ester-modified PEG, as described elsewhere [60]. Microchambers were fabricated using a spacer tape between glass slides and cover glasses, through which the sample could be imaged. Phase separation of synapsin-1 in the absence or presence of 0.5 mM SUVs was initiated by adding PEG-8000 at a final concentration of 4% ($w/v$) to reaction buffer (25 mM Tris-HCl, 150 mM NaCl, pH 7.4). Matched samples with and without N-terminally acetylated alpha-synuclein (200 μM), prepared in independent triplicates, were imaged using Zeiss LSM-880 laser scanning confocal microscope, with excitation wavelengths of 488 nm and 532 nm for eGFP and rhodamine channels, respectively. Detector gain was adjusted to minimize saturation of the rhodamine signal to allow quantitation.

Droplet images were analyzed using Fiji-ImageJ as follows. Droplets where rhodamine signal reaches saturation were manually removed. A Gaussian blur of 2 pixels was applied to the eGFP channel, followed by background subtraction using a rolling ball diameter of 40 pixels (5.27 μM), chosen to be larger than the size of the largest droplet so as to capture background features while excluding droplets. Thresholding of the eGFP channel was performed using the Li algorithm [61]. Mean rhodamine fluorescence of all particles of area greater than 0.2 square microns (to reduce noise from off-focus or very small droplets) within the eGFP-thresholded regions was measured. Mean rhodamine fluorescence outside of droplets was determined using an inverted threshold, and a partition coefficient for SUVs in each droplet was calculated as the ratio of mean rhodamine fluorescence inside and outside each droplet. To assess statistical significance, partition coefficient values were log-transformed, and outliers were removed based on interquartile range method. Upon assessing and verifying the near-normality of the transformed data using Shapiro–Wilk test, Student's t-test was performed to reject the null hypothesis that there was no significant difference between partition coefficient of SUVs inside synapsin-1 droplets in absence and presence of 200 μM alpha-synuclein. The difference between the two conditions remained highly significant ($p < 0.001$) when the data were analyzed without outlier removal.

## 3. Results

### *3.1. Linker Mutants Bias Linker Region Helicity in Micelle-Bound State*

We designed two conformationally selective mutants of alpha-synuclein in which we mutated residues 39 to 42 in the linker region from YVGS to either AAAE (3AE mutant) or GGGG (4G mutant) [62]. The mutants were designed to preferentially populate the extended-helix (3AE) or the broken-helix (4G) conformations (Figure 1). Alanine and glutamine have higher helical propensity than tyrosine, valine and serine, while glycine exhibits very low helical propensity [63,64]. We then investigated the structure of the 3AE and 4G mutants when bound to membrane-mimetic detergent micelles using solution-state NMR spectroscopy. $C_\alpha$ secondary shifts for the SDS micelle-bound linker region mutants show that the 3AE mutant features a higher helical propensity in the linker region, while the 4G mutant exhibits a more extensive break between helix-1 and helix-2 than the WT protein, with no effect observed outside the linker region (Figure 2A). Paramagnetic relaxation enhancement (PRE) measurements using MTSL labeling via an S9C mutation [25] were similar for the 3AE, 4G and WT proteins (Figure 2B), indicating proximity (within ~25 Å [65,66]) of the N- and C-termini of helix-1 and helix-2, respectively, for all three variants and suggesting that this proximity does not require a break in the helical structure, consistent with a previous report [67].

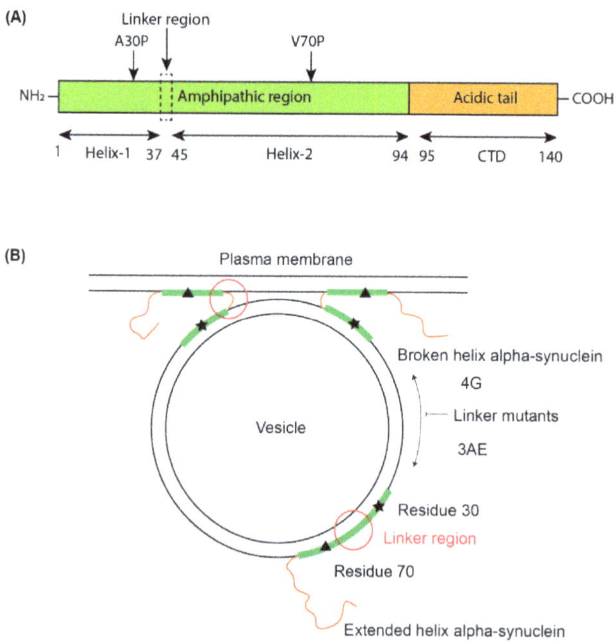


**Figure 1.** Schematic illustrating the design logic for alpha-synuclein linker mutants. (**A**) Domain structure of alpha-synuclein and location of mutations used in the study. (**B**) Proposed functional contexts of different membrane-bound conformations of alpha-synuclein and the effects of the two linker region mutations, 4G and 3AE, on these conformations. The helix-1 and helix-2 regions are depicted as green rectangles and the linker region in the broken-helix state and the disordered C-terminal tail as orange lines. The linker region is highlighted with a red circle in both conformations, and the positions of the A30P and V70P mutations are marked with a star and a triangle, respectively. The two linker regions mutations are expected to inhibit conformational exchange and bias the membrane-bound conformation of the protein towards the broken-helix (4G) or extended-helix (3AE) state.

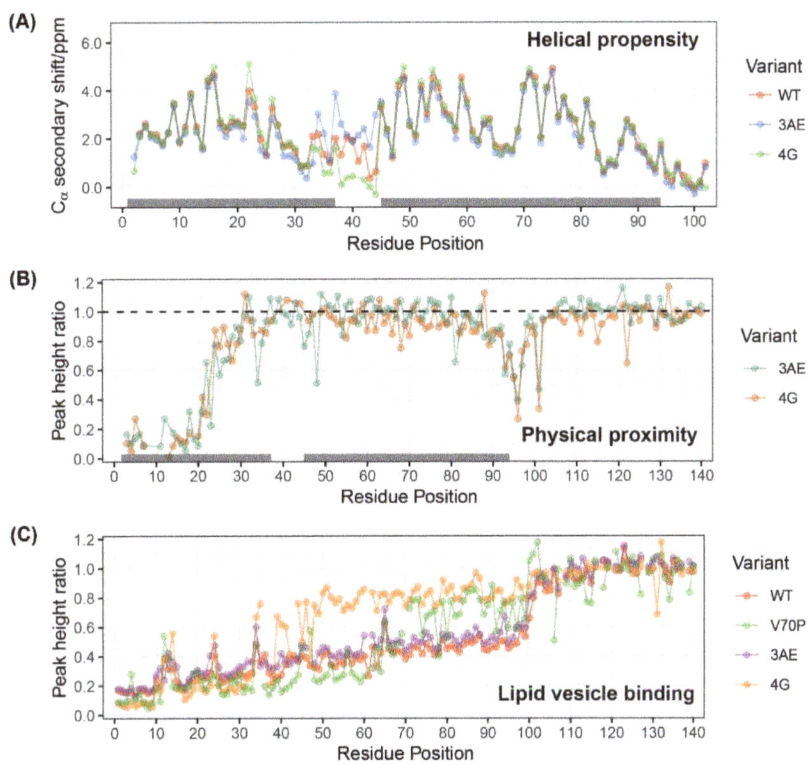


**Figure 2.** Effects of the linker region mutations on the micelle- and vesicle-bound states of alpha-synuclein. (**A**) NMR C-alpha secondary chemical shifts for micelle-bound WT alpha-synuclein and the 4G and 3AE mutants. Positive secondary shifts above ~1 PPM are indicative of significant helical propensity. Deuterated SDS concentration was 40 mM, and protein concentrations were 100–200 μM. Data were collected at 40 °C. (**B**) PRE in micelle-bound alpha-synuclein 4G and 3AE mutants labeled with a paramagnetic spin-label at position 9. SDS concentration was 40 mM, and protein concentrations were 100 μM. Data were collected at 40 °C. (**C**) Intensity ratios of signals from NMR $^{15}$N-$^{1}$H HSQC spectra of 50 μM alpha-synuclein variants obtained in the presence vs. the absence of SUVs at a total lipid concentration of 2.5 mM. Data were collected at 10 °C.

### *3.2. 4G Mutant Prevents Propagation of the Vesicle-Bound Extended-Helix Conformation*

Binding of alpha-synuclein to small unilamellar vesicles (SUVs) results in a fractional decrease of NMR signal intensities corresponding to the bound fraction for each residue involved in the interaction. Peak intensity ratios in the presence vs. absence of SUVs therefore report on the free state fraction for each position in the protein. Intensity ratio plots for 50 μM WT N-terminally acetylated alpha-synuclein in the presence of 2.5 mM lipid SUVs (Figure 2C, red) reveal that the bound fraction is maximal at the very N-terminus of the protein and gradually decreases towards the C-terminus of the lipid-binding domain, reflecting a consensus in the field that lipid binding proceeds from the N- to the C-terminus of the lipid-binding domain [68,69]. The 3AE mutant shows similar lipid-binding behavior to the WT protein (Figure 2C, purple). In contrast, the 4G mutant shows a distinct break in its lipid-binding profile, with residues 34–94 exhibiting higher intensity ratios compared to residues 1–33 (Figure 2C, orange). We previously observed a similar break in the vesicle-binding profiles of alpha-synuclein mutant V70P, which introduces a proline in the middle of helix-2. Interestingly, the intensity ratios at the very N-terminus are lower for the 4G and V70P (Figure 2C, green) mutants than for the WT or 3AE variants, suggesting that this apparently enhanced binding may be a common effect of truncating the extended-helix conformation. This effect could potentially result from a smaller binding footprint on the membrane surface, resulting in a greater number of available binding sites.

### 3.3. Linker Mutants Alter Effects of Alpha-Synuclein on Vesicle Release

We performed stimulated exocytosis assays in RBL-2H3 cells as described earlier [57]. We used RBL-2H3 rat basophilic leukemia cells as a model system to quantify $Ca^{2+}$-stimulated exocytosis of recycling endosomes labeled with the pH-sensitive fluorescent marker VAMP8-pHluorin. Exocytosis in these cells is stimulated by $Ca^{2+}$ release from the endoplasmic reticulum (ER), which can be triggered by signaling downstream of antigen receptors or by inhibition of the sarco/endoplasmic $Ca^{2+}$ ATPase (SERCA) using the SERCA inhibitor thapsigargin [70]. Cells were co-transfected with VAMP8-pHluorin and low (5 µg plasmid) or high (25 µg plasmid) levels of alpha-synuclein variants, stimulated with 250 nM thapsigargin and imaged for 400 s using a confocal microscope. Ammonium chloride was added at the end of the experiment to a final concentration of 50 mM to neutralize all vesicles in the cell interior and provide a measure of the total recycling vesicle pool. As we reported previously, low levels of WT alpha-synuclein lead to inhibition of vesicle release (Figure S1A), but high expression levels were associated with enhanced vesicle release (Figure S1B) compared to empty vector [57]. The 3AE mutant was less effective at inhibiting release at low levels and more effective at enhancing release at high levels compared to empty vector, consistent with its design goals of stabilizing the extended-helix state over the broken-helix state. However, despite favoring the broken-helix state (Figure 2A), the 4G mutant caused no inhibition of vesicle release at low expression levels, suggesting that the two flexibly coupled helices alone are insufficient for the inhibitory function of alpha-synuclein. Release at high expression levels of the 4G mutant was comparable to that for the WT protein, suggesting that the altered binding mode of this mutant was still somewhat effective at enhancing exocytosis.

We previously showed, using TIRF imaging of individual vesicle fusion events, that despite potentiation of vesicle release at high expression levels of alpha-synuclein, the rate of vesicle release is still lower than for the empty vector control, indicating that the inhibitory and potentiating activities of the protein are not mutually exclusive and instead operate in tandem [70]. In this case, the potentiating activity is better measured by comparing release at high levels to release at low levels, with the difference representing the isolated potentiating activity. Using this measure, we observe (Figure 3A) that the WT and 3AE variants have the greatest potentiation activity, followed by the 4G mutant. The V70P mutant exhibits release that is comparable to or slightly higher than that of the WT protein at high expression levels (Figure S1B), and we previously took this as evidence that the V70P mutant does not perturb enhancement of release [70]. However, when compared directly to its effects at low expression levels, it is evident that the V70P does not significantly enhance vesicle release. This analysis also confirmed the surprising result that the 4G mutant was able to sustain potentiation of exocytosis despite its abbreviated binding mode to lipid vesicles, while revealing that the V70P mutant, which exhibits a more extensive binding mode than the 4G mutant, does not potentiate vesicle release. These results suggest that binding interactions reflected in intensity ratio measurements may not capture all aspects of vesicle binding and led us to examine binding using additional assays. As an additional reference point, we examined the difference in release for low and high levels of the PD mutant, A30P. This mutant effectively inhibits release at low levels but showed no increase in release at higher levels (Figure 3A).

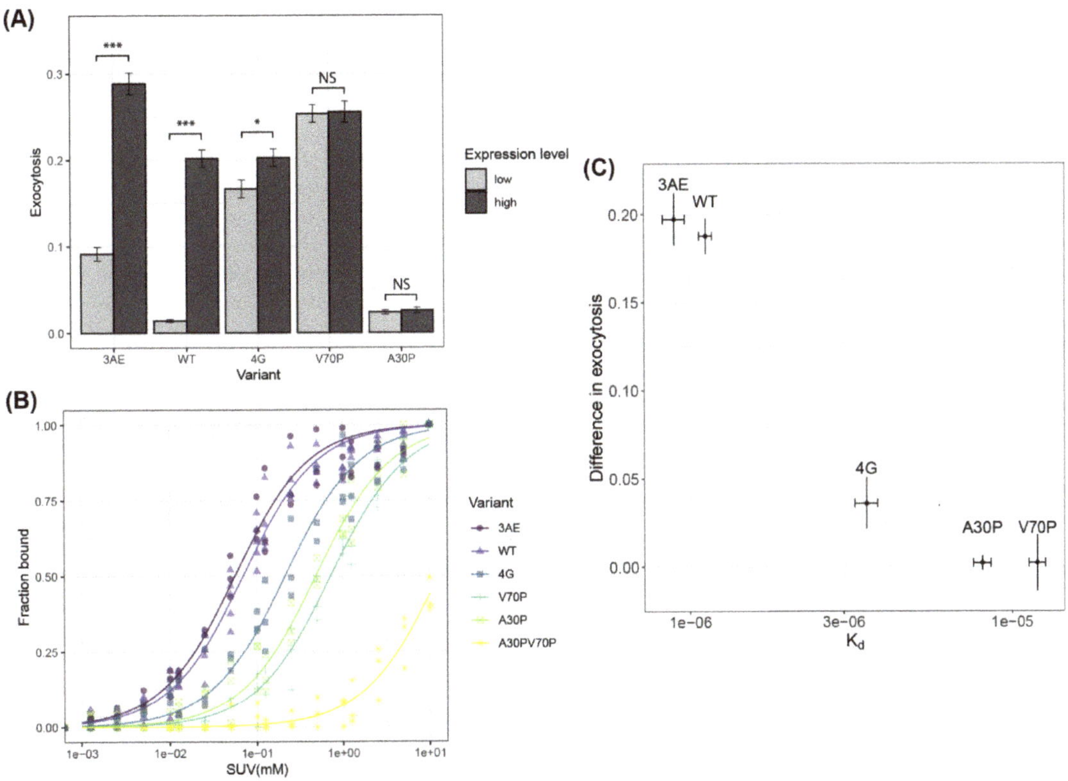


**Figure 3.** Functional assays and membrane affinity of alpha-synuclein variants. (**A**) Thapsigargin-stimulated exocytosis in RBL-2H3 cells transfected with low or high levels of WT or mutant alpha-synuclein measured using fluorescence of a VAMP8-pHlourin reporter. The difference in exocytosis levels between low and high expression levels represents the degree by which exocytosis is enhanced at high expression levels (*t*-test between high/low, *p* values *** < 0.001 < * < 0.05 < NS). (**B**) Lipid-binding curves for WT and mutant alpha-synuclein in an F4W mutant background measured by intrinsic tryptophan fluorescence as a function of increasing lipid concentrations. Protein concentrations were 0.1 µM, and lipid concentrations ranged from 1.25 µM to 10 mM. Data were acquired at a temperature of 22 °C (room temperature) by excitation at 280 nm and detection at 300–500 nm at 10 nm resolution, 60 flashes averaged and then baseline subtracted using a no-protein control, analyzed to extract bound fraction at every lipid/protein ratio and fit as described in the methods. Resulting fits are shown in solid lines. For each day of experiments with a new lipid vesicle preparation, WT data served as an internal control to account for variations. (**C**) Plot of membrane affinity derived from fits to the data in panel B vs. extent of enhancement of exocytosis derived from the data in panel A for WT and mutant alpha-synuclein.

### *3.4. Membrane Affinity for Isolated Vesicles Correlates with Potentiation of Exocytosis*

We previously argued that the potentiating activity of alpha-synuclein is derived from its ability to bind isolated vesicles. While lipid-binding profiles derived from NMR reveal changes in localized binding to lipid vesicles, they do not easily provide information on binding affinity since they are performed at relatively high protein and lipid concentrations. To investigate the correlation between membrane binding affinity and potentiating activity, we measured the affinity of our variants to lipid vesicles using fluorescence, following a previously reported protocol employing the introduction of a single tryptophan probe at position 4 (F4W mutation) [53] and relying on the environment-sensitive nature of tryptophan fluorescence emission spectra (Figure S2) [49,71] Notably, the F4W mutation was conclusively shown not to cause significant changes in the membrane affinity or secondary structure of vesicle-bound alpha-synuclein [53]. The sensitivity of this method allowed us to acquire data at very low (0.1 μM) protein concentrations, which enabled us to reach lipid/protein molar ratios of 100,000:1, at which point vesicle binding reached saturation for all of the mutants except one (Figure 3B and Figure S2B). Spectra at each lipid concentration were fit to a linear combination of the free-state and bound-state spectra yielding the fraction of protein in the bound state (Figure S3). Fits of the resulting binding curves to a quadratic bimolecular binding model between protein molecules and binding sites on lipid vesicles [53], in which the binding site concentration is related to the total lipid concentration by a proportionality constant, $B_{max}$ (maximum binding sites per lipid molecule; $1/B_{max}$ is then the minimum number of lipids per binding site) provided estimates of the dissociation constant, $K_d$. Notably, although $B_{max}$ can in principle be fit simultaneously with $K_d$, we could instead estimate $B_{max}$ from the fraction of bound protein determined by NMR studies in the saturation regime (see Section 2), where protein is in excess of binding sites (Figure 2C). In this regime, protein molecules will bind to all available sites (protein and lipid concentrations are well above $K_d$), so the presence of excess free protein indicates all sites are occupied. The amount of bound protein divided by the total lipid concentration thus provides an estimate of the number of lipids per binding sites. For N-terminally acetylated WT alpha-synuclein, we obtained a value of 59 lipid molecules per binding site, which is consistent with estimates from electron spin resonance (ESR) data using dimyristoyl phosphatidylglycerol (DMPG) membranes (36–100 lipids per binding site) [72] but significantly lower than that estimated for non-acetylated alpha-synuclein using circular dichroism spectroscopy (500 lipids per binding site) [53].

The $K_d$ values we obtained from fits of the fluorescence data (Table 1) indicate that WT alpha-synuclein and the 3AE mutant bind with the highest affinity, followed by 4G and then A30P and V70P. We also examined binding by the A30P/V70P double mutant, which showed weaker binding than either A30P or V70P alone and did not reach saturation at the highest lipid concentrations (Figure 3B and Figure S2B). The $K_d$ values show a correlation with the ability of the different variants to enhance vesicle release at high expression levels (Figure 3C), a result consistent with our proposal that binding to isolated vesicles is linked to the ability of alpha-synuclein to potentiate exocytosis. We previously reported that high expression of WT alpha-synuclein disperses mCherry-Rab11-labeled recycling endosomes from the endocytic recycling compartment (ERC) to the cell periphery [57]. We hypothesized that this redistribution leads to an increased abundance of vesicles readily available for fusion upon stimulation, contributing to the observed potentiation of exocytosis. Notably, the A30P mutant, which does not potentiate release at high expression levels, does not lead to a redistribution of mCherry-Rab11. Here, we examined the effects of the linker mutants on the distribution of fluorescently tagged Rab11. Both the 3AE and 4G variants showed a significant difference in exocytosis between high and low expression levels (Figure 3A), consistent with the ability of both mutants to enhance vesicle release. However, the extent of redistribution was similar for both the mutants and the WT protein (Figure S4), despite the decreased potentiating activity and vesicle affinity of the 4G mutant, indicating that RE redistribution is not linearly correlated with enhanced exocytosis or vesicle binding.

**Table 1.** $K_d$ values for the different mutants of alpha-synuclein extracted from fits to a biomolecular reaction between protein molecules and binding sites on the vesicle surface, where the total concentration of binding sites was taken to be $B_{max}$ times the lipid concentration, and $B_{max}$ was determined from NMR intensity ratios as described in the Section 2.

| Variant | $K_d$ | SEM | *p*-Value |
|---|---|---|---|
| **WT** | $1.12 \times 10^{-6}$ | $5.22 \times 10^{-8}$ | $7.98 \times 10^{-36}$ |
| **A30P** | $8.09 \times 10^{-6}$ | $5.01 \times 10^{-7}$ | $1.74 \times 10^{-21}$ |
| **V70P** | $1.21 \times 10^{-5}$ | $7.52 \times 10^{-7}$ | $2.36 \times 10^{-21}$ |
| **A30P/V70P** | $2.13 \times 10^{-4}$ | $1.41 \times 10^{-5}$ | $3.25 \times 10^{-20}$ |
| **3AE** | $8.96 \times 10^{-7}$ | $6.92 \times 10^{-8}$ | $1.39 \times 10^{-17}$ |
| **4G** | $3.54 \times 10^{-6}$ | $2.87 \times 10^{-7}$ | $9.22 \times 10^{-17}$ |

### *3.5. Alpha-Synuclein Can Disperse Vesicles from Condensates In Vitro*

Our investigations of the effects of alpha-synuclein on intracellular vesicle distributions were motivated by studies demonstrating that increasing levels of alpha-synuclein correlate with increasing dispersion of SVs from SV clusters in neurons [16,24]. Recent evidence suggests that SV clusters are formed by the sequestration of SVs in synapsin-1 phase-separated condensates [59]. We explored whether alpha-synuclein could either dissolve synapsin-1-eGFP condensates in vitro or disperse lipid vesicles from synapsin condensates as a potential mechanism by which it could disperse SV clusters at presynaptic nerve terminals. As previously reported, rhodamine-labeled SUVs were preferentially partitioned in the synapsin-1-eGFP phase-separated droplets formed in the presence of polyethylene glycol (PEG) (Figure S5A). The presence of 200 μM unlabeled alpha-synuclein did not eliminate synapsin droplets (Figure S5A) but did significantly decrease the partitioning of SUVs into the droplets (Figure S5A,B). This result is consistent with a recent report describing the interactions of alpha-synuclein with synapsin condensates [73] and could provide an explanation for the dispersal of SV clusters with increasing synuclein levels. In RBL-2H3 cells, we also observe that increasing levels of alpha-synuclein disperse recycling vesicles from intracellular stores, as described above, suggesting that a similar mechanism could be involved [57].

### *3.6. Bipartite Binding Contributes to Affinity of Alpha-Synuclein to Membranes*

The higher affinity of the 4G mutant for membranes compared with the V70P mutant is surprising, given that our NMR intensity ratio assay indicates that a larger portion of the protein interacts with membranes for the V70P variant (Figure 2C). To examine membrane binding more closely, we performed $^{15}$N-DEST NMR (Dark-state Exchange Saturation Transfer) experiments that probe the exchange between the invisible vesicle-bound state and the visible free state. These experiments are performed under conditions where only a small fraction of protein is bound at any one time, and unlike the intensity ratio assay, which reports on the populations of free protein, DEST probes transient interactions of individual residues with membranes, which result in a broadening of the corresponding DEST profiles [42]. As expected, DEST profiles of 100 μM N-terminally acetylated WT alpha-synuclein in the presence of 1 mM SUVs show broadening over the entire lipid-binding domain of the protein, while residues in the C-terminal tail that do not bind strongly to membranes show narrow DEST profiles and provide an internal control (Figure 4A). The 3AE mutant had a similar DEST profile as the WT, signifying that membrane binding modes of the protein are unperturbed for this mutant (Figure 4B). The V70P mutant, which showed disruption of the lipid interaction C-terminal to position 65 in the intensity ratio assay (Figure 2C), features broadening similar to the WT protein for residues 1–65 but features narrow DEST profiles C-terminal to position 65 that are similar to those for residues in the C-terminal tail (Figure 4C), indicating that there is no significant interaction with lipid membranes beyond this position. Interestingly, the 4G mutant exhibits DEST profiles broadened to a similar extent as for the WT protein both for residues 1–35 and for residues 50–94 (Figure 4D), which correspond roughly to the helix-1 and helix-2 regions of the broken-helix state. The presence of four glycine residues in the linker region of this mutant disallows helix propagation, and such a DEST profile thus strongly suggests independent binding of the helix-2 region to the membrane surface. This binding mode is apparently too transient to result in intensity decreases under the conditions used for the intensity ratio assay in Figure 2, but it does appear to contribute to membrane affinity, as reflected in the higher $K_d$ value of the 4G compared with the V70P mutant. Although the conformation of such a transiently bound state cannot be probed directly using these experiments, at higher lipid concentrations, binding of the helix-2 region of the 4G mutant can be observed using the intensity ratio assay (Figure S6). This suggests the possibility that when they are structurally decoupled, the helix-1 region of alpha-synuclein outcompetes the helix-2 region for membrane binding sites. This is consistent with a recent computational study that reports binding of the helix-2 region to membranes in both helical and disordered conformations with a lower affinity than the helix-1 region [74].

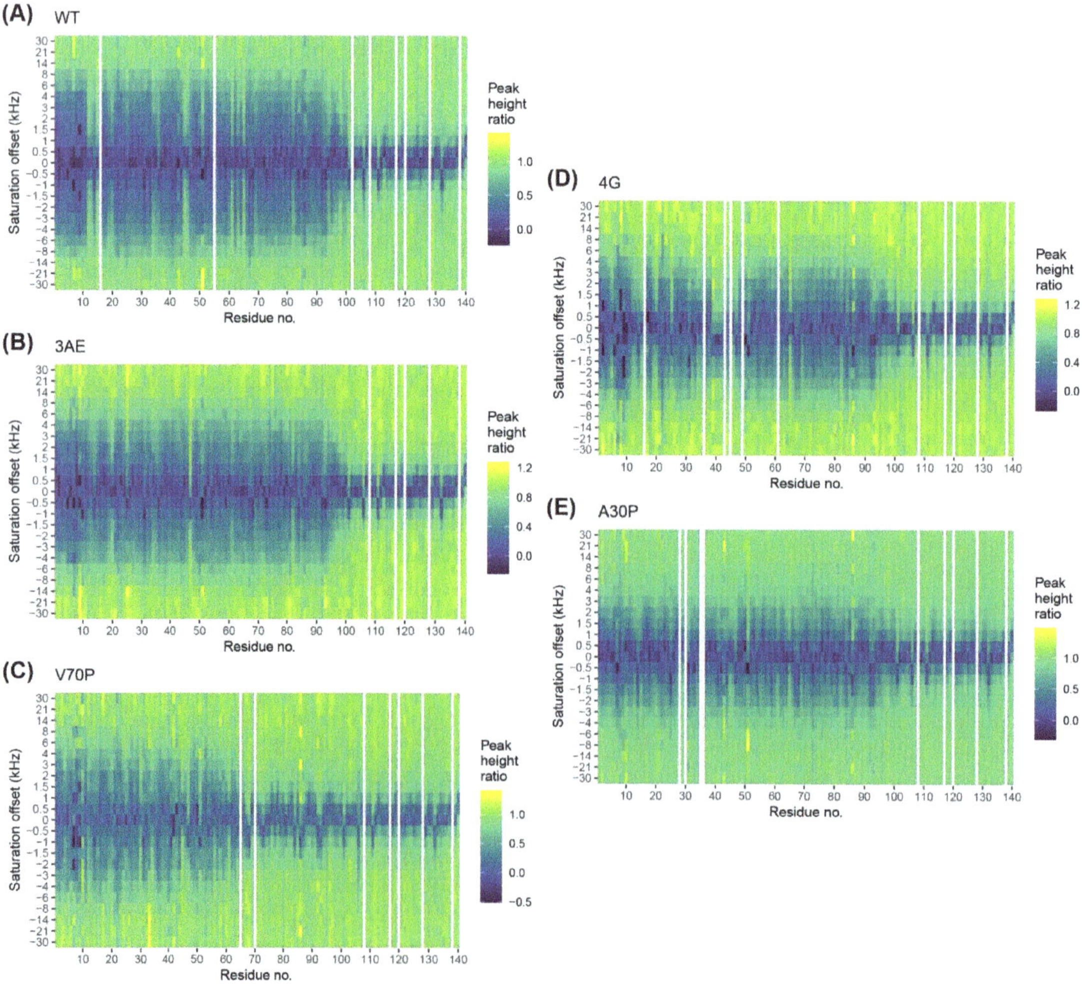


**Figure 4.** Alpha-synuclein membrane binding profiles. DEST intensity ratios as a function of saturation offset and residue number for 100 μM WT (**A**), 3AE (**B**), V70P (**C**), 4G (**D**) and A30P (**E**) alpha-synuclein with 1 mM 60:25:15 DOPC/DOPE/DOPS lipid SUVs at 13 °C, using a 700 MHz spectrometer and a saturation bandwidth of 400 Hz. Broad profiles indicate exchange with a slowly tumbling membrane-bound state while narrow profiles indicate less or no membrane interaction.

We also examined the DEST profiles of the PD mutant A30P in the presence of SUVs (Figure 4E). As previously reported, the DEST profiles are not as broad as those for the WT protein, consistent with a decrease in local binding to membranes throughout the protein sequence for this mutant. Furthermore, as for the 4G mutant but different from the V70P mutant, binding is evident beyond the site of mutation throughout the remainder of the lipid-binding domain. This observation was also interpreted previously as evidence for independent binding by the helix-2 region, but it is not clear to what extent the single proline mutation at position 30 decouples helix formation between the helix-1 and helix-2 regions, since the local helical structure of the micelle-bound A30P mutant is only perturbed for a few turns and resumes before the linker region [75]. Furthermore, the total helical content of this mutant when bound to vesicles at high lipid concentrations is only slightly decreased compared to the WT protein [76].

### *3.7. Quantitative Analysis of DEST Data Support Bipartite Binding Model*

Quantitative fits to DEST data can be used to extract information regarding the kinetics of membrane binding and release and the ensemble of membrane-bound conformations at the level of individual residues [42,77]. To apply the model developed by Fawzi et al., we first examined whether binding of alpha-synuclein to vesicles is consistent with a pseudo-first-order process. We showed that increases in $R_2$, which can be used as an estimate of the rate of membrane association, depend on lipid concentration but are relatively independent of protein concentration (Figure S7), confirming that binding can be modeled as a pseudo-first-order process. Next, we applied the DESTfit analysis package developed by Fawzi et al. to fit our DEST and $^{15}N$ $R_2$ relaxation data to different binding models [42,77]. Since the C-terminal tail does not bind to lipids, we only considered the lipid-binding domain (residue 1–98) for the fitting. The models, as described by Fawzi et al. [77], include a simple two-state model for membrane binding, where the protein exchanges between a single free state and a single bound state for each residue, and a pseudo-two-state model, where for each residue, bound-state conformations are divided into two subsets, those in which the residue is in direct contact with the vesicle surface or those where it is tethered to the vesicle surface by other nearby residues that are in direct contact (see Section 2). Notably, modeling multiple bound-state conformations allows for different bound-state relaxation properties, better accounting for variations in the shape of the DEST profiles.

The success of each model in fitting the data was assessed by its ability to reproduce the difference in $R_2$ measured in the absence and presence of vesicles. For the WT protein, the two-state model did not provide a good fit, while the pseudo-two-state model resulted in adequate fits over the entire sequence (Figure S8). This was also observed to be the case for the 3AE, the 4G and the A30P mutants. For each of these variants, the fit results suggest a significant population of direct contact binding modes for residues in the helix-2 region of the protein (Figure S9), consistent with an independent binding mode for this region. Interestingly, at the N-terminal region of each of these variants, the direct contact population becomes smaller, suggesting that tethered binding modes dominate in this region. The N-terminal region of alpha-synuclein is known to be the tightest binding region of the protein, as seen in our intensity ratio plots and as documented in other studies. Because of tight binding in this region, the local off-rate is expected to be quite slow, likely smaller than the longitudinal relaxation rate $R_1$, enabling only a fraction of direct contact binding events to contribute to the DEST profiles in this region. In addition, the tethered state R2 rates extracted from the model are much higher in this N-terminal region compared with the helix-2 region, consistent with tighter binding of this site even in tethered binding modes. Unlike for the other variants, the DEST data for the V70P mutant could be fit using the simpler two-state model, which does not require heterogeneous binding modes. This is consistent with the lack of independent binding of the helix-2 region of this mutant. The bound-state R2 values extracted from this model are consistent with the directly bound R2 values estimated for the other variants and an order of magnitude larger than those estimated for the tethered binding modes of the other variants, confirming that this binding mode corresponds to a direct contact mode.

## 4. Discussion

While the normal function(s) of alpha-synuclein remain incompletely understood, considerable evidence indicates that the protein can function in the regulation of the SV cycle by influencing endocytosis [78–81], intracellular vesicle pools [8,16,24,82] and exocytosis [16,17,83–86]. The structural underpinnings of these functions are poorly understood, despite considerable progress in delineating the different conformations that alpha-synuclein can adopt in vitro. We recently developed a structure–function assay for the effects of alpha-synuclein on vesicle exocytosis utilizing a model cell line, RBL-2H3, which provides facile stimulation of vesicle release that can be monitored conveniently via confocal fluorescence microscopy using the pHluorin assay [57]. Our studies revealed that low levels of alpha-synuclein expression are sufficient to inhibit the release of recycling endosomes triggered by thapsigargin or antigen treatment, while TIRF microscopy measurements of individual vesicle fusion events indicated that this inhibition was likely operating directly at the level of individual fusion events by reducing the probability of fusion. We identified a mutation, V70P, that abrogated the ability of alpha-synuclein to inhibit release in this assay and proposed that this mutation prevented the protein from bridging between the vesicle and plasma membranes via a broken-helix configuration that we posit is required for inhibitory activity [57]. In this conformation, we [25,87–90] and others [91] have proposed that the two helices of alpha-synuclein are thought to preferentially bind to either the vesicle membrane or inner plasma membrane, creating a membrane-bridging conformation.

Surprisingly, we observed that higher levels of alpha-synuclein expression lead to increased levels of vesicle release compared with empty vector controls [57]. Increased release was associated with a redistribution of mCherry-Rab11 staining, a marker for the endocytic recycling compartment from which recycling endosomes originate, from the interior of the cell to the vicinity of the plasma membrane. This suggested that increased release was associated with an increased pool of vesicles near the membrane. Supporting this hypothesis, the A30P mutant of alpha-synuclein, which was known to be defective in binding to isolated membrane vesicles, abrogated increased vesicle release at high expression levels and also eliminated the redistribution of Rab11 to the cell periphery. Since alpha-synuclein binds to isolated vesicles predominantly in an extended-helix conformation, we proposed that this binding mode is required for potentiation of vesicle release.

Although alpha-synuclein is a neuronal protein, our previous studies demonstrated that RBL-2H3 cells are advantageous for structure–function analyses of this protein because, as described above, they allow us to decouple the effects of alpha-synuclein on the fusion of docked vesicles from its effects in intracellular vesicle pools and distribution. These effects are difficult to deconvolute in neurons, where vesicles can only fuse at the active zone, whereas in our model cells vesicles can fuse anywhere on the plasma membrane. Here, we designed two new mutants of alpha-synuclein, designed to bias the membrane-bound conformation of the protein towards either the broken-helix state (4G mutant), which we posited should favor the inhibitory activity of the protein, or the extended-helix state (3AE mutant), which we posited should favor the potentiating activity of the protein. The 3AE mutant of alpha-synuclein indeed favors the extended-helix over the broken-helix state. As expected, this mutant does not inhibit vesicle release as effectively as the WT but is fully able to potentiate release. These results support the model in which the broken-helix state mediates direct inhibition of release, while the extended-helix state characteristic of binding to isolated vesicles mediates enhanced release. Surprisingly, despite favoring the broken-helix conformation, the 4G mutant was also able to potentiate vesicle release, although to a lesser extent than the WT and 3AE variants. We reasoned that the circumscribed binding mode of the 4G mutant to isolated vesicles was sufficient to confer a weak potentiating activity. However, the V70P mutant, which exhibits a vesicle binding mode that is more extensive than that of the 4G mutant but less than that of the WT protein, did not exhibit detectable enhancement of vesicle release. To investigate this apparent discrepancy, we measured the binding affinity of the different variants to vesicles using intrinsic tryptophan fluorescence spectroscopy. The results revealed that the 4G mutant in fact binds to vesicles more tightly that the V70P mutant, suggesting an explanation for its ability to enhance vesicle release. Indeed, a plot of the lipid binding affinity of the four different alpha-synuclein variants versus their potentiating activity indicates a correlation between these two properties.

We then investigated the basis for the higher affinity of the 4G mutant for vesicles, compared with the V70P mutant. Using DEST, we showed that the 4G mutant binds to vesicles using both the helix-1 and helix-2 regions. Binding in the helix-1 region occurs on a longer time scale and can be observed in equilibrium experiments, while binding by the helix-2 region is more transient. In contrast, the V70P mutant only features a single binding mode and is not capable of independent binding via its helix-2 region, presumably because the proline residue situated in the middle of helix-2 precludes this binding mode. Previous reports had also suggested that the helix-2 region of alpha-synuclein can bind to membranes independently based on DEST studies of the A30P mutant [92]. However, a single proline is not fully effective at interrupting helical structure, so the A30P mutation may not completely decouple helix propagation and membrane binding in the helix-2 region from helix-1. In contrast, propagation of helical structure across a four-glycine linker would be extremely unfavorable, so our observation of independent binding of the helix-2 region in the 4G mutant provides clear evidence of independent membrane binding in this region. A recent publication also concluded, based on a combination of biophysical measurements, that independent binding of the N-terminal region and the helix-2 regions of alpha-synuclein both contribute to its membrane binding affinity [69].

Surprisingly, despite creating a clear break between helix-1 and helix-2, the 4G mutant resulted in a loss of direct inhibition of vesicle release by alpha-synuclein. This result indicates that the ability to form two separate helices is not, by itself, sufficient for the inhibitory activity of alpha-synuclein. Further studies will be required to delineate what additional features are required for this activity. Since the 4G mutant effectively decouples propagation of helical structure across the linker region, it may be that some degree of structural coupling between the helix-1 and helix-2 regions is required for alpha-synuclein to directly inhibit vesicle release.

Inhibition of vesicle secretion by alpha-synuclein overexpression has been previously reported in neuronal cells and has been attributed in part to a reduction of intracellular vesicle pools [16]. Indeed, increasing levels of alpha-synuclein lead to a disruption of SV clustering [16,24]. In neurons, unlike in less specialized cells, vesicles must fuse with the plasma membrane at specialized sites known as active zones, which are closely apposed to SV clusters. Dispersion of vesicles from their clusters in such cells would be expected to lead to a reduction in vesicle release as vesicles are removed from the proximity of the active zones. Thus, while high levels of alpha-synuclein appear to cause a redistribution of vesicles out of internal pools in both neuronal cells and in RBL-2H3 cells, in the latter this leads to an increase in release as more vesicles are available at the outer cell membrane for fusion, while in neurons this leads to a decrease in release, as vesicles are mislocated away from the requisite sites of fusion.

The mechanisms by which alpha-synuclein may disperse vesicles from internal stores remains unclear, but the clustering of SVs was recently proposed to be mediated by the sequestration of vesicles in condensates formed by the presynaptic protein synapsin-1. We proposed previously that alpha-synuclein could interfere with this process either by interfering with synapsin-1 condensate formation or by release vesicles from such condensates [57]. Here, we investigated the effects of alpha-synuclein on synapsin-1 condensates with and without vesicles in vitro. Under our conditions, alpha-synuclein does not dissolve synapsin-1 condensates but does significantly reduce the degree to which fluorescently labeled vesicles co-localize with synapsin-1 droplets. Hence, alpha-synuclein appears to disperse vesicles from condensates, resulting in their release and mislocalization at presynaptic nerve terminals. A recent report also examined the interplay between alpha-synuclein and synapsin condensate formation [73], reporting that synuclein is recruited into synapsin condensates. This observation is complementary to, but consistent with, our own results, in that synuclein entry into synapsin condensates is likely required for its ability to release vesicles. This same study also reported that vesicles accelerate the formation of synapsin-vesicle condensates, while excess alpha-synuclein retards this process. Condensate formation was assessed by turbidity, and the vesicle content was not directly measured, but our results suggest that this effect of alpha-synuclein likely originates from a decrease in the number of vesicles within the forming condensates, leading to the reduced rate of condensate formation. Our observation that alpha-synuclein disperses vesicles from the endocytic recycling compartment led us to suggest that the mechanism involved may be similar and that the ERC may also be comprised, at least in parts, of protein-membrane condensates. Future work will be required to address this intriguing hypothesis.

**Supplementary Materials:** The following supporting information can be downloaded at: https://www.mdpi.com/article/10.3390/biom12121816/s1. Eleven supplementary figures with the following titles and legends.

Figure S1: Effects of alpha-synuclein variants on stimulated exocytosis at (A) low and (B) high expression levels. Thapsigargin-stimulated exocytosis of RBL-2H3 cells, quantified using VAMP8-pHluorin fluorescence time-course imaging of individual cells, for alpha-synuclein variants transfected at low (left panel) and high (right panel) concentrations. After removal of outliers (see Section 2), statistical significance was determined using one-way analysis of variance followed by pairwise p-values determined by Tukey's post hoc test. Individual data points used for statistical analysis are shown as blue circles; outlier values within the plotting range are shown as red circles. Significance levels: $p$ values *** < 0.001 < ** < 0.01 < * < 0.05 < NS. Error bars represent SEM;

Figure S2: Membrane binding monitored using intrinsic tryptophan fluorescence. (A) Alpha-synuclein extended-helix structure schematic showing location of a single tryptophan residue that was incorporated at position 4 (F4W mutation) of N-terminally acetylated alpha-synuclein variants to serve as an environment-sensitive fluorescence reporter. (B) Fluorescence spectra were collected at 10 nm resolution from proteins at 0.1 μM concentration with variable concentrations of lipid SUVs. After subtracting concentration-matched lipid-only spectra to account for scattering background, the background-subtracted spectra (filled points) were fit to a log-normal model [45–47] (solid lines). (C) Plots of peak position versus full width at half maxima for each fitted spectrum. The solid line represents the empirical linear correlation between peak position and peak width for zwitterionic tryptophan in solvents with decreasing degrees of hydrophobicity, plotted as $y = -156.7 + 0.624\,x$ [48]. Deviations from the line indicate heterogeneity in the hydrophobicity of the environment experienced by Trp probe, which may reflect a mixture of bound and unbound states;

Figure S3: Fitting of individual tryptophan fluorescence spectra as a linear combination of free-state and fully bound state spectra. Each background-subtracted spectrum at intermediate lipid concentrations (e.g., at 0.025 mM lipid, cyan spectrum, $S_i$) was fit as a linear combination of the free-state spectrum (red, $S_{free}$) and the fully bound-state spectrum at the highest lipid concentration (10 mM, green, $S_{bound}$), using the equation $\textit{Fitted } S_i = c_1 \cdot S_{free} + c_2 \cdot S_{bound}$. The bound fraction at every lipid concentration was determined using the fitted coefficients as $\frac{c_2}{c_1+c_2}$;

Figure S4: Redistribution of recycling endosomes at high expression levels of alpha-synuclein variants. mCherry fluorescence from mCherry-Rab11 was used as a measure of recycling endosomes. For individual cells, fluorescence integrated over a membrane-proximal zone, defined as a shell of ~800 nm thickness from the cell membrane, was divided by the total mCherry fluorescence to provide a measure of the fraction of membrane-proximal recycling endosomes. An increase in the membrane-proximal fraction in the presence of high levels of alpha-synuclein compared with control indicates a redistribution of recycling endosomes from the cell interior to the plasma membrane. The box represents 1st to 3rd quartile of the data, the midline represents the median value, and the cross represents the mean value. Significance levels: $p$ values *** < 0.001 < ** < 0.01 < * < 0.05 < NS;

Figure S5: Effect of wild-type alpha-synuclein on enrichment of small unilamellar vesicles in synapsin-1 phase-separated droplets: (A) Colocalization of rhodamine fluorescence from rhodamine-labeled SUVs (0.5 mM) with eGFP fluorescence from synapsin-1-eGFP phase-separated droplets in the absence (upper panels) and presence (lower panels) of N-terminally acetylated wild-type alpha-synuclein (200 μM) was used to assess the enrichment of vesicles in synapsin droplets. Representative micrographs are shown. All images were acquired using the same image acquisition and processing parameters. Scale bar: 10 μM. (B) Quantification of vesicle enrichment inside each droplet. Blue circles represent data points used for statistical analysis; red circles represent outlier values. In the box plot, the central line represents the median, the boundaries of the box represent the 25th and 75th percentile values and the ends of the vertical line represent the range of the data. Statistical significance was determined by t-test of normally distributed log-transformed data. *** $p < 0.001$;

Figure S6: SUV binding of N-terminally acetylated alpha-synuclein 4G mutant with increasing concentrations of lipid SUVs. The peak height ratio, representing the free fraction per residue, was determined by ratio of $^{1}$H-$^{15}$N HSQC resonances between samples with and without SUV. Data were collected at 50 μM protein concentration and 10 °C;

Figure S7: Dependence of N-terminally acetylated WT alpha-synuclein $\Delta R_2$ on lipid and protein concentration. (A) The difference in $R_2$ relaxation rates in presence versus absence of SUVs ($\Delta R_2$) for 100 μM N-terminally acetylated WT alpha-synuclein with different concentrations of lipid SUVs. (B) The difference in $R_2$ relaxation rates in presence versus absence of SUVs ($\Delta R_2$) for different concentrations of N-terminally acetylated WT alpha-synuclein with 1 mM lipid SUVs. SUV composition: DOPC/DOPE/DOPS = 60:25:15. Data were collected at 13 °C on a 700 MHz spectrometer;

Figure S8: Fits to DEST and $\Delta R_2$ data from wild-type alpha-synuclein using different models. (A) DEST profiles of select residues from different regions of the protein, measured at 700 MHz, at 175 and 400 Hz saturation bandwidths (red and blue circles, respectively). The dashed lines represent fits obtained using a two-state model, and the solid lines represent fits obtained using a pseudo-two-state model (see text). (B) Residue-specific $\Delta R_2$ values (black circles connected by dashed line) with fits obtained using a two-state model (blue circles) or a pseudo-two-state model (red circles). Error bars represent 1 standard deviation;

Figure S9: Residue-specific $K_3$ equilibrium constants for direct contact and tethered states. $K_3$ values were determined by simultaneous fitting of DEST and $R_2$ relaxation experiments for alpha-synuclein variants (WT, A30P, 3AE and 4G) that were best fit using a pseudo-two-state model (see text). Error bars represent 1 standard deviation;

Figure S10: Schematic for the two-state and pseudo-two-state models used to fit the DEST and $\Delta R_2$ data. (A) In the two-state model, binding of all residues occurs by conversion of a single molecular free state ensemble, $E^{m}_{visible}$, to a single molecular bound state ensemble, $E^{m}_{dark}$, and is dominated by same global $k^{app}_{on}$ and $k_{off}$ rate constants. (B) In the pseudo-two-state model, each residue sampling of a local free-state ensemble, $E^{res}_{visible}(i)$, transitions upon binding to an ensemble of states in which this residue is directly bound to the vesicle surface, $E^{res}_{contact}(i)$, or to an ensemble of states in which this residue is tethered to the surface by nearby directly bound residues, $E^{res}_{tethered}(i)$, via residue-specific on-rate constants $k^{app}_{1}(i)$ or $k^{app}_{2}(i)$, respectively (see [42,77]);

Figure S11: Protein expression levels in RBL-2H3 cells quantified using alpha-synuclein immunostaining under (A) conditions of high expression for the WT and linker mutants and (B) conditions of low and high expression for the 4G mutant. All data were collected using the same staining protocol and microscopy settings. Movies: Representative movies of exocytosis experiments for cells transfected with pcDNA, WT (low expression) or WT (high expression), 3AE (low expression), 3AE (high expression), 4G (low expression) or 4G (high expression) variants of alpha-synuclein.

**Author Contributions:** Conceptualization, T.D., D.S., B.A.B. and D.E.; data curation, T.D.; funding acquisition, K.Y.L., D.A.H., B.A.B. and D.E.; investigation, T.D., M.R., D.S., C.F. and P.C.; methodology, T.D., M.R. and D.E.; project administration, B.A.B. and D.E.; resources, P.C. and K.Y.L.; software, T.D.; supervision, K.Y.L., D.A.H., B.A.B. and D.E.; validation, T.D. and M.R.; visualization, T.D. and M.R.; writing—original draft, T.D.; writing—review and editing, T.D., M.R., P.C., B.A.B. and D.E. All authors have read and agreed to the published version of the manuscript.

**Funding:** Funding was provided by NIH grants R37AG019391 (D.E.), R35GM136686 (D.E.), R01GM117552 (B.A.B. and D.A.H.) and NSF grant MCB-1950525 (K.Y.L.). C.F. is grateful to the Carlos Chagas Foundation for Research Support of the State of Rio de Janeiro (FAPERJ) and the Brazilian National Council for Scientific and Technological Development (CNPq). D.E. is a member of the New York Structural Biology Center (NYSBC), which is supported by ORIP/NIH facility improvement grant CO6RR015495. The 700 MHz spectrometer was purchased with funds from NIH grant S10OD018509. Data collected using the 800 MHz Avance III spectrometer are supported by NIH grant S10OD016432. The 900 MHz NMR spectrometers were purchased with funds from NIH grant P41GM066354 and the New York State Assembly. Some of the work presented here was conducted at the Center on Macromolecular Dynamics by NMR Spectroscopy located at the New York Structural Biology Center, supported by NIH grant GM118302. The Weil Cornell NMR Core Facility is supported by grant S10OD016320 from NIH. Fluorescence imaging were carried out in the Cornell University Biotechnology Resource Center with funding for the Zeiss LSM 710 confocal microscope (NIH S10RR025502) and Zeiss Elyra microscope (NSF 1428922).

**Institutional Review Board Statement:** Not Applicable.

**Informed Consent Statement:** Not Applicable.

**Data Availability Statement:** NMR backbone chemical shift assignments for SDS micelle-bound alpha-synuclein 3AE and 4G variants have been deposited in the BMRB database (BMRB accession numbers 51632 and 51633, respectively). All other data are available upon request to David Eliezer (dae2005@med.conell.edu).

**Acknowledgments:** We are grateful to Shibani Bhattacharyya for her extensive support in performing NMR experiments at the New York Structural Biology Center (NYSBC).

**Conflicts of Interest:** The authors declare no conflict of interest.

## References

1. Feigin, V.L.; Nichols, E.; Alam, T.; Bannick, M.S.; Beghi, E.; Blake, N.; Culpepper, W.J.; Dorsey, E.R.; Elbaz, A.; Ellenbogen, R.G.; et al. Global, Regional, and National Burden of Neurological Disorders, 1990–2016: A Systematic Analysis for the Global Burden of Disease Study 2016. *Lancet Neurol.* **2019**, *18*, 459–480. [CrossRef] [PubMed]
2. Spillantini, M.G.; Schmidt, M.L.; Lee, V.M.-Y.; Trojanowski, J.Q.; Jakes, R.; Goedert, M. Alpha-Synuclein in Lewy Bodies. *Nature* **1997**, *388*, 839–840. [CrossRef] [PubMed]
3. Mahul-Mellier, A.-L.; Burtscher, J.; Maharjan, N.; Weerens, L.; Croisier, M.; Kuttler, F.; Leleu, M.; Knott, G.W.; Lashuel, H.A. The Process of Lewy Body Formation, Rather than Simply α-Synuclein Fibrillization, Is One of the Major Drivers of Neurodegeneration. *Proc. Natl. Acad. Sci. USA* **2020**, *117*, 4971–4982. [CrossRef] [PubMed]
4. Maroteaux, L.; Campanelli, J.T.; Scheller, R.H. Synuclein: A Neuron-Specific Protein Localized to the Nucleus and Presynaptic Nerve Terminal. *J. Neurosci.* **1988**, *8*, 2804–2815. [CrossRef] [PubMed]
5. George, J.M.; Jin, H.; Woods, W.S.; Clayton, D.F. Characterization of a Novel Protein Regulated during the Critical Period for Song Learning in the Zebra Finch. *Neuron* **1995**, *15*, 361–372. [CrossRef]
6. Liu, S.; Ninan, I.; Antonova, I.; Battaglia, F.; Trinchese, F.; Narasanna, A.; Kolodilov, N.; Dauer, W.; Hawkins, R.D.; Arancio, O. Alpha-Synuclein Produces a Long-Lasting Increase in Neurotransmitter Release. *EMBO J.* **2004**, *23*, 4506–4516. [CrossRef]
7. Gureviciene, I.; Gurevicius, K.; Tanila, H. Role of Alpha-Synuclein in Synaptic Glutamate Release. *Neurobiol. Dis.* **2007**, *28*, 83–89. [CrossRef]
8. Cabin, D.E.; Shimazu, K.; Murphy, D.; Cole, N.B.; Gottschalk, W.; McIlwain, K.L.; Orrison, B.; Chen, A.; Ellis, C.E.; Paylor, R.; et al. Synaptic Vesicle Depletion Correlates with Attenuated Synaptic Responses to Prolonged Repetitive Stimulation in Mice Lacking Alpha-Synuclein. *J. Neurosci.* **2002**, *22*, 8797–8807. [CrossRef]
9. Martín, E.D.; González-García, C.; Milán, M.; Fariñas, I.; Ceña, V. Stressor-Related Impairment of Synaptic Transmission in Hippocampal Slices from Alpha-Synuclein Knockout Mice. *Eur. J. Neurosci.* **2004**, *20*, 3085–3091. [CrossRef]
10. Abeliovich, A.; Schmitz, Y.; Fariñas, I.; Choi-Lundberg, D.; Ho, W.H.; Castillo, P.E.; Shinsky, N.; Verdugo, J.M.; Armanini, M.; Ryan, A.; et al. Mice Lacking Alpha-Synuclein Display Functional Deficits in the Nigrostriatal Dopamine System. *Neuron* **2000**, *25*, 239–252. [CrossRef]
11. Senior, S.L.; Ninkina, N.; Deacon, R.; Bannerman, D.; Buchman, V.L.; Cragg, S.J.; Wade-Martins, R. Increased Striatal Dopamine Release and Hyperdopaminergic-like Behaviour in Mice Lacking Both Alpha-Synuclein and Gamma-Synuclein. *Eur. J. Neurosci.* **2008**, *27*, 947–957. [CrossRef] [PubMed]
12. Yavich, L.; Tanila, H.; Vepsäläinen, S.; Jäkälä, P. Role of Alpha-Synuclein in Presynaptic Dopamine Recruitment. *J. Neurosci.* **2004**, *24*, 11165–11170. [CrossRef] [PubMed]
13. Watson, J.B.; Hatami, A.; David, H.; Masliah, E.; Roberts, K.; Evans, C.E.; Levine, M.S. Alterations in Corticostriatal Synaptic Plasticity in Mice Overexpressing Human Alpha-Synuclein. *Neuroscience* **2009**, *159*, 501–513. [CrossRef] [PubMed]

14. Chandra, S.; Fornai, F.; Kwon, H.-B.; Yazdani, U.; Atasoy, D.; Liu, X.; Hammer, R.E.; Battaglia, G.; German, D.C.; Castillo, P.E.; et al. Double-Knockout Mice for Alpha- and Beta-Synucleins: Effect on Synaptic Functions. *Proc. Natl. Acad. Sci. USA* **2004**, *101*, 14966–14971. [CrossRef]
15. Anwar, S.; Peters, O.; Millership, S.; Ninkina, N.; Doig, N.; Connor-Robson, N.; Threlfell, S.; Kooner, G.; Deacon, R.M.; Bannerman, D.M.; et al. Functional Alterations to the Nigrostriatal System in Mice Lacking All Three Members of the Synuclein Family. *J. Neurosci.* **2011**, *31*, 7264–7274. [CrossRef]
16. Nemani, V.M.; Lu, W.; Berge, V.; Nakamura, K.; Onoa, B.; Lee, M.K.; Chaudhry, F.A.; Nicoll, R.A.; Edwards, R.H. Increased Expression of α-Synuclein Reduces Neurotransmitter Release by Inhibiting Synaptic Vesicle Reclustering after Endocytosis. *Neuron* **2010**, *65*, 66–79. [CrossRef]
17. Wu, N.; Joshi, P.R.; Cepeda, C.; Masliah, E.; Levine, M.S. Alpha-Synuclein Overexpression in Mice Alters Synaptic Communication in the Corticostriatal Pathway. *J. Neurosci. Res.* **2010**, *88*, 1764–1776. [CrossRef]
18. Larsen, K.E.; Schmitz, Y.; Troyer, M.D.; Mosharov, E.; Dietrich, P.; Quazi, A.Z.; Savalle, M.; Nemani, V.; Chaudhry, F.A.; Edwards, R.H.; et al. Alpha-Synuclein Overexpression in PC12 and Chromaffin Cells Impairs Catecholamine Release by Interfering with a Late Step in Exocytosis. *J. Neurosci.* **2006**, *26*, 11915–11922. [CrossRef]
19. Scott, D.A.; Tabarean, I.; Tang, Y.; Cartier, A.; Masliah, E.; Roy, S. A Pathologic Cascade Leading to Synaptic Dysfunction in Alpha-Synuclein-Induced Neurodegeneration. *J. Neurosci.* **2010**, *30*, 8083–8095. [CrossRef]
20. Janezic, S.; Threlfell, S.; Dodson, P.D.; Dowie, M.J.; Taylor, T.N.; Potgieter, D.; Parkkinen, L.; Senior, S.L.; Anwar, S.; Ryan, B.; et al. Deficits in Dopaminergic Transmission Precede Neuron Loss and Dysfunction in a New Parkinson Model. *Proc. Natl. Acad. Sci. USA* **2013**, *110*, E4016-25. [CrossRef]
21. Sudhof, T.C.; Rizo, J. Synaptic Vesicle Exocytosis. *Cold Spring Harb Perspect. Biol.* **2011**, *3*, a005637. [CrossRef] [PubMed]
22. Rizzoli, S.O.; Betz, W.J. Synaptic Vesicle Pools. *Nat. Rev. Neurosci.* **2005**, *6*, 57–69. [CrossRef] [PubMed]
23. Alabi, A.A.; Tsien, R.W. Synaptic Vesicle Pools and Dynamics. *Cold Spring Harb Perspect. Biol.* **2012**, *4*, a013680. [CrossRef]
24. Vargas, K.J.; Schrod, N.; Davis, T.; Fernandez-Busnadiego, R.; Taguchi, Y.V.; Laugks, U.; Lucic, V.; Chandra, S.S. Synucleins Have Multiple Effects on Presynaptic Architecture. *Cell Rep.* **2017**, *18*, 161–173. [CrossRef]
25. Dikiy, I.; Fauvet, B.; Jovičić, A.; Mahul-Mellier, A.-L.; Desobry, C.; El-Turk, F.; Gitler, A.D.; Lashuel, H.A.; Eliezer, D. Semisynthetic and in Vitro Phosphorylation of Alpha-Synuclein at Y39 Promotes Functional Partly Helical Membrane-Bound States Resembling Those Induced by PD Mutations. *ACS Chem. Biol.* **2016**, *11*, 2428–2437. [CrossRef] [PubMed]
26. Imam, S.Z.; Zhou, Q.; Yamamoto, A.; Valente, A.J.; Ali, S.F.; Bains, M.; Roberts, J.L.; Kahle, P.J.; Clark, R.A.; Li, S. Novel Regulation of Parkin Function through C-Abl-Mediated Tyrosine Phosphorylation: Implications for Parkinson's Disease. *J. Neurosci.* **2011**, *31*, 157–163. [CrossRef]
27. Karuppagounder, S.S.; Brahmachari, S.; Lee, Y.; Dawson, V.L.; Dawson, T.M.; Ko, H.S. The C-Abl Inhibitor, Nilotinib, Protects Dopaminergic Neurons in a Preclinical Animal Model of Parkinson's Disease. *Sci. Rep.* **2014**, *4*, 4874. [CrossRef] [PubMed]
28. Ko, H.S.; Lee, Y.; Shin, J.H.; Karuppagounder, S.S.; Gadad, B.S.; Koleske, A.J.; Pletnikova, O.; Troncoso, J.C.; Dawson, V.L.; Dawson, T.M. Phosphorylation by the C-Abl Protein Tyrosine Kinase Inhibits Parkin's Ubiquitination and Protective Function. *Proc. Natl. Acad. Sci. USA* **2010**, *107*, 16691–16696. [CrossRef]
29. Weinreb, P.H.; Zhen, W.; Poon, A.W.; Conway, K.A.; Lansbury, P.T. NACP, a Protein Implicated in Alzheimer's Disease and Learning, Is Natively Unfolded. *Biochemistry* **1996**, *35*, 13709–13715. [CrossRef]
30. Eliezer, D.; Kutluay, E.; Bussell, R.; Browne, G. Conformational Properties of α-Synuclein in Its Free and Lipid-Associated States. *J. Mol. Biol.* **2001**, *307*, 1061–1073. [CrossRef]
31. Johnson, M.; Geeves, M.A.; Mulvihill, D.P. Production of Amino-Terminally Acetylated Recombinant Proteins in E. Coli. *Methods Mol. Biol.* **2013**, *981*, 193–200. [CrossRef]
32. Johnson, M.; Coulton, A.T.; Geeves, M.A.; Mulvihill, D.P. Targeted Amino-Terminal Acetylation of Recombinant Proteins in E. Coli. *PLoS ONE* **2010**, *5*. [CrossRef] [PubMed]
33. Narhi, L.; Wood, S.J.; Steavenson, S.; Jiang, Y.; Wu, G.M.; Anafi, D.; Kaufman, S.A.; Martin, F.; Sitney, K.; Denis, P.; et al. Both Familial Parkinson's Disease Mutations Accelerate α-Synuclein Aggregation. *J. Biol. Chem.* **1999**, *274*, 9843–9846. [CrossRef] [PubMed]
34. Coelho-Cerqueira, E.; Carmo-Gonçalves, P.; Sá Pinheiro, A.; Cortines, J.; Follmer, C. α-Synuclein as an Intrinsically Disordered Monomer - Fact or Artefact? *FEBS J.* **2013**, *280*, 4915–4927. [CrossRef] [PubMed]
35. Deutsch, J.W.; Kelly, R.B. Lipids of Synaptic Vesicles: Relevance to the Mechanism of Membrane Fusion. *Biochemistry* **1981**, *20*, 378–385. [CrossRef] [PubMed]
36. Takamori, S.; Holt, M.; Stenius, K.; Lemke, E.A.; Grønborg, M.; Riedel, D.; Urlaub, H.; Schenck, S.; Brügger, B.; Ringler, P.; et al. Molecular Anatomy of a Trafficking Organelle. *Cell* **2006**, *127*, 831–846. [CrossRef] [PubMed]
37. Breckenridge, W.C.C.; Morgan, I.G.G.; Zanetta, J.P.P.; Vincendon, G. Adult Rat Brain Synaptic Vesicles Ii. Lipid Composition. *Biochim. Biophys. Acta* **1973**, *320*, 681–686. [CrossRef]
38. Rouser, G.; Siakotos, A.N.; Fleischer, S. Quantitative Analysis of Phospholipids by Thin-Layer Chromatography and Phosphorus Analysis of Spots. *Lipids* **1966**, *1*, 85–86. [CrossRef]
39. Rouser, G.; Fleischer, S.; Yamamoto, A. Two Dimensional Thin Layer Chromatographic Separation of Polar Lipids and Determination of Phospholipids by Phosphorus Analysis of Spots. *Lipids* **1970**, *5*, 494–496. [CrossRef]

40. Delaglio, F.; Grzesiek, S.; Vuister, G.W.; Zhu, G.; Pfeifer, J.; Bax, A. NMRPipe: A Multidimensional Spectral Processing System Based on UNIX Pipes. *J. Biomol. NMR* **1995**, *6*, 277–293. [CrossRef]
41. Vranken, W.F.; Boucher, W.; Stevens, T.J.; Fogh, R.H.; Pajon, A.; Llinas, M.; Ulrich, E.L.; Markley, J.L.; Ionides, J.; Laue, E.D. The CCPN Data Model for NMR Spectroscopy: Development of a Software Pipeline. *Proteins* **2005**, *59*, 687–696. [CrossRef] [PubMed]
42. Fawzi, N.L.; Ying, J.; Torchia, D.A.; Clore, G.M. Probing Exchange Kinetics and Atomic Resolution Dynamics in High-Molecular-Weight Complexes Using Dark-State Exchange Saturation Transfer NMR Spectroscopy. *Nat. Protoc.* **2012**, *7*, 1523–1533. [CrossRef] [PubMed]
43. McConnell, H.M. Reaction Rates by Nuclear Magnetic Resonance. *J. Chem. Phys.* **1958**, *28*, 430–431. [CrossRef]
44. Helgstrand, M.; Härd, T.; Allard, P. Simulations of NMR Pulse Sequences during Equilibrium and Non-Equilibrium Chemical Exchange. *J. Biomol. NMR* **2000**, *18*, 49–63. [CrossRef] [PubMed]
45. Burstein, E.A.; Emelyanenko, V.I. Log-Normal Description of Fluorescence Spectra of Organic Fluorophores. *Photochem. Photobiol.* **1996**, *64*, 316–320. [CrossRef]
46. Burstein, E.A.; Abornev, S.M.; Reshetnyak, Y.K. Decomposition of Protein Tryptophan Fluorescence Spectra into Log-Normal Components. I. Decomposition Algorithms. *Biophys. J.* **2001**, *81*, 1699–1709. [CrossRef] [PubMed]
47. Siano, D.B.; Metzler, D.E. Band Shapes of the Electronic Spectra of Complex Molecules. *J. Chem. Phys.* **1969**, *51*, 1856–1861. [CrossRef]
48. Ladokhin, A.S.; Jayasinghe, S.; White, S.H. How to Measure and Analyze Tryptophan Fluorescence in Membranes Properly, and Why Bother? *Anal. Biochem.* **2000**, *285*, 235–245. [CrossRef]
49. Burstein, E.A.; Vedenkina, N.S.; Ivkova, M.N. Fluorescence and the Location of Tryptophan Residues in Protein Molecules. *Photochem. Photobiol.* **1973**, *18*, 263–279. [CrossRef]
50. Dikiy, I.; Eliezer, D. N-Terminal Acetylation Stabilizes N-Terminal Helicity in Lipid- and Micelle-Bound α-Synuclein and Increases Its Affinity for Physiological Membranes. *J. Biol. Chem.* **2014**, *289*, 3652–3665. [CrossRef]
51. Kang, L.; Moriarty, G.M.; Woods, L.A.; Ashcroft, A.E.; Radford, S.E.; Baum, J. N-Terminal Acetylation of α-Synuclein Induces Increased Transient Helical Propensity and Decreased Aggregation Rates in the Intrinsically Disordered Monomer. *Protein Sci.* **2012**, *21*, 911–917. [CrossRef] [PubMed]
52. Maltsev, A.S.; Ying, J.; Bax, A. Impact of N-Terminal Acetylation of α-Synuclein on Its Random Coil and Lipid Binding Properties. *Biochemistry* **2012**, *51*, 5004–5013. [CrossRef] [PubMed]
53. Pfefferkorn, C.M.; Lee, J.C. Tryptophan Probes at the Alpha-Synuclein and Membrane Interface. *J. Phys. Chem. B* **2010**, *114*, 4615–4622. [CrossRef]
54. Gosse, J.A.; Wagenknecht-Wiesner, A.; Holowka, D.; Baird, B. Transmembrane Sequences Are Determinants of Immunoreceptor Signaling. *J. Immunol.* **2005**, *175*, 2123–2131. [CrossRef] [PubMed]
55. Wilkes, M.M.; Wilson, J.D.; Baird, B.; Holowka, D. Activation of Cdc42 Is Necessary for Sustained Oscillations of Ca2+and PIP2 Stimulated by Antigen in RBL Mast Cells. *Biol. Open* **2014**, *3*, 700–710. [CrossRef] [PubMed]
56. Wilson, J.D.; Shelby, S.A.; Holowka, D.; Baird, B. Rab11 Regulates the Mast Cell Exocytic Response. *Traffic* **2016**, *17*, 1027–1041. [CrossRef] [PubMed]
57. Ramezani, M.; Wilkes, M.M.; Das, T.; Holowka, D.; Eliezer, D.; Baird, B. Regulation of Exocytosis and Mitochondrial Relocalization by Alpha-Synuclein in a Mammalian Cell Model. *NPJ Parkinsons Dis.* **2019**, *5*, 12. [CrossRef]
58. Schindelin, J.; Arganda-Carreras, I.; Frise, E.; Kaynig, V.; Longair, M.; Pietzsch, T.; Preibisch, S.; Rueden, C.; Saalfeld, S.; Schmid, B.; et al. Fiji: An Open-Source Platform for Biological-Image Analysis. *Nat. Methods* **2012**, 9. [CrossRef]
59. Milovanovic, D.; Wu, Y.; Bian, X.; De Camilli, P. A Liquid Phase of Synapsin and Lipid Vesicles. *Science* **2018**, *361*, 604–607. [CrossRef]
60. Roy, R.; Hohng, S.; Ha, T. A Practical Guide to Single-Molecule FRET. *Nat. Methods* **2008**, *5*, 507–516. [CrossRef]
61. Li, C.H.; Tam, P.K.S. An Iterative Algorithm for Minimum Cross Entropy Thresholding. *Pattern Recognit. Lett.* **1998**, *19*, 771–776. [CrossRef]
62. Snead, D. *Mechanisms of Inhibition of Synaptic Vesicle Exocytosis by Complexin*; Weill Medical College of Cornell University: New York, NY, USA, 2015.
63. Blaber, M.; Zhang, X.J.; Matthews, B.W. Structural Basis of Amino Acid Alpha Helix Propensity. *Science* **1993**, *260*, 1637–1640. [CrossRef] [PubMed]
64. Pace, C.N.; Scholtz, J.M. A Helix Propensity Scale Based on Experimental Studies of Peptides and Proteins. *Biophys. J.* **1998**, *75*, 422–427. [CrossRef]
65. Marius Clore, G.; Iwahara, J. Theory, Practice, and Applications of Paramagnetic Relaxation Enhancement for the Characterization of Transient Low-Population States of Biological Macromolecules and Their Complexes. *Chem. Rev.* **2009**, *109*, 4108–4139. [CrossRef] [PubMed]
66. Clore, G.M. *Practical Aspects of Paramagnetic Relaxation Enhancement in Biological Macromolecules*, 1st ed.; Qin, P.Z., Warncke, K., Eds.; Academic Press: Cambridge, MA, USA, 2015.
67. Rao, J.N.; Kim, Y.E.; Park, L.S.; Ulmer, T.S. Effect of Pseudorepeat Rearrangement on Alpha-Synuclein Misfolding, Vesicle Binding, and Micelle Binding. *J. Mol. Biol.* **2009**, *390*, 516–529. [CrossRef] [PubMed]
68. Fusco, G.; De Simone, A.; Gopinath, T.; Vostrikov, V.; Vendruscolo, M.; Dobson, C.M.; Veglia, G. Direct Observation of the Three Regions in α-Synuclein That Determine Its Membrane-Bound Behaviour. *Nat. Commun.* **2014**, *5*, 1–8. [CrossRef] [PubMed]

69. Cholak, E.; Bugge, K.; Khondker, A.; Gauger, K.; Pedraz-Cuesta, E.; Pedersen, M.E.; Bucciarelli, S.; Vestergaard, B.; Pedersen, S.F.; Rheinstädter, M.C.; et al. Avidity within the N-Terminal Anchor Drives α-Synuclein Membrane Interaction and Insertion. *FASEB J.* **2020**, 1–21. [CrossRef]
70. Lytton, J.; Westlin, M.; Hanley, M.R. Thapsigargin Inhibits the Sarcoplasmic or Endoplasmic Reticulum Ca-ATPase Family of Calcium Pumps. *J. Biol. Chem.* **1991**, *266*, 17067–17071. [CrossRef]
71. Chen, Y.; Barkley, M.D. Toward Understanding Tryptophan Fluorescence in Proteins. *Biochemistry* **1998**, *37*, 9976–9982. [CrossRef]
72. Ramakrishnan, M.; Jensen, P.H.; Marsh, D. Alpha-Synuclein Association with Phosphatidylglycerol Probed by Lipid Spin Labels. *Biochemistry* **2003**, *42*, 12919–12926. [CrossRef]
73. Hoffmann, C.; Sansevrino, R.; Morabito, G.; Logan, C.; Vabulas, R.M.; Ulusoy, A.; Ganzella, M.; Milovanovic, D. Synapsin Condensates Recruit Alpha-Synuclein. *J. Mol. Biol.* **2021**, *433*, 166961. [CrossRef] [PubMed]
74. Navarro-Paya, C.; Sanz-Hernandez, M.; De Simone, A. Plasticity of Membrane Binding by the Central Region of α-Synuclein. *Front. Mol. Biosci.* **2022**, *9*, 1–9. [CrossRef] [PubMed]
75. Ulmer, T.S.; Bax, A. Comparison of Structure and Dynamics of Micelle-Bound Human Alpha-Synuclein and Parkinson Disease Variants. *J. Biol. Chem.* **2005**, *280*, 43179–43187. [CrossRef] [PubMed]
76. Bussell, R.; Eliezer, D. Effects of Parkinson's Disease-Linked Mutations on the Structure of Lipid-Associated Alpha-Synuclein. *Biochemistry* **2004**, *43*, 4810–4818. [CrossRef] [PubMed]
77. Fawzi, N.L.; Ying, J.; Ghirlando, R.; Torchia, D.A.; Clore, G.M. Atomic-Resolution Dynamics on the Surface of Amyloid-β Protofibrils Probed by Solution NMR. *Nature* **2011**, *480*, 268–272. [CrossRef]
78. Busch, D.J.; Oliphint, P.A.; Walsh, R.B.; Banks, S.M.L.; Woods, W.S.; George, J.M.; Morgan, J.R. Acute Increase of α-Synuclein Inhibits Synaptic Vesicle Recycling Evoked during Intense Stimulation. *Mol. Biol. Cell* **2014**, *25*, 3926–3941. [CrossRef]
79. Ben Gedalya, T.; Loeb, V.; Israeli, E.; Altschuler, Y.; Selkoe, D.J.; Sharon, R. Alpha-Synuclein and Polyunsaturated Fatty Acids Promote Clathrin-Mediated Endocytosis and Synaptic Vesicle Recycling. *Traffic* **2009**, *10*, 218–234. [CrossRef]
80. Schechter, M.; Atias, M.; Abd Elhadi, S.; Davidi, D.; Gitler, D.; Sharon, R. α-Synuclein Facilitates Endocytosis by Elevating the Steady-State Levels of Phosphatidylinositol 4,5-Bisphosphate. *J. Biol. Chem.* **2020**, *295*, 18076–18090. [CrossRef]
81. Vargas, K.J.; Makani, S.; Davis, T.; Westphal, C.H.; Castillo, P.E.; Chandra, S.S. Synucleins Regulate the Kinetics of Synaptic Vesicle Endocytosis. *J. Neurosci.* **2014**, *34*, 9364–9376. [CrossRef]
82. Scott, D.; Roy, S. α-Synuclein Inhibits Intersynaptic Vesicle Mobility and Maintains Recycling-Pool Homeostasis. *J. Neurosci.* **2012**, *32*, 10129–10135. [CrossRef]
83. Burré, J.; Sharma, M.; Tsetsenis, T.; Buchman, V.; Etherton, M.R.; Südhof, T.C. Alpha-Synuclein Promotes SNARE-Complex Assembly in Vivo and in Vitro. *Science* **2010**, *329*, 1663–1667. [CrossRef] [PubMed]
84. Logan, T.; Bendor, J.; Toupin, C.; Thorn, K.; Edwards, R.H. α-Synuclein Promotes Dilation of the Exocytotic Fusion Pore. *Nat. Neurosci.* **2017**, *20*, 681–689. [CrossRef] [PubMed]
85. Darios, F.; Ruipérez, V.; López, I.; Villanueva, J.; Gutierrez, L.M.; Davletov, B. Alpha-Synuclein Sequesters Arachidonic Acid to Modulate SNARE-Mediated Exocytosis. *EMBO Rep.* **2010**, *11*, 528–533. [CrossRef] [PubMed]
86. Lai, Y.; Kim, S.; Varkey, J.; Lou, X.; Song, J.-K.; Diao, J.; Langen, R.; Shin, Y.-K. Nonaggregated α-Synuclein Influences SNARE-Dependent Vesicle Docking via Membrane Binding. *Biochemistry* **2014**, *53*, 3889–3896. [CrossRef]
87. Eliezer, D. Protein Folding and Aggregation in in Vitro Models of Parkinson's Disease: Structure and Function of α–Synuclein. In *Parkinson's Disease: Molecular and Therapeutic Insights from Model Systems*; Nass, R., Przedborski, S., Eds.; Academic Press: New York, NY, USA, 2008; pp. 575–595.
88. Georgieva, E.R.; Ramlall, T.F.; Borbat, P.P.; Freed, J.H.; Eliezer, D. The Lipid-Binding Domain of Wild Type and Mutant Alpha-Synuclein: Compactness and Interconversion between the Broken and Extended Helix Forms. *J. Biol. Chem.* **2010**, *285*, 28261–28274. [CrossRef]
89. Dikiy, I.; Eliezer, D. Folding and Misfolding of Alpha-Synuclein on Membranes. *Biochim. Biophys. Acta* **2012**, *1818*, 1013–1018. [CrossRef]
90. Snead, D.; Eliezer, D. A-Synuclein Function and Dysfunction on Cellular Membranes. *Exp. Neurobiol.* **2014**, *23*, 292–313. [CrossRef]
91. Man, W.K.; Tahirbegi, B.; Vrettas, M.D.; Preet, S.; Ying, L.; Vendruscolo, M.; De Simone, A.; Fusco, G. The Docking of Synaptic Vesicles on the Presynaptic Membrane Induced by α-Synuclein Is Modulated by Lipid Composition. *Nat. Commun.* **2021**, *12*, 927. [CrossRef]
92. Fusco, G.; Pape, T.; Stephens, A.D.; Mahou, P.; Costa, A.R.; Kaminski, C.F.; Kaminski Schierle, G.S.; Vendruscolo, M.; Veglia, G.; Dobson, C.M.; et al. Structural Basis of Synaptic Vesicle Assembly Promoted by α-Synuclein. *Nat. Commun.* **2016**, *7*, 12563. [CrossRef]

*biomolecules*

MDPI

*Article*

# The Difference in Structural States between Canonical Proteins and Their Isoforms Established by Proteome-Wide Bioinformatics Analysis

**Zarifa Osmanli [1,2], Theo Falgarone [1], Turkan Samadova [2], Gudrun Aldrian [1], Jeremy Leclercq [1], Ilham Shahmuradov [2] and Andrey V. Kajava [1,*]**

[1] CRBM, Université de Montpellier, CNRS, 1919 Route de Mende, CEDEX 5, 34293 Montpellier, France
[2] Institute of Biophysics, ANAS, Baku AZ1141, Azerbaijan
* Correspondence: andrey.kajava@crbm.cnrs.fr

**Abstract:** Alternative splicing is an important means of generating the protein diversity necessary for cellular functions. Hence, there is a growing interest in assessing the structural and functional impact of alternative protein isoforms. Typically, experimental studies are used to determine the structures of the canonical proteins ignoring the other isoforms. Therefore, there is still a large gap between abundant sequence information and meager structural data on these isoforms. During the last decade, significant progress has been achieved in the development of bioinformatics tools for structural and functional annotations of proteins. Moreover, the appearance of the AlphaFold program opened up the possibility to model a large number of high-confidence structures of the isoforms. In this study, using state-of-the-art tools, we performed in silico analysis of 58 eukaryotic proteomes. The evaluated structural states included structured domains, intrinsically disordered regions, aggregation-prone regions, and tandem repeats. Among other things, we found that the isoforms have fewer signal peptides, transmembrane regions, or tandem repeat regions in comparison with their canonical counterparts. This could change protein function and/or cellular localization. The AlphaFold modeling demonstrated that frequently isoforms, having differences with the canonical sequences, still can fold in similar structures though with significant structural rearrangements which can lead to changes of their functions. Based on the modeling, we suggested classification of the structural differences between canonical proteins and isoforms. Altogether, we can conclude that a majority of isoforms, similarly to the canonical proteins are under selective pressure for the functional roles.

**Keywords:** isoform; large-scale analysis; protein structure; AlphaFold; canonical protein

**Citation:** Osmanli, Z.; Falgarone, T.; Samadova, T.; Aldrian, G.; Leclercq, J.; Shahmuradov, I.; Kajava, A.V. The Difference in Structural States between Canonical Proteins and Their Isoforms Established by Proteome-Wide Bioinformatics Analysis. *Biomolecules* **2022**, *12*, 1610. https://doi.org/10.3390/biom12111610

Academic Editors: Stefania Brocca, Keith Dunker, Sonia Longhi and Prakash Kulkarni

Received: 15 September 2022
Accepted: 27 October 2022
Published: 1 November 2022



## 1. Introduction

Alternative splicing is one of the principal sources of structural and functional diversity in the proteomes of multicellular organisms. It is a process that may include or exclude particular exons of a multi-exonic gene from its processed messenger RNA. Different combinations of exons can produce multiple mRNA isoforms of a single gene. It is estimated that up to 95% of human multi-exonic genes are alternatively spliced [1,2]. The average number of splice variants per human gene is equal to four [3]. All this can drastically increase the number of different proteins in the proteome. Today, most genome-wide information about alternative splicing is generated on the nucleic acid level thanks to high-throughput data such as expressed sequence tags (ESTs) [4], microarrays [5], and RNA-seq data [6]. However, not all splicing variants are expressed as functional proteins. Although a very large number of alternatively spliced variants are detected in RNA-seq studies, large-scale mass spectrometry-based proteomics analyses detect only a small fraction of alternative isoforms on the protein level [7]. One of today's problems in this area is to establish the real number of splice variants that appear as functional proteins for each gene. In addition to the application of genome-wide mass spectrometry analyses, researchers pay special attention to the protein isoforms with the most cross-species conservation and those that are able to maintain protein structure integrity [1,8–10].

Although the way to obtain the exact set of real protein variants may take some time, the data already available thanks to a combination of approaches (proteomics, cross-species conservation, and 3D mapping) can be used for the subsequent structural and functional annotations. Today, high-quality collections of protein isoforms are stored in UniProt, NCBI RefSeq, Ensembl databanks [11–13], and in more specific ones such as APPRIS, ISOexpresso, and ASES [14–16].

Another important point is the existence of a single main protein isoform among several protein variants for each gene, which is called principal isoform or canonical protein. The canonical protein is identified by several criteria: experimental data on its functional role; data about its expression in different tissues of an organism; existence of the same combination of exons in orthologous proteins and in different curated databases. Although, in the annotated databases of proteomes [11–13] many canonical proteins are well distinguished from their isoforms, some of them are still poorly annotated.

Depending on the proteomes and quality of their annotation, the number of isoforms usually exceeds the amount of canonical proteins 2–3 times [11,17]. At the same time, if to compare the number of proteins with the available experimental structural information, the situation is opposite. Almost all proteins in the Protein Data Bank [18] are canonical. Thus, due to a large gap between abundant sequence information and meager structural data on the isoforms, there is a growing interest in assessing the structural states and functional roles of alternative protein isoforms. As we have already mentioned, the sequence data on the isoforms are abundant. Therefore, if we want to get a global view of the structural-functional difference between the canonical proteins and their isoforms, apparently, the most appropriate approach is bioinformatics rather than the time-consuming experimental methods. In line with this need, during the last decade, significant progress has been achieved in the development of bioinformatics tools for large-scale structural and functional annotations of proteins. In the early days of structural bioinformatics, the foremost efforts of researchers were devoted to proteins with globular 3D structures. However, today, it is becoming clear that non-globular protein regions, having either intrinsically disordered conformations, membrane domains, elongated structures with tandem repeats or being aggregation-prone also have important functional roles [19–21]. Thus, an accurate structural and functional prediction of protein molecule can only be achieved when accounting for all these structural states. Recently, in line with this need, we developed a computational pipeline called TAPASS, which was designed to do just that [20]. The TAPASS pipeline is using known cutting-edge predictors able to detect intrinsically disordered regions (IDRs), transmembrane regions, signal peptides, conserved structured domains, short linear motifs (SLiMs) and aggregation-prone regions in protein sequences. The main novelty of this tool is a more precise prediction of aggregation-prone regions by taking into consideration the other known or predicted structural states. Moreover, the appearance of the AlphaFold program [22] opened up the possibility to model a large number of high-confidence structures of the isoforms. This artificial intelligence program, in a short time, became the gold standard computational technique for prediction of the 3D structure of proteins based on their sequence thanks to its accuracy competitive with experimental structures in a majority of cases.

In this study, by taking advantage of these state-of-the-art bioinformatics tools, we systematically compared the structural states of canonical proteins and isoforms. The analysis was performed on a large scale using 58 eukaryotic proteomes and provided a global view of the prevalence of each of these types of structures in canonical and isoform sets. Moreover, in some cases, our analysis proposed functional implications caused by structural changes of the isoforms as well as the possibility of selective evolutionary pressure, to which they can be exposed for functional roles.

## 2. Materials and Methods

### *2.1. Construction of Datasets of Canonical Proteins and Their Isoforms*

#### 2.1.1. Main Dataset

Construction of properly divided large datasets of canonical proteins and their isoforms represents a challenge because some proteins are still poorly annotated. To obtain large subsets of canonical proteins and their isoforms, we retrieved corresponding sequences from reference proteomes of 58 eukaryotic species (Supplementary Table S1) by using July 2020 release of UniProt databank [11]. Our choice was justified by the fact that UniProt contains a large combined set of several databases. The UniProt uses the following criteria to identify the canonical proteins: experimental data on their functional role; data about their expression in different tissues of an organism; existence of the same combination of exons in orthologous proteins and in different curated databases (https://www.uniprot.org/help/canonical_and_isoforms (accessed on 25 August 2020)). First, we used option "Download all (FASTA (canonical & isoform)" to get 1,906,397 sequences including both canonical proteins and their isoforms. Second, we used "Download one protein sequence per gene" option to obtain a better-defined set of 1,244,044 canonical proteins. To avoid redundancy, we clustered the isoforms by CDhit [23] and removed the identical ones. This gave us 661,745 isoforms. Then we selected those isoform sequences, which had the same gene IDs as proteins from the canonical set and were highly similar BLAST (e-value < $10^{-35}$) with them [24]. As a result, we obtained a dataset of 263,475 canonical proteins and 565 942 isoforms, which was used in our analysis (Supplementary Table S2).

#### 2.1.2. Dataset of Proteins from Cancer-Related Genes with Well-Documented Expression Levels

Not all proteins from the UniProt databank have information about their expression level. Therefore, we built a smaller set of canonical proteins and corresponding isoforms of human cancer-related genes with well-documented expression levels in both 22 normal and cancer tissues. For this purpose, we used ISOexpresso database [15]. Our dataset contains 82 canonical and 166 isoform proteins, which were used for evaluation of the correlation between aggregation and expression level of proteins.

#### 2.1.3. Datasets for Estimation of the Structural Difference in Isoforms by Using AlphaFold Modeling

To evaluate the structural changes caused by the differences in the sequences (hereafter referred to as difference regions) of the corresponding canonical and isoform proteins, we used pairs of proteins with the difference regions inside well-conserved structured domains. For this purpose, we chose human proteins annotated in SwissProt [25] and having evidence of existence at the protein level (PE = 1). The conserved structural domains were detected by using HMM library of the CATH databank [26]. In the next step, we selected CATH domains that overlapped with the difference regions. A CATH domain found in a canonical protein may be shortened in the isoform so that the remaining domain is not able to fold. Therefore, we considered only isoforms where (1) the canonical CATH domain is shorter than 200 aa, and at least 70% of the domain remains in the isoform, or (2) the canonical domain is longer than 200 aa, and at least 50% of the domain remains in the isoform. For the modeling, we subsequently selected 168 canonical proteins with 223 corresponding isoforms where the difference regions were longer than 20 AA and located inside the CATH domains. Finally, to select the most conserved and studied domains, we ran the 168 canonical proteins by local BLASTP against PDB sequences from 7 species (*P. troglodytes*, *B. taurus*, *M. musculus*, *R. norvegicus*, *D. rerio*, *D. melanogaster*, *C. elegans*) and kept only those having more than 10 hits with e-value < $10^{-6}$. As a result, we obtained 53 canonical human proteins with 63 corresponding isoforms for the prediction by the AlphaFold program.

Subsequently, the 3D structures of the isoforms were predicted by AlphaFold Colab [27]. The structural models of the canonical proteins were obtained from the AlphaFold database (https://alphafold.com/download#proteomes-section (accessed on 10 May 2022)). The obtained structural models were analyzed by using PyMol [28].

*2.2. Bioinformatics Tools Used to Annotate Structural States of Proteins*

To annotate the structural states of proteins, we used the TAPASS pipeline, which includes several prediction tools. Structured domains were predicted by using HMM libraries (e-value < $10^{-3}$) of CATH. Intrinsically disordered regions were detected by IUPred [29] and an in-house BISMM filter, which chooses hydrophilic regions greater than 75% and proline-rich regions more than 25%. Signal peptide and transmembrane regions were predicted with SignalP and TMHMM, respectively [30,31]. The tool also predicts amyloidogenic regions (aggregation-prone motifs) by ArchCandy2.0 [32], TANGO [33], and PASTA 2.0 [34] with their default parameters. We detected short linear motifs (SLiMs) of degradation (degrons) by using motifs collected in the Eukaryotic Linear Motif (ELM) resource [35].

*2.3. Detection of Structural Changes in and around the Difference Regions*

All types of difference regions (insertion, deletion, non-identical, and mixed) can cause structural changes not only in the place of their location but also in the flanking regions with identical sequences. Most of the methods used in the TAPASS for structural annotation of canonical and isoform proteins detected these changes automatically. However, cases when deletions truncated CATH domains required additional rules (see Section 2.1.3). The application of these rules in our analysis affected the prediction of structured/unstructured regions and exposed aggregation-prone regions (EARs).

*2.4. Analysis of Tandem Repeats in Canonical Proteins and Isoforms*

Tandem repeat regions were identified by MetaRepeatFinder (MRF) (https://bioinfo.crbm.cnrs.fr/index.php?route=tools&tool=15 (accessed on 6 July 2022)) [36] tool in five proteomes (*H. sapiens*, *M. musculus*, *D. melanogaster*, *D. rerio*, *A. thaliana*). From several tandem repeat finders of MRF, we chose Regex, T-REKS [37], and TRUST [38], which are specialized in the detection of TRs with units of less than 3 residues, less than 15 residues, and more than 15 residues, respectively. As a result, the combination of these finders detects all types of tandem repeats. The overlap between the "difference" region and the TR region was counted if they had at least one common residue.

## 3. Results and Discussion

*3.1. Identification, Classification, and Distribution of Difference Regions*

Difference in the sequences of canonical proteins and their isoforms is quite specific in comparison with the differences between orthologous/paralogous proteins. Frequently, the differences between the orthologues represent point mutations and (or) short indels spread over the proteins. While canonical proteins and their isoforms always have a region(s) with identical sequences (corresponding to the same exons) and relatively long fragments where sequences can be completely different (Figure 1). To detect the difference regions, we generated pairwise alignments of canonical-isoform proteins by using Clustal Omega [39] and treated them by our in-house script (Supplementary Data S1).

We classified the differences between the canonical-isoform pairs into four groups choosing as a starting point canonical sequence: insertion, deletion, non-identical and mixed (Figure 1). The "non-identical" regions have different sequences of the same length. "Mixed" regions are those that have both amino acid substitutions and indels in the difference region. Sometimes, these regions also include identical regions shorter than 10aa.

The analysis showed that the "mixed" difference region is the most common case, followed by the deletions (Figure 1B). At the same time, a more detailed analysis of the "mixed" cases showed that it also contains a significant amount of deletions (68.6% of positions have deletions, 15.4% insertions, and 16% amino acids). Because of the frequent deletions, on average, the isoforms are shorter in length than canonical proteins (Figure 1C).

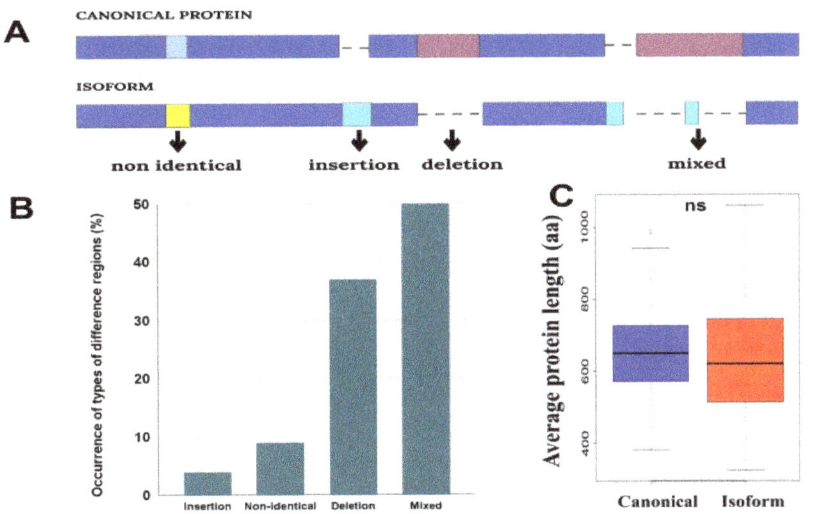


**Figure 1.** (**A**) Schematic representation of four groups of difference regions (dark blue and pink colors indicate identical and non-identical regions in the sequences, respectively). (**B**) Occurrence of types of the difference regions. (**C**) Distributions of the average length of canonical proteins and isoforms in proteomes. The distributions contain 58 points corresponding to the average length of each proteome. Here, ns means non-significant difference with $p$-value > 0.05.

### 3.2. Distribution of Structured and Unstructured Regions

Previous studies suggested that isoform proteins have a higher coverage of unstructured regions in comparison to canonical proteins [40–42]. This conclusion suggested a lower level of involvement of isoforms in functional activity than of canonical ones. We examined this conclusion by using our datasets and the TAPASS pipeline [20] (see Section 2.1.3). Our analysis showed that the proportion of proteins containing unstructured regions is slightly higher in the isoform set (Figure 2). The same tendency was observed when we compared the coverage of unstructured regions in proteins. At the same time, both of these differences were not statistically significant. Thus, our results do not confirm the previous conclusions about the higher number of unstructured residues in isoforms, rather suggesting that the canonical proteins and their isoforms have the same ratio of residues in structured/unstructured states. This also suggests that during evolution, isoforms preserve their structural domains, which play functional roles (Supplementary Table S3).

### 3.3. Changes in Subcellular Localization

To understand the functional role of a protein, it is important to know where it resides in the cell. There are a number of bioinformatics tools that can accurately predict the outcome of protein targeting in four major subcellular localizations: secreted proteins can be identified by SignalP [30], transmembrane regions (more exactly transmembrane helices) by TMHMM [31], nuclear proteins with nuclear localization signals can be found by regular expressions [35], and the remaining proteins as a rough approximation can be considered as cytosolic.

Our analysis of the proportion of proteins with signal peptide showed that it is significantly lower in isoforms than in canonical proteins (Figure 3A). It suggests that in some cases, the isoforms may maintain their globular functional domains but change their cellular localization from extracellular to cytosolic. A similar tendency was observed with the canonical proteins containing transmembrane helices (Figure 3B). Moreover, we found that the proportion of the nuclear localization signals in isoforms is significantly higher in comparison with canonical proteins. It indicates that isoforms are more often localized in the nucleus than canonical proteins (Figure 3C). The proportion of canonical proteins with transmembrane helices is higher than in isoforms, suggesting that a noticeable part of the isoforms loses their transmembrane localization. Parts of the difference regions that gain and lose signal peptides represent 2% and 4%, respectively. For the transmembrane helices, it is 2% and 7%. These changes may have important functional implications (Supplementary Table S3).

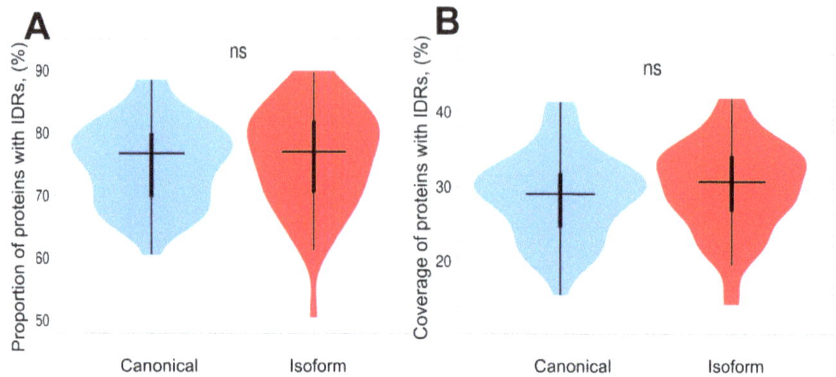


**Figure 2.** Violin plots of proportion and coverage of proteins containing IDRs in canonical and isoform proteins. The distributions contain 58 points corresponding to each proteome. (**A**) Proportion of proteins with IDRs in canonical proteins and isoforms. The difference between 2 sets is non-significant. (**B**) Coverage of IDRs in canonical proteins and isoforms. The coverage in isoforms is slightly higher; however, this difference is non-significant.

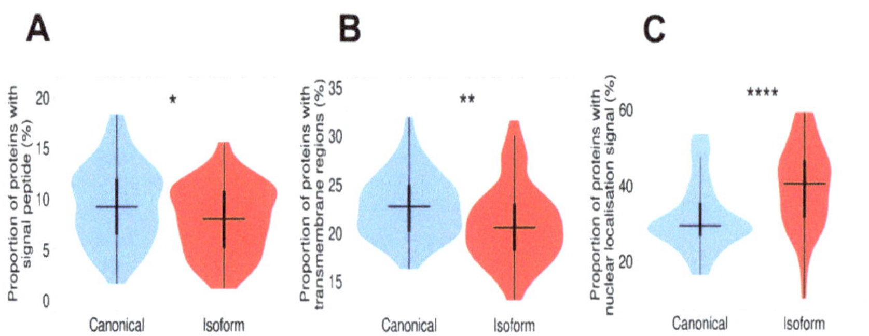


**Figure 3.** Difference in subcellular localization between canonical proteins and isoforms. (**A**) Proportion of proteins containing signal peptides. This value is significantly higher in canonical proteins than in isoforms. (**B**) Proportion of proteins containing transmembrane regions. The plot demonstrates a significant decrease in transmembrane proteins in the isoform set. (**C**) Proportion of proteins with nuclear localization signal. Isoforms have a remarkably high proportion of nuclear localization signals in comparison with canonical proteins. Signes *, **, **** mean significant differences with $p$-value < 0.05, $p$-value < 0.01, and $p$-value < 0.0001, respectively.

### 3.4. Proportion of Aggregation-Prone Regions

Proteins are usually soluble and easily degraded by proteases after having performed their functions. However, some of them, depending on the amino acid sequence and at certain conditions, can assemble into stable, protease-resistant aggregates. These aggregates are linked to serious diseases, which include, but are not limited to, Alzheimer's disease, Parkinson's disease, type II diabetes, and rheumatoid arthritis [43]. Moreover, protein aggregation can be "functional" and play a central role in liquid–liquid phase separation (LLPS), a process that leads to the formation of membrane-less organelles [44,45]. Several computational programs for the prediction of protein aggregation have been developed [46]. The most realistic evaluation of the aggregation potential requires the prediction of motifs located within unstructured regions and being aggregation-prone, which we call "Exposed Aggregation-prone Regions" (EARs) [20]. Here, we analyzed the EARs in canonical proteins and isoforms. Our interest in this analysis was also because, in general, canonical proteins have a higher level of cellular expression in comparison with their isoforms. It is reasonable to assume that to avoid aggregation, canonical proteins with a higher expression level may have a lower aggregation potential. The other reason for the higher aggregation potential of the isoforms may be the truncation of native globular domains and the unfolding of their remaining parts. For example, it was shown that the p53 isoform Δ133p53β, which is critical in promoting cancer activity, is regulated through an aggregation-dependent mechanism [41]. The analyses of the truncated DNA-binding domain of Δ133p53β suggest that its remaining part is most probably unfolded and contains the EARs.

We estimated an average aggregation potential of canonical proteins and isoforms by the proportion of EAR-containing proteins predicted by one of the predictors (ArchCandy, Pasta, Tango) in these two datasets. Our analysis revealed that the median value of proportion for isoforms with EARs is almost the same as for canonical proteins (Figure 4 and Supplementary Table S3).

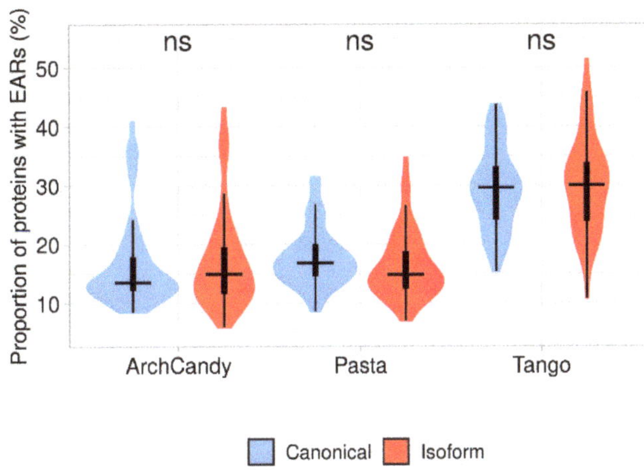


**Figure 4.** Proportion of EAR-containing proteins in canonical and isoform proteomes predicted by three tools (ArchCandy, Pasta, Tango). Differences between canonical proteins and isoforms are non-significant.

Although it is accepted that canonical proteins have higher expression levels than the isoforms [7,47], most proteins from our main dataset do not have reliable information about their expression level. Therefore, we also analyzed smaller sets with 82 canonical and 166 isoform proteins of human cancer genes with well-documented expression levels in normal and cancer tissues (Supplementary Tables S4 and S5). These sets were used for evaluation of the correlation between aggregation and expression level of the proteins. The results confirm that the average expression level of canonical proteins is significantly higher than that of their isoforms. We also compared the relationship between the expression level and aggregation potential of proteins in normal and cancer cells. The results of the analysis are shown in Figure 5. The expression of canonical proteins is higher in both normal and cancer cells. At the same time, the expression level of all proteins slightly decreases in cancer cells. We also found that the proteins with EARs are expressed less in both normal and cancer cells than the ones without EARs. These results are in agreement with the assumption that to avoid aggregation, proteins with a higher expression level may have a lower aggregation potential.

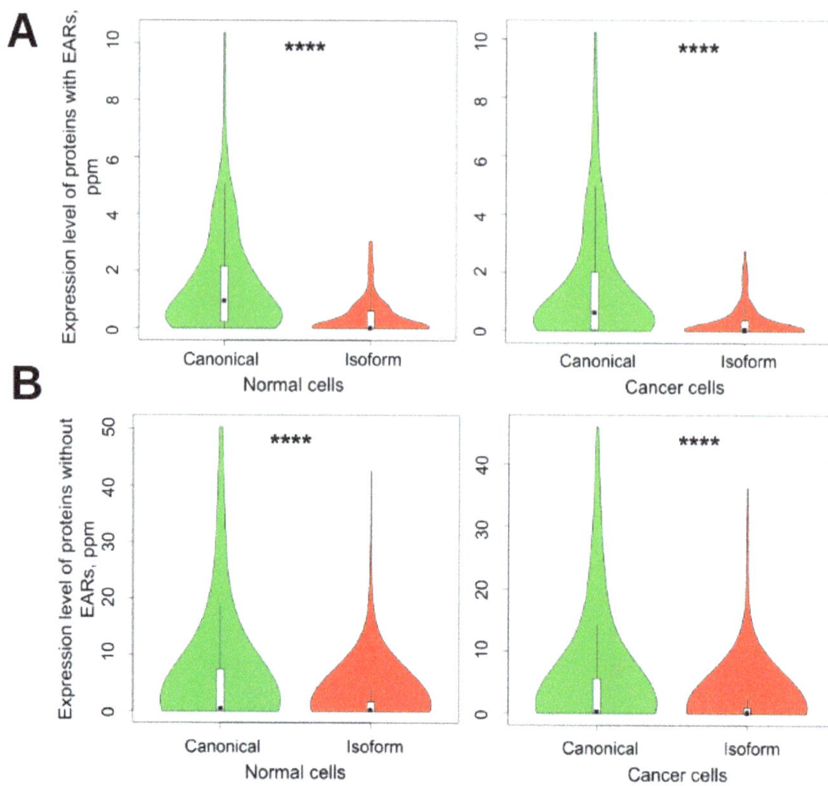


**Figure 5.** Violin plots of expression of canonical proteins and their isoforms in normal and cancer cells. (**A**) EAR-containing proteins and (**B**) non-EAR-containing proteins. EARs were predicted by using the ArchCandy program. Mean levels of expression for EAR-containing canonical proteins and isoforms in normal cells were 1.565 and 0.386, respectively, and in cancer cells, 1.490 and 0.306. For non-EAR-containing proteins, these values were 5.784, 1.773, and 4.984, 1.499, respectively. In accordance with *t*-test, all results were significant, with *p*-values of less than $10^{-13}$. **** means significant difference with *p*-value < 0.0001.

### 3.5. Canonical Proteins Have More Degradation Motifs Than Their Isoforms

The abundance of proteins in the cell mostly depends on the balance of two opposite processes: expression and degradation. In general, canonical proteins have a higher level of cellular expression in comparison with their isoforms. It was interesting to understand if there is any difference between these proteins in terms of their degradation. The experimental data on protein degradation is limited and controversial. We compared canonical and isoform proteins in silico by analyzing the occurrence of degron motifs by TAPASS [20]. The degrons are short linear motifs that increase the targeting of proteins for degradation [48,49]. We found that canonical proteins have a higher proportion of degrons in comparison to the isoforms and this difference is statistically significant (Figure 6 and Supplementary Table S6).

If the more frequent occurrence of degrons in the canonical proteins causes their higher degradation rate in comparison with the isoforms, this may decrease the difference in the abundance between canonical proteins and isoforms. In its turn, a similar level of abundance may explain almost the same proportion of the aggregation-prone proteins predicted (Figure 4) for the canonical and isoform sets.

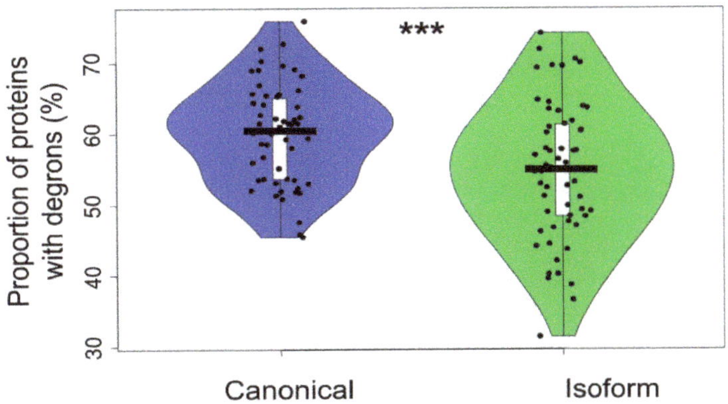


**Figure 6.** Proportion of canonical proteins and isoforms with degrons predicted by using SLiMs ($t$-test $p$-value = 0.00071). The distributions contain 58 points corresponding to each proteome. The proportion of degron-containing proteins is significantly higher in the canonical set than in the isoform one. Here, *** means significant difference with $p$-value < 0.001.

### 3.6. Occurrence of Tandem Repeats in Canonical Proteins and Isoforms

Many protein sequences contain arrays of repeats that are adjacent to each other [50,51] tandem repeats (TRs). *Several authors* have proposed that TRs might have evolved by exon duplication and rearrangement [52,53]. Therefore, it was interesting to get insight into the difference between canonical proteins and isoforms in these particular regions. We detected TRs in five well-annotated proteomes (*H. sapiens*, *M. musculus*, *D. melanogaster*, *D. rerio*, *A. thaliana*) by using MetaRepeatFinder (MRF) (https://bioinfo.crbm.cnrs.fr/index.php?route=tools&tool=15 (accessed on 6 July 2022)). These proteomes contain a total of 44,357 canonical proteins. We found that a large part (43%) of them contains at least one TR region, and each TR-containing protein has, on average, about two TR regions. A comparison of the occurrence of the TR regions in canonical proteins and isoforms revealed that isoforms have fewer TR regions than canonical proteins (0.5 vs. 0.81 TR region per protein) (Figure 7A). It is especially noticeable for TRs with a repeat length of 4–10 residues (Figure 7B). Partially, the decrease in TRs in the isoforms can be explained by the fact that among the differences between canonical proteins and isoforms, we predominantly observed deletions (see Section 3.1). It was interesting to study the relationship between the location of the TRs and the difference regions. Our analysis showed that among the difference regions detected in the aligned pairs, a significant part (35%) overlaps with TRs.

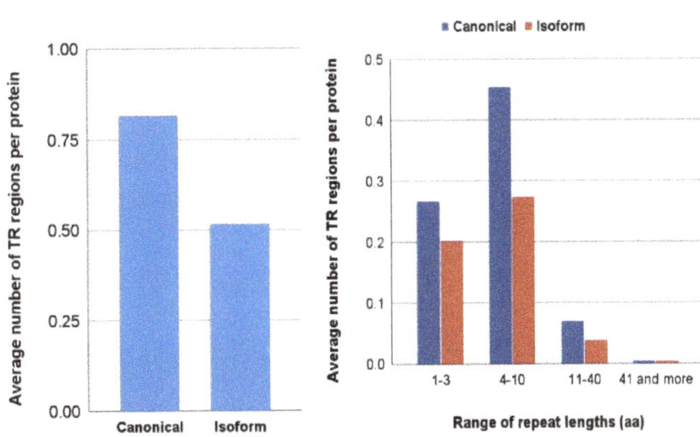


**Figure 7.** (**A**) Average number of tandem repeat regions determined per protein by MRF tool; (**B**) Distribution of proteins with tandem repeat by the length of their repetitive units.

### *3.7. Differences within the 3D Structures of Canonical Proteins and Isoforms Predicted by AlphaFold*

Our proteome-wide analysis provides a global view of the canonical-isoform protein difference. At the same time, it is also interesting to investigate these changes from within the 3D structures down to the atomic details. In orthologous and paralogous proteins, the difference in the amino acid sequences of more than 30% of identity may guarantee the same structural fold [54]. However, the character of the differences between canonical and isoform sequences is quite specific. They are identical at the location of the same exons; however, in the places of alternative splicing, they can have completely different sequences. This "mosaic" arrangement may trigger significant structural and functional changes.

Given the fact that almost all proteins with experimentally determined 3D structures are canonical, the comparison requires molecular modeling of isoform structures. Previously, this type of modeling of the isoform structures and their comparison with the structures of the corresponding canonical proteins was described for some particular proteins [10]. Today, with the development of an artificial intelligence program called AlphaFold [22], the scientific community got an opportunity to build high-quality structural models on a large scale. Here, we applied the AlphaFold program to obtain structural models of the isoform proteins. It was especially interesting to examine cases when the difference regions between the isoform and canonical proteins are conserved in several organisms and located within well-conserved structured domains. For the modeling, we used human proteins. To evaluate the cross-species conservation, we used seven species from the Animal Kingdom (*P. troglodytes*, *B. taurus*, *M. musculus*, *R. norvegicus*, *D. rerio*, *D. melanogaster*, *C. elegans*). We considered that AlphaFold structural models are reliable when their level of confidence (pLDDT) was higher than 70%, they did not have disallowed backbone conformations, and the inside residues of the structure were predominantly apolar and did not have charged residues, which were not involved in the ionic bonds. The detection of unstructured regions was based on criteria used in TAPASS [20]. Several isoforms had difference regions outside of the well-conserved structured domains, while inside these domains, they were identical to each other. Each group of these isoforms was reduced to one representative case. As a result, we compared the 3D structures of 50 canonical human proteins with 51 structural models of the corresponding isoforms predicted by AlphaFold. This allowed us to classify the 3D structure transformations into four subgroups.

#### 3.7.1. Exon Deletions with the Preservation of the Overall Structure

*Proteins with tandem repeats*

Though most of the selected proteins have globular structures, non-globular structures built of tandem repeats were found in 26% (13 of 51) of the cases. In the analyzed proteins with the difference regions inside of the complete structure, the most frequent situation is the deletion of one repetitive unit. As a rule, these changes (also with any integer number of the repeats) do not cause serious structural perturbations (Figure 8A). These cases are observed in proteins with tandem repeats from Class III, IV, and V [51,54,55]. In a few cases, the difference regions do not have an integer number of repeats. This could lead to structural changes if this difference is located in the middle of the repetitive structure. However, the isoform models showed that the change in the loop size between the repeats preserves the integrity of the whole structure (Supplementary Data S2 and Figure S1). In other such cases, these difference regions are located at the terminal parts of the repetitive domains with no effect on the overall structure (Supplementary Data S2 and Figure S1). The described structural changes preserve the overall structure by creating patches of new surfaces that can lead to the modification of protein functions.

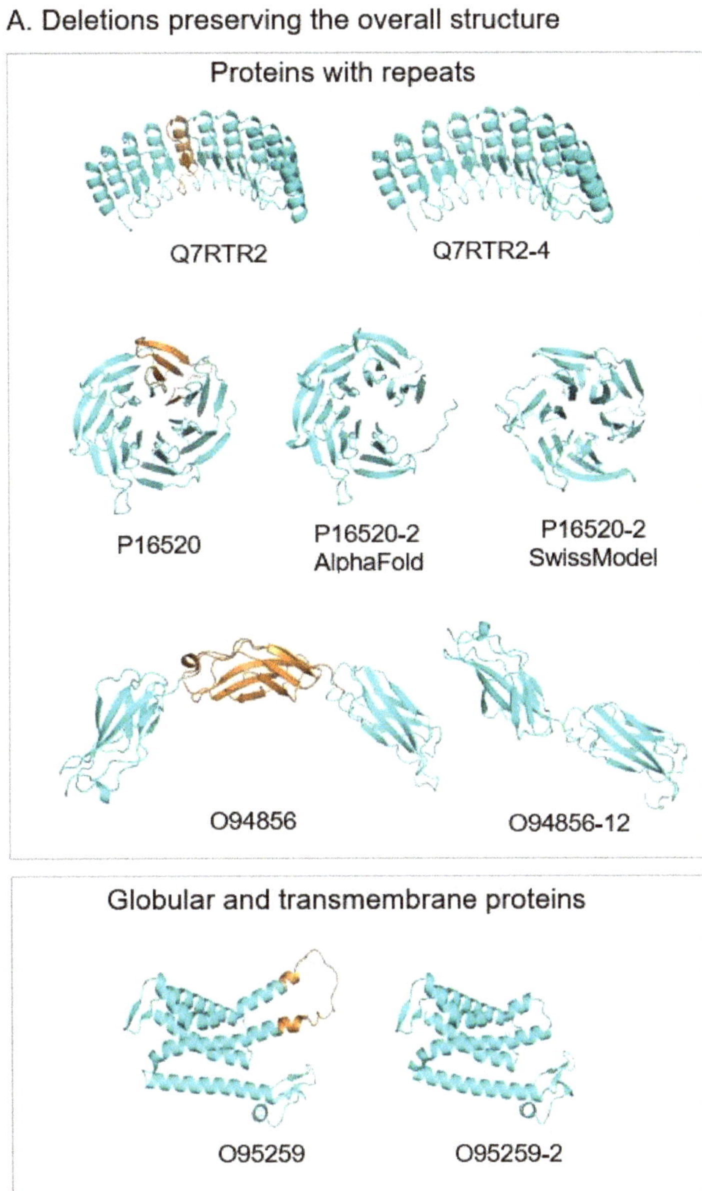


**Figure 8.** *Cont.*

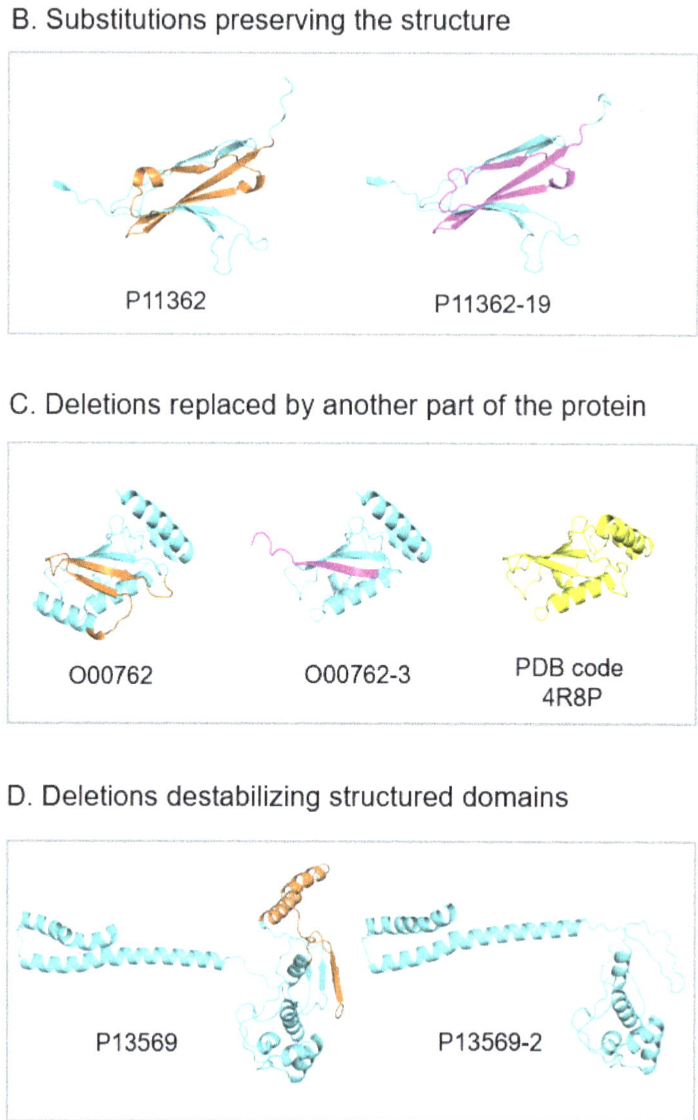


**Figure 8.** Ribbon representation of AlphaFold models of canonical proteins (left) and their isoforms (right). Fragments of canonical proteins deleted in the isoforms are in orange. Fragments of isoforms that substitute deleted fragments of the canonical proteins are in magenta. Representative structures of each subgroup from top to bottom are: (**A**). Deletions preserving the overall structure. Q7RTR2, LRR-protein of NLR family CARD domain-containing protein 3; P16520, 7-bladed beta-propeller of Guanine nucleotide-binding protein G(I)/G(S)/G(T) subunit beta-3. AlphaFold model of isoform represents 6-bladed structure with an open beta-propeller, SwissModel structure made based on the known 6-bladed structure (PDB code 1E1A) has closed beta-propeller; O94856, neurofascin; O95259, potassium voltage-gated channel subfamily H member 1; (**B**). Substitutions preserving the structure. P11362, fibroblast growth factor receptor 1; (**C**). Deletions replaced by another part of the protein. O00762, ubiquitin-conjugating enzyme E2 C, on the right, in yellow, the known crystal structure of ubiquitylation module similar to the truncated structure of the isoform in the center; (**D**). Deletions destabilizing structured domains. P13569, cystic fibrosis transmembrane conductance regulator.

*Globular proteins*

Among 51 analyzed pairs, there are 20 globular structures, representing 38% of the cases, with the deletions of exons in the middle of the structure. In most of these cases, the deletion does not lead to critical structural transformations (Figure 8A). In some cases, it makes shorter loops preserving α-helices or β-strands; sometimes, it removes one or several transmembrane helices. At the same time, these deletions can lead to changes in the binding properties of the isoforms and (or) changes in the oligomerization states of the protein [56].

### 3.7.2. Exon Substitutions That Preserve the 3D Structure

The other subgroup of four analyzed proteins (8% of the cases) is characterized by substitutions of exons. The size of the substituted exons is the same or almost the same, and the sequences of canonical and isoform variants are not identical but similar. AlphaFold suggests that the new exons of the isoforms fit the native structure. This does not change the overall structure but leads to local changes on the molecular surface. This can be a basis for the modification of protein functions [57] (Figure 8B).

### 3.7.3. Deletion That Is Substituted in the Structure by Another Part of the Molecule

We observed 6 of 51 cases (12%) where an exon deletion in the isoform removes a region that is critical for the structural integrity of the globular domain. In the AlphaFold model of the isoform, this part of the structure is filled by a new fragment, which, in the canonical protein, belongs to an unstructured region. This suggests that to provide structural diversity, proteins may have two or more neighboring regions. One is in the structure, and another is unstructured. If the first region is deleted in the isoform, the second one can dock into the structure, preserve it, and modify the function. (Figure 8C)

### 3.7.4. Deletions That Destabilize Structured Domains

We found eight cases (representing 16%) where exon deletions may destabilize the 3D structure of the isoforms. It mostly happened in large multi-domain proteins. We assigned these examples to a separate subgroup. In these structures, the domain, which may be destabilized by the deletion of a critical part, can be transformed into an unfolded linker connecting the other globular domains. Instead, in the canonical structure, these domains are connected by the structured domain (Figure 8D). In the case of canonical proteins with a single structured domain, the isoforms may represent intrinsically disordered proteins.

### 3.7.5. Limitations of AlphaFold in the Interpretation of the Conformational Changes

Our analysis revealed some limitations of AlphaFold modeling of the isoforms. For example, it is the case when we try to distinguish between isoforms with exon deletions, which preserve the overall structure, from the ones that destabilize it. In most of the cases, we could not base our decisions on the confidence score pLDDT for the reason that even structures, which missed a large part of the domain, frequently had pLDDT scores higher than 70%. These borderline cases were classified based on our visual analysis. In general, AlphaFold had a tendency to build isoform models that are very close to the canonical structures but with missing parts corresponding to the deleted exons. One of these examples is shown in Figure 8A, where an isoform of the canonical 7-bladed beta-propeller of guanine nucleotide-binding protein subunit beta-3 has six repetitive units. AlphaFold model of the isoform is almost identical to the canonical structure but misses one blade leading to the structure with an open beta-propeller. However, the SwissModel structure made based on the known 6-bladed structure (PDB code 1E1A) represents a closed 6-bladed beta-propeller. Such ambiguous cases cannot be resolved without experimental studies.

## 4. Conclusions

We took advantage of the progress achieved in the development of bioinformatics tools for large-scale structural annotations of proteins and examined the structural differences between canonical proteins and their isoforms. It became possible thanks to the TAPASS pipeline, which uses several state-of-the-art programs for the prediction of structured domains, unstructured regions, transmembrane regions, and aggregation-prone motifs [20]. Moreover, the availability of the AlphaFold program [22] opened up the possibility of modeling a large number of isoform structures. Altogether, our in silico analysis of 58 eukaryotic proteomes supported the concept that the majority of isoforms, similarly to the canonical proteins, are under selective pressure for functional roles. We also found that the proportions of proteins with a signal peptide and transmembrane helices are lower in isoforms than in canonical proteins. This suggested that some isoforms lose their transmembrane or extracellular localization and, eventually, their functional roles. At the same time, we did not observe significant differences between canonical proteins and their isoforms in the occurrence of unstructured regions or aggregation-prone motifs. Our modeling of the isoform structures demonstrated that the AlphaFold program is perfectly suitable for investigations of the structural differences of splicing variants at atomic details. It was shown that frequently the isoform sequences being different from the canonical ones still can fold in similar structures. At the same time, the isoforms may have significant structural rearrangements, which can lead to changes in their functions. We suggested the classification of the structural differences in the isoforms, which preserves the overall structure of the canonical proteins.

**Supplementary Materials:** The following supporting information can be downloaded at: https://mycore.core-cloud.net/index.php/s/VsMd0pthBJELe2h or https://drive.google.com/file/d/1XvQ2mzF1J54acRls2iGw7oqmO0gGIVkM/view?usp=sharing, Table S1: Species list; Table S2: Combined table of protein ids; Table S3: Combined results of canonical and isoform main sets by TAPASS tool. It includes predictions of amyloidogenicity by ArchCandy, Pasta, Tango, structured domains by CATH, unstructured regions by IUPred and BISMM filter, signal peptide by SignalP, transmembrane regions by TMHMM, degrons and nuclear localization signals by using SLiMs motifs; Table S4: Expression of canonical vs. isoforms in normal tissues; Table S5: Expression of canonical vs. isoforms in cancer tissues; Table S6: proportion of degron containing proteins; Figure S1: Gallery of AlphaFold models; Data S1: DReFinder script; Data S2: Atomic coordinates of AlphaFold models.

**Author Contributions:** Conceptualization, A.V.K., Z.O. and I.S.; methodology, Z.O., A.V.K., T.F. and J.L.; software, Z.O., T.F. and J.L.; validation, T.S., G.A. and Z.O.; data curation, T.S., G.A. and Z.O.; writing—original draft preparation, Z.O. and A.V.K.; writing—review and editing, A.V.K., Z.O., G.A., T.F. and I.S.; supervision, A.V.K. and I.S. All authors have read and agreed to the published version of the manuscript.

**Funding:** This research was funded by REFRACT project with Latin America in the RISE program (2018–2023) H2020-MSCA-RISE-2018 to A.V.K., Azerbaijan National Academy of Sciences, The Ministry of Science and Education of Azerbaijan to Z.O.

**Institutional Review Board Statement:** Not applicable.

**Informed Consent Statement:** Not applicable.

**Data Availability Statement:** Not applicable.

**Acknowledgments:** We thank Layla Hirsh and Nikola Arsic for discussion, careful reading of the manuscript, and valuable comments.

**Conflicts of Interest:** The authors declare no conflict of interest.

## References


1. Wang, E.T.; Sandberg, R.; Luo, S.; Khrebtukova, I.; Zhang, L.; Mayr, C.; Kingsmore, S.F.; Schroth, G.P.; Burge, C.B. Alternative isoform regulation in human tissue transcriptomes. *Nature* **2008**, *456*, 470–476. [CrossRef]
2. Pan, Q.; Shai, O.; Lee, L.J.; Frey, B.J.; Blencowe, B.J. Deep surveying of alternative splicing complexity in the human tran-scriptome by high-throughput sequencing. *Nat. Genet.* **2008**, *40*, 1413–1415. [CrossRef]
3. Melamud, E.; Moult, J. Structural implication of splicing stochastics. *Nucleic Acids Res.* **2009**, *37*, 4862–4872. [CrossRef]
4. Harrow, J.; Frankish, A.; Gonzalez, J.M.; Tapanari, E.; Diekhans, M.; Kokocinski, F.; Aken, B.L.; Barrell, D.; Zadissa, A.; Searle, S.; et al. GENCODE: The reference human genome annotation for The ENCODE Project. *Genome Res.* **2012**, *22*, 1760–1774. [CrossRef]
5. Sánchez-Pla, A.; Reverter, F.; de Villa, M.C.R.; Comabella, M. Transcriptomics: mRNA and alternative splicing. *J. Neuroimmunol.* **2012**, *248*, 23–31. [CrossRef]
6. Uhlén, M.; Fagerberg, L.; Hallström, B.M.; Lindskog, C.; Oksvold, P.; Mardinoglu, A.; Sivertsson, Å.; Kampf, C.; Sjöstedt, E.; Asplund, A.; et al. Proteomics. Tissue-Based Map of the Human Proteome. *Science* **2015**, *347*, 1260419. [CrossRef]
7. Tress, M.L.; Abascal, F.; Valencia, A. Alternative Splicing May Not Be the Key to Proteome Complexity. *Trends Biochem. Sci.* **2016**, *42*, 98–110. [CrossRef]
8. Savosina, P.; Karasev, D.; Veselovsky, A.; Miroshnichenko, Y.; Sobolev, B. Functional and structural features of proteins associated with alternative splicing. *Int. J. Biol. Macromol.* **2020**, *147*, 513–520. [CrossRef]
9. Hegyi, H.; Kalmár, L.; Horvath, T.; Tompa, P. Verification of alternative splicing variants based on domain integrity, truncation length and intrinsic protein disorder. *Nucleic Acids Res.* **2010**, *39*, 1208–1219. [CrossRef]
10. Birzele, F.; Csaba, G.; Zimmer, R. Alternative splicing and protein structure evolution. *Nucleic Acids Res.* **2007**, *36*, 550–558. [CrossRef]
11. The UniProt Consortium. UniProt: The universal protein knowledgebase in 2021. *Nucleic Acids Res.* **2021**, *49*, D480–D489. [CrossRef] [PubMed]
12. O'Leary, N.A.; Wright, M.W.; Brister, J.R.; Ciufo, S.; Haddad, D.; McVeigh, R.; Rajput, B.; Robbertse, B.; Smith-White, B.; Ako-Adjei, D.; et al. Reference sequence (RefSeq) database at NCBI: Current status, taxonomic expansion, and functional annotation. *Nucleic Acids Res.* **2016**, *44*, D733–D745. [CrossRef]
13. Cunningham, F.; Allen, E.J.; Allen, J.; Alvarez-Jarreta, J.; Amode, M.R.; Armean, I.M.; Austine-Orimoloye, O.; Azov, A.G.; Barnes, I.; Bennett, R.; et al. Ensembl 2022. *Nucleic Acids Res.* **2021**, *50*, D988–D995. [CrossRef]
14. Rodriguez, J.M.; Maietta, P.; Ezkurdia, I.; Pietrelli, A.; Wesselink, J.-J.; Lopez, G.; Valencia, A.; Tress, M.L. APPRIS: Annotation of principal and alternative splice isoforms. *Nucleic Acids Res.* **2012**, *41*, D110–D117. [CrossRef] [PubMed]
15. Yang, I.S.; Son, H.; Kim, S.; Kim, S. ISOexpresso: A web-based platform for isoform-level expression analysis in human cancer. *BMC Genom.* **2016**, *17*, 631. [CrossRef] [PubMed]
16. Zea, D.J.; Richard, H.; Laine, E. ASES: Visualizing evolutionary conservation of alternative splicing in proteins. *Bioinformatics* **2022**, *38*, 2615–2616. [CrossRef] [PubMed]
17. UniProt Consortium. UniProt: A worldwide hub of protein knowledge. *Nucleic Acids Res.* **2019**, *47*, D506–D515. [CrossRef]
18. Berman, H.M.; Westbrook, J.; Feng, Z.; Gilliland, G.; Bhat, T.N.; Weissig, H.; Shindyalov, I.N.; Bourne, P.E. The Protein Data Bank. *Nucleic Acids Res.* **2000**, *28*, 235–242. [CrossRef]
19. Uversky, V.N. Intrinsically Disordered Proteins and Their "Mysterious" (Meta)Physics. *Front. Phys.* **2019**, *7*, 10. [CrossRef]
20. Falgarone, T.; Villain, É.; Guettaf, A.; Leclercq, J.; Kajava, A.V. TAPASS: Tool for annotation of protein amyloidogenicity in the context of other structural states. *J. Struct. Biol.* **2022**, *214*, 107840. [CrossRef]
21. Uversky, V.N. Typical Functions of IDPs and IDPRs. In *Intrinsically Disordered Proteins*, 1st ed.; Gomes, G.M., Ed.; Springer: Cham, Switzerland, 2014; pp. 13–33. [CrossRef]
22. Jumper, J.; Evans, R.; Pritzel, A.; Green, T.; Figurnov, M.; Ronneberger, O.; Tunyasuvunakool, K.; Bates, R.; Žídek, A.; Potapenko, A.; et al. Highly accurate protein structure prediction with AlphaFold. *Nature* **2021**, *596*, 583–589. [CrossRef] [PubMed]
23. Fu, L.; Niu, B.; Zhu, Z.; Wu, S.; Li, W. CD-HIT: Accelerated for clustering the next-generation sequencing data. *Bioinformatics* **2012**, *28*, 3150–3152. [CrossRef] [PubMed]
24. Boratyn, G.M.; Schäffer, A.A.; Agarwala, R.; Altschul, S.F.; Lipman, D.J.; Madden, T.L. Domain enhanced lookup time accelerated BLAST. *Biol. Direct* **2012**, *7*, 12. [CrossRef]
25. Bairoch, A.; Apweiler, R. The SWISS-PROT protein sequence database and its supplement TrEMBL in 2000. *Nucleic Acids Res.* **2000**, *28*, 45–48. [CrossRef]
26. Sillitoe, I.; Bordin, N.; Dawson, N.; Waman, V.P.; Ashford, P.; Scholes, H.M.; Pang, C.S.M.; Woodridge, L.; Rauer, C.; Sen, N.; et al. CATH: Increased structural coverage of functional space. *Nucleic Acids Res.* **2020**, *49*, D266–D273. [CrossRef] [PubMed]
27. Mirdita, M.; Schütze, K.; Moriwaki, Y.; Heo, L.; Ovchinnikov, S.; Steinegger, M. ColabFold: Making protein folding accessible to all. *Nat. Methods* **2022**, *19*, 679–682. [CrossRef]
28. Schrödinger. *The PyMOL Molecular Graphics System*, Version 1.8; Schrödinger Technical: New York, NY, USA, 2015. Available online: http://www.pymol.org/pymol(accessed on 26 October 2022).
29. Mészáros, B.; Erdős, G.; Dosztányi, Z. IUPred2A: Context-dependent prediction of protein disorder as a function of redox state and protein binding. *Nucleic Acids Res.* **2018**, *46*, W329–W337. [CrossRef]
30. Petersen, T.N.; Brunak, S.; von Heijne, G.; Nielsen, H. SignalP 4.0: Discriminating signal peptides from transmembrane regions. *Nat. Methods* **2011**, *8*, 785–786. [CrossRef]

31. Krogh, A.; Larsson, B.; von Heijne, G.; Sonnhammer, E.L. Predicting transmembrane protein topology with a hidden markov model: Application to complete genomes. *J. Mol. Biol.* **2001**, *305*, 567–580. [CrossRef]
32. Ahmed, A.B.; Znassi, N.; Château, M.; Kajava, A.V. A structure-based approach to predict predisposition to amyloidosis. *Alzheimer's Dement.* **2014**, *11*, 681–690. [CrossRef]
33. Rousseau, F.; Schymkowitz, J.; Serrano, L. Protein aggregation and amyloidosis: Confusion of the kinds? *Curr. Opin. Struct. Biol.* **2006**, *16*, 118–126. [CrossRef] [PubMed]
34. Walsh, I.; Seno, F.; Tosatto, S.C.; Trovato, A. PASTA 2.0: An improved server for protein aggregation prediction. *Nucleic Acids Res.* **2014**, *42*, W301–W307. [CrossRef]
35. Kumar, M.; Michael, S.; Alvarado-Valverde, J.; Mészáros, B.; Sámano-Sánchez, H.; Zeke, A.; Dobson, L.; Lazar, T.; Örd, M.; Nagpal, A.; et al. The Eukaryotic Linear Motif resource: 2022 release. *Nucleic Acids Res.* **2021**, *50*, D497–D508. [CrossRef] [PubMed]
36. Richard, F.D.; Kajava, A.V. TRDistiller: A rapid filter for enrichment of sequence datasets with proteins containing tandem repeats. *J. Struct. Biol.* **2014**, *186*, 386–391. [CrossRef] [PubMed]
37. Szklarczyk, R.; Heringa, J. Tracking repeats using significance and transitivity. *Bioinformatics* **2004**, *20*, i311–i317. [CrossRef] [PubMed]
38. Jorda, J.; Kajava, A.V. T-REKS: Identification of Tandem REpeats in sequences with a K-meanS based algorithm. *Bioinformatics* **2009**, *25*, 2632–2638. [CrossRef] [PubMed]
39. Madeira, F.; Pearce, M.; Tivey, A.R.N.; Basutkar, P.; Lee, J.; Edbali, O.; Madhusoodanan, N.; Kolesnikov, A.; Lopez, R. Search and sequence analysis tools services from EMBL-EBI in 2022. *Nucleic Acids Res.* **2022**, *50*, W276–W279. [CrossRef]
40. Colak, R.; Kim, T.; Michaut, M.; Sun, M.; Irimia, M.; Bellay, J.; Myers, C.L.; Blencowe, B.J.; Kim, P.M. Distinct Types of Disorder in the Human Proteome: Functional Implications for Alternative Splicing. *PLOS Comput. Biol.* **2013**, *9*, e1003030. [CrossRef]
41. Arsic, N.; Slatter, T.; Gadea, G.; Villain, E.; Fournet, A.; Kazantseva, M.; Allemand, F.; Sibille, N.; Seveno, M.; de Rossi, S.; et al. Δ133p53β isoform pro-invasive activity is regulated through an aggregation-dependent mechanism in cancer cells. *Nat. Commun.* **2021**, *12*, 5463. [CrossRef]
42. Uversky, V.N.; Dunker, A.K. Understanding protein non-folding. *Biochim. Biophys. Acta (BBA)-Proteins Proteom.* **2010**, *1804*, 1231–1264. [CrossRef]
43. Pepys, M.B. Amyloidosis. *Annu. Rev. Med.* **2006**, *57*, 223–241. [CrossRef] [PubMed]
44. Tsang, B.; Pritišanac, I.; Scherer, S.W.; Moses, A.M.; Forman-Kay, J.D. Phase Separation as a Missing Mechanism for Interpretation of Disease Mutations. *Cell* **2020**, *183*, 1742–1756. [CrossRef] [PubMed]
45. Uversky, V.N. Protein intrinsic disorder-based liquid–liquid phase transitions in biological systems: Complex coacervates and membrane-less organelles. *Adv. Colloid Interface Sci.* **2017**, *239*, 97–114. [CrossRef] [PubMed]
46. Kotulska, M.; Wojciechowski, J.W. Bioinformatics Methods in Predicting Amyloid Propensity of Peptides and Proteins. In *Computer Simulations of Aggregation of Proteins and Peptides*, 1st ed.; Li, M.S., Kloczkowski, A., Cieplak, M., Kouza, M., Eds.; Methods in Molecular Biology, Humana: New York, NY, USA, 2022; Volume 2340, pp. 1–15.
47. Ezkurdia, I.; Rodriguez, J.M.; Pau, E.C.-D.S.; Vázquez, J.; Valencia, A.; Tress, M.L. Most Highly Expressed Protein-Coding Genes Have a Single Dominant Isoform. *J. Proteome Res.* **2015**, *14*, 1880–1887. [CrossRef] [PubMed]
48. Ravid, T.; Hochstrasser, M. Diversity of degradation signals in the ubiquitin–proteasome system. *Nat. Rev. Mol. Cell Biol.* **2008**, *9*, 679–689. [CrossRef]
49. Varshavsky, A. N-degron and C-degron pathways of protein degradation. *Proc. Natl. Acad. Sci. USA* **2019**, *116*, 358–366. [CrossRef]
50. A.Andradeab, M.; Iratxetaab, C.P.; Ponting, C. Protein Repeats: Structures, Functions, and Evolution. *J. Struct. Biol.* **2001**, *134*, 117–131. [CrossRef]
51. Kajava, A.V. Tandem repeats in proteins: From sequence to structure. *J. Struct. Biol.* **2011**, *179*, 279–288. [CrossRef]
52. Paladin, L.; Necci, M.; Piovesan, D.; Mier, P.; Andrade-Navarro, M.A.; Tosatto, S.C. A novel approach to investigate the evolution of structured tandem repeat protein families by exon duplication. *J. Struct. Biol.* **2020**, *212*, 107608. [CrossRef]
53. Liu, M.; Grigoriev, A. Protein domains correlate strongly with exons in multiple eukaryotic genomes—Evidence of exon shuffling? *Trends Genet.* **2004**, *20*, 399–403. [CrossRef]
54. Lesk, A.M.; Levitt, M.; Chothia, C. Alignment of the amino acid sequences of distantly related proteins using variable gap penalties. *Protein Eng. Des. Sel.* **1986**, *1*, 77–78. [CrossRef] [PubMed]
55. Paladin, L.; Bevilacqua, M.; Errigo, S.; Piovesan, D.; Mičetić, I.; Necci, M.; Monzon, A.M.; Fabre, M.L.; Lopez, J.L.; Nilsson, J.F.; et al. RepeatsDB in 2021: Improved data and extended classification for protein tandem repeat structures. *Nucleic Acids Res.* **2020**, *49*, D452–D457. [CrossRef] [PubMed]
56. Wise, H. The roles played by highly truncated splice variants of G protein-coupled receptors. *J. Mol. Signal.* **2012**, *7*, 13. [CrossRef]
57. Dardenne, E.; Pierredon, S.; Driouch, K.; Gratadou, L.; Lacroix-Triki, M.; Espinoza, M.P.; Zonta, E.; Germann, S.; Mortada, H.; Villemin, J.-P.; et al. Splicing switch of an epigenetic regulator by RNA helicases promotes tumor-cell invasiveness. *Nat. Struct. Mol. Biol.* **2012**, *19*, 1139–1146. [CrossRef] [PubMed]

*biomolecules*

MDPI

*Article*

# A Novel Tandem-Tag Purification Strategy for Challenging Disordered Proteins

Attila Mészáros [1,2,*], Kevin Muwonge [1,2], Steven Janvier [1,2], Junaid Ahmed [1,2] and Peter Tompa [1,2,3,*]

1 VIB-VUB Center for Structural Biology, Vlaams Instituut voor Biotechnologie (VIB), 1050 Brussels, Belgium
2 Structural Biology Brussels (SBB), Vrije Universiteit Brussel (VUB), 1050 Brussels, Belgium
3 Research Centre for Natural Sciences (RCNS), Institute of Enzymology, ELKH, 1117 Budapest, Hungary
* Correspondence: attila.meszaros@vub.be (A.M.); peter.tompa@vub.be (P.T.)

**Abstract:** Intrinsically disordered proteins (IDPs) lack well-defined 3D structures and can only be described as ensembles of different conformations. This high degree of flexibility allows them to interact promiscuously and makes them capable of fulfilling unique and versatile regulatory roles in cellular processes. These functional benefits make IDPs widespread in nature, existing in every living organism from bacteria and fungi to plants and animals. Due to their open and exposed structural state, IDPs are much more prone to proteolytic degradation than their globular counterparts. Therefore, the purification of recombinant IDPs requires extra care and caution, such as maintaining low temperature throughout the purification, the use of protease inhibitor cocktails and fast workflow. Even so, in the case of long IDP targets, the appearance of truncated by-products often seems unavoidable. The separation of these unwanted proteins can be very challenging due to their similarity to the parent target protein. Here, we describe a tandem-tag purification method that offers a remedy to this problem. It contains only common affinity-chromatography steps (HisTrap and Heparin) to ensure low cost, easy access and scaling-up for possible industrial use. The effectiveness of the method is demonstrated with four examples, Tau-441 and two of its fragments and the transactivation domain (AF1) of androgen receptor.

**Keywords:** intrinsically disordered proteins (IDPs); protein purification; affinity chromatography; Tau; androgen receptor (AF1)

**Citation:** Mészáros, A.; Muwonge, K.; Janvier, S.; Ahmed, J.; Tompa, P. A Novel Tandem-Tag Purification Strategy for Challenging Disordered Proteins. *Biomolecules* **2022**, *12*, 1566. https://doi.org/10.3390/biom12111566

Academic Editors: Prakash Kulkarni, Stefania Brocca, Keith Dunker and Sonia Longhi

Received: 27 September 2022
Accepted: 18 October 2022
Published: 26 October 2022



## 1. Introduction

The production of recombinant proteins is a crucial technique in both academic research and industrial applications [1]. In industry, such as the pharma sector, the use of biopharmaceuticals is becoming the dominant trend. In the last few years, close to 100 new biopharmaceuticals, the majority being recombinant proteins, have entered the market [2]. In molecular biology, recombinant protein purification is a vital technique for a broad range of applications, such as structural characterization by X-ray crystallography, nuclear magnetic resonance (NMR), small-angle X-ray scattering (SAXS) and cryo-electron microscopy (cryo-EM) [3–5]. Most of these require large quantities of protein with high quality, although cryo-EM is less stringent on sample requirement following its compatibility with protein purified from native samples [6]. Besides protein structure technologies, other in vitro biochemical studies and molecular biology applications also require protein of good quality and reasonable quantity. Therefore, significant effort is directed towards developing new and improved purification approaches [7–9].

Recombinant protein purification can be classified by the host organism in which the expression is performed [10]. Each system has its advantages and disadvantages, and the host of choice is often motivated by the downstream application of the purified protein [11,12]. Bacterial expression systems, for example, are well-established, easy to handle and relatively cheap [13]. However, they provide proteins without typical post-translational modifications (PTMs), which might be critical for the native, functional state of eukaryotic target proteins [14]. Furthermore, removing bacterial endotoxin from the sample can be cumbersome, compromising biotechnological applications [15]. Bacterial expression can also fail to produce soluble proteins, as many of them tend to form inclusion bodies (IBs) [16,17]. Eukaryotic systems, on the other hand, appear to be superior in producing soluble and active eukaryotic proteins [18–21]. However, their final yield tends to be lower, and the production process is more labor intensive, requiring special media and equipment, which may significantly increase the total cost of production. Another variation on the theme is the production of proteins in so-called cell-free systems. In this case, instead of using a host organism, an in vitro mixture is reconstituted for protein expression [22]. These systems are relatively costly and are not suited for high-level protein expression, even though they allow fast production and incorporation of special amino acids [23].

The most common host organism for recombinant protein production is *Escherichia coli* (*E. coli*) [24]. As an expression system, *E. coli* is easy to handle, cheap, has a high growth rate and usually produces large quantities of the desired recombinant protein [25]. As mentioned earlier, its major disadvantage is the lack of PTMs, which can result in improper folding and/or IB formation of expressed eukaryotic proteins. Furthermore, bacterial codon usage, which differs from that of eukaryotes, may also be a limiting factor. This is more evident in the case of human recombinant proteins expressed in bacteria. All in all, this system is still widely used due to its many advantages, as intense research is also being conducted to overcome its limitations [11,25–28]. For example, the introduction of solubility tag(s) or co-expression of molecular chaperones can significantly improve solubility of expressed proteins [29–32]. Codon bias can also be overcome via codon optimization, or by using special strains that contain transfer RNAs (tRNAs) at levels typical of eukaryotes [24,33]. Genetic modification of *E. coli* strains even allows the production of glycosylated antibodies [13,34,35].

Intrinsically disordered proteins (IDPs) are proteins that lack well-defined 3D structures [36]. Since their discovery, there has been a boom in studies highlighting their important roles in crucial cellular processes [37]. Due to their lack of well-defined 3D structures, IDPs are much more susceptible for proteolytic degradation [13,38]. For this reason, extra precaution needs to be taken during their purification, such as applying protease inhibitors, keeping the temperature low or optimizing for a very fast workflow. Despite all precautions, however, degradation still occurs most of the time. One approach to overcome the challenges of degradation is the development of tandem-tag based methods, applying different affinity tags on both termini of the protein of interest [39–43]. In such instances, however, only one tag is normally removed by targeted proteolytic cleavage, to ensure that the remaining tag can be used in subsequent applications such as pull-down or western-blot experiments [40–42]. In cases where both tags must be removed, this is usually achieved via two different proteases [43]. The other drawback of some tag combinations reported in the literature is that the necessary column and the elution reagents are either expensive or not easily available (e.g., FLAG-tag® or Twin-Strep-tag®) [39,41,42,44]. In addition, there are other plasmid constructs that contain two affinity tags for consecutive affinity steps to ensure higher purity, although in most cases the two tags are located at the same terminus [7].

Here, we describe a novel tandem-tag based method for IDP purification, in which both tags can be removed simultaneously. We demonstrate the versatility of the method on a few selected IDP examples with different charge properties.

## 2. Materials and Methods

### 2.1. Generation of pSUMO Plasmid

To generate the pSUMO plasmid for our method, we modified an existing one (pHYRSF53, a gift from Hideo Iwai, Addgene plasmid # 64696; http://n2t.net/addgene:64696, accessed on 25 September 2022; RRID: Addgene_64696) [45]. pHYRSF53 contains an N-terminal 6xHis-tag followed by a SUMO tag, and it was used as a backbone. First, a C-terminal DNA-binding domain (DBD) was added to the pHYRSF53 plasmid using the HiFi DNA Assembly Cloning kit (New England Biolabs (NEB), Ipswich, MA, USA), following the instructions of the manufacturer. The plasmid used as a template to generate the DBD fragment of androgen receptor was created in-house. The fragment was generated by polymerase chain reaction (PCR) using Q5 High-Fidelity DNA polymerase (NEB, Ipswich, MA, USA), following the manufacturer's instructions. After the successful insertion of the DBD fragment into pHYRSF53, we inserted a multiple cloning site (MCS) flanked by two TEV cleavage sites in between the two affinity tags. The MCS-TEV-site fragment was ordered as a single-stranded DNA fragment from Eurofins Genomics (Ebersberg, Germany). Transformants in NEB 5-alpha chemically competent bacterial cells were then selected using Luria Bertani (LB)-agar containing 50 µg/mL Kanamycin antibiotic (Duchefa Biochemie, Haarlem, The Netherlands). Single colonies were picked and grown overnight at 37 °C in LB broth media (Duchefa Biochemie, Haarlem, The Netherlands) supplemented with 50 µg/mL Kanamycin antibiotic (LB-Kanamycin). Plasmid DNA was then isolated from the liquid cultures using the MN-NucleoSpin Plasmid QuickPure kit (Fisher Scientific, Merelbeke, Belgium), and plasmid DNA sequences were verified by Sanger sequencing (Microsynth, Balgach, Switzerland).

### *2.2. Generation of pSUMO Expression Constructs*

#### 2.2.1. Generating pSUMO Constructs by HiFi Cloning (pSUMO-AF1, pSUMO-Tau-441)

We generated a pSUMO-AF1 plasmid for recombinant protein expression by inserting a DNA fragment coding for the activation function 1 (AF1) domain of Androgen receptor into the MCS of pSUMO plasmid by HiFi cloning. The AF1-coding DNA fragments were generated with Q5 High-Fidelity DNA polymerase, using a plasmid housing a coding sequence of full-length AF1 as a template (The plasmid containing AF1-coding sequence was also generated in-house). Similarly, the pSUMO-Tau-441 plasmid for recombinant protein expression was generated by inserting the full-length Tau-441-coding sequence into the pSUMO plasmid using the HiFi DNA Assembly Cloning kit.

In both cases, successful transformants in NEB 5-alpha bacterial cells were selected on LB-Kanamycin agar plates, and single colonies were then cultured in LB-Kanamycin liquid media for plasmid extraction. Plasmid DNA was isolated using MN-NucleoSpin Plasmid QuickPure kit, and plasmid DNA sequences were verified by Sanger sequencing.

#### 2.2.2. Generating pSUMO Constructs by Site-Directed Mutagenesis (pSUMO-Tau-NTMT, pSUMO-Tau-MTBR and pSUMO-AF1 (Only N-tag))

The pSUMO-Tau-NTMT plasmid was generated by site-directed mutagenesis of the pSUMO-Tau-441 construct via deletion of Tau-441's short C-terminal tail. The pSUMO-Tau-MTBR construct was then generated by further mutagenesis of pSUMO-Tau-NTMT plasmid, via deletion of the flexible N-terminal region of Tau-441. The pSUMO-AF1(only N-tag) construct was created by inserting a stop codon (TAG) at the end of the AF1 coding sequence, such that only the N-terminal 6xHis-SUMO-tag would be translated as a fusion to AF1 recombinant protein, without the C-terminal DBD-affinity tag.

In all cases, mutagenesis was performed using the Q5 High-Fidelity DNA polymerase kit following the manufacturer's instructions. After the mutagenesis PCR, nascent non-methylated DNA strands harboring respective DNA modifications were enriched by KLD (Kinase, Ligase and Dpn1) treatment (NEB, Ipswich, MA, USA). Successful transformants in NEB 5-alpha bacterial cells were selected on LB-Kanamycin agar plates, and single colonies were then cultured in LB-Kanamycin liquid media for plasmid extraction. Plasmid DNA was isolated using the MN-NucleoSpin Plasmid QuickPure kit, according to the manufacturer's instructions. Plasmid DNA sequences of pSUMO-Tau-NTMT, pSUMO-Tau-MTBR and pSUMO-AF1(only N-tag) were confirmed by Sanger sequencing before expression of recombinant proteins.

### 2.3. *Expression and Purification of pSUMO Constructs*

#### 2.3.1. Expression and Purification of pSUMO-AF1(Only N-tag)

As the pSUMO-AF1(only N-tag) construct was not codon optimized for *E. coli* expression, it was expressed in *E. coli* Rosetta 2 cells (Invitrogen, Waltham, MA, USA) to enhance expression of such a eukaryotic protein in a bacterial system. Cells were cultured at 37 °C in Terrific Broth (TB) that was prepared in-house and supplemented with 50 µg/mL of Kanamycin and 25 µg/mL of Chloramphenicol antibiotics (Duchefa Biochemie, Haarlem, The Netherlands). Recombinant protein expression was induced with 1 mM isopropyl β-D-1-thiogalactopyranoside (IPTG) (Sigma-Aldrich, St. Louis, USA) at an optical density ($OD_{600}$) of 1.2. Cells were then cultured for another 5 h at 30 °C, harvested by centrifugation (Avanti JXN-26, Beckman Coulter, CA, USA) at 5000 revolutions per minute (rpm) for 15 min, and bacterial pellets were stored at −80 °C awaiting protein purification.

To purify pSUMO-AF1, 1 L of bacterial pellet was resuspended in 75 mL of Lysis buffer, composed of 50 mM Tris (Sigma-Aldrich, St. Louis, USA), 250 mM NaCl (Sigma-Aldrich, St. Louis, USA), 25 mM Imidazole (Merck, Darmstadt, Germany), 10% Glycerol (VWR, Ohio, USA), 0.1% Triton X-100 (Sigma-Aldrich, St. Louis, USA), 0.5 mM tris(2-carboxyethyl) phosphine (TCEP) (VWR, Ohio, USA), 1 mM phenyl-methyl-sulfonyl fluoride (PMSF) (Sigma-Aldrich, St. Louis, USA), 2 tablets of Roche cOmplete protease inhibitor cocktail (Merck, Darmstadt, Germany), pH 8.0) using a glass homogenizer (Carl Roth, Karlsruhe, Germany). Cells were then lysed by sonication (5 s on/5 s off, 60% amplitude, 5 min), and the resulting bacterial lysate centrifuged for 1 h at 40,000× *g* (4 °C) in presence of 1 mg/mL DNase I (Sigma-Aldrich, St. Louis, USA). The lysate supernatant was filtered using a 0.45 µm syringe filter (Sarstedt, Nümbrecht, Germany) and every step was carefully performed on ice.

The filtered lysate was loaded onto a HisTrap-HP-5 mL column (Cytiva, Uppsala, Sweden) using an AKTA™ Pure chromatography system (GE Healthcare, Uppsala, Sweden) in a cooling cabinet. The HisTrap loading buffer (A) was 50 mM Tris, 250 mM NaCl, 25 mM imidazole, 0.5 mM TCEP, pH 8.0. After sample loading, a washing step was applied with buffer A to wash away unbound contaminants. For elution, a gradient was applied using an elution buffer (B) composed of 50 mM Tris, 250 mM NaCl, 500 mM Imidazole, 0.5 mM TCEP, pH 8.0. Elution fractions were analyzed by sodium dodecyl-sulfate polyacrylamide gel electrophoresis (SDS-PAGE), and selected fractions were pooled together in a sterile 50-mL Falcon tube (Greiner Bio-One, Vilvoorde, Belgium).

To the pooled fractions, 0.1 mg/mL of TEV protease (From a 10 mg/mL stock that was expressed and purified in-house) was added to cleave the N-terminal 6xHis-SUMO-tag. The digested protein mixture was dialyzed against buffer A in a cold-room maintained at 4 °C, and then taken through a Reverse HisTrap purification using HisTrap loading (A) and elution (B) buffers. The HisTrap flow-through was dialyzed against 50 mM Tris, 150 mM NaCl, 0.5 mM TCEP, pH 8.0, and concentrated by ultrafiltration (VivaSpin, 10 kDa MWCO, Sartorius, Stonehouse, UK) in preparation for size exclusion chromatography (SEC), which was performed using a Superdex XK 16/100 (200 pg) column (Cytiva, Uppsala, Sweden). After SEC, elution fractions were analyzed by SDS-PAGE, and selected fractions were pooled and concentrated as before. The protein concentration of the final product was determined by measuring the absorbance at 280 nm, using NanoDropTMOne (ThermoFisher Scientific, Waltham, MA, USA). The yield was calculated to 1 L of bacterial culture (0.8 mg protein/L).

#### 2.3.2. Expression and Purification of pSUMO-AF1

The pSUMO-AF1 construct was also not codon optimized for *E. coli* expression, and hence its recombinant expression was performed in *E. coli* Rosetta 2 cells. The cells were cultured at 37 °C in TB containing 50 µg/mL of Kanamycin and 25 µg/mL of Chloramphenicol antibiotics. Protein expression was induced with 1 mM IPTG at $OD_{600}$ = 1.2, and cells were cultured for another 5 h at 30 °C. Cells were then harvested by centrifugation at 5000 rpm for 15 min, and the bacterial pellets stored at −80 °C awaiting protein purification.

Just like pSUMO-AF1(only N-tag) purification, a 1 L pSUMO-AF1 bacterial pellet was resuspended in 75 mL of a Lysis buffer with identical composition. Resuspended cells were then lysed, centrifuged, and the lysate supernatant filtered as previously described.

The filtered lysate was loaded onto a HisTrap-HP-5 mL column equilibrated with HisTrap loading buffer (A1) composed of 50 mM Tris, 250 mM NaCl, 25 mM imidazole, 0.5 mM TCEP, pH 8.0. After sample loading, a high-salt ATP-wash buffer (A2) composed of 50 mM Tris, 500 mM NaCl, 500 mM KCl (VWR, Leuven, Belgium), 25 mM Imidazole, 10 mM ATP (Merck, Darmstadt, Germany), 20 mM $MgCl_2$ (Sigma-Aldrich, St. Louis, USA), 0.5 mM TCEP, pH 8.0, was applied to the column to wash away bacterial chaperone contaminants that co-purify with AF1. Bound proteins were eluted from the column by applying a gradient of elution buffer (B) composed of 50 mM Tris, 250 mM NaCl, 500 mM imidazole, 0.5 mM TCEP, pH 8.0. Elution fractions were analyzed by SDS-PAGE and selected fractions were pooled together in a sterile 50-mL Falcon tube.

Pooled fractions were dialyzed against Heparin loading buffer A (50 mM Tris, 5 mM TCEP, pH 8.0) and loaded onto a Heparin-HP-5 mL chromatography column (Cytiva, Uppsala, Sweden) equilibrated with the same buffer. Bound proteins were washed with Heparin buffer A and eluted using a gradient of Heparin Buffer B (50 mM Tris, 1 M NaCl, 5 mM TCEP, pH 8.0). After SDS-PAGE analysis of collected samples, selected fractions were pooled and mixed with 0.4 mg/mL TEV protease to simultaneously cleave both affinity tags. The protein mixture was then dialyzed overnight at 4 °C against HisTrap Buffer A1 and taken through a Reverse HisTrap purification using the same HisTrap buffers outlined above (except the buffer A2). The HisTrap flow-through was then dialyzed against Heparin Buffer A, concentrated by ultrafiltration (VivaSpin 10 kDa MWCO) and taken through a final Reverse Heparin purification step using buffers and Heparin chromatography column outlined above. The Reverse Heparin flow-through was further concentrated by ultrafiltration as before, and the protein concentration was determined by measuring the absorbance at 280 nm, using NanoDropTMOne (ThermoFisher Scientific, Waltham, MA, USA). The yield for 1 L bacterial culture was then calculated (4.6 mg protein/L).

#### 2.3.3. Expression and Purification of pSUMO-Tau-441

The pSUMO-Tau-441 construct was codon optimized for *E. coli* expression; therefore, the recombinant protein was expressed in *E. coli* BL21 Star™ (DE3) cells (NEB, Ipswich, MA, USA). Cells were cultured at 37 °C in TB supplemented with 50 μg/mL Kanamycin antibiotic. Protein expression was induced with 1 mM IPTG at $OD_{600}$ = 1.2, then cells were cultured for another 5 h at 30 °C. Bacterial cells were then harvested by centrifugation at 5000 rpm for 15 min and the bacterial pellets stored at −80 °C awaiting protein purification.

To purify pSUMO-Tau-441, bacterial pellet from one liter of medium was resuspended in 75 mL of Lysis buffer (50 mM HEPES (Sigma-Aldrich, St. Louis, MO, USA), 0.2 mM $MgCl_2$, 10% Glycerol, 0.1% Triton X-100, 5 mM TCEP, 1 mM PMSF, 2 tablets of Roche cOmplete protease inhibitor cocktail, pH 7.2) using a glass homogenizer. Cells were then lysed by sonication (5 s on/5 s off, 60% amplitude, 5 min), and bacterial lysate centrifuged for 1 h at 40,000× *g* (4 °C), in the presence of 1 mg/mL DNase I (Sigma-Aldrich, St. Louis, MO, USA). The lysate supernatant was filtered using a 0.45 μm syringe filter and every step was carefully performed on ice.

The filtered lysate was then loaded onto a Heparin-HP-5 mL chromatographic column equilibrated with Heparin buffer A (50 mM HEPES, 0.5 mM TCEP, pH 7.2). Bound proteins were washed with Heparin buffer A and eluted using a gradient of Heparin Buffer B (50 mM Tris, 1 M NaCl, 0.5 mM TCEP, pH 7.2). Elution fractions were analyzed by SDS-PAGE and selected fractions were pooled together in a 50-mL Falcon tube. After adding 25 mM imidazole to the pooled fractions, the protein solution was loaded onto a HisTrap-Hp-5 mL column equilibrated with HisTrap buffer A (50 mM HEPES, 250 mM NaCl, 25 mM imidazole, 0.5 mM TCEP, pH 7.2). Bound proteins were washed with HisTrap buffer A and eluted using a gradient of HisTrap Buffer B (50 mM HEPES, 250 mM NaCl, 500 mM imidazole, 0.5 mM TCEP, pH 7.2). After analyzing the fractions on SDS-PAGE, selected fractions were pooled together in a sterile 50-mL Falcon tube. From this point on, the purification followed the workflow employed for pSUMO-AF1, while using buffers outlined for pSUMO-Tau-441. The final yield was normalized to one liter of bacterial culture (2.5 mg protein/L).

2.3.4. Expression and Purification of pSUMO-Tau-NTMT

The pSUMO-Tau-NTMT recombinant protein was also expressed to high levels in *E. coli* BL21 Star™ (DE3) cells, as it was generated from the pSUMO-Tau-441 construct codon optimized for bacterial expression. Cells were cultured at 37 °C in TB supplemented with 50 μg/mL of Kanamycin antibiotic. Protein expression was induced with 1 mM IPTG at $OD_{600}$ = 1.2, and cells were cultured for another 5 h at 30 °C. Cells were then harvested by centrifugation at 5000 rpm for 15 min, and bacterial pellets were stored at −80 °C awaiting protein purification.

To purify pSUMO-Tau-NTMT, 1 L of bacterial pellet was resuspended in 50 mL of Lysis buffer (50 mM HEPES, 250 mM NaCl, 0.2 mM $MgCl_2$, 10% Glycerol, 0.1% Triton X-100, 0.5 mM TCEP, 1 mM PMSF, 1 tablet of Roche cOmplete protease inhibitor cocktail, pH 7.2) using a glass homogenizer. Cells were then lysed by sonication (5 s on/5 s off, 60% amplitude, 5 min) and the bacterial lysate was centrifuged for 1 h at 40,000× *g* (4 °C) in the presence of 1 mg/mL DNase I. The lysate supernatant was filtered using a 0.45 μm syringe filter, and every step was carefully performed on ice.

The purification of pSUMO-Tau-NTMT followed the workflow described for pSUMO-AF1 with minor differences highlighted below. HisTrap loading buffer (A) was composed of 50 mM HEPES, 250 mM NaCl, 25 mM imidazole, 0.5 mM TCEP, pH 7.2, whereas HisTrap elution buffer (B) was composed of 50 mM HEPES, 250 mM NaCl, 500 mM imidazole, 0.5 mM TCEP, pH 7.2. Similarly, the Heparin loading buffer (A) was HEPES-based at pH 7.2, with 250 mM NaCl and 1 mM TCEP. The Heparin elution buffer (B) contained additional salt (1 M NaCl) for gradient elution of proteins bound to the Heparin chromatography column. After the final purification step, the protein concentration of purified pSUMO-Tau-NTMT was determined by measuring absorbance at 205 nm (instead of 280 nm), considering that the Tau-NTMT polypeptide sequence lacked any Tryptophan residues, the major contributors to intrinsic fluorescence at 280 nm. The final yield was normalized to one liter of bacterial culture (1.8 mg protein/L).

2.3.5. Expression and Purification of pSUMO-Tau-MTBR

The shorter pSUMO-Tau-MTBR construct was expressed and purified in the same way as pSUMO-Tau-NTMT, with minor modifications. Despite attaining high-level expression of Tau-MTBR at 30 °C for 5 h when induced with 0.5 mM IPTG, a large portion of the expressed protein localized to bacterial inclusion bodies. This was probably because bacterial chaperones could not keep up with the translation machinery rapidly producing large amounts of aggregation-prone Tau-MTBR, resulting in the sequestration into inclusion bodies [40]. To overcome this, slow expression of pSUMO-Tau-MTBR was adopted where protein expression was induced by 0.5 mM IPTG, and bacteria further cultured overnight at 16 °C. This approach dramatically improved on the yield recovered in the soluble fraction of the bacterial lysate (Figure 6). In addition, unlike Tau-NTMT, purified Tau-MTBR was dialyzed and concentrated using 3 kDa MWCO dialysis membrane (Serva, Heidelberg, Germany) and VivaSpin ultrafiltration columns (Sartorius, Stonehouse, UK), respectively, due to the relatively smaller size of the cleaved final product (15.8 kDa). The final yield was normalized to one liter of bacterial culture (2.1 mg protein/L).

## *2.4. LC-MS/MS Analysis*

SDS-PAGE bands of interest were subjected to in-gel digestion with trypsin and ProteaseMax™ surfactant, both obtained from Promega (Maddison, WI, USA), following the manufacturer's instructions. Processed samples were then snap-frozen and stored at −80 °C awaiting further analysis.

LC-MS/MS analysis of the tryptic digests was performed by means of a Q-Exactive™ Focus Hybrid Orbitrap mass spectrometer equipped with a Thermo Scientific™ Vanquish™ ultra-high performance liquid chromatography system (Thermo Fisher Scientific, Waltham, MA, USA). Five microliters of the tryptic digests were injected and chromatographically separated by means of a 35-min linear gradient of 2–45% mobile phase B (mobile phase A: 0.1% formic acid, mobile phase B: 0.1% formic acid in acetonitrile) on an Acquity UPLC® CSH C18 column (2.1 × 150 mm, 1.7 μm) from Waters (Milford, MA, USA). The flow rate and column temperature were 0.3 mL/min and 45 °C, respectively. The Q-Exactive Focus, operating in data dependent acquisition mode (DDA), was set to perform a mass spectrometry (MS) scan (R= 70 000 at 200 m/z, AGC target 3.0 e6) from 375 to 1500 m/z, followed by HCD $MS^2$ spectra (R= 17 500 at 200 m/z, NCE =27%, AGC target = 1.0e5, max ion time = 50 ms) on the three most abundant precursors (quadrupole isolation width 1.4 m/z).

Data treatment and data analysis were performed by PEAKS studio 10.6 (Bioinformatics Solutions Inc., Waterloo, Canada). *De novo* sequencing and database searches (full Swiss-Prot database, downloaded 12 September 2022) were performed with a 10-ppm precursor mass tolerance, a 0.02 Da fragment tolerance and an FDR <0.1% on the peptide level. Oxidations of methionine were set as a variable modification, while the carbamidomethylation of cysteines was included as a fixed modification. Only fully tryptic peptides and a maximum of three trypsin mis-cleavages were allowed. Database searches were performed with and without the AA sequences of the respective DNA constructs present.

## 3. Results

### *3.1. Generation of pSUMO Plasmid with Tandem-Tags*

Generally, many expression plasmids contain one affinity tag fused to either the N- or C-terminus of the target protein for subsequent purification by affinity chromatography (Figure 1A). This is usually sufficient for many proteins. However, as we already mentioned in the Introduction, IDPs are prone to proteolytic degradation due to their lack of a stable structure [13,38]. This degradation leads to truncated products that still carry one of the affinity-tags; hence these unwanted products usually co-purify with the target protein. Therefore, extra purification steps such as ion-exchange (IEX) or size exclusion (SEC) chromatography are necessary to get rid of these degradation fragments. This extra step, however, can be challenging because of the high similarity and/or small size difference between the target protein and its truncated versions. As mentioned above, a second affinity step can isolate the intact protein from a heterogeneous mixture [39–43]. We generated a plasmid for bacterial expression (bearing the T7 promoter) containing two different affinity-tags at the N- and C-terminus of the target protein, respectively (Figure 1B). In between the tags and the gene of interest, TEV cleavage sites were cloned into the plasmid, to facilitate the simultaneous removal of both tags. On the N-terminus, a SUMO tag was inserted to ensure a high expression level and increased solubility of the target protein [46]. The SUMO tag itself does not participate in affinity purification, hence we combined it with a 6xHis-tag, the most commonly available tag for HisTrap purification. On the C-terminus, the DNA-binding domain (DBD) of androgen receptor was inserted. Heparin mimics DNA, therefore, DBDs can specifically bind to heparin columns [47]. However, heparin is not specific enough to enable a one-step purification, as it can also non-specifically bind other proteins. It can be applied, however, if there is another affinity step in the workflow [48]. These heparin columns have also been reported to behave as cation exchangers [49].

Upon partial proteolytic degradation, one of the two termini is truncated, which leads to the malfunction of the tag at the affected termini. By applying two subsequent affinity steps (HisTrap and Heparin), we could ensure that only the intact protein is isolated from a bacterial cell lysate. Following successful isolation, both tags can be simultaneously removed in a single TEV protease cleavage step (Figure 2A). As the TEV protease also carries a fused 6xHis-tag at its N-terminus, the free 6xHis-SUMO-tag, uncleaved target proteins and the protease can be simultaneously removed by a reverse HisTrap purification step (Figure 2B). Reverse Heparin purification acts as a final polishing step. Moreover, since the flow-through is collected in these last steps, non-specifically binding contaminants are also going to re-bind to the columns, further improving the overall purity of the final product.

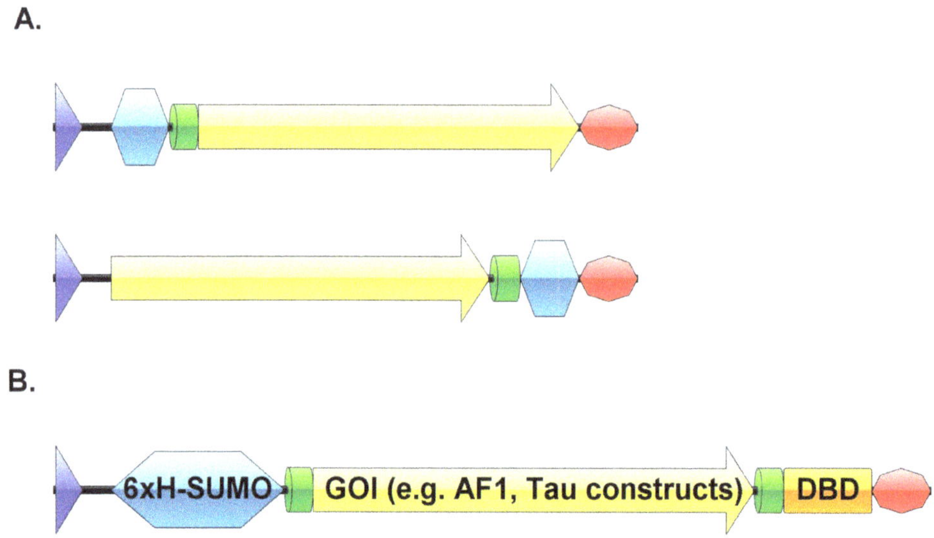


**Figure 1.** (**A**) Schematic representations of expression plasmids with either an N-terminal or a C-terminal affinity tag (blue—promoter region, silver blue—affinity tag, green—proteolytic cleavage site, yellow—gene of interest, orange—second affinity tag, red—terminator sequence). (**B**) Schematic representation of the pSUMO plasmid with tandem tags (blue—promoter region, silver blue—6xHis-SUMO-tag, green—TEV cleavage site, yellow—gene of interest, orange—DBD, red—terminator sequence). Created by IBS software [50].

### *3.2. Purification of pSUMO-AF1 and pSUMO-AF1(Only N-tag): Comparison of the Two Methods*

Androgen receptor is a transcription factor consisting of three domains: a globular DNA binding domain (DBD) sandwiched between a C-terminal ligand binding domain and a disordered N-terminal domain (NTD) [51]. The NTD contains an activation function domain (AF1) that is responsible for the recruitment of important cofactors, hence it has been studied extensively [52]. The existing purification strategy of AF1 recombinant protein is based on a single HisTrap via a 6xHis-tag followed by a SEC polishing step [53].

We compared our method with others applied for purifying the same protein. The purification of pSUMO-AF1(only N-tag) that contains only the N-terminal 6xHis-tag followed by the SUMO-tag relies on only one affinity purification step (HisTrap). This approach suffers from a high level of degradation by-products, which are very difficult to separate due to the resolution limit of the SEC columns (Supplementary Figure S1). Furthermore, as only a small portion of the fractions contains the intact protein, the yield is usually very low. In contrast, the purification of pSUMO-AF1 containing a DBD as an additional C-terminal tag for a second affinity purification (Heparin) resulted in recovery of intact protein, which was verified by MS (Supplementary Figure S10). Interestingly, after the first HisTrap, we observed two peaks. Analyzing the contents of these peak fractions by SDS-PAGE revealed that both peaks contained the desired product. However, one peak was almost entirely clean, containing only the intact protein with both tags, but the second peak also contained truncated degradation fragments (Supplementary Figure S2). As a precaution, these fractions were pooled separately for downstream purification steps (HisTrap pool I and II). The second affinity purification step (Heparin) did not improve on the purity of the first pool (HisTrap pool I), but it significantly improved the purity of

HisTrap pool II (Supplementary Figure S3). Nonetheless, we pooled interesting elution fractions separately for further purification steps (Heparin pool I and II). In the next step, both tags were removed by targeted proteolytic cleavage using TEV protease. The protease, uncleaved products and 6xHis-SUMO-tag were simultaneously removed in the reverse HisTrap step, and the fully cleaved product was recovered in the flow-through. As a final polishing step, another Heparin affinity purification (reverse Heparin) was performed. This served to remove the cleaved DBD-tag that bound to the Heparin column as the final product was collected in the flow-through (Figure 3). The final cleaved product was verified by MS (Supplementary Figure S11). The faint impurity in the final product was AF1 with uncleaved DBD that dimerized with fully cleaved AF1 and eluted together in the flow-through. This can be avoided by adding more reducing agent to the sample just before the reverse Heparin chromatography step as demonstrated later in the purification of Tau-MTBR.

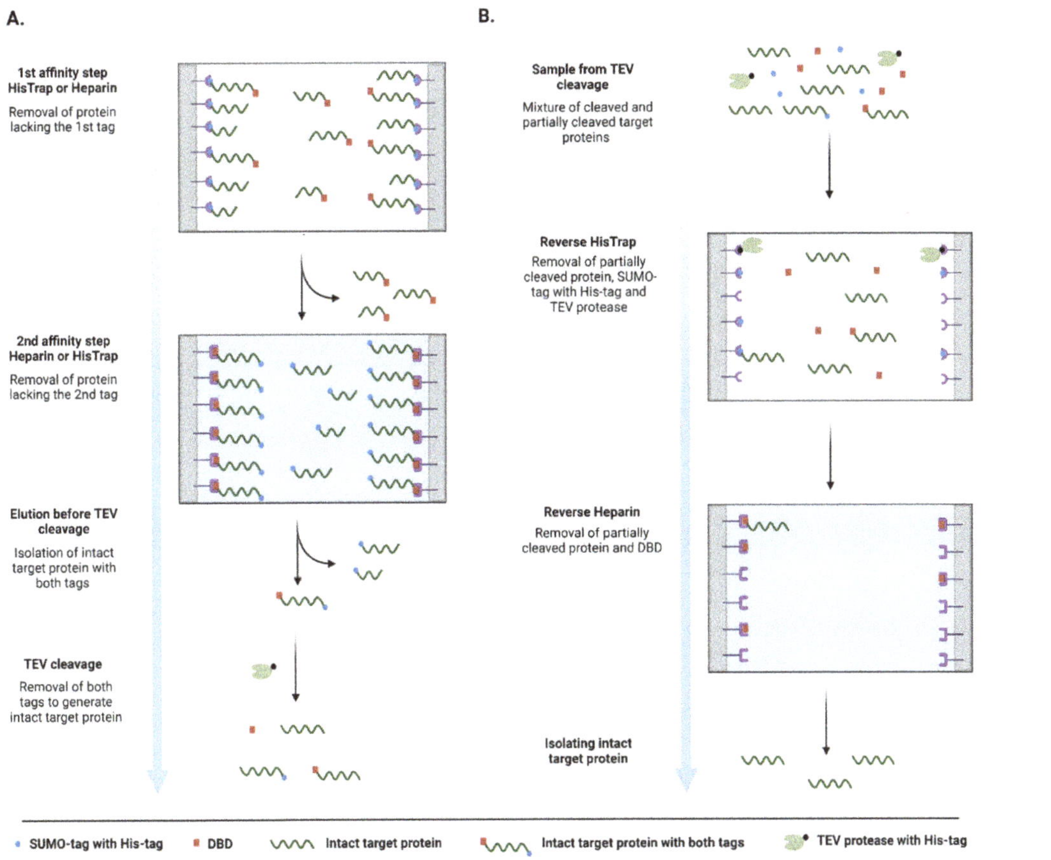


**Figure 2.** Schematic representation of the workflow of the tandem-tag purification method. (**A**) Representation the first two affinity steps before TEV cleavage. (**B**) Representation the reverse affinity steps after TEV cleavage. (Figure was created with BioRender.com).

It is important to note that, after performing all the purification steps, there was no difference in the purity between reverse Heparin pool I and II (Figure 3). This shows that the purification method worked equally well on the clean (HisTrap pool I) and dirty (HisTrap pool II) fractions of the first HisTrap step, demonstrating its separation capability and robustness.

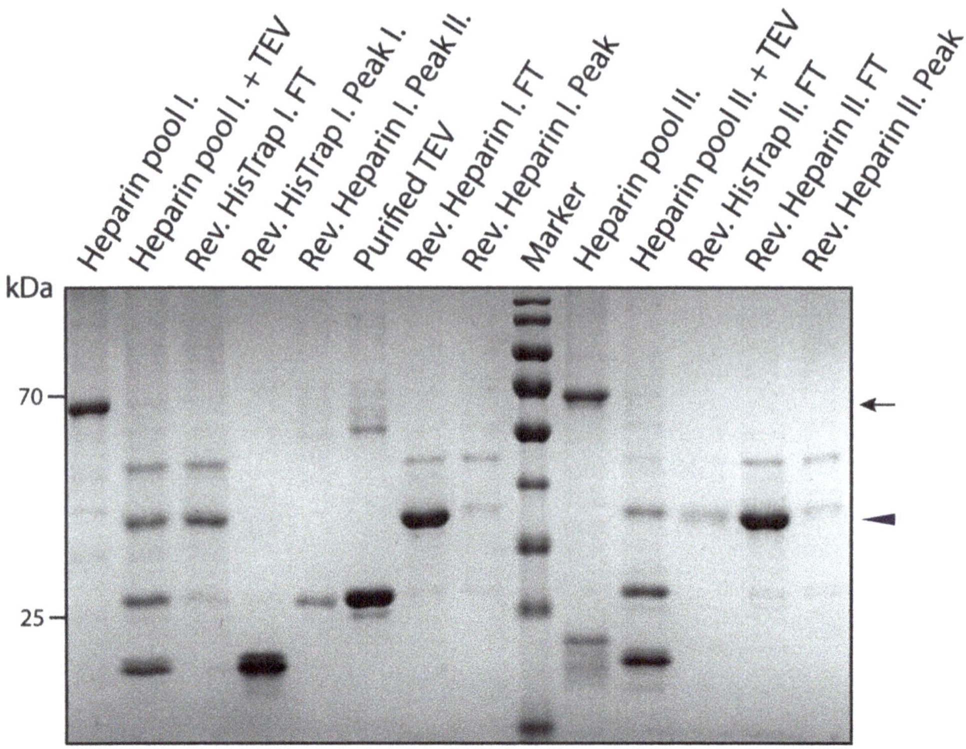


**Figure 3.** SDS-PAGE summarizing pSUMO-AF1 purification. The black arrow indicates intact AF1 having both affinity tags, while the black triangle indicates the cleaved final product. The lane 'Purified TEV' on the gel was loaded with a sample of TEV protease enzyme that was purified in-house and used in affinity-tag cleavage of all purified protein samples presented in this paper.

### *3.3. Purification of pSUMO-Tau Constructs*

#### 3.3.1. Purification of pSUMO-Tau-441

Microtubule-associated protein Tau has been implicated in various neurodegenerative diseases, such as Alzheimer's disease (AD) and frontotemporal dementia (FTD) [54]. It has multiple isoforms in the central nervous system, with the longest human isoform consisting of 441 amino acid residues, named Tau-441 (it is also known as htau40 and Tau-2N4R). Full-length Tau-441 is made up of a long flexible N-terminal region, a microtubule-binding region (MTBR) and a short C-terminal tail. Tau-441 is highly dynamic and behaves as an IDP in solution. Therefore, its structure has only been studied using NMR spectroscopy [54,55]. Recently, Tau-441 has been shown to undergo liquid-liquid phase separation (LLPS), which seems to play a key role in both its physiological functions and pathological aggregate formation [56–58]. Generally, LLPS and NMR studies require proteins of high purity and yield, which can only be satisfactorily produced in a bacterial expression system. Due to its intrinsically disordered nature, Tau-441 can withstand high temperature, thus the most common Tau purification methods in the literature include a boiling step on top of affinity chromatographic separation [57,58]. Nonetheless, the bacterial expression and purification of soluble and intact Tau-441, as well as its domain constructs, is not straightforward due to their open and exposed structural state; they are highly prone to proteolytic degradation. Moreover, they contain aggregation-prone motifs that can lead to solubility issues and sequestration into bacterial inclusion bodies [40].

To overcome these challenges, we applied our tandem-tag purification method to purify Tau-441 and two of its domains. Application of tandem tags can provide a remedy for truncated degradation products, while the N-terminal SUMO-tag also enhances solubility of the protein.

In the case of Tau-441, we changed the approach and decided to start the purification with Heparin chromatographic separation, as this method—unlike HisTrap—is not sensitive to high concentrations of reducing agents. It has been recently proposed that Tau molecules form higher-order oligomers via disulfide bonding of their cysteine residues, which can later lead to aggregate formation [59]. Therefore, we opted to use increased amounts of reducing agent in the lysis buffer (5 mM TCEP) to counter disulfide-bond formation among Tau monomers. As expected, Heparin affinity purification alone was not sufficient to produce a high purity sample, but it was able to enrich the target protein in separate fractions with significantly decreased levels of contamination (Supplementary Figure S4). Elution fractions that contained our protein of interest were pooled together for the next affinity purification step. Another advantage of this approach was that the elution buffer from the previous purification step (Heparin buffer B) was more compatible with the loading buffer of the next step (HisTrap buffer A). We could thus proceed without a buffer exchange of the Heparin pool sample (Supplementary Figure S5). The intact Tau-441 protein containing both tags was verified by MS (Supplementary Figure S12). In the next step, both tags were removed by TEV digestion and the method was continued the same way as described above for pSUMO-AF1. Interestingly, during the last reverse Heparin step, we isolated the final cleaved product in the elution fractions, rather than in the flow-through (Figure 4). This can be explained by the isoelectric point (pI) of the target protein (pI 7.85), coupled with the cation exchange behavior of Heparin columns. Successful cleavage of both affinity tags in the final product was also confirmed by MS analysis (Supplementary Figure S13).

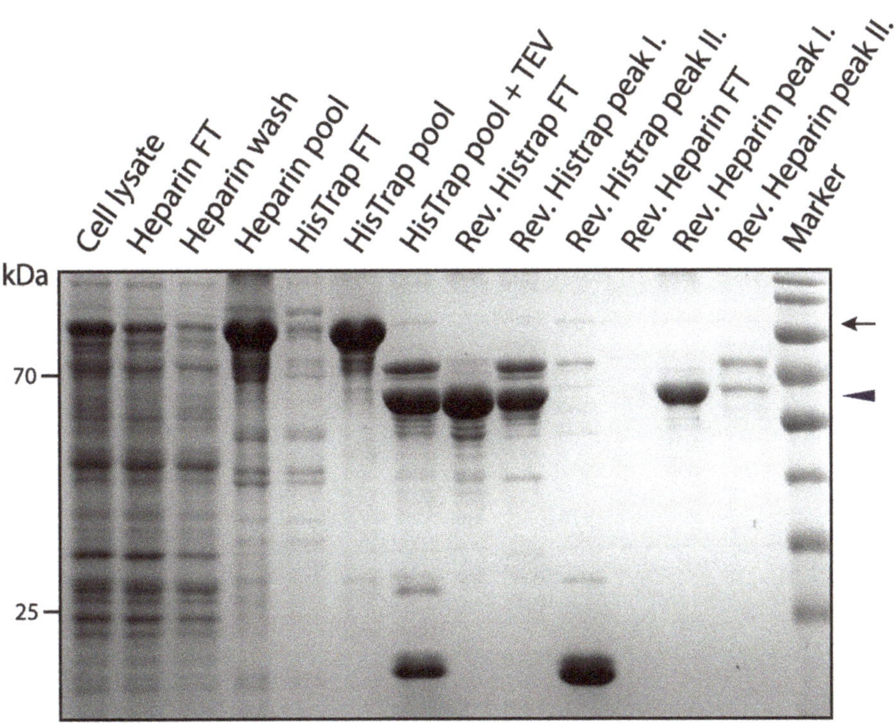


**Figure 4.** SDS-PAGE summarizing pSUMO-Tau-441 purification. The black arrow indicates intact Tau-441 having both affinity tags, while the black triangle indicates the cleaved final product.

3.3.2. Purification of pSUMO-Tau-NTMT

Biomolecular condensation of Tau has emerged as a crucial process in both its physiological microtubule-associated functions and pathological aggregation leading to neurodegeneration [56,57]. The N-terminal half of Tau-441 coupled to the microtubule-binding domain (Tau-NTMT; residues 1 to 372 of Tau-441) have been identified as the minimal construct driving its LLPS via intramolecular and intermolecular electrostatic interactions [56]. The purification of this construct from bacteria presents similar challenges as full-length Tau-441, considering that it only lacks a very short C-terminal tail, with its flexible N-terminal region and the aggregation-prone microtubule-binding region still present. We have attempted to purify this recombinant protein using our tandem-tag purification strategy following a similar workflow to that of pSUMO-Tau-441. Surprisingly, not much of the desired protein was recoverable when starting with Heparin affinity purification instead of the HisTrap purification (data not shown). This could be indicative of the lower binding capacity of the Heparin-HP-5 mL chromatographic column when compared to that of a HisTrap-HP-5 mL column, and not the binding affinity of the DBD-tag versus the 6xHis-tag to their respective affinity columns. It is for this reason that the purification strategy where the HisTrap preceded Heparin chromatographic separation, followed by affinity-tag cleavage using TEV protease and eventually separating the cleaved tags from purified Tau molecules, was employed for these two Tau domain constructs.

Considering that most of the expressed recombinant protein was in the soluble fraction of bacterial whole-cell lysate, pSUMO-Tau-NTMT was the most dominant protein bound to the HisTrap column in the first chromatographic step (Supplementary Figure S6). Therefore, all elution fractions containing pSUMO-Tau-NTMT were pooled, dialyzed against Heparin buffer A, and loaded onto the Heparin-HP-5 mL column. Here, pSUMO-Tau-NTMT was favored to bind to the Heparin column due to a higher salt concentration in the binding buffer (250 mM NaCl). On the other hand, the binding of residual bacterial contaminants and truncated degradation products was disfavored under these salt conditions (Supplementary Figure S7). The two affinity tags flanking our protein of interest were simultaneously cleaved via TEV protease digestion applied overnight at 4 °C. In the following morning, additional TEV protease was added to the dialyzed sample and the protein mixture was incubated on ice for another hour before reverse HisTrap chromatographic separation. The additional TEV protease ensured a maximal cleavage of pSUMO-Tau-NTMT, and hence a higher recovery of the cleaved full-length protein in the flow-through after the reverse HisTrap chromatography step (Figure 5). The 6xHis-SUMO-tag and TEV protease were retained on the HisTrap column (TEV protease also has an N-terminal 6xHis-tag), thereby facilitating their separation from our cleaved protein of interest. Reverse Heparin chromatography served as a final polishing step, also concentrating purified Tau-NTMT due to its ability to bind to the Heparin column, just like purified Tau-441 (Figure 5).

The final product, which eluted in the first peak of the reverse Heparin purification step, appeared to be an amalgamation of protein bands that were conjoined throughout the entire purification process (Figure 5). Our suspicion was that purified Tau tends to dimerize with both fully cleaved and partially cleaved Tau species (especially Tau with the DBD affinity tag still fused at the C-terminus). MS analysis confirmed our suspicion highlighting that the dominant gel band at 55 kilo Daltons (kDa) was pure Tau-NTMT with no affinity tags (Supplementary Figure S17), whereas the bands just above it—indicated with a red arrow on the gel in Figure 5—were Tau-NTMT-DBD with no 6xHis-SUMO affinity tag (Supplementary Figure S16). The high-molecular weight (HMW) bands—indicated at the top of the gel in Figure 5—were confirmed to be oligomers of pure Tau-NTMT by MS (Supplementary Figure S18). It is important to note that the purity of the final product could be drastically improved by addressing incomplete TEV digestion and dimerization/oligomerization of purified Tau monomers. This adjustment was made in the purification of Tau-MTBR (a more oligomerizing and aggregation-prone domain of Tau-441) and yielded satisfactory results, as described in the next section.

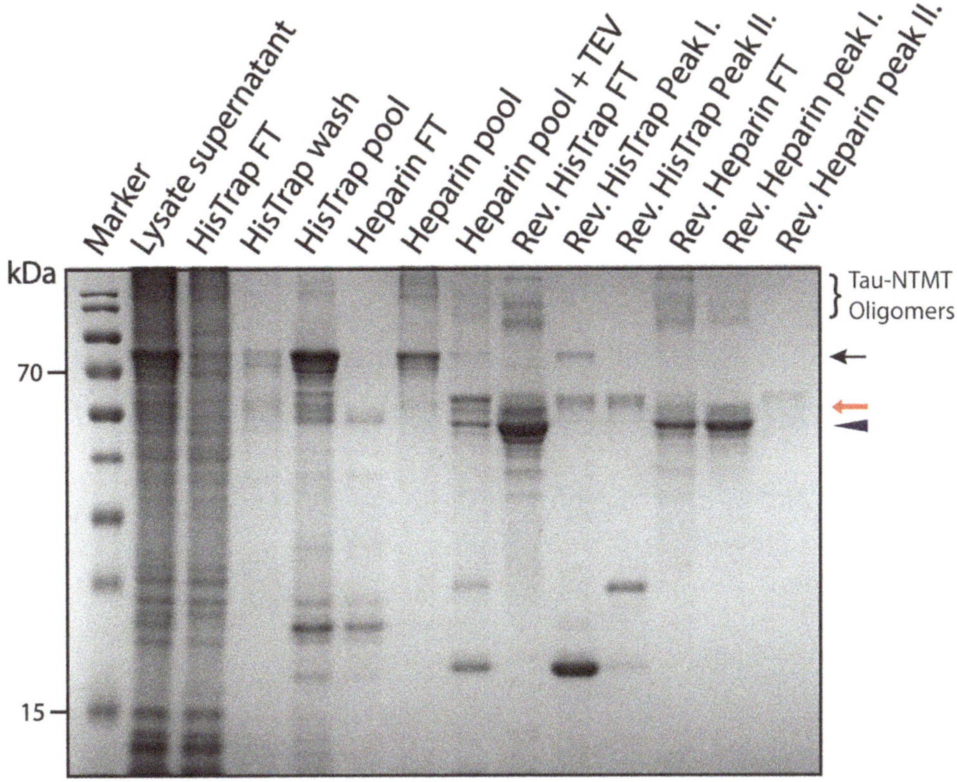


**Figure 5.** SDS-PAGE summarizing pSUMO-Tau-NTMT purification. The black arrow indicates intact Tau-NTMT with both affinity tags, the red arrow indicates Tau-NTMT with a DBD-affinity tag at its C-terminus and the black triangle indicates the fully cleaved Tau-NTMT product.

#### 3.3.3. Purification of pSUMO-Tau-MTBR

As a neuron-specific microtubule-associated protein, the primary functions of Tau include regulating microtubule dynamics, maintaining neuronal cytoskeletal integrity and facilitating both anterograde and retrograde axonal transport [60–62]. Tau's interaction with microtubules is mediated by its microtubule-binding domain (Tau-MTBR; residues 225–372 of Tau-441), which houses four pseudo-repeat sequences that bind to polymerized microtubule bundles with high affinity [63]. This mode of interaction allows the flexible N-terminal half to project away from bound microtubules and mediate microtubule spacing [64]. On the other hand, Tau-MTBR is usually at the center of both amyloid and amorphous aggregates in a range of diseases termed Tauopathies. The borders of the second and third pseudo-repeats house two highly hydrophobic hexapeptides, which form the core of Tau amyloids observed in patients with AD [65,66]. In other Tauopathies such as FTD, Tau forms amorphous aggregates causing neurodegeneration and dementia in affected individuals [67]. As a domain, Tau-MTBR does not phase separate on its own, rather it forms complex coacervates with polyanions such as RNA and heparin at favorable molar ratios [68].

The purification of pSUMO-Tau-MTBR presents a specific challenge, due to its known intrinsic aggregation propensity. In our hands, a solution of Tau-MTBR turned turbid in the dialysis bag upon incubation with TEV protease to cleave off the flanking affinity tags (including SUMO, which played a solubilizing role). This was indicative of LLPS among Tau-MTBR molecules with possibly nucleic acids, originally from the bacterial lysate, via charge interactions [58]. We found that performing the TEV cleavage in a buffer of higher salt (250 mM NaCl) and reducing agent (5 mM TCEP) concentrations prevented the protein solution from turning turbid. Coincidentally, like pSUMO-Tau-NTMT purification, high salt inhibited non-specific binding of bacterial contaminants to the HisTrap column and favored binding of our protein to the Heparin column rather than the 6xHis-SUMO-containing truncated fragments (Supplementary Figures S8, S9 and S19). The purification strategy where Heparin preceded HisTrap chromatographic separation still did not perform any better with this construct (data not shown), and for the same reason the approach fell short for pSUMO-Tau-NTMT. The slow expression of Tau-MTBR (16 °C, Overnight) facilitated its recovery into the soluble fraction of the bacterial cell lysate. This ensured more pSUMO-Tau-MTBR starting material was available in the lysate supernatant for the first purification step, fully saturating the HisTrap-HP-5 mL column with just one liter of bacterial lysate (Figure 6).

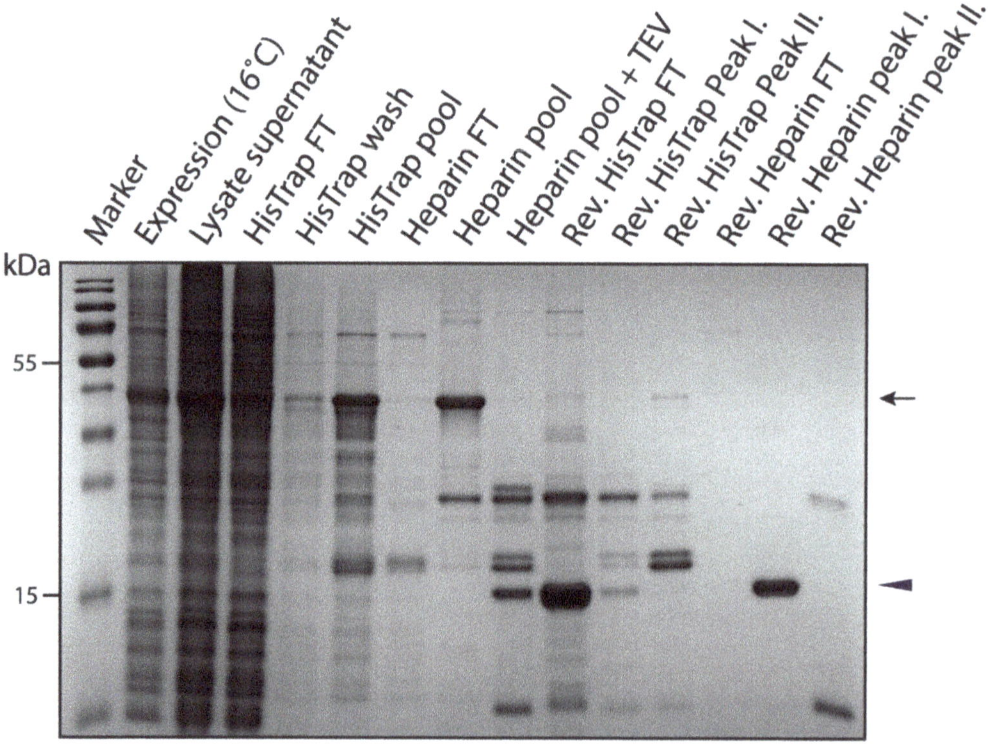


**Figure 6.** SDS-PAGE summarizing pSUMO-Tau-MTBR purification. The black arrow indicates intact Tau-MTBR having both affinity tags, while the black triangle indicates the cleaved final product.

To improve on the purity of the final product, extra TEV protease (0.2 mg/mL) was added to the Heparin pool fractions prior to overnight dialysis and extra reducing agent (10 mM TCEP) was added to the dialyzed flow-through sample from the reverse HisTrap prior to the final polishing step (Reverse Heparin). Extra TEV protease ensured the depletion of partially cleaved pSUMO-Tau-MTBR species before taking the protein solution through Reverse HisTrap and Heparin purification steps. The efficiency of TEV proteolytic cleavage can be appreciated in the digested sample, where the large majority of full-length pSUMO-Tau-MTBR was successfully cleaved (Figure 6). However, even with extra TEV protease, there was still a population of partially cleaved Tau-MTBR with a DBD-affinity tag at their C-terminus, which was confirmed by MS (Supplementary Figure S20). The addition of more TCEP to the protein solution before reverse Heparin chromatographic separation prevented oligomerization and ensured that the undesired truncations did not dimerize and elute with fully cleaved Tau-MTBR monomers, as it happened during the purification of Tau-NTMT and AF1. The complete separation of these two populations with extra TCEP resulted in elution of DBD-containing fragments together with free DBD-tag molecules in the second peak of Reverse Heparin purification step (Figure 6). Just as with Tau-441 and Tau-NTMT, fully cleaved Tau-MTBR also bound to the Heparin column, concentrating the final product in the process. Pure Tau-MTBR eluted in the first peak of the Reverse Heparin purification step, confirmed by MS analysis (Figure 6 and Supplementary Figure S21).

## 4. Discussion

Recombinant expression and purification of IDPs is a field where improvements are still needed. These proteins are often challenging to purify because of their unique characteristics. However, due to their crucial roles in many cellular processes and in LLPS, they are subject of immense interest, making their production necessary in high quantity and purity. Here, we have outlined a method to overcome one of the main hurdles in their preparation, i.e., truncated degradation contaminants. The combination of two affinity chromatography steps by using a tandem-tag bacterial expression system (pSUMO) enables their easy separation from degradation products. Moreover, the two chromatographic systems applied in the method (HisTrap and Heparin) are easy to use and commercially available from a variety of sources. The tags can be removed in one step and separation of the cleaved products from the final product, while polishing the sample in the process, happens at the same time without the need of size-exclusion chromatography (SEC). By eliminating the usual polishing SEC step, the method has a fast workflow that is crucial for IDPs and suitable for industrial scale-up, thereby widening the application scope.

The robustness and versatility of the method was shown through approaching challenging and important IDP examples. The presented examples have a vast pI range, from acidic (AF1) to basic (Tau-441, Tau-NTMT, Tau-MTBR). As a matter of fact, the Tau domain constructs have an increasing pI as the length decreases, yet the method still performs well independently of charge characteristics. When compared to the existing purifications in the literature, our method had a clear advantage in the examples shown. For instance, in case of AF1, the purity of the sample was substantially increased, with degradation contaminants falling almost below detection limit: only a small amount of partially cleaved protein could be detected due to the formation of mixed dimers between partially cleaved and fully cleaved AF1. This can be overcome by adding more reducing agent before the reverse Heparin purification step, as demonstrated in the case of Tau-MTBR. In addition to improving purity, we also achieved significantly increased yields. In the case of Tau and its domain constructs, the purity of the sample was as good or better than that reported in the literature, with practically undetectable degradation contaminants [69,70]. Our method also allows the elimination of the boiling step that is favored in the IDP field, but which can have adverse effects on sample quality (causing oxidation, deamidation, etc.), thereby compromising downstream experiments. In the case of Tau constructs, the yield was comparable or slightly better than that obtained with existing purification methods [69,70].

Another level of versatility of the method is changing the order of the affinity chromatography steps. By starting with HisTrap followed by Heparin, the method seems to be more general. However, starting with Heparin followed by HisTrap has the advantage of enabling the application of an increased amount of reducing agent during cell lysis. Furthermore, the elution buffer of the Heparin chromatography step is more compatible with the loading buffer of the HisTrap purification step, hence requiring less buffer exchange steps to ensure a faster workflow. Nonetheless, our experience with this system has shown us that it is best to optimize which order works best for a given target protein.

The findings in this study demonstrate that novel avenues of IDP purification are conceivable, overcoming the existing hurdles to ensure IDP products of high quantity and purity. To emphasize the generality of our method, we would like to produce labeled proteins for NMR experiments as a possible future application. For ongoing projects in our laboratory, we plan to purify labeled Tau-MTBR, alongside a few other IDPs, by making use of the presented method.

**Supplementary Materials:** The following supporting information can be downloaded at https://www.mdpi.com/article/10.3390/biom12111566/s1: Figures S1–S9: SDS-PAGE gel pictures for the various purifications, and Figures S10–S21: MS alignment pictures for the various purifications.

**Author Contributions:** A.M., J.A., K.M. and P.T. conceived the project and planned the experiments. A.M. and K.M. performed the cloning, and the protein purification experiments as well as prepared the MS samples. S.J. helped with MS sample preparation, performed the MS measurements, and analyzed the data. A.M. and K.M. wrote the manuscript with contribution from all the authors. P.T. corrected and edited the final version of the manuscript. All authors have read and agreed to the published version of the manuscript.

**Funding:** This work was supported by a VUB Strategic Research Program on Microfluidics (SRP51) at Vrije Universiteit Brussel (VUB, Brussels, Belgium, to P.T.), an EC H2020-WIDESPREAD-2020-5 Twinning grant (PhaseAGE, no. 952334, to P.T.), an EC H2020-MSCA-RISE Action grant (IDPfun, no. 778247, to P.T.), grants K124670 and K131702 from the National Research, Development and Innovation Office (NKFIH, Hungary, to P.T.) and FWO PhD fellowships in strategic basic research (FWOSB77, to J.A.) and fundamental research (11D0122N-BIO1, to K.M.).

**Institutional Review Board Statement:** Not applicable.

**Informed Consent Statement:** Not applicable.

**Data Availability Statement:** The data presented in this study are available upon request from the corresponding author.

**Conflicts of Interest:** The authors declare no conflict of interest.

## References

1. Wingfield, P.T. Overview of the Purification of Recombinant Proteins. *Curr. Protoc. Protein Sci.* **2015**, *80*, 6.1.1–6.1.35. [CrossRef] [PubMed]
2. Walsh, G. Biopharmaceutical Benchmarks 2018. *Nat. Biotechnol.* **2018**, *36*, 1136–1145. [CrossRef] [PubMed]
3. Kim, Y.; Bigelow, L.; Borovilos, M.; Dementieva, I.; Duggan, E.; Eschenfeldt, W.; Hatzos, C.; Joachimiak, G.; Li, H.; Maltseva, N.; et al. Chapter 3. High-Throughput Protein Purification for X-Ray Crystallography and NMR. *Adv. Protein Chem. Struct. Biol.* **2008**, *75*, 85–105. [PubMed]
4. Edwards, A.M.; Arrowsmith, C.H.; Christendat, D.; Dharamsi, A.; Friesen, J.D.; Greenblatt, J.F.; Vedadi, M. Protein Production: Feeding the Crystallographers and NMR Spectroscopists. *Nat. Struct. Biol.* **2000**, *7*, 970–972. [CrossRef] [PubMed]
5. Hura, G.L.; Menon, A.L.; Hammel, M.; Rambo, R.P.; Poole, F.L., 2nd; Tsutakawa, S.E.; Jenney, F.E., Jr.; Classen, S.; Frankel, K.A.; Hopkins, R.C.; et al. Robust, High-Throughput Solution Structural Analyses by Small Angle X-Ray Scattering (SAXS). *Nat. Methods* **2009**, *6*, 606–612. [CrossRef]
6. Stark, H.; Chari, A. Sample Preparation of Biological Macromolecular Assemblies for the Determination of High-Resolution Structures by Cryo-Electron Microscopy. *Microscopy* **2016**, *65*, 23–34. [CrossRef]
7. Li, Y. Commonly Used Tag Combinations for Tandem Affinity Purification. *Biotechnol. Appl. Biochem.* **2010**, *55*, 73–83. [CrossRef]
8. Schmidt, T.G.M.; Skerra, A. The Strep-Tag System for One-Step Purification and High-Affinity Detection or Capturing of Proteins. *Nat. Protoc.* **2007**, *2*, 1528–1535. [CrossRef]
9. Kronqvist, N.; Sarr, M.; Lindqvist, A.; Nordling, K.; Otikovs, M.; Venturi, L.; Pioselli, B.; Purhonen, P.; Landreh, M.; Biverstål, H.; et al. Efficient Protein Production Inspired by How Spiders Make Silk. *Nat. Commun.* **2017**, *8*, 15504. [CrossRef]
10. Tripathi, N.K.; Shrivastava, A. Recent Developments in Bioprocessing of Recombinant Proteins: Expression Hosts and Process Development. *Front. Bioeng. Biotechnol.* **2019**, *7*, 420. [CrossRef]
11. Demain, A.L.; Vaishnav, P. Production of Recombinant Proteins by Microbes and Higher Organisms. *Biotechnol. Adv.* **2009**, *27*, 297–306. [CrossRef]
12. Adrio, J.-L.; Demain, A.L. Recombinant Organisms for Production of Industrial Products. *Bioeng. Bugs* **2010**, *1*, 116–131. [CrossRef]
13. Gupta, S.K.; Shukla, P. Advanced Technologies for Improved Expression of Recombinant Proteins in Bacteria: Perspectives and Applications. *Crit. Rev. Biotechnol.* **2016**, *36*, 1089–1098. [CrossRef]
14. Ferrer-Miralles, N.; Domingo-Espín, J.; Corchero, J.L.; Vázquez, E.; Villaverde, A. Microbial Factories for Recombinant Pharmaceuticals. *Microb. Cell Fact.* **2009**, *8*, 17. [CrossRef]

15. Mamat, U.; Wilke, K.; Bramhill, D.; Schromm, A.B.; Lindner, B.; Kohl, T.A.; Corchero, J.L.; Villaverde, A.; Schaffer, L.; Head, S.R.; et al. Detoxifying *Escherichia coli* for Endotoxin-Free Production of Recombinant Proteins. *Microb. Cell Fact.* **2015**, *14*, 57. [CrossRef]
16. Carrió, M.M.; Villaverde, A. Protein Aggregation as Bacterial Inclusion Bodies Is Reversible. *FEBS Lett.* **2001**, *489*, 29–33. [CrossRef]
17. Carrió, M.M.; Villaverde, A. Construction and Deconstruction of Bacterial Inclusion Bodies. *J. Biotechnol.* **2002**, *96*, 3–12. [CrossRef]
18. Owczarek, B.; Gerszberg, A.; Hnatuszko-Konka, K. A Brief Reminder of Systems of Production and Chromatography-Based Recovery of Recombinant Protein Biopharmaceuticals. *Biomed Res. Int.* **2019**, *2019*, 4216060. [CrossRef]
19. Fletcher, E.; Krivoruchko, A.; Nielsen, J. Industrial Systems Biology and Its Impact on Synthetic Biology of Yeast Cell Factories. *Biotechnol. Bioeng.* **2016**, *113*, 1164–1170. [CrossRef]
20. McKenzie, E.A.; Abbott, W.M. Expression of Recombinant Proteins in Insect and Mammalian Cells. *Methods* **2018**, *147*, 40–49. [CrossRef]
21. Puetz, J.; Wurm, F.M. Recombinant Proteins for Industrial versus Pharmaceutical Purposes: A Review of Process and Pricing. *Processes* **2019**, *7*, 476. [CrossRef]
22. Bernhard, F.; Tozawa, Y. Cell-Free Expression–Making a Mark. *Curr. Opin. Struct. Biol.* **2013**, *23*, 374–380. [CrossRef] [PubMed]
23. Silverman, A.D.; Karim, A.S.; Jewett, M.C. Cell-Free Gene Expression: An Expanded Repertoire of Applications. *Nat. Rev. Genet.* **2020**, *21*, 151–170. [CrossRef] [PubMed]
24. Rosano, G.L.; Ceccarelli, E.A. Recombinant Protein Expression in *Escherichia coli*: Advances and Challenges. *Front. Microbiol.* **2014**, *5*, 172. [CrossRef] [PubMed]
25. Sezonov, G.; Joseleau-Petit, D.; D'Ari, R. *Escherichia coli* Physiology in Luria-Bertani Broth. *J. Bacteriol.* **2007**, *189*, 8746–8749. [CrossRef]
26. Sahdev, S.; Khattar, S.K.; Saini, K.S. Production of Active Eukaryotic Proteins through Bacterial Expression Systems: A Review of the Existing Biotechnology Strategies. *Mol. Cell. Biochem.* **2008**, *307*, 249–264. [CrossRef]
27. Gopal, G.J.; Kumar, A. Strategies for the Production of Recombinant Protein in *Escherichia coli*. *Protein J.* **2013**, *32*, 419–425. [CrossRef]
28. Rosano, G.L.; Morales, E.S.; Ceccarelli, E.A. New Tools for Recombinant Protein Production in *Escherichia coli*: A 5-Year Update. *Protein Sci.* **2019**, *28*, 1412–1422. [CrossRef]
29. Liu, M.; Wang, B.; Wang, F.; Yang, Z.; Gao, D.; Zhang, C.; Ma, L.; Yu, X. Soluble Expression of Single-Chain Variable Fragment (scFv) in *Escherichia coli* Using Superfolder Green Fluorescent Protein as Fusion Partner. *Appl. Microbiol. Biotechnol.* **2019**, *103*, 6071–6079. [CrossRef]
30. Paraskevopoulou, V.; Falcone, F.H. Polyionic Tags as Enhancers of Protein Solubility in Recombinant Protein Expression. *Microorganisms* **2018**, *6*, 47. [CrossRef]
31. De Marco, A. Protocol for Preparing Proteins with Improved Solubility by Co-Expressing with Molecular Chaperones in *Escherichia coli*. *Nat. Protoc.* **2007**, *2*, 2632–2639. [CrossRef]
32. Jo, B.H. An Intrinsically Disordered Peptide Tag That Confers an Unusual Solubility to Aggregation-Prone Proteins. *Appl. Environ. Microbiol.* **2022**, *88*, e0009722. [CrossRef]
33. Gupta, S.K.; Dangi, A.K.; Smita, M.; Dwivedi, S.; Shukla, P. Effectual Bioprocess Development for Protein Production. In *Applied Microbiology and Bioengineering*; Elsevier: Amsterdam, The Netherlands, 2019; pp. 203–227. ISBN 9780128154076.
34. Wacker, M.; Linton, D.; Hitchen, P.G.; Nita-Lazar, M.; Haslam, S.M.; North, S.J.; Panico, M.; Morris, H.R.; Dell, A.; Wren, B.W.; et al. N-Linked Glycosylation in Campylobacter Jejuni and Its Functional Transfer into *E. coli*. *Science* **2002**, *298*, 1790–1793. [CrossRef]
35. Valderrama-Rincon, J.D.; Fisher, A.C.; Merritt, J.H.; Fan, Y.-Y.; Reading, C.A.; Chhiba, K.; Heiss, C.; Azadi, P.; Aebi, M.; DeLisa, M.P. An Engineered Eukaryotic Protein Glycosylation Pathway in *Escherichia coli*. *Nat. Chem. Biol.* **2012**, *8*, 434–436. [CrossRef]
36. Tompa, P. Intrinsically Disordered Proteins: A 10-Year Recap. *Trends Biochem. Sci.* **2012**, *37*, 509–516. [CrossRef]
37. Uversky, V.N. Introduction to Intrinsically Disordered Proteins (IDPs). *Chem. Rev.* **2014**, *114*, 6557–6560. [CrossRef]
38. Suskiewicz, M.J.; Sussman, J.L.; Silman, I.; Shaul, Y. Context-Dependent Resistance to Proteolysis of Intrinsically Disordered Proteins. *Protein Sci.* **2011**, *20*, 1285–1297. [CrossRef]
39. Hammarberg, B.; Nygren, P.A.; Holmgren, E.; Elmblad, A.; Tally, M.; Hellman, U.; Moks, T.; Uhlén, M. Dual Affinity Fusion Approach and Its Use to Express Recombinant Human Insulin-like Growth Factor II. *Proc. Natl. Acad. Sci. USA* **1989**, *86*, 4367–4371. [CrossRef]
40. McInnes, J.; Zhou, L.; Verstreken, A.P. Purification of Soluble Recombinant Human Tau Protein from Bacteria Using Double-Tag Affinity Purification. *Bio. Protoc.* **2018**, *8*, e3043. [CrossRef]
41. Ortega, C.; Prieto, D.; Abreu, C.; Oppezzo, P.; Correa, A. Multi-Compartment and Multi-Host Vector Suite for Recombinant Protein Expression and Purification. *Front. Microbiol.* **2018**, *9*, 1384. [CrossRef]
42. Bekesi, A.; Abdellaoui, S.; Holroyd, N.; Van Delm, W.; Pardon, E.; Pauwels, J.; Gevaert, K.; Steyaert, J.; Derveaux, S.; Borysik, A.; et al. Challenges in the Structural-Functional Characterization of Multidomain, Partially Disordered Proteins CBP and p300: Preparing Native Proteins and Developing Nanobody Tools. *Methods Enzymol.* **2018**, *611*, 607–675. [PubMed]
43. Huang, L.; Qu, X.; Chen, Y.; Xu, W.; Huang, C. Sandwiched-Fusion Strategy Facilitates Recombinant Production of Small Labile Proteins. *Protein Sci.* **2021**, *30*, 650–662. [CrossRef] [PubMed]

44. Einhauer, A.; Jungbauer, A. The FLAG Peptide, a Versatile Fusion Tag for the Purification of Recombinant Proteins. *J. Biochem. Biophys. Methods* **2001**, *49*, 455–465. [CrossRef]
45. Guerrero, F.; Ciragan, A.; Iwaï, H. Tandem SUMO Fusion Vectors for Improving Soluble Protein Expression and Purification. *Protein Expr. Purif.* **2015**, *116*, 42–49. [CrossRef] [PubMed]
46. Young, C.L.; Britton, Z.T.; Robinson, A.S. Recombinant Protein Expression and Purification: A Comprehensive Review of Affinity Tags and Microbial Applications. *Biotechnol. J.* **2012**, *7*, 620–634. [CrossRef]
47. Siri, A.; Balza, E.; Carnemolla, B.; Castellani, P.; Borsi, L.; Zardi, L. DNA-Binding Domains of Human Plasma Fibronectin. pH and Calcium Ion Modulation of Fibronectin Binding to DNA and Heparin. *Eur. J. Biochem.* **1986**, *154*, 533–538. [CrossRef]
48. Bolten, S.N.; Rinas, U.; Scheper, T. Heparin: Role in Protein Purification and Substitution with Animal-Component Free Material. *Appl. Microbiol. Biotechnol.* **2018**, *102*, 8647–8660. [CrossRef]
49. Staby, A.; Sand, M.-B.; Hansen, R.G.; Jacobsen, J.H.; Andersen, L.A.; Gerstenberg, M.; Bruus, U.K.; Jensen, I.H. Comparison of Chromatographic Ion-Exchange Resins IV. Strong and Weak Cation-Exchange Resins and Heparin Resins. *J. Chromatogr. A* **2005**, *1069*, 65–77. [CrossRef]
50. Liu, W.; Xie, Y.; Ma, J.; Luo, X.; Nie, P.; Zuo, Z.; Lahrmann, U.; Zhao, Q.; Zheng, Y.; Zhao, Y.; et al. IBS: An Illustrator for the Presentation and Visualization of Biological Sequences. *Bioinformatics* **2015**, *31*, 3359–3361. [CrossRef]
51. Davey, R.A.; Grossmann, M. Androgen Receptor Structure, Function and Biology: From Bench to Bedside. *Clin. Biochem. Rev.* **2016**, *37*, 3–15.
52. Monaghan, A.E.; McEwan, I.J. A Sting in the Tail: The N-Terminal Domain of the Androgen Receptor as a Drug Target. *Asian J. Androl.* **2016**, *18*, 687–694.
53. Reid, J.; Kelly, S.M.; Watt, K.; Price, N.C.; McEwan, I.J. Conformational Analysis of the Androgen Receptor Amino-Terminal Domain Involved in Transactivation. Influence of Structure-Stabilizing Solutes and Protein-Protein Interactions. *J. Biol. Chem.* **2002**, *277*, 20079–20086. [CrossRef]
54. Iqbal, K.; Liu, F.; Gong, C.-X. Tau and Neurodegenerative Disease: The Story so Far. *Nat. Rev. Neurol.* **2016**, *12*, 15–27. [CrossRef]
55. Mukrasch, M.D.; Bibow, S.; Korukottu, J.; Jeganathan, S.; Biernat, J.; Griesinger, C.; Mandelkow, E.; Zweckstetter, M. Structural Polymorphism of 441-Residue Tau at Single Residue Resolution. *PLoS Biol.* **2009**, *7*, e34. [CrossRef]
56. Kanaan, N.M.; Hamel, C.; Grabinski, T.; Combs, B. Liquid-Liquid Phase Separation Induces Pathogenic Tau Conformations in vitro. *Nat. Commun.* **2020**, *11*, 2809. [CrossRef]
57. Wegmann, S.; Eftekharzadeh, B.; Tepper, K.; Zoltowska, K.M.; Bennett, R.E.; Dujardin, S.; Laskowski, P.R.; MacKenzie, D.; Kamath, T.; Commins, C.; et al. Tau Protein Liquid-Liquid Phase Separation Can Initiate Tau Aggregation. *EMBO J.* **2018**, *37*, e98049. [CrossRef]
58. Hochmair, J.; Exner, C.; Franck, M.; Dominguez-Baquero, A.; Diez, L.; Brognaro, H.; Kraushar, M.L.; Mielke, T.; Radbruch, H.; Kaniyappan, S.; et al. Molecular Crowding and RNA Synergize to Promote Phase Separation, Microtubule Interaction, and Seeding of Tau Condensates. *EMBO J.* **2022**, *41*, e108882. [CrossRef]
59. Soeda, Y.; Yoshikawa, M.; Almeida, O.F.X.; Sumioka, A.; Maeda, S.; Osada, H.; Kondoh, Y.; Saito, A.; Miyasaka, T.; Kimura, T.; et al. Toxic Tau Oligomer Formation Blocked by Capping of Cysteine Residues with 1,2-Dihydroxybenzene Groups. *Nat. Commun.* **2015**, *6*, 10216. [CrossRef]
60. Morris, S.L.; Tsai, M.-Y.; Aloe, S.; Bechberger, K.; König, S.; Morfini, G.; Brady, S.T. Defined Tau Phosphospecies Differentially Inhibit Fast Axonal Transport Through Activation of Two Independent Signaling Pathways. *Front. Mol. Neurosci.* **2020**, *13*, 610037. [CrossRef]
61. Venkatramani, A.; Panda, D. Regulation of Neuronal Microtubule Dynamics by Tau: Implications for Tauopathies. *Int. J. Biol. Macromol.* **2019**, *133*, 473–483. [CrossRef]
62. Barbier, P.; Zejneli, O.; Martinho, M.; Lasorsa, A.; Belle, V.; Smet-Nocca, C.; Tsvetkov, P.O.; Devred, F.; Landrieu, I. Role of Tau as a Microtubule-Associated Protein: Structural and Functional Aspects. *Front. Aging Neurosci.* **2019**, *11*, 204. [CrossRef] [PubMed]
63. Kadavath, H.; Hofele, R.V.; Biernat, J.; Kumar, S.; Tepper, K.; Urlaub, H.; Mandelkow, E.; Zweckstetter, M. Tau Stabilizes Microtubules by Binding at the Interface between Tubulin Heterodimers. *Proc. Natl. Acad. Sci. USA* **2015**, *112*, 7501–7506. [CrossRef] [PubMed]
64. Frappier, T.F.; Georgieff, I.S.; Brown, K.; Shelanski, M.L. Tau Regulation of Microtubule-Microtubule Spacing and Bundling. *J. Neurochem.* **1994**, *63*, 2288–2294. [CrossRef] [PubMed]
65. Gulisano, W.; Maugeri, D.; Baltrons, M.A.; Fà, M.; Amato, A.; Palmeri, A.; D'Adamio, L.; Grassi, C.; Devanand, D.P.; Honig, L.S.; et al. Role of Amyloid-β and Tau Proteins in Alzheimer's Disease: Confuting the Amyloid Cascade. *J. Alzheimers's Dis.* **2018**, *64*, S611–S631. [CrossRef] [PubMed]
66. Von Bergen, M.; Barghorn, S.; Li, L.; Marx, A.; Biernat, J.; Mandelkow, E.M.; Mandelkow, E. Mutations of Tau Protein in Frontotemporal Dementia Promote Aggregation of Paired Helical Filaments by Enhancing Local Beta-Structure. *J. Biol. Chem.* **2001**, *276*, 48165–48174. [CrossRef]
67. Lin, L.-C.; Nana, A.L.; Hepker, M.; Hwang, J.-H.L.; Gaus, S.E.; Spina, S.; Cosme, C.G.; Gan, L.; Grinberg, L.T.; Geschwind, D.H.; et al. Preferential Tau Aggregation in von Economo Neurons and Fork Cells in Frontotemporal Lobar Degeneration with Specific MAPT Variants. *Acta Neuropathol. Commun.* **2019**, *7*, 159. [CrossRef]

68. Parolini, F.; Tira, R.; Barracchia, C.G.; Munari, F.; Capaldi, S.; D'Onofrio, M.; Assfalg, M. Ubiquitination of Alzheimer's-Related Tau Protein Affects Liquid-Liquid Phase Separation in a Site- and Cofactor-Dependent Manner. *Int. J. Biol. Macromol.* **2022**, *201*, 173–181. [CrossRef]
69. Ferrari, L.; Rüdiger, S.G.D. Recombinant Production and Purification of the Human Protein Tau. *Protein Eng. Des. Sel.* **2018**, *31*, 447–455. [CrossRef]
70. Barghorn, S.; Biernat, J.; Mandelkow, E. Purification of Recombinant Tau Protein and Preparation of Alzheimer-Paired Helical Filaments in Vitro. *Methods Mol. Biol.* **2005**, *299*, 35–51.

*biomolecules*

MDPI

*Article*

# Sequence Properties of an Intramolecular Interaction that Inhibits p53 DNA Binding

**Emily Gregory and Gary W. Daughdrill** *

Department of Cell Biology, Microbiology, and Molecular Biology, University of South Florida, Tampa, FL 33620, USA
* Correspondence: gdaughdrill@usf.edu; Tel.: +1-813-974-2503

**Abstract:** An intramolecular interaction between the p53 transactivation and DNA binding domains inhibits DNA binding. To study this autoinhibition, we used a fragment of p53, referred to as ND WT, containing the N-terminal transactivation domains (TAD1 and TAD2), a proline rich region (PRR), and the DNA binding domain (DBD). We mutated acidic, nonpolar, and aromatic amino acids in TAD2 to disrupt the interaction with DBD and measured the effects on DNA binding affinity at different ionic strengths using fluorescence anisotropy. We observed a large increase in DNA binding affinity for the mutants consistent with reduced autoinhibition. The $\Delta\Delta G$ between DBD and ND WT for binding a consensus DNA sequence is $-3.0$ kcal/mol at physiological ionic strength. $\Delta\Delta G$ increased to $-1.03$ kcal/mol when acidic residues in TAD2 were changed to alanine (ND DE) and to $-1.13$ kcal/mol when all the nonpolar residues, including W53/F54, were changed to alanine (ND NP). These results indicate there is some cooperation between acidic, nonpolar, and aromatic residues from TAD2 to inhibit DNA binding. The dependence of DNA binding affinity on ionic strength was used to predict excess counterion release for binding both consensus and scrambled DNA sequences, which was smaller for ND WT and ND NP with consensus DNA and smaller for scrambled DNA overall. Using size exclusion chromatography, we show that the ND mutants have similar Stokes radii to ND WT suggesting the mutants disrupt autoinhibition without changing the global structure.

**Keywords:** tumor suppressor p53; intrinsically disordered proteins; intramolecular interaction; salt-dependent binding affinity; counterion condensation theory; DNA binding; fluorescence anisotropy; van't Hoff; hydrodynamic radius

**Citation:** Gregory, E.; Daughdrill, G.W. Sequence Properties of an Intramolecular Interaction that Inhibits p53 DNA Binding. *Biomolecules* **2022**, *12*, 1558. https://doi.org/10.3390/biom12111558

Academic Editors: Prakash Kulkarni, Stefania Brocca, Keith Dunker and Sonia Longhi

Received: 25 September 2022
Accepted: 18 October 2022
Published: 25 October 2022



## 1. Introduction

In response to cellular stress, the p53 tumor suppressor binds promoter response element DNA, activating transcription by recruiting the general transcription machinery [1–3]. Transcribed genes control cell fate decisions including cell cycle arrest, senescence, and apoptosis [4]. It is the most frequently mutated gene found in cancer, and mutations that interfere with DNA binding are found in a large subset of solid tumors [5–7]. p53 DNA binding and transcriptional activation is regulated by posttranslational modification, accumulation level, and association with other cellular factors [3,6–10]. p53 binds a 20 base pair DNA sequence consisting of two inverted repeats with the degenerate consensus sequence RRRCWWGYYY, where R is A/G, W is A/T, and Y is C/T [11,12]. p53 binds DNA as a homodimer to one 10 base pair repeat. The binding of a dimer to one repeat recruits a second dimer to the second repeat in a highly cooperative manner [13–15] and this homotetramer is the functional form of p53 [16,17]. Binding affinity of p53 to promoter DNA correlates with transactivation of genes, with dissociation constants ($K_D$) ranging over three orders of magnitude, from low nanomolar to low micromolar, and binding affinity is higher for promoters that control cell cycle arrest and lower for promoters that control apoptosis [12,17–21]. Like most DNA-binding proteins, p53 binds both specific and nonspecific DNA [8,13].

p53's DNA binding affinity is regulated by an autoinhibitory intramolecular interaction between the disordered N-terminal transactivation domain (TAD) and the ordered DNA binding domain (DBD), resulting in a lowered DNA binding affinity and an increase in specificity for target DNA [22,23]. As shown in Figure 1a,b, the domain structure of p53 is defined as a TAD that is further divided into TAD1 (1–39) and TAD2 (40–60), followed by a proline rich region (PRR, residues 61–93), a DNA-binding or Core domain (94–292), a linker (293–322), a tetramerization domain (TET, residues 323–355), and regulatory domain (REG, residues 356–393).

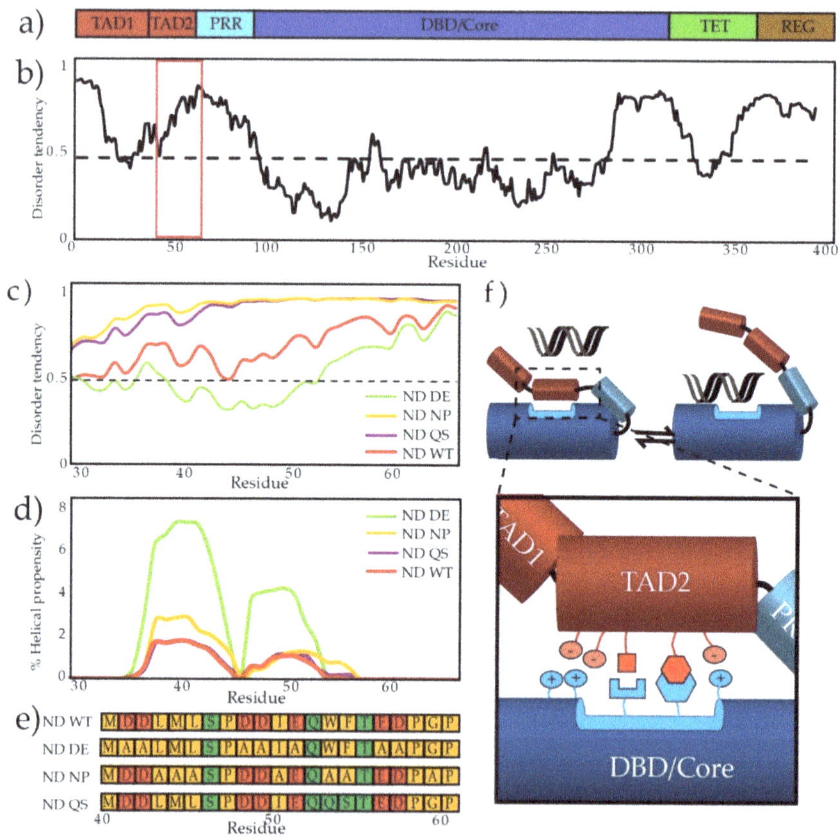


**Figure 1.** p53's disordered TAD2 interacts with DBD. (**a**) A domain map shows p53's domains. (**b**) IUPRED plot of full length p53 WT predicts regions of disorder based on sequence. The red box defines the region containing TAD2. (**c**) Inset from red box in (**b**) of IUPRED plot of a region containing TAD2 compares the disorder prediction of the wild type TAD2 and three mutants, where residues above the 0.5 line are predicted to be disordered. (**d**) Agadir prediction of helical propensity of the TAD2 region using wild type TAD2 and three mutants. (**e**) TAD2 sequences of the WT and mutants used in this study; red boxes indicate negatively charged residues, green boxes indicate polar residues, and gold boxes indicate nonpolar residues. (**f**) TAD2 interacts with DBD, inhibiting DNA binding by a combination a charge-based and specific interactions.

The p53 intramolecular interaction primarily involves TAD2 and PRR [22] with a small contribution from TAD1 [24]. TAD2 is acidic and phosphorylation of TAD2 modulates DNA binding affinity [24]; additionally, the intramolecular interaction is neutralized at high salt concentrations [23] which suggests a strong electrostatic component. However, NMR data implicate several of TAD2's noncharged residues in the interaction, pointing to a more complicated mechanism than merely the attraction of a negatively charged TAD with a positively charged DBD [22]. It is notable that the TAD2-DBD interaction does not confer a stable secondary structure to TAD2; it remains disordered even when bound to DBD [22]. The persistent disorder of the bound state is common in intrinsically disordered regions (IDRs) and is thought to decrease the entropic penalty of association [25,26]. A dynamic bound state is also observed in other IDRs that autoinhibit DNA binding like Ets-1 and HMGB1 [27,28]. The interaction between TAD2 and DBD is too weak *in trans* to measure by ITC. NMR data suggests a $K_D$ in the micromolar to millimolar range [22]. Despite this weak interaction, when TAD2 is tethered to DBD by PRR, the free energy of binding to a consensus DNA sequence is decreased by 3 kcal/mol. The tethering of TAD to the DBD increases the effective concentration as is seen in other examples of disordered regulatory regions, where *in trans* binding affinities range from the low micromolar to the low millimolar and yet have large effects on DNA binding affinity [29,30]. There is some evidence from our group and others that TAD2 from one subunit of the dimer contacts the DBD from the other subunit [22,24] and the intramolecular interaction we observed for a p53 monomer may become intermolecular in the dimer.

IDRs are enriched in transcription factors [31–33], and their contribution to promoter selection is increasingly recognized [33–36]. The mechanism IDRs use to inhibit DNA often appears to rely on negatively charged residues in the IDR screening the DNA binding pocket. This can result from a disordered acidic domain interacting with a positively charged DNA-binding domain, as observed for the FOXO transcription factors [36], RFX1 [37], the HMG box family member UBF [38], HMGB1 [27], RBBP1 [30], the Sox transcription factors [39], and p53 [22,23]. IDRs can also inhibit DNA binding when phosphorylated, as seen for B-Myb [40] while Ets-1 uses a combination of phosphorylated serines and aromatic residues to tune inhibition [28,41].

Because the intramolecular interaction between TAD2 and DBD is weak *in trans*, we assess the interaction *in cis* using DNA binding. We introduce mutations to TAD2 that are predicted to weaken the intramolecular interaction and lead to increased DNA binding affinity. Because TAD2 lacks secondary structure in its apo and DBD-bound states, we used predictive tools to assess our designed mutants. An IUPRED plot predicts changes to disorder of TAD2 (Figure 1c) [42] and the Agadir plot of helical propensity (Figure 1d) [43] shows predicted changes for the mutants. This study uses DBD (94–312), a fragment containing the N-terminus and DBD (ND; 1–312), and mutants of ND (shown in Figure 1e) with substitutions where 7 acidic residues were changed to alanine (ND DE), where 7 nonpolar residues were changed to alanine (ND NP), and where W53/F54 were changed to QS (ND QS). Figure 1f shows a model of the TAD2-DBD interaction with an emphasis on charged and nonpolar interactions. Because the interaction is dynamic, there not a single structure that corresponds to the autoinhibited state. However, we assume charge-charge and nonpolar-nonpolar interactions occur even if there is multivalency [44]. Using high and low affinity DNA sequences, we compare the ability of the TAD2 mutants to inhibit DNA binding across a range of ionic strength (IS) with the expectation that electrostatic features of the TAD2-DBD interaction will be more sensitive to changes in salt concentration than nonelectrostatic features.

## 2. Materials and Methods

### 2.1. Protein Expression

Synthetic cDNA fragments of p53 (Genscript, Piscataway, NY, USA) were ligated into the pGEX-6P-1 plasmid (Sigma-Aldrich, Burlington, MA, USA) using BamHI and EcoRI restriction sites. cDNA for the ND DE and ND NP mutants were synthesized and the ND QS mutant was generated by site-directed mutagenesis starting with ND WT using Agilent's Quikchange II protocol and kit (Santa Clara, CA, USA). All p53 fragments contain four stabilizing mutations in DBD: M133A, V203A, N239Y, and N268D [45]. Plasmids were transformed and expressed in BL21 (DE3) *E. coli* using minimal media at 37 °C to an O.D. of 0.5 at which point the media was supplemented with 20 μM $ZnCl_2$, allowed to cool to

15 °C, and induced with 1 mM IPTG for 20 h. Cells were centrifuged at 7168 rcf for 5 min and frozen at −80 °C. To purify protein, one liter of pelleted cells was resuspended in 25 mL lysis buffer containing 50 mM Tris (pH 7.4), 500 mM NaCl, 2 mM DTT, 0.02% $NaN_3$ and a fresh tablet of Pierce EDTA-free protease inhibitor (Thermo Fisher, Waltham, MA, USA). Cells were lysed via French press at approximately 1000 psi and centrifuged at 38,000 rcf for one hour. The supernatant was passed through a GST Fast-Flow Sepharose column (Cytiva, Marlboro, MA, USA) and eluted with 10 mM reduced glutathione. The eluted fractions containing the GST-tagged ND fragments were pooled and incubated with a 1:100 ratio of the HRV3C protease overnight at 4 °C to cleave the GST tag. The cleaved GST tag was removed by passing the mixture over another GST column. Following separation of p53 and the GST tag, fragments containing the TAD were dialyzed into a low-salt buffer and passed through a Q Sepharose High Performance anion exchange column (Cytiva), eluted in buffer containing 20 mM Tris at a pH of 7–8 depending on isoelectric point of the protein, 0–1 M NaCl, 2 mM DTT, and 0.02% $NaN_3$. All fragments were analyzed using polyacrylamide gel electrophoresis and protein samples were pooled and concentrated 25–50 µM and loaded on a 16/600 mm Superdex 75 column (Cytiva) in a buffer composed of 50 mM $NaH_2PO_4$ (pH 7), 300 mM NaCl, 1 mM DTT, and 0.02% $NaN_3$. Protein purity was evaluated via SDS-PAGE and concentration assessed using a Nanodrop 1000 Spectrometer (Thermo Fisher).

### 2.2. Preparation of DNA

HPLC-purified, 6-Carboxyfluorescein (6-FAM) tagged DNA was obtained from IDTDNA (Coralville, IA, USA) as single strands. Double-stranded DNA was annealed by boiling at 95 °C for 10 minutes and allowing to cool to room temperature. The sequences used are as follows: consensus 5′ AGACATGCCTAGACATGCCT and scrambled 5′ TGCCGATCAAAAC-CGATTCG. Annealing was confirmed using nondenaturing gel electrophoresis.

### 2.3. Fluorescence Anisotropy

Purified samples of DBD, ND WT, ND DE, ND NP, and ND QS were concentrated to 20–200 µM depending on the IS of the buffer and co-dialyzed with DNA twice against a buffer containing 10 mM $NaH_2PO_4$ (pH 7.4), 30–200 mM NaCl, 5 mM DTT, 0.02% $NaN_3$, and 0.01% Triton-X 100 for a total dilution factor of $1 \times 10^6$. 10 nM labeled DNA was aliquoted into Corning™ COSTAR 96-Well Solid Black Polystyrene Microplates (Thermo Fisher) and protein samples were added at increasing concentrations from 1 nM to saturation at 20–100 µM for a total volume of 100 µL. Fluorescence was measured using a Synergy H1 microplate reader from Biotek (Winooski, VT, USA) at 25 °C, and at 1.5° increments from 21–37 °C for van't Hoff analysis. Excitation and emission wavelengths were 485 nm and 528 nm, respectively, with a sample height of 7 cm, gain of 50, and shake and delay steps of 30 s and 20 s, respectively.

Binding affinities were estimated using a cooperative binding model for p53's interaction with consensus DNA as described previously [13] where p53 is evaluated as a dimer:

$$\Delta A = \frac{[p]^2}{K_D + [p]^2} \tag{1}$$

Where ΔA is the normalized anisotropy change, [p] is p53 dimer concentration. Binding affinity to scrambled DNA was calculated using a one-to-one binding model [46]:

$$\Delta A = \frac{[p] + [DNA] + K_D - \sqrt{([p] + [DNA] + K_D)^2 - 4[p][DNA]}}{2[DNA]} \tag{2}$$

The Hill coefficient was evaluated using the following equation [46]:

$$\Delta A = \frac{[p]^h / K_D{}^h}{1 + [p]^h / K_D{}^h} \quad (3)$$

where $h$ is the Hill coefficient, indicating the cooperativity of the binding event where 1 is a noncooperative event and greater than 1 is a cooperative event.

Enthalpy and entropy estimates were calculated from van't Hoff plots. These were generated by measuring anisotropy at physiological IS across a range of temperatures as previously described [47].

*2.4. Estimating Counterion Release*

The counterion condensation theory developed by Record and colleagues expands on the polyelectrolyte theory [48] to estimate ionic contacts and excess ion release for protein-nucleic acid binding [49] using the following relationship:

$$\log(K_A) = \log(K'_A) - N^*\log[\text{Salt}] \quad (4)$$

Where $K_A$ is the association constant, $K'_A$ is the nonelectrostatic component of binding, and $N$*log[Salt] is the electrostatic component of binding. $N$ is the slope of a double log plot of $K_A$ versus [Salt]. In this theory the electrostatic component of binding refers to the positive entropy associated with ion release [49,50]. It is unclear if this approach can quantitatively discriminate the salt-dependent entropic component of binding from other components, but we think it provides a useful qualitative segregation of components of binding affinity [51,52]. Because of this we refer to it these as the salt-dependent and salt-independent components of binding rather than as the electrostatic and nonelectrostatic components. The salt-independent component is inferred from the y-intercept of a log($K_A$) vs. log[Salt]. The slope of this plot, $N$, is further defined as:

$$N = Z\Psi + \beta \quad (5)$$

where Z is the number of protein-DNA backbone contacts made, Ψ is the fractional number of ions bound by phosphate, 0.7 for short oligonucleotides [53], and β is the number of excess ions released from protein. Our study utilizes only NaCl as the salt. Studies have found that variation of the monovalent cation, which is condensed around and ultimately released from DNA, is unimportant in evaluating ion release [50,54] although introduction of a divalent cation can have complicated effects on apparent ion release [55]. Variation of the anion may affect apparent ion release; however, the change in apparent ion release based on anion identity may reflect on the size of the anion or its relative attraction to water versus the protein side chains and thus varying the anion is not predicted to reveal additional information about the protein's DNA binding interface [50,56,57].

A reevaluation of the theory by Manning and colleagues resulted in the following relationship [58]:

$$\log(K_A) = \log(K_0) + \log V + 0.513Z - 0.434 - Z^*\log[\text{Salt}] \quad (6)$$

where $K_A$ is the association constant, $K_0$ is the salt-independent component of binding, V is the reaction volume, and Z represents the number of charged molecules associated with the binding event, which is interchangeable with $N$ from Equation (2).

Both these approaches use the section of a double log plot where log($K_A$) versus log[Salt] becomes linear, a range that is uniquely determined for a given protein. In this case, while fluorescence anisotropy was conducted on DBD and ND WT over an IS range of 15–225 mM, Supplementary Table S1 , the double log plot is linear in the 125–225 mM range. Thus, fluorescence anisotropy was only conducted on ND mutants in the 85–225 mM range and these were evaluated using the counterion condensation theory from 125–225 mM IS.

### 2.5. Size Exclusion Chromatography

Stokes radii ($R_H$) of the p53 fragments were determined using size exclusion chromatography (SEC). The Cytiva Gel Filtration Calibration Kit LMW was used to generate a calibration curve in a buffer of 50 mM $NaH_2PO_4$, pH 7.4, 300 mM NaCl, 0.02% $NaN_3$ using a HiLoad 16/600 mm Superdex 75 column (Cytiva, Marlboro, MA, USA) at 4 °C. A high ionic strength buffer was used to reduce binding to the sephadex beads and decrease line broadening. The elution volume of each protein was taken as the average of three injections, each of which contained 0.6–0.8 mg/mL of protein. The peak elution volume is used to find the partition coefficient, $K_{av}$:

$$K_{av} = (V_t - V_o)/(V_c - V_o) \quad (7)$$

Where $V_c$ is the total column volume, $V_o$ is the void volume, and $V_t$ is the elution volume. A plot of log($K_{av}$) versus the known $R_H$ of calibration kit standards generates a trendline from which $R_H$ of an unknown protein can be estimated [59,60]. Error of $R_H$ values is determined by the average of three runs where the resolution of the elution volume is 0.02 mL. We acknowledge previous work by Langridge and Whitten showing that the hydrodynamic radius of TAD increases with decreasing temperature [61].

## 3. Results

### 3.1. Salt Dependence of p53 DBD Binding DNA

We conducted binding experiments using fluorescence anisotropy in buffers with IS ranging from 15–225 mM. We used two DNA sequences. One is a high affinity sequence taken from a consensus promoter sequence [62], which we refer to as consensus DNA. The other is a scrambled version of this sequence that maintains the same GC content and is used as a representative of nontarget DNA. Figure 2 shows the normalized anisotropy values of fluorescently labeledDNA plotted as a function of DBD concentration. Dashed lines show the fit to a cooperative binding model in the case of consensus DNA (Figure 2a), and a single-site binding model was used to fit the data for scrambled DNA (Figure 2b). Both models assume p53 binds DNA as a dimer of dimers [13]. As salt concentration increases, binding affinity of DBD to DNA decreases. This is in accordance with observations of p53 specifically [13] and of DNA-binding proteins in general [63,64]. Hill coefficients are approximately 1.8 for DBD binding to consensus DNA and 1 for binding to scrambled DNA. This supports previous studies showing that p53 binds its target DNA in a cooperative manner and nontarget DNA in a noncooperative manner [13]. We observed the same trend in cooperativity when ND WT and the mutants bind to DNA, but $K_D$ values are 5–200 times larger (Table S1 and Figure S1). At 125 mM IS the $K_D$ for DBD binding consensus DNA was 0.9 ± 0.07 nM and at 225 mM IS $K_D$ was 104.5 ± 5 nM. For binding to scrambled DNA, $K_D$ ranges from 89.1 ± 5 nM to 1388 ± 44 nM over the same range of IS. These results are in the same range as previously observed binding affinities of DBD to DNA [22,65]. Similar trends are observed for ND WT, for which fluorescence anisotropy curves across a range of IS are shown in Figure S2. The $K_D$ for ND WT binding to consensus DNA ranges from 43 ± 3.4 nM to 3861 ± 40 nM and binding to scrambled DNA ranges from 193 ± 8.2 nM to 3705 ± 230 nM. See Table S1 for full range of values. Error bars in Figure 2a,b represent the standard deviation of three measurements at each IS and the fitting errors presented in Table S2 are the standard error of estimate.

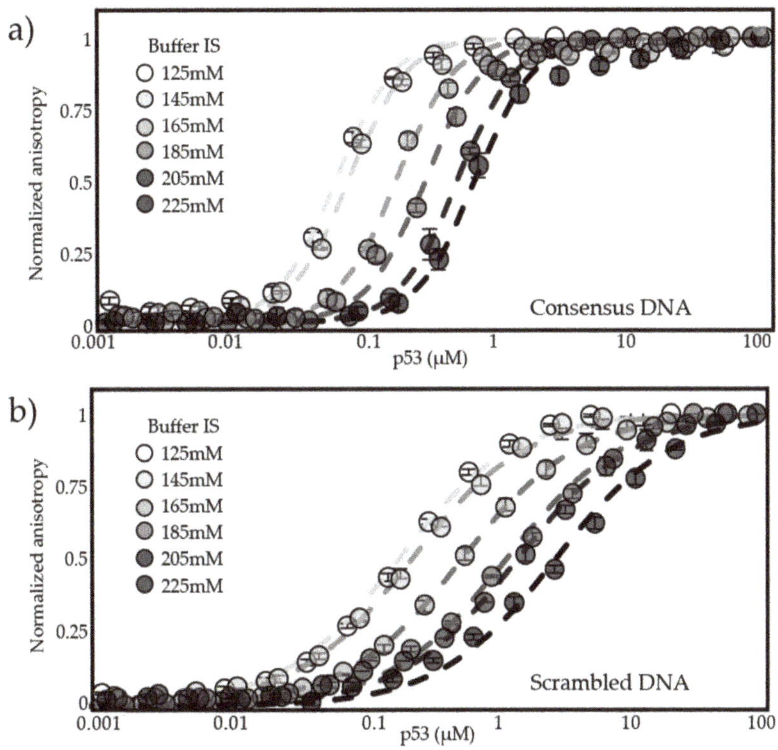


**Figure 2.** DBD binds DNA across IS. Fluorescence anisotropy plots show the change in signal from a fluorescently tagged DNA fragment as protein is added: an increase in the concentration of p53 needed to achieve saturation when DNA concentration is kept stable as buffer salt concentration increases. (**a**) fluorescence anisotropy plots of DBD bound to consensus DNA at 125–225 mM IS; (**b**) fluorescence anisotropy plots of DBD bound to scrambled DNA at 125–225 mM IS.

*3.2. DBD, ND, and ND Mutants Binding to Consensus and Scrambled DNA at Physiological IS*

To determine the contributions of charged and nonpolar interactions between TAD2 and DBD in the autoinhibition of DNA binding we designed three mutants where all aspartic and glutamic acid residues in TAD2 were changed to alanine (ND DE), where all the nonpolar residues from TAD2, including W53 and F54, were changed to alanine (ND NP), and where W53 and F54, were changed to glutamine and serine (ND QS) (See Figure 1e for sequences). The ND QS mutant is based on an early study of p53 in which this mutation inhibited transactivation and apoptosis by inhibiting interactions with multiple domains of CBP/p300 [66–68]. A decrease in the intramolecular interaction should lead to increased DNA binding affinity. Figure 3a shows the binding curves of fluorescence anisotropy experiments for DBD, ND WT, and the ND mutants at physiological IS (145 mM). The ND mutants have a binding affinity for consensus DNA that is closer to DBD than ND WT, indicating all the mutants disrupt the intramolecular interaction between TAD2 and DBD. ND DE and ND NP have similar binding affinities to one another for consensus and scrambled DNA, increasing the free energy of binding for consensus DNA relative to ND WT by −1.99 and −1.89 kcal/mol, respectively (Table 1). The ND QS mutant has DNA binding affinity between ND NP and ND WT and increases the free energy of consensus DNA binding by −1.49 kcal/mol relative to ND WT.

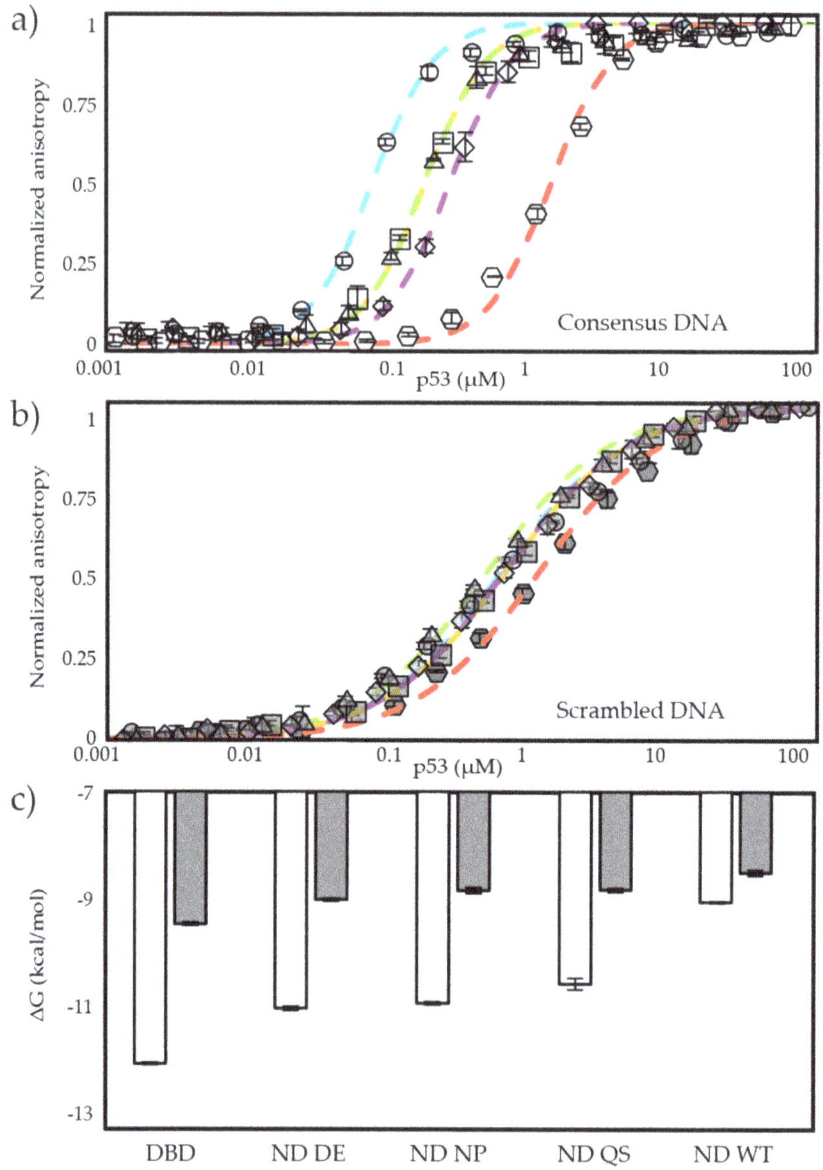


**Figure 3.** Binding of DBD and ND fragments to consensus and scrambled DNA at physiological IS (145 mM). (**a**) Fluorescence anisotropy plots of p53 constructs binding consensus DNA, where ○ is DBD, ⬡ is ND WT, △ is ND DE, □ is ND NP, ◇ is ND QS, (**b**) p53 constructs binding scrambled DNA, where ● is DBD, ⬢ is ND WT, ▲ is ND DE, DNA, ■ is ND NP, ◆ is ND QS, (**c**) ΔG of all fragments with consensus and scrambled DNA. Each data set represents three titrations.

**Table 1.** ΔΔG (column-row) in kcal/mol at physiological IS.

| | DBD | ND DE | ND NP | ND QS | ND WT |
|---|---|---|---|---|---|
| **Consensus DNA** | | | | | |
| **DBD** | 0.00 | 1.03 | 1.13 | 1.49 | 3.02 |
| **ND DE** | −1.03 | 0.00 | 0.10 | 0.46 | 1.99 |
| **ND NP** | −1.13 | −0.10 | 0.00 | 0.36 | 1.89 |
| **ND QS** | −1.49. | −0.46 | −0.36 | 0.00 | 1.53 |
| **ND WT** | −3.02 | −1.99 | −1.89 | −1.53 | 0.00 |
| **Scrambled DNA** | | | | | |
| | **DBD** | **ND DE** | **ND NP** | **ND QS** | **ND WT** |
| **DBD** | 0.00 | 0.46 | 0.64 | 0.63 | 0.97 |
| **ND DE** | −0.46 | 0.00 | 0.18 | 0.17 | 0.51 |
| **ND NP** | −0.64 | −0.18 | 0.00 | −0.01 | 0.32 |
| **ND QS** | −0.63 | −0.17 | 0.01 | 0.00 | 0.33 |
| **ND WT** | −0.97 | −0.51 | −0.32 | −0.33 | 0.00 |

Figure 3b shows that binding affinities for the ND mutants with scrambled DNA are in a similar order as we observe for consensus DNA. ND DE increases the free energy of binding by −0.51 kcal/mol relative to ND WT; ND NP and ND QS both increase free energy of binding by −0.32 and −0.33 kcal/mol, respectively (Table 1 and Figure 3c). The ND fragments binding consensus DNA have a ΔΔG with DBD ranging from −1.03 kcal/mol to −3.02 kcal/mol. ND fragments binding scrambled DNA have a ΔΔG with DBD ranging from −0.46 kcal/mol to −1.04 kcal/mol. Similar to DBD, the ND fragments show cooperative binding to consensus DNA and noncooperative binding to scrambled DNA, as seen in Figure S1 where consensus DNA binding data points match a fit line with a Hill coefficient of 2 and scrambled DNA binding data points match a fit line with a Hill coefficient of 1. Thus, the intramolecular interaction does not affect cooperativity of DBD on target DNA. In summary, we find that introduction of mutations to TAD2 decreases the intramolecular interaction and increases DNA binding affinity. We find the ND DE mutant has the largest change in autoinhibition, followed by ND NP, and then ND QS.

### *3.3. Effects of IS on Binding Specificity of DBD, ND WT, and the ND Mutants*

Binding specificity is commonly estimated as $\Delta G_{specific} - \Delta G_{nonspecific}$ [69,70]. Figure 4 shows the ΔΔG values for DBD and ND WT at 55–225 mM IS, and the ND mutants at 85–225 mM IS. Below physiological ionic strength, ND WT has greater specificity than DBD for consensus DNA than scrambled DNA as evidenced by the larger negative ΔΔG; however, this trend reverses between 85–125 mM IS. Figure 4 also shows that at higher IS, ND NP has a similar binding specificity to DBD and the binding specificity for ND DE closer to ND WT. This is interesting because we expect the nonpolar interactions between TAD2 and DBD to be more specific than the charged interactions and our data shows that removing them increases DNA binding specificity while removing the charged interactions between TAD2 and DBD reduces specificity. We think ND DE has lower binding specificity because the strength of the hydrophobic effect between nonpolar residues in TAD2 and DBD becomes stronger at higher IS [71,72]. In contrast, ΔΔG for ND NP tracks with DBD at higher salt concentrations, indicating that the acidic residues in TAD2 are responsible for inhibiting binding to nonspecific DNA. We expect residues W53 and F54 in TAD2 to play a role in forming specific interactions with DBD but introduction of Q53/S54 reduces DNA binding specificity, suggesting the introduction of these amino acids, and not removal of W53/F54, is driving this effect. The ND WT fragment used in this study lacks the tetramerization domain and only enhances DNA binding specificity at low ionic strength even though it shows strong inhibition of DNA binding and maintains binding cooperativity for specific DNA up to 225 mM IS. As shown in Figure 4, the DBD can bind

DNA specifically in the absence of TAD2 and the TET, and Figure 3c shows that ND WT inhibits binding to either consensus or scrambled DNA by a similar amount.

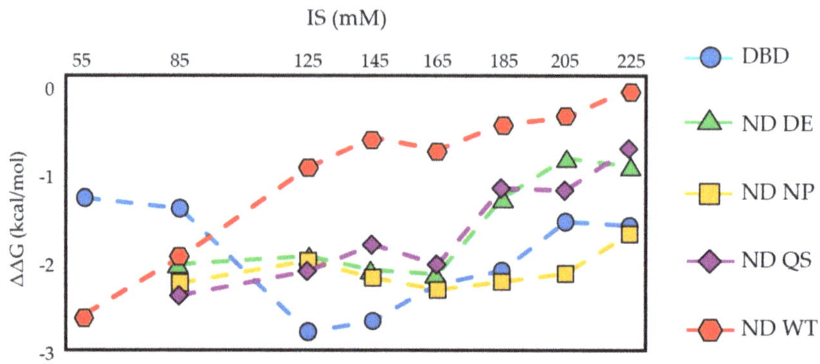


**Figure 4.** Binding specificity of DBD, ND WT, and ND mutants. For each p53 fragment, $\Delta\Delta G = \Delta G_{consensus} - \Delta G_{scrambled}$ at a given IS indicates binding specificity.

In our previous work we showed the intramolecular interaction between TAD2 and DBD in monomeric p53 became intermolecular when the tetramerization domain (TET) was present [22]. In a related study, Wright and colleagues showed that adding TAD2 to a p53 fragment containing the DBD and TET enhances DNA binding specificity by inhibiting binding to nonspecific DNA but has no effect on binding to specific DNA [23]. The binding studies by Wright and colleagues were conducted at an IS close to 165 mM using similar specific and nonspecific sequences to ours. Using full length p53 with and without TAD2, their $K_D$ ratio for binding was 1 for specific DNA and 5.7 for nonspecific DNA. By comparison our $K_D$ ratios for ND WT and DBD binding to specific and nonspecific DNA are 70 and 5.3, respectively. Taken together these data suggest that inhibition of DNA binding to both specific and nonspecific sequences is driven by the intramolecular interaction between TAD2 and DBD and specificity enhancement depends on this interaction becoming intermolecular when p53 is tetrameric. As mentioned, we think addition of the tetramerization domain reduces the hydrophobic effect between TAD2 and DBD and this could be happening due differences in the way TAD2 interacts with DBD when the intramolecular interaction becomes intermolecular.

### 3.4. Estimating Ion Release Using Counterion Condensation Theory

To assess the sensitivity of the TAD2-DBD interaction to IS, we conducted fluorescence anisotropy binding experiments on ND WT and the ND mutants from 125–225 mM IS. Figures 5 and 6 show the linear region of log($K_A$) versus log[Salt] plots. Figure 5a shows that the binding of consensus DNA to DBD is tighter than to ND WT at every IS and that the presence of TAD2 in ND WT inhibits DNA binding at a level that corresponds to increasing IS by 70–80 mM for DBD. Binding of DBD and ND WT to scrambled DNA (Figure 5b) shows a similar trend in affinity where the inhibition of DNA binding by TAD2 corresponds to an increased IS of 40–60 mM for DBD.

Counterion condensation theory proposes that ions are uniformly condensed on DNA at a concentration that is relatively independent of buffer conditions or the type of protein binding. When a positively charged protein binds DNA, a number of counterions equivalent (or fractionally equivalent) to the number of nonspecific ionic contacts made between the protein and DNA backbone are released into solution [48]. The oligolysine model developed by Record and colleagues as an extension of the counterion condensation theory predicts that the observed decrease in DNA binding affinity as salt concentration increases can be used to estimate the number of these nonspecific ionic contacts [49,73]. In Equation (5), the slope (*N*) of the double log plots in Figures 5 and 6 is proportional to the fractional number of counterions released from the DNA backbone (Ψ), approximately 0.7 per phosphate contact for short oligonucleotides [53], and any excess ions released from the protein (β). According to this theory, a smaller slope corresponds to release of fewer ions, whether they originate from backbone phosphates or from protein. As shown in Table 2, DBD has a larger slope than ND WT when binding consensus DNA, corresponding to greater predicted ion release.

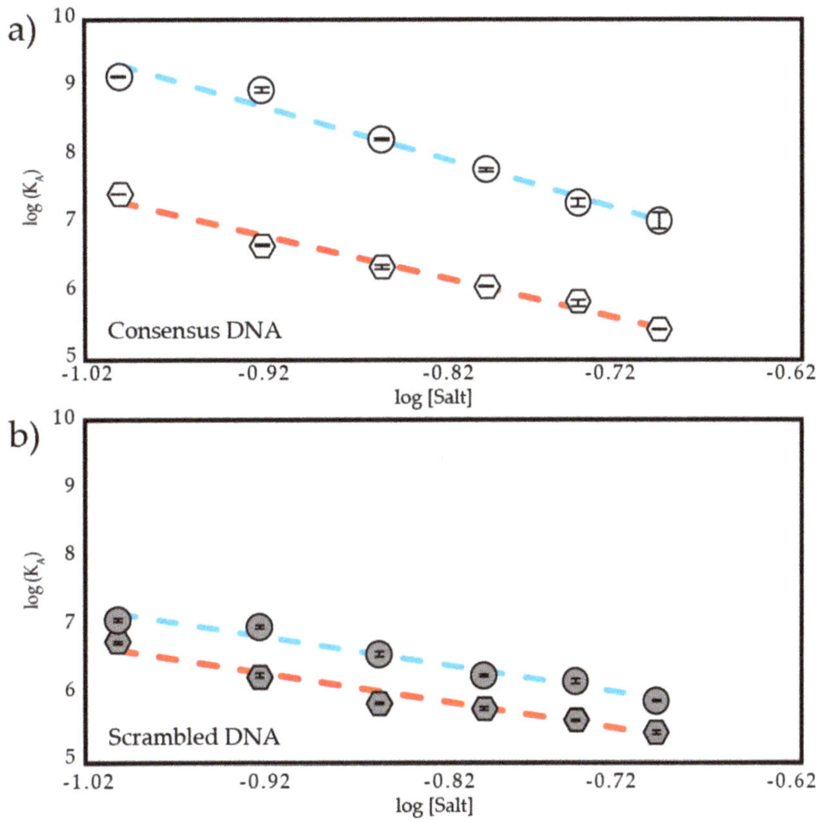


**Figure 5.** Salt-dependent binding affinity of DBD and ND WT. Plot of log ($K_A$) vs. log [Salt] from 125–225 mM IS of (**a**) DBD and ND WT binding to consensus DNA where ○ is DBD, ○ is ND WT, (**b**) DBD and ND WT binding to scrambled DNA where ● is DBD, ● is ND WT. $R^2$ values for all fit lines are between 0.96 and 0.99.

Crystallographic studies show five DNA backbone contacts made by DBD when bound to the p21 promoter [74,75]. We assume the same number of DNA backbone contacts are made by DBD to consensus DNA because our consensus sequence is similar to the p21 sequence. We also assume ND WT and ND mutants make the same number of contacts as DBD because TAD2 does not interact with DNA [22] or affect binding cooperativity according to the Hill plots in Figure S1. The difference in the slopes between DBD and ND WT when binding consensus DNA corresponds to a difference in the predicted release of excess ions when binding DNA (Table 2) where DBD is predicted to release 3.9 excess ions and ND WT is predicted to release 2.5 excess ions. This small difference in ion release corresponds to a difference in salt sensitivity where DBD experiences a 117-fold increase in $K_D$ versus ND WT's 86-fold increase in $K_D$ over this range of IS. We also observe that inhibition of DNA binding is greater for ND WT as IS decreases, indicating a stronger intramolecular interaction at lower salt concentrations. A similar divergence of salt-dependent binding affinity was seen in a previous study of an autoinhibitory IDR-DBD interaction [27], in which the addition of an acidic domain lowered both DNA binding affinity and changed the slope of its double log plot. By contrast, ND WT binding to scrambled DNA has a slope similar to that of DBD (Table 2). We assume the same number of backbone contacts are made when p53 binds a nontarget sequence as is suggested by structures of low affinity p53-DNA complexes [75]. Assuming five backbone contacts, the slopes of ND WT and DBD when binding scrambled DNA correspond to predicted excess ion release of 0.7 and 0.6, respectively.

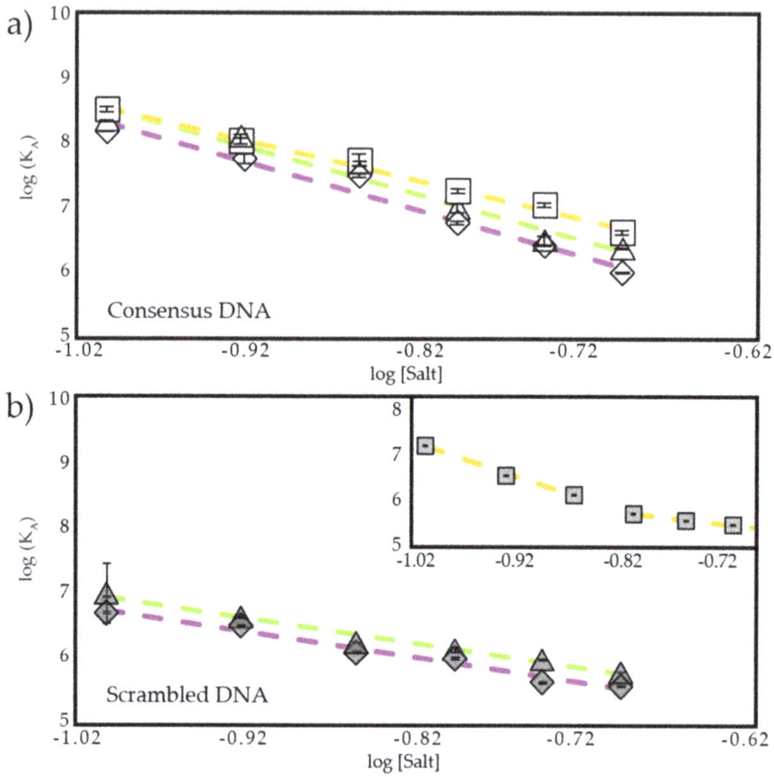


**Figure 6.** Salt-dependent binding affinity of ND mutants. Plot of log ($K_A$) vs. log[Salt] from 125–225 mM IS of (**a**) ND mutants binding consensus DNA, where △ is ND DE, □ is ND NP, ◇ is ND QS (**b**) ND mutants binding scrambled DNA, where ▲ is ND DE, ◆ is ND QS. Inset shows ND NP binding scrambled DNA, ■. $R^2$ values for all fit lines are between 0.96 and 0.99.

**Table 2.** Slope of log ($K_A$) versus log [Salt] predicts ion release.

| | Slope, *N*, with Consensus DNA | Predicted Excess Ions released | Slope, *N*, with Scrambled DNA | Predicted Excess Ions released |
|---|---|---|---|---|
| **DBD** | −7.39 | 3.9 | −4.09 | 0.6 |
| **ND DE** | −7.08 | 3.6 | −3.89 | 0.5 |
| **ND NP** | −5.94 | 2.4 | −6.91, −2.35 | 3.4, 0.0 |
| **ND QS** | −7.16 | 3.6 | −3.90 | 0.4 |
| **ND WT** | −5.99 | 2.5 | −4.15 | 0.7 |

Figure 6a shows ND DE, ND NP, and ND QS bind consensus DNA more tightly than ND WT (also see Table S1). Slope values for ND DE and ND QS are close to DBD, while ND NP has a slope close to ND WT (Table 2). From these results we can make three conclusions:

(1) ion release after removal of acidic residues (ND DE) is similar to ion release of DBD, (2) removal of several nonpolar residues in TAD2, including W53 and F54, (ND NP) has no effect on ion release relative to ND WT, and (3) introduction of Q53 and S54, not removal of W53 and F54, is responsible for changes in ion release of ND QS. The first two conclusions were expected and the third suggests the Q53/S54 mutant may do more than interfere with binding to CBP.

When binding scrambled DNA, the slopes are similar for DBD, ND WT, and the ND DE and ND QS mutants (Figure 6b). We predict that ND DE and ND QS release 0.5 and 0.4 excess ions, respectively, when binding scrambled DNA, similar to DBD and ND WT. ND NP does not have a single linear slope over the 125–225 mM range when binding scrambled DNA. Instead, it appears to have a linear portion at 125–165 mM IS with a slope of −6.91 and another linear portion at 185–225 mM IS with a slope of −2.35 as shown in the inset in Figure 6b. Slopes and estimated excess ion release from these two states are shown in Table 2 to be different from each other and from other p53 fragments. This suggests to us that ND NP binds scrambled DNA in multiple states.

According to the oligolysine model, ΔG of binding can be separated into electrostatic and nonelectrostatic components, where the slopes of the plots in Figures 5 and 6 multiplied by log[Salt] is the salt-dependent entropy due to ions being released into solution from the phosphate backbone [49,58]. As shown in Figure 7 and Table S3, the salt-dependent entropy is predicted to be the energetic driver of the p53 fragments binding to consensus DNA, ranging from 68–85% of the total energy. However, in an earlier binding study from our group at an IS of 85 mM using isothermal titration calorimetry we observed a large entropic penalty for DBD binding consensus DNA and a smaller penalty for ND WT and both had a large enthalpy change upon binding [22]. Van't Hoff plots using temperature-dependent fluorescence anisotropy data also predict a large enthalpic component of binding (Figure S4 and Table S5) [76]. This suggests to us that for p53 the salt-dependent component of binding is not just made up of an entropic contribution from ion release. According to the Record model, the salt-dependent and independent contributions to binding free energy for DBD are predicted to be −9.30 kcal/mol and −2.77 kcal/mol, respectively, and for ND WT they are −7.55 kcal/mol and −1.50 kcal/mol, respectively. For all the fragments except ND NP, a smaller contribution for binding to scrambled DNA comes from the salt-dependent component. For DBD, the salt-dependent and independent components of binding to scrambled DNA are −5.14 kcal/mol and −4.32 kcal/mol, respectively, and for ND WT are −5.22 kcal/mol and −3.28 kcal/mol, respectively. An analysis of these components using Manning's model, Equation (6), also predicts that salt-dependent entropy is a larger component of binding to consensus DNA than to scrambled DNA (Table S4 and Figure S3).

Salt-dependent ion release is one of several mechanisms that proteins use to achieve specificity in DNA binding. Studies have characterized systems in which the salt-dependent component of binding is higher for specific than nonspecific DNA binding [77], in which the salt-dependent component is similar for specific and nonspecific DNA binding [50,78], in which the salt-dependent component is lower for specific than for nonspecific DNA binding [57,79,80], in which the salt-dependent component is relatively low for both specific and nonspecific binding [47,81,82], and in which the salt-dependent component follows no clear trend between specific and nonspecific DNA binding [83,84]. It appears that our p53 fragments utilize salt-dependent components of the interaction for specific binding to a greater degree than the salt-independent components, and this trend is reversed for nonspecific DNA. Our mutants also follow this trend, with the exception of ND NP, which may switch between two modes depending on the IS.

In summary, using the salt-dependent component of binding, we find that predicted excess ion release upon protein-DNA binding is greater when our p53 fragments binding consensus DNA than scrambled DNA. Whereas excess ion release varies by fragment when binding consensus DNA, it is similar between all fragments when binding scrambled DNA excepting ND NP. This salt-dependent component comprises a variable amount of the free energy of binding for each fragment and generally comprises a greater amount of the free energy of binding for consensus DNA than scrambled DNA.

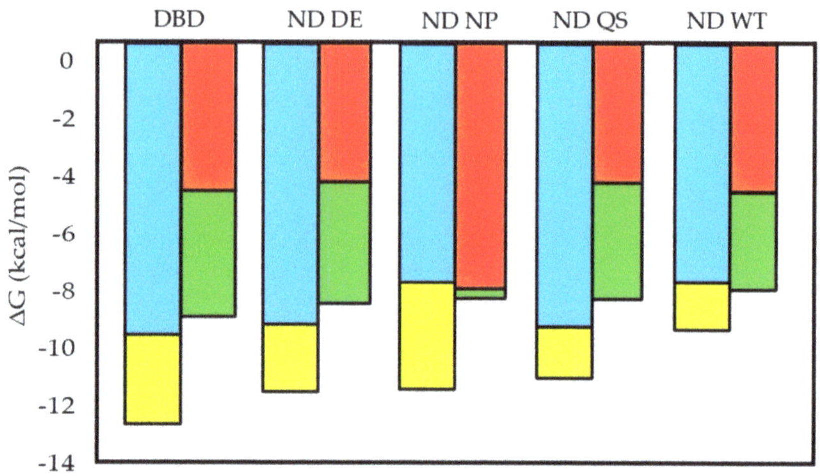


**Figure 7.** Salt-dependent and salt-independent components of Gibbs free energy at 145 mM IS from Record's model. Free energy is apportioned into categories by assuming complete inhibition of salt-dependent components at 1M NaCl so the remainder of free energy is salt-independent. Slopes of double log plot slopes are used to estimate binding affinity at 1M NaCl, where □ is the salt-dependent component and □ is the salt-independent component for consensus DNA and □ is the salt-dependent component and □ is the salt-independent component for scrambled DNA.

### *3.5. The Intramolecular Interaction Affects Stokes Radius and Apparent Molecular Weight*

Using size exclusion chromatography (SEC) at high IS (410 mM), the elution volumes of p53 constructs were compared to elution volumes of known standards (see methods) to determine their Stokes radii and apparent molecular weights. As shown in Figure 8, ND mutant constructs elute at a lower volume than ND WT, which elutes at a lower volume than DBD. As shown in Table 3, we find the Stokes radius of DBD to be 2.74 ± 0.004 nm, in agreement with a previously published Stokes radius of the same DBD fragment using dynamic light scattering (2.74 nm) [85], whereas the radius of ND WT was found to be 3.55 ± 0.004 nm. The change in radius with the tethered TAD is relatively small given that p53 residues 1–93, including TAD1, TAD2, and PRR, has a Stokes radius of 3.5 nm at 5 °C [61]. ND WT appears to be more compact than predicted for 93 disordered residues attached to 219 ordered residues, but the ND WT is more expanded than predicted for a folded protein of the same number of residues (2.51 ± 0.59 nm) [86]. Estimating the hydrodynamic radius of a protein containing both ordered and disordered sections is an ongoing challenge [86,87]. Both DBD and ND WT have an apparent molecular weight greater than their actual molecular weight, as shown in Table 3. For DBD this is likely due to a disordered segment near the C-terminus from residues 292–312 (**PDB 4HJE**) [75]. ND WT and the ND mutants have apparent molecular weights almost twice as large as their actual molecular weights using this technique.

We observe a small decrease in the elution volume of the ND mutants relative to ND WT, but it is larger than the resolution error of the volume measurement (+/−0.02 mL). Small changes in Stokes radii are evidence the mutants do not disrupt the global structure of ND, which was unexpected given the increase in DNA binding affinity of the mutants relative to ND WT. We suspect maintenance of the global structure is being driven by the PRR and will test this hypothesis in the future. We also conducted SEC on ND WT at 150 mM IS to test for changes in elution volume at low IS and compared this result to the elution volume at 410 mM IS. Shown in Figure S5, ND WT's elution volume varies between these two conditions by <0.2 mL, a difference that corresponds to an approximately 0.03 nm difference in Stokes radius and less than 1 kDa difference in apparent molecular weight.

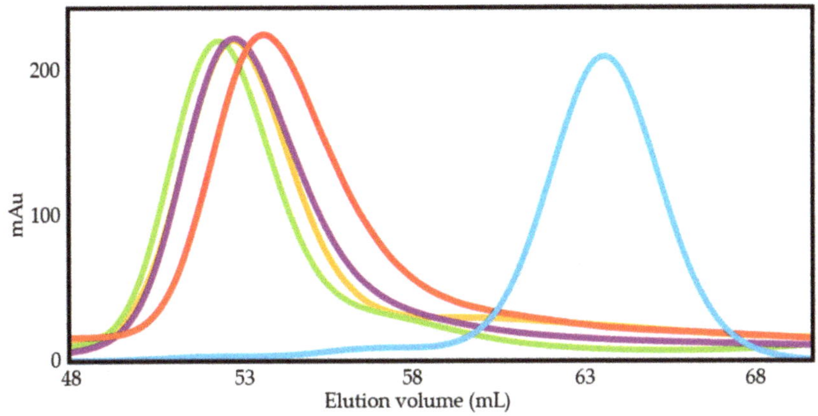


**Figure 8.** Size exclusion chromatography is used to compare p53 constructs. Elution profiles of p53 constructs where lower elution volume indicates a larger hydrodynamic radius: — DBD, — ND DE, — ND NP, — ND QS, — ND WT.

**Table 3.** Stokes radii and apparent molecular weights of p53 constructs assessed by SEC.

| | Stokes Radius (nm) | Elution Volume (mL) | Apparent Molecular Weight (kDa) | Actual Molecular Weight (kDa) |
|---|---|---|---|---|
| **DBD** | 2.74 ± 0.004 | 63.78 ± 0.05 | 34.76 ± 0.13 | 24.55 |
| **ND DE** | 3.71 ± 0.001 | 52.41 ±0.01 | 67.89 ± 0.07 | 34.23 |
| **ND NP** | 3.65 ± 0.004 | 52.90 ± 0.04 | 65.98 ± 0.17 | 34.13 |
| **ND QS** | 3.65 ± 0.004 | 52.90 ± 0.02 | 65.98 ± 0.17 | 34.45 |
| **ND WT** | 3.55 ± 0.004 | 53.82 ± 0.04 | 62.46 ± 0.19 | 34.57 |

## 4. Discussion

We find that the intramolecular interaction between the TAD2 and DBD domains of p53 is disrupted by mutations targeting multiple types of interactions. Alanine substitutions of TAD2's negatively charged residues, ND DE, increased consensus DNA free energy of binding by −1.99 kcal/mol relative to ND WT, suggesting that electrostatics play a large role in the intramolecular interaction and autoinhibition of DNA binding. Alanine substitutions of nonpolar residues, ND NP, increased DNA free energy of binding by −1.89 kcal/mol, suggesting a nonelectrostatic component. A targeted substitution of W53/F54 to Q53/S54, ND QS, chosen because of its established ability to disrupt other important TAD2 interactions [66–68], increases consensus DNA free energy of binding by −1.49 kcal/mol. The sum of the effects of the ND DE and ND NP mutants on the autoinhibition of DNA binding is 1 kcal/mol greater than the effect of ND WT. This indicates some cooperativity between the acidic, nonpolar, and aromatic residues of TAD2 to inhibit DNA binding.

A previous analysis of transcription factor-DNA complexes using the counterion condensation theory, notably HMG boxes and homeodomains, showed the salt-dependent component of binding was similar for specific and nonspecific DNA and the salt-independent components, attributed to hydrogen bonds and van der Waals interactions, were the drivers of specificity [50]. By contrast, our study shows that p53 has a larger salt-dependent component of binding for consensus DNA versus scrambled DNA; according to the counterion condensation theory, this represents a dependency on entropy derived from ion release when p53 binds consensus DNA that is not present when it binds the scrambled DNA sequence. Critiques of the counterion condensation theory have noted that ion release is not the only energetic component of the salt-dependent binding affinity, nor is the salt-dependent component entirely entropic [51,52,84,88]. Our data is discussed in the context of entropy derived from predicted ion release; however, our van't Hoff data and previous ITC data [22] suggests a large enthalpic component, meaning the difference we see is a combination of ion release and other energetic components that drive specificity.

Our results show how the presence of TAD2 decreases the apparent number of ions released by DBD when binding consensus DNA. We propose that the interactions between the positively charged residues in the DNA binding pocket and the negatively charged residues of TAD2 reduce the need for ionic interactions between those same positive charges of DBD and negatively charged solutes. This conclusion is consistent with the differences in ion release we see between the ND DE and ND NP mutants. The ND DE mutant releases almost the same number of ions as DBD. By eliminating the negative charges of TAD2 we have eliminated the intramolecular screening and now ions from the solute reestablish their positions around the positively charged amino acids of the DBD. The ND NP mutant has the negatively charged residues of TAD2 present, and the ion release is almost identical to that of ND WT. Thus, we show that the differences in ion release between DBD and ND WT are primarily moderated by negatively charged residues in TAD2. We also think the differences in the salt dependence of DNA binding between DBD and ND WT could be relevant for p53 function. Prior to DNA damage TAD1 is primarily responsible for the interaction with MDM2 that leads to p53 degradation [89]. However, following DNA damage, posttranslational modifications regulate numerous interactions between TAD2 and other cofactors [68,90–92]. It is reasonable to expect these other interactions will compete with the autoinhibitory function of TAD2, resulting in increased DNA binding.

**Supplementary Materials:** Supporting information can be downloaded at: https://www.mdpi.com/article/10.3390/biom12111558/s1: Table S1: $K_D$s of p53 constructst binding consensus and scrambled DNA at various ionic strengths; Table S2: Standard error of estimate; Table S3: Percentage of Gibbs free energy originating from salt-dependent and salt-independent components estimated using Record's interpretation of the counterion condensation model; Table S4: Percentage of Gibbs free energy originating from salt-dependent and salt-independent components estimated using Manning's interpretation of the counterion condensation model; Table S5: Van't Hoff plot-derived thermodynamics; Figure S1: Hill Coefficients of p53 constructs; Figure S2: ND WT binds DNA across ionic strengths; Figure S3: Salt-dependent and salt-independent components of Gibbs free energy at physiological ionic strength using Manning's model; Figure S4: Van't Hoff plots of DBD and ND with consensus and scrambled DNA; Figure S6: Separation of ND WT in seize exclusion column in low and high ionic strength buffer. References [76,93–98] are cited in Supplementary Materials.

**Author Contributions:** Conceptualization, E.G. and G.W.D.; methodology, G.W.D.; validation, G.W.D.; formal analysis, E.G. and G.W.D.; investigation, E.G.; resources, G.W.D.; data curation, E.G.; writing—original draft preparation, E.G.; writing—review and editing, E.G. and G.W.D.; visualization, E.G.; supervision, G.W.D.; project administration, G.W.D.; funding acquisition, G.W.D. All authors have read and agreed to the published version of the manuscript.

**Funding:** This research was funded by NIH grants R01CA141244, R01GM115556, and Florida Department of Health Project 20B17.

**Institutional Review board Statement:** Not applicable.

**Informed Consent Statement:** Not applicable.

**Data Availability Statement:** Not applicable.

**Acknowledgments:** The authors acknowledge support from the lab of Jianfeng Cai with the University of South Florida Chemistry Department for the use of and training with their equipment to generate fluorescence anisotropy data.

**Conflicts of Interest:** The authors declare no conflict of interest. The funders had no role in the design of the study; in the collection, analyses, or interpretation of data; in the writing of the manuscript; or in the decision to publish the results.

## References


1. Tishler, R.B.; Calderwood, S.K.; Coleman, C.N.; Price, B.D. Increases in sequence specific DNA binding by p53 following treatment with chemotherapeutic and DNA damaging agents. *Cancer Res.* **1993**, *53* (Suppl. 10), 2212–2216. [PubMed]
2. Zhan, Q.; Carrier, F.; Fornace, A.J., Jr. Induction of cellular p53 activity by DNA-damaging agents and growth arrest. *Mol. Cell. Biol.* **1993**, *13*, 4242–4250. [PubMed]
3. Beckerman, R.; Prives, C. Transcriptional regulation by p53. *Cold Spring Harb. Perspect. Biol.* **2010**, *2*, a000935. [CrossRef]
4. Picksley, S.M.; Lane, D.P. p53 and Rb: Their cellular roles. *Curr. Opin. Cell Biol.* **1994**, *6*, 853–858. [CrossRef]
5. Hainaut, P.; Hollstein, M. p53 and human cancer: The first ten thousand mutations. *Adv. Cancer Res.* **2000**, *77*, 81–137. [PubMed]
6. Hupp, T.R.; Meek, D.; Midgley, C.; Lane, D. Regulation of the Specific DNA-Binding Function of P53. *Cell* **1992**, *71*, 875–886. [CrossRef]
7. Kruse, J.P.; Gu, W. Modes of p53 Regulation. *Cell* **2009**, *137*, 609–622. [CrossRef]
8. Arbely, E.; Natan, E.; Brandt, T.; Allen, M.D.; Veprintsev, D.B.; Robinson, C.V.; Chin, J.W.; Joerger, A.C.; Fersht, A.R. Acetylation of lysine 120 of p53 endows DNA-binding specificity at effective physiological salt concentration. *Proc. Natl. Acad. Sci. USA* **2011**, *108*, 8251–8256. [CrossRef]
9. Shaw, P.; Freeman, J.; Bovey, R.; Iggo, R. Regulation of specific DNA binding by p53: Evidence for a role for O-glycosylation and charged residues at the carboxy-terminus. *Oncogene* **1996**, *12*, 921–930.
10. Riley, T.; Sontag, E.; Chen, P.A.; Levine, A. Transcriptional control of human p53-regulated genes. *Nat. Rev. Mol. Cell Biol.* **2008**, *9*, 402–412. [CrossRef]
11. El-Deiry, W.S.; Kern, S.E.; Pietenpol, J.A.; Kinzler, K.W.; Vogelstein, B. Definition of a consensus binding site for p53. *Nat. Genet.* **1992**, *1*, 45–49. [CrossRef] [PubMed]
12. Menendez, D.; Inga, A.; Resnick, M.A. The expanding universe of p53 targets. *Nat. Rev. Cancer* **2009**, *9*, 724–737. [CrossRef] [PubMed]
13. Weinberg, R.L.; Veprintsev, D.B.; Fersht, A.R. Cooperative binding of tetrameric p53 to DNA. *J. Mol. Biol.* **2004**, *341*, 1145–1159. [CrossRef] [PubMed]
14. Balagurumoorthy, P.; Sakamoto, H.; Lewis, M.S.; Zambrano, N.; Clore, G.M.; Gronenborn, A.M.; Appella, E.; Harrington, E.R. 4 P53 DNA-Binding Domain Peptides Bind Natural P53-Response Elements and Bend the DNA. *Proc. Natl. Acad. Sci. USA* **1995**, *92*, 8591–8595. [CrossRef]
15. McLure, K.G.; Lee, P.W.K. How p53 binds DNA as a tetramer. *EMBO J.* **1998**, *17*, 3342–3350. [CrossRef]
16. Jordan, J.J.; Menendez, D.; Inga, A.; Nourredine, M.; Bell, D.; Resnick, M.A. Noncanonical DNA motifs as transactivation targets by wild type and mutant p53. *PLoS Genet.* **2008**, *4*, e1000104. [CrossRef]
17. Weinberg, R.L.; Veprintsev, D.B.; Bycroft, M.; Fersht, A.R. Comparative binding of p53 to its promoter and DNA recognition elements. *J. Mol. Biol.* **2005**, *348*, 589–596. [CrossRef]
18. Vousden, K.H.; Lu, X. Live or let die: The cell's response to p53. *Nat. Rev. Cancer* **2002**, *2*, 594–604. [CrossRef]
19. Chen, X.; Ko, L.J.; Jayaraman, L.; Prives, C. p53 levels, functional domains, and DNA damage determine the extent of the apoptotic response of tumor cells. *Genes Dev.* **1996**, *10*, 2438–2451. [CrossRef]
20. Senitzki, A.; Safieh, J.; Sharma, V.; Golovenko, D.; Danin-Poleg, Y.; Inga, A.; Haran, E.T. The complex architecture of p53 binding sites. *Nucleic Acids Res.* **2021**, *49*, 1364–1382. [CrossRef]
21. Szak, S.T.; Mays, D.; Pietenpol, J.A. Kinetics of p53 binding to promoter sites in vivo. *Mol. Cell. Biol.* **2001**, *21*, 3375–3386. [CrossRef] [PubMed]
22. He, F.; Borcherds, W.; Song, T.; Wei, X.; Das, M.; Chen, L.; Daughdrill, G.W.; Chen, J. Interaction between p53 N terminus and core domain regulates specific and nonspecific DNA binding. *Proc. Natl. Acad. Sci. USA* **2019**, *116*, 8859–8868. [CrossRef] [PubMed]
23. Krois, A.S.; Dyson, H.J.; Wright, P.E. Long-range regulation of p53 DNA binding by its intrinsically disordered N-terminal transactivation domain. *Proc. Natl. Acad. Sci. USA* **2018**, *115*, E11302–E11310. [CrossRef]
24. Sun, X.; Dyson, H.J.; Wright, P.E. A phosphorylation-dependent switch in the disordered p53 transactivation domain regulates DNA binding. *Proc. Natl. Acad. Sci. USA* **2021**, *118*, e2021456118. [CrossRef]
25. Tompa, P.; Fuxreiter, M. Fuzzy complexes: Polymorphism and structural disorder in protein-protein interactions. *Trends Biochem. Sci.* **2008**, *33*, 2–8. [CrossRef] [PubMed]
26. Arbesu, M.; Iruela, G.; Fuentes, H.; Teixeira, J.M.C.; Pons, M. Intramolecular Fuzzy Interactions Involving Intrinsically Disordered Domains. *Front. Mol. Biosci.* **2018**, *5*, 39. [CrossRef]
27. Wang, X.; Greenblatt, H.M.; Bigman, L.S.; Yu, B.; Pletka, C.C.; Levy, Y.; Iwahara, J. Dynamic Autoinhibition of the HMGB1 Protein via Electrostatic Fuzzy Interactions of Intrinsically Disordered Regions. *J. Mol. Biol.* **2021**, *433*, 167122. [CrossRef]
28. Desjardins, G.; Meeker, C.A.; Bhachech, N.; Currie, S.L.; Okon, M.; Graves, B.J.; McIntosh, L.P. Synergy of aromatic residues and phosphoserines within the intrinsically disordered DNA-binding inhibitory elements of the Ets-1 transcription factor. *Proc. Natl. Acad. Sci. USA* **2014**, *111*, 11019–11024. [CrossRef]
29. Bourgeois, B.; Gui, T.; Hoogeboom, D.; Hocking, H.G.; Richter, G.; Spreitzer, E.; Viertler, M.; Richter, K.; Madl, T.; Burgering, B.M. Multiple regulatory intrinsically disordered motifs control FOXO4 transcription factor binding and function. *Cell Rep.* **2021**, *36*, 109446. [CrossRef]
30. Gong, W.B.; Liang, Q.; Tong, Y.; Perrett, S.; Feng, Y. Structural Insight into Chromatin Recognition by Multiple Domains of the Tumor Suppressor RBBP1. *J. Mol. Biol.* **2021**, *433*, 167224. [CrossRef]

31. Liu, J.; Perumal, N.B.; Oldfield, C.J.; Su, E.W.; Uversky, V.N.; Dunker, A.K. Intrinsic disorder in transcription factors. *Biochemistry* **2006**, *45*, 6873–6888. [CrossRef] [PubMed]
32. Minezaki, Y.; Homma, K.; Kinjo, A.R.; Nishikawa, K. Human transcription factors contain a high fraction of intrinsically disordered regions essential for transcriptional regulation. *J. Mol. Biol.* **2006**, *359*, 1137–1149. [CrossRef]
33. Singh, G.P.; Dash, D. Intrinsic disorder in yeast transcriptional regulatory network. *Proteins* **2007**, *68*, 602–605. [CrossRef] [PubMed]
34. Liu, Y.; Matthews, K.S.; Bondos, S.E. Multiple intrinsically disordered sequences alter DNA binding by the homeodomain of the Drosophila hox protein ultrabithorax. *J. Biol. Chem.* **2008**, *283*, 20874–20887. [CrossRef]
35. Brodsky, S.; Matthews, K.S.; Bondos, S.E. Intrinsically Disordered Regions Direct Transcription Factor In Vivo Binding Specificity. *Mol. Cell* **2020**, *79*, 459–471.e4. [CrossRef] [PubMed]
36. Spreitzer, E.; Alderson, T.R.; Bourgeois, B.; Eggenreich, L.; Habacher, H.; Bramerdorfer, G.; Pritišanac, I.; Sánchez-Murcia, P.A.; Madl, T. FOXO transcription factors differ in their dynamics and intra/intermolecular interactions. *Curr. Res. Struct. Biol.* **2022**, *4*, 118–133. [CrossRef]
37. Katan-Khaykovich, Y.; Shaul, Y. Nuclear import and DNA-binding activity of RFX1. Evidence for an autoinhibitory mechanism. *Eur. J. Biochem.* **2001**, *268*, 3108–3116. [CrossRef]
38. Ueshima, S.; Nagata, K.; Okuwaki, M. Internal Associations of the Acidic Region of Upstream Binding Factor Control Its Nucleolar Localization. *Mol. Cell. Biol.* **2017**, *37*, e00218-17. [CrossRef]
39. Wiebe, M.S.; Nowling, T.K.; Rizzino, A. Identification of novel domains within Sox-2 and Sox-11 involved in autoinhibition of DNA binding and partnership specificity. *J. Biol. Chem.* **2003**, *278*, 17901–17911. [CrossRef]
40. Wijeratne, T.U.; Guiley, K.Z.; Lee, H.-W.; Müller, G.A.; Rubin, S.M. Cyclin-dependent kinase-mediated phosphorylation and the negative regulatory domain of transcription factor B-Myb modulate its DNA binding. *J. Biol. Chem.* **2022**, *298*, 102319. [CrossRef]
41. Ning, S.; Chao, H.-J.; Li, S.; Zhou, R.; Zou, L.; Zhang, X.; Liu, J.; Yan, D.; Duan, M. The auto-inhibition mechanism of transcription factor Ets-1 induced by phosphorylation on the intrinsically disordered region. *Comput. Struct. Biotechnol. J.* **2022**, *20*, 1132–1141. [CrossRef] [PubMed]
42. Dosztanyi, Z.; Csizmok, V.; Tompa, P.; Simon, I. IUPred: Web server for the prediction of intrinsically unstructured regions of proteins based on estimated energy content. *Bioinformatics* **2005**, *21*, 3433–3434. [CrossRef] [PubMed]
43. Munoz, V.; Serrano, L. Elucidating the folding problem of helical peptides using empirical parameters. *Nat. Struct. Biol.* **1994**, *1*, 399–409. [CrossRef] [PubMed]
44. Mittag, T.; Orlicky, S.; Choy, W.-Y.; Tang, X.; Lin, H.; Sicheri, F.; Kay, L.E.; Tyers, M.; Forman-Kay, J.D. Dynamic equilibrium engagement of a polyvalent ligand with a single-site receptor. *Proc. Natl. Acad. Sci. USA* **2008**, *105*, 17772–17777. [CrossRef]
45. Nikolova, P.V.; Henckel, J.; Lane, D.; Fersht, A.R. Semirational design of active tumor suppressor p53 DNA binding domain with enhanced stability. *Proc. Natl. Acad. Sci. USA* **1998**, *95*, 14675–14680. [CrossRef]
46. LiCata, V.J.; Wowor, A.J. Applications of fluorescence anisotropy to the study of protein-DNA interactions. *Methods Cell Biol.* **2008**, *84*, 243–262.
47. Zou, Y.; Bassett, H.; Walker, R.; Bishop, A.; Amin, S.; Geacintov, N.E.; Van Houten, B. Hydrophobic forces dominate the thermodynamic characteristics of UvrA-DNA damage interactions. *J. Mol. Biol.* **1998**, *281*, 107–119. [CrossRef]
48. Manning, G.S. Limiting Laws and Counterion Condensation in Polyelectrolyte Solutions. 1. Colligative properties. *J. Chem. Phys.* **1969**, *51*, 922–934. [CrossRef]
49. Record, M.T., Jr.; Lohman, M.L.; De Haseth, P. Ion effects on ligand-nucleic acid interactions. *J. Mol. Biol.* **1976**, *107*, 145–158. [CrossRef]
50. Privalov, P.L.; Dragan, A.I.; Crane-Robinson, C. Interpreting protein/DNA interactions: Distinguishing specific from non-specific and electrostatic from non-electrostatic components. *Nucleic Acids Res.* **2011**, *39*, 2483–2491. [CrossRef]
51. Fogolari, F.; Elcock, A.; Esposito, G.; Viglino, P.; Briggs, J.; McCammon, J. Electrostatic effects in homeodomain-DNA interactions. *J. Mol. Biol.* **1997**, *267*, 368–381. [CrossRef] [PubMed]
52. Sharp, K.A.; Friedman, R.A.; Misra, V.; Hecht, J.; Honig, B. Salt Effects on Polyelectrolyte-Ligand Binding—Comparison of Poisson-Boltzmann, and Limiting Law Counterion Binding Models. *Biopolymers* **1995**, *36*, 245–262. [CrossRef] [PubMed]
53. Olmsted, M.C.; Bond, J.; Anderson, C.; Record, M. Grand-Canonical Monte-Carlo Molecular and Thermodynamic Predictions of Ion Effects on Binding of an Oligocation (L(8+)) to the Center of DNA Oligomers. *Biophys. J.* **1995**, *68*, 634–647. [CrossRef]
54. Ha, J.H.; Capp, M.W.; Hohenwalter, M.D.; Baskerville, M.; Record, M.T., Jr. Thermodynamic stoichiometries of participation of water, cations and anions in specific and non-specific binding of lac repressor to DNA. Possible thermodynamic origins of the "glutamate effect" on protein-DNA interactions. *J. Mol. Biol.* **1992**, *228*, 252–264. [CrossRef]
55. Misra, V.K.; Hecht, J.L.; Yang, A.S.; Honig, B. Electrostatic contributions to the binding free energy of the lambdacI repressor to DNA. *Biophys. J.* **1998**, *75*, 2262–2273. [CrossRef]
56. Grucza, R.A.; Bradshaw, J.M.; Mitaxov, V.; Waksman, G. Role of electrostatic interactions in SH2 domain recognition: Salt-dependence of tyrosyl-phosphorylated peptide binding to the tandem SH2 domain of the Syk kinase and the single SH2 domain of the Src kinase. *Biochemistry* **2000**, *39*, 10072–10081. [CrossRef]
57. Cravens, S.L.; Hobson, M.; Stivers, J.T. Electrostatic properties of complexes along a DNA glycosylase damage search pathway. *Biochemistry* **2014**, *53*, 7680–7692. [CrossRef]

58. Fenley, M.O.; Russo, C.; Manning, G.S. Theoretical assessment of the oligolysine model for ionic interactions in protein-DNA complexes. *J. Phys. Chem. B* **2011**, *115*, 9864–9872. [CrossRef]
59. Kunji, E.R.; Harding, M.; Butler, P.J.G.; Akamine, P. Determination of the molecular mass and dimensions of membrane proteins by size exclusion chromatography. *Methods* **2008**, *46*, 62–72. [CrossRef]
60. Rodbard, D.; Chrambach, A. Unified theory for gel electrophoresis and gel filtration. *Proc. Natl. Acad. Sci. USA* **1970**, *65*, 970–977. [CrossRef]
61. Langridge, T.D.; Tarver, M.J.; Whitten, S.T. Temperature effects on the hydrodynamic radius of the intrinsically disordered N-terminal region of the p53 protein. *Proteins* **2014**, *82*, 668–678. [CrossRef] [PubMed]
62. Wang, Y.; Schwedes, J.F.; Parks, D.; Mann, K.; Tegtmeyer, P. Interaction of p53 with its consensus DNA-binding site. *Mol. Cell. Biol.* **1995**, *15*, 2157–2165. [CrossRef]
63. Misra, V.K.; Honig, B. On the magnitude of the electrostatic contribution to ligand-DNA interactions. *Proc. Natl. Acad. Sci. USA* **1995**, *92*, 4691–4695. [CrossRef] [PubMed]
64. Record, M.T., Jr.; Ha, J.H.; Fisher, M.A. Analysis of equilibrium and kinetic measurements to determine thermodynamic origins of stability and specificity and mechanism of formation of site-specific complexes between proteins and helical DNA. *Methods Enzym.* **1991**, *208*, 291–343.
65. Beno, I.; Rosenthal, K.; Levitine, M.; Shaulov, L.; Haran, T.E. Sequence-dependent cooperative binding of p53 to DNA targets and its relationship to the structural properties of the DNA targets. *Nucleic Acids Res.* **2011**, *39*, 1919–1932. [CrossRef]
66. Zhu, J.; Zhou, W.; Jiang, J.; Chen, X. Identification of a novel p53 functional domain that is necessary for mediating apoptosis. *J. Biol. Chem.* **1998**, *273*, 13030–13036. [CrossRef]
67. Miller Jenkins, L.M.; Feng, H.; Durell, S.R.; Tagad, H.D.; Mazur, S.J.; Tropea, J.E.; Bai, Y.; Appella, E. Characterization of the p300 Taz2-p53 TAD2 complex and comparison with the p300 Taz2-p53 TAD1 complex. *Biochemistry* **2015**, *54*, 2001–2010. [CrossRef] [PubMed]
68. Teufel, D.P.; Freund, S.M.; Bycroft, M.; Fersht, A.R. Four domains of p300 each bind tightly to a sequence spanning both transactivation subdomains of p53. *Proc. Natl. Acad. Sci. USA* **2007**, *104*, 7009–7014. [CrossRef] [PubMed]
69. Von Hippel, P.H.; Berg, O.G. On the specificity of DNA-protein interactions. *Proc. Natl. Acad. Sci. USA* **1986**, *83*, 1608–1612. [CrossRef]
70. Eaton, B.E.; Gold, L.; Zichi, D.A. Let's get specific: The relationship between specificity and affinity. *Chem. Biol.* **1995**, *2*, 633–638. [CrossRef]
71. Chen, J.; Sun, Y. Modeling of the salt effects on hydrophobic adsorption equilibrium of protein. *J. Chromatogr. A* **2003**, *992*, 29–40. [CrossRef]
72. Zangi, R.; Hagen, M.; Berne, B.J. Effect of ions on the hydrophobic interaction between two plates. *J. Am. Chem. Soc.* **2007**, *129*, 4678–4686. [CrossRef] [PubMed]
73. Fried, M.G.; Stickle, D.F. Ion-exchange reactions of proteins during DNA binding. *Eur. J. Biochem.* **1993**, *218*, 469–475. [CrossRef] [PubMed]
74. Kitayner, M.; Rozenberg, H.; Kessler, N.; Rabinovich, D.; Shaulov, L.; Haran, T.E.; Shakked, Z. Structural basis of DNA recognition by p53 tetramers. *Mol. Cell* **2006**, *22*, 741–753. [CrossRef] [PubMed]
75. Chen, Y.; Zhang, X.; Machado, A.C.D.; Ding, Y.; Chen, Z.; Qin, P.Z.; Rohs, R.; Chen, L. Structure of p53 binding to the BAX response element reveals DNA unwinding and compression to accommodate base-pair insertion. *Nucleic Acids Res.* **2013**, *41*, 8368–8376. [CrossRef]
76. Zhukov, A.; Karlsson, R. Statistical aspects of van't Hoff analysis: A simulation study. *J. Mol. Recognit.* **2007**, *20*, 379–385. [CrossRef]
77. Moraitis, M.I.; Xu, H.; Matthews, K.S. Ion concentration and temperature dependence of DNA binding: Comparison of PurR and LacI repressor proteins. *Biochemistry* **2001**, *40*, 8109–8117. [CrossRef]
78. Oda, M.; Furukawa, K.; Ogata, K.; Sarai, A.; Nakamura, H. Thermodynamics of specific and non-specific DNA binding by the c-Myb DNA-binding domain. *J. Mol. Biol.* **1998**, *276*, 571–590. [CrossRef]
79. Winter, R.B.; von Hippel, P.H. Diffusion-driven mechanisms of protein translocation on nucleic acids. 2. The Escherichia coli repressor–operator interaction: Equilibrium measurements. *Biochemistry* **1981**, *20*, 6948–6960. [CrossRef]
80. DeHaseth, P.L.; Lohman, T.M.; Record, M.T., Jr. Nonspecific interaction of lac repressor with DNA: An association reaction driven by counterion release. *Biochemistry* **1977**, *16*, 4783–4790. [CrossRef]
81. Chakraborty, M.; Sengupta, A.; Bhattacharya, D.; Banerjee, S.; Chakrabarti, A. DNA binding domain of RFX5: Interactions with X-box DNA and RFXANK. *Biochim. Biophys. Acta* **2010**, *1804*, 2016–2024. [CrossRef]
82. Poon, G.M.; Gross, P.; Macgregor, R.B., Jr. The sequence-specific association of the ETS domain of murine PU.1 with DNA exhibits unusual energetics. *Biochemistry* **2002**, *41*, 2361–2371. [CrossRef] [PubMed]
83. Koblan, K.S.; Ackers, G.K. Site-specific enthalpic regulation of DNA transcription at bacteriophage lambda OR. *Biochemistry* **1992**, *31*, 57–65. [CrossRef] [PubMed]
84. Misra, V.K.; Hecht, J.L.; Sharp, K.A.; Friedman, R.A.; Honig, B. Salt effects on protein-DNA interactions. The lambda cI repressor and EcoRI endonuclease. *J. Mol. Biol.* **1994**, *238*, 264–280. [CrossRef] [PubMed]

85. Klein, C.; Georges, G.; Künkele, K.-P.; Huber, R.; Engh, R.A.; Hansen, S. High thermostability and lack of cooperative DNA binding distinguish the p63 core domain from the homologous tumor suppressor p53. *J. Biol. Chem.* **2001**, *276*, 37390–37401. [CrossRef]
86. Wilkins, D.K.; Grimshaw, S.B.; Receveur, V.; Dobson, C.M.; Jones, J.A.; Smith, L.J. Hydrodynamic radii of native and denatured proteins measured by pulse field gradient NMR techniques. *Biochemistry* **1999**, *38*, 16424–16431. [CrossRef]
87. Marsh, J.A.; Forman-Kay, J.D. Sequence Determinants of Compaction in Intrinsically Disordered Proteins. *Biophys. J.* **2010**, *98*, 2383–2390. [CrossRef]
88. Yu, B.; Pettitt, B.M.; Iwahara, J. Dynamics of Ionic Interactions at Protein-Nucleic Acid Interfaces. *Acc. Chem. Res.* **2020**, *53*, 1802–1810. [CrossRef]
89. Kussie, P.H.; Gorina, S.; Marechal, V.; Elenbaas, B.; Moreau, J.; Levine, A.J.; Pavletich, N.P. Structure of the MDM2 oncoprotein bound to the p53 tumor suppressor transactivation domain. *Science* **1996**, *274*, 948–953. [CrossRef]
90. Ferreon, J.C.; Lee, C.W.; Arai, M.; Martinez-Yamout, M.A.; Dyson, H.J.; Wright, P.E. Cooperative regulation of p53 by modulation of ternary complex formation with CBP/p300 and HDM2. *Proc. Natl. Acad. Sci. USA* **2009**, *106*, 6591–6596. [CrossRef]
91. Di Lello, P.; Jenkins, L.M.M.; Jones, T.N.; Nguyen, B.D.; Hara, T.; Yamaguchi, H.; Dikeakos, J.D.; Appella, E.; Legault, P.; Omichinski, J.G. Structure of the Tfb1/p53 complex: Insights into the interaction between the p62/Tfb1 subunit of TFIIH and the activation domain of p53. *Mol. Cell* **2006**, *22*, 731–740. [CrossRef] [PubMed]
92. Zhao, L.; Ouyang, Y.; Li, Q.; Zhang, Z. Modulation of p53 N-terminal transactivation domain 2 conformation ensemble and kinetics by phosphorylation. *J. Biomol. Struct. Dyn.* **2020**, *38*, 2613–2623. [CrossRef] [PubMed]
93. Liu, Y.; Sturtevant, J.M. Significant discrepancies between van't Hoff and calorimetric enthalpies. II. *Protein Sci.* **1995**, *4*, 2559–2561.
94. Datta, K.; Wowor, A.J.; Richard, A.J.; LiCata, V.J. Temperature dependence and thermodynamics of Klenow polymerase binding to primed-template DNA. *Biophys. J.* **2006**, *90*, 1739–1751.
95. Demir, O.; Ieong, P.U.; Amaro, R.E. Full-length p53 tetramer bound to DNA and its quaternary dynamics. *Oncogene* **2007**, *36*, 1451–1460.
96. Lambrughi, M.; De Gioia, L.; Gervasio, F.L.; Lindorff-Larsen, K.; Nussinov, R.; Urani, C.; Bruschi, M.; Papaleo, E. DNA-binding protects p53 from interactions with cofactors involved in transcription-independent functions. *Nucleic Acids Res.* **2016**, *44*, 9096–9109.
97. Melero, R.; Rajagopalan, S.; Lázaro, M.; Joerger, A.C.; Brandt, T.; Veprintsev, D.B.; Lasso, G.; Gil, D.; Scheres, S.H.; Carazo, J.M. Electron microscopy studies on the quaternary structure of p53 reveal different binding modes for p53 tetramers in complex with DNA. *Proc. Natl. Acad. Sci. USA* **2011**, *108*, 557–562.
98. Takeda, Y.; Ross, P.D.; Mudd, C.P. Thermodynamics of Cro protein-DNA interactions. *Proc. Natl. Acad. Sci. USA* **1992**, *89*, 8180–8184.

biomolecules

MDPI

*Article*

# Quantifying Coexistence Concentrations in Multi-Component Phase-Separating Systems Using Analytical HPLC

Anne Bremer [1,†], Ammon E. Posey [2,†], Madeleine B. Borgia [1], Wade M. Borcherds [1], Mina Farag [2], Rohit V. Pappu [2,*] and Tanja Mittag [1,*]

[1] Department of Structural Biology, St. Jude Children's Research Hospital, 262 Danny Thomas Place, Memphis, TN 38105, USA
[2] Department of Biomedical Engineering, Center for Biomolecular Condensates (CBC), James McKelvey School of Engineering, Washington University in St. Louis, St. Louis, MO 63130, USA
* Correspondence: pappu@wustl.edu (R.V.P.); tanja.mittag@stjude.org (T.M.)
† These authors contributed equally to this work.

**Abstract:** Over the last decade, evidence has accumulated to suggest that numerous instances of cellular compartmentalization can be explained by the phenomenon of phase separation. This is a process by which a macromolecular solution separates spontaneously into dense and dilute coexisting phases. Semi-quantitative, in vitro approaches for measuring phase boundaries have proven very useful in determining some key features of biomolecular condensates, but these methods often lack the precision necessary for generating quantitative models. Therefore, there is a clear need for techniques that allow quantitation of coexisting dilute and dense phase concentrations of phase-separating biomolecules, especially in systems with more than one type of macromolecule. Here, we report the design and deployment of analytical High-Performance Liquid Chromatography (HPLC) for in vitro separation and quantification of distinct biomolecules that allows us to measure dilute and dense phase concentrations needed to reconstruct coexistence curves in multicomponent mixtures. This approach is label-free, detects lower amounts of material than is accessible with classic UV-spectrophotometers, is applicable to a broad range of macromolecules of interest, is a semi-high-throughput technique, and if needed, the macromolecules can be recovered for further use. The approach promises to provide quantitative insights into the balance of homotypic and heterotypic interactions in multicomponent phase-separating systems.

**Keywords:** phase separation; biomolecular condensates; coexistence line

**Citation:** Bremer, A.; Posey, A.E.; Borgia, M.B.; Borcherds, W.M.; Farag, M.; Pappu, R.V.; Mittag, T. Quantifying Coexistence Concentrations in Multi-Component Phase-Separating Systems Using Analytical HPLC. *Biomolecules* **2022**, *12*, 1480. https://doi.org/10.3390/biom12101480

Academic Editors: Prakash Kulkarni, Stefania Brocca, Keith Dunker and Sonia Longhi

Received: 11 September 2022
Accepted: 3 October 2022
Published: 14 October 2022



## 1. Introduction

Phase separation is a biophysical process in which a macromolecular solution separates spontaneously into two coexisting phases. The two phases are a dense phase enriched in macromolecules with a macromolecular concentration $c_{dense}$, and a dilute phase deficient in macromolecules with a macromolecular concentration $c_{sat}$. In a binary mixture comprising macromolecules of a specific type dissolved in a complex solvent, the separation into two coexisting phases occurs at and above a system-specific threshold macromolecular concentration denoted as $c_{sat}$. At concentrations below $c_{sat}$, the solution is stable as a one-phase system. Over the last decade, evidence has accumulated that phase separation, which is a segregative transition, can be used to explain the observed compartmentalization of cellular matter [1–4]. Phase separation is thought to contribute, at least in part [5], to the formation of membraneless compartments known as biomolecular condensates [2]. These include condensates such as nucleoli [6,7] in the nucleus, and stress granules [8–10] and P bodies [11] in the cytoplasm. Phase separation has also been implicated in the formation of DNA repair foci [12,13], transcription centers [14–16] and membrane receptor clusters [17,18].

The ubiquitous roles invoked for phase separation in cells suggest that its dysregulation can result in disease states [19,20]. Indeed, evidence is accumulating that cancer pathogenesis can be mediated by either the abrogation of functional condensates through mutations of scaffolding molecules [21], or by the creation of aberrant condensates through the fusion of phase separating molecules with effector domains via chromosomal translocations [22–24]. Aberrant maturation of condensates is also thought to underlie the pathogenesis of a spectrum of neurodegenerative diseases [25–30].

Understanding the mechanisms of phase separation and quantitative comparisons of the system-specific driving forces for phase separation requires the quantitative characterization of phase-separating systems. Of particular interest is the contribution of multiple macromolecular components to phase separation [31], and how ligands influence the phase behavior [32,33]. Semi-quantitative methods for mapping phase diagrams include titrating a series of input concentrations of the constituent components and determining the presence or absence of phase separation via microscopy or turbidity measurements. This creates a grid, that yields an approximation of the low concentration arm of a phase boundary, where the number of input concentrations that are titrated and the concentration ranges they help explore will determine the accuracy and resolution of the inferred phase boundary [21,26,34]. These semi-quantitative phase boundaries have proven very useful in determining some key features of biomolecular condensates [9]. However, these methods often lack the precision necessary to be used for generating quantitative models [35,36]. Such physics-based models can provide insight into the underlying interactions by extracting thermodynamic parameters such as the critical temperature, free energy of mixing, and interaction strengths [35–39]. In addition, accurately determining the coexisting dilute and dense phase concentrations in multi-component systems provides information on the underlying contributions from homo- vs. heterotypic interactions [31,40]. For example, while two ligands may have similar observable effects on a condensate in terms of their partition coefficients, the underlying molecular mechanisms and effects on the driving force for phase separation may be very different [32,33]. Only with quantitative measurements over a range of input concentrations is it possible to distinguish underlying mechanisms of the modulation or regulation of phase behavior. Hence, there is a clear need for techniques that allow quantitation of coexisting dilute and dense phase concentrations of phase-separating biomolecules, especially in systems with more than one type of macromolecule. While quantitative methods have been developed for single-component systems [41,42], accurate determination of the coexistence concentrations of all species in multi-component systems is still a challenge.

Here, we report the development and deployment of analytical High-Performance Liquid Chromatography (HPLC) to separate and quantify distinct biomolecules and thereby access dilute and dense phase concentrations needed to reconstruct coexistence curves in multicomponent mixtures in vitro. This approach proves to be suitable because: (i) it can separate several components in multi-component systems, (ii) it is a label-free technique, (iii) through the choice of a suitable input volume it can frequently detect lower amounts of material than can be accessed using classic UV-spectrophotometers, (iv) it is applicable to a broad range of macromolecules of interest (e.g., nucleic acids, proteins, sugars), (v) it is a semi-high-throughput technique in that ca. 200 samples can be queued for measurement with the instrumentation used in this work, and finally (vi) if needed, the macromolecules can also be recovered for further use by coupling the HPLC instrument to a fraction collector.

To illustrate the utility of the HPLC approach, we deploy it to determine the saturation concentrations for a single-component phase-separating system and show their agreement with previously measured coexistence concentrations determined via classic UV-spectroscopy measurements. We then demonstrate that this method enables the determination of concentrations of each of the components in the dilute and dense phases of a two-component phase-separating system. This, as will be discussed in detail in a separate contribution [43], can provide access to information regarding the slopes of tie lines. The

HPLC method that we introduce here has the potential to provide quantitative insights into the interplay between homo- and heterotypic interactions in multi-component systems and help further our understanding of the driving forces for phase separation in biochemical reconstitutions of the phase behaviors of complex macromolecular mixtures.

## 2. Results

### 2.1. Rationale for the Proposed Approach

Accurate dilute and dense phase concentrations in single-component phase-separating systems can be determined by absorbance measurements of the dilute phase after centrifugation to pellet the dense phase [35,36,41]. If the saturation concentration is low, or if samples have low extinction coefficients, the absorbance values may be below the reliable detection limit of the spectrophotometer. Furthermore, in multi-component phase-separating systems, it is non-trivial to quantify the concentrations of individual components. If the sample contains more than one protein, UV absorption at 280 nm cannot be used to separate the contributions from the two proteins. Likewise, in systems containing protein and nucleic acids, the overlapping absorption spectra cannot be reliably deconvoluted. Labelling components with spectroscopically resolvable fluorophores and determining their concentrations in the dense phase from the fluorescence intensity is an option, but many controls are needed for accurate measurements [35]. In turn, fluorophores may influence the phase behavior. Large fluorophores, like GFP, may have a dramatic effect on the solubility [44,45], but even small fluorophores may be able to perturb phase behavior as they are often charged and have aromatic moieties which are key determinants of phase behaviors of many phase-separating biomolecules [10,31,35–37,39,44,46].

As mentioned previously, it is possible to generate semi-quantitative estimates of the locations of phase boundaries by titrating the components and then using turbidity or microscopy to determine the presence/absence of droplets. These methods, however, are semi-quantitative, only yielding estimates of saturation concentrations because concentrations are evaluated in a stepwise fashion. They also do not provide access to tie lines. For microscopy, there is the additional concern that interactions (or lack thereof) of condensates with the slide surface can interfere with observation, and surfaces may need to be functionalized and optimized separately for mutants of the biomolecules. Given these complications, using a label-free method that relies on the intrinsic properties of the native molecules would be preferable.

### 2.2. Details of the Approach

Combining the established approach of separating dilute and dense phases via centrifugation with the use of analytical HPLC to separate and quantify sample components yields a semi-high-throughput, robust, and highly quantitative method (Figure 1). First, a column needs to be selected and tested to confirm that it can separate the components (see "General considerations for implementation of the method"). Then, separate standard curves for each component are determined by making several injections of known concentration ($c_A$) and volume ($V_A$) and integrating the peak from the chromatogram ($I_A$). This generates a standard curve for each species (example given in Figure 2A), which is fit to Equation (1)

$$I_A = slope \times n_A + intercept \tag{1}$$

to determine the slope and intercept for the given component. Here, $n_A = c_A \times V_A$, the amount of biomolecule in moles. This yields the input concentration $c_A$ for a given volume $V_A$. With the resulting standard curve, $c_{\text{sat}}$ and $c_{\text{dense}}$ for the component can be determined from injections of appropriate samples as follows, and schematically shown in Figure 1. After preparation and incubation of the phase-separating sample, the dilute and dense phases are separated by centrifugation. A sample of the dilute phase is removed, injected onto the HPLC and eluted with the same gradient as used for the calibration measurements. The amount of each component present in the sample is quantified by integration of the relevant peak ($I_A$) such that the initial sample concentration of each component ($c_A$) is computed based on their individual standard curves and the volume of sample injected ($V_A$). Dense phase concentrations are determined in the same way; this requires the preparation of a dense phase sample that is large enough to remove a defined volume, which is then diluted appropriately for HPLC processing.

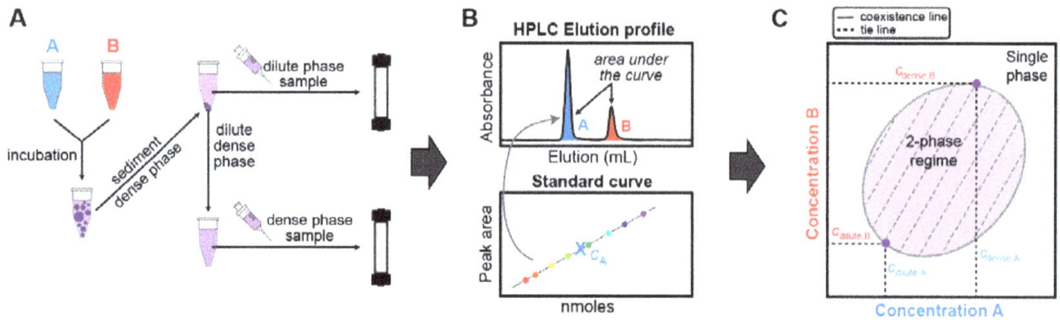


**Figure 1.** Schematic overview of the workflow used in this study to reconstruct phase boundaries via analytical HPLC. (**A**) Biomolecules A and B undergo phase separation when present at suitable concentrations and molar ratios. The sample is incubated for equilibration. Separation of the dense and dilute phases is achieved via centrifugation. Known volumes of dilute and dense phase are each separately injected onto the HPLC column and eluted with an appropriate method. (**B**) HPLC elution profile for a sample containing biomolecules A and B. Peaks are integrated and the concentrations of biomolecules A and B are determined using a standard curve. (**C**) The coexistence line and tie lines of biomolecules A and B can be reconstructed from the dilute and dense phase concentrations extracted from the elution profiles. The tie line connects the coexisting dilute and dense phase concentrations. A tie line or tie simplex is defined by its slope, and it identifies the components whose concentrations need to be constrained relative to one another to yield the concentrations of the coexisting phases.

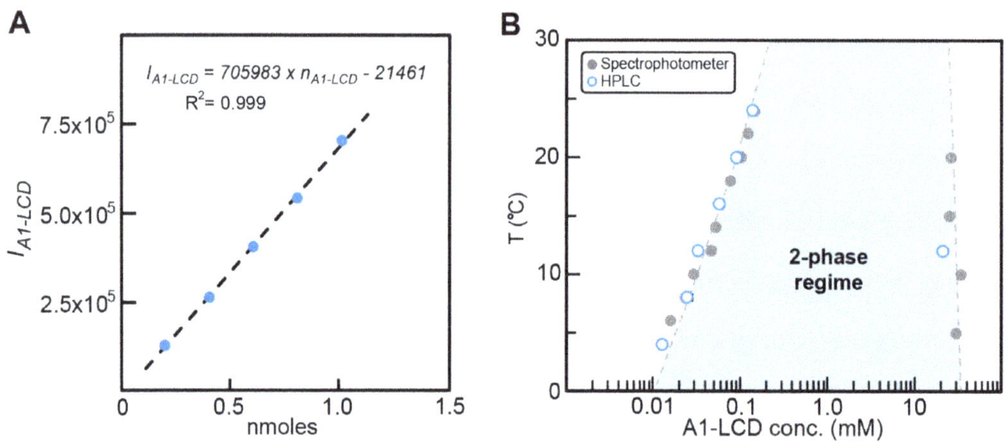


**Figure 2.** Determination of $c_{sat}$ and $c_{dense}$ by analytical HPLC. (**A**) The standard curve for A1-LCD was determined by plotting the area under the HPLC elution peak ($I_{A1\text{-}LCD}$) from injections of different amounts of A1-LCD, $n_{A1\text{-}LCD}$ in nmoles. (**B**) Comparison of $c_{sat}$ and $c_{dense}$ values obtained via measurements using HPLC vs values from UV absorption measurements by spectrophotometer for A1-LCD as a function of temperature [36]. The dashed line represents a fit of the Flory Huggins equation to the spectrophotometer data. The shaded area thus represents the 2-phase regime determined by the coexistence concentrations and other extant data [35,36].

### 2.3. Validation of the Method with a Single-Component Phase-Separating System

To test the capabilities of analytical HPLC for determining coexisting dilute and dense phase concentrations, we determined the coexisting concentrations of phase-separated samples of the low-complexity domain (LCD) of hnRNPA1, hereafter referred to as A1-LCD. We previously determined the concentrations of coexisting phases by UV absorption [35,41]. The standard curve shows a linear relationship between input sample amount and peak area allowing the use of linear regression analysis to determine sample concentrations (Figure 2A). In all cases, $c_{sat}$ and $c_{dense}$ that we estimated from the HPLC chromatograms agree with the previously reported values (Figure 2B) [35]. Thus, the method appears to be suitable for the quantitative determination of coexisting dilute and dense phase concentrations of biomolecules.

### 2.4. Application to Multi-Component Phase Separation

Having established the functionality of analytical HPLC in a single-component phase-separating system, we next employed the approach to quantify the saturation concentration of a Gcn4 construct (for details see Methods) in the presence of polyethylene glycol with an average molecular weight of 8 kDa, referred to hereafter as PEG8000. The absorbance spectrum for PEG8000 partially overlaps that of Gcn4 at 280 nm, and thus its detection must be performed separately from Gcn4 in order to determine the dilute phase concentration of Gcn4. Before using HPLC to quantify the concentrations of multiple components in a sample, it needs to be confirmed that the components elute separately and with sufficient resolution. This was shown to be case, thus allowing for accurate assessments of the concentration of the Gcn4 component. We quantified the $c_{sat}$ values of Gcn4 as a function of different input concentrations of PEG8000. As expected, the concentration of Gcn4 in the coexistent dilute phase decreases with increasing concentrations of PEG8000 (Figure 3A). This implies that for the Gcn4 system, PEG8000 behaves mostly like a crowder that enhances the driving forces for phase separation through depletion mediated attractions, which refers to the enhanced inter-Gcn4 attractions that most likely arises from exclusion of the crowder from the dense phase.

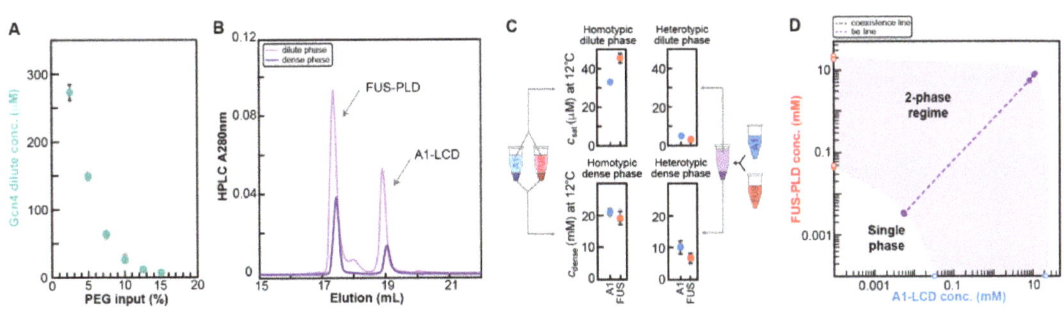


**Figure 3.** Determination of dilute and dense phase concentrations for multi-component systems. (**A**) Dilute phase concentrations of Gcn4 in the presence of increasing concentrations of PEG8000. Individual measurements are shown in grey, with the average shown in green. At least three replicates per sample were measured and error bars represent the standard deviation. (**B**) HPLC chromatogram showing the elution profile of samples of the dilute and dense phase for the A1-LCD/FUS-PLD mixture. The dense phase sample was diluted prior to injection. For a comparison of absorbance at 230 and 280 nm, please see Figure S2. (**C**) $c_{sat}$ and $c_{dense}$ for A1-LCD and FUS-PLD for the case of their homotypic or heterotypic phase separation. (**D**) Data points from C are shown in a 2D phase diagram. The tie line between coexisting dilute and dense phase concentrations in the heterotypic system is shown as dashed line. The 2-phase regime is approximated as shaded area as expected from few presented data points.

Next, we investigated a two-protein system consisting of A1-LCD and the FUS prion-like domain (FUS-PLD); the two proteins can phase separate on their own. However, they also form a single dense phase when mixed. The mixture of the A1-LCD and FUS-PLD is a ternary system comprising two macromolecules in a solvent. To study phase separation in this mixture, we need to be able to measure the concentrations of both macromolecules in the coexisting dense and dilute phases in the scenario where the ternary system separates into two coexisting phases. We quantified the dilute and dense phase concentrations for both protein components at a single mixing ratio and concentration. The chromatogram in Figure 3B shows that A1-LCD and FUS-PLD elute separately, allowing for integration of the peaks and determination of dilute and dense phase concentrations of each component; the results are shown in Figure 3C. The saturation concentrations of separate A1-LCD and FUS-PLD solutions are higher than the coexisting dilute phase concentrations of each of the components in the mixture. In fact, even the sum of the dilute phase concentrations in the mixture is lower than the $c_{sat}$ values measured for either system in a binary mixture comprising just one type of macromolecule and the solvent. We also determined dense phase concentrations for the A1-LCD/FUS-PLD mixture and the tie line (Figure 3C,D), which provides insights into contributions from homo- and heterotypic interactions to phase separation [40]. A detailed discussion of the complex phase behaviors of A1-LCD and FUS-PLD mixtures will be presented elsewhere [43].

We have demonstrated that determination of dilute and dense phase concentrations of coexisting phases can be achieved by integrating peaks in HPLC chromatograms. The reliability of the measurements is established via favorable comparisons to estimates obtained using measurements based on established techniques [41]. We can go beyond the study of binary mixtures comprising just one type of macromolecule and leverage the ability to separate a system of multiple components using HPLC to determine concentrations of more than one type of macromolecule in coexisting phases. This information gives access to coexisting dilute and dense phase concentrations in a multi-component phase boundary, from which tie lines can be determined. This provides powerful information that is difficult to procure in other ways and can be used to dissect contributions from homo- and heterotypic interactions to phase separation.

### 2.5. General Considerations for Implementation of the Method

To employ and adapt the approach described in this study, the following points need to be considered:

(a) Column: Columns to achieve separation include normal-phase, reverse-phase, ion exchange and size exclusion columns, which are readily available for HPLC systems. The work presented here used C4 or C18 (ReproSil Gold 200; Dr. Maisch) reverse-phase columns.

(b) Mobile phase: The mobile phase solvents used are primarily dictated by the column. The typical chromatographic buffers used for size exclusion and ion exchange are aqueous buffers, while RP-HPLC uses a gradient of organic solvents in water. However, also within the remit of RP-HPLC, there are different options possible for the organic solvent, including acetonitrile, methanol, and tetrahydrofuran. The solvents used must be miscible with water and of HPLC-grade quality to minimize their contribution to the absorbance signals measured. In this work, gradients used involved the mixing of $H_2O$ + 0.1% TFA (trifluoroacetic acid) with pure acetonitrile. 0.1% TFA yields a pH of 2.1 ensuring full ionization of analytes and acts as a weak ion-pairing agent thereby conferring more uniform binding of each analyte. This yields sharper peaks and more reproducible elution profiles. Use of TFA in just water, and not in the acetonitrile, is employed as it effectively adds an ion exchange component to the RP-HPLC separation and can result in better peak separation. Of note, the low pH results in the denaturation of protein structure. In cases where it is desirable to recover the components, reverse-phase columns are only suitable if the macromolecules readily refold.

(c) Gradient: Optimization of the gradient is required to obtain sufficient separation between eluting species. It is important to consider sufficient equilibration time given the column volume if step changes are made at any point in the overall gradient run. The appropriate combination of points a, b and c is key to a successful use of the HPLC methodology and likely requires iteration for optimization based on the types of samples that are being studied.
(d) Detection: HPLC systems may have different detection capabilities ranging from a single absorbance wavelength to setups with photodiode array detectors providing absorbance spectra rather than single wavelengths, or even fluorescence detectors. This work made use of an HPLC system with a dual-selectable wavelength detector. The selection of wavelengths to be monitored will depend on the macromolecule of interest. Typical choices include 280 nm for proteins containing aromatic residues, 260 nm for nucleic acids and 230 to 215 nm for proteins lacking aromatic residues. The monitored wavelength should also avoid interference from solvent components.
(e) Column loading: The range of volumes that can be injected onto the column will depend on the system at hand. Injection of accurate volumes, a prerequisite for accurate determination of coexistence concentrations, is most easily achieved with an autoinjector. Further, the amount of macromolecule of interest in the sample should yield an absorbance signal in the linear range of the detector as confirmed through the standard curve. The amounts for which this can be achieved will vary based on the extinction coefficient of the molecule, the wavelength being monitored, and the width of the elution peak, which can be optimized by solvent choice and gradient properties. A further consideration is that loading of high concentrations of some buffer components such as glycerol or PEG can lead to contamination and ultimately damage the column.
(f) Washing: It is good practice to perform wash programs/cycles between batches of samples to ensure that the column remains in good working order and is frequently cleaned. This avoids material or contamination from previous runs interfering with following measurements.
(g) Tests: Routine running of blank injections using the method gradient is valuable to check that sample material has not been retained on the column. Retention in the column can lead to subsequent elution that interferes with the quantitation of components in injected samples.
(h) Sample recovery: If the HPLC system is coupled to a fraction collector, the eluted peaks can be collected to recover sample components. In the case of RP-HPLC, these fractions are best dried on a speed-vac and then resuspended in the buffer of choice. Keep in mind however, that as RP-HPLC denatures the protein, structured proteins need to be refolded. *Considerations regarding handling of dense phase:*
(i) Viscosity of dense phase: The dense phase is highly viscous and needs to be carefully pipetted. We recommend the use of a positive displacement pipette to minimize errors and achieve accurate volumes (see also [41]. The variability in the measured dense phase concentrations is higher than the measured dilute phase concentrations as can be seen in Figures 2B and 3C,D, but the percentage errors are relatively small. Compared to error sources in other approaches for determining dense phase concentrations, e.g., microscopic determination of fluorescence intensity in the dense phase, the error contribution from pipetting the viscous dense phase is relatively small and manageable. Several replicate measurements should be performed to get a sense of their precision.
(j) Sample requirements: The required biomolecule amounts to generate sufficient dilute phase for detection depend almost exclusively on the extinction coefficient of the biomolecule. Dense phase requirements can be more limiting. We typically remove 2 μL of dense phase for dilution and subsequent injection into the HPLC. The amount of protein needed to generate a slightly larger volume of dense phase depends on the dilute vs. dense phase concentrations and the concentration of the stock solution. If we, e.g., consider the hnRNPA1 LCD (Figure 2B) with dilute and dense phase

concentrations at 20 °C of ~100 µM and ~20 mM, a stock solution used to generate a dense phase sample could be 100 µL of a 1 mM protein. Induction of phase separation (by addition of NaCl to 150 mM final concentration in this experiment) would result in approximately 95.5 µL of 100 µM dilute phase and 4.5 µL of 20 mM dense phase. Notably, the resulting dense phase volume is not only determined by the total amount of protein but also by how far above the saturation concentration the preparation starts, with higher concentrations capturing a larger fraction of protein in the dense phase. Less concentrated dense phases require substantially lower protein amounts.

## 3. Materials and Methods

### *3.1. Details of Protein Constructs*

Three different proteins, namely A1-LCD and FUS-PLD, which are prion-like disordered domains of the Fused in Sarcoma (F), Ewing Sarcoma (E), and Taf15 (T) (i.e., FET) family of proteins, and a short version of the canonical yeast transcription factor Gcn4, were expressed in *E. coli* and purified. A1-LCD was expressed as detailed in reference [8], and the same cloning, expression and purification strategy was employed for $FUS^{1-214}$ (UniProt: P35637) and Gcn4. The variant of Gcn4 spans the central activation domain (residues 101–141) from *S. cerevisiae* (UniProt: P03069) connected by a short $(GS)_4$-linker to the DNA-binding domain of Gcn4 (residues 222–281).

### *3.2. Phase Separation Assay*

Phase separation of A1-LCD and FUS-PLD, respectively, was induced by adding NaCl to 150 mM in 20 mM HEPES (pH 7.0). Phase separation of Gcn4 was induced by titrating PEG8000 from 2.5% to 15% in 20 mM HEPES (pH 7.3), 150 mM potassium acetate, 2 mM DTT. For the multi-component A1-LCD/FUS-PLD system, 1.1 mg/mL A1-LCD was mixed with 1.1 mg/mL FUS-PLD in 20 mM HEPES (pH 7.0), 150 mM NaCl. The samples were incubated at the desired temperature for 20 min, then centrifuged at this temperature for 5 min at 12,000 rpm to separate the dilute and dense phases. Known amounts of dilute and dense phases were removed. The dense phase volume was diluted into a defined volume of 6 M GdmHCl as needed. Aliquots of the separated phases were then applied to the HPLC to determine the concentrations.

### *3.3. HPLC instrumentation, Columns, and Solvents*

The dilute ($c_{sat}$) and dense phase ($c_{dense}$) concentrations were determined on a HPLC instrument with UV/Vis Detector. Samples were run on a Waters HPLC system with an Autosampler (Waters 2707), a Binary HPLC Pump (Waters 1525) and a dual-channel UV/Visible Detector (Waters 2489). The wavelengths monitored were 280 nm and 230 nm. Monitored wavelengths should be chosen to avoid any interference from solvent components. ReproSil Gold 200 (5 µm, 250 mm × 4.6 mm; Dr. Maisch, Germany) columns were used; C18 for protein only samples, and C4 for Gcn4 + PEG8000 samples. The solvents used were $H_2O$ + 0.1% TFA (Sigma-Aldrich, Saint Louis, MO, USA) and acetonitrile (Alfa Aesar, Haverhill, MA, USA).

### *3.4. Calibration of Concentration Measurements by HPLC*

For each protein, a standard curve was measured by injecting 5–6 different volumes ($V_A$) of solution in buffer with known concentration ($c_A$). The integral of the elution peak ($I_A$) was obtained with the built-in Waters Empower HPLC-software. A plot of $I_A$ vs. $n_A$, where $n_A = c_A \times V_A$, yields the line which was fit with Equation (1) (Figure 2A) to obtain the *slope* and *intercept*. The resulting standard curve enables determination of the concentration of samples with known injection volumes and resulting peak integrals.

### *3.5. Determination of the Dilute ($c_{sat}$) and Dense Phase ($c_{dense}$) Concentration Using HPLC*

Quantitation of $c_{sat}$ was achieved by injecting known volumes of each dilute phase sample onto the HPLC. Dense phase concentrations were assessed by dilution of 2 mL of dense phase, obtained using a positive displacement pipette, with 6 M GdmHCl before loading onto the HPLC. Dilution facilitates complete loading and the denaturant prevents precipitation that may occur upon dilution with incompatible buffers. For all resulting chromatograms the amounts of the relevant components were calculated from their respective standard curves, allowing the reconstruction of the coexistence line. For all data points presented, at least three replicates were measured and averaged. The resulting values were compared to dilute and dense phase concentrations determined by UV absorption on a spectrophotometer that we have previously reported [35].

## 4. Discussion

To address the need for quantitative methods to measure phase boundaries in multi-component systems, we present an analytical HPLC method to separate and quantify multiple components in coexisting dilute and dense phases. We tested the accuracy of the HPLC method by reproducing measured dilute and dense phase concentrations of coexisting phases for the binary mixture comprising a single type of macromolecule namely, A1-LCD. We then deployed the method to study a ternary system with two macromolecular components namely, Gcn4 and PEG8000 system. We used this system because the two macromolecular components have overlapping absorption spectra. We showed that the HPLC based approach enabled the separation of both macromolecules, thus allowing us to quantify the concentration of Gcn4 in the coexisting dilute phase as function of the concentration of PEG8000. The methodology was then used to determine all four coexisting concentrations for the two-component A1-LCD/FUS-PLD phase-separating system. The four concentrations are the individual dilute phase concentrations of A1-LCD and FUS-PLD, and their coexisting dense phase concentrations. These data provide direct access to tie lines, and with additional input concentrations we can map the full coexistence curve, and use information regarding the shapes of these curves as well as the slopes of tie lines to uncover the interplay between homo- vs. heterotypic interactions—a topic that we will analyze and discuss elsewhere [43].

The basic methodology described here should be able to determine the concentrations of as many species as one can resolve on the chosen column. For increasingly complex systems, not all species may be resolvable on a single column, and future development of the methodology will center around using parallel columns with different chemistries to resolve a larger number of species. This would allow for the accurate determination of coexisting concentrations for increasingly complex systems.

For protein-RNA mixtures, additional challenges include high apparent affinities in the nanomolar or sub-nanomolar range, and therefore, they may remain bound to one another even during the HPLC run. Further, the dilute phase will likely comprise a mixture of bound and unbound species as defined by a binding polynomial. Separating the bound and unbound species would provide a fuller species characterization and is a challenge to be addressed that will also be highly relevant for systems that form pre-percolation clusters [47]. Combining the HPLC methodology with other approaches is likely to be promising in this regard.

Overall, the HPLC methodology reported here enables label-free, quantitative measurements of coexisting concentrations in complex systems at semi-high throughput. The method has the potential to further our understanding of the contributions of homotypic and heterotypic interactions and how they are encoded in the sequence of biomacromolecules.

**Supplementary Materials:** The following supporting information can be downloaded at: https://www.mdpi.com/article/10.3390/biom12101480/s1, Figure S1. HPLC elution profile of (A) A1-LCD, (B) FUS-PLD and (C) Gcn4 were used to generate respective standard curves (D) and (E), where the ordinate is the area under the HPLC elution peak, and the abscissa is n, given in nmoles. For the A1-LCD standard curve, see Figure 2; Figure S2. HPLC elution profile of A1-LCD and FUS-PLD monitored at 230 and 280 nm.

**Author Contributions:** A.E.P. and R.V.P. conceived of the method. A.B., A.E.P. and M.B.B. designed the implementation. A.B. prototyped the measurements and received assistance with troubleshooting from M.B.B., W.M.B. and T.M. A.B., M.B.B., W.M.B. and T.M. analyzed the data, and wrote the first draft of the manuscript. M.F. provided inspiration for studying the mixtures of A1-LCD and FUS-PLD based on his unpublished simulation results. All others read and edited the manuscript. T.M. and R.V.P. secured funding. All authors have read and agreed to the published version of the manuscript.

**Funding:** T.M. and R.V.P. acknowledge funding by the St. Jude Children's Research Hospital Research Collaborative on the Biology and Biophysics of RNP granules, and by NIH grant R01NS121114. T.M. acknowledges funding by the American Lebanese Syrian Associated Charities.

**Institutional Review Board Statement:** Not applicable.

**Informed Consent Statement:** Not applicable.

**Data Availability Statement:** All data are available upon request.

**Conflicts of Interest:** Tanja Mittag is a member of the Scientific Advisory Board of Faze Medicines. Rohit Pappu is a member of the Scientific Advisory Board of Dewpoint Therapeutics. The work reported here was not influenced by these affiliations.

## References

1. Shin, Y.; Brangwynne, C.P. Liquid phase condensation in cell physiology and disease. *Science* **2017**, *357*. [CrossRef] [PubMed]
2. Banani, S.F.; Lee, H.O.; Hyman, A.A.; Rosen, M.K. Biomolecular condensates: Organizers of cellular biochemistry. *Nat. Rev. Mol. Cell Biol.* **2017**, *18*, 285–298.
3. Holehouse, A.S.; Pappu, R.V. Functional Implications of Intracellular Phase Transitions. *Biochemistry* **2018**, *57*, 2415–2423. [CrossRef] [PubMed]
4. Alberti, S.; Gladfelter, A.; Mittag, T. Considerations and Challenges in Studying Liquid-Liquid Phase Separation and Biomolecular Condensates. *Cell* **2019**, *176*, 419–434. [CrossRef] [PubMed]
5. Mittag, T.; Pappu, R.V. A conceptual framework for understanding phase separation and addressing open questions and challenges. *Mol. Cell* **2022**, *82*, 2201–2214. [CrossRef]
6. Brangwynne, C.P.; Mitchison, T.J.; Hyman, A.A. Active liquid-like behavior of nucleoli determines their size and shape in Xenopus laevis oocytes. *Proc. Natl. Acad. Sci. USA* **2011**, *108*, 4334–4339. [CrossRef] [PubMed]
7. Feric, M.; Vaidya, N.; Harmon, T.S.; Mitrea, D.M.; Zhu, L.; Richardson, T.M.; Kriwacki, R.W.; Pappu, R.V.; Brangwynne, C.P. Coexisting Liquid Phases Underlie Nucleolar Subcompartments. *Cell* **2016**, *165*, 1686–1697. [CrossRef]
8. Sanders, D.W.; Kedersha, N.; Lee, D.S.W.; Strom, A.R.; Drake, V.; Riback, J.A.; Bracha, D.; Eeftens, J.M.; Iwanicki, A.; Wang, A.; et al. Competing Protein-RNA Interaction Networks Control Multiphase Intracellular Organization. *Cell* **2020**, *181*, 306–324.e28. [CrossRef]
9. Yang, P.; Mathieu, C.; Kolaitis, R.M.; Zhang, P.; Messing, J.; Yurtsever, U.; Yang, Z.; Wu, J.; Li, Y.; Pan, Q.; et al. G3BP1 Is a Tunable Switch that Triggers Phase Separation to Assemble Stress Granules. *Cell* **2020**, *181*, 325–345.e28. [CrossRef]
10. Guillen-Boixet, J.; Kopach, A.; Holehouse, A.S.; Wittmann, S.; Jahnel, M.; Schlussler, R.; Kim, K.; Trussina, I.; Wang, J.; Mateju, D.; et al. RNA-Induced Conformational Switching and Clustering of G3BP Drive Stress Granule Assembly by Condensation. *Cell* **2020**, *181*, 346–361.e17. [CrossRef] [PubMed]
11. Xing, W.; Muhlrad, D.; Parker, R.; Rosen, M.K. A quantitative inventory of yeast P body proteins reveals principles of composition and specificity. *Elife* **2020**, *9*, e56525. [CrossRef] [PubMed]
12. Kilic, S.; Lezaja, A.; Gatti, M.; Bianco, E.; Michelena, J.; Imhof, R.; Altmeyer, M. Phase separation of 53BP1 determines liquid-like behavior of DNA repair compartments. *EMBO J.* **2019**, *38*, e101379. [CrossRef]
13. Levone, B.R.; Lenzken, S.C.; Antonaci, M.; Maiser, A.; Rapp, A.; Conte, F.; Reber, S.; Mechtersheimer, J.; Ronchi, A.E.; Muhlemann, O.; et al. FUS-dependent liquid-liquid phase separation is important for DNA repair initiation. *J. Cell Biol.* **2021**, *220*. [CrossRef] [PubMed]
14. Boija, A.; Klein, I.A.; Sabari, B.R.; Dall'Agnese, A.; Coffey, E.L.; Zamudio, A.V.; Li, C.H.; Shrinivas, K.; Manteiga, J.C.; Hannett, N.M.; et al. Transcription Factors Activate Genes through the Phase-Separation Capacity of Their Activation Domains. *Cell* **2018**, *175*, 1842–1855.e16. [CrossRef] [PubMed]
15. Cho, W.K.; Spille, J.H.; Hecht, M.; Lee, C.; Li, C.; Grube, V.; Cisse, I.I. Mediator and RNA polymerase II clusters associate in transcription-dependent condensates. *Science* **2018**, *361*, 412–415. [CrossRef]
16. Boehning, M.; Dugast-Darzacq, C.; Rankovic, M.; Hansen, A.S.; Yu, T.; Marie-Nelly, H.; McSwiggen, D.T.; Kokic, G.; Dailey, G.M.; Cramer, P.; et al. RNA polymerase II clustering through carboxy-terminal domain phase separation. *Nat. Struct. Mol. Biol.* **2018**, *25*, 833–840. [CrossRef]
17. Su, X.; Ditlev, J.A.; Hui, E.; Xing, W.; Banjade, S.; Okrut, J.; King, D.S.; Taunton, J.; Rosen, M.K.; Vale, R.D. Phase separation of signaling molecules promotes T cell receptor signal transduction. *Science* **2016**, *352*, 595–599. [CrossRef]

18. Case, L.B.; Zhang, X.; Ditlev, J.A.; Rosen, M.K. Stoichiometry controls activity of phase-separated clusters of actin signaling proteins. *Science* **2019**, *363*, 1093–1097. [CrossRef]
19. Alberti, S.; Dormann, D. Liquid-Liquid Phase Separation in Disease. *Annu. Rev. Genet.* **2019**, *53*, 171–194. [CrossRef]
20. Boija, A.; Klein, I.A.; Young, R.A. Biomolecular Condensates and Cancer. *Cancer Cell* **2021**, *39*, 174–192. [CrossRef]
21. Bouchard, J.J.; Otero, J.H.; Scott, D.C.; Szulc, E.; Martin, E.W.; Sabri, N.; Granata, D.; Marzahn, M.R.; Lindorff-Larsen, K.; Salvatella, X.; et al. Cancer Mutations of the Tumor Suppressor SPOP Disrupt the Formation of Active, Phase-Separated Compartments. *Mol. Cell* **2018**, *72*, 19–36.e18. [CrossRef] [PubMed]
22. Chandra, B.; Michmerhuizen, N.L.; Shirnekhi, H.K.; Tripathi, S.; Pioso, B.J.; Baggett, D.W.; Mitrea, D.M.; Iacobucci, I.; White, M.R.; Chen, J.; et al. Phase Separation Mediates NUP98 Fusion Oncoprotein Leukemic Transformation. *Cancer Discov.* **2022**, *12*, 1152–1169. [CrossRef] [PubMed]
23. Ahn, J.H.; Davis, E.S.; Daugird, T.A.; Zhao, S.; Quiroga, I.Y.; Uryu, H.; Li, J.; Storey, A.J.; Tsai, Y.H.; Keeley, D.P.; et al. Phase separation drives aberrant chromatin looping and cancer development. *Nature* **2021**, *595*, 591–595. [CrossRef] [PubMed]
24. Tulpule, A.; Guan, J.; Neel, D.S.; Allegakoen, H.R.; Lin, Y.P.; Brown, D.; Chou, Y.T.; Heslin, A.; Chatterjee, N.; Perati, S.; et al. Kinase-mediated RAS signaling via membraneless cytoplasmic protein granules. *Cell* **2021**, *184*, 2649–2664.e18.
25. Mackenzie, I.R.; Nicholson, A.M.; Sarkar, M.; Messing, J.; Purice, M.D.; Pottier, C.; Annu, K.; Baker, M.; Perkerson, R.B.; Kurti, A.; et al. TIA1 Mutations in Amyotrophic Lateral Sclerosis and Frontotemporal Dementia Promote Phase Separation and Alter Stress Granule Dynamics. *Neuron* **2017**, *95*, 808–816.e9. [CrossRef]
26. White, M.R.; Mitrea, D.M.; Zhang, P.; Stanley, C.B.; Cassidy, D.E.; Nourse, A.; Phillips, A.H.; Tolbert, M.; Taylor, J.P.; Kriwacki, R.W. C9orf72 Poly(PR) Dipeptide Repeats Disturb Biomolecular Phase Separation and Disrupt Nucleolar Function. *Mol. Cell* **2019**, *74*, 713–728.e6. [CrossRef]
27. Zhang, P.; Fan, B.; Yang, P.; Temirov, J.; Messing, J.; Kim, H.J.; Taylor, J.P. Chronic optogenetic induction of stress granules is cytotoxic and reveals the evolution of ALS-FTD pathology. *eLife* **2019**, *8*, e39578. [CrossRef]
28. Fernandopulle, M.; Wang, G.; Nixon-Abell, J.; Qamar, S.; Balaji, V.; Morihara, R.; St George-Hyslop, P.H. Inherited and Sporadic Amyotrophic Lateral Sclerosis and Fronto-Temporal Lobar Degenerations arising from Pathological Condensates of Phase Separating Proteins. *Hum. Mol. Genet.* **2019**, *28*, R187–R196. [CrossRef]
29. Nedelsky, N.B.; Taylor, J.P. Pathological phase transitions in ALS-FTD impair dynamic RNA-protein granules. *Rna* **2022**, *28*, 97–113. [CrossRef]
30. Mathieu, C.; Pappu, R.V.; Taylor, J.P. Beyond aggregation: Pathological phase transitions in neurodegenerative disease. *Science* **2020**, *370*, 56–60. [CrossRef]
31. Choi, J.M.; Holehouse, A.S.; Pappu, R.V. Physical Principles Underlying the Complex Biology of Intracellular Phase Transitions. *Annu. Rev. Biophys.* **2020**, *49*, 107–133. [CrossRef] [PubMed]
32. Ruff, K.M.; Dar, F.; Pappu, R.V. Polyphasic linkage and the impact of ligand binding on the regulation of biomolecular condensates. *Biophys. Rev.* **2021**, *2*, 021302. [CrossRef] [PubMed]
33. Ruff, K.M.; Dar, F.; Pappu, R.V. Ligand effects on phase separation of multivalent macromolecules. *Proc. Natl. Acad. Sci. USA* **2021**, *118*, e2017184118. [CrossRef] [PubMed]
34. Li, P.; Banjade, S.; Cheng, H.C.; Kim, S.; Chen, B.; Guo, L.; Llaguno, M.; Hollingsworth, J.V.; King, D.S.; Banani, S.F.; et al. Phase transitions in the assembly of multivalent signalling proteins. *Nature* **2012**, *483*, 336–340. [CrossRef]
35. Martin, E.W.; Holehouse, A.S.; Peran, I.; Farag, M.; Incicco, J.J.; Bremer, A.; Grace, C.R.; Soranno, A.; Pappu, R.V.; Mittag, T. Valence and patterning of aromatic residues determine the phase behavior of prion-like domains. *Science* **2020**, *367*, 694–699. [CrossRef]
36. Bremer, A.; Farag, M.; Borcherds, W.M.; Peran, I.; Martin, E.W.; Pappu, R.V.; Mittag, T. Deciphering how naturally occurring sequence features impact the phase behaviours of disordered prion-like domains. *Nat. Chem.* **2022**, *14*, 196–207. [CrossRef]
37. Brady, J.P.; Farber, P.J.; Sekhar, A.; Lin, Y.H.; Huang, R.; Bah, A.; Nott, T.J.; Chan, H.S.; Baldwin, A.J.; Forman-Kay, J.D.; et al. Structural and hydrodynamic properties of an intrinsically disordered region of a germ cell-specific protein on phase separation. *Proc. Natl. Acad. Sci. USA* **2017**, *114*, E8194–E8203. [CrossRef]
38. Wei, M.T.; Elbaum-Garfinkle, S.; Holehouse, A.S.; Chen, C.C.; Feric, M.; Arnold, C.B.; Priestley, R.D.; Pappu, R.V.; Brangwynne, C.P. Phase behaviour of disordered proteins underlying low density and high permeability of liquid organelles. *Nat. Chem.* **2017**, *9*, 1118–1125. [CrossRef]
39. Wang, J.; Choi, J.M.; Holehouse, A.S.; Lee, H.O.; Zhang, X.; Jahnel, M.; Maharana, S.; Lemaitre, R.; Pozniakovsky, A.; Drechsel, D.; et al. A Molecular Grammar Governing the Driving Forces for Phase Separation of Prion-like RNA Binding Proteins. *Cell* **2018**, *174*, 688–699.e16. [CrossRef]
40. Choi, J.-M.; Dar, F.; Pappu, R.V. LASSI: A lattice model for simulating phase transitions of multivalent proteins. *PLoS Ccomput. Biol.* **2019**, *15*, e1007028. [CrossRef]
41. Milkovic, N.M.; Mittag, T. Determination of Protein Phase Diagrams by Centrifugation. *Methods Mol. Biol.* **2020**, *2141*, 685–702. [PubMed]
42. Peran, I.; Martin, E.W.; Mittag, T. Walking Along a Protein Phase Diagram to Determine Coexistence Points by Static Light Scattering. *Methods Mol. Biol.* **2020**, *2141*, 715–730. [PubMed]
43. Farag, M.; Bremer, A.; Borcherds, W.M.; Mittag, T.; Pappu, R.V. Electrostatic interactions contribute to the cooperative interplay in co-phase separation of mixtures of prion-like low complexity domains. 2022; *in preparation*.

44. Pak, C.W.; Kosno, M.; Holehouse, A.S.; Padrick, S.B.; Mittal, A.; Ali, R.; Yunus, A.A.; Liu, D.R.; Pappu, R.V.; Rosen, M.K. Sequence Determinants of Intracellular Phase Separation by Complex Coacervation of a Disordered Protein. *Mol. Cell* **2016**, *63*, 72–85. [CrossRef] [PubMed]
45. Martin, E.W.; Thomasen, F.E.; Milkovic, N.M.; Cuneo, M.J.; Grace, C.R.; Nourse, A.; Lindorff-Larsen, K.; Mittag, T. Interplay of folded domains and the disordered low-complexity domain in mediating hnRNPA1 phase separation. *Nucleic Acids Res.* **2021**, *49*, 2931–2945. [CrossRef]
46. Boeynaems, S.; Holehouse, A.S.; Weinhardt, V.; Kovacs, D.; Van Lindt, J.; Larabell, C.; Van Den Bosch, L.; Das, R.; Tompa, P.S.; Pappu, R.V.; et al. Spontaneous driving forces give rise to protein-RNA condensates with coexisting phases and complex material properties. *Proc. Natl. Acad. Sci. USA* **2019**, *116*, 7889–7898. [CrossRef]
47. Kar, M.; Dar, F.; Welsh, T.J.; Vogel, L.T.; Kuhnemuth, R.; Majumdar, A.; Krainer, G.; Franzmann, T.M.; Alberti, S.; Seidel, C.A.M.; et al. Phase-separating RNA-binding proteins form heterogeneous distributions of clusters in subsaturated solutions. *Proc. Natl. Acad. Sci. USA* **2022**, *119*, e2202222119. [CrossRef]

*biomolecules*

MDPI

*Article*

# Digging into the 3D Structure Predictions of AlphaFold2 with Low Confidence: Disorder and Beyond

**Apolline Bruley, Jean-Paul Mornon, Elodie Duprat *,† and Isabelle Callebaut *,†**

Sorbonne Université, Muséum National d'Histoire Naturelle, UMR CNRS 7590, Institut de Minéralogie, de Physique des Matériaux et de Cosmochimie, IMPMC, 75005 Paris, France
* Correspondence: elodie.duprat@sorbonne-universite.fr (E.D.); isabelle.callebaut@sorbonne-universite.fr (I.C.)
† These authors contributed equally to this work.

**Abstract:** AlphaFold2 (AF2) has created a breakthrough in biology by providing three-dimensional structure models for whole-proteome sequences, with unprecedented levels of accuracy. In addition, the AF2 pLDDT score, related to the model confidence, has been shown to provide a good measure of residue-wise disorder. Here, we combined AF2 predictions with pyHCA, a tool we previously developed to identify foldable segments and estimate their order/disorder ratio, from a single protein sequence. We focused our analysis on the AF2 predictions available for 21 reference proteomes (AFDB v1), in particular on their long foldable segments (>30 amino acids) that exhibit characteristics of soluble domains, as estimated by pyHCA. Among these segments, we provided a global analysis of those with very low pLDDT values along their entire length and compared their characteristics to those of segments with very high pLDDT values. We highlighted cases containing conditional order, as well as cases that could form well-folded structures but escape the AF2 prediction due to a shallow multiple sequence alignment and/or undocumented structure or fold. AF2 and pyHCA can therefore be advantageously combined to unravel cryptic structural features in whole proteomes and to refine predictions for different flavors of disorder.

**Keywords:** long foldable segments; pyHCA; soluble domains; protein sequence; conditional order; hidden order; dark proteomes; intrinsically disordered domains

**Citation:** Bruley, A.; Mornon, J.-P.; Duprat, E.; Callebaut, I. Digging into the 3D Structure Predictions of AlphaFold2 with Low Confidence: Disorder and Beyond. *Biomolecules* **2022**, *12*, 1467. https://doi.org/10.3390/biom12101467

Academic Editors: Stefania Brocca, Keith Dunker, Sonia Longhi and Prakash Kulkarni

Received: 14 September 2022
Accepted: 5 October 2022
Published: 13 October 2022



## 1. Introduction

AlphaFold2 [1] and RoseTTAfold [2] have recently achieved an impressive breakthrough in the field of structural biology, providing accurate models of three-dimensional (3D) structures of proteins based on only knowledge of their amino acid sequences alone. Based on deep-learning techniques, they take advantage of the vast existing knowledge of protein sequences and 3D structures, recently expanded through environmental genomics and structural genomics approaches. In particular, they extensively used evolutionary information to detect co-variation of residues (or correlated mutations), the underlying idea being that residues that have co-evolved are close in 3D space. The first version of the AlphaFold2 database (AFDB v1) [3] included predictions for a very large part of proteomes from 21 widely studied organisms. It has been extended to provide open access to over 200 million predictions, covering nearly every organism with protein sequence data. This provides the scientific community with a wealth of knowledge, which could accelerate the understanding of protein structure-function relationships and have a profound impact on many areas of biology, including human health and the environment.

Several studies have already been conducted to estimate the extent to which AlphaFold2 (AF2) improves the coverage in structural biology, as well as to analyze its current advantages and limitations (e.g., [4–10]). One striking feature of AF2 is that it provides a per-residue metric, reflecting confidence in the structural assignment (predicted local distance difference test (pLDDT)) [1]. High values of pLDDT are observed for folded domains, contrasting with low values typically associated with linkers and unstructured or disordered regions [11]. The relevance of pLDDT as a predictor of disorder has been supported on the CAID benchmark dataset [12] and compared to other state-of-the-art disorder predictors, such as SPOT-Disorder2 or IUPred2 [1,4,11].

At least two questions need to be considered when focusing on very low confidence regions (pLDDT < 50) in AF2 predictions, which are assumed to be globally disordered. The first question is whether it is possible to reveal conditional order within these intrinsically disordered regions (IDRs), from amino acid sequence information alone. Such IDRs may be involved in molecular recognition, to which hydrophobic interactions make major contributions [13,14]. In many cases, these regions undergo a disorder-to-order transition (induced folding) to a more structured state upon binding with a partner [15]. High-resolution multi-dimensional NMR studies have demonstrated that such IDRs, ranging in length from 10 up to 70 amino acids and referred to over time by different names ([16,17], molecular recognition elements [18], primary contact sites [19], preformed structural elements [20], pre-structured motifs [21]), can be pre-populated by transient local structural elements, presaging the target-bound conformation [21]. The plasticity of these IDRs can allow for a range of secondary structures in the bound state, as shown by the example of the p53 tumor suppressor protein [22]. Some IDRs are also able to retain a significant degree of structural heterogeneity in the bound states [23], leading to the definition of fuzzy complexes [24,25]. Some IDRs involved in molecular recognition consist of or incorporate short linear motifs (SLiMs), i.e., short conserved sequences, which enable low affinity, transient, and conditional interactions and are often located within disordered regions [26]. Specifying the structural unit in which these short interacting motifs are embedded should inform on the global features of the interaction (such as affinity, specificity, fuzziness). Regarding conditional order, another question to consider is whether it is possible to identify, within these very low confidence AF2 predictions, longer binding IDRs that meet the definition of intrinsically disordered domains (IDD) [27–29], which must be stabilized by a partner within protein complexes to adopt a stable fold?

The second issue related to very low pLDDT regions is to evaluate whether some might not be disordered and might still adopt a well-folded 3D structure, but AF2 cannot predict it (what we call "hidden order"). This hypothesis is conceivable as the co-evolutionary information, necessary to predict inter-residue contacts, is lacking for some protein sequences. These proteins, not predicted as disordered, escape any annotation coming from sequence or structure databases and constitute the dark proteome [30,31]. They still represent 10% of the human proteome after annotation with the AlphaFold2 predictions [7].

We recently analyzed the pLDDT values observed for the AF2 3D structure predictions on the 21 reference proteomes (AFDB v1) in light of another metric, called the HCA score (Bruley et al. [32]). The HCA score is based on Hydrophobic Cluster Analysis (HCA), a two-dimensional approach allowing the analysis of the content of an amino acid sequence in regular secondary structures (see [33] for a recent review of the methodology). Indeed, the hydrophobic clusters defined by this approach mainly correspond to the positions of regular secondary structures constituting the building blocks of folded domains [34–36]. The analysis of the composition of a sequence in hydrophobic clusters thus provides information on its architecture in domains and the disorder/order content of the delineated domains. A tool has been developed to automatically partition protein sequences into foldable segments based on a measure of hydrophobic cluster density [37]. The calculation of an HCA score provides information about the composition of the sequence in clusters and hydrophobic amino acids within the clusters, which thus reflects the overall order/disorder ratio of the foldable segments [32].Using this HCA score, we disentangled different types of disorder and appreciate disorder-to-order continuum. While residues with low-pLDDT values were enriched in non-foldable segments, a significant portion of foldable segments with HCA scores typical of well-folded domains also had low mean pLDDT values in AF2 3D structure predictions. This suggests that these regions carry specific functional information (corresponding to the two cases mentioned above) that remains unraveled by AF2 (Bruley et al. [32]).

Here, we further explored the source of this apparent inconsistency between foldability and low confidence AF2 prediction, which is widely assimilated to disorder in the literature. To this end, we analyzed, from the same 21 reference proteomes, the long soluble-like foldable segments as defined by the pyHCA tool, whose residues all have a very low AF2 pLDDT value (hereafter referred to as full-VL segments). Particular focus was on segments of length > 30 amino acids, which corresponds to the minimum length considered for globular domains [38]. Moreover, this minimal length excludes a large number of short motifs (SLiMs) undergoing induced folding and which are otherwise associated with higher HCA score values (Bruley et al. [32]). To analyze these long, soluble-like full-VL segments, we considered four features related to their amino acid sequences and 3D structures, as predicted by AF2. We described these 3D structure models by the proportion of residues involved in a regular secondary structure (RSS) and by the proportion of residues accessible to the solvent. In addition, we described the protein sequences on which these predictions were based, by the proportion of residues predicted as disordered by IUPred2 and by the average number of homologs per residue as found in the large environmental BFD database. The latter feature allowed us to consider co-evolution information, essential for the reliable prediction of amino acid contacts by AF2. We compared these features to their distribution for long soluble-like foldable segments whose residues all have a very high pLDDT values (hereafter referred to as full-VH segments), for the 21 proteomes included in AFDB v1.

## 2. Material and Methods

### 2.1. Proteomes from AlphaFold Protein Structure Database v1

Amino acid (aa) sequences and predicted 3D structures were downloaded from the AlphaFold Protein Structure database (AFDB) v1 ([3], https://alphafold.ebi.ac.uk, accessed on 21 July 2021) for the 21 reference model-organism proteomes. The per-residue model confidence values (pLDDT) were extracted from the 3D coordinate files (B-factor column in PDB format).

### 2.2. Delineation of Soluble-Like Foldable Segments within Protein Sequences

The *segment* function of the pyHCA tool (provided at https://github.com/DarkVador-HCA/pyHCA, accessed on 14 September 2022) was used to automatically delineate foldable segments (FS), i.e., segments with a high density of hydrophobic clusters (HC), as defined by the Hydrophobic Cluster Analysis (HCA) [33]. HC consist of strong hydrophobic amino acids (V,I,L,M,F,Y,W) and are separated from each other by at least four other amino acids or a proline. For FS delineation, cysteine (C) is integrated into the hydrophobic alphabet and HC consist of only one or two consecutive hydrophobic amino acid(s) are not considered, as they are mainly associated with coils [36].

The HCA score, which measures the density of hydrophobic clusters and strong hydrophobic amino acids of foldable segments (Bruley et al. [32]), was calculated using the *segment* function (pyHCA tool). Soluble-like segments were defined according to an HCA score value between $-1$ and 3.5.

### 2.3. Description of Sequence and Structural Features

Our final dataset consisted of proteins from AFDB v1, encompassing at least one long (>30 a.a.), globular soluble-like ($-1 \leq$ HCA score $\leq 3.5$) foldable segments, entirely made of residues with very low (VL) or very high (VH) 3D prediction confidence (pLDDT $\leq 50$ and pLDDT $> 90$, respectively). We considered four different features to characterize the amino acid sequence and AF2 3D models of these segments, as detailed below. For each feature, we defined a threshold value based on the distribution of these full-VH segments and delimiting an interval encompassing at least 95% of them. These threshold values were further used for the dataset description by binary trees.

#### 2.3.1. Per-Residue Disorder Prediction

Disorder was predicted using the IUPred2A [39] *long* disorder predictor on the whole protein sequences. IUPred2A calculates a per-residue score between 0 and 1 that reflects the estimated stabilizing effect of other residues on each residue of one amino acid sequence. The coverage of the FS by disorder was then calculated (in percentage of the segment length), considered as disordered amino acids having a score above 0.5. The coverage threshold was set to 33.4%. The number of FS with a value below this threshold were as follows: 14,077 segments over 30,644 and 10,827 segments over 11,395 in case of full-VL and full-VH, respectively.

#### 2.3.2. Known Homologs

The multiple sequence alignments used to build the AlphaFold2 models were not provided in AFDB repositories. Therefore, a search for known homologs in the reduced Big Fantastic Database (BFD) was performed using *jackhmmer* (from HMMER 3.3.2 [40], http://hmmer.org/, accessed on 14 September 2022). The parameters (e-value threshold of 0.0001, 1 iteration) were those used by AF2 in the similarity search step. The Big Fantastic Database (BFD) [1] (https://bfd.mmseqs.com, accessed on 24 May 2022) is a database containing 2.5 billion clustered protein sequences. It is the most comprehensive database used by AF2 in order to build multiple sequence alignments, gathering sequences from genomic and metagenomic databases (UniprotKB [41] and metaclust [42] and datasets assembled with Plass [43]). The reduced version of BFD contains only representative sequences of each cluster (65,984,053 sequences). This one was downloaded following the recommendations given on the AF2 github (https://github.com/deepmind/alphafold, accessed on 24 May 2022). In this work, the sequence similarity search was performed on the whole protein sequences. The number of aligned sequences per FS position was then calculated and averaged over the length of the FS. The mean number threshold was set to 23.5 BFD homologs per segment residue. The number of FS with a value above this threshold were as follows: 3347 segments over 30,644 and 10,829 segments over 11,395 in case of full-VL and full-VH, respectively.

#### 2.3.3. Secondary Structure Assignment

Secondary structures were assigned from the coordinates of the AF2 3D structure models (PDB files, full-length proteins) using the DSSP program [44] available in the biopython module v1.78 for python v3.6.3. All amino acids found in alpha helices (encoded as "H" in DSSP), 3–10 helices ("G"), Pi helices ("I"), strands ("E"), and isolated beta-bridge residues ("B") were considered to participate in regular secondary structures (RSS). The percentage of the FS residues participating in a RSS was then calculated. The number of FS with at least 1 RSS were as follows: 10,993 segments out of 30,644 and 11,393 segments out of 11,395 in case of full-VL and full-VH, respectively.

#### 2.3.4. Solvent Accessibility

Using the same module, the residues relative accessible surface area was calculated. This value was obtained by normalizing the residue accessible surface area (ASA) by the maximum ASA for the residue, computed on Gly-X-Gly tripeptides (where X is the residue of interest). By default, DSSP referred to the Sander and Rost scale for maximum ASA values per residue [45]. We considered a residue to be solvent accessible if the relative ASA was above 0.36 (based on Rost and Sander [45]). The percentage of accessible residues was calculated on each FS. The feature threshold was set to 82.9%. The number of FS with a value below this threshold were as follows: 1908 segments over 30,644 and 10,823 segments over 11,395 in case of full-VL and full-VH, respectively.

2.3.5. 3D Structure Comparison

The Dali server ([46], http://ekhidna2.biocenter.helsinki.fi/dali, accessed on 14 September 2022) was used to compare the AF2 3D structure models of the foldable segments with PDB experimental 3D structures.

2.3.6. Figure Creation

3D structures were visualized with the UCSF Chimera software [47]. HCA plots were drawn using the DrawHCA program (http://osbornite.impmc.upmc.fr/hca/hca-seq.html, accessed on 14 September 2022). Hydrophobic clusters (HC) affinities for RSS were extracted from HCDB [36]. Binary tree diagrams were created using the R package *ggparty* (https://github.com/martin-borkovec/ggparty, accessed on 14 September 2022).

## 3. Results

### *3.1. General Features of Full-VL and Full-VH Segments from AFDB v1*

Figure 1 illustrates the technical flow used in this study to extract 30,644 full-VL and 11,395 full-VH long soluble-like foldable segments from AFDB v1 using the pyHCA tool.

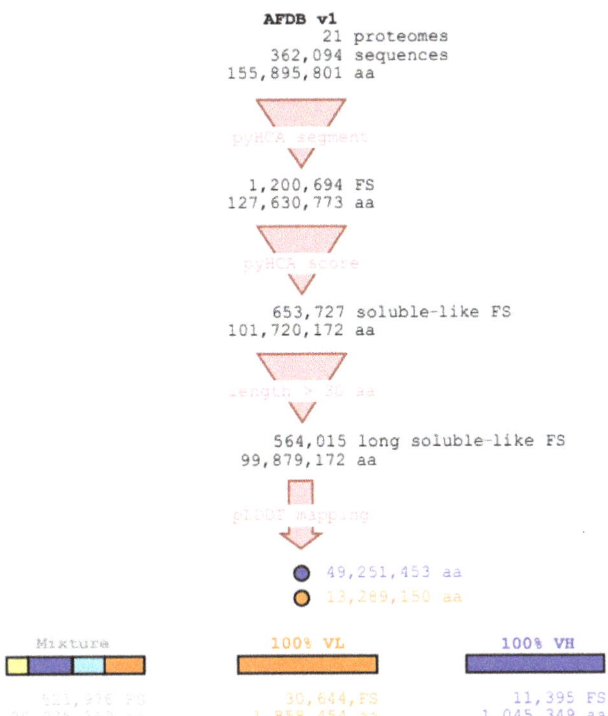


**Figure 1.** The technical flow for definition of the long soluble-like full-VH and full-VL foldable segments from AFDB v1 by using the pyHCA tool. The number of foldable segments (FS) and the number of residues (aa) are indicated at each step of the flow. The dataset further analyzed in this study consists of the full-VL and full-VH segments. For quantitative details about each of the 21 proteomes, see Supplementary Table S1.

Details for each of the 21 proteomes are given in Supplementary Table S1. Most of the residues in AFDB v1 (64.1%) are included in long soluble-like foldable segments (from 55.5% up to 73.9% in the proteomes of *Leishmania infantum* and *E. coli*, respectively). These segments are mainly composed of residues with a very high pLDDT value (49.3% VH, 13.3%

VL). This trend is also observed for each proteome, except for *Plasmodium falciparum* (23.8% VH, 40.4% VL). However, the set of the full-VL segments is larger than the set of full-VH segments, both in the number of segments and in the number of residues (Figure 1). This trend is observed for each of the 17 eukaryotic proteomes, where at least 9.1% VL residues included in a long soluble-like foldable segment are part of a full-VL segment (up to 22.4% and 24.0% for *Leishmania infantum* and *Plasmodium falciparum*, respectively). On the contrary, less than 6.3% of the VL residues included in a long soluble-like foldable segment are part of a full-VL segment for prokaryotic proteomes, where only a few cases of full-VL segments were found (1, 8, 14, and 29 segments for the archaeon *Methanocaldococcus jannaschii* and the bacteria *Escherichia coli*, *Staphylococcus aureus*, and *Mycobacterium tuberculosis* respectively). Furthermore, for eukaryotic proteomes, less than 2.6% of the VH residues included in a long soluble-like foldable segment are part of a full-VH segment (from 3.7% up to 8.1% for the prokaryotic proteomes). In AFDB v1, the mean length of full-VL segments (60.7 amino acids) is smaller than the mean length of full-VH segments (91.7 aa). This trend is observed for each of the 21 proteomes.

Figure 2 illustrates the technical flow used in this study for the description of the AF2 3D models and protein sequences for the full-VL and full-VH segment datasets. We described each segment by four quantitative features and explored their distribution for each dataset.

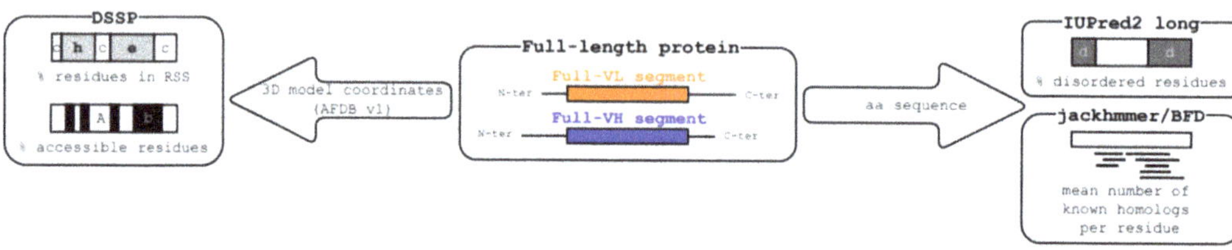


**Figure 2.** The technical flow for feature description of the segment dataset. Each AFDB v1 full-length protein comprising at least one full-VL or one full-VH long soluble-like foldable segment was analyzed by different tools (DSSP on the 3D coordinates, IUPred2 *long*, and *jackhmmer* on the amino acid sequence) allowing for calculation of four quantitative features describing each segment. Labels used for the different tools are: (i) for DSSP secondary structure assignment: h, helix; e, strand (extended); c, coil; (ii) for DSSP solvent accessibility: A, accessible, b, buried; (iii) for IUPred2 long: d, disorder.

### 3.2. Full-VH Segments

Figure 3 depicts the classification of the 11,395 full-VH segments using a binary tree based on the features used to describe the 3D models and the amino acid sequences (see Figure 2 and Section 2 for details). Representative examples of the different categories are shown in Figure 4.

Quantitative thresholds were defined for each feature based on 95% full-VH, except for the proportion of segment residues participating in a RSS, as assigned by DSSP from the AF2 3D models (see Section 2 for details). For this 3D feature, we considered two classes of segments based on the presence/absence of RSS. All the long soluble-like foldable segments whose residues all have a very high pLDDT value (full-VH segments) are associated with the presence of RSS, except for two cases, corresponding to thrombospondin (TSP) repeats (Figure 4e). As observed in the experimental 3D structures that can serve as templates for homology modeling (pdb entries 1yo8 and 3fby), TSP repeats are folded domains with calcium ions bound into the core through acidic (aspartate) residues. The foldable segments delineated here contain conserved cysteine residues that form interdomain disulfide-bridges, providing tight interactions in the wire architecture typical of the TSP-2 signature domain [48].

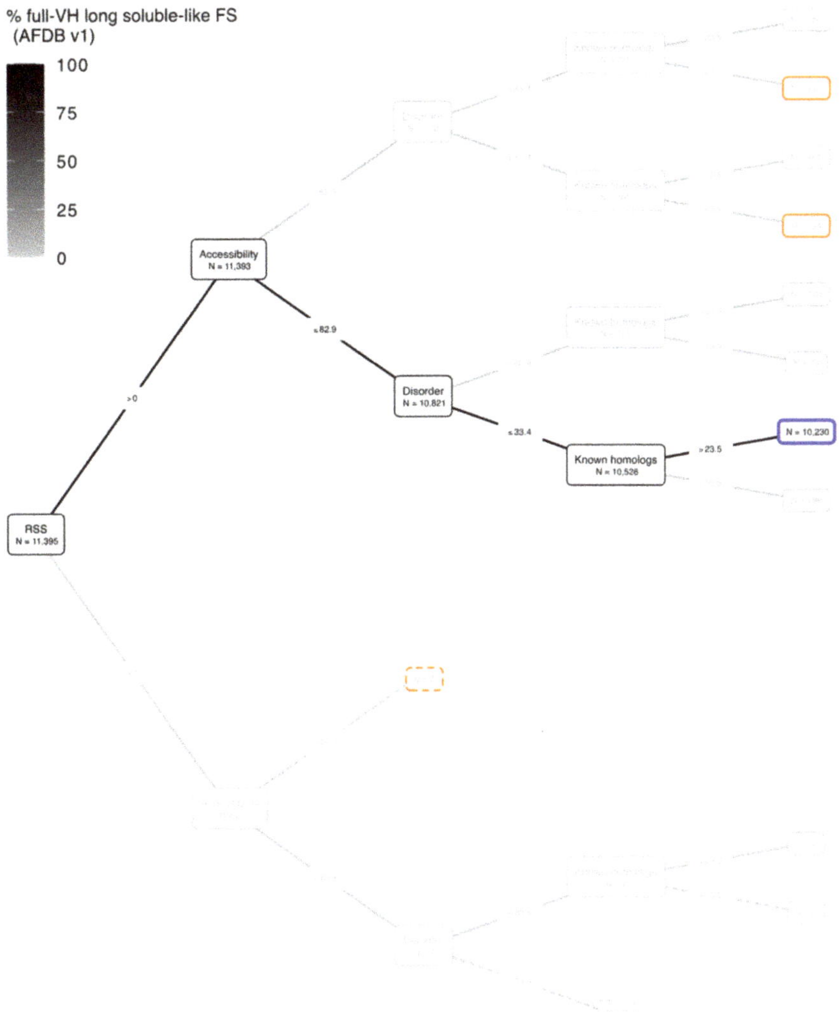


**Figure 3.** Binary tree diagram of the full-VH segments according to the feature thresholds. The four levels of the tree (from the root on the left to the last internal nodes on the right) correspond to the four features describing the segments (see Figure 2 for the technical flow), as follows: percentage of segment residues participating in a regular secondary structure (RSS), percentage of segment residues accessible to the solvent (Accessibility), percentage of segment residues predicted to be disordered (Disorder), the mean number of BFD homologs per segment residue (Known homologs). The binary conditions based on each feature threshold are indicated on the edges of the tree (for details, see Section 2). The number of foldable segments with a given feature below or above each threshold is indicated in the internal and terminal nodes. The total number of full-VH segments is indicated within the root node. The terminal nodes corresponding to the most abundant subsets of full-VH segments (this figure) and full-VL segments (Section 3.3) are highlighted in blue and orange, respectively. For quantitative details about each of the 21 proteomes, see Supplementary Figure S1.

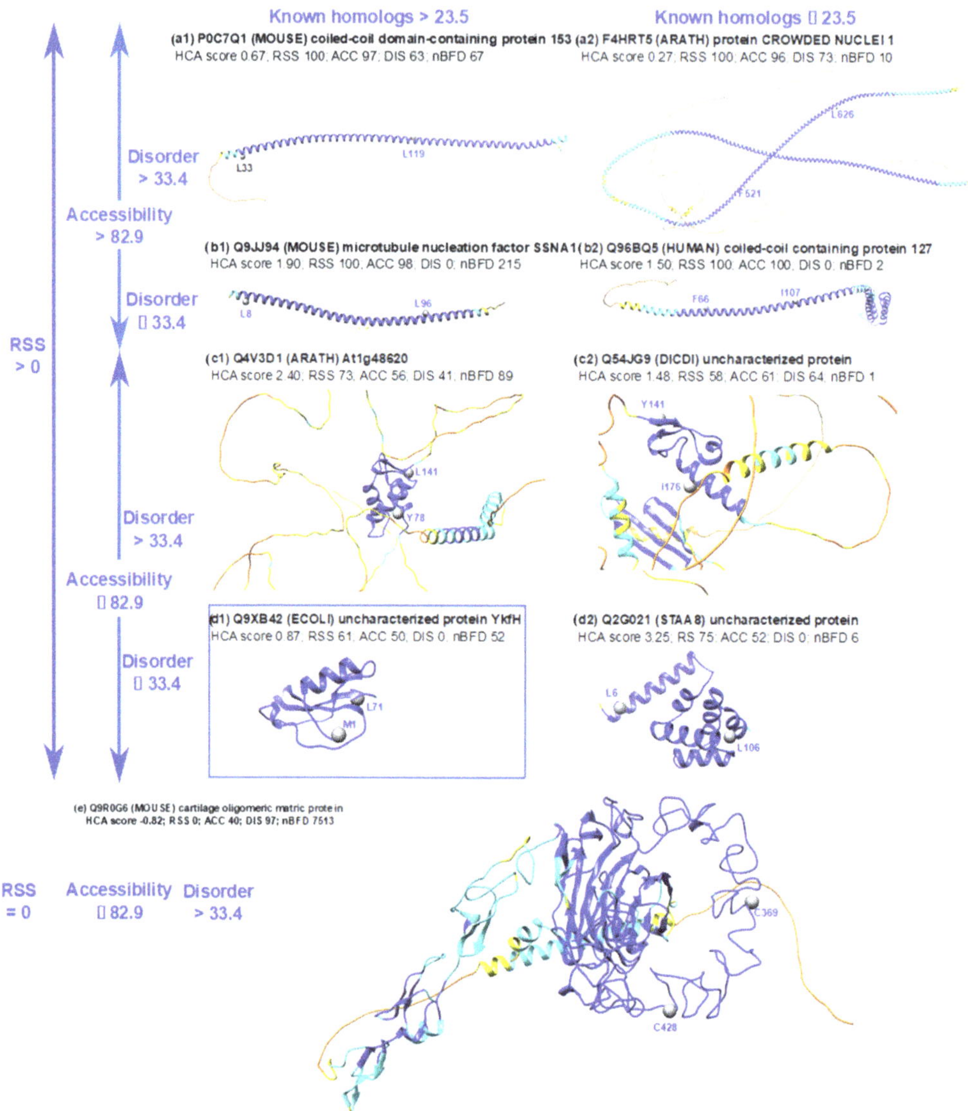


**Figure 4.** Examples of full-VH soluble-like foldable segments, distinguished according to the four features. The examples were extracted from the binary tree diagram shown in Figure 3. The AF2 3D structure models are colored according to pLDDT values, with the positions of the first and last amino acids of the full-VH soluble-like foldable segments indicated. The corresponding HCA score values are also reported, as well as those of the four features. The example extracted from the most populated leaf in Figure 3 is boxed in blue. HCA plots of the corresponding sequences are illustrated in Supplementary Figure S2. Subfigures a show examples with RSS > 0, Accessibility > 82.9, Disorder > 33.4 and BFD homologs per position > 23.5 (**a1**) and ≤ 23.5 (**a2**). Subfigures b show examples with RSS > 0, Accessibility > 82.9, Disorder ≤ 33.4 and BFD homologs per position > 23.5 (**b1**) and ≤ 23.5 (**b2**). Subfigures c show examples with RSS > 0, Accessibility ≤ 82.9, Disorder > 33.4 and BFD homologs per position > 23.5 (**c1**) and ≤ 23.5 (**c2**). Subfigures d show examples with RSS > 0, Accessibility ≤ 82.9, Disorder ≤ 33.4 and BFD homologs per position > 23.5 (**d1**) and ≤ 23.5 (**d2**). Subfigure (**e**) shows one of the two similar cases with RSS = 0, Accessibility ≤ 82.9, Disorder > 33.4.

The most abundant category of full-VH long soluble-like foldable segments (10,230 full-VH segments over 11,395, boxed in blue in Figures 3 and 4(d1)) corresponds to folded domains with low predicted disorder and a high number of BFD homologs. Domains were considered as folded as they contain RSS assembled together and have relative low solvent accessibility due to the involvement of a large number of amino acids in a hydrophobic core. Supplementary Figure S2 provides details of the HCA plots of the foldable segments whose 3D structures are shown in Figure 4. The folded domains contain ~1/3 strong hydrophobic amino acids distributed in clusters, which correspond to the positions of RSS. A significant number of cases also exist with a smaller number of BFD homologs (296 segments, Figures 3 and 4(d2)). Here, the consideration of experimental 3D structures as templates can explain the accurate AF2 prediction (pdb:1sed for the example shown in Figure 4(d2)). Other interesting cases are those of folded domains corresponding to sequences predicted to be disordered for a large part, but which are clearly not (Figure 4(c1,c2) corresponding to histone fold, 29% identity with pdb 2lso-A, and to a case with no obvious similarity with known 3D structures, respectively). Finally, the cases of accurate AF2 predictions associated with models globally accessible to the solvent concern long helices, typical of coiled-coil assembly, whose sequences are predicted as disordered or not (Figure 4(a1,a2,b1,b2)). When no experimental 3D structure is available, the AF2 prediction is supported by a sufficiently informative periodic pattern and self-organizing structure, regardless of the number of BFD homologs.

### 3.3. *Full-VL Segments*

Figure 5 shows the binary tree diagram of full-VL, long soluble-like foldable segments, according to the same threshold values as in Figure 3 for full-VH segments. The full-VL segments are much more dispersed across the different categories than the full-VH segments (see boxes in Figures 3 and 5). Four categories are populated by at least 10% of the full-VL segments. In contrast, there was only one in category in this case for full-VH segment, including 90% of them. Another notable point is that the mean values of the four features (RSS, Accessibility, Disorder, Known homologs) differ significantly between full-VH and full-VL segments, even when considering a same binary class (Figure 6). In particular, (i) full-VL segments with at least one RSS contain on average fewer residues participating in a RSS than similar full-VH segments (Figure 6a); (ii) full-VL segments with accessibility less than 82.9% are more accessible to solvent than similar full-VH segments (Figure 6b); (iii) full-VL segments with disorder less than 33.4% are predicted to be more disordered than similar full-VH segments (Figure 6c); finally, the full-VL segments with at least 23.5 known homologs per site in BFD have fewer homologs than similar full-VH segments (Figure 6d).

#### 3.3.1. Full-VL Segments with AF2 Well-folded Models

The category that is most populated for full-VH segments, i.e., 3D models with low solvent accessibility and tight contacts between the RSS, accounts for a substantial number of full-VL cases, although not predominant (293 segments: Figure 5, blue box). These AF2 predictions correspond to well-folded 3D structures, as illustrated with the yeast uncharacterized protein YBR032W (UniProt P38223, Figure 7b, blue box). This was predicted as an alpha + beta fold, but no significant structural similarity could be detected in the PDB database by the Dali server.

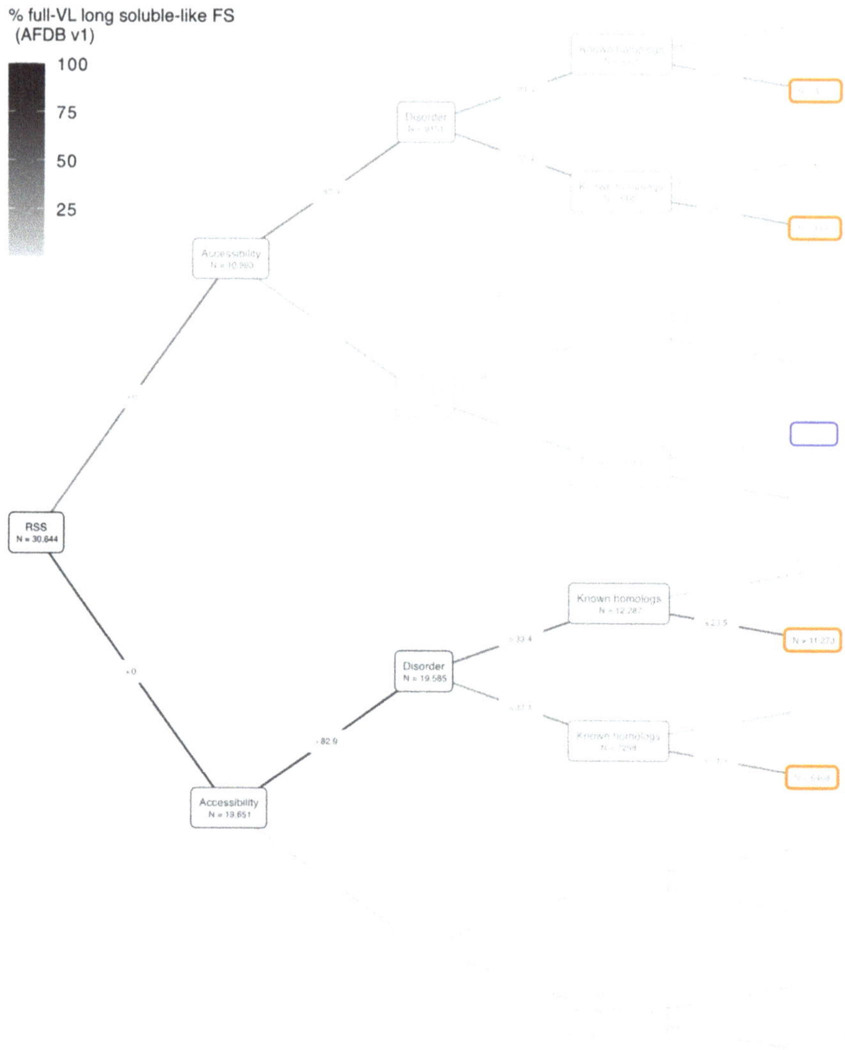


**Figure 5.** Binary tree diagram of the full-VL segments according to the feature thresholds. The four levels of the tree (from the root on the left to the last internal nodes on the right) correspond to the four features describing the segments (see Figure 2 for technical flow), as follows: percentage of segment residues participating in a regular secondary structure (RSS), percentage of segment residues accessible to the solvent (Accessibility), percentage of segment residues predicted to be disordered (Disorder), mean number of BFD homologs per segment residue (Known homologs). The binary conditions based on each feature threshold are indicated on the edges of the tree (for details, see Section 2). The number of foldable segments with a given feature below or above each threshold is indicated within the internal and terminal nodes. The total number of full-VL segments is indicated in the root node. The terminal nodes corresponding to the most abundant subsets of full-VL segments (this figure) and full-VH segments (Figure 3) are highlighted in orange and blue, respectively. For quantitative details about each of the 21 proteomes, see Supplementary Figure S3.

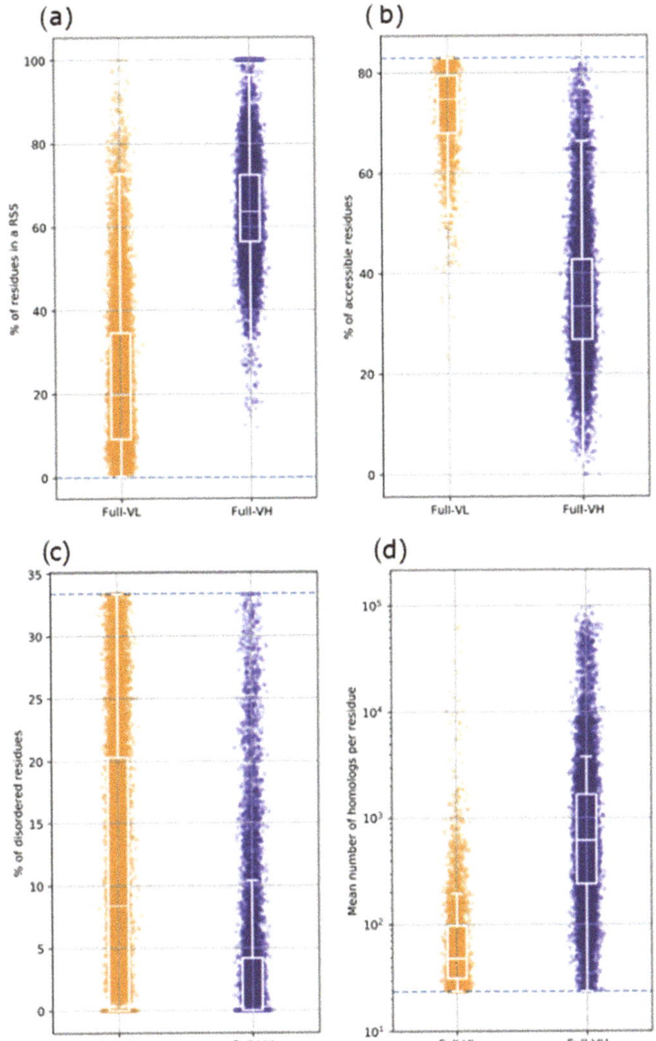


**Figure 6.** Distribution of features of 3D models and amino acid sequences for full-VH (blue) and full-VL (orange) long soluble-like foldable segments from AFDB v1 (21 proteomes). For the structural feature corresponding to the percentage of segment residues participating in a regular secondary structure (RSS) (**a**), only segments with at least 1 RSS as assigned by DSSP from the full-length protein 3D coordinates are shown (see Section 2 for quantitative details). For each feature in (**b–d**), the blue dashed line indicates the threshold value defined based on 95% of the full-VH segments (see Section 2 for details). For both full-VL and full-VH segments, only values falling in these intervals are shown.

Such AF2 predictions cannot be reported with high confidence for several reasons. They could correspond to the adopted structures, but represent novel folds, with amino acid contacts not yet described in the folds used for the AF2 machine learning step and insufficient depth of the multiple sequence alignment. Conversely, RSS could also be misassembled or insufficiently relative to what is happening in the actual structure.

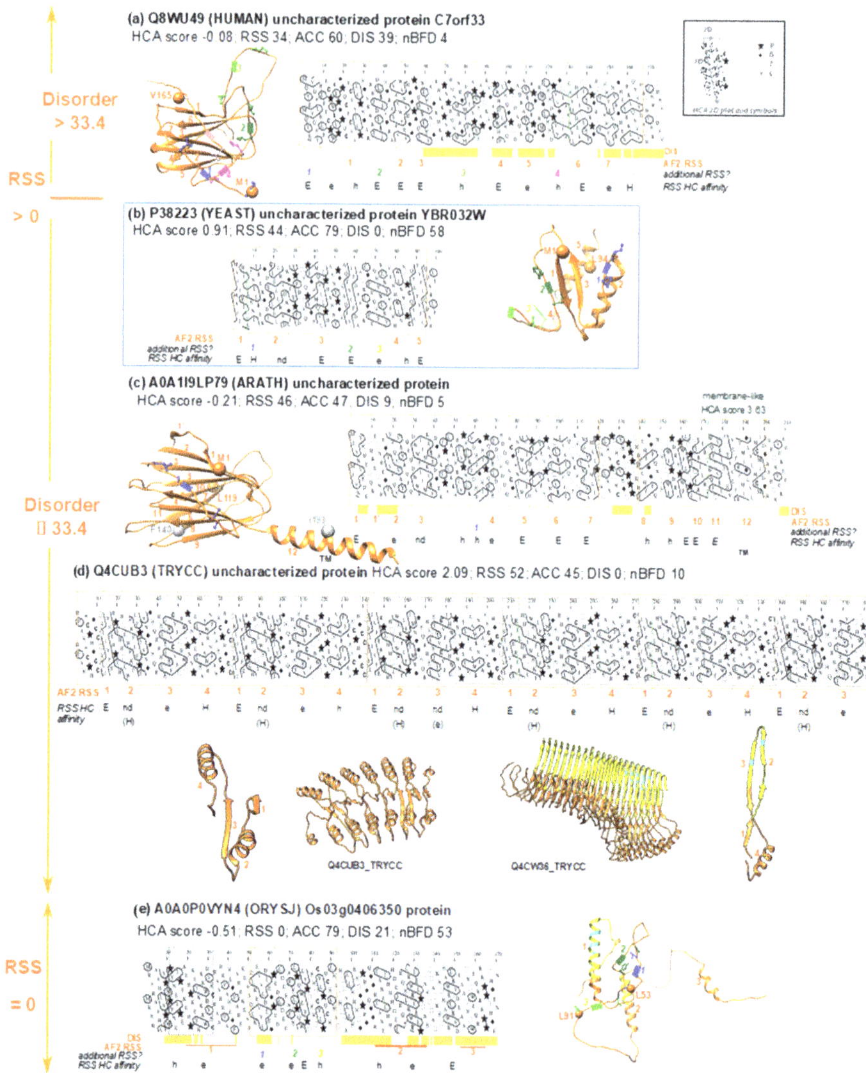


**Figure 7.** Examples of full-VL soluble-like foldable segments corresponding to folded AF2 predictions. Examples were extracted from the binary tree diagram shown in Figure 5. AF2 3D structure models, colored according to the pLDDT values, are shown, along with the positions of the first and last amino acids of the full-VL soluble-like foldable segments (orange balls). The values of the four features are indicated, along with the HCA scores. HCA plots of sequences of the full-VL soluble-like foldable segments are also shown (orange, dashed boxes). How to read sequences (1D) and secondary structures (2D) is shown in the inset, as well as the special symbols used to designate four amino acids with respect to their particular structural behavior. Regular secondary structures (RSS), as observed in the AF2 3D structure models, are designated with orange numbers, which are also reported below the HCA plot in order to indicate the correspondence with hydrophobic clusters. RSS predicted only according to the presence of hydrophobic clusters are reported in other colors, and their positions are indicated on the AF2 3D structure models (with the first and last amino acids shown in atomic details). The hydrophobic cluster affinities for RSSs, calculated using only the binary pattern information, are

indicated, as extracted from HCDB v2 [36]. The upper (H,E) and lower (h,e) cases stand for strong and weak preferences, respectively. H stands for alpha-helix, E for beta-strand. Nd stands for hydrophobic clusters for which there are insufficient statistics in HCDB for the assignment of RSS affinity. TM stands for Transmembrane. IUPred2 long disorder predictions (DIS) are indicated in orange. Hydrophobic clusters corresponding to two successive regular secondary structures are broken down into their components (vertical lines). The sequence repeat in panel (**d**) is boxed on the HCA plot, whereas the basic unit of the repeat was extracted from the 3D structures (shown at the left and right ends). The 3D structure at right illustrates the AF2 prediction for a member of the same family as the protein sequence shown on the left. The blue box corresponds to the sequence included in the leaf that is the most populated in the full-VH tree shown in Figure 3. Subfigures (**a**) to (**d**) correspond to examples with RSS > 0, disorder ≥ 33.4 (**a**) or disorder < 33.4 (**b**–**d**). Subfigure (**e**) corresponds to an example without RSS.

Logically, about five times as many cases are found with a low number of BFD homologs (1274 segments: Figure 5). This reinforces the observation that while assigning a low confidence score, AF2 can propose models even when little evolutionary information is available (Figure 7c,d). A first example (UniProt A0A1I9LP79, Figure 7c) corresponds to an uncharacterized protein from *Arabidopsis thaliana*, whose 3D structure is predicted as a 12-stranded beta-sandwich. A Dali-server search in the PDB database revealed multiple hits with similar structures but with a lower strand content (Z-scores up to 6.6 and sequence identities below 15% (e.g., pdb:4q7g-A, Z-score of 6.6, 8% identity)). Examination of the HCA plot indicated that all the hydrophobic clusters match the regular secondary structures predicted by AF2. This suggests that the basic secondary structure elements are indeed present in the proposed model, arranged correctly or not. However, no conclusion can be drawn in the absence of a sufficient number of homologs (mean BFD homologs per position: 5.08, mean sequence identity > 60%). A second intriguing example is an uncharacterized protein from *Trypanosoma cruzi* (Q4CUB3), consisting of a repeated motif of 70 amino acids (mean BFD homologs per position: 10.16, identity > 80%) (Figure 7d). This is predicted to form a repeated beta-alpha-beta-alpha motif, with the two helices arranged on either side of a central beta sheet of parallel beta-strands, forming an elongated structure with a continuous hydrophobic core. A Dali search revealed structural alignments with different tandem-repeat structures (Z-scores up to 4.4, with sequences identity below 10%), belonging to distinct structural families (armadillo repeats (pdb:6dee-A, Z-score: 4.3, 7% identity), right-handed beta-helix (pdb:5zru-A, Z-score: 4.1, 3% identity; 1bhe-A, Z-score: 4.1, 5% identity), heat repeats (pdb:5loi-A, Z-score: 4.0, 9% identity)). In addition, AF2 predictions made for some homologous sequences correspond to a different repeat fold, always predicted with a low to very low level of accuracy (e.g., Q4CW36_TRYCC, >80% mean identity on the repeated sequences, AF2 prediction corresponding to a right-handed beta helix, at right on Figure 7d). This suggests that this repeat module may correspond to a novel 3D structure, which deserves to be explored experimentally.

A third example (UniProt Q8WU49, Figure 7a) illustrates a case containing amino acids predicted to be disordered, in contrast to the former. It corresponds to the uncharacterized human protein C7orf33, which is taxonomically restricted to primates (mean BFD homologs per position: 4.49, mean identity 76%). The 3D structure predicted by AF2 corresponds to a beta-sandwich, with seven strands. A Dali search yielded many results with similar structures (Z-scores up to 5.9 and sequence identities below 15% (e.g., pdb:6eon-A, Z-score 5.7, 8% identity)). Examination of the HCA plot indicated that not all the hydrophobic clusters present in the sequence correspond to the regular secondary structures predicted by AF2. Instead, there are at least five hydrophobic clusters that correspond in the AF2 model to large, unstructured coils. Many of these clusters have strong affinity for the extended (beta-strand) state, as deduced from our hydrophobic cluster dictionary [36]. This suggests that the 3D structure of this protein could incorporate these clusters as additional regular secondary structures. Alternatively, as part of this sequence is predicted to disordered by IUPred2, it is also possible that this sequence corresponds to a disordered compact domain, helping to maintain a metastable/transient interface for target recognition, as discussed for the C-terminal domain of protein 4.1G [49].

A last category of full-VL, long soluble-like foldable segments with poor solvent accessibility are the cases without RSS. Most of these cases correspond to unfolded segments in contact with other, well-folded protein regions under consideration, making them comparable to the principal category described below. However, a few cases correspond to segments that show a tendency to form a hydrophobic core without the presence of true secondary structures (see for instance the case of a protein from *Oryza sativa* in Figure 7e).

These examples indicate that such foldable domains, with very low AF2 pLDDT values but a presence of regular secondary structures interacting with each other, may correspond to original, well-folded structures. These are thus prime targets for experimental investigation, especially in the absence of sufficiently divergent homologous sequences. These include tandem repeats, which are relatively poorly represented in the PDB compared to other folds [50].

#### 3.3.2. Full-VL Segments with AF2 Unfolded Models

The most abundant category of the full-VL segments (orange boxes in Figure 5) corresponds to unfolded 3D models (encompassing more than 82.9% residues considered solvent accessible by DSSP). These are predicted as disordered or not by IUPred2 and have a low number of known homologs in BFD. This supports the general observation that VL residues are mostly associated with disorder, as no or very few unassembled RSS can be predicted by AF2. The fact that cases with few BFD homologs are about ten times more numerous than cases with a high number of BFD homologs supports the assignment of these segments to the "disorder" category, because IDR sequences are known to be less conserved. However, the HCA score values and the content in hydrophobic clusters suggest that these segments contain conditional order. Nevertheless, it cannot be ruled out that AF2 fails to predict RSS that can assemble into stable, well-folded 3D structures due to the lack of evolutionary information (or, for cases with a high number of BFD homologs, to insufficient depth of multiple sequence alignments). Such cases are referred to as "hidden" (unconditional) order. These hypotheses of conditional or unconditional order cannot be unequivocally demonstrated without the use of experimentation. Nevertheless, we give below some examples supported by experiments that confirm these hypotheses.

The first category (conditional order) is further supported by the fact that some instances are annotated in the DisProt database (Figure 8, green box). This is illustrated by a first example (Figure 8c) corresponding to a foldable segment of the mouse glucocorticoid receptor (GCR, UniProt P06537), including its core transactivation domain (DisProt DP00030, 94.2% identity with human GCR). This domain is intrinsically disordered but forms three helices that are ~30% pre-populated [51]. These three helices correspond to the positions of hydrophobic clusters on the HCA plot.

A second example (Figure 8d) is the foldable segment of the human sodium/hydrogen exchanger 1 (SLC9A1, UniProt P19634), located in its intrinsically disordered intracellular distal tail (aa 686–815, DisProt DP01241). NMR performed on two distant homologs suggested the presence of transient secondary structures and a role in molecular recognition [52]. This role was further supported by a point mutation introduced in the region that disrupts the putative binding feature and impairs trafficking to the plasma membrane [52]. These secondary structures correspond to the positions of hydrophobic clusters on the HCA plots, the first one belonging to the foldable segment described here.

A third example (Figure 8e) is the foldable domain present in the middle of the regulatory (R) region of mouse Cystic Fibrosis Transmembrane conductance Regulator (CFTR, UniProt P26361), a chloride channel belonging to the ABC transporter superfamily (DisProt DP00012, 64% identity with human CFTR). The entire R region of CFTR is a well-known example of an intrinsically disordered sequence whose phosphorylation regulates channel activity [53]. The R region has been shown to interact with the nucleotide-binding domain 1 (NBD1) via multiple transient helices [54]. One of them is included in the foldable region considered here, which is located in the middle of the R region, while the two N- and C-terminal part of the R domain are embedded in the foldable segments of the preceding (Nucleotide Binding Domain 1) and succeeding (Membrane-Spanning Domain 1) folded domains, respectively.

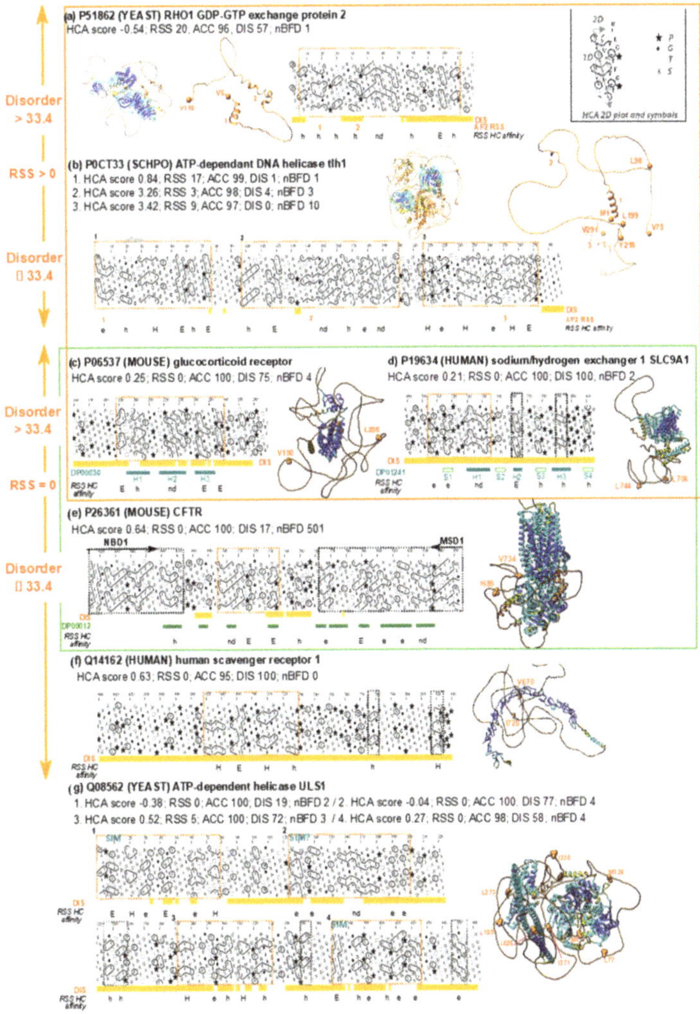


**Figure 8.** Examples of full-VL soluble-like foldable segments corresponding to unfolded AF2 predictions. See legend in Figure 7. The green box illustrates cases of disordered sequences with transient regular structures (highlighted in green below the HCA plot), documented in the DisProt database (accession number in green at left). SIM stands for Sumo-Interacting Motif. Orange boxes correspond to sequences included in the most populated leaf of the full-VL tree shown in Figure 5. Subfigures a and b correspond to examples with RSS coverage (predicted by AF2) > 0 and disorder coverage (IUPred2 predictions) > 33.4 (**a**) and $\leq$ 33.4 (**b**). Subfigures c to f correspond to examples RSS coverage = 0 and disorder coverage > 33.4 (**c**,**d**) and $\leq$ 33.4 (**e**,**f**). Subfigure (**g**) corresponds to an example with multiple full-VL soluble-like segments, some of which including SLiMs.

Only a small fraction of the foldable segments corresponding to such AF2 predictions (i.e., low pLDDT values, no regular secondary structures, HCA scores typical of folded, soluble domains and high IUPred2 coverage) correspond to sequences included in DisProt, with experimental evidence of conditional disorder. This suggests that the remaining segments, which are numerous, may be interesting targets for experimental studies. One such example is the human scavenger receptor F member 1 (SCARF1 (SREC_HUMAN), UniProt Q14162, Figure 8f), which plays a key role in the binding and endocytosis of endogenous and exogenous ligand. The importance of SCARF1 in immunological processes was demonstrated using a SCARF1-deficient mice model, which developed systemic lupus erythematosus-like autoimmune disease [55]. A foldable segment (aa 670–728, HCA score: 0.62, IUPred2 coverage 100%) with three hydrophobic clusters typical of an alpha-beta-alpha motif can be found in its large, otherwise, intrinsically disordered cytoplasmic domain (Figure 8d), for which a role in signaling has been suggested but this function has yet to be elucidated [55]. The foldable segment highlighted here is a good candidate for further exploration of conditional order, even though this remains to be supported at the experimental level.

Short linear motifs (SLiMs) [56] are a priori excluded from this study because their lengths are below the threshold fixed here (30 amino acids) and as they often contain only a single hydrophobic cluster [30]. Such cases are associated with higher HCA scores (Bruley et al. [32]). However, some SLiMs can be embedded in larger foldable segments [30], allowing their detection in the present dataset. This is illustrated by four foldable segments detected in the N-terminal region of yeast ULS1 (UniProt Q08562, Figure 8g). This ATP-dependent helicase is required for end-joining inhibition at telomeres and interacting with the silencing regulator Sir4 [57]. SUMO-interacting motifs can be found within the first and fourth foldable segments, while a third can be suspected in the second foldable segment. The advantage of the HCA-based approach is to propose a prediction, through the boundaries of the foldable domains, of the structurally coherent neighborhood of the interacting modules, and thus highlight the sequences that confer flexibility, adaptability, and dynamic character to the IDRs.

Finally, we also observed cases of full-VL, long soluble-like foldable segments with RSS but accessible to solvent (Figure 8a,b). These can be compared to the most populated category without RSS, corresponding to either possible conditional or hidden order. Consideration of disorder predictions can help to distinguish between the different categories.

## 4. Discussion

It is now widely accepted that the low confidence structural predictions of AF2 correspond mainly to disorder [1,4,11]. In agreement with other investigations [4,5], we have recently shown that a large fraction of these sequences are indeed included in non-foldable segments as defined by pyHCA, which can therefore be considered as "full disorder" [32]. However, a substantial part of sequences with very low confidence scores in AF2 also belongs to foldable segments, in particular, those with a density in hydrophobic clusters typical of soluble domains. This led us to further study their structural characteristics, with respect to the type of order they might contain. The non-foldability/foldability of sequences is estimated by pyHCA from the sole information of a single amino acid sequence, independently of the existence of homologs, whose consideration is one of the pillars of AF2 efficiency.

The key lesson that can be drawn from our study is that the long foldable segments predicted as unfolded by AF2 with very low confidence scores (represented in the form of full-length spaghetti, like those of non-foldable segments), in fact most likely contain either conditional order or hidden, non-conditional order.

Conditional order (or disorder) can be considered as a consequence of the marginal stability of the folded state, making us aware that structure can be determined by both the sequence and the environment [58]. Here, we specifically addressed the issue of intrinsically disordered domains (IDDs), since we only considered long segments (>30 amino acids) that, moreover, are likely to correspond to homogeneous structural units, according to the definition of foldable segments. Shorter foldable segments, including a large part of MoRFs, belong to another category, characterized by higher HCA scores [32], which was not explored here. It should be noted that short linear motifs (SLiMs) can be embedded in larger foldable segments, constituting the structural unit that can modulate their interaction properties. For instance, the study of CBP interaction domain (CID) of the p160 transcriptional co-activator NCOA3 revealed that its flanking regions promote binding through short-lived, non-specific hydrophobic contacts with the partner [59]. These hydrophobic contacts are provided by hydrophobic clusters that are part of the foldable segments in which CID is included.

A recent study has shown that AF2 predicts 60% of the conditional order with high accuracy, capturing the folded state [5]. This reinforces the assumption that low scoring corresponds to full disorder. Our study provides a refined analysis and new insights for additional conditional order unidentified by AF2, which represent interesting targets worth investigating an experimental level.

The long, soluble-like full-VL foldable segments studied here may correspond to (i) cases of induced folding without the formation of a folded domain, resulting from the interaction of individual regular secondary structures with a partner, (ii) cases where a folded 3D structure is formed, dependent on the partner to be induced/stabilized, (iii) cases where a folded 3D structure is stably formed, independent of the environment (what we designate as unconditional, hidden order). This unconditional order remains completely invisible in AF2 predictions, presumably due to the lack of homologs or insufficient depth of the multiple sequence alignments used in the machine learning process.

While cases of conditional order can be supported by taking into account the DisProt database, this is not obvious for cases of hidden, non-conditional order. These indeed correspond to the unknow part of the proteomes (also described as dark proteomes). However, the HCA characteristics of these foldable segments with unfolded AF2 models (Figure 8) are comparable to those of well-folded AF2 models from the full-VL (Figure 7) and full-VH (Figure 4) categories. This supports the hypothesis that these foldable segments are still unexplored reservoirs of well-folded 3D structures. Whether these sequences correspond to true orphans, or at least taxonomically restricted genes, or whether they share distant relationships that cannot be detected by current homology detection methods is a difficult question to answer. It requires in particular novel methods going beyond sequence similarities. Recent developments for the detection of distant homologs (e.g., [60]) but also for 3D structure prediction from single protein sequences without known homologs (e.g., [61], based on the protein language model) will thus open new perspectives to decipher these cases.

The distinction between conditional and hidden, non-conditional order is not straightforward, but can be guided by taking into account current disorder predictors, in particular integrating more information on the amino acid composition. Useful information could also be given by the hydrophobic cluster composition (e.g., based on the HCA toolkit), as well as by sequences linking the hydrophobic clusters, which correspond mainly to loops.

Several hypotheses can explain the low confidence scores associated with the folded AF2 model segments. First, the proposed 3D structures should be adopted but are not yet validated by AF2 due to either: original folds/structures, the lack of representation in the databases used for learning, or an insufficient amount of homologous sequences to validate the predicted contacts. This hypothesis was recently supported in particular by Sen and colleagues [62], showing lower AF2 pLDDT values for models of sequences corresponding to unassigned domains, compared to those corresponding to CATH or Pfam entries.

Second, the proposed 3D structures should not be adopted, due to incorrect RSS assembly, with sometimes some RSSs not yet well predicted. Nevertheless, the signature of folding is there and thus, given that these proteins are largely uncharacterized, they constitute interesting targets for experimental validation, and characterization of new functions. Among these uncharacterized sequences are de novo gene candidates, as illustrated with the yeast YBR032W protein in Figure 7b [63]. Other cases are protein repeats, which are widespread periodic units involved in a wide range of functions but are generally difficult to predict due to artifacts resulting from inherent translational symmetry [64]. At the protein level, the structural mechanisms of orphan gene emergence remain to be understood. A fine-grained exploration of foldable segments within the expanding reported cases in eukaryotic proteomes (e.g., Drosophila [65], Oryza [66], Yeast [63]) would shed light on a still open debate related to the suggested disordered nature of de novo proteins, as a first structural intermediate after gene birth (e.g., [67–71]).

**Supplementary Materials:** The following supporting information can be downloaded at: https://www.mdpi.com/article/10.3390/biom12101467/s1, Figure S1: Full-VH binary tree diagrams per proteome; Figure S2: HCA plots of the sequences of full-VH soluble-like foldable segments; Figure S3: Full-VL binary tree diagrams per proteome; Table S1: Distribution of VH and VL residues within the long soluble-like foldable segments of the AFDB v1 dataset (21 proteomes).

**Author Contributions:** Conceptualization, E.D. and I.C.; methodology, A.B., E.D. and I.C.; software, A.B.; validation, A.B., E.D. and I.C.; formal analysis, A.B., E.D. and I.C.; investigation, A.B., J.-P.M., E.D. and I.C.; resources, A.B.; data curation, A.B., E.D. and I.C.; writing—original draft preparation, E.D., I.C.; writing—review and editing, A.B., J.-P.M., E.D. and I.C.; visualization, A.B., E.D., I.C.; supervision, E.D. and I.C.; project administration, I.C.; funding acquisition, E.D. and I.C. All authors have read and agreed to the published version of the manuscript.

**Funding:** A.B. was supported by the PhD program of Doctoral School "Complexité du Vivant" (ED515, Sorbonne Université). This work was supported by the French National Research Agency (PHOSTORE: ANR-19-CE01-0005 and APOTHESIS: ANR-21-CE12-0021).

**Institutional Review Board Statement:** Not applicable.

**Informed Consent Statement:** Not applicable.

**Data Availability Statement:** The data that support the findings of the present study are available upon request from the corresponding authors E.D. and I.C.

**Acknowledgments:** Analyses were processed with the support of the computer cluster "Plateforme Calcul Intensif Algorithmique" (UMS2700-PCIA) of the Muséum National d'Histoire Naturelle (MNHN).

**Conflicts of Interest:** The authors declare no conflict of interest.

## References

1. Jumper, J.; Evans, R.; Pritzel, A.; Green, T.; Figurnov, M.; Ronneberger, O.; Tunyasuvunakool, K.; Bates, R.; Žídek, A.; Potapenko, A.; et al. Highly accurate protein structure prediction with AlphaFold. *Nature* **2021**, *596*, 583–589. [CrossRef] [PubMed]
2. Baek, M.; DiMaio, F.; Anishchenko, I.; Dauparas, J.; Ovchinnikov, S.; Lee, G.R.; Wang, J.; Cong, Q.; Kinch, L.N.; Schaeffer, R.D.; et al. Accurate prediction of protein structures and interactions using a three-track neural network. *Science* **2021**, *373*, 871–876. [CrossRef] [PubMed]
3. Varadi, M.; Anyango, S.; Deshpande, M.; Nair, S.; Natassia, C.; Yordanova, G.; Yuan, D.; Stroe, O.; Wood, G.; Laydon, A.; et al. AlphaFold Protein Structure Database: Massively expanding the structural coverage of protein-sequence space with high-accuracy models. *Nucleic Acids Res.* **2022**, *50*, D439–D444. [CrossRef] [PubMed]
4. Akdel, M.; Pires, D.E.V.; Porta Pardo, E.; Jänes, J.; Zalevsky, A.O.; Mészáros, B.; Bryant, P.; Good, L.L.; Laskowski, R.A.; Pozzati, G.; et al. A structural biology community assessment of AlphaFold 2 applications. *bioRxiv* **2021**. [CrossRef]
5. Alderson, T.R.; Pritišanac, I.; Moses, A.M.; Forman-Kay, J.D. Systematic identification of conditionally folded intrinsically disordered regions by AlphaFold2. *bioRxiv* **2022**. [CrossRef]
6. Binder, J.L.; Berendzen, J.; Stevens, A.O.; He, Y.; Wang, J.; Dokholyan, N.V.; Oprea, T.I. AlphaFold illuminates half of the dark human proteins. *Curr. Opin. Struct. Biol.* **2022**, *74*, 102372. [CrossRef] [PubMed]
7. Porta-Pardo, E.; Ruiz-Serra, V.; Valentini, S.; Valencia, A. The structural coverage of the human proteome before and after AlphaFold. *PLoS Comput. Biol.* **2022**, *18*, e1009818. [CrossRef]
8. Ruff, K.M.; Pappu, R.V. AlphaFold and Implications for Intrinsically Disordered Proteins. *J. Mol. Biol.* **2021**, *433*, 167208. [CrossRef] [PubMed]
9. Tang, Q.-Y.; Ren, W.; Wang, J.; Kaneko, K. The Statistical Trends of Protein Evolution: A Lesson from AlphaFold Database. *bioRxiv* **2022**. [CrossRef]
10. Wilson, C.J.; Choy, W.Y.; Karttunen, M. AlphaFold2: A Role for Disordered Protein/Region Prediction? *Int. J. Mol. Sci.* **2022**, *23*, 4591. [CrossRef]
11. Tunyasuvunakool, K.; Adler, J.; Wu, Z.; Green, T.; Zielinski, M.; Žídek, A.; Bridgland, A.; Cowie, A.; Meyer, C.; Laydon, A.; et al. Highly accurate protein structure prediction for the human proteome. *Nature* **2021**, *596*, 590–596. [CrossRef]

12. Necci, M.; Piovesan, D.; CAID Predictors; DisProt Curators; Tosatto, S.C.E. Critical assessment of protein intrinsic disorder prediction. *Nat. Methods* **2021**, *18*, 472–481. [CrossRef] [PubMed]
13. Van der Lee, R.; Buljan, M.; Lang, B.; Weatheritt, R.J.; Daughdrill, G.W.; Dunker, A.K.; Fuxreiter, M.; Gough, J.; Gsponer, J.; Jones, D.T.; et al. Classification of intrinsically disordered regions and proteins. *Chem. Rev.* **2014**, *114*, 6589–6631. [CrossRef] [PubMed]
14. Morris, O.M.; Torpey, J.H.; Isaacson, R.L. Intrinsically disordered proteins: Modes of binding with emphasis on disordered domains. *Open Biol.* **2021**, *11*, 210222. [CrossRef] [PubMed]
15. Wright, P.E.; Dyson, H.J. Linking folding and binding. *Curr. Opin. Struct. Biol.* **2009**, *19*, 31–38. [CrossRef] [PubMed]
16. Mohan, A.; Oldfield, C.J.; Radivojac, P.; Vacic, V.; Cortese, M.S.; Dunker, A.K.; Uversky, V.N. Analysis of molecular recognition features (MoRFs). *J. Mol. Biol.* **2006**, *362*, 1043–1059. [CrossRef]
17. Yan, J.; Dunker, A.K.; Uversky, V.N.; Kurgan, L. Molecular recognition features (MoRFs) in three domains of life. *Mol. Biosyst.* **2016**, *12*, 697–710. [CrossRef] [PubMed]
18. Oldfield, C.J.; Cheng, Y.; Cortese, M.S.; Romero, P.; Uversky, V.N.; Dunker, A.K. Coupled folding and binding with alpha-helix-forming molecular recognition elements. *Biochemistry* **2005**, *44*, 12454–12470. [CrossRef]
19. Csizmók, V.; Bokor, M.; Bánki, P.; Klement, E.; Medzihradszky, K.F.; Friedrich, P.; Tompa, K.; Tompa, P. Primary contact sites in intrinsically unstructured proteins: The case of calpastatin and microtubule-associated protein 2. *Biochemistry* **2005**, *44*, 3955–3964. [CrossRef]
20. Fuxreiter, M.; Simon, I.; Friedrich, P.; Tompa, P. Preformed structural elements feature in partner recognition by intrinsically unstructured proteins. *J. Mol. Biol.* **2004**, *338*, 1015–1026. [CrossRef] [PubMed]
21. Lee, S.H.; Kim, D.H.; Han, J.J.; Cha, E.J.; Lim, J.E.; Cho, Y.J.; Lee, C.; Han, K.H. Understanding pre-structured motifs (PreSMos) in intrinsically unfolded proteins. *Curr. Protein Pept. Sci.* **2012**, *13*, 34–54. [CrossRef] [PubMed]
22. Watson, M.; Stott, K. Disordered domains in chromatin-binding proteins. *Essays Biochem.* **2019**, *63*, 147–156. [CrossRef] [PubMed]
23. Borgia, A.; Borgia, M.B.; Bugge, K.; Kissling, V.M.; Heidarsson, P.O.; Fernandes, C.B.; Sottini, A.; Soranno, A.; Buholzer, K.J.; Nettels, D.; et al. Extreme disorder in an ultrahigh-affinity protein complex. *Nature* **2018**, *555*, 61–66. [CrossRef] [PubMed]
24. Tompa, P.; Fuxreiter, M. Fuzzy complexes: Polymorphism and structural disorder in protein–protein interactions. *Trends Biochem. Sci.* **2008**, *33*, 2–8. [CrossRef]
25. Sharma, R.; Raduly, Z.; Miskei, M.; Fuxreiter, M. Fuzzy complexes: Specific binding without complete folding. *FEBS Lett.* **2015**, *589*, 2533–2542. [CrossRef]
26. Davey, N.E.; Van Roey, K.; Weatheritt, R.J.; Toedt, G.; Uyar, B.; Altenberg, B.; Budd, A.; Diella, F.; Dinkel, H.; Gibson, T.J. Attributes of short linear motifs. *Mol. Biosyst.* **2012**, *8*, 268–281. [CrossRef]
27. Tompa, P.; Fuxreiter, M.; Oldfield, C.J.; Simon, I.; Dunker, A.K.; Uversky, V.N. Close encounters of the third kind: Disordered domains and the interactions of proteins. *Bioessays* **2009**, *31*, 328–335. [CrossRef]
28. Williams, R.W.; Xue, B.; Uversky, V.N.; Dunker, A.K. Distribution and cluster analysis of predicted intrinsically disordered protein Pfam domains. *Intrinsically Disord. Proteins* **2013**, *1*, e25724. [CrossRef]
29. Zhou, J.; Oldfield, C.J.; Yan, W.; Shen, B.; Dunker, A.K. Intrinsically disordered domains: Sequence ➔ disorder ➔ function relationships. *Protein Sci.* **2019**, *28*, 1652–1663. [CrossRef]
30. Bitard-Feildel, T.; Callebaut, I. Exploring the dark foldable proteome by considering hydrophobic amino acids topology. *Sci. Rep.* **2017**, *7*, 41425. [CrossRef]
31. Perdigão, N.; Heinrich, J.; Stolte, C.; Sabir, K.S.; Buckley, M.J.; Tabor, B.; Signal, B.; Gloss, B.S.; Hammang, C.J.; Rost, B.; et al. Unexpected features of the dark proteome. *Proc. Natl. Acad. Sci. USA* **2015**, *112*, 15898–15903. [CrossRef] [PubMed]
32. Bruley, A.; Bitard-Feildel, T.; Callebaut, I.; Duprat, E. A sequence-based foldability score combined with AlphaFold2 predictions to disentangle the protein order/disorder continuum. *Proteins* **2022**. *in revision*. [CrossRef]
33. Bitard-Feildel, T.; Lamiable, A.; Mornon, J.-P.; Callebaut, I. Order in disorder as observed by the "Hydrophobic Cluster Analysis" of protein sequences. *Proteomics* **2018**, *18*, e1800054. [CrossRef] [PubMed]
34. Callebaut, I.; Labesse, G.; Durand, P.; Poupon, A.; Canard, L.; Chomilier, J.; Henrissat, B.; Mornon, J.-P. Deciphering protein sequence information through hydrophobic cluster analysis (HCA): Current status and perspectives. *Cell Mol. Life Sci.* **1997**, *53*, 621–645. [CrossRef]
35. Eudes, R.; Le Tuan, K.; Delettré, J.; Mornon, J.-P.; Callebaut, I. A generalized analysis of hydrophobic and loop clusters within globular protein sequences. *BMC Struct. Biol.* **2007**, *7*, 2. [CrossRef]
36. Lamiable, A.; Bitard-Feildel, T.; Rebehmed, J.; Quintus, F.; Schoentgen, F.; Mornon, J.P.; Callebaut, I. A topology-based investigation of protein interaction sites using Hydrophobic Cluster Analysis. *Biochimie* **2019**, *167*, 68–80. [CrossRef]
37. Faure, G.; Callebaut, I. Comprehensive repertoire of foldable regions within whole genomes. *PLOS Comput. Biol.* **2013**, *9*, e1003280. [CrossRef]
38. Linding, R.; Russell, R.B.; Neduva, V.; Gibson, T.J. GlobPlot: Exploring protein sequences for globularity and disorder. *Nucleic Acids Res.* **2003**, *31*, 3701–3708. [CrossRef]
39. Mészáros, B.; Erdos, G.; Dosztányi, Z. IUPred2A: Context-dependent prediction of protein disorder as a function of redox state and protein binding. *Nucleic Acids Res.* **2018**, *46*, W329–W337. [CrossRef]
40. Eddy, S. Accelerated Profile HMM Searches. *PLoS Comput. Biol.* **2011**, *7*, e1002195. [CrossRef]
41. The UniProt Consortium. UniProt: The universal protein knowledgebase in 2021. *Nucleic Acids Res.* **2021**, *49*, D480–D489. [CrossRef] [PubMed]

42. Steinegger, M.; Söding, J. Clustering huge protein sequence sets in linear time. *Nat. Commun.* **2018**, *9*, 2542. [CrossRef] [PubMed]
43. Steinegger, M.; Mirdita, M.; Söding, J. Protein-level assembly increases protein sequence recovery from metagenomic samples manyfold. *Nat. Methods* **2019**, *16*, 603–606. [CrossRef] [PubMed]
44. Kabsch, W.; Sander, C. Dictionary of protein secondary structure: Pattern recognition of hydrogen-bonded and geometrical features. *Biopolymers* **1983**, *22*, 2577–2637. [CrossRef]
45. Rost, B.; Sander, C. Conservation and prediction of solvent accessibility in protein families. *Proteins* **1994**, *20*, 216–226. [CrossRef]
46. Holm, L. Dali server: Structural unification of protein families. *Nucleic Acids Res.* **2022**, *50*, W210–W215. [CrossRef]
47. Pettersen, E.F.; Goddard, T.D.; Huang, C.C.; Couch, G.S.; Greenblatt, D.M.; Meng, E.C.; Ferrin, T.E. UCSF Chimera—A visualization system for exploratory research and analysis. *J. Comput. Chem.* **2004**, *25*, 1605–1612. [CrossRef]
48. Carlson, C.B.; Bernstein, D.A.; Annis, D.S.; Misenheimer, T.M.; Hannah, B.L.; Mosher, D.F.; Keck, J.L. Structure of the calcium-rich signature domain of human thrombospondin-2. *Nat. Struct. Mol. Biol.* **2005**, *12*, 910–914. [CrossRef]
49. Wang, D.; Wu, S.; Wang, D.; Song, X.; Yang, M.; Zhang, W.; Huang, S.; Weng, J.; Liu, Z.; Wang, W. The importance of the compact disordered state in the fuzzy interactions between intrinsically disordered proteins. *Chem. Sci.* **2022**, *13*, 2363–2377. [CrossRef]
50. Kajava, A.V. Tandem repeats in proteins: From sequence to structure. *J. Struct. Biol.* **2012**, *179*, 279–288. [CrossRef]
51. Kim, D.-H.; Wright, A.; Han, K.-H. An NMR study on the intrinsically disordered core transactivation domain of human glucocorticoid receptor. *BMB Rep.* **2017**, *50*, 522–527. [CrossRef] [PubMed]
52. Nørholm, A.B.; Hendus-Altenburger, R.; Bjerre, G.; Kjaergaard, M.; Pedersen, S.F.; Kragelund, B.B. The intracellular distal tail of the Na+/H+ exchanger NHE1 is intrinsically disordered: Implications for NHE1 trafficking. *Biochemistry* **2011**, *50*, 3469–3480. [CrossRef] [PubMed]
53. Ostedgaard, L.S.; Baldursson, O.; Vermeer, D.W.; Welsh, M.J.; Robertson, A.D. A functional R domain from cystic fibrosis transmembrane conductance regulator is predominantly unstructured in solution. *Proc. Natl. Acad. Sci. USA* **2000**, *97*, 5657–5662. [CrossRef]
54. Baker, J.M.R.; Hudson, R.P.; Kanelis, V.; Choy, W.-Y.; Thibodeau, P.H.; Thomas, P.J.; Forman-Kay, J.D. CFTR regulatory region interacts with NBD1 predominantly via multiple transient helices. *Nat. Struct. Mol. Biol.* **2007**, *14*, 738–745. [CrossRef]
55. Patten, D.A. SCARF1: A multifaceted, yet largely understudied, scavenger receptor. *Inflamm. Res.* **2018**, *67*, 627–632. [CrossRef]
56. Weatheritt, R.J.; Luck, K.; Petsalaki, E.; Davey, N.E.; Gibson, T.J. The identification of short linear motif-mediated interfaces within the human interactome. *Bioinformatics* **2012**, *28*, 976–982. [CrossRef]
57. Lescasse, R.; Pobiega, S.; Callebaut, I.; Marcand, S. End-joining inhibition at telomeres requires the translocase and polySUMO-dependent ubiquitin ligase Uls1. *EMBO J.* **2013**, *32*, 805–815. [CrossRef] [PubMed]
58. Hausrath, A.C.; Kingston, R.L. Conditionally disordered proteins: Bringing the environment back into the fold. *Cell Mol. Life Sci.* **2017**, *74*, 3149–3162. [CrossRef] [PubMed]
59. Karlsson, E.; Schnatwinkel, J.; Paissoni, C.; Andersson, E.; Herrmann, C.; Camilloni, C.; Jemth, P. Disordered Regions Flanking the Binding Interface Modulate Affinity between CBP and NCOA. *J. Mol. Biol.* **2022**, *434*, 167643. [CrossRef] [PubMed]
60. Schütze, K.; Heinzinger, M.; Steinegger, M.; Rost, B. Nearest neighbor search on embeddings rapidly identifies distant protein relations. *bioRxiv* **2022**. [CrossRef]
61. Chowdhury, R.; Bouatta, N.; Biswas, S.; Rochereau, C.; Church, G.M.; Sorger, P.K.; AlQuraishi, M. Single-sequence protein structure prediction using language models from deep learning. *bioRxiv* **2021**. [CrossRef]
62. Sen, N.; Anishchenko, I.; Bordin, N.; Sillitoe, I.; Velankar, S.; Baker, D.; Orengo, C. Characterizing and explaining the impact of disease-associated mutations in proteins without known structures or structural homologs. *Brief. Bioinform.* **2022**, *23*, bbac187. [CrossRef] [PubMed]
63. Vakirlis, N.; Hebert, A.S.; Opulente, D.A.; Achaz, G.; Hittinger, C.T.; Fischer, G.; Coon, J.J.; Lafontaine, I. A Molecular Portrait of De Novo Genes in Yeasts. *Mol. Biol. Evol.* **2017**, *35*, 631–645. [CrossRef] [PubMed]
64. Espada, R.; Parra, R.G.; Mora, T.; Walczak, A.M.; Ferreiro, D.U. Capturing coevolutionary signals inrepeat proteins. *BMC Bioinform.* **2015**, *16*, 207. [CrossRef] [PubMed]
65. Heames, B.; Schmitz, J.; Bornberg-Bauer, E. A Continuum of Evolving De Novo Genes Drives Protein-Coding Novelty in Drosophila. *J. Mol. Evol.* **2020**, *88*, 382–398. [CrossRef]
66. Zhang, T.; Faraggi, E.; Li, Z.; Zhou, Y. Intrinsically semi-disordered state and its role in induced folding and protein aggregation. *Cell Biochem. Biophys.* **2013**, *67*, 1193–1205. [CrossRef] [PubMed]
67. Carvunis, A.R.; Rolland, T.; Wapinski, I.; Calderwood, M.A.; Yildirim, M.A.; Simonis, N.; Charloteaux, B.; Hidalgo, C.A.; Barbette, J.; Santhanam, B.; et al. Proto-genes and de novo gene birth. *Nature* **2012**, *487*, 370–374. [CrossRef]
68. Vakirlis, N.; Acar, O.; Hsu, B.; Castilho Coelho, N.; Van Oss, S.B.; Wacholder, A.; Medetgul-Ernar, K.; Bowman, R.W., 2nd; Hines, C.P.; Iannotta, J.; et al. De novo emergence of adaptive membrane proteins from thymine-rich genomic sequences. *Nat. Commun.* **2020**, *11*, 781. [CrossRef]
69. Wilson, B.A.; Foy, S.G.; Neme, R.; Masel, J. Young genes are highly disordered as predicted by the preadaptation hypothesis of de novo gene birth. *Nat. Ecol. Evol.* **2017**, *1*, 0146. [CrossRef]
70. Bitard-Feildel, T.; Heberlein, M.; Bornberg-Bauer, E.; Callebaut, I. Detection of orphan domains in Drosophila using "hydrophobic cluster analysis". *Biochimie* **2015**, *119*, 244–253. [CrossRef] [PubMed]
71. Bungard, D.; Copple, J.S.; Yan, J.; Chhun, J.J.; Kumirov, V.K.; Foy, S.G.; Masel, J.; Wysocki, V.H.; Cordes, M.H.J. Foldability of a Natural De Novo Evolved Protein. *Structure* **2017**, *25*, 1687–1696. [CrossRef]

*biomolecules*

MDPI

*Article*

# Deciphering the Alphabet of Disorder—Glu and Asp Act Differently on Local but Not Global Properties

**Mette Ahrensback Roesgaard †, Jeppe E. Lundsgaard †, Estella A. Newcombe, Nina L. Jacobsen, Francesco Pesce, Emil E. Tranchant, Søren Lindemose, Andreas Prestel, Rasmus Hartmann-Petersen, Kresten Lindorff-Larsen * and Birthe B. Kragelund ***

Linderstrøm-Lang Centre for Protein Science, Department of Biology, University of Copenhagen, Ole Maaløes Vej 5, 2200 Copenhagen, Denmark

* Correspondence: lindorff@bio.ku.dk (K.L.-L.); bbk@bio.ku.dk (B.B.K.)

† These authors contributed equally to this work.

**Abstract:** Compared to folded proteins, the sequences of intrinsically disordered proteins (IDPs) are enriched in polar and charged amino acids. Glutamate is one of the most enriched amino acids in IDPs, while the chemically similar amino acid aspartate is less enriched. So far, the underlying functional differences between glutamates and aspartates in IDPs remain poorly understood. In this study, we examine the differential effects of aspartate and glutamates in IDPs by comparing the function and conformational ensemble of glutamate and aspartate variants of the disordered protein Dss1, using a range of assays, including interaction studies, nuclear magnetic resonance spectroscopy, small-angle X-ray scattering and molecular dynamics simulation. First, we analyze the sequences of the rapidly growing database of experimentally verified IDPs (DisProt) and show that glutamate enrichment is not caused by a taxonomy bias in IDPs. From analyses of local and global structural properties as well as cell growth and protein-protein interactions using a model acidic IDP from yeast and three Glu/Asp variants, we find that while the Glu/Asp variants support similar function and global dimensions, the variants differ in their binding affinities and population of local transient structural elements. We speculate that these local structural differences may play roles in functional diversity, where glutamates can support increased helicity, important for folding and binding, while aspartates support extended structures and form helical caps, as well as playing more relevant roles in, e.g., transactivation domains and ion-binding.

**Keywords:** Dss1; intrinsically disordered protein; IDPs; molecular dynamics; NMR; sequence composition; SAXS

**Citation:** Roesgaard, M.A.; Lundsgaard, J.E.; Newcombe, E.A.; Jacobsen, N.L.; Pesce, F.; Tranchant, E.E.; Lindemose, S.; Prestel, A.; Hartmann-Petersen, R.; Lindorff-Larsen, K.; et al. Deciphering the Alphabet of Disorder—Glu and Asp Act Differently on Local but Not Global Properties. *Biomolecules* **2022**, *12*, 1426. https://doi.org/10.3390/biom12101426

Academic Editors: Prakash Kulkarni, Stefania Brocca, Keith Dunker and Sonia Longhi

Received: 16 August 2022

Accepted: 28 September 2022

Published: 4 October 2022



## 1. Introduction

Intrinsically disordered proteins (IDPs) are involved in various cellular processes, including cell cycle regulation, cellular signaling and protein degradation. The malfunctioning or aggregation of IDPs can cause diseases such as cancer, diabetes, Parkinson's disease and Alzheimer's disease [1,2]. IDPs are characterized by not adapting one specific spatial structure, instead fluctuating between a large number of conformations, distinguishing them from folded proteins. While the sequence-structure-function paradigm is by now well established for folded proteins, the link between sequence and function for IDPs is still poorly understood [3]. Understanding the relationship between sequence, conformational ensemble and function is important for understanding the structural basis of fundamental processes in life and for reaching treatment options for complex diseases.

The pioneering bioinformatics work in the early 2000s recognized IDPs as a separate class of proteins and showed that the amino acid composition of IDPs was distinct from that of folded proteins. The most significant characteristics were a low content of hydrophobic residues and a high content of charged and polar residues [4,5]. Some amino acids were found to be significantly more enriched or depleted in IDPs [6], and subsequent studies by Uversky et al. from 2007 and 2013 [7,8] found similar enrichment. Glutamate is one of the most enriched amino acids in IDPs, while aspartate, which differs from glutamate only by having one less methylene group in the sidechain (Figure 1A) was surprisingly found not to be substantially enriched. The same can be observed for the two similar amino acids, glutamine and asparagine, where glutamine is more enriched in IDPs than asparagine, with no explanation for these differences having been provided. One possible explanation for the difference in the enrichment of glutamate and aspartate could be that the two amino acids have distinct disorder-promoting properties; however, the difference could also be related to differences in the behavior of aspartate and glutamate in the reference set of folded proteins. The extra methylene group in glutamate compared to aspartate results in a larger conformational space, which could favor solvent-exposed conformations over buried conformations. Glutamate is also more frequently observed in α helices than aspartate [9] and is more helix-stabilizing, because the carboxyl group is further from the backbone and thus imposes fewer restraints on the conformational space of the residues in the helix [10]. Transient helicity is frequently observed in IDPs, and a larger population of transient helicity in the free state has been linked to faster binding to a target [11,12], which could be a structural and functional explanation of the observed enrichment of glutamate. However, the analysis of the sequence composition of IDPs could be biased by factors that are not related to structure. Most of the proteins in the intrinsic disorder database DisProt [13–15] are eukaryotic, and more than a third of the proteins are human, while the Protein Data Bank (PDB) [16] contains many bacterial proteins, as of October 2022. A bias in sequence composition when comparing DisProt to the PDB could therefore be because the sequence composition from eukaryotes differs from bacteria or because the human protein sequence composition is distinct from the average sequence composition. Further, since the early studies by Uversky and co-workers [7,8], the number of experimentally verified disordered sequences deposited in DisProt has increased five times, as of July 2022.

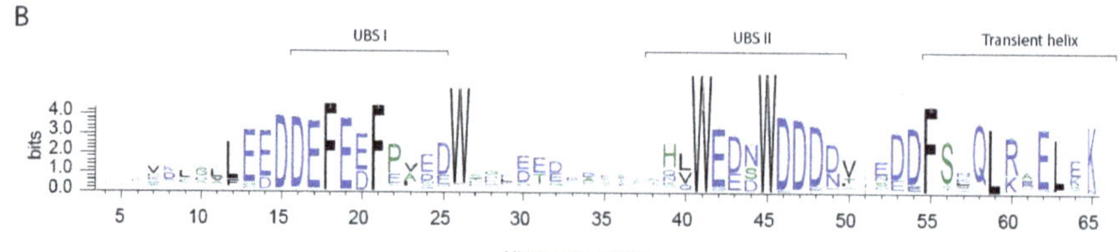


**Figure 1.** Glutamate and aspartate have similar chemical properties. (**A**) Chemical structure of the amino acids, glutamate and aspartate, (**B**) graphical representation of multiple sequence alignment of the Pfam [17] Dss1/Sem1 family (PF05160), only showing the positions that are present in *S. Pombe* Dss1, with information content on the y-axis (for the full MSA, see Figure S1). The sequence logo was made using WebLogo 3 [18].

In this work, we revisited the sequence enrichment profile of IDPs using a larger dataset of disordered proteins and investigated if compositional bias arising from other factors than structural properties exists, including species variability. The analyses confirmed the previous observation of a greater enrichment of glutamates than aspartates in

IDPs, and we did not find significant differences across species. To address if glutamate is more disorder-promoting than aspartate, we designed Glu/Asp variants of a model IDP. Deleted in split hand/split foot 1 (Dss1) from *S. pombe* was chosen because it is a well-studied IDP with a high content of negatively charged residues, many of which are fully conserved, showing preferences for either Glu or Asp, as shown in Figure 1B. Dss1 is known to have multiple interaction partners [19] and is found in various complexes, including the 26S proteasome [20,21]. Dss1 adapts different conformations when bound in different complexes but retains disordered regions upon binding [22]. The C-terminal region adapts an α-helical conformation when bound to, e.g., the proteasome and the T-REX complex, and it also exists transiently when free in solution [23]. In the proteasome, Dss1 functions as a ubiquitin-receptor, binding polyubiquitylated substrates destined for degradation. Thus, while Dss1 binds mono-ubiquitin, it can also bind chains of ubiquitin, exploiting two disordered ubiquitin-binding motifs [24] (UBS) I from D38–D49 and UBS II from D16–N25, with UBS I has the strongest affinity for ubiquitin [23]. A transiently formed C-terminal helix in Dss1 interacts dynamically with UBS I, and this fold-back structure may function to shield the binding site and thereby regulate binding partner interaction [19]. From the alignments, the sequences of UBS I, UBS II and the C-terminal helix are highly conserved within the family (Figure 1B). By designing variants of Dss1 containing only aspartate, only glutamate or Glu/Asp swaps, we addressed the impact of Glu/Asp variants on the function of Dss1 both by growth assays in yeast and by measuring ubiquitin binding. Additionally, we investigated the conformational ensembles of the Dss1 Glu/Asp variants using nuclear magnetic resonance (NMR) spectroscopy, small-angle X-ray scattering (SAXS) and molecular dynamics (MD) simulation. We found that Glu/Asp variants do not impair yeast growth and that the proteins have similar global dimensions and bind ubiquitin, but the specific Glu/Asp pattern in UBS I influences the binding affinity. Additionally, the length and population of the transient C-terminal helix was found to depend on the presence of N-terminal helix-capping aspartates and the presence of a glutamate in the center of the helix. We concluded that glutamate and aspartate confer different functional and structural properties in an IDP acting on a local scale, which may contribute to the observed higher enrichment of glutamate in IDPs.

## 2. Materials and Methods

Composition profiles were constructed using the web tool developed by Uversky et al., Composition Profiler [8], using 50,000 boot strap iterations and a statistical significance value of 0.05. The query dataset consisted of all non-redundant, non-ambiguous and non-obsolete sequences from DisProt v. 9.0.1 [13–15], and for the background dataset, the reviewed part of the database UniProt-UniParc (release 2022_02) [25] was used to obtain a non-redundant set of protein sequences with natural origin from the Protein Data Bank (PDB) [16].

We designed three variants of *S. pombe* Dss1 consisting of a variant with all aspartates substituted for glutamate (All-E), all glutamates substituted for aspartate (All-D), all glutamates substituted for aspartate and all aspartates substituted for glutamate (Swap) and included the wildtype (WT) for comparison. All variants carried a substitution of asparagine for cysteine at the C-terminus, and codons were optimized for *E. coli* expression. Additionally, peptides corresponding to the transient helical region of Dss1 (residues 51–69) with substitutions at key positions were designed and purchased from Pepscan (The Netherlands).

### *2.1. Yeast Strains and Techniques*

The *dss1Δ* strain has been described before [23]. The pDUAL vector [26] was used for the expression of $dss1^+$ and the *dss1* variants carrying N-terminal HFG (6His, Flag, green fluorescent protein (GFP)) tags. Cloning and mutagenesis were performed using Genscript. The yeast strains were transformed using lithium acetate [27]. Growth assays were performed on Edinburgh minimal media (EMM2) (San Diego, CA, USA) as described previously [23]. The preparation of cell lysate samples for SDS-PAGE was performed using trichloroacetic acid and glass beads as described previously [23]. The samples were separated by SDS-PAGE on 12.5% acrylamide gels and transferred to 0.2 μm pore-size nitrocellulose membranes (Advantec, Tokyo, Japan). The antibody was anti-GFP (1:1000, Chromotek, Planegg, Germany, Cat# 3H9). Secondary horseradish-peroxidase-conjugated antibodies were from Dako Cytomation. Equal loading was checked using stain-free imaging with 0.5% trichloroethanol (Sigma, St. Louis, MO, USA).

### 2.2. Protein Purification

All four variants of *S. pombe* Dss1 were designed to encode an N-terminal $His_6$-SUMO tag to be cleaved with ubiquitin-like protein protease 1 (ULP1) following initial purification with a nickel column [28]. All four variants were purified with isotope labeling, as described in previous work [29], resulting in lyophilized pure protein, ready for resuspension in the buffer of choice. $His_6$-SUMO ubiquitin was purified for protein interaction studies with Dss1, similarly to what has been described in previous work [30].

### 2.3. NMR Assignment

Assignments of Dss1 variants were carried out from a series of $^1$H-$^{15}$N HSQC, HN-CACB, HN(CO)CACB, HNCO, HN(CA)CO and HN(CA)NNH spectra, as described previously for WT Dss1 [19]. Spectra were recorded either on a Bruker Avance III HD 750 MHz spectrometer, with a Bruker proton-optimized triple-resonance 5mm TCI cryoprobe, or on a Bruker Avance Neo 800 MHz spectrometer, with a Bruker 5 mm CPTXO cryoprobe. Spectra were processed in TopSpin version 3.6.2 (Bruker, Fällanden, Switzerland), transformed using qMDD version 3.2 [8] and processed through nmrDraw version 9.9 [31]. Analysis was carried out using CcpNmr Analysis version 2.5.0 [32]. Samples were transferred to single-use LabScape Essence 5 mm NMR Tubes (Bruker, Switzerland), and all spectra were recorded at 10 °C, unless noted otherwise.

### 2.4. Secondary Chemical Shift (SCS) Analysis

Random coil chemical shifts were computed for variants using a predictor based on previous work [33,34]. Conditions (10 °C and pH 7.4) were specified as input together with the sequence. From the observed assigned chemical shifts ($\delta_{observed}$) and the predicted random coil chemical shifts ($\delta_{random\ coil}$), the secondary chemical shifts ($SCS$ or $\Delta\delta$) for different nuclei were calculated [35]:

$$SCS = \Delta\delta = \delta_{observed} - \delta_{random\ coil} \quad (1)$$

The resulting nuclei-specific datasets were used as an indication of local and structural properties in the different Dss1 variants.

### 2.5. Ubiquitin Binding

The Dss1 variants and ubiquitin were brought to the same buffer solution, with a final concentration of 20 mM $Na_2HPO_4/NaH_2PO_4$ and 150 mM NaCl at pH 7.4; 10% ($v/v$) $D_2O$, 1% ($v/v$) sodium 2,2-dimethyl-2-silapentane-5-sulfonate (DSS) and 5 mM dithiothreitol (DTT) were added, and the pH was readjusted to 7.4. Ubiquitin and the Dss1 variant were mixed, resulting in 1:1, 1:3, 1:6, 1:9, 1:20 and 1:40 molar equivalents of ubiquitin, keeping the Dss1 variant concentration at 50 μM in all titration points. The $^1$H-1D and $^1$H-$^{15}$N heteronuclear single quantum coherence (HSQC) spectra were recorded for all titration points as well as for a sample of 50 μM of a Dss1 variant with no ubiquitin present. The recorded spectra were referenced in TopSpin (version 3.6) and analyzed in CcpNmr Analysis (version 2.5.2) over the titration series. Chemical shift perturbations (CSPs) for the $^1$H-$^{15}$N HSQC spectra were then exported, as calculated by [36]:

$$CSP = \sqrt{(\Delta\delta_H)^2 + (0.1 \cdot \Delta\delta_N)^2} \quad (2)$$

with $\Delta\delta_H$ representing the perturbation in the hydrogen dimension and $\Delta\delta_N$ the perturbation in the nitrogen dimension. CSPs for residues T39 and L40 of the Dss1 WT were used to fit and derive the maximum perturbation expected at the saturation of binding for the WT, $\Delta\delta_{max}$, as well as the dissociation constant, $K_D$, based on the following relationship [36]:

$$CSP = \Delta\delta_{max}\frac{[P_0]\cdot[L_0]\cdot K_D - \sqrt{([P_0]\cdot[L_0]\cdot K_D)^2 - 4\cdot[P_0]\cdot[L_0]}}{2\cdot[P_0]} \quad (3)$$

with $[P_0]$ and $[L_0]$ being the concentrations of Dss1 variant and ubiquitin, respectively. Then, using Equation (3) and in a global fit keeping the $\Delta\delta_{max}$ fixed to the values found for the WT protein, the CSPs of T39 and L40 were fitted to derive the relative dissociation constant, $K_D$, for each variant compared to WT Dss1. This relative $K_D$ determination rather than absolute determination was performed as the ubiquitin titrations did not reach saturation, and intermediate exchange was observed in the Dss1 WT NMR experiments.

### 2.6. Small-Angle X-ray Scattering (SAXS)

SAXS data were collected at the DIAMOND beamline B21, London, UK, using a mono-chromatic ($\lambda$ = 0.9524 Å) beam operating with a flux of 2 × 104 photons/s. The detector was an EigerX 4M (Dectris). The detector to sample distance was set to 3.7 m. Samples were placed in a Ø = 1.5 mm capillary at 288 K during data acquisition. The SAXS intensity profiles of the four proteins were measured at a temperature of 15 °C and a protein concentration of 3 mg/mL (WT, All-E) or 2 mg/mL (All-D, Swap). The average $R_g$ was calculated from the SAXS profiles using ATSAS 3.0.1 [37], using the Guinier approximation with a $q_{\max}$ corresponding to $q_{\max} * R_g = 0.9$, which is commonly used for IDPs [38].

### 2.7. Diffusion-Ordered NMR Spectroscopy

Diffusion-ordered spectroscopy (DOSY) experiments [39] were performed on the four Dss1 variants (50 µM) to determine hydrodynamic radii ($R_H$). The buffer was 20 mM $Na_2HPO_4/NaH_2PO_4$, 150 mM NaCl and 5 mM DTT, pH 7.4, 10% ($v/v$) $D_2O$ and 0.25 mM DSS. Translational diffusion constants were calculated by fitting the peak intensity decay within the methyl region (0.85–0.9 ppm), which was compared to the diffusion constant of the internal reference 1,4-dioxane (0.02% $v/v$) to estimate each protein's $R_H$ as described [35]. Spectra were recorded (a total of 16 scans) on a Bruker Avance Neo 800 MHz spectrometer, with a Bruker carbon/nitrogen-optimized triple-resonance NMR observe cryoprobe with Z-field gradient over gradients strengths from 2 to 98% and using a diffusion time ($\Delta$) of 200 ms and a gradient length of 3 ms ($\delta$). Diffusion constants were fitted in GraphPad Prism v9.2.0.

### 2.8. Dss1 Peptide Assignment

The NMR resonances of Dss1 peptides ($Dss1_{51-69}$) were assigned using spectra recorded via Bruker Topspin v3.6.2 on a Bruker 800 MHz spectrometer equipped with a cryogenic probe and Z-field gradient using natural isotope abundance (peptide concentration 1.2 mM, 20 mM $Na_2HPO_4/NaH_2PO_4$, 150 mM NaCl, 2% ($v/v$) trifluoroethanol, 10% ($v/v$) $D_2O$ and 0.25 mM DSS; pH 7.4), acquiring TOCSY, ROESY, $^{15}N$-HSQC and $^{13}C$-HSQC for manual assignments. Spectra were transformed and referenced using TopSpin v3.6.2 (Bruker, Switzerland) before being analyzed in CCPN Analysis v2.5 [32]. The fractional helicity of the peptides was calculated by averaging the Cα chemical shifts of residues D/E54 to G67, both included, and using 3.1 and 3.8 ppm as average reference chemical shifts for a fully formed helix (max *helix*), with the expression [40–42]:

$$\frac{\sum_i(\Delta\delta\, C_i^\alpha)}{\text{max}helix} \quad (4)$$

*2.9. Molecular Dynamics Simulations*

Molecular dynamics simulations were performed with GROMACS v. 2019.6 [43–46] and plumed v. 2.6.1. [47–49] with the force field a99SB-*disp* [50]. Parallel bias metadynamics [51] with well-tempered bias potentials [52] was used for all simulations. The backbone dihedral angles and the $R_g$ of the $C^\alpha$ atoms were chosen as collective variables for biasing the simulations of full-length Dss1. The $R_g$ was biased within an interval of 1.3–4.0 nm. Only the backbone dihedral angles were biased for simulations of the Dss1 C-terminal-region peptides. We used multiple walkers [53] with 20 replicas per variant for the simulation of full-length Dss1. Gaussian "hills" were added to the bias potential at a frequency of 400 fs and a width determined using diffusion-based adaptive gaussians [54]. The bias factor was set to 32. A dodecahedral box was used, and periodic boundary conditions were applied. We used the a99SB-*disp* water model and added ions to a concentration of 150 mM. Simulations were run at a temperature of 283 K. Energy minimization was performed using a steepest descend algorithm followed by a conjugate gradient algorithm. The first equilibration was run for 2 ns with position restraints. The second equilibration was run for 2 ns without position restraints with pressure coupling using the Berendsen barostat. Constraints on bonds were applied with the LINCS algorithm [55]. For the equilibrations, constraints were applied to all bonds, and for the production, constraints were applied for bonds to hydrogen atoms. A cut-off of 0.9 nm was used for non-bonded interactions, and PME [56] was used for electrostatic interactions. A leap-frog integrator was used for the equilibration and production runs. The Parrinello-Rahman barostat was used in the production runs, and the velocity-rescale thermostat [57] was used for the equilibrations and production. Simulations of full-length Dss1 and variants were run for 13 µs (0.65 µs per replica) and the peptides for 8 µs with a timestep of 2 fs. We discarded the first 0.15 µs of each replica in the full-length Dss1 and 2 µs of the peptide simulations as equilibration, after calculating the accumulated bias for the entire trajectory, as the bias hereafter is considered static, allowing for reconstruction of the un-biased probability distribution using reweighting [58]. The reweighted and un-biased trajectories were used for the subsequent data analysis. Errors on the $R_g$ were estimated using block averaging [59,60]. The block size used for error estimation of the average $R_g$ of each simulation was the smallest block size in the plateaued region for the block error analysis of the free energy surface as a function of the $R_g$.

Theoretical small-angle X-ray scattering (SAXS) profiles for each of the full-length protein simulations were calculated for every 50th frame (0.5 ns) with Pepsi-SAXS [61]. The theoretical profiles were compared to experimental SAXS profiles by reduced $\chi 2$ statistics. To account for uncertainty of the experimental errors, experimental errors were rescaled using a correction factor calculated with BayesApp v. 1.0 [62,63].

The secondary structure content of the MD-simulated conformations of full-length Dss1 and the helix region was calculated using the Dictionary of Secondary Structure of Proteins (DSSP) algorithm [64] using DSSP v. 2.2.1. DSSP assigns a secondary structure to a protein from the coordinates of the backbone atoms based on the possible hydrogen bonding patterns. Hydrogen bonding is defined by electrostatic interaction energy, and the cut-off is set as high, allowing the algorithm to pick up on hydrogen bonds that deviate from the ideal length and angle. Conformations classified by DSSP as $\alpha$ helix, $3_{10}$ helix or $\pi$ helix were considered helical. DSSP was applied to all frames in the trajectories.

Contact maps for all simulations were calculated with the python package MDTraj [65]. Contacts between residues were defined with a distance cut-off of 8.5 Å between C$\alpha$ atoms and calculated for every 10th frame in the trajectory. The contact maps were averaged and weighted by the metadynamics bias associated with each frame to arrive at a contact map representing the weighted fraction of simulation time that each residue pair was in contact. The differences in the contact maps between the wildtype and the variants were calculated as the log ratio of the fraction of contacts for each residue pair in the variant to those of the wildtype.

## 3. Results

### 3.1. The Alphabet of Disorder

We first examined how much the observed enrichment and depletion of the different types of amino acids in IDPs relative to a background of folded proteins depends on the chosen background dataset. Since, here, we define the enrichment of amino acids in the IDP sequence composition profile as the difference to the background normalized to the frequency of the amino acid in the background dataset, variations within the background dataset will affect the calculated enrichment. We examined three background datasets with different criteria for folded proteins: (1) the standard Composition Profiler folded protein dataset with high-quality X-ray structures, (2) a dataset with X-ray structures with low B-factors and thus, perhaps, less dynamic proteins, and (3) a broadly defined dataset including proteins with lower resolution (Figure S2). We found variations in amino acid frequencies in the three reference sets, and thus, the enrichment profile was dependent on the chosen background dataset, in particular for amino acids of low frequency. We decided that a broader definition of folded proteins would give us a more representative sequence profile for IDPs, because restricting the folded dataset to high-resolution globular proteins would also include the enrichment of amino acids in these structures (Figure S3). We chose a set of non-redundant and naturally occurring proteins that had an entry in the PDB as the best representation of folded proteins, which would include more diverse proteins than the standard Composition Profiler folded background dataset.

Next, we explored the composition profile in the DisProt database using the selected background data. Here, we found that the recent growth of DisProt and the use of the larger and more diversely defined folded protein dataset available in 2022 resulted in some differences compared to the earlier enrichment profile described by Uversky et al. 2013 [7] (Figure 2A). First, the main characteristics of the IDP sequence composition stands: a depletion in hydrophobic amino acids, an enrichment in polar and charged amino acids and an enrichment in the structure-disrupting amino acid proline. However, the most enriched and most depleted amino acids were now less extremely enriched or depleted. Glutamate thus appeared to be much less enriched compared to the original profile. A few amino acids shifted from being depleted or slightly enriched to being more enriched in IDPs, including asparagine, threonine, glycine, and aspartate. From a structural viewpoint, the enrichment of glycine in IDPs can be explained by the rotational freedom from the lack of a sidechain, allowing for a larger conformational space. Although not as pronounced as previously observed, we still observed a greater enrichment of glutamate and glutamine compared to the similar amino acids, aspartate and asparagine with a one-carbon-shorter sidechain.

Next, we investigated whether the observed differences could be explained by species-specific amino acid frequencies, as DisProt mainly contains eukaryotic proteins and sequences mostly from humans. To remove this potential bias, we created sequence profiles containing only eukaryotic or human sequences in both the folded and the disordered datasets. For all three sets, we observed a similar IDP composition profile (Figure 2B). However, we found a difference in the enrichment of glutamine in the species-specific profiles, indicating that the glutamine enrichment in IDPs might partly be explained by a depletion of glutamine in prokaryotes compared to eukaryotes. There was no substantial difference in the enrichment of glutamate and aspartate in the species-specific composition profiles, and we could thus not explain the difference in glutamate and aspartate enrichment by a species bias.

### 3.2. Functional Effect of Aspartate and Glutamate in Dss1

To investigate whether this apparent bias towards glutamate in IDPs would relate to functional effects, we used the small acidic IDP from *S. pombe*, Dss1, which is a component of several different protein complexes [19,22], including the 26S proteasome [66–68], and which can bind both mono- and poly-ubiquitin. Dss1 is overall highly negatively charged (−18) with a distributed content of both glutamates (9) and aspartates (14), totaling 23 negative charges. We designed three variants with different Glu/Asp ratios: an All-E variant, where all 23 acidic residues were glutamates, an All-D variant where all were aspartates and a Swap variant where we exchanged glutamate for aspartate and *vice versa* (Swap). Together with the wildtype (WT) protein, we first assessed the functional effect of the aspartate-glutamate substitutions.

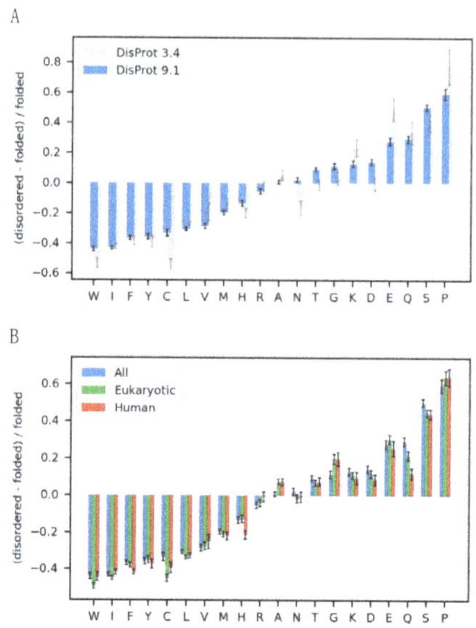


**Figure 2.** The disorder alphabet revisited. (**A**) Sequence enrichment profile of IDPs with the enrichment of amino acids in disordered proteins from the newest (9.1) and the older version (3.4) of DisProt [13–15] compared to folded proteins, defined, respectively, as non-redundant sequences in the PDB or the standard Composition Profiler dataset PDBselect25. (**B**) Sequence enrichment profiles for eukaryotic and human IDPs from DisProt v. 9.1 using a background set of non-redundant sequences in the PDB. Error bars show the boot strap confidence intervals with a statistical significance value of 0.05.

### 3.2.1. The Glu/Asp Variants Are Functional in Vivo

To investigate whether the Glu/Asp variants of Dss1 retained function, we tested the ability of overexpressed GFP-tagged versions of the Dss1 variants to rescue the temperature-sensitive growth defect of a Dss1 knockout strain (*dss1*Δ). First, we tested the expression of the recombinant Dss1 variants by analyzing whole-cell extracts via SDS-PAGE and Western blotting. This revealed that all variants were expressed at roughly equal levels (Figure 3A). As shown before [23], the *dss1* null mutant is viable at 29 °C, but unable to form colonies at 37 °C (Figure 3B). In all cases, we observed that the overexpression of the recombinant *dss1* variants suppressed this temperature-sensitive growth defect as efficiently as WT (Figure 3B). We conclude, therefore, that any changes in the conformational ensembles of the variants are too subtle to substantially impair the Dss1 function relevant to this phenotype in vivo. We note, however, that this phenotype complementation assay may not be sensitive enough to capture small effects and that the temperature-sensitive phenotype of the *dss1*Δ strain is primarily linked to a lack of Dss1 incorporation in the 26S proteasome [23,66–68]. The assay therefore does not report on the other cellular functions of Dss1.

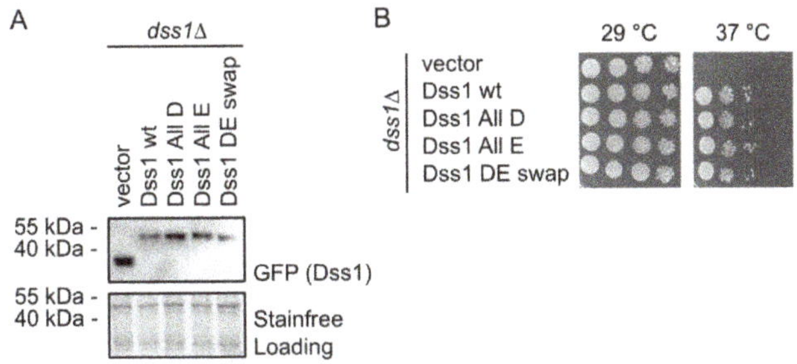


**Figure 3.** Growth effects of Asp and Glu substitutions in Dss1. (**A**) *S. pombe* cells deleted for Dss1 (*dss1*Δ) were transformed to express GFP-tagged wildtype (wt) Dss1 and the indicated Dss1 variants. Whole-cell lysates were then compared via SDS-PAGE and Western blotting using antibodies to GFP. Stain-free labeling was used as a loading control. (**B**) The growth of the strains from (**A**) was compared by serial dilution and incubation on solid media at 29 °C and 37 °C. Note that the temperature-sensitive growth defect of the *dss1*Δ strain is rescued by all Dss1 variants.

3.2.2. Ubiquitin Binding Affinity, but Not Binding Ability, Depends on Glutamate

We then used NMR spectroscopy to assess if all Dss1 variants could bind to ubiquitin. This was carried out by quantifying changes in the chemical shifts (chemical shift perturbation, CSP analysis) in a $^{15}$N-HSQC NMR spectrum after the addition of 40 molar excess of mono-ubiquitin (Figure 4); the chemical shifts of each variant were first assigned using sets of triple-resonance 3D NMR spectra. Overall, the same residues in the variants were affected by the addition of ubiquitin, confirming the binding of ubiquitin to all four variants (Figure 4A). However, binding to UBS I in Dss1 led to the disappearance of peaks in the spectra, mainly of the WT, but also to a much lesser degree in the All-E variant, indicating exchange between free and bound states on an intermediate NMR time scale.

The broadening or loss of signals makes it difficult to quantify binding affinity; thus, in a recent study, we titrated $^{15}$N-ubiquitin with unlabeled Dss1 to circumvent the loss of signal intensity to quantify the affinity of WT Dss1 for mono-ubiquitin giving a $K_d$ of 380 μM [24]. The titration of the variants with ubiquitin into 40 molar excess showed smaller CSPs than WT Dss1 (Figure 4B and Figure S4), suggesting weaker affinities and a smaller population of the bound state. However, since ubiquitin is known to form dimers at mM concentrations [69], we were unable to reach saturation. Instead, here, we determined the fold-change in affinity from global fitting to the CSPs of all variants using data from the same residue, either T39 or L40 (Figure 4B). All variants bound mono-ubiquitin 3.5–7.5-fold weaker than WT Dss1, with the Swap variant having the lowest affinity. The sequence of UBS I in WT contains a central glutamate and is flanked by two aspartates on each side. The WT and All-E variant both have this central glutamate in common, while the two other variants have an aspartate in this position. Thus, the overall weaker affinity suggests a combined preference for glutamate in one specific position within the UBS I motif with flanking aspartates for increased affinity. However, the weaker affinity of the variants for mono-ubiquitin may be rescued to some extent when binding to ubiquitin chains, due to avidity and a local concentration effect, thus possibly explaining the lack of a phenotype in yeast.

*3.3. Global Compaction Does Depend on Glu vs. Asp Ratio*

To assess chain compaction of the four variants, we determined the hydrodynamic radius, $R_H$, of the four Glu/Asp Dss1 variants via pulse field gradient NMR and the radius of gyration, $R_g$, via SAXS. In parallel, we performed all-atom molecular dynamics (MD) simulations of each of the four Dss1 variants, applying enhanced sampling techniques to push the simulations to explore more conformations in the ensemble (Figure 5). We used parallel bias metadynamics [51] and chose the $R_g$ and the dihedral angles as collective variables to increase the sampling of backbone conformation without directly biasing the simulations towards a helical conformation. We calculated the average $R_g$ from the MD simulations of each variant by calculating the $R_g$ from the coordinates of the atoms in each frame. We then compared the dimensions of the four variants (Figure 5B).

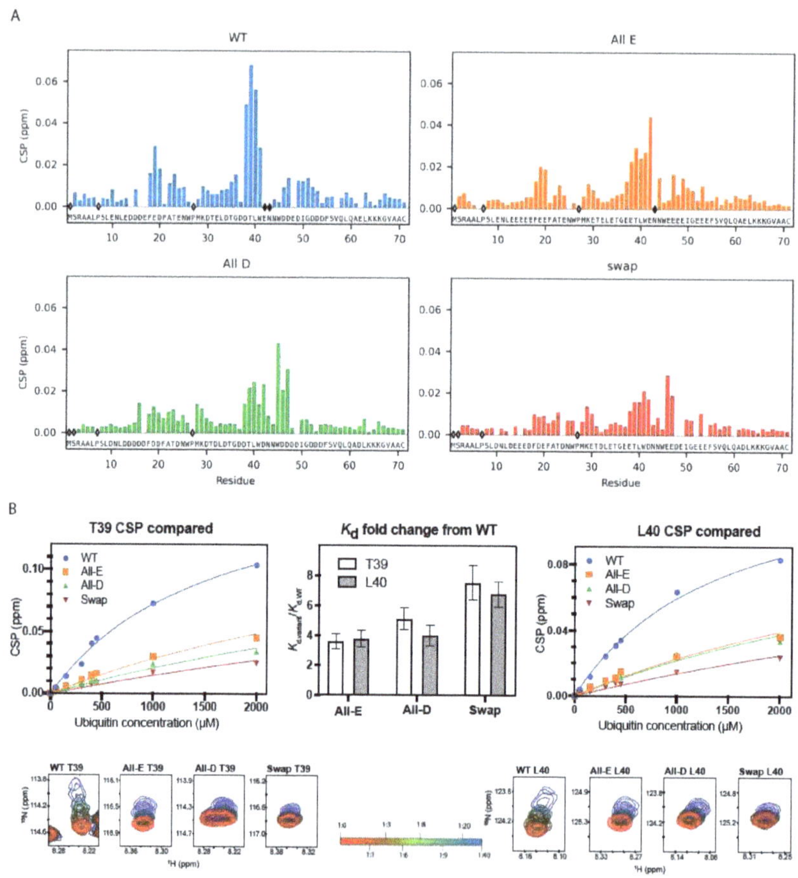


**Figure 4.** Ubiquitin binding ability is supported by both glutamate and aspartate. (**A**) CSP for Dss1 WT and Glu/Asp variants (each at 50 μM concentration) binding to a 40 times molar excess of ubiquitin (2 mM). Empty diamonds indicate non-assigned residues, and filled diamonds indicate disappearance of peaks upon ubiquitin addition. (**B**) CSP of Dss1 WT and Glu/Asp variants for residue T39 (left) and L40 (right) as a function of ubiquitin concentration, fitted to derive the relative affinities for each variant. The middle bar plot shows the fold-change in affinity for the three variants (compared to WT Dss1) derived from the fits to either T39 or L40. The bottom figures show the HSQC peaks for T39 (left) and L40 (right) during ubiquitin titration (from 1:1 to 1:40) indicated by the red-green-blue color change, respectively.

First, we found that the global dimensions of Dss1 as extracted via NMR diffusion and SAXS were overall relatively similar for all Glu/Asp variants of Dss1, although we note that the $R_H$ for the All-E and Swap variants were slightly larger than that of WT Dss1 (by 17% and 13%, respectively). Second, we compared the $R_g$ calculated from the MD ensembles to the $R_g$ measured via SAXS and found no substantial differences between the variants nor between the simulation or experiment, except for the simulation of the Swap variant, which showed a slightly more expanded chain by MD. Additionally, we calculated the theoretical SAXS intensity profiles from the MD ensembles using Pepsi-SAXS and compared the profiles directly to the experimental SAXS intensity profiles, which showed similar agreement between the simulation and experiment (Figure S5).

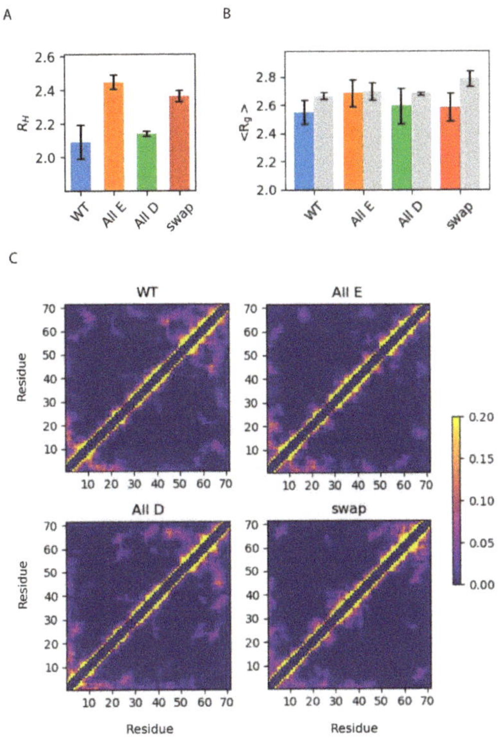


**Figure 5.** Global chain properties of Dss1 WT, All-E, All-D and Swap. (**A**) The hydrodynamic radius (nm) determined via diffusion NMR; (**B**) average radius of gyration (nm) determined experimentally via SAXS (color) or via MD simulations (gray); (**C**) contact maps showing the frequency of contacts between each of the 71 residues in the four Dss1 variants in the MD simulations, where yellow corresponds to an interaction found > 20%.

From the simulations, we also calculated the average number of contacts between $C^{\alpha}$ residues within the protein (Figure 5C) and were able to observe contacts between the C-terminal region and the UBS I, in agreement with conclusions from a study using paramagnetic relaxation enhancement NMR [19]. In our MD simulations, we observed that this interaction primarily took place between the last part of the region that also samples helical structures, where there are three consecutive lysines, and the UBS I. We also observed that interactions between the C-terminal region and the UBS I were more frequent in the WT and All-D, but nearly absent in the All-E variant. This implies that the aspartates on both sides of the binding site facilitate this interaction, which could, as proposed in [19], be a mechanism to regulate the accessibility of the binding site. While the effects are small, we also note that both the $R_H$ and $R_g$ values suggest that the WT is the most compact variant.

### 3.4. Local Structural Changes in Dss1 Depending on Glu/Asp Variants

Since the global properties of Dss1 appeared to be mostly indifferent to either glutamate or aspartate, but with changes in contacts between the C-terminal and the UBS I, we asked if the glutamate bias would be explained by effects on local structure formation in Dss1. To answer this question, we analyzed the secondary chemical shifts (SCS) of the $C^{\alpha}$ and $C^{\beta}$ atoms in all four Dss1 variants (Figure 6). Additionally, we analyzed the local structure in the MD ensembles by calculating the most likely hydrogen-bond patterning based on the distances between atoms in each frame with DSSP and simulated the C-terminal region from residue 50 to 71 of the four proteins alone, allowing us to run longer simulations and observe the formation and unfolding of the helix in a single trajectory.

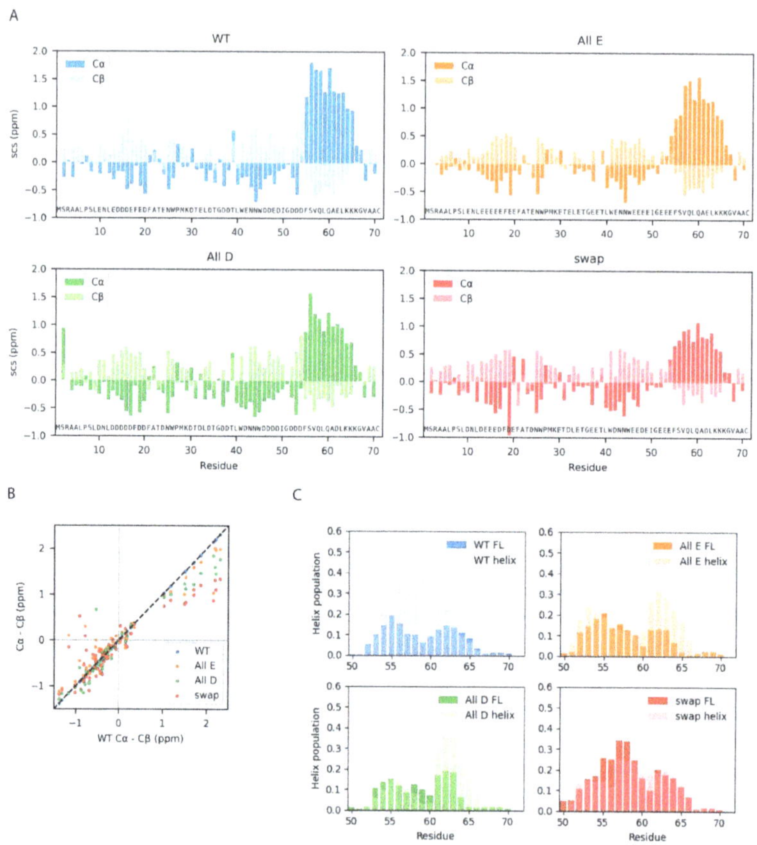


**Figure 6.** Local chain properties of Dss1 WT, All-E, All-D and Swap. (**A**) NMR secondary chemical shifts ($C^{\alpha}$ and $C^{\beta}$) for Dss1 wildtype and the three variants; (**B**) correlation between SCSs of WT Dss1 and the three variants; (**C**) helix population of the C-terminal region from MD simulations of full-length (FL) Dss1 overlayed with the helix population from the simulations of the helix region alone.

All four Dss1 variants showed the formation of a transient C-terminal α helix, which was evident both from NMR SCSs and from the MD simulations of both full-length Dss1 and the C-terminal helical region (Figure 6). From the NMR secondary chemical shifts, we found that the largest difference between the Glu/Asp variants and WT Dss1 was in the population of the transient C-terminal α helix, where all variants have a smaller α helix population compared to wildtype variants, as well as in a region capping the N-terminal side of the helix (Figure 6). Here, WT and All-D had a short stretch of negative $C^{\alpha}$ SCSs indicating an extended structure, which was absent in the Swap and All-E variants. The extended structures of the disordered ubiquitin binding sites were maintained, but it appeared that consecutive aspartates increased the extendedness compared to consecutive glutamates. Thus, locally, there appears to be an effect of Glu/Asp variation where aspartate may better support the local structure of extended characters.

While the MD simulations do capture the formation and unfolding of the C-terminal transiently populated α helix in all variants, we did not observe a larger population in WT

Dss1. In the simulations of the C-terminal regions, we observed that the formation and unfolding of the α helix is a slow process even when applying a bias against already visited conformations, where the process takes around 1 μs (Figures S6 and S7); we note here that the bias that is used to enhance sampling means that the kinetics and mechanism of helix formation/breaking is perturbed. Possibly, the simulations of full-length Dss1 of 10 μs do not therefore capture the population of the C-terminal α helix precisely. However, the populations of the C-terminal α helix are similar in both the simulations of the C-terminal peptides and the full-length Dss1 variants, indicating that the true population can be expected to be in between these populations.

In the MD simulations, UBS I appears to be transiently helical (Figure S8). As this is not observed in the NMR SCS analysis, it is perhaps a result of remaining force field inaccuracies, for example, related to the solubility of hydrophobic amino acids [50]. When the helix is formed, two tryptophans are positioned across from each other, and their hydrophobic interaction is slightly stronger, thus perhaps overstabilizing the helical conformation. In the C-terminal transiently helical region, there are no tryptophans, and thus, we did not expect this issue to have a major impact on the helix population in this region.

In the simulations, we observed that the full eleven-residue C-terminal helix rarely forms, while either one or the other half of the helix forms more frequently (Figures S6–S8). While this could be because of the slow formation of the full helix, it could also describe the same transient helicity as the NMR secondary chemical shifts, as these represent a bulk average and could thus have contributions from conformations with either end of the helix formed. We did not observe a direct effect on helix population of the helix-capping residues being either glutamate or aspartate in the MD simulations but observed that glutamates (residue 52–54) positioned before the helix in the All-E and Swap variants are more frequently helical at these positions than the aspartates in the wildtype and All-D variants. This might indicate that glutamate supports a helix conformation better than aspartate, or that aspartate is more frequently found in a helix-capping conformation and thus does not have a helical geometry. We also note that in the simulations, in some cases, we observed a small dip in the average fraction of helical structure near the middle of the helix; this did not appear to be observed in the experiments. A more direct comparison, however, would require better methods for calculating small changes in secondary chemical shifts from simulations.

The most pronounced effect of Glu/Asp variations on the local structure was observed for the helix population, which changed by just minor alterations to the amino acid composition. We therefore decided to examine specific single-amino-acid substitutions that we hypothesized would either increase or decrease the helix population. To be able to capture these minor changes in population in an otherwise disordered chain, we analyzed the effects using peptides corresponding to the helical 19-residue C-terminal region of Dss1, $_{51}$GDDDFSVQLQAELKKKGVA$_{69}$. From the MD simulations of the full-length Dss1 proteins, we observed that the lysines K65 and K66 often formed salt bridges with E62 in the WT and All-E variants more frequently than the aspartate E62D in the Swap and All-D variants, which formed salt bridges with K64 more often (Figure S9). When the residues are in an α-helical conformation, the sidechain of D62 will be in proximity of the residues K65 and K66, while K64 would be on the other side of the helix. Salt bridges between E62 and K65 and K66 thus likely form helix-stabilizing interactions, while conformations with salt bridges between E62 and K64 are unlikely to be α-helical. We thus speculated that the interaction between K65 and E62 could stabilize the helix, and that the sidechain of aspartate is likely too short to support this interaction. The NMR chemical shifts indicate the possible helix-capping function of D52–D54, which could also stabilize the helix. This would explain why the population of the helix is largest in the WT, as it both has a glutamate at position 62 and aspartates at positions 52–54. Based on these observations, we designed the following peptides (residue 51–69): WT (12% helicity predicted by Agadir) and two helix-modulating D/E-Swap variants, D54E (26% predicted helicity) and E62D (9% predicted helicity). Using NMR spectroscopy, we extracted helicity from the SCS for the $C^{\alpha}$

atoms (Figure 7, see Materials and Methods). Here, we found that although not reaching the full effect, the predicted effects of the substitutions were captured experimentally, with the D54E variant increasing in helicity (from 27.7% to 29.3%), mostly at the substitution site, and the E62D losing helicity (from 27.7% to 20%). Thus, in the peptides, the D54E gained 6% in helicity and the E62D lost 28%. This suggests that the substitution of aspartate for glutamate in the *N*-terminal of the helix increases helicity and removing a glutamate in the middle of the helix destabilizes it. Thus, these data support that glutamate is preferred for the stabilization of this transient helix.

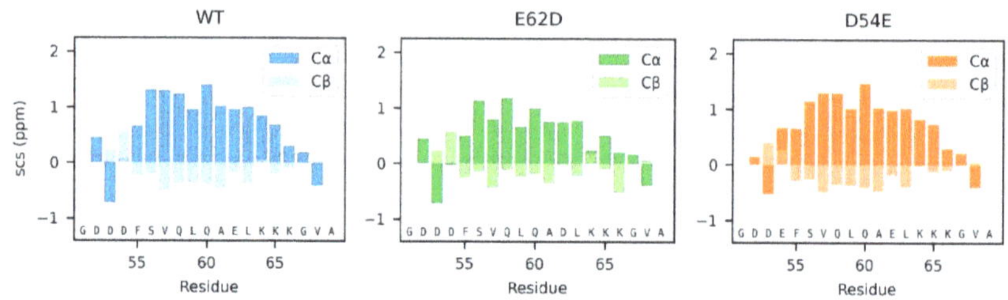


**Figure 7.** Local effects of Asp and Glu in helix stabilization. The SCS of the Cα and Cβ atoms of the three peptides corresponding to WT, D54E and E62D in the context of $Dss1_{51-69}$.

## 4. Discussion

The enrichment of glutamate over aspartate in IDPs has been known since the early 2000s, but to our knowledge, no systematic attempt to explain this bias has been performed. Since then, current databases on IDPs have expanded, and additional proteomes have become available, enabling us to revisit the basis for this bias. Using a disorder dataset containing five times as many sequences and comparing to a more diverse dataset of folded proteins, we found that while glutamate is indeed enriched in IDPs, the difference is less pronounced than originally found. We also exclude the possibility that this difference is due to an underlying bias caused by the dominance of human proteins in the disorder database.

Using yeast Dss1 as a model IDP, we sought a functional explanation for the enrichment of glutamate over aspartate. We found, however, that all tested Dss1 variants were able to complement the growth defect of a *dss1* deletion mutant, irrespective of the type of anionic sidechain. Since previous studies have shown that the temperature-sensitive phenotype of the *dss1*Δ strain is tightly connected with the incorporation of Dss1 into the 26S proteasome [23,66–68], it is likely that the structural and dynamic effects of the Asp/Glu substitutions are not sufficiently pronounced enough to disrupt its interaction with the 26S proteasome in vivo. However, we note that the Dss1 variants were GFP-tagged and overexpressed, which could mask subtle effects. In addition, Dss1 is also involved in other cellular processes and is often found as a subunit in larger complexes [19]. Since the effects of these functions of Dss1 and more subtle effects on proteasome incorporation may not have been captured by our growth assays, we cannot rule out the concept that some Dss1 functions are not affected by Asp/Glu substitutions.

As also suggested by the growth assays, we observed that all variants were capable of binding mono-ubiquitin. Although exchanging the anionic amino acids did not impair the ability of Dss1 to bind to ubiquitin, variants bound mono-ubiquitin 3.5–7.5-fold weaker. Since Dss1 prefers ubiquitin chains over mono-ubiquitin, avidity in binding may rescue some of this effect, and hence may not therefore lead to any phenotype. Additionally, stronger binding to ubiquitin appears to be a combined result of expanding the binding region from the flanking aspartates and optimizing affinity in UBS I by the glutamate. In a study on Dss1, Schenstrøm et al. found that the C-terminal region of Dss1 bends back and shields the UBS I and suggested that this may be a mechanism for regulating ubiquitin binding [19]. In our MD simulations, we were able to observe this interaction for all but the All-E variant. Interestingly, we found that the more aspartates the variants contain in the

UBS I, the more frequent are the interactions between the C-terminal region and the UBS I. Similarly, in a recent study, Zeng et al. found that aspartate in IDPs forms hydrogen bonds more frequently than glutamate, likely stabilizing observed local chain compaction [70].

No large differences in $R_H$ or average $R_g$ in the simulated conformational ensemble and SAXS experiment were observed, and glutamate was therefore not found to promote substantially more expanded ensembles than aspartate. Only minor differences in local structure were observed via NMR, where consecutive aspartates better support extended structures than consecutive glutamates, and the positional effects of having a Glu or an Asp can influence the population of transient helices. Although these small population changes were detectable via NMR, the effects were likely too small to be manifested and detectable in the $R_H$ or $R_g$ measurements. Likely, for Dss1, which is already highly charged, changing the type of anionic sidechain will have little effect on the net charge per residue and hence may not affect chain collapse [71].

Instead, we observed that local structural effects and binding strength could be modulated by exchanging Glu/Asp preferences. For Dss1, the transient helix in the C-terminal region was populated differently in the Glu/Asp variants. Using peptide variants, we could establish that glutamate stabilized the transient helix both when positioned near the N-terminus and when inserted into the helix at positions that enable salt bridge formation with positively charged sidechains positioned three residues C-terminally of it. This is consistent with previous work on folded proteins, where glutamate is more frequently observed in $\alpha$ helices than aspartate [9] and where glutamate in the center of a helix is generally more stabilizing than aspartate, because the carboxyl group is more distant to the backbone. Thus, glutamate imposes less restraints on the conformational space of the residues in the helix [10].

Our work has explored a potential link between a Glu/Asp bias in IDPs, local conformational preferences and functional effects. A question remains as to when and why evolution would favor glutamate over aspartate in IDPs. Recent work has shown that higher helicity in the free state of an IDP may lead to higher affinity for a partner protein [12,72]. Combined with the preference for glutamate seen here to stabilize transient helices, this could suggest that glutamate would be a preferred residue for highly populated transient helices in IDPs. The preformation of highly populated helices could be important in folding-upon-binding reactions, where increased helicity has been shown to affect affinity through effects on both *on*- and *off*-rate constants [12,72]. However, aspartate is a known helix N-capping residue [73], and has been shown in several IDPs to initiate helices, even at several positions within the same helical stretch [74,75], but a quantitative comparison of the two amino acids for this property has to our knowledge not been performed. Finally, in a recent work studying the interactions between IDPs and calcium ions, a preference for aspartate over glutamate in the so-called Escaliber-like motif was noted [29]. Thus, another possible reason for suppressing the use of aspartate in IDPs would be to minimize the binding of divalent cations.

While our results point towards a bias arising from the function of the anionic amino acids in IDPs, the difference in the enrichment of glutamate and aspartate could also arise through other multiple local effects, explanations to which need exploring. The subtle differences in the choice of amino acid at specific positions may be able to shift the equilibrium populations of the conformational ensemble in IDPs and thus impact their function and interactome.

## 5. Conclusions

Here, we addressed the compositional bias in IDPs, which have preference for glutamates over aspartates, a phenomenon pointed out already in the early 2000s [4,6–8]. We found the dimension of the disordered Dss1 is largely indifferent to the difference between these anionic amino acids, whereas highly local effects both on the populations of transient structures as well as on binding affinity were seen. We hypothesize that stabilizing local transient helix structures through capping effects and intra-helix salt bridges, as well as adding binding strength through the additional methylene group, may be important reasons for the preference of glutamates over aspartates in IDPs. Finally, functional biases towards glutamates in regions undergoing helical folding-and-binding and towards aspartates in transactivation domains and calcium-binding regions are likely just some of several functional reasons for the selection of glutamate or aspartate in specific IDPs.

**Supplementary Materials:** The following supporting information can be downloaded at: https://www.mdpi.com/article/10.3390/biom12101426/s1, Table S1: Sequences of S. pombe Dss1 wildtype and Glu/Asp variants; Figure S1: MSA logo for the Dss1_Sem1 family; Figure S2: Influence of background dataset on composition profile; Figure S3: Composition profile of high-resolution X-ray crystal structures; Figure S4: NMR titrations of Dss1 variants with ubiquitin; Figure S5: Comparing experimental and theoretical SAXS profiles; Figure S6: Helix formation time series, peptides; Figure S7: Helix formation time series; Figure S8: Helix population simulations; Figure S9: Salt bridge frequency simulations.

**Author Contributions:** Conceptualization, K.L.-L. and B.B.K.; methodology, B.B.K., K.L.-L., R.H.-P. and F.P.; validation, M.A.R., J.E.L., E.A.N., N.L.J., E.E.T. and R.H.-P.; formal analyses, M.A.R. and J.E.L.; investigation, M.A.R., J.E.L., E.A.N., N.L.J. and S.L.; resources, B.B.K., K.L.-L. and R.H.-P.; data curation, K.L.-L., R.H.-P., E.A.N., A.P., F.P. and B.B.K.; writing—original draft preparation, M.A.R. and B.B.K.; writing—review and editing, B.B.K., R.H.-P. and K.L.-L.; visualization, M.A.R., J.E.L., E.E.T. and R.H.-P.; supervision, B.B.K., K.L.-L., R.H.-P. and F.P.; project administration, B.B.K. and K.L.-L.; funding acquisition, B.B.K., K.L.-L. and R.H.-P. All authors have read and agreed to the published version of the manuscript.

**Funding:** This research was funded by the Novo Nordisk Foundation, grant number NNF18OC0033926 (to B.B.K. and R.H.P.) and the Lundbeck Foundation BRAINSTRUC initiative, grant number R155-2015-2666 (to K.L-L. and B.B.K.). E.A.N. has received funding from the European Union's Horizon 2020 research and innovation programme under the Marie Sklodowska-Curie grant agreement No. 101023654. The NMR spectra were recorded at cOpenNMR, an infrastructure initiative supported by the Novo Nordisk Foundation grant number NNF18OC0032996.

**Institutional Review Board Statement:** Not applicable.

**Informed Consent Statement:** Not applicable.

**Data Availability Statement:** MD simulations and SAXS data will be available via https://github.com/KULL-Centre/_2022_Roesgaard-Lundsgaard_DSS1, accessed on 16 August 2022. Chemical shifts for all four Dss1 variants are available at BMRB, IDs:27618, 51551, 51552 and 51557.

**Acknowledgments:** Jesper E. Dreier is thanked for the initial growth assays, and we thank Signe A. Sjørup for the technical assistance. This work was carried out using computational resources at Computerome 2.0 and the core facility for biocomputing at the Department of Biology. We thank the beamline scientists at the DIAMOND light source, UK, for help with SAXS data acquisition.

**Conflicts of Interest:** The authors declare no conflict of interest.

## References

1. Wright, P.E.; Dyson, H.J. Intrinsically disordered proteins in cellular signalling and regulation. *Nat. Rev. Mol. Cell Biol.* **2015**, *16*, 18–29. [CrossRef] [PubMed]
2. Martinelli, A.H.S.; Lopes, F.C.; John, E.B.O.; Carlini, C.R.; Ligabue-Braun, R. Modulation of Disordered Proteins with a Focus on Neurodegenerative Diseases and Other Pathologies. *Int. J. Mol. Sci.* **2019**, *20*, 1322. [CrossRef] [PubMed]
3. Uversky, V.N. Intrinsic Disorder, Protein-Protein Interactions, and Disease. *Adv. Protein Chem. Struct. Biol.* **2018**, *110*, 85–121. [CrossRef]
4. Tompa, P. Intrinsically unstructured proteins. *Trends Biochem. Sci.* **2002**, *27*, 527–533. [CrossRef]
5. Dunker, A.K.; Lawson, J.D.; Brown, C.J.; Williams, R.M.; Romero, P.; Oh, J.S.; Oldfield, C.J.; Campen, A.M.; Ratliff, C.M.; Hipps, K.W.; et al. Intrinsically disordered protein. *J. Mol. Graph Model* **2001**, *19*, 26–59. [CrossRef]
6. Williams, R.M.; Obradovi, Z.; Mathura, V.; Braun, W.; Garner, E.C.; Young, J.; Takayama, S.; Brown, C.J.; Dunker, A.K. The protein non-folding problem: Amino acid determinants of intrinsic order and disorder. *Pac. Symp. Biocomput.* **2001**, *2000*, 89–100. [CrossRef]
7. Uversky, V.N. The alphabet of intrinsic disorder: II. Various roles of glutamic acid in ordered and intrinsically disordered proteins. *IDP* **2013**, *1*, e24684. [CrossRef]

8. Vacic, V.; Uversky, V.N.; Dunker, A.K.; Lonardi, S. Composition Profiler: A tool for discovery and visualization of amino acid composition differences. *BMC Bioinform.* **2007**, *8*, 211. [CrossRef]
9. Nagano, K. Logical analysis of the mechanism of protein folding. I. Predictions of helices, loops and beta-structures from primary structure. *J. Mol. Biol.* **1973**, *75*, 401–420. [CrossRef]
10. Pace, C.N.; Scholtz, J.M. A helix propensity scale based on experimental studies of peptides and proteins. *Biophys. J.* **1998**, *75*, 422–427. [CrossRef]
11. Daughdrill, G.W. Disorder for Dummies: Functional Mutagenesis of Transient Helical Segments in Disordered Proteins. *Methods. Mol. Biol.* **2020**, *2141*, 3–20. [CrossRef] [PubMed]
12. Iesmantavicius, V.; Dogan, J.; Jemth, P.; Teilum, K.; Kjaergaard, M. Helical propensity in an intrinsically disordered protein accelerates ligand binding. *Angew. Chem. Int. Ed. Engl.* **2014**, *53*, 1548–1551. [CrossRef] [PubMed]
13. Quaglia, F.; Meszaros, B.; Salladini, E.; Hatos, A.; Pancsa, R.; Chemes, L.B.; Pajkos, M.; Lazar, T.; Pena-Diaz, S.; Santos, J.; et al. DisProt in 2022: Improved quality and accessibility of protein intrinsic disorder annotation. *Nucleic Acids Res.* **2022**, *50*, D480–D487. [CrossRef] [PubMed]
14. Hatos, A.; Hajdu-Soltesz, B.; Monzon, A.M.; Palopoli, N.; Alvarez, L.; Aykac-Fas, B.; Bassot, C.; Benitez, G.I.; Bevilacqua, M.; Chasapi, A.; et al. DisProt: Intrinsic protein disorder annotation in 2020. *Nucleic Acids Res.* **2020**, *48*, D269–D276. [CrossRef]
15. Piovesan, D.; Tabaro, F.; Micetic, I.; Necci, M.; Quaglia, F.; Oldfield, C.J.; Aspromonte, M.C.; Davey, N.E.; Davidovic, R.; Dosztanyi, Z.; et al. DisProt 7.0: A major update of the database of disordered proteins. *Nucleic Acids Res.* **2017**, *45*, D219–D227. [CrossRef]
16. Berman, H.M.; Westbrook, J.; Feng, Z.; Gilliland, G.; Bhat, T.N.; Weissig, H.; Shindyalov, I.N.; Bourne, P.E. The Protein Data Bank. *Nucleic Acids Res.* **2000**, *28*, 235–242. [CrossRef] [PubMed]
17. Mistry, J.; Chuguransky, S.; Williams, L.; Qureshi, M.; Salazar, G.A.; Sonnhammer, E.L.L.; Tosatto, S.C.E.; Paladin, L.; Raj, S.; Richardson, L.J.; et al. Pfam: The protein families database in 2021. *Nucleic Acids Res.* **2021**, *49*, D412–D419. [CrossRef]
18. Crooks, G.E.; Hon, G.; Chandonia, J.M.; Brenner, S.E. WebLogo: A sequence logo generator. *Genome. Res.* **2004**, *14*, 1188–1190. [CrossRef]
19. Schenstrom, S.M.; Rebula, C.A.; Tatham, M.H.; Hendus-Altenburger, R.; Jourdain, I.; Hay, R.T.; Kragelund, B.B.; Hartmann-Petersen, R. Expanded Interactome of the Intrinsically Disordered Protein Dss1. *Cell. Rep.* **2018**, *25*, 862–870. [CrossRef]
20. Ellisdon, A.M.; Dimitrova, L.; Hurt, E.; Stewart, M. Structural basis for the assembly and nucleic acid binding of the TREX-2 transcription-export complex. *Nat. Struct. Mol. Biol.* **2012**, *19*, 328–336. [CrossRef]
21. Sone, T.; Saeki, Y.; Toh-e, A.; Yokosawa, H. Sem1p is a novel subunit of the 26 S proteasome from Saccharomyces cerevisiae. *J. Biol. Chem.* **2004**, *279*, 28807–28816. [CrossRef] [PubMed]
22. Kragelund, B.B.; Schenstrom, S.M.; Rebula, C.A.; Panse, V.G.; Hartmann-Petersen, R. DSS1/Sem1, a Multifunctional and Intrinsically Disordered Protein. *Trends Biochem. Sci.* **2016**, *41*, 446–459. [CrossRef] [PubMed]
23. Paraskevopoulos, K.; Kriegenburg, F.; Tatham, M.H.; Rosner, H.I.; Medina, B.; Larsen, I.B.; Brandstrup, R.; Hardwick, K.G.; Hay, R.T.; Kragelund, B.B.; et al. Dss1 is a 26S proteasome ubiquitin receptor. *Mol. Cell.* **2014**, *56*, 453–461. [CrossRef] [PubMed]
24. Dreier, J.E.; Prestel, A.; Martins, J.M.; Brondum, S.S.; Nielsen, O.; Garbers, A.E.; Suga, H.; Boomsma, W.; Rogers, J.M.; Hartmann-Petersen, R.; et al. A context-dependent and disordered ubiquitin-binding motif. *Cell. Mol. Life Sci.* **2022**, *79*, 484. [CrossRef] [PubMed]
25. UniProt, C. UniProt: The universal protein knowledgebase in 2021. *Nucleic Acids Res.* **2021**, *49*, D480–D489. [CrossRef]
26. Matsuyama, A.; Shirai, A.; Yashiroda, Y.; Kamata, A.; Horinouchi, S.; Yoshida, M. pDUAL, a multipurpose, multicopy vector capable of chromosomal integration in fission yeast. *Yeast* **2004**, *21*, 1289–1305. [CrossRef] [PubMed]
27. Suga, M.; Hatakeyama, T. A rapid and simple procedure for high-efficiency lithium acetate transformation of cryopreserved Schizosaccharomyces pombe cells. *Yeast* **2005**, *22*, 799–804. [CrossRef] [PubMed]
28. Pedersen, C.P.; Seiffert, P.; Brakti, I.; Bugge, K. Production of Intrinsically Disordered Proteins for Biophysical Studies: Tips and Tricks. *Methods Mol. Biol.* **2020**, *2141*, 195–209. [CrossRef] [PubMed]
29. Newcombe, E.A.; Fernandes, C.B.; Lundsgaard, J.E.; Brakti, I.; Lindorff-Larsen, K.; Langkilde, A.E.; Skriver, K.; Kragelund, B.B. Insight into Calcium-Binding Motifs of Intrinsically Disordered Proteins. *Biomolecules* **2021**, *11*, 1173. [CrossRef] [PubMed]
30. Ruidiaz, S.F.; Dreier, J.E.; Hartmann-Petersen, R.; Kragelund, B.B. The disordered PCI-binding human proteins CSNAP and DSS1 have diverged in structure and function. *Protein Sci.* **2021**, *30*, 2069–2082. [CrossRef] [PubMed]
31. Delaglio, F.; Grzesiek, S.; Vuister, G.W.; Zhu, G.; Pfeifer, J.; Bax, A. NMRPipe: A multidimensional spectral processing system based on UNIX pipes. *J. Biomol. NMR* **1995**, *6*, 277–293. [CrossRef] [PubMed]
32. Vranken, W.F.; Boucher, W.; Stevens, T.J.; Fogh, R.H.; Pajon, A.; Llinas, M.; Ulrich, E.L.; Markley, J.L.; Ionides, J.; Laue, E.D. The CCPN data model for NMR spectroscopy: Development of a software pipeline. *Proteins* **2005**, *59*, 687–696. [CrossRef] [PubMed]
33. Kjaergaard, M.; Poulsen, F.M. Sequence correction of random coil chemical shifts: Correlation between neighbor correction factors and changes in the Ramachandran distribution. *J. Biomol. NMR* **2011**, *50*, 157–165. [CrossRef] [PubMed]
34. Kjaergaard, M.; Brander, S.; Poulsen, F.M. Random coil chemical shift for intrinsically disordered proteins: Effects of temperature and pH. *J. Biomol. NMR* **2011**, *49*, 139–149. [CrossRef] [PubMed]
35. Prestel, A.; Bugge, K.; Staby, L.; Hendus-Altenburger, R.; Kragelund, B.B. Characterization of Dynamic IDP Complexes by NMR Spectroscopy. *Methods Enzymol.* **2018**, *611*, 193–226. [CrossRef] [PubMed]

36. Teilum, K.; Kunze, M.B.; Erlendsson, S.; Kragelund, B.B. (S)Pinning down protein interactions by NMR. *Protein Sci.* **2017**, *26*, 436–451. [CrossRef]
37. Franke, D.; Petoukhov, M.V.; Konarev, P.V.; Panjkovich, A.; Tuukkanen, A.; Mertens, H.D.T.; Kikhney, A.G.; Hajizadeh, N.R.; Franklin, J.M.; Jeffries, C.M.; et al. ATSAS 2.8: A comprehensive data analysis suite for small-angle scattering from macromolecular solutions. *J. Appl. Crystallogr.* **2017**, *50*, 1212–1225. [CrossRef]
38. Borgia, A.; Zheng, W.; Buholzer, K.; Borgia, M.B.; Schuler, A.; Hofmann, H.; Soranno, A.; Nettels, D.; Gast, K.; Grishaev, A.; et al. Consistent View of Polypeptide Chain Expansion in Chemical Denaturants from Multiple Experimental Methods. *J. Am. Chem. Soc.* **2016**, *138*, 11714–11726. [CrossRef]
39. Wu, D.H.; Chen, A.D.; Johnson, C.S. Johnson. An Improved Diffusion-Ordered Spectroscopy Experiment Incorporating Bipolar-Gradient Pulses. *J. Magn. Reson. Ser. A* **1995**, *115*, 260–264. [CrossRef]
40. Dyson, H.J.; Wright, P.E. Insights into the structure and dynamics of unfolded proteins from nuclear magnetic resonance. *Adv. Protein Chem.* **2002**, *62*, 311–340. [CrossRef]
41. Mielke, S.P.; Krishnan, V.V. Characterization of protein secondary structure from NMR chemical shifts. *Prog. Nucl. Magn. Reson. Spectrosc* **2009**, *54*, 141–165. [CrossRef] [PubMed]
42. Spera, S.; Ikura, M.; Bax, A. Measurement of the exchange rates of rapidly exchanging amide protons: Application to the study of calmodulin and its complex with a myosin light chain kinase fragment. *J. Biomol. NMR* **1991**, *1*, 155–165. [CrossRef] [PubMed]
43. Pronk, S.; Pall, S.; Schulz, R.; Larsson, P.; Bjelkmar, P.; Apostolov, R.; Shirts, M.R.; Smith, J.C.; Kasson, P.M.; van der Spoel, D.; et al. GROMACS 4.5: A high-throughput and highly parallel open source molecular simulation toolkit. *Bioinformatics* **2013**, *29*, 845–854. [CrossRef] [PubMed]
44. Van Der Spoel, D.; Lindahl, E.; Hess, B.; Groenhof, G.; Mark, A.E.; Berendsen, H.J. GROMACS: Fast, flexible, and free. *J. Comput. Chem.* **2005**, *26*, 1701–1718. [CrossRef] [PubMed]
45. Hess, B.; Kutzner, C.; van der Spoel, D.; Lindahl, E. GROMACS 4: Algorithms for Highly Efficient, Load-Balanced, and Scalable Molecular Simulation. *J. Chem. Theory. Comput.* **2008**, *4*, 435–447. [CrossRef] [PubMed]
46. Lindahl, E.; Abraham, M.J.; Hess, B.; van der Spoel, D. *GROMACS 2019.6 Source Code*. Available online: https://zenodo.org/record/3685922#.YdzzHfnYsdU(accessed on 16 August 2022).
47. Consortium, P. Promoting transparency and reproducibility in enhanced molecular simulations. *Nat. Methods* **2019**, *16*, 670–673. [CrossRef]
48. Tribello, G.A.; Bonomi, M.; Branduardi, D.; Camilloni, C.; Bussi, G. PLUMED 2: New feathers for an old bird. *Comput. Phys. Commun.* **2014**, *185*, 604–613. [CrossRef]
49. Bonomi, M.; Branduardi, D.; Bussi, G.; Camilloni, C.; Provasi, D.; Raiteri, P.; Donadio, D.; Marinelli, F.; Pietrucci, F.; Broglia, R.A.; et al. PLUMED: A portable plugin for free-energy calculations with molecular dynamics. *Comput. Phys. Commun.* **2009**, *180*, 1961–1972. [CrossRef]
50. Robustelli, P.; Piana, S.; Shaw, D.E. Developing a molecular dynamics force field for both folded and disordered protein states. *Proc. Natl. Acad. Sci. USA* **2018**, *115*, E4758–E4766. [CrossRef] [PubMed]
51. Pfaendtner, J.; Bonomi, M. Efficient Sampling of High-Dimensional Free-Energy Landscapes with Parallel Bias Metadynamics. *J. Chem. Theory Comput.* **2015**, *11*, 5062–5067. [CrossRef] [PubMed]
52. Barducci, A.; Bussi, G.; Parrinello, M. Well-tempered metadynamics: A smoothly converging and tunable free-energy method. *Phys. Rev. Lett.* **2008**, *100*, 020603. [CrossRef]
53. Raiteri, P.; Laio, A.; Gervasio, F.L.; Micheletti, C.; Parrinello, M. Efficient reconstruction of complex free energy landscapes by multiple walkers metadynamics. *J. Chem. Phys. B* **2006**, *110*, 3533–3539. [CrossRef] [PubMed]
54. Branduardi, D.; Bussi, G.; Parrinello, M. Metadynamics with Adaptive Gaussians. *J. Chem. Theory. Comput.* **2012**, *8*, 2247–2254. [CrossRef] [PubMed]
55. Hess, B.; Bekker, H.; Berendsen, H.; Fraaije, J. LINCS: A Linear Constraint Solver for molecular simulations. *J. Chem. Phys.* **1998**, *18*, 1463–1472. [CrossRef]
56. Essmann, U.; Perera, L.; Berkowitz, M.L.; Darden, T.; Lee, H.; Pedersen, L.G. A smooth particle mesh Ewald method. *J. Chem. Phys.* **1995**, *103*, 8577–8593. [CrossRef]
57. Bussi, G.; Donadio, D.; Parrinello, M. Canonical sampling through velocity rescaling. *J. Chem. Phys.* **2007**, *126*, 014101. [CrossRef] [PubMed]
58. Torrie, G.M.; Valleau, J.P. Nonphysical sampling distributions in Monte Carlo free-energy estimation: Umbrella sampling. *J. Comp. Phy.* **1977**, *23*, 187–199. [CrossRef]
59. Flyvbjerg, H. Error estimates on averages of correlated data. In *Advances in Computer Simulation*; Springer: Berlin/Heidelberg, Germany, 1998; pp. 88–103.
60. Bussi, G.; Tribello, G.A. Analyzing and Biasing Simulations with PLUMED. *Methods Mol. Biol.* **2019**, *2022*, 529–578. [CrossRef]
61. Grudinin, S.; Garkavenko, M.; Kazennov, A. Pepsi-SAXS: An adaptive method for rapid and accurate computation of small-angle X-ray scattering profiles. *Acta Crystallogr. D Struct. Biol.* **2017**, *73*, 449–464. [CrossRef] [PubMed]
62. Larsen, A.H.; Pedersen, M.C. Experimental noise in small-angle scattering can be assessed using the Bayesian indirect Fourier transformation. *J. Appl. Crystallogr.* **2021**, *54*, 1281–1289. [CrossRef]
63. Hansen, S. BayesApp: A web site for indirect transformation of small-angle scattering data. *J. Appl. Crystallogr.* **2012**, *45*, 566–567. [CrossRef]

64. Kabsch, W.; Sander, C. Dictionary of protein secondary structure: Pattern recognition of hydrogen-bonded and geometrical features. *Biopolymers* **1983**, *22*, 2577–2637. [CrossRef]
65. McGibbon, R.T.; Beauchamp, K.A.; Harrigan, M.P.; Klein, C.; Swails, J.M.; Hernandez, C.X.; Schwantes, C.R.; Wang, L.P.; Lane, T.J.; Pande, V.S. MDTraj: A Modern Open Library for the Analysis of Molecular Dynamics Trajectories. *Biophys J.* **2015**, *109*, 1528–1532. [CrossRef] [PubMed]
66. Tomko, R.J., Jr.; Hochstrasser, M. The intrinsically disordered Sem1 protein functions as a molecular tether during proteasome lid biogenesis. *Mol. Cell.* **2014**, *53*, 433–443. [CrossRef]
67. Funakoshi, M.; Li, X.; Velichutina, I.; Hochstrasser, M.; Kobayashi, H. Sem1, the yeast ortholog of a human BRCA2-binding protein, is a component of the proteasome regulatory particle that enhances proteasome stability. *J. Cell. Sci.* **2004**, *117*, 6447–6454. [CrossRef] [PubMed]
68. Josse, L.; Harley, M.E.; Pires, I.M.; Hughes, D.A. Fission yeast Dss1 associates with the proteasome and is required for efficient ubiquitin-dependent proteolysis. *Biochem. J.* **2006**, *393*, 303–309. [CrossRef] [PubMed]
69. Liu, Z.; Zhang, W.P.; Xing, Q.; Ren, X.; Liu, M.; Tang, C. Noncovalent dimerization of ubiquitin. *Angew. Chem. Int. Ed. Engl.* **2012**, *51*, 469–472. [CrossRef]
70. Zeng, X.; Ruff, K.M.; Pappu, R.V. Competing interactions give rise to two-state behavior and switch-like transitions in charge-rich intrinsically disordered proteins. *Proc. Natl. Acad. Sci. USA* **2022**, *119*, e2200559119. [CrossRef] [PubMed]
71. Bremer, A.; Farag, M.; Borcherds, W.M.; Peran, I.; Martin, E.W.; Pappu, R.V.; Mittag, T. Deciphering how naturally occurring sequence features impact the phase behaviours of disordered prion-like domains. *Nat. Chem.* **2022**, *14*, 196–207. [CrossRef]
72. Crabtree, M.D.; Borcherds, W.; Poosapati, A.; Shammas, S.L.; Daughdrill, G.W.; Clarke, J. Conserved Helix-Flanking Prolines Modulate Intrinsically Disordered Protein:Target Affinity by Altering the Lifetime of the Bound Complex. *Biochemistry* **2017**, *56*, 2379–2384. [CrossRef] [PubMed]
73. Doig, A.J.; Baldwin, R.L. N- and C-capping preferences for all 20 amino acids in alpha-helical peptides. *Protein Sci.* **1995**, *4*, 1325–1336. [CrossRef] [PubMed]
74. Wells, M.; Tidow, H.; Rutherford, T.J.; Markwick, P.; Jensen, M.R.; Mylonas, E.; Svergun, D.I.; Blackledge, M.; Fersht, A.R. Structure of tumor suppressor p53 and its intrinsically disordered N-terminal transactivation domain. *Proc. Natl. Acad. Sci. USA* **2008**, *105*, 5762–5767. [CrossRef] [PubMed]
75. Jensen, M.R.; Communie, G.; Ribeiro, E.A., Jr.; Martinez, N.; Desfosses, A.; Salmon, L.; Mollica, L.; Gabel, F.; Jamin, M.; Longhi, S.; et al. Intrinsic disorder in measles virus nucleocapsids. *Proc. Natl. Acad. Sci. USA* **2011**, *108*, 9839–9844. [CrossRef] [PubMed]

*biomolecules*

*Article*

# The Ni(II)-Binding Activity of the Intrinsically Disordered Region of Human NDRG1, a Protein Involved in Cancer Development

**Ylenia Beniamino [1], Vittoria Cenni [2], Mario Piccioli [3], Stefano Ciurli [1,*] and Barbara Zambelli [1,*]**

1 Laboratory of Bioinorganic Chemistry, Department of Pharmacy and Biotechnology, University of Bologna, Viale Giuseppe Fanin 40, 40127 Bologna, Italy
2 CNR Institute of Molecular Genetics "Luigi-Luca Cavalli-Sforza" Unit of Bologna, Via di Barbiano 1/10, 40136 Bologna, Italy
3 Department of Chemistry, Center for Magnetic Resonance, University of Florence, 50121 Florence, Italy
* Correspondence: stefano.ciurli@unibo.it (S.C.); barbara.zambelli@unibo.it (B.Z.); Tel.: +38-051-2096204 (S.C.); +38-051-2096233 (B.Z.)

**Abstract:** Nickel exposure is associated with tumors of the respiratory tract such as lung and nasal cancers, acting through still-uncharacterized mechanisms. Understanding the molecular basis of nickel-induced carcinogenesis requires unraveling the mode and the effects of Ni(II) binding to its intracellular targets. A possible Ni(II)-binding protein and a potential focus for cancer treatment is *h*NDRG1, a protein induced by Ni(II) through the hypoxia response pathway, whose expression correlates with higher cancer aggressiveness and resistance to chemotherapy in lung tissue. The protein sequence contains a unique C-terminal sequence of 83 residues (*h*NDRG1*C), featuring a three-times-repeated decapeptide, involved in metal binding, lipid interaction and post-translational phosphorylation. In the present work, the biochemical and biophysical characterization of unmodified *h*NDRG1*C was performed. Bioinformatic analysis assigned it to the family of the intrinsically disordered regions and the absence of secondary and tertiary structure was experimentally proven by circular dichroism and NMR. Isothermal titration calorimetry revealed the occurrence of a Ni(II)-binding event with micromolar affinity. Detailed information on the Ni(II)-binding site and on the residues involved was obtained in an extensive NMR study, revealing an octahedral paramagnetic metal coordination that does not cause any major change of the protein backbone, which is coherent with CD analysis. *h*NDRG1*C was found in a monomeric form by light-scattering experiments, while the full-length *h*NDRG1 monomer was found in equilibrium between the dimer and tetramer, both in solution and in human cell lines. The results are the first essential step for understanding the cellular function of *h*NDRG1*C at the molecular level, with potential future applications to clarify its role and the role of Ni(II) in cancer development.

**Keywords:** nickel; intrinsically disordered regions; lung cancer; nmr; isothermal titration calorimetry; circular dichroism; light scattering

**Citation:** Beniamino, Y.; Cenni, V.; Piccioli, M.; Ciurli, S.; Zambelli, B. The Ni(II)-Binding Activity of the Intrinsically Disordered Region of Human NDRG1, a Protein Involved in Cancer Development. *Biomolecules* **2022**, *12*, 1272. https://doi.org/10.3390/biom12091272

Academic Editors: Prakash Kulkarni, Stefania Brocca, Keith Dunker and Sonia Longhi

Received: 29 July 2022
Accepted: 5 September 2022
Published: 9 September 2022



## 1. Introduction

Lung cancer is the first cause of tumor-related death worldwide [1]. The major risk factor is cigarette smoke, but exposition to air pollution also significantly increases tumor incidence: in particular, according to the World Health Organization (WHO), inhalation of fine dusts is responsible for more than three million deaths every year, mostly due to cancer [2]. One of the dangerous elements present both in cigarette smoke and in fine dusts is nickel. In 1990, compounds containing this metal were classified as class I carcinogens by WHO [3]. Moreover, occupational exposure to nickel compounds in nickel mines or refineries increases nasal and lung cancer [4]. Understanding the mechanisms of nickel-induced carcinogenesis is therefore of social importance, especially considering the ever-growing level of air pollution [5], and would allow for new drugs to be found to prevent or to cure lung cancer.

Exposure to nickel activates the cell hypoxia response, a tumorigenic state that allows the transformed cells to grow faster due to increased blood supply, promoting the transcription of several genes from the hypoxia-recognizing elements (HREs) [6]: Ni(II) ions likely substitute Fe(II) in the non-heme metal center of asparaginyl and prolyl hydroxylases [6]. These enzymes hydroxylate the regulatory subunit of the hypoxia-inducing factor 1 (HIF-1) [7], named HIF-1$\alpha$, driving its ubiquitination and rapid proteasomal degradation [8]. Their inhibition, caused by metal substitution, stabilizes HIF-1$\alpha$, which translocates into the nucleus, where it interacts with HIF-1$\beta$ and promotes HIF-dependent transcription from HREs [6]. Imitation of the state of hypoxia by nickel compounds is likely involved in the nickel-induced carcinogenic transformation, as it may select cells with altered energy metabolism, changed growth control and/or resistance to apoptosis.

Among Ni(II)- and hypoxia-induced genes, the one known as human N-myc downstream-regulated gene 1 (*h*NDRG1) is particularly noteworthy [9]. Its transcription is repressed by the C-Myc and N-myc oncogenes, often upregulated in cancers [10]. As Myc overexpression is linked to cell proliferation and metastasis, the genes repressed by Myc, including that coding for *h*NDRG1, are believed to regulate tumor progression [11]. The *h*NDRG1 protein is predominantly cytosolic [12], but it translocates into the nucleus in response to DNA damage in different cell lines, and it was also found in mitochondria and membranes, suggesting a function in stress response and DNA repair [13,14]. A large bulk of literature proposes for *h*NDRG1 a multiplicity of functions, including embryogenesis and development, cell growth and differentiation, lipid biosynthesis and myelination, stress responses, vesicle sorting and trafficking, and immunity [15], rendering it a central regulator of cellular biochemistry.

*h*NDRG1 is a key regulator of multiple signaling pathways that modulate tumor progression [14]. Interestingly, the protein shows a strictly tissue-specific pleiotropic activity in carcinogenesis: it works as metastasis suppressor in brain [16], breast [17], colon [18], glioma [19] and prostate cancers [14], as it suppresses the TGF-$\beta$/SMAD pathway by reducing the expression of the downstream targets SMAD2 and SMAD3 [20], as well as the Wnt/$\beta$-catenin pathway by interacting with the Wnt co-receptor LRP6 [21]. On the other hand, its overexpression in cervical [22], hepatocellular [23], renal [24] and lung [25] cancers is associated with poor prognosis and higher tumor aggressiveness [26], making this protein a negative prognostic biomarker for these tumors. The molecular role of *h*NDRG1 in worsening the tumor outcomes in these tissues is not known.

In the non-small-cell lung carcinoma (NSCLC), higher expression of *h*NDRG1 correlates with higher cancer aggressiveness and resistance to chemotherapy, especially with cisplatin medication [25]. It also favors stem-like properties in tumor-initiating cells (TICs) by interacting with the Skp2 kinase and preventing the degradation of C-Myc through Skp2-mediated ubiquitination [27]. For these reasons, *h*NDRG1 has been proposed as a possible target for both treating tumor aggressiveness and modulating the cellular effects of nickel compounds in lung cells [28]. Interestingly, *h*NDRG1 showed Ni(II)-binding properties [29], suggesting a possible link between the oncologic potential of nickel for lung cells and the tumor response. A function for this protein as a modulator of nickel-dependent toxicity, through Ni(II) chelation and detoxification, has been proposed [30].

Despite the large number of studies that revealed the multiplicity of cellular pathways regulated by *h*NDRG1, its mechanism of action in the cellular machinery remains elusive. Understanding the precise function that this protein plays in the carcinogenic process at the cellular, molecular and structural level is necessary to design potential drugs that modulate or inhibit its specific role in tumors.

*h*NDRG1 belongs to the *h*NDRG family, which contains four orthologues, namely *h*NDRG1, *h*NDRG2, *h*NDRG3 and *h*NDRG4. All these proteins are encoded by genes repressed by Myc and share approximately 60% sequence identity, with a nonenzymatic $\alpha/\beta$ hydrolase-fold usually linked to protein scaffold or regulation of protein–protein interactions [31,32]. The crystallographic structures of *h*NDRG2, *h*NDRG3 and more recently *h*NDRG1, all artificially truncated at the N- and C-termini to favor crystallization, were determined [33–35]. They all contain a common α/β hydrolase fold made of six β-strands surrounded by eight α-helices.

Like other *h*NDRG proteins, *h*NDRG1 contains a N-terminal flexible region of 32 residues, not present in the crystal structures due to the truncation, which is quite conserved between *h*NDRG1 and *h*NDRG3. This sequence contains several hydrophobic residues and likely folds as an α-helix at the side of the α/β hydrolase domain as a regulatory lid, as suggested by CD and SAXS [35]. This region is post-translationally modified by SUMOylation at Arg14 [36] and by truncation in a *h*NDRG1 isoform found in different cell types [37,38].

Distinctively among other *h*NDRG proteins, *h*NDRG1 features a C-terminal sequence of 83 residues, rich in charged residues and featuring a three-times-repeated decapeptide GTRSRSHTSE. This region, here called *h*NDRG1*C ($^{311}$GMGYMPSASMTRLMRSRTASGSSVTSLDG TRSRSHTSEGTRSRSHTSEGTRSRSHTSEAHLDITPNSGAAGNSAGPKSMEVSC$^{394}$) binds to transition metal ions such as Ni(II) [30,39], Zn(II) and Cu(II) [40], as well as Mn(II) and Co(II) [40], and it is responsible for the conformational change observed in vitro for *h*NDRG1 upon interaction with lipids [35]. This sequence is modified both in vitro and in vivo through Ser-Thr phosphorylation performed by different kinases, such as calmoduline kinase-II, protein kinase A (PKA), serum- and glucocorticoid-induced kinase 1 (SGK-1) and glycogen synthase kinase 3 beta (GSK3beta) [41,42]. This post-translational modification likely regulates the biological functions of the *h*NDRG1, determining its cell-cycle-dependent protein localization and protein–protein interactions [38,43–46]. Phosphorylation of Ser330 and Thr346 by SGK1 is related to the suppression of the NF-kB signaling pathway and of the CXC cytokine production in pancreatic cancer cells [45]. The SGK1-mediated additional phosphorylation of Thr356 and Thr366, located—together with Thr346—in the decapeptide sequences and found in the liver, lung, spleen and skeletal muscle of mice, primes the further action of GSK3, which could then phosphorylate Ser342, Ser352 and Ser362 in the three repeated regions [42]. All these observations pinpoint an important role of this unique C-terminal region for the *h*NDRG1 characteristic cellular function.

In the present work, the structural, biochemical and biophysical characterization of the C-terminal 83-residues portion of *h*NDRG1 (here named *h*NDRG1*C) is reported. Analysis of the sequence assigned it to the family of the intrinsically disordered regions (IDRs). The protein was expressed and purified from *Escherichia coli*, and experiments of isothermal titration calorimetry, light scattering and circular dichroism were carried out to establish its metal-binding activity, as well as secondary and quaternary structure, which were then compared with those observed for the native *h*NDRG1. A thorough analysis of the spectroscopic fingerprint of $^{1}$H- and $^{13}$C-detected NMR spectra provided detailed information on the effect of pH and Ni(II) binding onto the structure of the protein backbone and side chains. The biophysical data were integrated with the analysis of the Ni(II)-induced expression, subcellular localization and oligomeric states of *h*NDRG1 natively expressed in two cell lines, one deriving from lung adenocarcinoma. The results are discussed considering the possible role of *h*NDRG1*C in the Ni(II)-driven lung cancer progression.

## 2. Materials and Methods

### 2.1. Gene Cloning

The gene coding for the full-length human NDRG1 (*h*NDRG1, Uniprot code: Q92597-1), 1197 bp, was commercially synthesized and cloned into the pEX-K4 subcloning vector by Eurofins, introducing the recognition sites for NdeI and BamHI restriction enzymes at the 5′ and 3′ positions, respectively. The *pEX-K4: hNDRG1* construct, purified from *Escherichia coli* XL10-Gold Ultracompetent Cells (Agilent, Santa Clara, CA, USA), was double-digested with the restriction enzymes NdeI and BamHI (Fermentas, Waltham, MA, USA). The DNA fragment, corresponding to the *hNDRG1* gene, was purified from a 1% (*w*/*v*) agarose gel and ligated, using T4 DNA ligase (Promega, Madison, WI, USA), into a modified *pET15b* expression vector (5.7 kb), previously digested with the same restriction enzymes, containing the sequence for a N-terminal Strep-tag (MASWSHPQFEKGAENLYFQGH) [47]. The purified construct *pET15b:hNDRG*1 was analyzed by restriction analysis and sequenced at both strands.

The C-terminal sequence of the protein (*h*NDRG1*C) was PCR-amplified from the *pET15b:hNDRG1* using the Easy-A High-Fidelity PCR Cloning Enzyme (Agilent, Santa Clara, CA, USA) and the primer pairs *h*NDRG_*C _F (5′-CAACATATGGGCTATATGC CGAGCG-3′)/T7-terminator_R (5′-GCTAGTTATTGCTCAGCGG-3′). The purified PCR product was double-digested with NdeI and BamHI restriction enzymes (Fermentas, Waltham, MA, USA) and cloned into the modified *pET15b* expression vector described above [47]. The resulting construct was analyzed by restriction analysis and sequenced on both strands. Subsequently, the obtained *pET15b:hNDRG1*C* was double-digested with NcoI and BamHI restriction enzymes (Fermentas, Waltham, MA, USA). The *Strep-hNDRG1*C* gene was cloned into the *pETZZ_1a* expression vector [48]. This construct coded for *h*NDRG1*C fused to a N-terminal His tag, an IgG-binding domain ZZ (ZZ-tag) and a Strep tag. The entire fusion protein could be excised by TEV cleavage between the StrepTag and *h*NDRG1*C, leaving a Gly-His dipeptide at the N-terminal region of the protein. Subsequent removal of the non-native His residue was performed by site-directed mutagenesis using the Quikchange mutagenesis kit (Agilent, Santa Clara, CA, USA) following the manufacturer's instructions, to yield the wild-type sequence $^{311}$GMGYMPSASMTRLMRSRTASGSSVTSLDGTRSRSHTSEGTRSRSHTSEGTRSRSHTSEA HLDITPNSGAAGNSAGPKSMEVSC$^{394}$.

### 2.2. Protein Expression and Purification

Large-scale expression of the full-length protein *h*NDRG1 followed transformation of *E. coli* BL21-CodonPlus (DE3) RiL competent cells (Agilent, Santa Clara, CA, USA) with *pET15b:hNDRG1*, and was achieved in 1 L of autoinduction medium [49] (10 g $L^{-1}$ triptone, 5 g $L^{-1}$ yeast extract, 5 g $L^{-1}$ glycerol, 3.3 g $L^{-1}$ $(NH_4)_2SO_4$, 6.8 g $L^{-1}$ $KH_2PO_4$, 7.1 g $L^{-1}$ $Na_2HPO_4$, 0.120 g $L^{-1}$ $MgSO_4$, 0.5 g $L^{-1}$ glucose, 2 g $L^{-1}$ lactose), supplemented with 100 µg/mL of ampicillin and 34 µg/mL of chloramphenicol. The cells were grown at 37 °C for 4 h with vigorous stirring; then, the temperature was reduced to 26 °C and the expression was carried out for 18 h. The cells were harvested by centrifugation at 6000× *g* for 20 min at 4 °C and the cellular pellet was resuspended in 60 mL of the lysis buffer containing 50 mM Tris-HCl buffer pH 8, 150 mM NaCl, 1 mM EDTA, 5 mM DTT, 20 µg/mL DNaseI, 10 mM $MgCl_2$. A protease inhibitor cocktail of 1 mM PMSF (Sigma-Aldrich, Saint Louis, MO, USA), 1 mM benzamidine hydrochloride (Acros Organics, Geel, Belgium), 5 µg/mL pepstatin A (VWR) was added. The bacterial cells were disrupted by two passages through a French pressure cell (SLM Aminco) at 20,000 pounds/square inch. The soluble fraction, obtained after removal of the precipitated material by centrifugation at 25,000× *g* for 20 min at 4 °C, was loaded onto a StrepTrap HP 5 mL column (GE Healthcare, Chicago, IL, USA), pre-equilibrated with 50 mM Tris-HCl pH 8, 150 mM NaCl, 1 mM EDTA and 5 mM DTT. The same buffer was used to wash out the unbound material until the absorbance at 280 nm returned to baseline. The protein was eluted with 30 mL of the elution buffer containing 2.5 mM D-desthiobiotin (IBA-Lifesciences, Göttingen, Germany), and the fractions containing *h*NDRG1 were combined and concentrated using 10-kDa cut-off membrane ultra-filtration units (Millipore, Burlington, MA, USA) before the final polishing step obtained on a Superdex 200 16/60 column equilibrated with 20 mM HEPES buffer at pH 7.5, containing 150 mM NaCl and 1 mM TCEP (working buffer). The final yield of the protein was 6–8 mg $L^{-1}$ of initial culture. The purity of the purified protein was verified by SDS-PAGE using NuPAGE 4–12% Bis–tris acrylamide gels and the fractions containing *h*NDRG1 were concentrated and stored at −80 °C. The final protein concentration, referred to the monomeric form of the protein, was determined using as molar extinction coefficient at 280 nm the theoretical value 38,890 $M^{-1}$ $cm^{-1}$ calculated using the ProtParam website (https://web.expasy.org/protparam/, accessed on 17 June 2022).

Large-scale expression of the C-terminal peptide *h*NDRG1*C (residues 312–394) followed transformation of *E. coli* BL21-CodonPlus (DE3) RiL competent cells (Agilent, Santa Clara, CA, USA) with *pETZZ1a:hNDRG1*C*, and was achieved in 1–2 L of lisogeny broth (LB) at 37 °C, supplemented with 30 µg/mL kanamicin and 34 µg/mL chrolamphenicol with vigorous stirring. Protein expression was induced with 0.5 mM IPTG at $OD_{600}$ = 0.6 and performed at 26 °C for 18 h. To produce labeled *h*NDRG1*C, bacterial cells were centrifuged before induction and transferred in one-fourth of the initial volume using 2x M9 minimal medium (1.26 g $L^{-1}$ $(NH_4)_2SO_4$, 12 g $L^{-1}$ $Na_2HPO_4$, 6 g $L^{-1}$ $KH_2PO_4$, 1 g $L^{-1}$ NaCl, 4 g $L^{-1}$ glucose, 0.240 g $L^{-1}$ $MgSO_4$), containing either $^{15}N$ or $^{15}N/^{13}C$ isotopes, and induced with 0.5 mM IPTG [50,51], as above. Cells were harvested by centrifugation at 6000× *g* for 20 min at 4 °C, resuspended in 30 mL of 50 mM Tris HCl buffer pH 7.6, 300 mM NaCl, 20 mM imidazole, 20 µg/mL DNase I and 10 mM $MgCl_2$; and a protease inhibitor cocktail of 1 mM PMSF (Sigma-Aldrich, Saint Louis, MO, USA), 1 mM benzamidine hydrochloride (Acros Organics, Geel, Belgium) and 5 µg/mL pepstatin A (VWR). The bacterial cells were disrupted by two passages through a French pressure cell (SLM Aminco, Urbana, IL, USA) at 20,000 pounds/square inch and the lysate was centrifugated at 76,000× *g* for 20 min at 4 °C. The clarified fraction was loaded onto a HisTrap HP 5 mL column (GE Healthcare, Chicago, IL, USA), pre-equilibrated with 25 mL of 50 mM TrisHCl pH 7.6 containing 300 mM NaCl and 20 mM imidazole. The column was washed with the same buffer until the baseline was stable and the protein was eluted by a linear gradient from 20 to 500 mM imidazole. The fractions containing the His-Tag/ZZ-tag/Strep-Tag fusion polypeptide were collected and incubated with a 1:50 protease:protein ratio with TEV protease for 3 h at room temperature. Subsequently, the protein was loaded onto a HiPrep 16/60 desalting column pre-equilibrated with 20 mM Tris-HCl pH 7.5, 2 mM EDTA and 1 mM DTT. *h*NDRG1*C was separated by the His-Tag/ZZ-tag/Strep-Tag peptide using a SP-sepharose cation exchange chromatography XK 16/10 column, equilibrated with the same buffer, and eluted by a linear gradient from 0 to 1 M NaCl. *h*NDRG1*C eluted at 240–280 mM NaCl. Fractions containing the protein were pooled, concentrated with 3 kDa MWCO Centricon ultra-filtration units (Millipore, Burlington, MA, USA) and further purified into a Superdex 75 XK 16/60 column (GE Healthcare, Chicago, IL, USA) equilibrated with the working buffer. Protein quantification was performed by amino acid analysis (Alphalyse, Odense, Denmark). The obtained extinction coefficient at 280 nm was 2450 $cm^{-1}$ $M^{-1}$. The final yield of the purified protein was 5–10 mg $L^{-1}$ of initial culture.

The purity of both *h*NDRG1 and *h*NDRG1*C, as well as the molecular mass of the isolated variants under denaturing conditions, were estimated by SDS-PAGE using XCell SureLockTM Mini-Cell Electrophoresis System (Thermo Fisher Scientific, Waltham, MA, USA) apparatus and NuPAGE 4–12% or 12% Bis–tris acrylamide gels, stained using ProBlue Safe stain (Giotto Biotech, Firenze, Italy). The absence of any metal bound to the purified proteins was confirmed by ICP-ES, using a procedure previously described [52].

### 2.3. Isothermal Titration Calorimetry

Ni(II) binding to *h*NDRG1*C (90–130 µM) was investigated at 25 °C using a high-sensitivity VP-ITC microcalorimeter (MicroCal, Northhampton, MA, USA). The protein solution in the working buffer was loaded into the sample cell (cell volume = 1.4093 mL) and 29 injections of a 10 µL solution containing $NiSO_4$ (2 mM) were added using a computer-controlled 310-µL microsyringe. Intervals of 300 s were applied between the injections to allow the system to reach thermal equilibrium after each addition. The heat of dilution was negligible, as verified by control experiments performed titrating Ni(II) over the buffer, under the same experimental conditions.

The data were processed using the Origin software package (MicroCal, Northhampton, MA, USA) and fitted using a nonlinear least-squares minimization algorithm to theoretical curves corresponding to different binding models. $\Delta H$ (reaction enthalpy change), $K_A$

(binding affinity constant), *n* (number of binding sites) and *N* (binding stoichiometry) were the fitting parameters. The $\chi^2$ parameter was used to establish the best fit. The reaction entropy was calculated using the equations: $\Delta G = -RT \ln K_a$ ($R$ =1.9872 cal mol$^{-1}$ K$^{-1}$, $T$ = 298 K) and $\Delta G = \Delta H—T\Delta S$. The values obtained for $\Delta H$ and $\Delta S$ are apparent, and include contributions not only from metal binding, but also from associated events such as protonation/deprotonation of the amino acid residues involved in the binding and consequent change in the buffer ionization state.

### 2.4. Circular Dichroism Spectroscopy

The secondary structures of *h*NDRG1 (30 μM) and *h*NDRG1*C (200 μM) were estimated by recording circular dichroism (CD) spectra in the working buffer using a JASCO J-810 spectropolarimeter flushed with $N_2$ and a cuvette with 0.1 cm path length. Experiments were conducted in the absence and in the presence of different concentrations of $NiSO_4$. Ten spectra were registered at 25 °C from 260 nm to 190 nm at 0.2 nm intervals and averaged to achieve an appropriate signal-to-noise-ratio. The secondary structure compositions of *h*NDRG1 and *h*NDRG1*C were quantitatively evaluated using Best Structure Selection (BeStSel) [53].

### 2.5. NMR Spectroscopy

NMR experiments were carried out using 0.2–0.3 mL samples of 0.8–1.0 mM purified U-$^{15}$N or U-$^{13}$C,$^{15}$N *h*NDRG1*C in the working buffer at pH 6.5 containing 10% $D_2O$, in 3-mm NMR tubes. Standard $^1$H-detected protein NMR spectra for the assignment of nuclei belonging to backbone [$^1$H-$^{15}$N BEST-TROSY, $^1$H-$^{13}$C HSQC, HNCO, HNcaCO, HNCA, HNCACB, CBCAcoNH, HNcoCACB, HBHANH and HBHAcoNH], aliphatic side-chains [$^1$H-$^{13}$C HSQC, hCCH-TOCSY, HCcH-TOCSY and CcoNH] and aromatic side chains [2D $^1$H-$^1$H TOCSY and hbCBcgcdHD (CBHD)] were collected at 298 K on a Bruker AVANCE NEO/III spectrometer operating at 28.2 T (1200.73 MHz $^1$H Larmor frequency) and equipped with a 3 mm triple-resonance inverse TCI z-gradient cryoprobe. $^{13}$C-detected NMR spectra (CON, hCACO and hCBCACO) [54–56] were acquired at 298 K using a 16.4 T Bruker AVANCE NEO spectrometer operating at 16.4 T (700.06 MHz $^1$H Larmor frequency), equipped with a 3 mm TXO cryoprobe optimized for $^{13}$C direct detection. Proton chemical shifts were referenced to 2,2-dimethyl-2-silapentane-5-sulfonic acid sodium salt (DSS), while the $^{13}$C and $^{15}$N chemical shifts were referenced indirectly to DSS, using the ratios of the gyromagnetic constants. Table S1 reports the NMR spectra acquisition parameters.

All NMR spectra were processed using NMRpipe [57] and the SMILE (Sparse Multidimensional Iterative Lineshape-Enhanced) [58] reconstruction algorithm plug-in module implemented in NMRpipe, for both non-uniformly sampled (NUS) and conventional NMR spectra. Table S1 reports the details of all acquired NMR spectra. Analysis and assignments of the 2D and 3D data sets were carried out using NMRFAM-SPARKY [59] and POKY. [60] The assignment process was facilitated by using the PINE [61,62] server for initial automated assignments before completing the assignments manually. LACS (Linear Analysis of Chemical Shifts) [63] was used to validate the final assignment. The assignment was deposited in the Biological Magnetic Resonance Bank (BMRB) with the accession code 50803.

$^1$H-NMR experiments tailored for the identification of hyperfine shifted and fast relaxing signals [64] were performed on an AVANCE 600 Bruker NMR spectrometer equipped with a room-temperature 5 mm $^1$H selective probe and operating at 600.13 MHz $^1$H Larmor frequency, without gradients. Spectra were collected using a short $^1$H pulse (3 μs, corresponding to a ca. 35° pulse), in order to excite the large spectral window (640 kHz) needed for the experiments. A 200 ms presaturation pulse was applied to suppress the water signal. A total of 32 K data points were acquired over 46 ms. The number of scans acquired ranged from 200 K to 400 K, corresponding to 18–36 h of experimental time. Prior to Fourier transform, FIDs were multiplied by a cosine-square weighting function followed by a 40 Hz Lorentzian line broadening. Phase and baseline correction were performed manually.

### 2.6. Light Scattering

The oligomeric properties and the hydrodynamic radii of *h*NDRG1 and *h*NDRG1*C in the absence and in the presence of Ni(II) were determined combining size-exclusion chromatography (SEC) with multiple-angle light scattering (MALS) and quasi-elastic light scattering (QELS). *h*NDRG1 (300 μL, 140 μM) and *h*NDRG1*C (300 μL, 1.2 mM) in the working buffer were loaded onto a Superdex 200 10/300 GL column (GE Healthcare, Chicago, IL, USA) (*h*NDRG1) or a Superdex 75 10/300 GL column (GE Healthcare, Chicago, IL, USA) (*h*NDRG1*C) equilibrated with the same buffer, in the absence or in the presence of equimolar concentrations of Ni(II). $NiSO_4$ (28 μM for *h*NDRG1 and 240 μM for *h*NDRG1*C) was also added to the working buffer for the experiments in the presence of Ni(II). Elution was carried out at room temperature with a flow rate of 0.5 mL/min. The column was connected downstream to a multiple-angle laser light (690 nm)-scattering DAWN EOS photometer (Wyatt Technology) and to a quasi-elastic light-scattering apparatus (Wyatt QELS). The used value for the specific refractive index increment (dn/dc) was 0.185 mL/g [65]. The value of 1.331 for the solvent refractive index was determined using the refractive index detector. The latter was used also to determine the concentration of the protein while eluting from the chromatographic column. The weight-average molecular masses were determined from MALS measurements across the entire elution profile, in intervals of 0.2 s, using the ASTRA software (Wyatt Technology). A Rayleigh–Debye–Gans light-scattering model was used to determine the molecular weight (MW), using a Zimm plot. The uncertainties on MW are a measure of the statistical consistency of the MALS data, obtained combining the standard deviations calculated for each slice in the analyzed peaks. Data analysis was performed using Astra version 5.3.4 following the manufacturer's instructions.

### 2.7. Cultures and Cellular Treatments

Human A549 cell line derived from lung carcinoma was a gift from Dr A. Tesei (Istituto Scientifico Romagnolo per lo Studio e la Cura dei Tumori, Meldola, Italy). HeLa cells were obtained from Istituto Ortopedico Rizzoli, Bologna Italy. Cells were cultured in Dulbecco's Modified Eagle Medium (DMEM) GlutaMAX (Gibco, ThermoFisher Scientific, Monza, Italy) supplemented with 10% of heat-inactivated fetal bovine serum (FBS, Gibco). Cells were maintained in a humidified atmosphere with 5% $CO_2$ at 37 °C and subcultured twice a week. Where indicated, cells were treated with $NiSO_4$.

### 2.8. Preparation of Protein Extracts

Total lysates were prepared in SDS lysis buffer (20 mM Tris-HCl, pH 7.5, 1% SDS, 1 mM $Na_3VO_4$, 1 mM PMSF, 5% β-mercaptoethanol and protease inhibitors). Nuclear extracts were prepared as follows: cells were trypsinized, collected and resuspended in hypotonic buffer (10 mM Tris-HCl pH 7.8, 5 mM $MgCl_2$). Then, 0.2% Triton X-100 was added. Cells were sheared through a 22-G needle. Nuclei were recovered by centrifugation and lysed in 20 mM Tris-HCl 7.0, 1% NP-40, 150 mM NaCl, 10% glycerol, 10 mM EDTA, 20 mM NAF, 5 mM $Na_4P_2O_7$, 1 mM $Na_3VO_4$, 1 mM PMSF and protease inhibitors and cleared by centrifugation. Protein amount was evaluated by Bradford colorimetric assay. Equal amounts of protein lysates were resolved by SDS-PAGE. Nondenaturing conditions were maintained by resuspending lysates in native sample buffer (0.3 M Tris-HCl pH 6.8, 0.03% bromophenol blue and 50% glycerol); denaturing conditions were achieved after boiling samples resuspended in full Laemli sample buffer (0.3 M Tris-HCl pH 6.8, 0.03% bromophenol blue, 9% sodium dodecyl sulfate (SDS), 9% β-mercaptoethanol, 50% glycerol). Samples were transferred to nitrocellulose membrane overnight at 4 °C. Incubation with primary antibodies was performed for the indicated time. Bands were revealed using the Amersham ECL detection system and analyzed with ImageJ (National Institute of Health, Bethesda, MD, USA). Purity of nuclei was analyzed by immunoblot detection of β-tubulin. Antibodies used were anti-NDRG1 (1:500), anti-Actin (1:1000), anti-Lamin A/C (1:200), all from Santa Cruz (Santa Cruz Biotechnology, DBA Italia SRL, Segrate, Italy); anti-β-Catenin

(1:3000), α-tubulin (1:2000) and anti-GAPDH (1:8000) from Merck (Merck Life Science S.r.l., Milan, Italy); anti-p21 (1:1000) (Thermo-Fisher Scientific).

## 3. Results

### *3.1. Analysis of Protein Disorder*

Analysis of the canonical sequence of *h*NDRG1 (UniProt accession number Q92597) using disorder predictors in the D2P2 database [66] highlighted a differentiation in disorder content along the protein sequence (Figure 1A) that concerns the three domains previously reported [35]: indeed, disorder is prevalent in the short N-terminal sequence (residues 1–30) and in the unique C-terminal domain of 83 residues (*h*NDRG1*C, residues 312–394), which contains several predicted phosphorylation sites, as well as a three-times-repeated His-containing decapeptide with Ni(II)-binding properties (Figure 1B) [30,39]. On the other hand, the α/β hydrolase domain (residues 30–312) is predicted to be well-folded (Figure 1B).

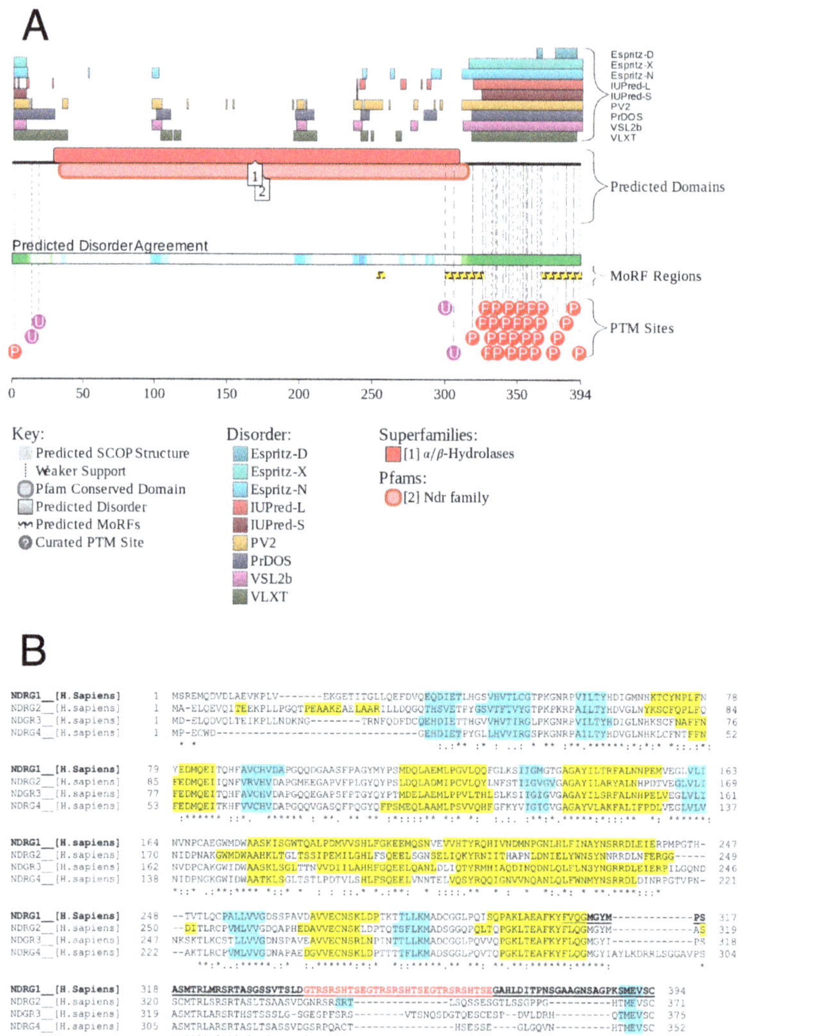


**Figure 1.** Analysis of the sequence of *h*NDRG1. (**A**) Disordered regions of the sequence of *h*NDRG1 as predicted by the D2P2 server (http://d2p2.pro/, accessed on 29 January 2021). The predicted

disordered regions (top), folded domains (middle), and disorder consensus (bottom) are indicated by bars over the residue numbers. The sites with predicted post-translational modifications are also indicated. (**B**) Prediction of the secondary structure content of *h*NDRG1 using the software JPred [67]: predicted a-helices are indicated in yellow, b-strands are in cyan. *h*NDRG1*C sequence is underlined and the repeated decapeptide is shown as red text.

According to the BioGRID database (https://thebiogrid.org, accessed on 29 January 2021), *h*NDRG1 interacts with as many as 140 proteins in the cell [68]. Consistently, a visualization of the interactivity of *h*NDRG1 by the STRING computational platform [69], which produces a network of predicted associations for a particular group of proteins, shows that this protein is part of a massive interactome (Figure S1A). The intrinsically disordered nature of the C-terminal region might be indicative of the observed promiscuity of *h*NDRG1, reported to be involved in a plethora of different cellular metabolisms [15]. This observation is supported by a prediction of two molecular recognition sequences (MORF, Figure 1A) [70] at the beginning and at the end of the protein C-terminus, as well as by the output of the Anchor software [71], which identifies a long disordered-based interaction sequence that includes *h*NDRG1*C (residues 308–394, Figure 1B). Coherently, the charge–hydropathy (CH) plot [72], which analyzes the protein properties in terms of the mean net absolute charge vs. mean hydrophobicity, shows that while the full-length protein falls in the region of the plot typical of polypeptides with a well-folded behavior, *h*NDRG1*C resides in the disordered region of the plot, confirming its propensity for an intrinsically disordered behavior (Figure S1B).

### 3.2. Protein Expression and Purification

The canonical sequence coding for *h*NDRG1 was obtained by retrotranslating its amino acid sequence and cloned into a modified *pET15b* expression vector. The obtained construct was used to overproduce a N-terminal Strep-tagged protein (45.28 kDa) in *E. coli* BL21-CodonPlus (DE3) RiL competent cells using an autoinduction medium enriched with glucose and lactose. The expressed polypeptide accumulated in the soluble fraction of the cellular extract and was purified using a Strep-tactin affinity column, followed by size-exclusion chromatography. The identity and purity of the protein was confirmed by gel electrophoresis (Figure S2A) and by mass spectrometry, which evidenced the cleavage of the N-terminal methionine after protein production. In the SDS-PAGE, the presence of a band at ca. 85–90 kDa, coexisting with that expected at 45 kDa and corresponding to the protein monomer, suggests a possible dimerization of the protein sample (Figure S2A). This is sometimes visible in SDS-PAGE, when denaturing conditions used (in this case, heating the sample at 90 °C for 3 min) are not sufficient to fully dissociate the dimer.

The nucleotide sequence coding for *h*NDRG1*C was initially cloned into a modified *pET15b* vector, but the expression of the Strep-tagged protein was insignificant. Thus, a different expression approach, involving the N-terminal tagging of *h*NDRG1*C with the modified immunoglobulin-binding domain of protein A of *Staphylococcus aureus* (ZZ-tag), fused with a His-rich sequence at the N-terminus, was applied [48]. This tag allowed for the improvement of yields, solubility and conformational stability of the expressed protein. The obtained construct was used to overproduce His-ZZ-Strep-*h*NDRG1*C (27.98 kDa) in *E. coli* BL21-CodonPlus (DE3) RiL cells by induction with IPTG. The polypeptide was purified from the soluble fraction of the cellular extract with a Ni(II)-based affinity chromatography, followed by TEV protease cleavage and by cation-exchange chromatography. High-purity *h*NDRG1*C (8.83 kDa) was obtained using size-exclusion chromatography (SEC). The identity and purity of the protein were confirmed by gel electrophoresis (Figure S2B) and by mass spectrometry.

### 3.3. Isothermal Titration Calorimetry

Titration of Ni(II) on *h*NDRG1*C (Figure 2A) produces a binding isotherm with a single inflection point, which was fitted using a single site model. The binding parameters, obtained from the fit of four binding isotherms (Figure 2B) and by averaging the results of the fits, indicate that the protein presents a single binding event ($n$ = 1.4 ± 0.5) with affinity in the micromolar range ($K_A$ = 1.4 ± 0.3 × $10^4$; $K_D$ = 70 ± 1 μM) and enthalpically driven (ΔH = −9 ± 6 kcal $mol^{-1}$ and ΔS = −19 ± 13 cal $mol^{-1}$ $K^{-1}$).

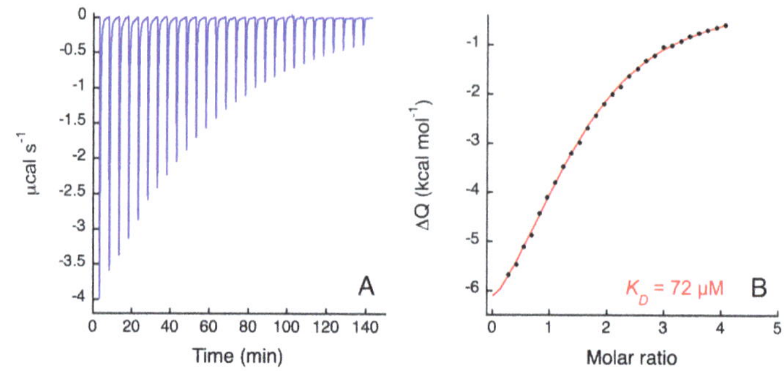


**Figure 2.** ITC titration data for the binding of $NiSO_4$ to *h*NDRG1*C. (**A**) Raw titration data represent the thermal effect of 27 × 10 μL injections of Ni(II) onto or *h*NDRG1*C solution at pH 7.5. (**B**) Normalized heat of reaction data for the binding events of Ni(II) to *h*NDRG1*C were obtained integrating the raw data. The solid line represents the best fit of the integrated data, obtained by a nonlinear least-squares procedure, as described in the text. The calculated dissociation constant is indicated.

### 3.4. Circular Dichroism

The CD spectrum of *h*NDRG1*C is typical of an intrinsically disordered protein, with a pronounced negative peak around 198 nm and a quantitative analysis of the spectrum confirming the low amount of secondary structure, with 6.6% α-helices, 26.3% β-strands and 67% unordered (Figure 3). The addition of Ni(II) does not significantly influence the protein secondary structure (Figure 3). The CD spectrum of *h*NDRG1 in the presence and in the absence of Ni(II), shown in Figure S3 for comparison with the one of *h*NDRG1*C, is very similar to that reported previously [35], independently from the presence of Ni(II) ions. Quantitative evaluation of secondary structures, using BestSel, indicated a marked prevalence of α-helical content (25%) and a minor number of β-strands (19%), well in accordance with the secondary structure content extracted from the crystal structure (25% α-helix and 12.5% β-strands) [35].

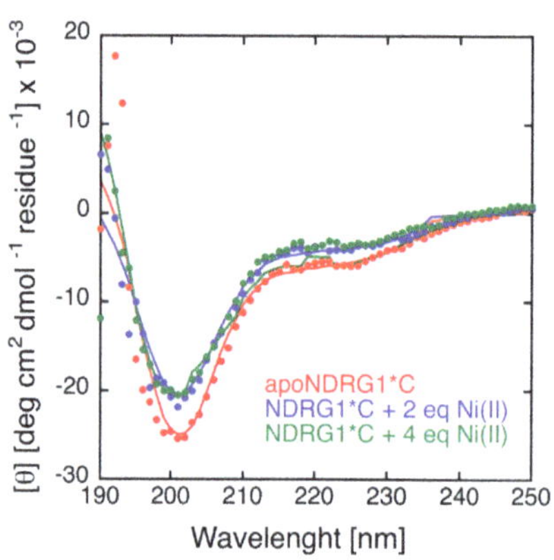


**Figure 3.** Far-UV CD spectra of *h*NDRG1*C in the absence and in the presence of increasing concentrations of Ni(II).

### 3.5. NMR Spectroscopy on hNDRG1*C

In order to establish the molecular structural details of *h*NDRG1*C in solution at the atomic/molecular level and the effects of Ni(II) binding, high-resolution nuclear magnetic resonance spectroscopy (NMR) was extensively applied. The full signal assignment was first achieved at pH 6.5, followed by the analysis of the effect of solution pH on the spectra and finally by the investigation of the modifications induced by Ni(II) binding at pH 7.5.

#### 3.5.1. $^{1}H$, $^{13}C$ and $^{15}N$ Signal Assignment

Two-dimensional and three-dimensional high-resolution solution NMR spectra of *h*NDRG1*C were acquired using data recorded at 1.2 GHz $^{1}H$ Larmor frequency, with the aim of assigning the backbone and side-chain signals of the protein. In the following description, the residues are numbered according to the sequence of the full *h*NDRG1 protein, from Gly311 to Cys394. Initial attempts to record $^{1}H$,$^{15}N$ HSQC spectra at the pH of the working buffer (7.5) resulted in a significantly lower number of signals than expected, suggesting the presence of exchange phenomena involving the amide NH protons at this pH. On the other hand, lowering the pH at 6.5 resulted in a significant improvement of the spectra; therefore, the NMR signal assignment was initially carried out at pH 6.5.

The BEST-TROSY $^{1}H$-$^{15}N$ HSQC spectrum of hNDRG1*C at pH 6.5 (Figure 4A) features a narrow $^{1}H$ chemical shift dispersion (8.0–8.6 ppm) typically seen for intrinsically disordered proteins (IDPs). Outside this region, the signals of amide $NH_2$ protons of the side chains of the two asparagine residues Asn377 and Asn383 could also be detected. A total of 76 amide NH signals were assigned in the $^{1}H$-$^{15}N$ HSQC spectrum. The N-terminal is not observed because of the fast exchange with water, while the NH resonance of Met312 is not detected probably because its NMR signal is broadened beyond detection due to conformational exchange phenomena occurring with rates comparable to the frequency differences among the different conformers. In addition, Pro316, Pro376 and Pro387 are not observable in the $^{1}H$,$^{15}N$ HSQC spectrum, but the signals of CB and CG of these residues were obtained using triple-resonance experiments; the conformation of these residues was determined by calculating the difference between proline CA and CG chemical shifts [73,74], which indicated that all proline residues are linked to the preceding amino acid by a peptide bond in the *trans* conformation. The $^{13}C$ chemical shift of the CB nucleus of Cys394, the last protein residue in the sequence and the only cysteine residue in the sequence, is 29.06 ppm; considering that this value depends on the redox state of the terminal S atom, being <32 ppm for the reduced thiol state and >35 ppm for the oxidized disulfide state [75], it can be concluded that Cys394 is in the Cys-SH state, preventing the formation of higher-order aggregates by disulfide bridges; this is consistent with the observation that the signals display peak widths and intensities that are invariant upon dilution, as well as with the results of light-scattering measurements.

The essential absence of significant secondary structures in *h*NDRG1*C was confirmed by the inspection of the deviations of the chemical shifts of backbone nuclei from their predicted random coil chemical shifts (RCCS), carried out using CheZOD [76] and CheSPI [77], specific algorithms to quantify the statistical composition of structural states in IDPs (http://www.protein-nmr.org, accessed on 25 February 2022). This analysis indicated that NDRG1*C is largely unfolded, with the notable exception of the $^{318}$ASMTRLMRS$^{327}$R region near the N-terminus, which features a small α-helix propensity, consistent with a transient helical character (Figure 5).

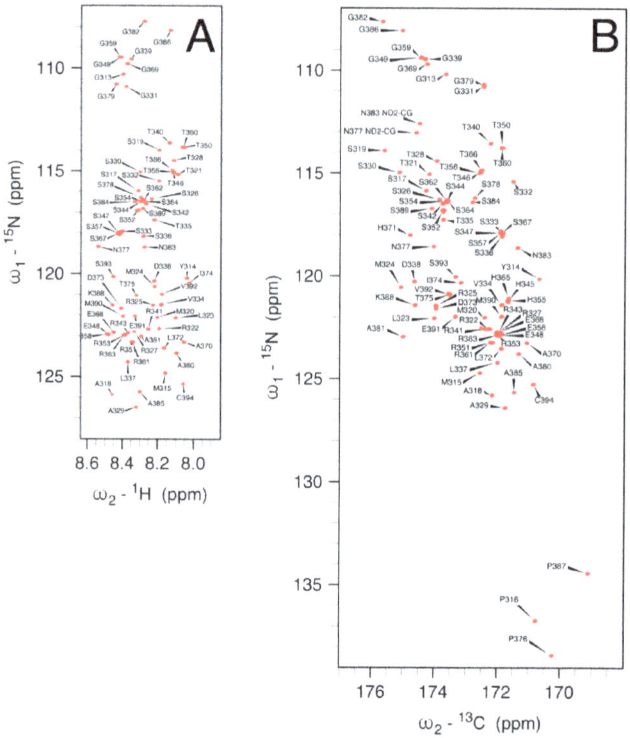

**Figure 4.** (**A**) 1.2 GHz (28.2 T) 2D $^1H,^{15}N$ BEST-TROSY spectrum and (**B**) 700 MHz (16.4 T) 2D CON spectrum obtained by $^{13}C$ direct detection, acquired at 298 K on samples of $^{13}C,^{15}N$ labeled *h*NDRG1-C, at pH 6.5. In the CON spectrum, the resonances are labeled according to the $^{15}N$ frequency. Assigned peaks are labeled.

Notably, the peptide NH signals for all histidine residues are missing from the $^1H,^{15}N$ HSQC spectrum: it has been shown that amide proton exchange with water significantly increases in the presence of a protonated histidine imidazole ring [78], suggesting that the side chains of His345, His355, His365 and His371 are in the imidazolium form at the solution pH of 6.5. On the other hand, the $^{13}C$-detected CON spectrum, which correlates the peptide $^{15}N$ nucleus of each residue to the carbonyl $^{13}C$ nucleus of the preceding amino acid independently of the amide $^1H$ exchange phenomena, allows the observation of the $^{15}N$ signals of all histidines, together with those of Pro316, Pro376 and Pro387 (Figure 4B). The assignment of the side-chain $^{13}C$ nuclei of each residue, performed using triple-resonance experiments, allowed for the full assignment of the $^{13}C$-detected CACO (Figure 6) spectrum, which additionally revealed the signals for the side-chain carboxyl and amide carbons of Asp338, Asp373, Glu348, Glu358, Glu368, Asn377 and Asn383; for the latter two residues, the side-chain amide $^{15}N$ nuclei could also be assigned in the CON spectrum (Figure 4B).

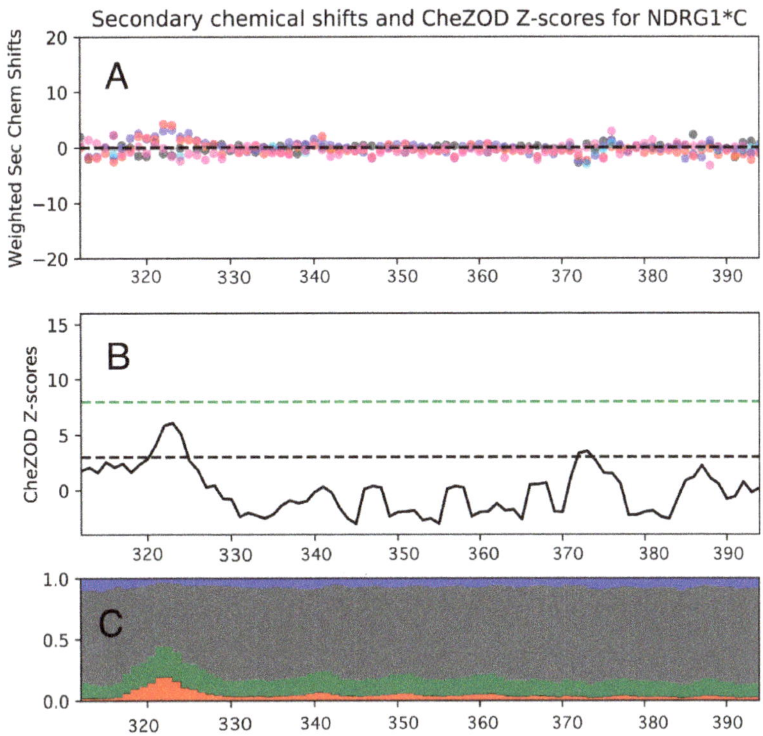


**Figure 5.** (**A**) Weighted secondary chemical shifts, (**B**) CheZOD Z-scores and (**C**) stacked bar plot of CheSPI populations of "extended" (blue), "helical" (red), "turn" (green) and "non-folded" (grey) local structures calculated for *h*NDRG1*C.

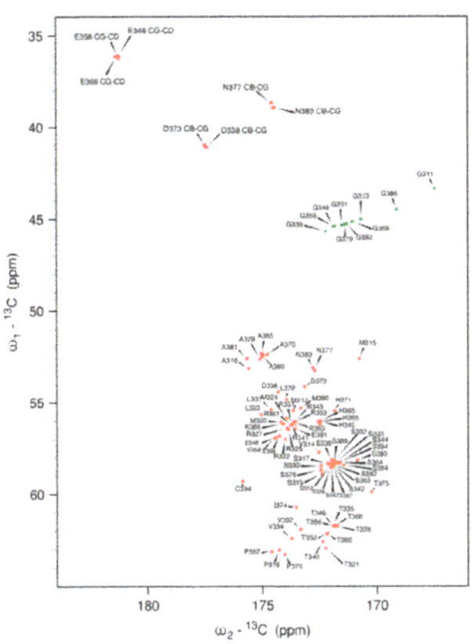


**Figure 6.** 700 MHz (16.4 T) 2D CACO spectrum obtained by $^{13}$C direct detection, acquired at 298 K on samples of $^{13}$C,$^{15}$N labeled *h*NDRG1-C, at pH 6.5. The resonances are labeled according to the $^{13}$C frequencies of the CO and CA nuclei of each amino acid, except for the signals for the side chains of Asp and Glu residues, for which the CB-CG and CG-CD are explicitly indicated.

Considering the potential importance of the histidine residues in Ni(II) binding, a special effort was dedicated to the full assignment of the side chains of these four amino acids using a combination of NMR spectra. The general scheme for this task is illustrated in Figure S4, and the full assignment is reported in Table 1. Considering the characteristic chemical shifts of the CD2 and CE1 nuclei for the histidine imidazole ring in the neutral (113.6 and 135.5 ppm) and doubly protonated cationic (119.4 and 136.3 ppm) states [79–81], the chemical shift values for all histidine residues (CE1 ~138 ppm and CD2 ~120 ppm) support the doubly protonated state of all imidazole rings at pH 6.5. This was confirmed by the patterns observed in the 2J $^1$H,$^{15}$N-HMQC spectra, which revealed chemical shifts for ND1 and NE1 nuclei (in the 180–210 ppm range), typical for doubly protonated His residues, [79–81] and is consistent with a low Ni(II)-binding capability of *h*NDRG1*C at pH 6.5, due to this unfavorable protonation state of the His residues.

**Table 1.** Chemical shift (ppm) signal assignment of the histidine residues in hNDRG1*C at pH 6.5 and pH 7.5.

| | His335 | | His345 | | His355 | | His371 | |
|---|---|---|---|---|---|---|---|---|
| | *pH 6.5* | *pH 7.5* | *pH 6.5* | *pH 7.5* | *pH 6.5* | *pH 7.5* | *pH 6.5* | *pH 7.5* |
| **(H)N** | 121.16 | 121.89 | 121.29 | 121.89 | 121.29 | 121.78 | 118.11 | 118.43 |
| **C** | 175.24 | 175.69 | 175.24 | 175.69 | 175.20 | 175.60 | 174.54 | 175.14 |
| **CA** | 56.13 | 56.67 | 56.13 | 56.67 | 56.26 | 56.65 | 55.56 | 56.20 |
| **CB** | 30.14 | 31.03 | 30.14 | 31.03 | 30.07 | 30.95 | 29.42 | 30.40 |
| **HD2** | 7.13 | 6.99 | 7.13 | 6.99 | 7.13 | 6.99 | 7.16 | 6.99 |
| **CD2** | 119.97 | 119.86 | 119.97 | 119.86 | 119.97 | 119.86 | 120.0 | 119.86 |
| **HE1** | 8.05 | 7.75 | 8.05 | 7.75 | 8.08 | 7.75 | 8.23 | 7.83 |
| **CE1** | 137.97 | 138.66 | 137.97 | 138.66 | 137.90 | 138.66 | 137.5 | 138.51 |

#### 3.5.2. Effect of pH on the NMR Spectra

Considering that Ni(II) binding would be more physiologically significant to explore at pH higher than 6.5, the effect of pH on the NMR spectra of *h*NDRG1*C was monitored by recording the $^1$H,$^{15}$N HSQC, CON and CACO fingerprint spectra in the 6.5–7.5 pH range with steps of 0.25 pH units, allowing us to assign the NMR signals observed at pH 7.5. Similarly, the $^1$H,$^{13}$C HSQC in the aromatic region was monitored in the same pH range to assign the observed signals to the non-exchangeable protons of the histidine side chains.

The intrinsically disordered behavior of the protein is not affected by pH in the 6.5–7.5 interval. The $^1$H,$^{15}$N HSQC spectrum at pH 7.5 is essentially obliterated, consistently with the increased rate of hydrogen exchange (HX) at higher pH, except for signals related to Tyr314, Met315, Leu323, Asp338, Ala370, Leu372, Asp373, Ile374, Thr375, Gly386, Lys388, Glu391, Val392, Ser393 and Cys394; while the chemical shifts of most of these signals are not affected by the pH change, the signals of Leu372 and Asp373 are clearly perturbed, which is a phenomenon that can be rationalized by considering that in the protein sequence, they are in the vicinity of His371, a residue that could be involved in a protonation/deprotonation event in the explored pH range.

This observation reflects different hydrogen-exchange (HX) behaviors along the protein backbone and could be used to gain information about the structure and conformational dynamics of *h*NDRG1*C. The HX rate is determined by a number of factors, including solvent shielding caused by the presence of folded segments, H-bond formation, amino acid sequence composition, temperature and ionic strength [82]. The presence of structural elements and H-bond formation will slow HX, and this is expressed by the so-called protection factor (PF = $k_{\mathrm{intr}}/k_{\mathrm{obs}}$), namely the ratio between the intrinsic HX rate in a random coil model ($k_{\mathrm{intr}}$) and the HX rate measured for the protein ($k_{\mathrm{obs}}$) [83]. The values of $k_{\mathrm{intr}}$, which depend on the protein sequence, are normally not available, but can be calculated by considering the identity of the side chains bracketing each of the amide hydrogens in the sequence [84,85] using SPHERE (https://protocol.fccc.edu/research/labs/roder/sphere/sphere.html, accessed on 25 February

2022). Another crucial parameter that influences the HX rates is the local electrostatic potential at all backbone amide positions along the chain [82]. Plots of (i) the amide NH signal intensities at pH 6.5, (ii) the $k_{intr}$ values calculated using SPHERE, as well as (iii) the electrostatic potential and (iv) the protection factor computed using a recently proposed approach [82], all calculated at pH 6.5 (Figure S5), suggest that the observed smaller effect of pH on the intensity of signals at the C-terminal portion of the protein vs. the N-terminal region is not due to an increased propensity towards a more structured ensemble, but rather to a decrease in the positive electrostatic potential, observed for the first ca. 65–70 residues in the sequence, to smaller and even negative values in the last 15–20 residues of the sequence. The consequence of this trend is a decrease in the local hydroxide ion concentration in the region of the protein that features a smaller electrostatic potential, which in turn induces a decreased HX rate and an increase in the signal intensity at higher pH.

On the other hand, in the CON and CACO spectra, no significant modification of the signals intensities is observed in the explored pH range, consistently with the independence of these signals on amide proton solvent exchange phenomena. In these spectra, the largest chemical shift perturbations affect the signals of the histidine residues, indicating that the deprotonation events occurring upon raising the pH from 6.5 to 7.5 only have local effects, without significant modification of the overall conformational space occupied by the protein. To analyze the effect of pH on the histidine side chains, the dependence of the $^{1}$H chemical shifts of the HE1 signals in the $^{13}$C HSQC spectra of *h*NDRG1*C in the aromatic region as a function of pH in the 6.5–7.5 range was investigated. The corresponding $pK_a$ values for the four histidine residues were obtained from simultaneous nonlinear fits of the chemical shifts of HE1 and HD2 protons of each imidazole ring to the following one-ionization Equation (1):

$$\delta_{obs} = \frac{[H^+]\cdot\delta_{HisH} + K\cdot\delta_{His}}{[H^+] + K} \quad (1)$$

where $d_{obs}$ is the observed experimental chemical shift, $d_{HisH}$ and $d_{His}$ are the chemical shifts of the protonated and neutral forms of the histidine imidazole, and $K$ is the dissociation constant for the ionization equilibrium. The estimated $pK_a$ values (Figure S6) are similar for all histidines (6.65 for His335 and His345, 6.72 for His355, and 6.84 for His371), with a significant fraction (10–20%) of protonated states still maintained at pH 7.5 in all cases.

3.5.3. Ni(II) Binding to NDRG1*C by NMR Spectroscopy

The obtained assignment of NMR spectra at pH 7.5 was then used to monitor the effects of Ni(II) binding onto *h*NDRG1*C. The $^{1}$H-$^{13}$C HSQC spectra in the aromatic region, where the signals of the CE1-HE1 and CD2-HD2 could be monitored as a function of the Ni(II)/protein ratio in the 0 to 3 equivalents (an upper limit known from calorimetry to saturate the binding equilibrium), indicated the progressive and concomitant disappearance of the signals of His371, His345, His355 and His365 without any detectable selectivity (Figure 7). The CON (Figure S7) and CACO (Figure S8) spectra further revealed that several additional signals disappear upon adding Ni(II) to the polypeptide solution, while those signals that are still visible do not modify their chemical shift. This phenomenon indicates that the protein does not undergo any observable change in the backbone folding upon Ni(II) binding. A large portion of the signals centered around the repeated decapeptide decrease their intensity upon Ni(II) binding, and in particular the N, CO and CA signals of all histidine residues are absent in the spectrum of the Ni-bound protein (Figure 8). Characteristically, the N, CO and CA signals of the residues that precede and follow the histidines are also canceled, suggesting the presence of bound paramagnetic high-spin S = 1 Ni(II) $d^8$ ions hexacoordinated in (pseudo)octahedral geometry. In addition, the signals relating the side-chain carbonyl C atoms of Asp338, Glu348, Glu358, Glu368 and Asn377 disappear from the CON and CACO spectra, while the corresponding signals for Asn383 are not affected by Ni(II) binding. Finally, the N, CO and CA signals of the C-terminal Cys394 are completely obliterated upon Ni(II) binding, which is an indication that its thiolate group and/or the carboxylate C-terminus are involved in Ni(II) binding. Overall, the picture that can be drawn from the NMR spectra analysis indicates the involvement of the side chains of Asp338, His345, Glu348, His355, Glu358, His365, Glu368, His371, Asp373, Asn377 and Cys394 in the uptake of Ni(II) by *h*NDRG1*C. In the following sequence, the residues observed to bind Ni(II) are shown in red, while the decapeptides are underlined:

$^{311}$GMGYMPSASMTRLMRSRTASGSSVTSLDGTRSRSHTSEGTRSRSHTSEGTRSRSHTSEGAHLDITPNSGAAGNSAGPKSMEVSC$^{394}$.

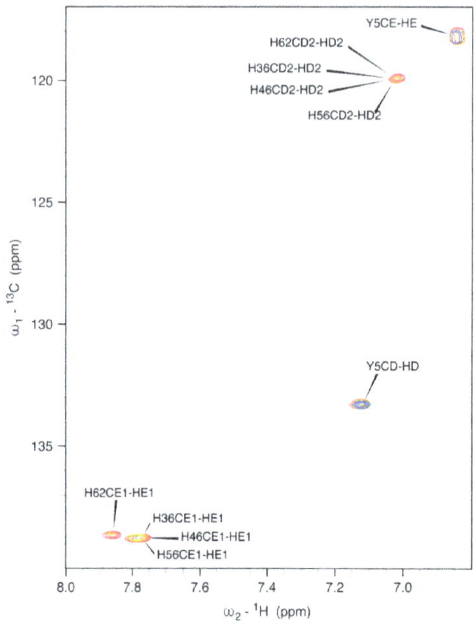


**Figure 7.** A 700 MHz (16.4 T) $^1$H,$^{13}$C HSQC spectrum of *h*NDRG1*C in the aromatic region, highlighting the signals for the side-chain imidazole rings of His345, His355, His365 and His371, as well as Tyr314, at pH 7.5, as a function of incremental addition of Ni(II) (red: 0 eq; orange: 1 eq; cyan: 2 eq; blue: 3 eq).

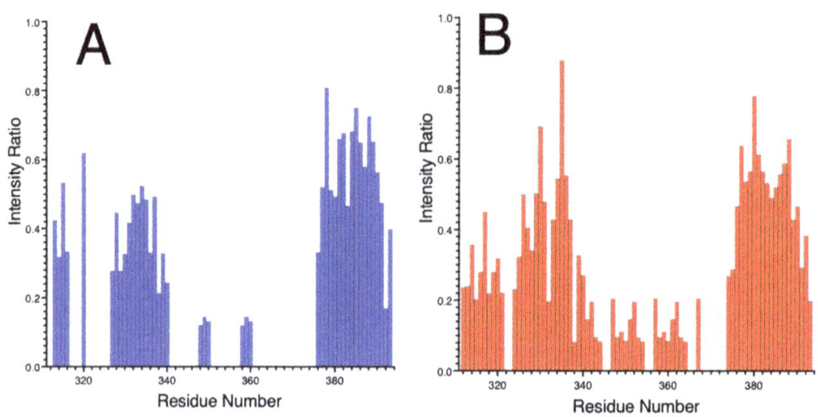


**Figure 8.** Intensity ratios of the peaks in the (**A**) CON spectrum and (**B**) CACO spectrum obtained at 700 MHz (16.4 T) by $^{13}$C direct detection at pH 7.5 in the absence and presence of 3 eq. Ni(II).

3.5.4. Effects of Ni(II) Binding by Paramagnetic NMR

The interaction between *h*NDRG1*C and Ni(II) was then investigated using $^1$H NMR spectra tailored for the observation of signals of residues bound to the Ni(II) center and affected by its paramagnetism. The spectrum of *h*NDRG1*C in the presence of four equivalents of Ni(II) at pH 7.5 and 298 K, (Figure 9) contains five hyperfine-shifted signals (A-E) in the range from +130 to −10 ppm, outside the diamagnetic region, arising from contact and pseudo-contact shifts involving high-spin (S = 1) Ni(II) centers. The chemical shifts, line widths and the Curie-type temperature dependence of the chemical shifts are consistent with the presence of a single paramagnetic Ni(II) center with S = 1 in octahedral coordination [64,86]. Signal B, at 68 ppm, disappears in $D_2O$, indicating that it belongs to an exchangeable proton; these features suggest that it belongs to either Hε2 or Hδ1 of Ni-bound histidine residues [64,86–88]. Signals C and D belong to non-exchangeable protons; their chemical shift is typical for imidazole Hε1 and Hδ2 histidine protons [64,86–88]. Signal A disappears by lowering the temperature to 288 K: its chemical shift is consistent with Hβ protons of Ni(II)-bound cysteine residues [64,89,90], suggesting the involvement of the C-terminal Cys394 thiol. The fact that its intensity varies with temperature is interpreted as indicating the presence of exchange phenomena by which Ni(II) is binding to different sites with equilibria that shift according to the available kinetic energy of the system.

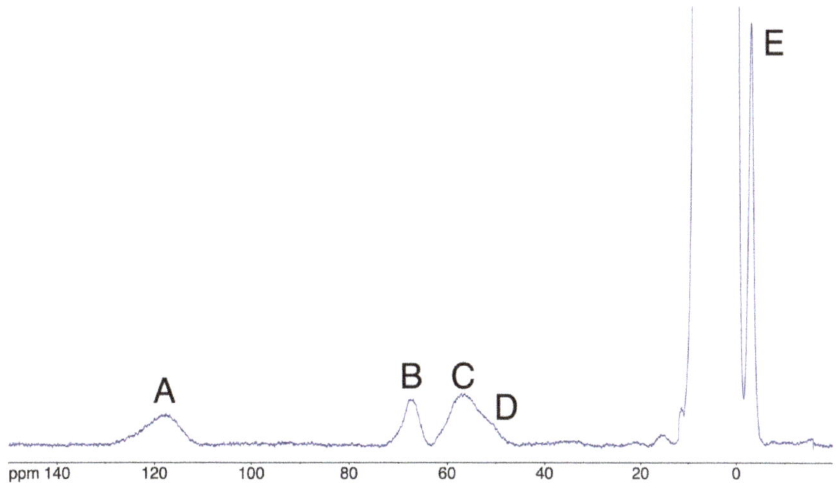


**Figure 9.** $^1$H NMR spectrum of NDRG1*C at pH 7.5 and 298 K in the presence of 4 equiv. Ni(II). Signals A, B, C, D and E are relative to nuclei sensing the hyperfine shift (contact and pseudo-contact) due to the presence of the paramagnetic Ni(II) S = 1 ion bound to *h*NDRG1-C.

### 3.6. *Light Scattering*

The hydrodynamic and oligomeric properties of *h*NDRG1 and *h*NDRG1*C in solution were determined using multiple-angle light scattering (MALS) and quasi-elastic light scattering (QELS) in line with a size-exclusion chromatography column (SEC). Elution of *h*NDRG1 occurred in three different peaks, corresponding to different oligomeric states of the protein in solution (Figure 10A). The calculated molar masses and hydrodynamic radii were MW = 150 kDa and $R_h$ = 5.3 nm, respectively, for the first eluted species. The obtained MW value is intermediate between the molar mass of the tetramer (180 kDa) and that of the trimer (135 kDa). As this species is largely superimposed to the oligomeric form with lower molar mass, an underestimation of the calculated molecular weight is expected, as also supported by the profile shown by the dots, each representing the molar mass calculated for every slice under the peaks. This observation strongly suggests that the first eluted species is indeed the tetrameric form. The second eluting peak features MW = 90.5 kDa and $R_h$ = 4 nm, corresponding to the values expected for the protein dimer. The last eluting peak is associated to MW = 47.4 kDa and $R_h$ = 2.4 nm, corresponding to the expected values for the protein monomer (theoretical MW = 45 kDa). The oligomeric equilibrium observed in solution is not significantly altered in the presence of Ni(II) (Figure 10A). No significant change in the elution volume was observed when performing the SEC experiment in the presence of 10 mM DTT, excluding that the higher oligomeric states are formed by covalent disulfide bonds.

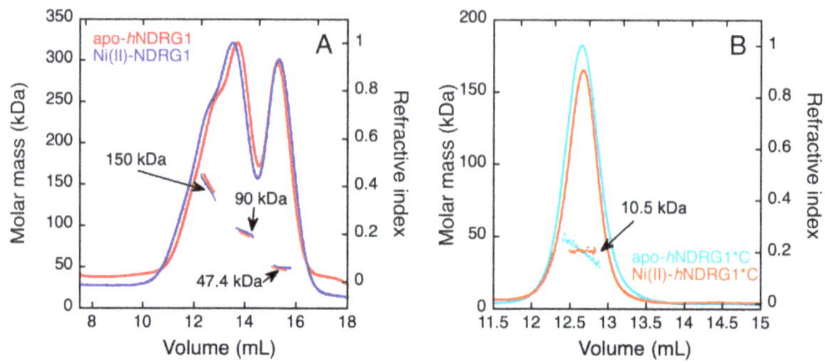


**Figure 10.** Molar mass and hydrodynamic radius determined by static and dynamic light scattering in line with a size-exclusion chromatography column. The chromatogram represents the trace of refractive index detector (lines) and the weight-averaged molar mass distribution, calculated on the eluting species, is represented as dots. The profile of *h*NDRG1 (**A**) and of *h*NDRG1*C (**B**) is reported in the absence and in the presence of four equivalents of Ni(II).

Differently from the full-length protein, elution of *h*NDRG1*C from the SEC-MALS-QELS flow occurs as a unique peak with the characteristic of the monomer (MW = 10.5 kDa, $R_h$ = 1.7 nm; theoretical MW = 8.6), both in the absence and in the presence of Ni(II) (Figure 10B), indicating that the full protein sequence is necessary to reach a multimeric form.

### 3.7. *In-Cell Experiments*

In order to confirm the physiological relevance of *h*NDRG1 oligomeric equilibrium observed in solution, the expression profile and oligomeric state of endogenous *h*NDRG1 was assayed in two human cellular lines, Hela and A549—the latter derived from human lung carcinoma—in the absence and in the presence of $NiSO_4$. At the end of the metal exposure, cells were lysed under nondenaturing conditions to preserve the stability of the *h*NDRG1 oligomers, and cellular lysates were resolved on SDS-PAGE under both denaturing and reducing conditions and nondenaturing conditions. For HeLa cells (Figure 11A), two bands at ca. 50 kDa and at ca.100 kDa, likely corresponding to the monomeric and the dimeric forms of the protein, were easily detectable. Of note, it is possible that all the faint bands around 50 kDa-band corresponded to different phosphorylation states of *h*NDRG1 [42] or of an N-terminal truncated form [38]. At longer exposure of the film, a faint band at ca. 200 kDa corresponding to the tetrameric form of *h*NDRG1 was also detectable. When HeLa cells were cultured with Ni(II), an increase in the level of expression of the monomeric form of *h*NDRG1 and a corresponding decrease in the band corresponding to the dimer are visible, suggesting that the presence of the metal ion shifts the equilibrium toward the low-molecular-weight states.

Differently from HeLa cells, no dimeric form of *h*NDRG1 was detected in A549 cells under these conditions, while the monomeric and the tetrameric forms were well-visible both in the absence and in the presence of Ni(II) (Figure 11B). Similarly to what was observed for the HeLa cells, addition of Ni(II) caused a marked expression of the monomeric specie. No form of *h*NDRG1 was found in the nucleus, both in the absence or in the presence of Ni(II), indicating that in this cellular line and under the experimental conditions the localization of *h*NDRG1 is cytoplasmatic. As previously reported, Ni(II) exposure increased the expression of b-catenin [91] (Figure 11B).

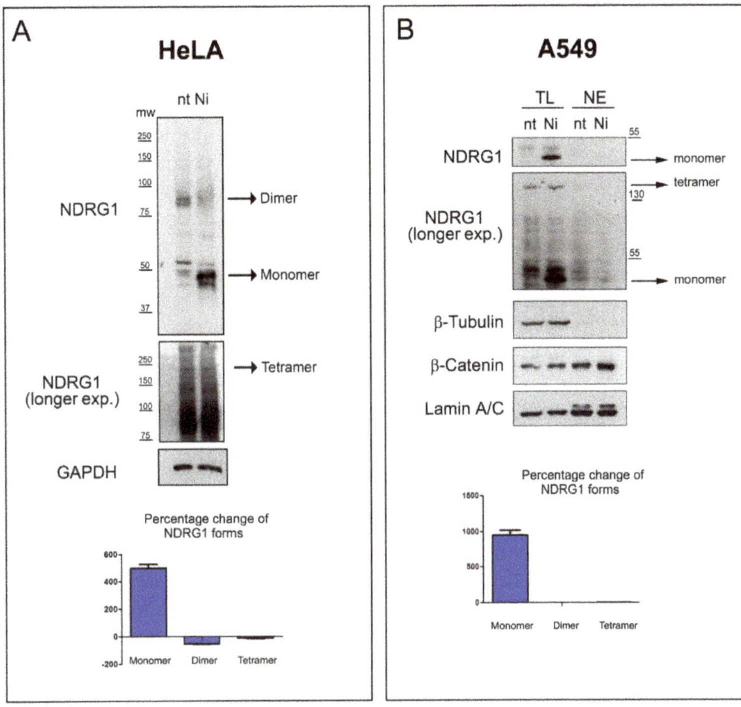


**Figure 11.** Nickel exposure promotes the expression of the monomeric form of *h*NDRG1. (**A**) HeLa cells were treated with 1 mM $NiSO_4$ for 24 h or left untreated (nt) and lysed under nondenaturing conditions, and lysates were then resolved by an SDS-PAGE under nondenaturing conditions. Corresponding filters were incubated with an antibody against the N-terminal region of NDRG1. GAPDH was used as equal loading marker. Histogram indicates the fold variation of the indicated forms of NDRG1 after nickel exposure compared to the untreated counterparts. (**B**) Lung carcinoma-derived A549 cells were treated with 1 mM $NiSO_4$ for two days (Ni) or left untreated (nt). Total lysates (TL) and nuclear extracts (NE) were analyzed in SDS-PAGE and the amount of NDRG1 was evaluated by Western blot analysis. β-tubulin and lamin A/C indicated the purity of nuclear extraction and equal loading. Histogram indicates the fold variation of the indicated forms of *h*NDRG1 after nickel exposure compared to the untreated counterparts. Western blots represent the most representative images of four repetitions of the same experiment.

## 4. Discussion

Nickel is an essential element for unicellular organisms and plants, being responsible for the catalytic activity of several enzymes, many involved in bacterial pathogenesis [92], and being tightly regulated intracellularly [93]. For humans, nickel is considered a dangerous metal ion, responsible of several pathologies such as immunotoxicity and cancer [94]. The carcinogenic effect of nickel for the respiratory tissues has been observed for more than thirty years, but the molecular mechanisms that cause nickel-driven carcinogenesis are still unclear [95]. Ni(II)-induced cellular damage occurs mainly through epigenetic mechanisms [5]. One of the possible pathways is the ability of Ni(II) to substitute cognate Fe(II) ions in metal-binding enzymes responsible for a balanced epigenetic landscape and for the regulation of gene expression [96]. Understanding how Ni(II) binds its intracellular targets is thus a necessary step to unravel the molecular basis of its carcinogen effects and to develop antitumoral drugs for detoxifying it.

One of the promising intracellular targets for Ni(II) is *h*NDRG1 [12,28], a protein that is induced by Ni(II) through the hypoxia response pathway, showing an oncogenic effect in lung carcinomas and responding to iron-chelation therapy [15]. *h*NDRG1 contains a unique C-terminal region (*h*NDRG1*C) of 83 residues reported to be very flexible and able to bind Ni(II) [39]. The intrinsically disordered behavior of this protein portion is reflected in its primary structure, which shows high content of charged and hydrophilic residues and low abundance of hydrophobic amino acids. It is known that IDRs, identified in all living organisms, play important roles in the regulation of cellular metabolisms and gene expression, usually present very large interactomes and are linked to the progress of several diseases such as cancer [97–99]. Head and neck cancer cells transfected with the *h*NDRG1 gene truncated in the sequence coding for the C-terminal 338–394 residues showed remarkable lower migration and invasion abilities, as compared to the same cells transfected with the full-length *h*NDRG1 gene, indicating that the C-terminal IDR plays a crucial role for facilitating cell motility, a hallmark for cancer metastasis and progression [46]. In addition, deletion of this sequence abolished *h*NDRG1 nuclear translocation, which was reported to promote motility [46], also supporting the role of *h*NDRG1*C for promoting the carcinogenic process. In the present work, we characterized the flexible behavior of *h*NDRG1*C and studied its Ni(II) binding activity.

The protocol for the overexpression and purification to homogeneity of the wild-type *h*NDRG1*C, initially fused to a N-terminal ZZ-tag and subsequently cleaved, is reported. The absence of stable secondary and tertiary structures was experimentally proven using CD and NMR spectroscopies. In particular, the far-UV CD spectra were quantitatively analyzed and are typical of a highly flexible polypeptide with almost 70% of the protein structure attributed to random coil conformation. The $^{1}H$ $^{15}N$ HSQC spectra of *h*NDRG1*C are characteristic of an intrinsically disordered protein, with low signal dispersion in the $^{1}H$ dimension. This observation is maintained from pH 6.5 to 7.5 and confirms that this region is dominated by random coil conformations, lacking a well-defined structure, as predicted by the in silico disorder prediction analysis. Assignment of the NMR signals, initially performed in the $^{1}H$,$^{15}N$-HSQC spectrum at pH 6.5 then translated to the more physiological pH 7.5, and analysis of the secondary structure propensity from the chemical shift analysis, confirmed the prevalence of random coil structures, with a small helical propensity in the N-terminal region.

In solution, *h*NDRG1*C behaves as a monomer. Differently, SEC-MALS data show that the full-length *h*NDRG1 exists in solution in equilibrium between three oligomeric forms: tetramer, dimer and monomer. These forms were also identified for the natively expressed protein in lung adenocarcinoma and in HeLa cells, implying that the oligomeric states observed in solutions are not an artifact of the experimental conditions, such as the high protein concentration. The homologous protein *h*NDRG3 was reported to form dimers in solution [34], while *h*NDRG2 was observed as a monomer [33]. A similar SEC-MALS analysis on the full-length and truncated variants of *h*NDRG1 showed that these proteins eluted mostly in a single peak corresponding to the monomeric form, while a minor peak, attributed to a dimer, was observed for two truncated variants [35]. No tetrameric form was observed under these conditions, likely because the protein amount injected was ca. 20 times lower as compared to the present work (100 μg in [35] vs. 1.9 mg in this work), which implies that a significantly lower protein concentration was attained in the SEC column. This observation is consistent with a concentration-dependent oligomeric equilibrium.

In the hypothesis that Ni(II) ions exert a physiological or pathological cellular role through binding to *h*NDRG1, Ni(II)-binding capacity of this protein was previously studied using ITC, showing a single binding event with $K_D$ at ca. 100 μM at pH 7.0 [35]. A similar binding event was observed for a truncated variant lacking the C-terminal region, leading the authors of this past study to conclude that the Ni(II)-binding site of *h*NDRG1 was located in the globular $\alpha/\beta$ hydrolase-like domain and not in the C-terminal region [35]. This observation disagrees with previously reported experiments, which indicated that the three-fold-repeated decapeptide (3R-motif) contained in *h*NDRG1*C is able to bind Ni(II) in a square planar diamagnetic coordination using the His imidazole ring and three amides from the protein backbone [30,39]. Indeed, the data on the full-length protein could have been affected by the presence, in the recombinant *h*NDRG1, of a non-native His2 residue, deriving from the cloning procedure and the subsequent TEV protease cleavage of the N-terminal His-tag used for the purification [35]. This residue typically forms a Ni-hook that is known to have substantial Ni(II) binding affinity, forming a non-native metal-binding site that could shield or alter the physiological Ni(II)-binding site in *h*NDRG1, located in *h*NDRG1*C [100]. On the other hand, previous studies on the Ni(II)-binding activity on the 3R-motif did not consider the entire *h*NDRG1*C region; rather, they only analyzed the repeated sequence containing three histidine residues, corresponding to His335, His345 and His355 [30,39]. Notably, these studies reported the ability of the 3R-motif to bind three Ni(II) ions, with each decapeptide repeat being the minimum motif for Ni(II) binding, implying that there is not any chelation effect [30,39]. The results of these studies are affected by the absence, in the investigated peptides, of an additional histidine (His371) and a C-terminal Cys394, potential Ni(II) binding residues that are instead present in the primary structure of *h*NDRG1*C.

In the present study, the Ni(II)-binding capacity of the *entire* *h*NDRG1*C was confirmed and investigated using ITC and NMR. ITC data, obtained by averaging the thermodynamic parameters derived from four independent experiments, clearly show that *h*NDRG1 is able to bind 1–2 Ni(II) ions in a single binding event with mild affinity ($K_D$ ca. 70 μM). The difficulty in establishing the exact stoichiometry using ITC is most likely due to the absence, in the protein primary structure, of Trp residues and to the presence of very few aromatic residues (one Tyr and three Phe), which makes significant the relative error in estimating protein concentration by absorbance at 280 nm. The binding is enthalpically driven and shows negative entropy, suggesting that Ni(II) binding induces some conformational rearrangement. CD and NMR spectroscopies, however, did not reveal any major change in the protein backbone upon Ni(II) binding, nor any acquisition of structure of the disordered protein region, which was previously suggested [35]. The content of Ni(II) in lung tissue, measured as 20 ng/g and 8–330 ng/g of wet tissue in non-occupationally exposed subjects [101,102], corresponds to an approximative Ni(II) intracellular concentration ranging from 80 nM to 1.25 μM. Higher concentrations are possible for people inhaling nickel compounds from pollution, cigarette smoke or occupational exposure. Thus, the affinity measured for *h*NDGR1*C might be significant, especially under pathological conditions.

Ni(II) binding results in the disappearance of several signals in the diamagnetic region of the NMR spectra, assigned to the side-chain and backbone nuclei of the four histidine residues found in the protein sequence, as well as to residues immediately preceding and following them; moreover, the NMR signals of nuclei belonging to the side chains of one aspartate, three glutamates and one asparagine residue, as well as those of the terminal single cysteine, also disappear upon Ni(II) binding. This observation suggests that Ni(II) is bound to *h*NDRG1*C in an octahedral or square-pyramidal geometry, resulting in a paramagnetic metal site. This conclusion was confirmed by the observation of large hyperfine shifts in NMR spectra tailored for the detection of fast-relaxing proton signals. The large number of residues experiencing the paramagnetic effect of the bound metal ion suggests the coexistence of different metal-bound conformation in solution separated by a relatively flat energy landscape and possibly undergoing intramolecule metal transfer between different binding sites. This situation has been observed as typical for IDPs or IDRs, often showing a low affinity and highly dynamic binding sites for metal ions [103,104].

In addition to the binding of Ni(II) [30,39], a direct interaction of the 3R-peptide has been observed with Cu(II) [40], Zn(II) [105], Co(II) and Mn(II) [106]. In all cases, the binding center of the peptide fragment is associated with histidine and glutamate residues. Even though these studies clearly indicate the ability of the C-terminal region of *h*NDRG1 to interact with diverse divalent metal ions, they cannot be used to compare the results with the Ni(II)-binding affinity found here for *h*NDRG1*C, due to the difference in primary structure between the 3R-motif and the *h*NDRG1*C sequence, as described above. In addition, while Ni(II) and Co(II) induce the expression of *h*NDRG1 [107], as well as Fe(II) chelation [108], no in vivo effect has been reported for Cu(II), Mn(II) and Zn(II). Interestingly, a previous study showed that Ni(II) and Co(II) were able to affect the stability of the isolated full-length *h*NDRG1, while Fe(II), Fe(III) and Mg(II) did not show any effect and the isolated protein precipitated in the presence of Zn(II) [35].

It is known that *h*NDRG1*C is phosphorylated in vivo on serine or threonine residues, this post-translational modification strongly determining its ability to promote nuclear localization of *h*NDRG1 and cell migration [46]. It is likely that this important change in protein functionality reflects a significant modification of protein folding and/or interactions such as Ni(II)-binding affinity that can be affected by the presence of covalently bound phosphate groups. This observation can give reason of the relatively low Ni(II)-binding affinity reported for nonphosphorylated *h*NDRG1*C and *h*NDRG1 [35]. The study of the conformational changes associated to *h*NDRG1*C phosphorylation is currently under development in our laboratory.

**Supplementary Materials:** The following supporting information can be downloaded at: https://www.mdpi.com/article/10.3390/biom12091272/s1.

**Author Contributions:** Conceptualization, S.C. and B.Z.; data curation, Y.B., V.C., M.P., S.C. and B.Z.; formal analysis, V.C., M.P., S.C. and B.Z.; funding acquisition, S.C. and B.Z.; investigation, Y.B., S.C. and B.Z.; methodology, Y.B.; project administration, B.Z.; supervision, B.Z.; writing—original draft, S.C. and B.Z. All authors have read and agreed to the published version of the manuscript.

**Funding:** This work was funded by CARISBO foundation, grant 2019.0558. The authors are grateful to CIRMMP (University of Firenze, Italy) that provided access to the NMR facility, which was funded by the INSTRUCT-Eric (PID: 12704).

**Acknowledgments:** Fabio Calogiuri and Massimo Lucci are duly acknowledged for their support in collecting all NMR spectra at CERM—Centre for Magnetic Resonance—University of Firenze, Italy. The authors are grateful to Frans Mulder for insightful advice in the analysis of NMR chemical shifts.

**Conflicts of Interest:** The authors declare no conflict of interest.

## References

1. Siegel, R.L.; Miller, K.D.; Fuchs, H.E.; Jemal, A. Cancer Statistics, 2021. *CA Cancer J. Clin.* **2021**, *71*, 7–33. [CrossRef] [PubMed]
2. IARC. *Outdoor Air Pollution*; International Angency for Research on Cancer: Lyon, France, 2016; Volume 109. Available online: https://publications.iarc.fr/Book-And-Report-Series/Iarc-Monographs-On-The-Identification-Of-Carcinogenic-Hazards-To-Humans/Outdoor-Air-Pollution-2015 (accessed on 31 August 2022).
3. IARC. Nickel and nickel compounds. In *Arsenic, Metals, Fibres, and Dusts*; IARC: Lyon, France, 2012; Volume 100C.
4. Zambelli, B.; Uversky, V.N.; Ciurli, S. Nickel impact on human health: An intrinsic disorder perspective. *Biochim. Biophys. Acta* **2016**, *1864*, 1714–1731. [CrossRef] [PubMed]
5. Chen, Q.Y.; Brocato, J.; Laulicht, F.; Costa, M. Mechanisms of Nickel Carcinogenesis. *Molec. Integr. Toxicol.* **2017**, *15*, 181–197. [CrossRef]
6. Maxwell, P.; Salnikow, K. HIF-1: An oxygen and metal responsive transcription factor. *Cancer Biol. Ther.* **2004**, *3*, 29–35. [CrossRef]
7. Noguchi, C.T. Is there something fishy about EPO? *Blood* **2004**, *104*, 1238. [CrossRef]
8. Salceda, S.; Caro, J. Hypoxia-inducible factor 1alpha (HIF-1alpha) protein is rapidly degraded by the ubiquitin-proteasome system under normoxic conditions. Its stabilization by hypoxia depends on redox-induced changes. *J. Biol. Chem.* **1997**, *272*, 22642–22647. [CrossRef]
9. Wang, Q.; Li, L.H.; Gao, G.D.; Wang, G.; Qu, L.; Li, J.G.; Wang, C.M. HIF-1alpha up-regulates NDRG1 expression through binding to NDRG1 promoter, leading to proliferation of lung cancer A549 cells. *Mol. Biol. Rep.* **2013**, *40*, 3723–3729. [CrossRef]
10. Zhang, J.; Chen, S.; Zhang, W.; Zhang, J.; Liu, X.; Shi, H.; Che, H.; Wang, W.; Li, F.; Yao, L. Human differentiation-related gene NDRG1 is a Myc downstream-regulated gene that is repressed by Myc on the core promoter region. *Gene* **2008**, *417*, 5–12. [CrossRef]
11. Vervoorts, J.; Luscher-Firzlaff, J.; Luscher, B. The ins and outs of MYC regulation by posttranslational mechanisms. *J. Biol. Chem.* **2006**, *281*, 34725–34729. [CrossRef]
12. Geleta, B.; Makonnen, E. N-myc downstream regulated gene (NDRG):role in cancer metastasis suppression and as drug target in cancer therapeutics. *J. Cancer Sci. Ther.* **2016**, *8*, 154–159. [CrossRef]

13. Lachat, P.; Shaw, P.; Gebhard, S.; van Belzen, N.; Chaubert, P.; Bosman, F.T. Expression of NDRG1, a differentiation-related gene, in human tissues. *Histochem. Cell Biol.* **2002**, *118*, 399–408. [CrossRef]
14. Sun, J.; Zhang, D.; Bae, D.H.; Sahni, S.; Jansson, P.; Zheng, Y.; Zhao, Q.; Yue, F.; Zheng, M.; Kovacevic, Z.; et al. Metastasis suppressor, NDRG1, mediates its activity through signaling pathways and molecular motors. *Carcinogenesis* **2013**, *34*, 1943–1954. [CrossRef]
15. Fang, B.A.; Kovacevic, Z.; Park, K.C.; Kalinowski, D.S.; Jansson, P.J.; Lane, D.J.; Sahni, S.; Richardson, D.R. Molecular functions of the iron-regulated metastasis suppressor, NDRG1, and its potential as a molecular target for cancer therapy. *Biochim. Biophys. Acta* **2014**, *1845*, 1–19. [CrossRef]
16. Said, H.M.; Safari, R.; Al-Kafaji, G.; Ernestus, R.I.; Lohr, M.; Katzer, A.; Flentje, M.; Hagemann, C. Time- and oxygen-dependent expression and regulation of NDRG1 in human brain cancer cells. *Oncol. Rep.* **2017**, *37*, 3625–3634. [CrossRef]
17. Ring, B.Z.; Seitz, R.S.; Beck, R.; Shasteen, W.J.; Tarr, S.M.; Cheang, M.C.; Yoder, B.J.; Budd, G.T.; Nielsen, T.O.; Hicks, D.G.; et al. Novel prognostic immunohistochemical biomarker panel for estrogen receptor-positive breast cancer. *J. Clin. Oncol.* **2006**, *24*, 3039–3047. [CrossRef]
18. Strzelczyk, B.; Szulc, A.; Rzepko, R.; Kitowska, A.; Skokowski, J.; Szutowicz, A.; Pawelczyk, T. Identification of high-risk stage II colorectal tumors by combined analysis of the NDRG1 gene expression and the depth of tumor invasion. *Ann. Surg Oncol.* **2009**, *16*, 1287–1294. [CrossRef]
19. Sun, B.; Chu, D.; Li, W.; Chu, X.; Li, Y.; Wei, D.; Li, H. Decreased expression of NDRG1 in glioma is related to tumor progression and survival of patients. *J. Neurooncol.* **2009**, *94*, 213–219. [CrossRef]
20. Chen, Z.; Zhang, D.; Yue, F.; Zheng, M.; Kovacevic, Z.; Richardson, D.R. The iron chelators Dp44mT and DFO inhibit TGF-beta-induced epithelial-mesenchymal transition via up-regulation of N-Myc downstream-regulated gene 1 (NDRG1). *J. Biol. Chem.* **2012**, *287*, 17016–17028. [CrossRef]
21. Liu, W.; Xing, F.; Iiizumi-Gairani, M.; Okuda, H.; Watabe, M.; Pai, S.K.; Pandey, P.R.; Hirota, S.; Kobayashi, A.; Mo, Y.Y.; et al. N-myc downstream regulated gene 1 modulates Wnt-beta-catenin signalling and pleiotropically suppresses metastasis. *EMBO Mol. Med.* **2012**, *4*, 93–108. [CrossRef]
22. Nishio, S.; Ushijima, K.; Tsuda, N.; Takemoto, S.; Kawano, K.; Yamaguchi, T.; Nishida, N.; Kakuma, T.; Tsuda, H.; Kasamatsu, T.; et al. Cap43/NDRG1/Drg-1 is a molecular target for angiogenesis and a prognostic indicator in cervical adenocarcinoma. *Cancer Lett.* **2008**, *264*, 36–43. [CrossRef]
23. Chua, M.S.; Sun, H.; Cheung, S.T.; Mason, V.; Higgins, J.; Ross, D.T.; Fan, S.T.; So, S. Overexpression of NDRG1 is an indicator of poor prognosis in hepatocellular carcinoma. *Mod. Pathol.* **2007**, *20*, 76–83. [CrossRef]
24. Masuda, K.; Ono, M.; Okamoto, M.; Morikawa, W.; Otsubo, M.; Migita, T.; Tsuneyoshi, M.; Okuda, H.; Shuin, T.; Naito, S.; et al. Downregulation of Cap43 gene by von Hippel-Lindau tumor suppressor protein in human renal cancer cells. *Int. J. Cancer* **2003**, *105*, 803–810. [CrossRef]
25. Azuma, K.; Kawahara, A.; Hattori, S.; Taira, T.; Tsurutani, J.; Watari, K.; Shibata, T.; Murakami, Y.; Takamori, S.; Ono, M.; et al. NDRG1/Cap43/Drg-1 may predict tumor angiogenesis and poor outcome in patients with lung cancer. *J. Thorac. Oncol.* **2012**, *7*, 779–789. [CrossRef]
26. Dai, T.; Dai, Y.; Murata, Y.; Husni, R.E.; Nakano, N.; Sakashita, S.; Noguchi, M. The prognostic significance of N-myc downregulated gene 1 in lung adenocarcinoma. *Pathol. Int.* **2018**, *68*, 224–231. [CrossRef]
27. Wang, H.; Li, W.; Xu, J.; Zhang, T.; Zuo, D.; Zhou, Z.; Lin, B.; Wang, G.; Wang, Z.; Sun, W.; et al. NDRG1 inhibition sensitizes osteosarcoma cells to combretastatin A-4 through targeting autophagy. *Cell Death Dis.* **2017**, *8*, e3048. [CrossRef]
28. Bae, D.H.; Jansson, P.J.; Huang, M.L.; Kovacevic, Z.; Kalinowski, D.; Lee, C.S.; Sahni, S.; Richardson, D.R. The role of NDRG1 in the pathology and potential treatment of human cancers. *J. Clin. Pathol.* **2013**, *66*, 911–917. [CrossRef]
29. Zoroddu, M.A.; Kowalik-Jankowska, T.; Kozlowski, H.; Salnikow, K.; Costa, M. Ni(II) and Cu(II) binding with a 14-aminoacid sequence of Cap43 protein, TRSRSHTSEGTRSR. *J. Inorg. Biochem.* **2001**, *84*, 47–54. [CrossRef]
30. Zoroddu, M.A.; Peana, M.; Medici, S.; Anedda, R. An NMR study on nickel binding sites in Cap43 protein fragments. *Dalton. Trans.* **2009**, *28*, 5523–5534. [CrossRef]
31. Shimono, A.; Okuda, T.; Kondoh, H. N-myc-dependent repression of ndr1, a gene identified by direct subtraction of whole mouse embryo cDNAs between wild type and N-myc mutant. *Mech. Dev.* **1999**, *83*, 39–52. [CrossRef]
32. Shaw, E.; McCue, L.A.; Lawrence, C.E.; Dordick, J.S. Identification of a novel class in the alpha/beta hydrolase fold superfamily: The N-myc differentiation-related proteins. *Proteins 2* **2002**, *47*, 163–168. [CrossRef]
33. Hwang, J.; Kim, Y.; Kang, H.B.; Jaroszewski, L.; Deacon, A.M.; Lee, H.; Choi, W.C.; Kim, K.J.; Kim, C.H.; Kang, B.S.; et al. Crystal structure of the human N-Myc downstream-regulated gene 2 protein provides insight into its role as a tumor suppressor. *J. Biol. Chem.* **2011**, *286*, 12450–12460. [CrossRef]
34. Kim, K.R.; Kim, K.A.; Park, J.S.; Jang, J.Y.; Choi, Y.; Lee, H.H.; Lee, D.C.; Park, K.C.; Yeom, Y.I.; Kim, H.J.; et al. Structural and Biophysical Analyses of Human N-Myc Downstream-Regulated Gene 3 (NDRG3) Protein. *Biomolecules* **2020**, *10*, 90. [CrossRef]
35. Mustonen, V.; Muruganandam, G.; Loris, R.; Kursula, P.; Ruskamo, S. Crystal and solution structure of NDRG1, a membrane-binding protein linked to myelination and tumour suppression. *FEBS J.* **2020**, *288*, 3507–3529. [CrossRef]
36. Lee, J.E.; Kim, J.H. SUMO modification regulates the protein stability of NDRG1. *Biochem. Biophys. Res. Commun.* **2015**, *459*, 161–165. [CrossRef]

37. Ghalayini, M.K.; Dong, Q.; Richardson, D.R.; Assinder, S.J. Proteolytic cleavage and truncation of NDRG1 in human prostate cancer cells, but not normal prostate epithelial cells. *Biosci. Rep.* **2013**, *33*, e00042. [CrossRef]
38. Park, K.C.; Menezes, S.V.; Kalinowski, D.S.; Sahni, S.; Jansson, P.J.; Kovacevic, Z.; Richardson, D.R. Identification of differential phosphorylation and sub-cellular localization of the metastasis suppressor, NDRG1. *Biochim. Biophys. Acta Mol. Basis. Dis.* **2018**, *1864*, 2644–2663. [CrossRef]
39. Zoroddu, M.A.; Peana, M.; Kowalik-Jankowska, T.; Kozlowski, H.; Costa, M. Nickel(II) binding to Cap43 protein fragments. *J. Inorg. Biochem.* **2004**, *98*, 931–939. [CrossRef]
40. Zoroddu, M.A.; Kowalik-Jankowska, T.; Medici, S.; Peana, M.; Kozlowski, H. Copper(II) binding to Cap43 protein fragments. *Dalton Trans.* **2008**, 6127–6134. [CrossRef]
41. Sugiki, T.; Murakami, M.; Taketomi, Y.; Kikuchi-Yanoshita, R.; Kudo, I. N-myc downregulated gene 1 is a phosphorylated protein in mast cells. *Biol. Pharm. Bull.* **2004**, *27*, 624–627. [CrossRef]
42. Murray, J.T.; Campbell, D.G.; Morrice, N.; Auld, G.C.; Shpiro, N.; Marquez, R.; Peggie, M.; Bain, J.; Bloomberg, G.B.; Grahammer, F.; et al. Exploitation of KESTREL to identify NDRG family members as physiological substrates for SGK1 and GSK3. *Biochem. J.* **2004**, *384*, 477–488. [CrossRef]
43. McCaig, C.; Potter, L.; Abramczyk, O.; Murray, J.T. Phosphorylation of NDRG1 is temporally and spatially controlled during the cell cycle. *Biochem. Biophys. Res. Commun.* **2011**, *411*, 227–234. [CrossRef] [PubMed]
44. Banz, V.M.; Medova, M.; Keogh, A.; Furer, C.; Zimmer, Y.; Candinas, D.; Stroka, D. Hsp90 transcriptionally and post-translationally regulates the expression of NDRG1 and maintains the stability of its modifying kinase GSK3beta. *Biochim. Biophys. Acta* **2009**, *1793*, 1597–1603. [CrossRef] [PubMed]
45. Murakami, Y.; Hosoi, F.; Izumi, H.; Maruyama, Y.; Ureshino, H.; Watari, K.; Kohno, K.; Kuwano, M.; Ono, M. Identification of sites subjected to serine/threonine phosphorylation by SGK1 affecting N-myc downstream-regulated gene 1 (NDRG1)/Cap43-dependent suppression of angiogenic CXC chemokine expression in human pancreatic cancer cells. *Biochem. Biophys. Res. Commun.* **2010**, *396*, 376–381. [CrossRef]
46. You, G.R.; Chang, J.T.; Li, H.F.; Cheng, A.J. Multifaceted and Intricate Oncogenic Mechanisms of NDRG1 in Head and Neck Cancer Depend on Its C-Terminal 3R-Motif. *Cells* **2022**, *11*, 1581. [CrossRef]
47. Miraula, M.; Ciurli, S.; Zambelli, B. Intrinsic disorder and metal binding in UreG proteins from Archae hyperthermophiles: GTPase enzymes involved in the activation of Ni(II) dependent urease. *J. Biol. Inorg. Chem.* **2015**, *20*, 739–755. [CrossRef]
48. Bogomolovas, J.; Simon, B.; Sattler, M.; Stier, G. Screening of fusion partners for high yield expression and purification of bioactive viscotoxins. *Protein Expr. Purif.* **2009**, *64*, 16–23. [CrossRef] [PubMed]
49. Studier, F.W. Protein production by auto-induction in high density shaking cultures. *Protein Expr Purif* **2005**, *41*, 207–234. [CrossRef] [PubMed]
50. Azatian, S.B.; Kaur, N.; Latham, M.P. Increasing the buffering capacity of minimal media leads to higher protein yield. *J. Biomol. NMR* **2019**, *73*, 11–17. [CrossRef]
51. Marley, J.; Lu, M.; Bracken, C. A method for efficient isotopic labeling of recombinant proteins. *J. Biomol. NMR* **2001**, *20*, 71–75. [CrossRef] [PubMed]
52. Stola, M.; Musiani, F.; Mangani, S.; Turano, P.; Safarov, N.; Zambelli, B.; Ciurli, S. The nickel site of Bacillus pasteurii UreE, a urease metallo-chaperone, as revealed by metal-binding studies and X-ray absorption spectroscopy. *Biochemistry* **2006**, *45*, 6495–6509. [CrossRef]
53. Micsonai, A.; Wien, F.; Bulyaki, E.; Kun, J.; Moussong, E.; Lee, Y.H.; Goto, Y.; Refregiers, M.; Kardos, J. BeStSel: A web server for accurate protein secondary structure prediction and fold recognition from the circular dichroism spectra. *Nucleic Acids Res.* **2018**, *46*, W315–W322. [CrossRef] [PubMed]
54. Bermel, W.; Bertini, I.; Csizmok, V.; Felli, I.C.; Pierattelli, R.; Tompa, P. H-start for exclusively heteronuclear NMR spectroscopy: The case of intrinsically disordered proteins. *J. Magn. Reson.* **2009**, *198*, 275–281. [CrossRef]
55. Bermel, W.; Bertini, I.; Felli, I.C.; Pierattelli, R. Speeding up (13)C direct detection biomolecular NMR spectroscopy. *J. Am. Chem. Soc.* **2009**, *131*, 15339–15345. [CrossRef]
56. Felli, I.C.; Pierattelli, R. Novel methods based on (13)C detection to study intrinsically disordered proteins. *J. Magn. Reson.* **2014**, *241*, 115–125. [CrossRef] [PubMed]
57. Delaglio, F.; Grzesiek, S.; Vuister, G.W.; Zhu, G.; Pfeifer, J.; Bax, A. NMRPipe: A multidimensional spectral processing system based on UNIX pipes. *J. Biomol. NMR* **1995**, *6*, 277–293. [CrossRef] [PubMed]
58. Ying, J.; Delaglio, F.; Torchia, D.A.; Bax, A. Sparse multidimensional iterative lineshape-enhanced (SMILE) reconstruction of both non-uniformly sampled and conventional NMR data. *J. Biomol. NMR* **2017**, *68*, 101–118. [CrossRef]
59. Lee, W.; Tonelli, M.; Markley, J.L. NMRFAM-SPARKY: Enhanced software for biomolecular NMR spectroscopy. *Bioinformatics* **2015**, *31*, 1325–1327. [CrossRef]
60. Lee, W.; Rahimi, M.; Lee, Y.; Chiu, A. POKY: A software suite for multidimensional NMR and 3D structure calculation of biomolecules. *Bioinformatics* **2021**, *37*, 3041–3042. [CrossRef]
61. Lee, W.; Westler, W.M.; Bahrami, A.; Eghbalnia, H.R.; Markley, J.L. PINE-SPARKY: Graphical interface for evaluating automated probabilistic peak assignments in protein NMR spectroscopy. *Bioinformatics* **2009**, *25*, 2085–2087. [CrossRef]
62. Bahrami, A.; Assadi, A.H.; Markley, J.L.; Eghbalnia, H.R. Probabilistic interaction network of evidence algorithm and its application to complete labeling of peak lists from protein NMR spectroscopy. *PLoS Comput. Biol.* **2009**, *5*, e1000307. [CrossRef]

63. Wang, L.; Eghbalnia, H.R.; Bahrami, A.; Markley, J.L. Linear analysis of carbon-13 chemical shift differences and its application to the detection and correction of errors in referencing and spin system identifications. *J. Biomol. NMR* **2005**, *32*, 13–22. [CrossRef] [PubMed]
64. Banci, L.; Piccioli, M. Cobalt (II)- and Nickel (II)-Substituted Proteins. In *Encyclopedia of Magnetic Resonance*; John Wiley & Sons, Ltd.: New York, NY, USA, 1996; pp. 1365–1378.
65. Charlwood, P.A. Partial Specific Volumes of Proteins in Relation to Composition and Environment. *J. Am. Chem. Soc.* **1957**, *79*, 776–781. [CrossRef]
66. Oates, M.E.; Romero, P.; Ishida, T.; Ghalwash, M.; Mizianty, M.J.; Xue, B.; Dosztanyi, Z.; Uversky, V.N.; Obradovic, Z.; Kurgan, L.; et al. D(2)P(2): Database of disordered protein predictions. *Nucleic Acids Res.* **2013**, *41*, D508–D516. [CrossRef] [PubMed]
67. Drozdetskiy, A.; Cole, C.; Procter, J.; Barton, G.J. JPred4: A protein secondary structure prediction server. *Nucleic Acids Res.* **2015**, *43*, W389–W394. [CrossRef]
68. Oughtred, R.; Rust, J.; Chang, C.; Breitkreutz, B.J.; Stark, C.; Willems, A.; Boucher, L.; Leung, G.; Kolas, N.; Zhang, F.; et al. The BioGRID database: A comprehensive biomedical resource of curated protein, genetic, and chemical interactions. *Protein. Sci.* **2021**, *30*, 187–200. [CrossRef]
69. Szklarczyk, D.; Franceschini, A.; Kuhn, M.; Simonovic, M.; Roth, A.; Minguez, P.; Doerks, T.; Stark, M.; Muller, J.; Bork, P.; et al. The STRING database in 2011: Functional interaction networks of proteins, globally integrated and scored. *Nucleic Acids Res.* **2011**, *39*, D561–D568. [CrossRef] [PubMed]
70. Mohan, A.; Oldfield, C.J.; Radivojac, P.; Vacic, V.; Cortese, M.S.; Dunker, A.K.; Uversky, V.N. Analysis of molecular recognition features (MoRFs). *J. Mol. Biol.* **2006**, *362*, 1043–1059. [CrossRef] [PubMed]
71. Dosztanyi, Z.; Meszaros, B.; Simon, I. ANCHOR: Web server for predicting protein binding regions in disordered proteins. *Bioinformatics* **2009**, *25*, 2745–2746. [CrossRef] [PubMed]
72. Uversky, V.N.; Gillespie, J.R.; Fink, A.L. Why are "natively unfolded" proteins unstructured under physiologic conditions? *Proteins* **2000**, *41*, 415–427. [CrossRef]
73. Schubert, M.; Labudde, D.; Oschkinat, H.; Schmieder, P. A software tool for the prediction of Xaa-Pro peptide bond conformations in proteins based on 13C chemical shift statistics. *J. Biomol. NMR* **2002**, *24*, 149–154. [CrossRef]
74. Shen, Y.; Bax, A. Prediction of Xaa-Pro peptide bond conformation from sequence and chemical shifts. *J. Biomol. NMR* **2010**, *46*, 199–204. [CrossRef]
75. Sharma, D.; Rajarathnam, K. 13C NMR chemical shifts can predict disulfide bond formation. *J. Biomol. NMR* **2000**, *18*, 165–171. [CrossRef] [PubMed]
76. Nielsen, J.T.; Mulder, F.A. There is Diversity in Disorder-"In all Chaos there is a Cosmos, in all Disorder a Secret Order". *Front. Mol. Biosci.* **2016**, *3*, 4. [CrossRef] [PubMed]
77. Nielsen, J.T.; Mulder, F.A.A. CheSPI: Chemical shift secondary structure population inference. *J. Biomol. NMR* **2021**, *75*, 273–291. [CrossRef]
78. Molday, R.S.; Englander, S.W.; Kallen, R.G. Primary structure effects on peptide group hydrogen exchange. *Biochemistry* **1972**, *11*, 150–158. [CrossRef] [PubMed]
79. Pelton, J.G.; Torchia, D.A.; Meadow, N.D.; Roseman, S. Tautomeric states of the active-site histidines of phosphorylated and unphosphorylated IIIGlc, a signal-transducing protein from Escherichia coli, using two-dimensional heteronuclear NMR techniques. *Protein Sci.* **1993**, *2*, 543–558. [CrossRef]
80. Li, S.; Hong, M. Protonation, tautomerization, and rotameric structure of histidine: A comprehensive study by magic-angle-spinning solid-state NMR. *J. Am. Chem. Soc.* **2011**, *133*, 1534–1544. [CrossRef]
81. Platzer, G.; Okon, M.; McIntosh, L.P. pH-dependent random coil 1H, 13C, and 15N chemical shifts of the ionizable amino acids: A guide for protein pKa measurements. *J. Biomol. NMR* **2014**, *60*, 109–129. [CrossRef]
82. Dass, R.; Corliano, E.; Mulder, F.A.A. The contribut.tion of electrostatics to hydrogen exchange in the unfolded protein state. *Biophys. J.* **2021**, *120*, 4107–4114. [CrossRef]
83. Englander, S.W.; Kallenbach, N.R. Hydrogen exchange and structural dynamics of proteins and nucleic acids. *Q. Rev. Biophys.* **1983**, *16*, 521–655. [CrossRef]
84. Bai, Y.; Milne, J.S.; Mayne, L.; Englander, S.W. Primary structure effects on peptide group hydrogen exchange. *Proteins* **1993**, *17*, 75–86. [CrossRef] [PubMed]
85. Connelly, G.P.; Bai, Y.; Jeng, M.F.; Englander, S.W. Isotope effects in peptide group hydrogen exchange. *Proteins* **1993**, *17*, 87–92. [CrossRef] [PubMed]
86. Spronk, C.; Zerko, S.; Gorka, M.; Kozminski, W.; Bardiaux, B.; Zambelli, B.; Musiani, F.; Piccioli, M.; Basak, P.; Blum, F.C.; et al. Structure and dynamics of Helicobacter pylori nickel-chaperone HypA: An integrated approach using NMR spectroscopy, functional assays and computational tools. *J. Biol. Inorg. Chem.* **2018**, *23*, 1309–1330. [CrossRef]
87. Ming, L.-J.; Banci, L.; Luchinat, C.; Bertini, I.; Valentine, J.S. Characterization of copper-nickel and silvernickel bovine superoxide dismutase by 1H NMR spectroscopy. *Inorg. Chem.* **1988**, *27*, 4458–4463. [CrossRef]
88. Donaire, A.; Salgado, J.; Moratal, J.M. Determination of the magnetic axes of cobalt(II) and nickel(II) azurins from 1H NMR data: Influence of the metal and axial ligands on the origin of magnetic anisotropy in blue copper proteins. *Biochemistry* **1998**, *37*, 8659–8673. [CrossRef]

89. Salgado, J.; Kalverda, A.P.; Diederix, R.E.; Canters, G.W.; Moratal, J.M.; Lawler, A.T.; Dennison, C. Paramagnetic NMR investigations of Co(II) and Ni(II) amicyanin. *J. Biol. Inorg. Chem.* **1999**, *4*, 457–467. [CrossRef] [PubMed]
90. Goodfellow, B.J.; Duarte, I.C.; Macedo, A.L.; Volkman, B.F.; Nunes, S.G.; Moura, I.; Markley, J.L.; Moura, J.J. An NMR structural study of nickel-substituted rubredoxin. *J. Biol. Inorg. Chem.* **2010**, *15*, 409–420. [CrossRef] [PubMed]
91. Jose, C.C.; Jagannathan, L.; Tanwar, V.S.; Zhang, X.; Zang, C.; Cuddapah, S. Nickel exposure induces persistent mesenchymal phenotype in human lung epithelial cells through epigenetic activation of ZEB1. *Mol. Carcinog.* **2018**, *57*, 794–806. [CrossRef] [PubMed]
92. Maroney, M.J.; Ciurli, S. Bioinorganic Chemistry of Nickel. *Inorganics* **2019**, *7*, 131. [CrossRef]
93. Musiani, F.; Zambelli, B.; Bazzani, M.; Mazzei, L.; Ciurli, S. Nickel-responsive transcriptional regulators. *Metallomics* **2015**, *7*, 1305–1318. [CrossRef]
94. Zambelli, B.; Ciurli, S. Nickel and human health. *Met. Ions Life Sci.* **2013**, *13*, 321–357. [CrossRef] [PubMed]
95. Zhu, Y.; Costa, M. Metals and molecular carcinogenesis. *Carcinogenesis* **2020**, *41*, 1161–1172. [CrossRef] [PubMed]
96. Guo, H.; Liu, H.; Wu, H.; Cui, H.; Fang, J.; Zuo, Z.; Deng, J.; Li, Y.; Wang, X.; Zhao, L. Nickel Carcinogenesis Mechanism: DNA Damage. *Int. J. Mol. Sci.* **2019**, *20*, 4690. [CrossRef] [PubMed]
97. Patil, A.; Nakamura, H. Disordered domains and high surface charge confer hubs with the ability to interact with multiple proteins in interaction networks. *FEBS Lett.* **2006**, *580*, 2041–2045. [CrossRef]
98. Choudhary, S.; Lopus, M.; Hosur, R.V. Targeting disorders in unstructured and structured proteins in various diseases. *Biophys. Chem.* **2022**, *281*, 106742. [CrossRef]
99. Zou, H.; Pan, T.; Gao, Y.; Chen, R.; Li, S.; Guo, J.; Tian, Z.; Xu, G.; Xu, J.; Ma, Y.; et al. Pan-cancer assessment of mutational landscape in intrinsically disordered hotspots reveals potential driver genes. *Nucleic Acids Res.* **2022**, *50*, e49. [CrossRef]
100. Basak, P.; Zambelli, B.; Cabelli, D.E.; Ciurli, S.; Maroney, M.J. Pro5 is not essential for the formation of 'Ni-hook' in nickel superoxide dismutase. *J. Inorg. Biochem.* **2022**, *234*, 111858. [CrossRef]
101. Rezuke, W.N.; Knight, J.A.; Sunderman, F.W., Jr. Reference values for nickel concentrations in human tissues and bile. *Am. J. Ind. Med.* **1987**, *11*, 419–426. [CrossRef]
102. Dudek-Adamska, D.; Lech, T.; Konopka, T.; Koscielniak, P. Nickel Content in Human Internal Organs. *Biol. Trace Elem. Res.* **2021**, *199*, 2138–2144. [CrossRef]
103. Faller, P.; Hureau, C.; La Penna, G. Metal ions and intrinsically disordered proteins and peptides: From Cu/Zn amyloid-beta to general principles. *Acc. Chem. Res.* **2014**, *47*, 2252–2259. [CrossRef]
104. Pontoriero, L.; Schiavina, M.; Murrali, M.G.; Pierattelli, R.; Felli, I.C. Monitoring the Interaction of alpha-Synuclein with Calcium Ions through Exclusively Heteronuclear Nuclear Magnetic Resonance Experiments. *Angew. Chem. Int. Ed. Engl.* **2020**, *59*, 18537–18545. [CrossRef] [PubMed]
105. Zoroddu, M.A.; Medici, S.; Peana, M.; Anedda, R. NMR studies of zinc binding in a multi-histidinic peptide fragment. *Dalton Trans.* **2010**, *39*, 1282–1294. [CrossRef] [PubMed]
106. Peana, M.; Medici, S.; Nurchi, V.M.; Crisponi, G.; Lachowicz, J.I.; Zoroddu, M.A. Manganese and cobalt binding in a multi-histidinic fragment. *Dalton Trans.* **2013**, *42*, 16293–16301. [CrossRef]
107. Salnikow, K.; Su, W.; Blagosklonny, M.V.; Costa, M. Carcinogenic metals induce hypoxia-inducible factor-stimulated transcription by reactive oxygen species-independent mechanism. *Cancer Res.* **2000**, *60*, 3375–3378. [PubMed]
108. Le, N.T.; Richardson, D.R. Iron chelators with high antiproliferative activity up-regulate the expression of a growth inhibitory and metastasis suppressor gene: A link between iron metabolism and proliferation. *Blood* **2004**, *104*, 2967–2975. [CrossRef]

*biomolecules*

MDPI

*Article*

# An Interpretable Machine-Learning Algorithm to Predict Disordered Protein Phase Separation Based on Biophysical Interactions

**Hao Cai [1], Robert M. Vernon [1] and Julie D. Forman-Kay [1,2,*]**

[1] Molecular Medicine Program, Hospital for Sick Children, Toronto, ON M5G 0A4, Canada
[2] Department of Biochemistry, University of Toronto, Toronto, ON M5S 1A8, Canada
* Correspondence: forman@sickkids.ca

**Abstract:** Protein phase separation is increasingly understood to be an important mechanism of biological organization and biomaterial formation. Intrinsically disordered protein regions (IDRs) are often significant drivers of protein phase separation. A number of protein phase-separation-prediction algorithms are available, with many being specific for particular classes of proteins and others providing results that are not amenable to the interpretation of the contributing biophysical interactions. Here, we describe LLPhyScore, a new predictor of IDR-driven phase separation, based on a broad set of physical interactions or features. LLPhyScore uses sequence-based statistics from the RCSB PDB database of folded structures for these interactions, and is trained on a manually curated set of phase-separation-driving proteins with different negative training sets including the PDB and human proteome. Competitive training for a variety of physical chemical interactions shows the greatest contribution of solvent contacts, disorder, hydrogen bonds, pi–pi contacts, and kinked beta-structures to the score, with electrostatics, cation–pi contacts, and the absence of a helical secondary structure also contributing. LLPhyScore has strong phase-separation-prediction recall statistics and enables a breakdown of the contribution from each physical feature to a sequence's phase-separation propensity, while recognizing the interdependence of many of these features. The tool should be a valuable resource for guiding experiments and providing hypotheses for protein function in normal and pathological states, as well as for understanding how specificity emerges in defining individual biomolecular condensates.

**Keywords:** biomolecular condensates; machine learning; predictor; physical interactions; intrinsically disordered proteins; phase separation

**Citation:** Cai, H.; Vernon, R.M.; Forman-Kay, J.D. An Interpretable Machine-Learning Algorithm to Predict Disordered Protein Phase Separation Based on Biophysical Interactions. *Biomolecules* **2022**, *12*, 1131. https://doi.org/10.3390/biom12081131

Academic Editors: Prakash Kulkarni, Stefania Brocca, Keith Dunker and Sonia Longhi

Received: 6 July 2022
Accepted: 12 August 2022
Published: 17 August 2022



## 1. Introduction

Protein phase separation has recently been recognized as an important mechanism of compartmentalization in cells contributing to the formation of biomolecular condensates [1,2]. Liquid–liquid phase separation (LLPS) is not the only physical phenomenon that can contribute to the formation of these condensates, with these including sol-gel transitions and phase separation coupled to percolation (PSCP) [1,3,4]. Here we use the term "phase separation" as an imprecise shorthand for these mechanisms that rely on exchanging multivalent interactions [5] that give rise to biomolecular condensates. Biomolecular condensates are found in a wide range of biological contexts, including intracellular condensates and membraneless organelles [6,7] such as signaling puncta [8,9], nuclear pores [10], transcription centers [11], and mRNA transport granules [12–14], as well as extracellular biological materials such as those in elastin [15–17], mussel foot [18,19], and squid beak [20–23]. Biomolecular condensates are also implicated in pathological aggregation (e.g., ALS [24] and Alzheimer's disease [25]).

The physical mechanistic understanding of protein phase separation in all its complexity is challenged due to the richness and versatility of its driving forces. Phase separation can be affected by a large set of sequence-dependent factors, with a significant role of intrinsically disordered protein regions (IDRs) in many cases. For phase separation driven by IDRs, numerous weak interaction forces have been highlighted to contribute, including electrostatic interactions [26–28], pi–pi stacking [29–31], cation–pi interactions [19,26,32], and hydrogen bonding [33,34], with multiple forces often implicated as being seen in low-complexity aromatic-rich kinked segments (LARKS) [33], which exhibit kinked-beta-backbone hydrogen bonding and aromatic sidechain interactions. In elastin and elastin-like peptides, the hydrophobic effect is important for phase separation [35,36]. For phase separation driven by folded domains, specific sequence motifs, SLiMs [37], and their cognate folded binding domains are key; while these are an important driver of biological phase separation, our focus here is on IDR-driven phase separation.

Since most of the physicochemical factors that facilitate phase separation are sequence-dependent, there have been numerous efforts to use statistical learning to draw physical insights from known phase-separating sequences, i.e., to predict whether a sequence will undergo phase separation by comparing it against tested sequences, as previously summarized in a 2019 review [38]. However, the algorithms mentioned in that review focus on specific categories of condensates or biophysical features, and can only predict a subset of phase-separating proteins with high confidence. There is a high level of correlation among biophysical features, e.g., pi–pi and solvent interactions [29], electrostatic interactions and hydrophobic interactions [28,39], but none of these algorithms can estimate phase-separation propensities based on all of these physical forces, limiting the overall predictive capability of these "first-generation" predictors. In subsequent work [40], a machine-learning-based prediction tool (PSPredictor) that uses word2vec sequence encoding and the Gradient Boosting Decision Tree (GBDT) model outperformed all the "first-generation predictors" and achieved a 96% prediction accuracy. However, because of the design of word2vec encoding [41], its prediction results cannot provide quantitative information about the contributions from different driving forces, and therefore it lacks clear physical interpretability. Recently, a number of additional tools have been developed to quantify phase-separation propensity. One of these, PSPer, focuses on the prediction of prion-like RNA-binding proteins that phase separate using a Hidden Markov Model (HMM) [42]. PSPer showed good predictability (0.87 Spearman correlation score between its output and the critical concentration of FUS-like proteins); however, it has limited ability to predict phase-separating proteins that are not RNA-binding. Another, ParSe, combines two physical features, the hydrodynamic size of monomeric proteins and the beta-turn propensity estimated from polymer models, to predict phase-separation propensity; however, it only uses the composition and not the residue context when making predictions [43]. A third, PSAP, uses the compositional bias of phase-separating proteins and sequence-based biochemical features to train random-forest classifier with a 0.89 AUROC (area under the receiver operating characteristics curve), yet also lacks residue context in the prediction [44].

A major issue in developing a phase-separation predictor is the selection of a negative training set. Most recently developed predictors use sequences of the folded proteins in the RCSB Protein Data Bank (PDB) [45] as the negative set [29,40]; however, this leads to a bias towards a final classification algorithm that distinguishes between intrinsically disordered proteins/regions and folded proteins, since most proteins that are found to phase separate are IDPs or have IDRs. This classification does not identify the driving forces of phase separation, however, since many IDPs/IDRs are not phase-separating. In addition, many proteins phase separate during crystallization [46]. While most of these proteins do not contain IDRs and thus likely do not phase separate due to the sequence features of IDRs within their sequences, the PDB is not an optimal phase-separation-negative set for training a predictor of IDR phase separation. To avoid the issues with the PDB, in other cases the human proteome was chosen as the negative benchmark, bringing in higher structural complexity [33,47,48]. Another computational approach that has been developed to predict the propensity to phase separate, FuzDrop, has estimated that up to 40% of the human proteome can potentially undergo phase separation under certain conditions [49].

Therefore, it is clear that training a phase-separation-prediction algorithm on negative datasets such as the human proteome or PDB could include many false negatives, leading to significant challenges.

In the present work, we based our strategy on the idea that a combination of multiple different physical interactions drives phase separation, and developed a machine-learning-based predictor (LLPhyScore) that predicts based on a set of phase-separation-related physical interactions or features. While "LLPhyScore" was named by combining the acronym "LLPS" and "physical feature-based scoring", the tool is not only focused on "liquid-liquid phase separation" but is intended as a general predictor of phase separation by various mechanisms that rely on exchanging multi-valent interactions within IDRs. We adapted the constrained training approach from our previous work on PScore [29] that focused on planar pi–pi interactions and extended it to a total of 16 (8 pairs) of physical measurements or features. The eight general features are not independent but are often discussed as separate terms. Our predictor development process was divided into two stages. In the first stage, we acquired sequence-based statistics (contact frequency/number of atoms/structure probability) from the PDB database of folded structures for each physical feature/interaction. We divided these observations by distinct residue pairs with varying sequence separation and developed a statistical method to predict the expected physical-feature values given a protein sequence. In the second stage, we trained the predictor to rank sequences by the weighted combination of the expected physical-feature values. During the predictor training, we used a genetic algorithm to optimize (i) the number of physical features to utilize in our final algorithm, (ii) the direction of contribution to the score (sign) of each feature, and (iii) the weights of each physical feature chosen for the final algorithm. The predictive model is a three-layer "neural network"-like model that infers the statistics of the input sequence based on physical features, residue types and residue counts and positions. The training revealed the better-appreciated importance of pi contacts and disorder, but also the less well-appreciated significance of solvent interactions, hydrogen bonds and kinked beta-structure. In order to address the "imperfect negative dataset" issue, we used three different negative training sets: the PDB, a curated human proteome, and a mixture of both the PDB and a curated human proteome, and examined their impact on the final predictor's performance. The final predictor (LLPhyScore) achieved excellent predictive power (AUROC of 0.978) and demonstrated significant physical interpretability by providing a breakdown of the contribution from each physical feature to a sequence's propensity to phase separate.

## 2. Methods

### 2.1. Data Preparation

The overall data preparation workflow is shown Figure 1 and includes the following four parts:

(1) The curation of phase-separation-positive (PS-positive) sequences: In this paper, we defined PS-positive proteins as proteins that can undergo phase separation on their own in vitro. We noticed that in several recently published phase-separation sequence databases, including LLPSDB [50], PhaSepDB [51] and PhaSePro [52], two main issues exist: (i) Many phase-separated systems are multi-component, comprised of "scaffold" proteins that are PS-positive and "client" proteins that are phase-separation-negative (PS-negative) on their own. However, "client" proteins were often mislabeled as PS-positive. (ii) There were many sequence errors (e.g., missing fluorescence tags; incorrect species; mishandled mutations and cleavages). To tackle these issues, we screened 142 papers (Supplementary Table S1) from July 2013 to January 2019, excluded sequences that can only undergo phase separation with DNA/RNA/other proteins from our positive set, and manually extracted 565 sequences (see Supplementary File S1) as our PS-positive set (workflow shown in Figure 1). Then, we used LLPSDB and PhaSepPro to cross-check the sequences.

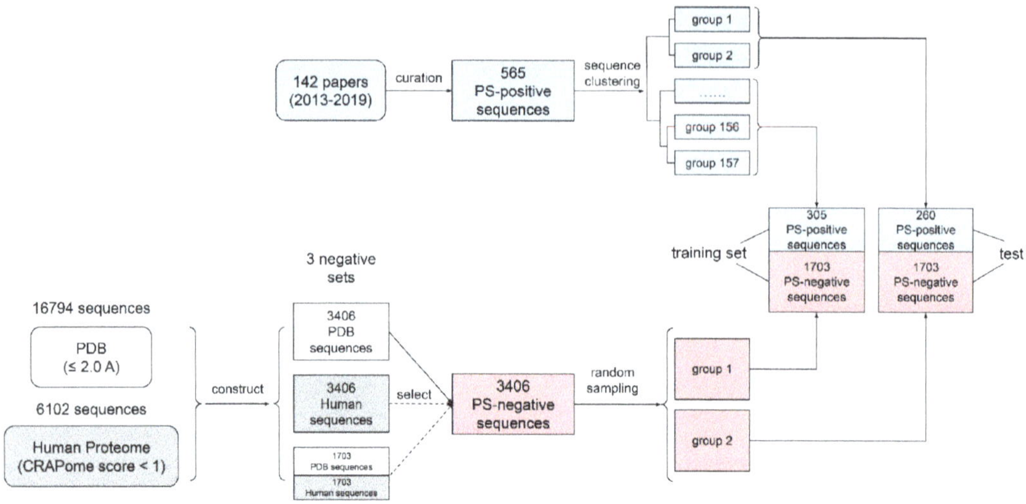


**Figure 1.** Data curation workflow. A schematic diagram of how data for training were obtained and processed.

(2) The clustering of PS-positive sequences (for train–test split): A common practice used in other related work [29,40,48] is to split the training and test data in order to use previously discovered sequences for training and newly found sequences for testing. However, we noticed that there are many similar sequences reported at different times (e.g., sequences from a family that was worked on by the same lab in many years, or different mutants of the same wild type), so performing a time-based split will cause greater bias as we are training and testing similar samples on the algorithm. This issue would be even more problematic considering the limited sample size for PS-positive proteins reported at the start of our work (<1000 samples). Therefore, before splitting the training and test set, we applied a hierarchical clustering to 565 PS-positive sequences, and obtained 157 sequence groups, as shown in Supplementary File S2 and Figure S1, where sequences within the same group have a pairwise similarity of higher than 50%. The subsequent train–test split was then conducted based on sequence groups instead of individual sequences, so that training and testing set proteins are derived from separate sequence groups.

(3) PS-negative training sets: We created two negative sequence databases: (i) the PDB sequence database, from which we collected 16,794 sequences (see Supplementary File S3) from high-resolution ($\leq$2.0 A) structures in the PDB, and (ii) the curated human proteome sequence database, from which we collected 20,380 human proteome sequences (see Supplementary File S4) from Uniprot and removed sequences with either null values or high values (top 20%) in CRAPome [38,53]. We chose CRAPome as the method of filtering out phase-separation-prone sequences because it is an empirical measurement, rather than a prediction, of non-specific interactions in human proteins [53]. This resulted in a "clean" human set of 6102 sequences (Supplementary Table S2). Supplementary Table S3 contains the CRAPome (along with final LLPhyScore) scores for all of the human sequences, including both those within the curated negative training set and those not in the curated list. It is worth noting that false negatives existed in both the PDB and curated human negative sequence data sets. While we attempted to minimize false negatives, both the PDB and curated human sets were compromised to an unknown degree. Certainly, there were fewer positives in these sets than in the known PS-positive sequences, but perfect discrimination is likely impossible because the training sets are not gold-standard truths, and the percentage of human-proteome- and PDB-derived sequences that undergo phase separation is unknown.

Since the positive sample size was much smaller than the negative sample size, we then randomly selected 3406 sequences each from the (i) PDB sequence database and (ii) curated human sequence database, and constructed two negative sets: (a) the PDB set (3406 sequences) and (b) the curated human set (3406 sequences). Finally, we mixed (a) and (b) at a 1:1 ratio and built (c), a mixture of the PDB and the curated human set of 3406 sequences (randomly selecting 1703 from PDB and 1703 from human). During the initial predictor training, the PDB set was used as the main set for determining both the "signs" of the features and the number of features to retain; then, all three sets were used to optimize the "weights" of the features and to compare the three final predictors' performances with each other and other predictors.

(4) The construction of the training/test/evaluation sets: For the training of the predictor models and the optimization of the model parameters, we initially constructed training and test sets by adopting a 70–30% train–test split ratio for the PS-positive and negative sample sets in steps (1) to (3). For the positive samples, random sampling was conducted at the clustered group level until >30% of sequences went to the test set. However, due to the existence of large sequence groups (30–50 sequences), the end result was actually close to an even ratio with 305 sequences in the training set and 260 in the test set, as shown in Figure 1. Therefore, we then used an even ratio between the training and test sets for the negative samples, where the random split was conducted at the sequence level, given that the issue with similar sequences did not affect the negative set.

For the evaluation of the final models' performances trained on different negative databases (PDB, human, human + PDB) on a defined dataset, we constructed evaluation set 1, composed of the entire PS-positive set (565 sequences) and the entire PDB proteome (16,794); for the comparison of our predictor's performance with other state-of-the-art phase-separation-prediction algorithms, we constructed evaluation set 2, composed of the entire PS-positive set (565 sequences) and the entire human proteome (20,380 sequences).

For more details on the constructed training, test and evaluation sets, see Supplementary Table S4.

### 2.2. Construction of Physical-Feature Collection in LLPhyScore

We made two core assumptions in this work to develop a sequence-based predictive algorithm: (i) phase separation is driven by multiple physical forces and structural factors; (ii) for any phase-separated system, these forces and features together build up the system's free energy to drive the phase transition. Then, we constructed a set of 16 (8 pairs) sequence-based, phase-separation-related physical features, including weak interactions and structural patterns, as described below. More details on each of these can be found in the Technical Methods and Supplementary Table S5. The motivation for our design derives from the focus of much of the phase-separation literature on protein–protein interactions, often ignoring protein–water interactions (see below), and the assumptions that one or a few certain specific physical or chemical interactions are dominant contributors or that some are not important (e.g., h-bonding, kinked beta). We also initially hypothesized that proteins found in distinct biomolecular condensates would use specific types of physical "interactions" based on our definitions as a way to generate specific condensates.

Protein–water interactions. As pointed out by others in the field [36,54], protein–water interactions represent a largely overlooked driving force in phase separation because of its synergistic nature with other interactions such as pi–pi, hydrogen-bonding and electrostatic interactions [29,36,54]. Here, we considered it as a separate force/feature and explored its role in phase separation. We defined protein–water interaction by contacts, and measured solvation contacts and hydrophobic contacts using two inversely correlated terms, a residue–water interaction count and a residue–carbon interaction count. The frequency measurement followed the same protocol as for pi–pi interactions in PScore [29].

Helices and strands. While most phase-separating proteins contain IDRs that play a significant role in driving phase separation, in some cases [33,37,55], these IDRs transiently exhibit a folded structure (either helices or beta-structures with varied dynamics and sizes) that can play a critical role. Here, we used the DSSP program to assign the secondary structure [56] and enable helices (H) and strands (E) to be considered as contributing features. Disorder was categorized as a separate feature, because most reported phase-separating systems are IDR-driven, and the statistics are highly skewed towards disorder, which could be detrimental to the algorithm training. Boolean values (true or false) instead of frequency were utilized for helices and strands.

"Long-range" and "short-range" disorder. Due to the large difference in structural context between short (<5 residues long) and long (>15 residues long) disordered regions [57,58], disorder was divided into these two categories. Here we defined the presence or absence of disorder as Boolean values (true or false), and measured disorder based on the lack of helix or strand DSSP assignment of consecutive residues in a sequence.

Long-range and short-range electrostatic interactions. Electrostatic interactions have been established as another important driving force for phase separation, especially for highly charged sequences in complex coacervation systems, such as for the tau protein [59]. Here we defined electrostatic interaction using coulombic interaction energy with atomic partial charges taken from the Talaris2014 force field [60], dividing the interaction energies by the sequence separation of the involved atom pairs into short-range (<5 residues apart) and long-range ($\geq$5 residues apart). We note that complex coacervation will not be predicted as the approach is based on the phase separation of a single protein.

Long-range and short-range hydrogen bonds. Hydrogen bonding was found in some cases to co-exist with other driving forces, including pi–pi contacts [21] and protein–solvent interactions [61]. In this work, we considered it as a separate force and explored its role in phase separation. We used the PHENIX software suite [62] to identify OH-N hydrogen bonds and measured inter-residue hydrogen-bond interaction counts in short-range (<5 residues apart) and long-range ($\geq$5 residues apart) contexts.

Long-range and short-range pi–pi interactions. We utilized our previous approach from the PScore phase-separation predictor based on planar pi–pi contacts [29], determining the contact frequency for residue pairs in the context of short-range (<5 residues apart) and long-range ($\geq$5 residues apart) interactions.

Long-range and short-range cation–pi interactions. Cation–pi interactions were found to have a specific residue-type preference among the cations arginine and lysine and the aromatics phenylalanine, tyrosine and histidine, and the substitution of preferred residues in certain systems cause drastic change in phase-separation behavior [63]. In order to crudely estimate the potential cation–pi interactions, we adapted the electrostatic potential by adding partial negative charges above and below the planes of aromatic ring systems, balanced it with an in-plane positive charge, and then calculated the change relative to our standard electrostatic term. These measurements were again split into short-range (<5 residues apart) and long-range ($\geq$5 residues apart).

Kinked beta-strands (K-Beta). It has been observed that specific sequences from some phase-separating proteins can form fibrils of kinked beta-strands, with beta-strand hydrogen bonding occurring without extended backbone torsion angles and forming fibrils similar to amyloids [33]. The prediction of this feature has previously been performed by the energetic assessment of a sequence's ability to adopt the topology found in these fibril structures [64], and we created an analogous classification strategy by identifying sequences in the PDB that were similar or dissimilar (measured by backbone RMSD) to these kinked beta-strands [65]. Two Boolean metrics, K-Beta similarity and K-Beta non-similarity, were determined from RMSD values after the structural superposition calculations.

Based on the above 16 (8 pairs) features, we designed a sequence-representation system (See Technical Methods and Figure 2) to convert a sequence into inferred residue-level feature values (frequencies/numbers/Booleans). Note that many of these features are highly interdependent, particularly protein–water interactions with all of the others, cation–pi with pi–pi and electrostatics, and kinked beta with hydrogen bonds and pi–pi [33]. In addition, the role of residue-type preferences, which are also terms that are fit during training (including counts and positions), cannot easily be deconvoluted from these features.

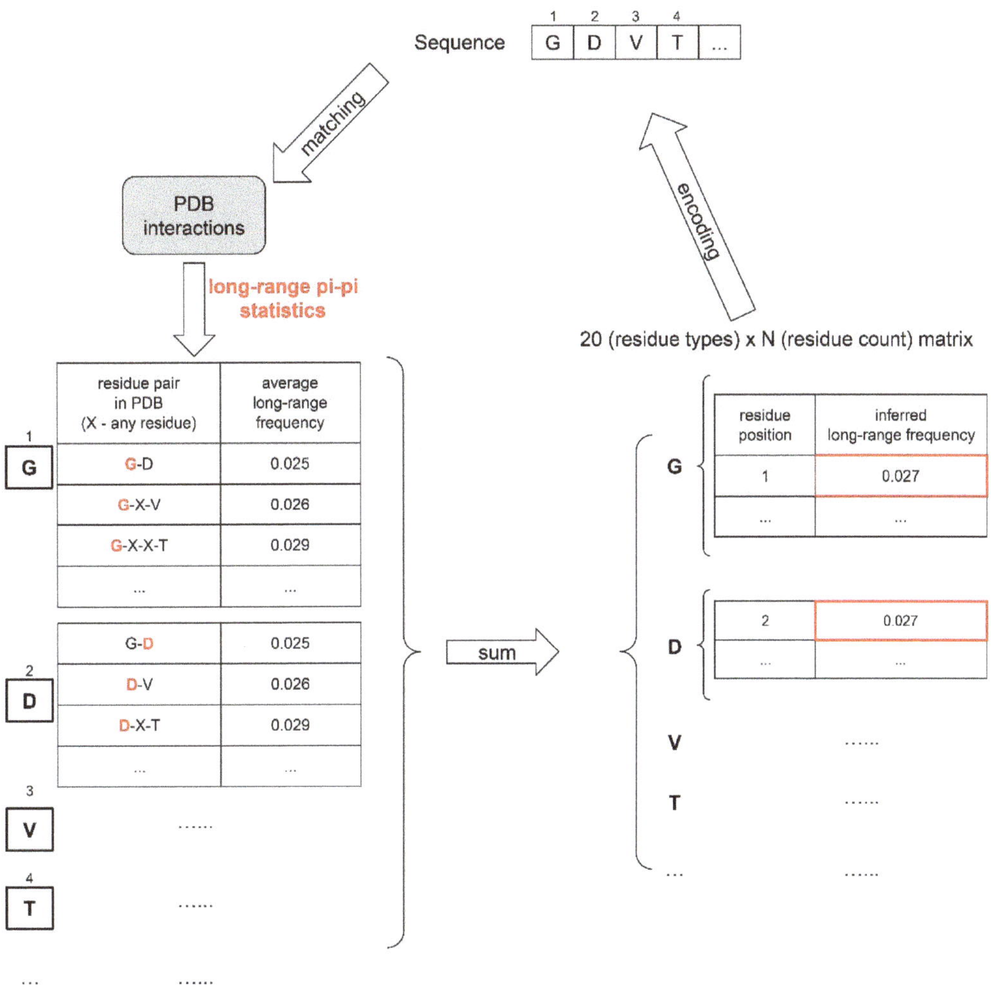


**Figure 2.** Physical-interaction- and structure-based feature extraction. An example is given of the feature representation of sequences for the sequence "GDVT" converted to the pi–pi (long-range) feature matrix.

## 3. Results and Discussion

### *3.1. Predictor Training*

The concept of "predictor training" in this work means: (i) for a specific sequence, the algorithm outputs a summed score calculated by a weighted combination of the expected physical-feature values, and (ii) during the predictor training, we optimized the combination of physical features, as well as the "weight" for each feature. The workflow of the predictor training is shown in Figure 3.

The predictor training has three outcomes, described here:

(1) "Signs" of features were determined using individual feature training. Some features in our list were positively correlated with the performance of the developing predictor, while other features were negatively correlated. Therefore, before combining the 16 features, we first trained each feature individually and let the algorithm decide the "direction" (positive or negative) of its correlation with performance (measured by AUROC). As shown in Figure 4 and Supplementary Table S6, the features that were found to correlate negatively

were protein–carbon interactions, the helical secondary structure, long-range electrostatic interactions, both short- and long-range cation–pi interactions and the kinked-beta (K-Beta) non-similarity. While the negative correlation for protein–carbon interactions and K-Beta non-similarity are consistent with an understanding that these features do not contribute to phase separation, in general these results are not simply interpretable as contributing positively or negatively to phase separation. This is particularly the case for electrostatic interactions including cation–pi, as it is not clear how the predictor deals with locally repulsive electrostatic interactions (clustered charges) that may favorably interact over longer ranges with oppositely charged clusters, or how well our crude estimate of cation–pi interactions works. Certainly, complex coacervation was not predicted as this tool was limited to homotypic phase separation, i.e., involving a single protein sequence.

(2). The number of features to include was determined using competitive feature training. After determining the "signs" of the features and applying them, we combined 16 features and allowed them to "compete" with each other through "competitive" training, then ranked their importance based on the final contribution (positive or negative) of each feature to distinguishing phase-separating from non-phase-separating proteins, as shown in Figure 5. While all 16 features achieved an average z-score greater than 1.5, the average z-scores for protein–water, protein–carbon, long-range hydrogen-bond and long-range pi–pi interactions were larger than 3.0, and those for disorder (within both short and long segments) and kinked-beta similarity were larger than 2.5. While the competitive training approach suggests the ability to quantitatively compare the significance of these physical interactions in phase separation over the input positive set, the interdependence of the terms and the convolution with the residue-type preference makes this comparison much more qualitative. We then came up with three different combinations of features according to the ranking, combining the top 8 or top 12 features based on ranking or combining all 16 features, in order to identify the minimal number of features that provides both good performance and physical interpretability. We conducted competitive training on each of the 8-, 12-, and 16-feature algorithms and assessed their performance. As shown in Supplementary Table S7, the combinations of 12 and 16 features did not demonstrate better performance than the combination of 8 features. To avoid overtraining, we chose the 8-feature combination in the final predictor training, with the weights of the smaller number of terms from training (see "(3)" below) reflecting the contributions of the features that were dropped. Thus, the choice of eight features cannot be interpreted as these features being the only ones that physically contribute to phase separation or that the resulting predictor ignores the contribution of those features. Cation–pi interactions are a clear example of this, as they are represented in the 8-feature predictor as a combination of residue-type preference, electrostatics and pi–pi interactions, even though they are not discretely represented as their own term.

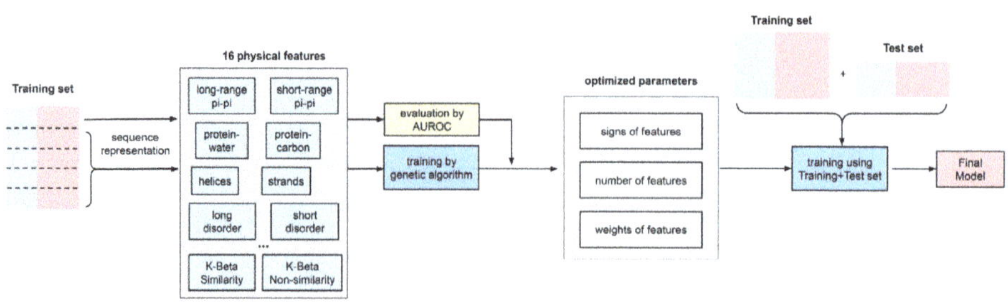


**Figure 3.** Predictor training workflow. A schematic diagram of the steps in training is shown.

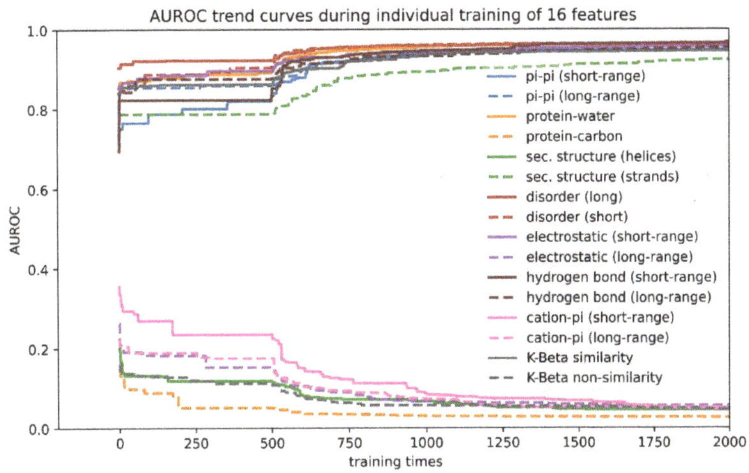


**Figure 4.** Direction of correlation of features with performance of the developing phase-separation predictor. Training curves of 16 features to reveal the direction of correlation of each feature with score. Features that rise towards AUROC = 1.0 have "positive" features; features that decline towards AUROC = 0.0 have "negative" signs.

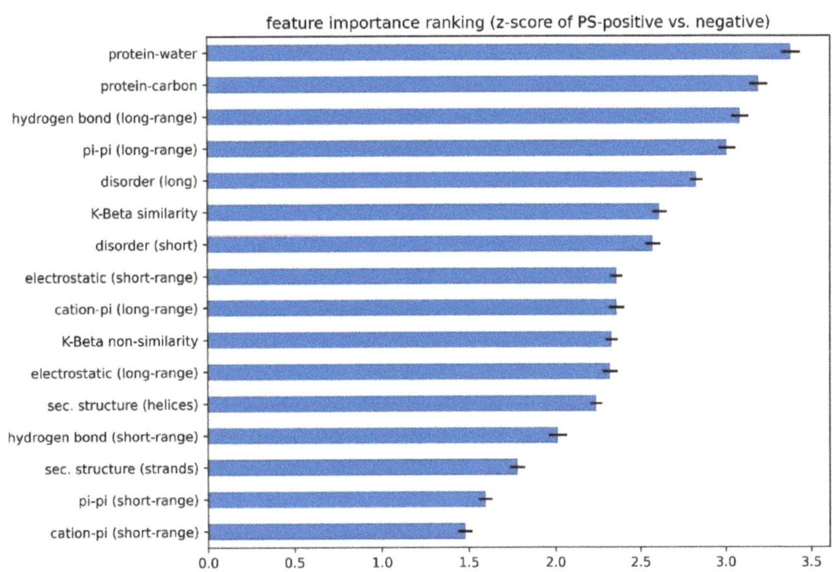


**Figure 5.** Ranking of the importance of features to discrimination in the developing phase-separation predictor between PS-positive and PS-negative sequences. The z-score of PS-positive sequences' individual feature values against the mean PS-negative sequences' values is shown.

(3) The "weights" of features in the final predictor were determined using competitive feature training on the entire dataset. We built the final predictor ("LLPhyScore") and optimized the "weights" for the chosen eight features with their respective signs by competitive training on training set 1 and tested the model performance on test set 1 (Supplementary Table S4). We chose AUROC as the model-performance metric, which was 0.969 for training and 0.942 for the test (Supplementary Table S7) with the PDB as the negative set. This indicates that minimal overtraining occurred during the "weights" optimization. Then, we

trained the "weights" again on training set 1 + test set 1 to yield the final predictor called "LLPhyScore-PDB model" based on its use of the PDB as a negative set. The LLPhyScore–PDB model achieved an AUROC value of 0.978 (Supplementary Figure S2) on evaluation set 1 (including all PS-positive sequences and the full PDB proteome, Supplementary Table S4) and good separation between positives and negatives (Supplementary Figures S3 and S4).

### 3.2. Model Performance Comparison against Different Negative Training Sets

As noted previously, there is no perfect negative sample set for phase-separation-predictor development; therefore, after we trained the LLPhyScore-PDB model (on training set 1 + test set 1), we also trained the LLPhyScore-Human model (on training set 2 + test set 2) and the LLPhyScore-Human + PDB model (on training set 3 + test set 3), and evaluated the three final models using both evaluation set 1 (all PS-positive sequences and full PDB proteome) and evaluation set 2 (all PS-positive sequences and full human proteome). The results shown in Figure 6 and Supplementary Figures S2–S4 indicate that the PDB model showed the best performance on evaluation set 1 against the PDB (AUROC of 0.978), but the worst performance on evaluation set 2 against the human proteome (AUROC of 0.824); the human model showed the best performance on evaluation set 2 (AUROC of 0.941) and the worst performance on evaluation set 1 (AUROC of 0.908). This indicates that the negative training set of different models had a significant impact on the final model performance. The model using only folded proteins from the PDB as the negative training sequences tended to have less power to generalize on evaluation set 2 (including the full human proteome), which contained many disordered regions. On the other hand, the model only using human proteins as the negative training sequences still had a strong ability to discriminate most PS-positive sequences from PDB sequences in evaluation set 1. This is also reflected by the fact that the human + PDB model showed a more balanced result for evaluation set 1 (AUROC of 0.947) and evaluation set 2 (AUROC of 0.933). Together, these results support the use of the curated human proteome as a negative set, alone or with the PDB, and our choice of the human + PDB model as the optimal model.

### 3.3. Predictor Validation

To validate the final predictors' performances, we constructed three sets of baselines. (1) Instead of providing PDB-based physical-feature values to the genetic algorithm, we provided random values from a normal distribution N(0, 1) in the weight-training step. (2) Instead of providing sequence-based physical-feature values, we provided random values from the distribution of residue-specific physical-feature values. (3) Instead of optimizing 20 weights for 20 residue types for each physical feature, we optimized 1 weight for all 20 residue types for each physical feature (removing residue specificity) during training. As shown in Figure 7 for the human + PDB model and Supplementary Figure S5 for all three models, baselines 1 and 2 showed a very high training AUROC but a low test AUROC, whereas the final models had both high training and test AUROCs. This suggests that the final predictors' good performances did not result from overtraining the genetic algorithm, which was the case for baselines 1 and 2. The comparison between baseline 3 and the final models also suggests that it is important to have residue specificity in our model for good prediction performance.

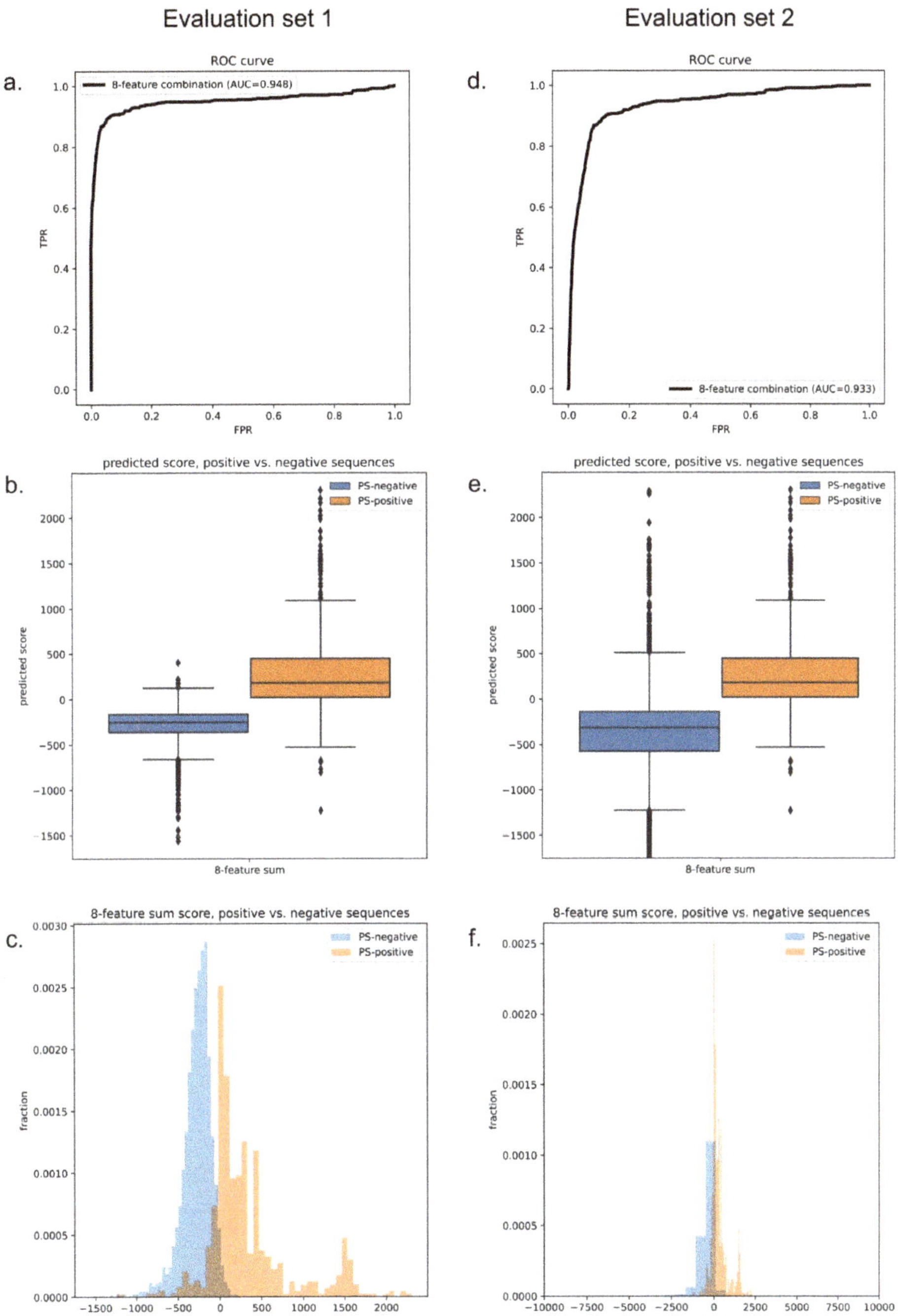


**Figure 6.** Final predictor of model performance. Performance plots of the final human + PDB model on evaluation set 1 (left, PS-positive sequences and the entire PDB proteome) and evaluation set 2 (right, PS-positive sequences and the entire human proteome). (**a**,**d**) ROC curves. (**b**,**e**) Predicted score boxplots of positive vs. negative sequences. (**c**,**f**) Distribution histograms of positive vs. negative sequences.

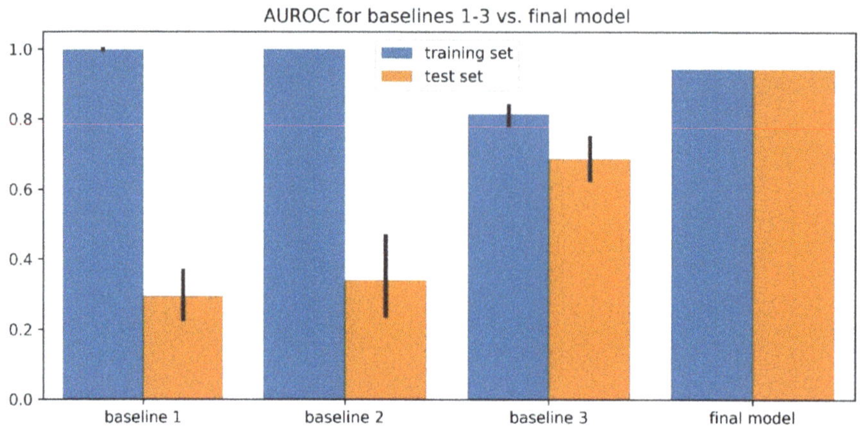


**Figure 7.** Comparison of three training baselines and the final human + PDB predictor model for validation. Baseline 1 was created by providing random values from a normal distribution N(0, 1) in the weight-training step instead of providing PDB-based physical-feature values to the genetic algorithm. Baseline 2 was created by providing random values from the distribution of residue-specific physical-feature values instead of providing sequence-based physical-feature values. Baseline 3 was created by optimizing 1 weight for 20 residue types for each physical feature (removing residue specificity) during training instead of optimizing 20 weights for 20 residue types for each physical feature.

### *3.4. Comparison of Prediction Using Eight Features or Single Features*

To test whether a combination of eight features can outperform the prediction using a single feature, we extracted from the three final models each of the feature components as one-feature predictors and evaluated these one-feature predictors on evaluation set 1. As shown in Figure 8a and Supplementary Figure S6, the receiver operating curves (ROCs) of one-feature predictors were outperformed by the eight-feature predictors. We also plotted Venn diagrams showing their recalled sequences at a confidence threshold that returns 2% of the PDB as a positive result (chosen based on the methods described in previous work [32,38,40]) as shown in Figure 8b and Supplementary Figure S7. We observed that each of the one-feature predictors missed a number of sequences (48–350 sequences) that were captured by the eight-feature models. This result supports our underlying assumption that phase separation is driven by a combination of different physical features, and that driving forces for different sequences can vary.

### *3.5. Comparison between LLPhyScore and Other Phase-Separation Predictors*

We compared the performance of our predictor (LLPhyScore, three final models) with PSPredictor, as well as two first-generation predictors, PScore and catGRANULE, in Figure 9. The comparison was conducted on both evaluation set 1 (PS-positive sequences and the entire PDB proteome) and evaluation set 2 (PS-positive sequences and the entire human proteome).

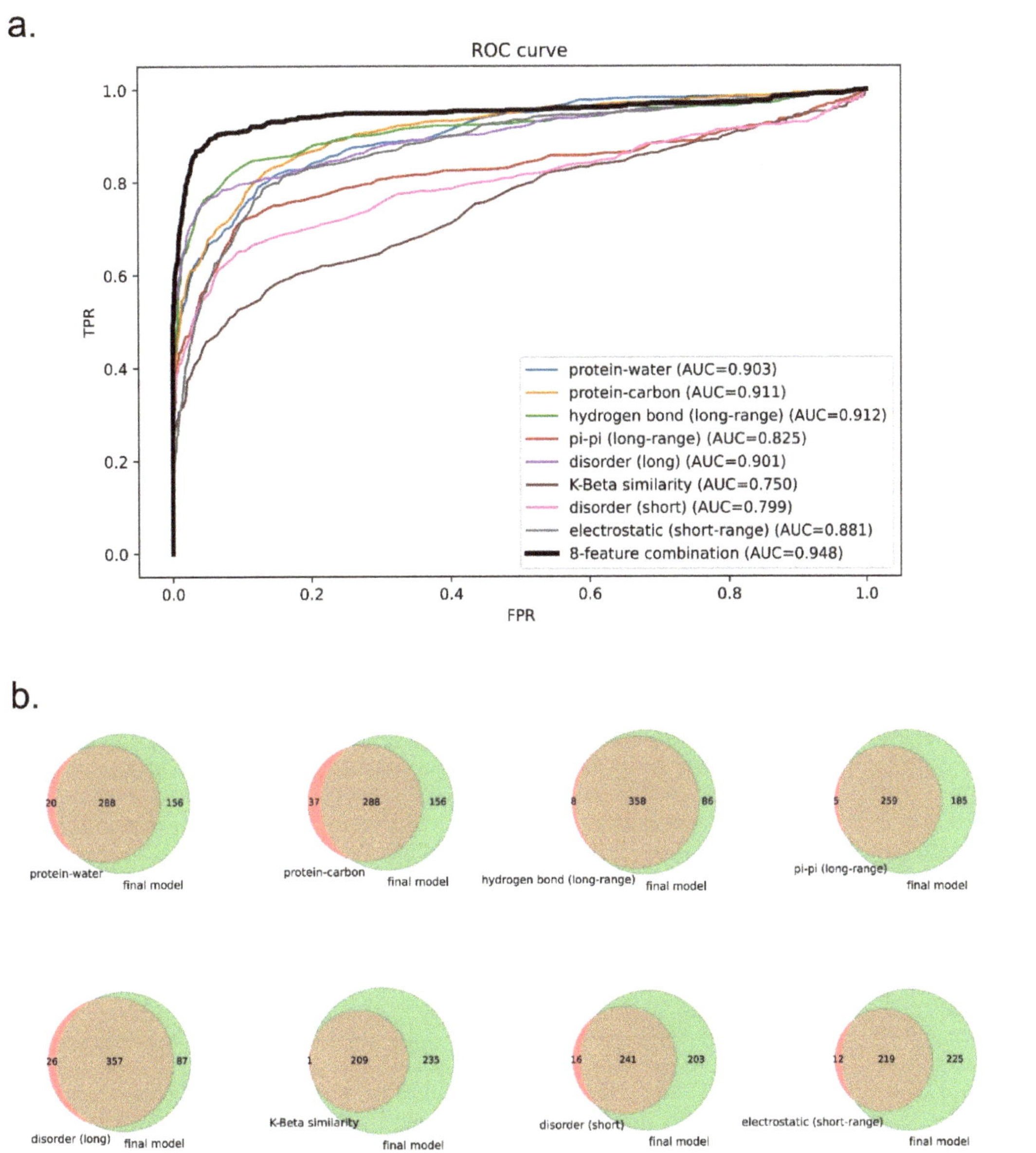


**Figure 8.** Comparison of the performance of predictors trained on eight features vs. one feature for the human + PDB model. (**a**) ROC curves of one-feature predictors vs. the eight-feature predictor. (**b**) Venn diagrams showing the coverage overlaps of PS-positive sequences by one-feature predictors vs. the eight-feature predictor at a confidence threshold that returns 2% of the PDB.

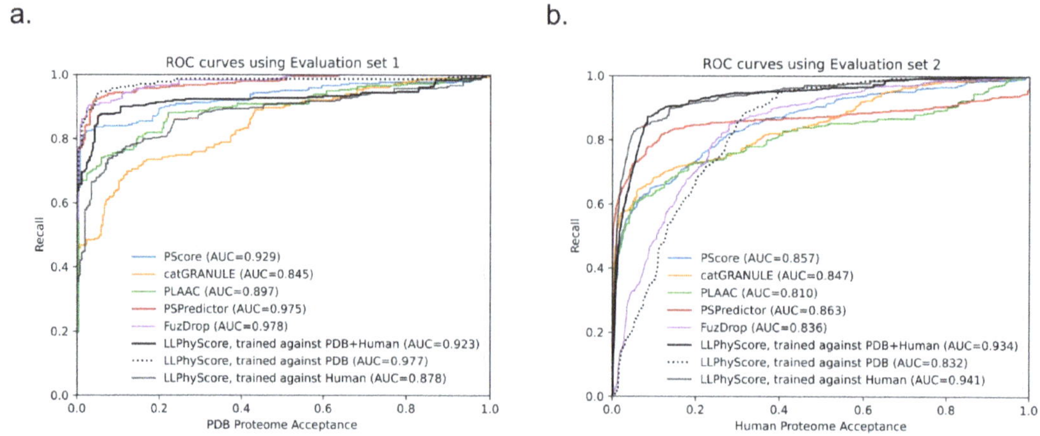


**Figure 9.** Comparison of LLPhyScore (three models) with other phase-separation predictors. Relationship between percent recall and total percentage of (**a**) evaluation set 1 and (**b**) evaluation set 2 accepted at the given thresholds for PScore, catGRANULE, PLAAC, PSPredictor, FuzDrop and LLPhyScore.

We can see that the LLPhyScore-PDB model showed the best performance on evaluation set 1 and even slightly outperformed PSPredictor, which was trained against 5258 sequences from the PDB. The LLPhyScore-PDB model also showed a better AUROC than PScore, which is based solely on planar pi–pi interactions. The LLPhyScore-Human + PDB model showed a slightly decreased performance on evaluation set 1 compared to the LLPhyScore-PDB model; however, it was still better than all of the first-generation predictors. The LLPhyScore-Human model showed a comparable performance to PLAAC.

However, on evaluation set 2, the LLPhyScore-PDB model did not show better recall statistics than the other first-generation predictors until a 30% acceptance threshold, as shown in Figure 9b. This is in line with the estimate of up to 40% of the human proteome driving phase separation [49], and could be considered support for an estimate of at least 30% of the proteome being involved in phase separation. On the other hand, the LLPhyScore-Human model and LLPhyScore-Human + PDB model both showed good performance on evaluation set 2, indicating that, by mixing the human and PDB sequences, the training algorithm can optimize PS-positive sequences from both negative sets. We also see (Figure 9a,b) that the LLPhyScore-PDB model showed comparable recall trends with FuzDrop. As a phase-separation predictor also based on biophysical principles combined with statistical training, FuzDrop uses a protein's binding entropy as the target function. The fact that the LLPhyScore-PDB model and FuzDrop showed similar statistics supports the utility of approaches directly addressing the biophysical features and energetic driving forces underlying the formation of condensates.

### *3.6. Feature-Based Breakdown of Scores for Different Sequences*

To further explore the general expectation that the phase separation of different sequences can be driven by different physical features, we clustered PS-positive sequences based on their single-feature scores after normalization. As shown in Figure 10 and Supplementary Figures S8 and S9, FUS, Nup98, an elastin-like peptide (ELP), and MEG-3 were categorized into different clusters, which demonstrates the ability of LLPhyScore to treat different types of sequences, although most proteins were not clearly distinguishable. This underscores the interdependence of many of the physical features. For the LLPhyScore-Human + PDB model, the breakdown of the scores (Figure 10) shows that Nup98 has high scores for protein–carbon interactions but low scores for disorder, pi–pi interactions, and K-beta, whereas for FUS, the scores are high for most of the features.

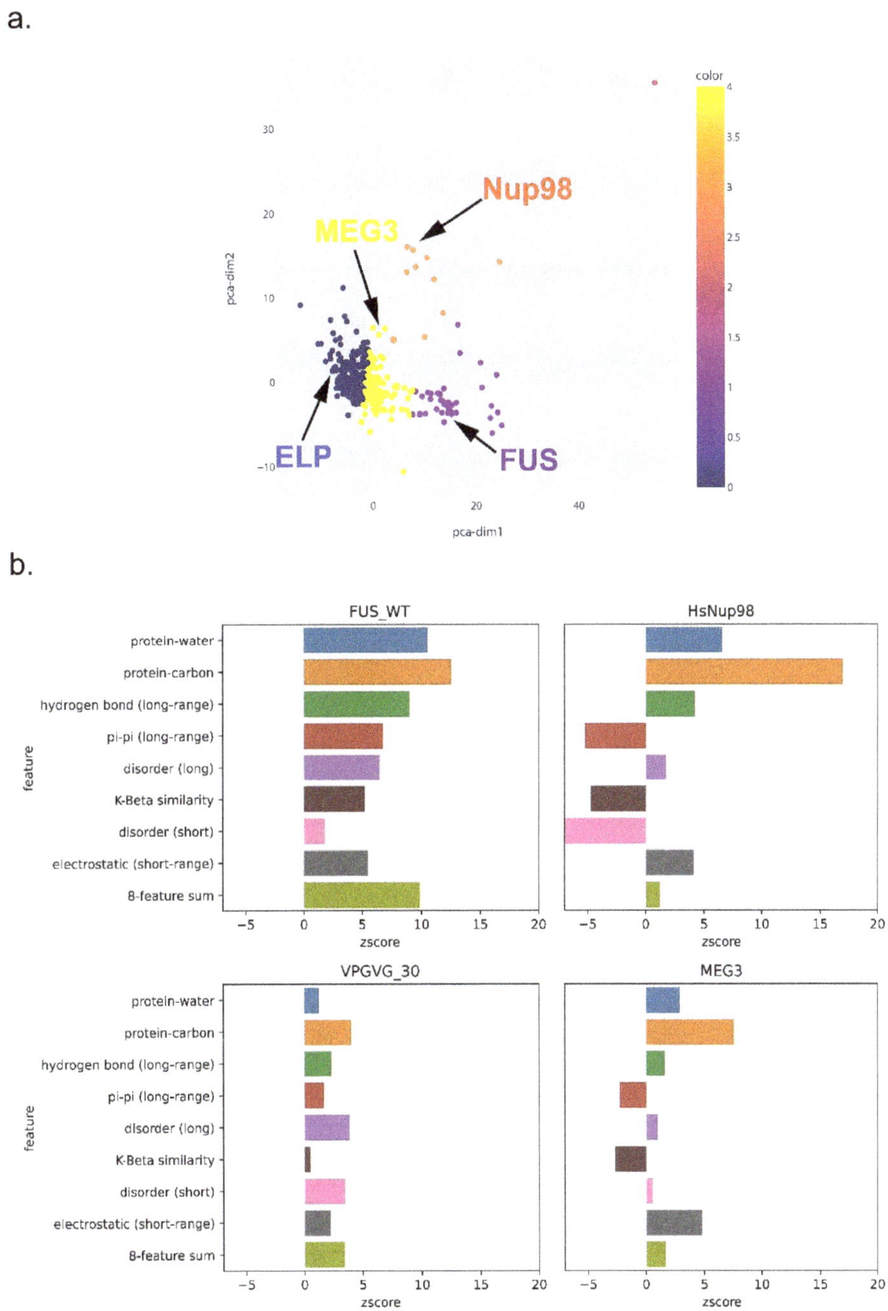


**Figure 10.** Feature-score-based clustering for PS-positive proteins for the human + PDB model. (**a**) Plot of two abstracted dimensions for clustering based on feature z-scores, showing the separation of different types of phase-separating sequences. (**b**) The score breakdown of four example sequences from four distinct clusters in (**a**): FUS (human), Nup98 (human), elastin-like peptide (ELP, VPGVG_30, 30 repeats of VPGVG) and MEG-3 (*C. elegans*).

### 3.7. Gene Ontology Term Enrichment

To explore our hypothesis that different biomolecular condensates would include proteins driven by similar features, we analyzed the enrichment of GO terms for human proteins in the top 10% (high confidence threshold) of scores from the LLPhyScore models predicted by eight single features as well as the combination of eight features in the final predictors. As shown in Figure 11, Supplementary Figures S10 and S11 and Supplementary Table S8, most GO terms identified by first-generation predictors [38] and by PSPredictor [40] were also enriched in sequences identified by LLPhyScore, such as extracellular matrix and nuclear body. For certain annotations associated with phase separation such as cytoplasmic stress granule, postsynaptic density and transcription factor complex, we observed differences depending on which features were utilized, which suggests that, for different biomolecular condensates with different functional roles for phase separation, the features linked to phase separation are also different, and are rooted in their sequence-specific biophysical landscape.

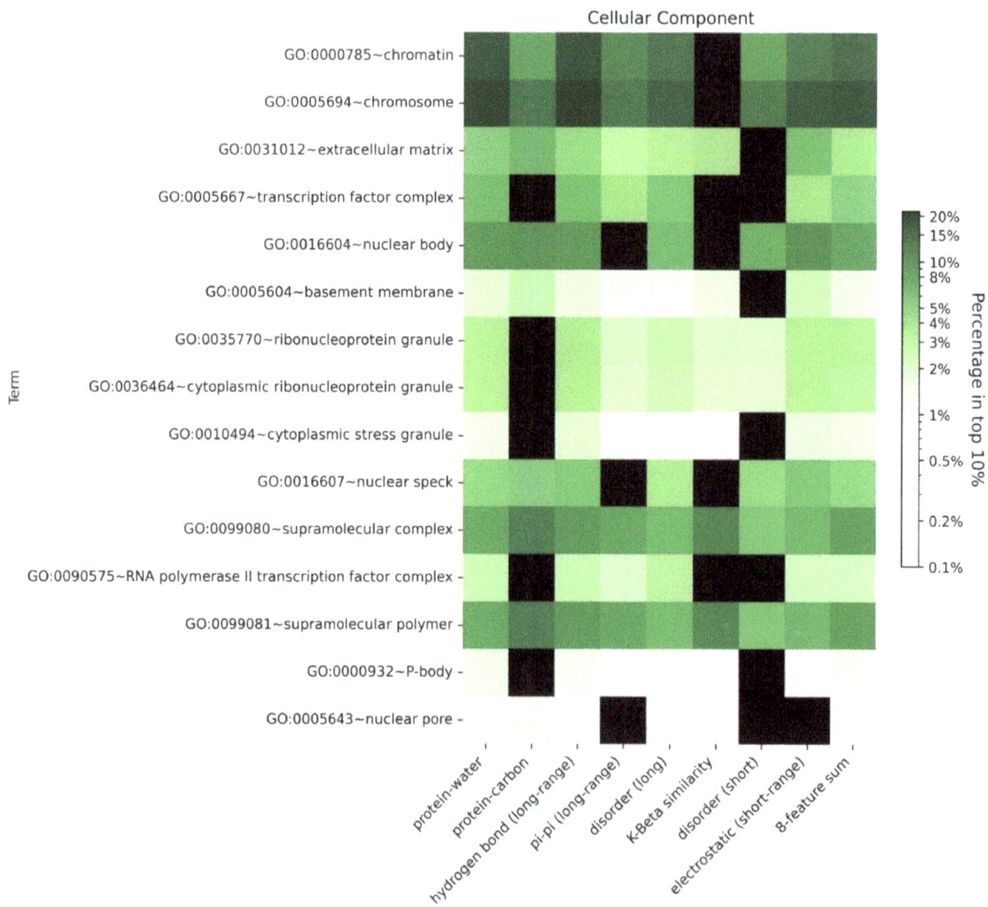


**Figure 11.** Enrichment heatmap by GO functional annotations for different features for the human + PDB model. Heatmap showing the enrichment of the proteins with a given functional annotation that fall under a 10% confidence threshold for each single-feature score and the eight-feature sum score. The color gradient shows the natural logarithm of the enrichment percentage. The black boxes indicate that no proteins in this GO term are within the top 10% of the corresponding score type.

### 3.8. Physical Insights into Phase Separation Based on LLPhyScores of the PDB Set

The assessment of the physical basis of the LLPhyScore predictions is complicated not only by the interdependence of the features but also by the detailed choices made during the training process, where the weight given to a feature by the final model not only reflects that feature but the full sequence context of the residue including residue-type preferences. Therefore, to assess how scores relate to the physical features for which we trained, we applied the predictor to the sequences of known structures in order to assess phase-separation scores by directly comparing them to "true" measurements of sequences in observed structural contexts. For this, we scored each amino acid independently, comparing the physical features associated with being in the top 50% of scores against the overall distribution for that residue type. Figure 12 shows high score enrichment statistics for a variety of physical features, including secondary structure (a), short-range pi–pi interactions (b), kinked-beta similarity and dissimilarity (c), disorder (d), short- and long-range electrostatics (d,e), and local water/carbon contacts (f,g).

For the protein sequences found in our PDB set, the predictor generally assigned low scores to structures that can satisfy their interactions locally. Helical residues that fully satisfy their backbone hydrogen bonds typically had low scores (Figure 12a), as did residues with stabilizing charge interactions found between nearby local residues (Figure 12e). Notably, charge interactions between non-local residues (Figure 12f) had above-average scores, consistent with the known effects of blocks of like charges in driving phase separation [26,66,67]. For short-range electrostatics (Figure 12e), attractive interactions (negative numbers) were not favorable and repulsive interactions (positive numbers) were, with long-range electrostatics (Figure 12f) generally flipping this relationship, which is consistent with the idea of locally self-satisfied interactions not being favorable.

For secondary structure, there appeared to be three categories of effects based on backbone hydrogen-bonding satisfaction and torsion-angle regularity. Fully self-satisfied structures, specifically helices, had the lowest scores. Ordered but not necessarily locally-satisfied structures, which include beta-strands as well as $3_{10}$ helices (often associated with short helices [68,69] and capping motifs [70]), had intermediate scores. Irregular secondary structures, including elements with defined hydrogen-bonding patterns (turns, bulges, and bent/kinked strands), as well as solvent-bound loops, had the highest scores. In general, the ability to form hydrogen bonds with a solvent was consistently associated with higher scores, as was the lack of a repetitive ordered structure. In this analysis, bent and twisted strands typically scored better than fully disordered residues, especially for proline, suggesting that the availability of backbone hydrogen bonding plays a role, and not just the lack of structure.

The differences between disorder prediction and phase-separation prediction are further defined in Figure 12d. In general, disordered residues were more likely to score high, with long stretches of disorder scoring higher than short disordered loops. However, the majority of this bias results from hydrophobic or aromatic residues, specifically V, L, I, F, H, Y, and W. This is consistent with disorder on its own being insufficient for phase separation, with disorder that forces hydrophobic and aromatic residues into contact with the solvent supporting phase separation.

This indirect solvent relationship can also be directly observed by the measurement of solvent interactions and overall burial, as shown in Figure 12g,h. In general, residues with a high number of observed water contacts had higher scores, and residues with a high degree of burial (assessed by the number of carbon contacts) had lower scores. However, this trend was more pronounced for hydrophobic residues and was not observed for polar or negative residues (N, Q, E and D). This may be expected given that the hydrophobic effect is driven by the solvent, with the energy associated with a reduction in solvation driving hydrophobic residues together (i.e., solvent relationships are what makes hydrophobics sticky). In this context, we observed that hydrophobic residues that were forced to be in contact with the solvent by their local sequence context were predicted to contribute to phase separation.

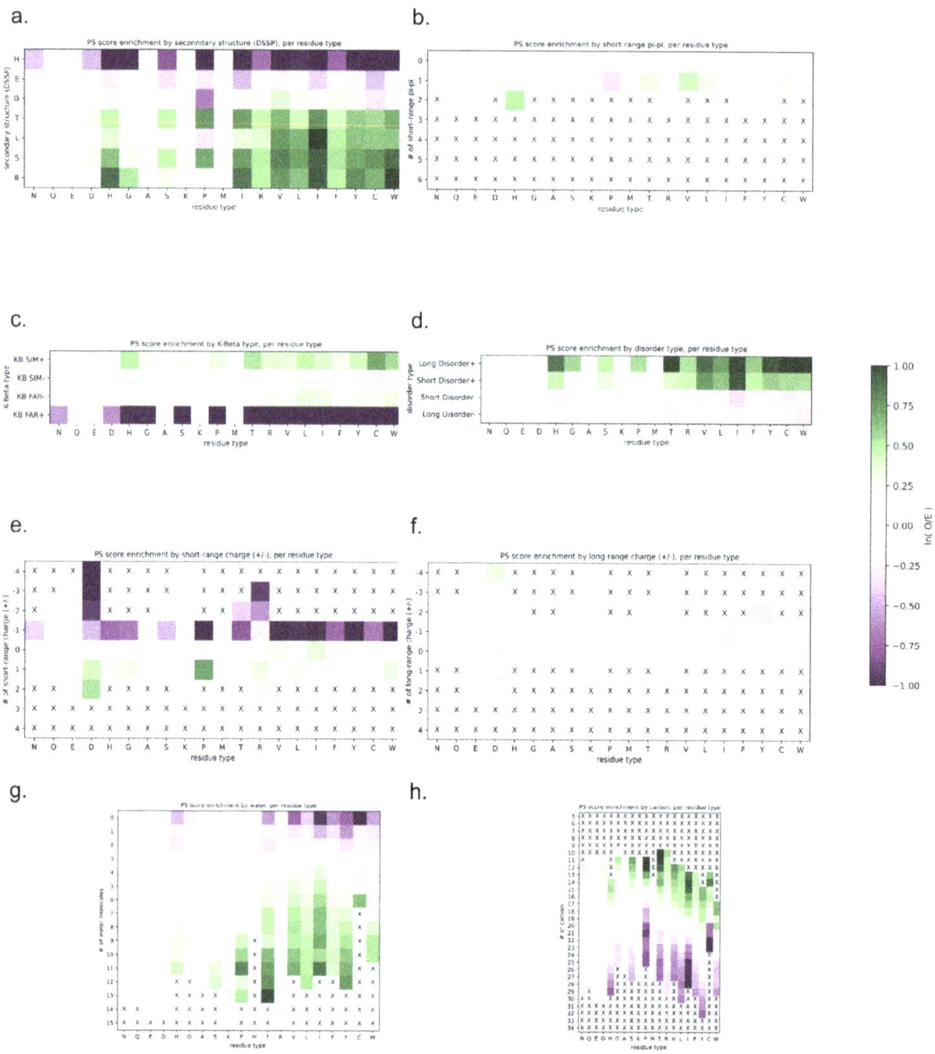


**Figure 12.** LLPhyScore score enrichment by eight selected physical features for the PDB proteome, per residue type, for the human + PDB model. Heatmaps show the score enrichment in PDB protein sequences by each feature's discrete values, normalized to each residue type. The color gradient shows the natural logarithm of the observed over expected ratio. Enrichment for (**a**) secondary structure (H, alpha-helix; E, beta-sheet; G, $3_{10}$ helix; T, hydrogen-bonded turn; L, loop; S, bend; B, single-pair beta-sheet), (**b**) short-range pi–pi, (**c**) K-beta, (**d**) disorder, (**e**) short-range electrostatic, (**f**) long-range electrostatic, (**g**) protein–water and (**h**) protein–carbon. The color bar for all heatmaps is shown at the right.

The notion that sequences that force solvation are prone to phase separation matches the observations for the secondary structure. We note that while extended beta-sheets can often exclude solvent, by forming flat planar interactions with other sheets, kinked beta-strands cannot. Figure 12c shows that sequences with high structural similarity to kinked beta-structures had higher scores, especially for hydrophobics and aromatics.

Together, our analyses of the LLPhyScores for the PDB structures supports the view that disorder itself does not drive phase separation, but locally unsatisfied sequences that are constrained in their ability to exclude the solvent, including those that can adopt an irregular or kinked beta-structure to contribute backbone hydrogen bonds, do drive phase separation. These results may contribute to the current discussion of the role of sequences with the propensity to form a kinked beta-structure in protein phase separation [33,64].

### 3.9. High-Scoring Structures in the PDB Trend towards Disorder

The protein structure databank is often used as a negative set when training phase-separation classifiers, but this is not a ground truth, and the true fraction of proteins with structures present in the PDB that are also capable of driving phase separation is unknown. To demonstrate this issue, we scored the set of PDB reference sequences used in this study and observed that the highest-scoring proteins were not random; the score selected for proteins with significant disorder relative to the average structure found within the PDB (Figure 13). This was true using multiple definitions of disorder: (i) of the highest-scoring 1% (N = 167), 99 had more than 50% of the reference sequence missing from the density (Figure 13a), and (ii) for the residues that were found within the density (Figure 13b), 128 of these proteins had more than 50% of the residues in secondary-structure classes other than helix and strand, with 62 of these having more than 50% of their residues in contiguous loop/turn/random coil segments of four or more residues in length.

The highest-scoring sequences for LLPhyScore in the PDB depart significantly from the expectation of well-ordered folded domains, and their function is unlikely to be defined simply by their ability to form the state observed within these structures. These results, in addition to describing physical features that are correlated with phase separation, highlight the need for a biophysically defined empirical negative set for future work in training phase-separation classifiers.

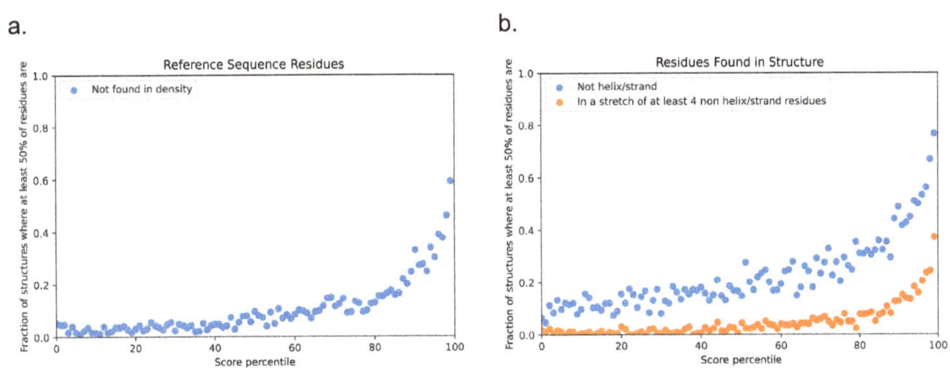


**Figure 13.** Disordered character of PDB sequences according to the LLPhyScore of chain reference sequences. Panel (**a**) shows the fraction of proteins in each percentile bin of LLPhyScore for which more than 50% of the reference sequence is missing from density (protein sequence that does not show up in the structure). Panel (**b**) shows the disordered/irregular structural character of residues that are within the density in the structure, with blue showing the fraction of proteins in each percentile bin for which more than 50% of the observed residues have a DSSP assignment other than helix or strand, and orange shows the fraction for which more than 50% of such residues are found in stretches of at least four residues in length with no helical or sheet structure.

## 4. Conclusions

In this work, we demonstrated the utility of combining different physicochemical interactions as driving forces in the prediction of protein phase separation. We addressed the issue of the "imperfect negative training set" by training three predictor models on three different negative sets and compared their performances. We optimized the combination of physical features in the final predictor models and achieved a superior performance over first-generation predictors. Importantly, our predictors enable a physical interpretability that is not possible with another comprehensive predictor, PSPredictor. Our results are consistent with the understanding that phase separation is driven by a combination of inter-related physical factors, including protein–water interactions, pi–pi contacts, disorder, hydrogen bonding—such as in the context of a kinked beta-structure—and electrostatics. By clustering sequences based on their physical-feature scores, we can differentiate some phase-separating sequences by their contributing driving forces, suggesting one contributor to the basis for specificity in the formation of the large number of unique biomolecular condensates found in biology. However, we found that many proteins used combinations of most or all of the features, reflecting their highly interdependent nature. We also observed that almost all the features were correlated with protein–water interactions. Therefore, the idea of protein–protein interactions driving phase separation themselves is simplistic, and for biomolecular condensates there is likely always a three-way interaction involving two or more protein groups and water. LLPhyScore should be a useful tool for the protein phase-separation field to provide hypotheses regarding key interactions driving phase separation, as well as for screening proteins that may play important biological roles in the context of biomolecular condensates.

## 5. Technical Methods

### 5.1. Curation of PS-Positive Sequences

We performed a search on PubMed for all papers published from July 2013 to January 2019 that contained keywords related to phase separation ("phase separation", "liquid condensates", "membraneless organelles", etc.), and manually screened 142 papers out of 689 articles that described in vitro phase-separating systems. Then, we extracted all the sequences from the papers (main content/supplementary information/Uniprot entry) that had clear evidence of phase separation on their own (either a detailed phase diagram, or mentioned as "phase separation positive" in the text) in the content. A total of 565 sequences were extracted and were checked twice (Supplementary Table S1, Supplementary File S1).

### 5.2. Clustering of PS-Positive Sequences

The clustering of positive sequences was performed by hierarchical clustering (shown in Supplementary File S2 and Supplementary Figure S1). First, a 20 × 20 dipeptide count number grid was calculated for each sequence, with each number being the number of a residue pair (e.g., AG) in the sequence (Equation (1)). Then, a Jaccard similarity value was calculated for any two sequences by dividing the overlap of the union of two 20 × 20 grids (Equation (2)). If two sequences had different lengths, then a sliding window of the smaller length was applied to the longer sequence, and the highest similarity value calculated for all sliding windows was kept. Finally, we used the hierarchical clustering package in Python scikit-learn [71] to conduct the clustering for 565 sequences. A cutoff similarity threshold of 0.5 was chosen.

$$A, B = \begin{bmatrix} N_{(Ala-Ala)} & \cdots & N_{(Ala-Val)} \\ \vdots & \ddots & \vdots \\ N_{(Val-Ala)} & \cdots & N_{(Val-Val)} \end{bmatrix} = (a_{ij}) \in \mathbb{R}^{20\times 20}, (b_{ij}) \in \mathbb{R}^{20\times 20} \quad (1)$$

$$J(A,B) = \frac{|A \cap B|}{|A \cup B|} = \sum_{i=1}^{20}\sum_{j=1}^{20} \min(a_{ij}, b_{ij}) \Big/ \sum_{i=1}^{20}\sum_{j=1}^{20} \max(a_{ij}, b_{ij}) \quad (2)$$

### 5.3. Preparation of PS-Negative Sequences

Two PS-negative sequence databases were prepared. First, 16,794 sequences were collected from high-resolution (≤2.0 A) structures in the PDB as the first negative sequence database ("PDB base"; Supplementary File S3); Second, 20,380 human proteome sequences were collected from Uniprot [72] (Supplementary File S4), then we used their CRAPome scores calculated in Vernon et al. [38] as a filter for PS-positive sequences. Sequences with either null values or high values (top 20%) in CRAPome were removed, resulting in a "clean" human set of 6102 sequences ("Human base"; Supplementary Table S2). The CRAPome-filtered curated human proteome set should have fewer positives than the uncurated human proteome, with final LLPhyScores shown for these two sets in Supplementary Figure S12, demonstrating an overall shift to negative scores for the curated sequences.

From these two PS-negative sequence databases, three negative sets were prepared: (a) a PDB set, including 3406 sequences randomly selected from the PDB base; (b) a human set, including 3406 sequences randomly selected from the human base; (c) a human + PDB set, including 1703 sequences randomly selected from the PDB set and 1703 sequences randomly selected from the human set.

### 5.4. Construction of Training/Test/Evaluation Datasets

The construction of the training set and test set began with PS-positive sequences. Random sampling was conducted on 565 PS-positive sequences at the clustered group level, with 305 sequences assigned to the training set and 260 sequences assigned to the test set. Then, a 50–50% split ratio was applied to three PS-negative sets at the sequence level, with 1703 sequences from each set assigned to the training set, and another 1703 sequences assigned to the test set. A total of three training–test set pairs were constructed accordingly.

Two evaluation sets were constructed. (1) The entire PS-positive set (565 sequences) + the entire PDB base (16,794); (2) The entire PS-positive set (565 sequences) + the entire human base (20,380 sequences).

For more details, see Supplementary Table S1.

### 5.5. Physical-Feature-Based Sequence Representation

Eight different pairs of general phase-separation-driving factors were defined to represent a protein sequence, resulting in a total of 16 physical features, as summarized in Supplementary Table S5. For each of these features, its sequence-based statistics (contact frequency/number of atoms/structure probability) in the PDB were acquired by mining the structures of folded proteins in the PDB. The observations were split by distinct residue pairs with varying sequence separations, leading to a database of "feature values", with each "feature value" being an empirical, per amino acid energy potential corresponding to the frequencies of specific contact types in the PDB. Then, for a given input sequence, inferred "feature values" for each residue of this sequence were obtained by matching its residue pair and sequence context to the "feature value database". For example, the short-range pi–pi contact frequency for valine in the tripeptide valine–glycine–tryptophan can be inferred by taking the average short-range pi–pi contact frequency for the residue pair valine–glycine with 0 separation and valine–tryptophan with 1-residue separation (see also Figure 2).

Specific definitions for each of these are as follows:

Pi–pi Contacts. Pi–pi contacts were defined using the method in Vernon et al. [29], and then divided into short-range and long-range by sequence separation. Less than five residues apart was defined as short-range, and greater than or equal to 5 residues apart was defined as long-range.

Hydrogen Bonding Terms. Structures were probed for OH-N hydrogen bonds using PHENIX [62], with the following commands used to extract hydrogen-bond information.

Phenix.reduce -Quiet -FLIP [pdb file] > /PHENIX_ALL/PHENIXL.pdb

Phenix.probe "NITROGEN,OXYGEN,HYDROGEN" -Quiet -ONEDOTeach -NOCLASHOUT -SUMMARY -NOVDWOUT. /PHENIX_ALL/PHENIXL.pdb | grep greentint > /N17.PHENIX/HLIST.txt

Bonds were than classified as short-range and long-range by sequence separation (short-range $< 5$, long-range $\geq 5$).

Water/Carbon Contact Counts. Water and carbon counts were calculated only for the subset of proteins in our training set that had a total number of water molecules greater than the number of protein residues. This captured almost all of the models with a resolution $\leq 1.8$ but removed lower-resolution models. Counts were measured for residues in their crystallographic context (measurement includes atoms from symmetry partners).

Secondary Structure. The DSSP letter code was used for secondary-structure assignments, with H/G used for helix, E for strand, and all others binned to loop.

Disorder. For identifying disordered residues, a DSSP assignment of "not G/H/E" over a span of at least 3 residues was used to classify residues as loops. These loop residues were then assigned as short disorder if they fell within 3 residues of G/H/E and as long disorder if they did not.

Charge. PHENIX (via the phenix.reduce command) was used to complete the PDB structures by adding hydrogen atoms, and charge interactions were calculated using the following pseudocode, with partial charges taken from the Talaris energy function [60].

q1 = partial_charge for atom X of amino acid 1
q2 = partial_charge for atom Y of amino acid 2
absF = 330.0 * abs(q1*q2)/(distance**2)
if q1*q2 < 0.0: absF * = −1.0
if SequenceSeparation ≥ 10: add absF to electrostatic (long-range)
if SequenceSeparation < 10: add absF to electrostatic (short-range)
Final per-residue values were then binned as follows:
bin = np.clip(int(round(residue_value/16.0)), −9, 9)

Cation–Pi. We recalculated the electrostatic scores after adding arbitrary partial charges to the surfaces of aromatic rings, with a partial charge value of −0.05 added 0.85 Å above and below the plane of the ring for each atom, counterbalanced by a partial charge of 0.1 at the atom. The cation–pi score was then taken from the difference between this modified score and the unmodified electrostatic score.

Kinked Beta. Superpositions to kinked beta-fibrils were made for chain A in each of 5 structures, PDB IDs 6bwz, 6bxv, 6bxx, 6bzm, and 6bzp. The full chain of each was superimposed to every overlapping window (same number of residues as the chain with none missing) in our PDB training set, and kinked-beta similarity was measured for each individual PDB residue by taking the minimum CA-RMSD over all of the measurements the residue was involved in. Residues were then classified as K-Beta similar if the minimum CA-RMSD was under 1.0 Å and as K-Beta dissimilar if it was over 2.0 Å.

These 16 physical features were converted to an inferred feature statistics value for every sequence with representation at the residue level and sequence level. At the residue level, each amino acid was represented by $16 \times 3$ numbers describing the impact of each of the 16 biophysical forces on each residue: (1) the amino acid position number, (2) the score from the comparison to the upper feature value threshold ($W_U$) and (3) the score from the comparison to the lower feature value threshold ($W_L$).

Inferred feature statistics for a protein sequence were based on $16 \times 20 \times N$ matrices, based on three components in the sequence representation, which function as 3 layers of our machine-learning model architecture. (i) A sequence is characterized by 16 physical features acting on each residue. (ii) The impact of each physical feature is dependent on residue type, represented by 20 residue-type groups. (iii) $N$ is the number of residues of a specific type within the sequence, with $z$ being the position (or index, see below).

Thus, the inferred feature statistical values are determined by translating protein sequences into $16 \times 20 \times N$ matrices (See Equation (3) and Figure 2).

$$S = En(seq)[x][y][z] \tag{3}$$

where

$$x \in 16\ features,$$

$$y \in 20\ residues,$$

$$z \in N\ residue\ positions,$$

$$S\text{—}inferred\ feature\ statistics\ value\ from\ PDB.$$

### 5.6. Predictor Training

Predictor training had the following steps: (1) For each physical feature and each residue type, we set a upper and lower threshold ("weight") for its inferred feature value, thereby constructing a 16 × 20 × 2 (each feature has two weight values: upper and lower threshold) array (Equation (4)). (2) We initialized the "sum feature score" for each physical score to 0. (3) For each residue in a sequence, if its feature score was higher than the upper threshold, we considered this residue as "abnormally active" in terms of this physical feature, and rewarded the corresponding "sum feature score" by adding 1 to it; if its feature score was lower than the lower threshold, then we considered it as "abnormally inactive" in terms of this physical feature, and penalized the corresponding "sum feature score" by subtracting 1 from it; if its feature score was between the upper and lower thresholds, then we considered it to be "within normal range", and did nothing (Equations (5) and (6)). (4) By optimizing the AUROC score function (Equation (7)) for each feature, we found the best feature combination and the best weight that maximized the gap of the sum feature score(s) between PS-positive sequences and PS-negative sequences (Equation (8)). (5) By summing "sum feature scores" and training the weights of features using a genetic algorithm, we calculated a "total sum probability" for any sequence, which was the final estimate of its phase-separation ability (Equation (9)).

$$W = Th[x][y] = \begin{pmatrix} W_U \\ W_L \end{pmatrix} \tag{4}$$

where

$$x \in 16\ features,$$

$$y \in 20\ residues,$$

$$W_U\text{—}upper\ feature\ value\ threshold,$$

$$W_L\text{—}lower\ feature\ value\ threshold.$$

$$f(x, seq, W) = \sum_{y=1}^{20} \sum_{z=1}^{N} P(En(seq)[x][y][z], W) \tag{5}$$

$$P(S, W) = \sum((S > W_U \rightarrow 1) + (S < W_L \rightarrow -1)) \tag{6}$$

where

$$x \in 16\ features,$$

$$y \in 20\ residues,$$

$$z \in N\ residue\ positions,$$

$$S\text{—}Inferred\ feature\ values,$$

$$W\text{—}Weights\ for\ inferred\ feature\ values.$$

$$AUC(f, X, W) = \sum_{x \in X} \frac{\sum_{seq_1 \in D^{PDB}} \sum_{seq_2 \in D^{PS}} (f(x, seq_1, W) < f(x, seq_2, W) \rightarrow 1)}{|D^{PDB}| \cdot |D^{PS}|} \tag{7}$$

$$W_{optimum} = \arg\max_{X,W} AUC(f, X, W) \tag{8}$$

where

$$X\text{—}selected\ feature\ combination,$$

$$x \in X\ features,$$

$$f\text{—}feature\ score\ function\ (Equation\ (5)),$$

$$D^{PDB}\text{-}set\ of\ sequences\ from\ PDB\ (negative\ samples),$$

$$D^{PS}\text{-}set\ of\ sequences\ that\ are\ PS-positive\ (positive\ samples).$$

$$Pred(seq) = \sum_{x \in X} f(x, seq, W_{optimum}) \quad (9)$$

where

$$seq\text{—}input\ sequence,$$

$$X\text{—}selected\ feature\ combination,$$

$$x \in X\ features,$$

$$f\text{—}feature\ score\ function\ (Equation\ (5)),$$

$$W_{optimum}\text{-}optimized\ weights\ in\ feature\ score\ function.$$

The optimization process for parameters in a predictive algorithm is called "training". In this work, the training of the phase-separation predictor had two parts: (i) training of the upper and lower weights of "binary feature score" for 16 features × 20 residue types; (ii) training of the combination of features to include in the final predictor. Numerically, the number of parameters trained was 16 × 20 × 2 weights = 640 weights (16 biophysical forces; 20 residue types; two weights; $W_U$ (upper threshold); $W_L$ (lower threshold)). Another "hyperparameter" being trained here was the selection of biophysical forces to include, with only 8 out of the 16 biophysical forces ultimately being used to avoid overfitting (requiring only 320 weights). The data used for the initial training were from the sequences of the PS-positive proteins in the training set that were separated from the test set (565–260 sequences) and the PS-negative sequences (1703 from either the PDB, human or human + PDB). The data used for training the "final models" included all 565 sequences of the PS-positive proteins and 3406 sequences from either the PDB, human, or PDB + human PS-negative sets.

This training was conducted on the positive and negative training datasets using a genetic algorithm. Specifically, we randomly generated an initial set of 640 weights, and then, for each iteration, we randomly picked a subset of these 640 weights to change and accepted the changes that improved the behavior (loss function based on "genetic operators"). We performed many iterations until a fixed number of generations was reached. The loss function was the AUROC curve (area under the receiver operating characteristic curve) as described above (Equation (7)); the performance of the predictor was then evaluated using the test set as well as by comparison against the baseline models.

Importantly, we used a genetic algorithm to optimize the weights (parameters) with the overall architecture being a 3-layer "neural network"-like predictive model with a non-convex loss function. For more details on implementation of training and prediction, please see https://github.com/julie-forman-kay-lab/LLPhyScore (accessed on 1 July 2022).

### 5.7. Proteome Analysis

Human proteins with scores in the top 10% of the human proteome using 8 predicted single-feature scores as well as the final predictor (8-feature sum score) were separately uploaded to DAVID 6.7 (https://david-d.ncifcrf.gov/ (accessed on 1 July 2022)) [73]. The enrichments of biological process, cellular component, and molecular function GO terms were analyzed for the proteins, with their respective $p$-values (EASE score) obtained. The resulting GO term enrichments were compared against the results in Vernon et al. [29], Vernon et al. [38], and Chu et al. [40].

**Supplementary Materials:** The following supporting information can be downloaded at: https://www.mdpi.com/article/10.3390/biom12081131/s1, This article contains supplementary material. Supplementary Figures S1–S12 and Supplementary Tables S4–S7 are included in the Supplementary Material document. Additional Supplementary Tables and Files are provided as separate files: One attached Excel file contains, on separate tabs, Tables S1–S3 and S8–S11. Table S1. Detailed information of 565 PS-positive sequences with PMID of each sequence's paper. Table S2. Uniprot IDs of 6102 sequences from human proteome that represent the negative training set using CRAPome as filtering method. Table S3. LLPhyScore and CRAPome scores for all human sequences, including both those within the curated negative training set and those not in the curated list. Table S8. (A). GO enrichment analysis for PDB + human model. (B). GO enrichment analysis for PDB model. (C). GO enrichment analysis for human model. Table S9. Uniprot IDs of 3406 sequences from PDB base. Table S10. Uniprot IDs of 3406 sequences randomly selected from human base in Table S2. Table S11. Uniprot IDs of 6812 sequences from PDB + human base. File S1. 565 PS-positive sequences (fasta file). File S2. Hierarchical clustering dendrogram of PS-positive sequences (pdf file). File S3. 16,794 PDB sequences (fasta file). File S4. 20,380 human sequences (fasta file). The software for running LLPhyScore and more details on the training are provided in the following GitHub: https://github.com/julie-forman-kay-lab/LLPhyScore (accessed on 1 July 2022).

**Author Contributions:** Conceptualization, R.M.V., H.C. and J.D.F.-K.; Methodology, R.M.V. and H.C.; Software, H.C.; Validation, R.M.V. and H.C.; Formal Analysis, R.M.V. and H.C.; Investigation, H.C. and R.M.V.; Resources, J.D.F.-K.; Data Curation, H.C.; Writing—Original Draft Preparation, H.C. and R.M.V.; Writing—Review & Editing, J.D.F.-K., R.M.V. and H.C.; Visualization, H.C. and R.M.V.; Supervision, R.M.V. and J.D.F.-K.; Project Administration, J.D.F.-K.; Funding Acquisition, J.D.F.-K. All authors have read and agreed to the published version of the manuscript.

**Funding:** The work was supported by funds to J.D.F.-K. from the Natural Sciences and Engineering Research Council of Canada (NSERC, 2016-06718), the Canadian Institutes of Health Research (CIHR, FND-148375) and the Canada Research Chairs Program.

**Institutional Review Board Statement:** Not applicable.

**Informed Consent Statement:** Not applicable.

**Data Availability Statement:** Please see https://github.com/julie-forman-kay-lab/LLPhyScore (accessed on 1 July 2022) for details of implementation of training and prediction as well as to utilize the code. Data from analysis regarding functional annotations were separately uploaded to DAVID 6.7 (https://david-d.ncifcrf.gov/ (accessed on 1 July 2022)) [73].

**Acknowledgments:** We sincerely thank Zhuqing Zhang and Jianfeng Pei for providing PSPredictor scores and Monika Fuxreiter and Michele Vendruscolo for providing FuzDrop scores, in both cases for the human proteome and PDB sequences. The authors acknowledge Alan Moses and Andrew Chong for critical reading of the manuscript.

**Conflicts of Interest:** J.D.F.-K. is an advisor for Faze Medicines. The authors declare that this affiliation has not influenced the work reported here in any way.

## References

1. Banani, S.F.; Lee, H.O.; Hyman, A.A.; Rosen, M.K. Biomolecular condensates: Organizers of cellular biochemistry. *Nat. Rev. Mol. Cell Biol.* **2017**, *18*, 285–298. [CrossRef] [PubMed]
2. Li, P.; Banjade, S.; Cheng, H.-C.; Kim, S.; Chen, B.; Guo, L.; Llaguno, M.; Hollingsworth, J.V.; King, D.S.; Banani, S.F. Phase transitions in the assembly of multivalent signalling proteins. *Nature* **2012**, *483*, 336–340. [CrossRef] [PubMed]
3. Weber, S.C. Evidence for and against liquid-liquid phase separation in the nucleus. *Non-Coding RNA* **2019**, *5*, 50.
4. Mittag, T.; Pappu, R.V. A conceptual framework for understanding phase separation and addressing open questions and challenges. *Mol. Cell* **2022**, *82*, 2201–2214. [CrossRef]
5. Harmon, T.S.; Holehouse, A.S.; Rosen, M.K.; Pappu, R.V. Intrinsically disordered linkers determine the interplay between phase separation and gelation in multivalent proteins. *eLife* **2017**, *6*, e30294. [CrossRef] [PubMed]
6. Hyman, A.A.; Brangwynne, C.P. Beyond Stereospecificity: Liquids and Mesoscale Organization of Cytoplasm. *Dev. Cell* **2011**, *21*, 14–16. [CrossRef] [PubMed]
7. Mitrea, D.M.; Kriwacki, R.W. Phase separation in biology; functional organization of a higher order. *Cell Commun. Signal.* **2016**, *14*, 1. [CrossRef]

8. Su, X.; Ditlev, J.A.; Hui, E.; Xing, W.; Banjade, S.; Okrut, J.; King, D.S.; Taunton, J.; Rosen, M.K.; Vale, R.D. Phase separation of signaling molecules promotes T cell receptor signal transduction. *Science* **2016**, *352*, 595–599. [CrossRef] [PubMed]
9. Chong, P.A.; Forman-Kay, J.D. Liquid–liquid phase separation in cellular signaling systems. *Curr. Opin. Struct. Biol.* **2016**, *41*, 180–186. [CrossRef]
10. Frey, S.; Richter, R.P.; Görlich, D. FG-Rich Repeats of Nuclear Pore Proteins Form a Three-Dimensional Meshwork with Hydrogel-Like Properties. *Science* **2006**, *314*, 815–817. [CrossRef]
11. Hnisz, D.; Shrinivas, K.; Young, R.A.; Chakraborty, A.K.; Sharp, P.A. A Phase Separation Model for Transcriptional Control. *Cell* **2017**, *169*, 13–23. [CrossRef] [PubMed]
12. Al-Husini, N.; Tomares, D.T.; Bitar, O.; Childers, W.S.; Schrader, J.M. α-Proteobacterial RNA Degradosomes Assemble Liquid-Liquid Phase-Separated RNP Bodies. *Mol. Cell* **2018**, *71*, 1027–1039.e14. [CrossRef] [PubMed]
13. Sfakianos, A.P.; Whitmarsh, A.J.; Ashe, M.P. Ribonucleoprotein bodies are phased in. *Biochem. Soc. Trans.* **2016**, *44*, 1411–1416. [CrossRef] [PubMed]
14. Brangwynne, C.P.; Eckmann, C.R.; Courson, D.S.; Rybarska, A.; Hoege, C.; Gharakhani, J.; Jülicher, F.; Hyman, A.A. Germline P Granules Are Liquid Droplets That Localize by Controlled Dissolution/Condensation. *Science* **2009**, *324*, 1729–1732. [CrossRef] [PubMed]
15. Muiznieks, L.D.; Sharpe, S.; Pomès, R.; Keeley, F.W. Role of Liquid–Liquid Phase Separation in Assembly of Elastin and Other Extracellular Matrix Proteins. *J. Mol. Biol.* **2018**, *430*, 4741–4753. [CrossRef]
16. Bellingham, C.M.; Woodhouse, K.A.; Robson, P.; Rothstein, S.J.; Keeley, F.W. Self-aggregation characteristics of recombinantly expressed human elastin polypeptides. *Biochim. Biophys. Acta (BBA)-Protein Struct. Mol. Enzymol.* **2001**, *1550*, 6–19. [CrossRef]
17. Reichheld, S.E.; Muiznieks, L.D.; Keeley, F.W.; Sharpe, S. Direct observation of structure and dynamics during phase separation of an elastomeric protein. *Proc. Natl. Acad. Sci. USA* **2017**, *114*, E4408–E4415. [CrossRef]
18. Wei, W.; Petrone, L.; Tan, Y.; Cai, H.; Israelachvili, J.N.; Miserez, A.; Waite, J.H. An Underwater Surface-Drying Peptide Inspired by a Mussel Adhesive Protein. *Adv. Funct. Mater.* **2016**, *26*, 3496–3507. [CrossRef]
19. Kim, S.; Huang, J.; Lee, Y.; Dutta, S.; Yoo, H.Y.; Jung, Y.M.; Jho, Y.; Zeng, H.; Hwang, D.S. Complexation and coacervation of like-charged polyelectrolytes inspired by mussels. *Proc. Natl. Acad. Sci. USA* **2016**, *113*, E847–E853. [CrossRef]
20. Le Ferrand, H.; Duchamp, M.; Gabryelczyk, B.; Cai, H.; Miserez, A. Time-Resolved Observations of Liquid–Liquid Phase Separation at the Nanoscale Using in Situ Liquid Transmission Electron Microscopy. *J. Am. Chem. Soc.* **2019**, *141*, 7202–7210. [CrossRef]
21. Gabryelczyk, B.; Cai, H.; Shi, X.; Sun, Y.; Swinkels, P.J.M.; Salentinig, S.; Pervushin, K.; Miserez, A. Hydrogen bond guidance and aromatic stacking drive liquid-liquid phase separation of intrinsically disordered histidine-rich peptides. *Nat. Commun.* **2019**, *10*, 5465. [CrossRef] [PubMed]
22. Cai, H.; Gabryelczyk, B.; Manimekalai, M.S.S.; Grüber, G.; Salentinig, S.; Miserez, A. Self-coacervation of modular squid beak proteins—A comparative study. *Soft Matter* **2017**, *13*, 7740–7752. [CrossRef] [PubMed]
23. Tan, Y.; Hoon, S.; Guerette, P.A.; Wei, W.; Ghadban, A.; Hao, C.; Miserez, A.; Waite, J.H. Infiltration of chitin by protein coacervates defines the squid beak mechanical gradient. *Nat. Chem. Biol.* **2015**, *11*, 488–495. [CrossRef] [PubMed]
24. Conicella, A.E.; Zerze, G.H.; Mittal, J.; Fawzi, N.L. ALS Mutations Disrupt Phase Separation Mediated by α-Helical Structure in the TDP-43 Low-Complexity C-Terminal Domain. *Structure* **2016**, *24*, 1537–1549. [CrossRef] [PubMed]
25. Ambadipudi, S.; Biernat, J.; Riedel, D.; Mandelkow, E.; Zweckstetter, M. Liquid–liquid phase separation of the microtubule-binding repeats of the Alzheimer-related protein Tau. *Nat. Commun.* **2017**, *8*, 275. [CrossRef] [PubMed]
26. Nott, T.J.; Petsalaki, E.; Farber, P.; Jervis, D.; Fussner, E.; Plochowietz, A.; Craggs, T.D.; Bazett-Jones, D.P.; Pawson, T.; Forman-Kay, J.D.; et al. Phase Transition of a Disordered Nuage Protein Generates Environmentally Responsive Membraneless Organelles. *Mol. Cell* **2015**, *57*, 936–947. [CrossRef] [PubMed]
27. Lin, Y.-H.; Forman-Kay, J.D.; Chan, H.S. Sequence-Specific Polyampholyte Phase Separation in Membraneless Organelles. *Phys. Rev. Lett.* **2016**, *117*, 178101. [CrossRef]
28. Pak, C.W.; Kosno, M.; Holehouse, A.S.; Padrick, S.B.; Mittal, A.; Ali, R.; Yunus, A.A.; Liu, D.R.; Pappu, R.V.; Rosen, M.K. Sequence Determinants of Intracellular Phase Separation by Complex Coacervation of a Disordered Protein. *Mol. Cell* **2016**, *63*, 72–85. [CrossRef]
29. Vernon, R.M.; Chong, P.A.; Tsang, B.; Kim, T.H.; Bah, A.; Farber, P.; Lin, H.; Forman-Kay, J.D. Pi-Pi contacts are an overlooked protein feature relevant to phase separation. *eLife* **2018**, *7*, e31486. [CrossRef]
30. Quiroz, F.G.; Chilkoti, A. Sequence heuristics to encode phase behaviour in intrinsically disordered protein polymers. *Nat. Mater.* **2015**, *14*, 1164–1171. [CrossRef]
31. Brangwynne, C.P.; Tompa, P.; Pappu, R.V. Polymer physics of intracellular phase transitions. *Nat. Phys.* **2015**, *11*, 899–904. [CrossRef]
32. Sherrill, C.D. Energy Component Analysis of π Interactions. *Acc. Chem. Res.* **2013**, *46*, 1020–1028. [CrossRef] [PubMed]
33. Hughes, M.P.; Sawaya, M.R.; Boyer, D.R.; Goldschmidt, L.; Rodriguez, J.A.; Cascio, D.; Chong, L.; Gonen, T.; Eisenberg, D.S. Atomic structures of low-complexity protein segments reveal kinked β sheets that assemble networks. *Science* **2018**, *359*, 698–701. [CrossRef] [PubMed]
34. Kato, M.; Han, T.W.; Xie, S.; Shi, K.; Du, X.; Wu, L.C.; Mirzaei, H.; Goldsmith, E.J.; Longgood, J.; Pei, J.; et al. Cell-free Formation of RNA Granules: Low Complexity Sequence Domains Form Dynamic Fibers within Hydrogels. *Cell* **2012**, *149*, 753–767. [CrossRef]

35. Yeo, G.C.; Keeley, F.W.; Weiss, A.S. Coacervation of tropoelastin. *Adv. Colloid Interface Sci.* **2011**, *167*, 94–103. [CrossRef]
36. Zaslavsky, B.Y.; Uversky, V.N. In aqua veritas: The indispensable yet mostly ignored role of water in phase separation and membrane-less organelles. *Biochemistry* **2018**, *57*, 2437–2451. [CrossRef]
37. Mittag, T.; Parker, R. Multiple modes of protein–protein interactions promote RNP granule assembly. *J. Mol. Biol.* **2018**, *430*, 4636–4649. [CrossRef]
38. Vernon, R.M.; Forman-Kay, J.D. First-generation predictors of biological protein phase separation. *Curr. Opin. Struct. Biol.* **2019**, *58*, 88–96. [CrossRef]
39. Boeynaems, S.; Alberti, S.; Fawzi, N.L.; Mittag, T.; Polymenidou, M.; Rousseau, F.; Schymkowitz, J.; Shorter, J.; Wolozin, B.; Van Den Bosch, L. Protein phase separation: A new phase in cell biology. *Trends Cell Biol.* **2018**, *28*, 420–435. [CrossRef]
40. Chu, X.; Sun, T.; Li, Q.; Xu, Y.; Zhang, Z.; Lai, L.; Pei, J. Prediction of liquid–liquid phase separating proteins using machine learning. *BMC Bioinform.* **2022**, *23*, 72. [CrossRef]
41. Mikolov, T.; Sutskever, I.; Chen, K.; Corrado, G.S.; Dean, J. Distributed representations of words and phrases and their compositionality. In Proceedings of the NIPS'13: 26th International Conference on Neural Information Processing Systems, Lake Tahoe, NV, USA, 5–10 December 2013; pp. 3111–3119.
42. Orlando, G.; Raimondi, D.; Tabaro, F.; Codicè, F.; Moreau, Y.; Vranken, W.F. Computational identification of prion-like RNA-binding proteins that form liquid phase-separated condensates. *Bioinformatics* **2019**, *35*, 4617–4623. [CrossRef] [PubMed]
43. Paiz, E.A.; Allen, J.H.; Correia, J.J.; Fitzkee, N.C.; Hough, L.E.; Whitten, S.T. Beta turn propensity and a model polymer scaling exponent identify intrinsically disordered phase-separating proteins. *J. Biol. Chem.* **2021**, *297*, 101343. [CrossRef] [PubMed]
44. van Mierlo, G.; Jansen, J.R.; Wang, J.; Poser, I.; van Heeringen, S.J.; Vermeulen, M. Predicting protein condensate formation using machine learning. *Cell Rep.* **2021**, *34*, 108705. [CrossRef]
45. Berman, H.M.; Westbrook, J.; Feng, Z.; Gilliland, G.; Bhat, T.N.; Weissig, H.; Shindyalov, I.N.; Bourne, P.E. The protein data bank. *Nucleic Acids Res.* **2000**, *28*, 235–242. [CrossRef]
46. Benvenuti, M.; Mangani, S. Crystallization of soluble proteins in vapor diffusion for X-ray crystallography. *Nat. Protoc.* **2007**, *2*, 1633–1651. [CrossRef] [PubMed]
47. Alberti, S.; Halfmann, R.; King, O.; Kapila, A.; Lindquist, S. A Systematic Survey Identifies Prions and Illuminates Sequence Features of Prionogenic Proteins. *Cell* **2009**, *137*, 146–158. [CrossRef]
48. Bolognesi, B.; Lorenzo Gotor, N.; Dhar, R.; Cirillo, D.; Baldrighi, M.; Tartaglia, G.G.; Lehner, B. A Concentration-Dependent Liquid Phase Separation Can Cause Toxicity upon Increased Protein Expression. *Cell Rep.* **2016**, *16*, 222–231. [CrossRef]
49. Hardenberg, M.; Horvath, A.; Ambrus, V.; Fuxreiter, M.; Vendruscolo, M. Widespread occurrence of the droplet state of proteins in the human proteome. *Proc. Natl. Acad. Sci. USA* **2020**, *117*, 33254–33262. [CrossRef]
50. Li, Q.; Peng, X.; Li, Y.; Tang, W.; Zhu, J.; Huang, J.; Qi, Y.; Zhang, Z. LLPSDB: A database of proteins undergoing liquid–liquid phase separation in vitro. *Nucleic Acids Res.* **2019**, *48*, D320–D327. [CrossRef]
51. You, K.; Huang, Q.; Yu, C.; Shen, B.; Sevilla, C.; Shi, M.; Hermjakob, H.; Chen, Y.; Li, T. PhaSepDB: A database of liquid–liquid phase separation related proteins. *Nucleic Acids Res.* **2019**, *48*, D354–D359. [CrossRef]
52. Mészáros, B.; Erdős, G.; Szabó, B.; Schád, É.; Tantos, Á.; Abukhairan, R.; Horváth, T.; Murvai, N.; Kovács, O.P.; Kovács, M.; et al. PhaSePro: The database of proteins driving liquid–liquid phase separation. *Nucleic Acids Res.* **2019**, *48*, D360–D367. [CrossRef] [PubMed]
53. Mellacheruvu, D.; Wright, Z.; Couzens, A.L.; Lambert, J.-P.; St-Denis, N.A.; Li, T.; Miteva, Y.V.; Hauri, S.; Sardiu, M.E.; Low, T.Y.; et al. The CRAPome: A contaminant repository for affinity purification–mass spectrometry data. *Nat. Methods* **2013**, *10*, 730–736. [CrossRef] [PubMed]
54. Ribeiro, S.S.; Samanta, N.; Ebbinghaus, S.; Marcos, J.C. The synergic effect of water and biomolecules in intracellular phase separation. *Nat. Rev. Chem.* **2019**, *3*, 552–561. [CrossRef]
55. Conicella, A.E.; Dignon, G.L.; Zerze, G.H.; Schmidt, H.B.; Alexandra, M.; Kim, Y.C.; Rohatgi, R.; Ayala, Y.M.; Mittal, J.; Fawzi, N.L. TDP-43 α-helical structure tunes liquid–liquid phase separation and function. *Proc. Natl. Acad. Sci. USA* **2020**, *117*, 5883–5894. [CrossRef] [PubMed]
56. Frishman, D.; Argos, P. Knowledge-based protein secondary structure assignment. *Proteins Struct. Funct. Bioinform.* **1995**, *23*, 566–579. [CrossRef]
57. Walsh, I.; Giollo, M.; Di Domenico, T.; Ferrari, C.; Zimmermann, O.; Tosatto, S.C. Comprehensive large-scale assessment of intrinsic protein disorder. *Bioinformatics* **2015**, *31*, 201–208. [CrossRef]
58. Mohan, A.; Uversky, V.N.; Radivojac, P. Influence of sequence changes and environment on intrinsically disordered proteins. *PLoS Comput. Biol.* **2009**, *5*, e1000497. [CrossRef]
59. Boyko, S.; Qi, X.; Chen, T.-H.; Surewicz, K.; Surewicz, W.K. Liquid–liquid phase separation of tau protein: The crucial role of electrostatic interactions. *J. Biol. Chem.* **2019**, *294*, 11054–11059. [CrossRef]
60. O'Meara, M.J.; Leaver-Fay, A.; Tyka, M.D.; Stein, A.; Houlihan, K.; DiMaio, F.; Bradley, P.; Kortemme, T.; Baker, D.; Snoeyink, J. Combined covalent-electrostatic model of hydrogen bonding improves structure prediction with Rosetta. *J. Chem. Theory Comput.* **2015**, *11*, 609–622. [CrossRef]
61. Murthy, A.C.; Dignon, G.L.; Kan, Y.; Zerze, G.H.; Parekh, S.H.; Mittal, J.; Fawzi, N.L. Molecular interactions underlying liquid−liquid phase separation of the FUS low-complexity domain. *Nat. Struct. Mol. Biol.* **2019**, *26*, 637–648. [CrossRef]

62. Adams, P.D.; Afonine, P.V.; Bunkóczi, G.; Chen, V.B.; Davis, I.W.; Echols, N.; Headd, J.J.; Hung, L.-W.; Kapral, G.J.; Grosse-Kunstleve, R.W. PHENIX: A comprehensive Python-based system for macromolecular structure solution. *Acta Crystallogr. Sect. D Biol. Crystallogr.* **2010**, *66*, 213–221. [CrossRef] [PubMed]
63. Wang, J.; Choi, J.-M.; Holehouse, A.S.; Lee, H.O.; Zhang, X.; Jahnel, M.; Maharana, S.; Lemaitre, R.; Pozniakovsky, A.; Drechsel, D.; et al. A Molecular Grammar Governing the Driving Forces for Phase Separation of Prion-like RNA Binding Proteins. *Cell* **2018**, *174*, 688–699.e16. [CrossRef] [PubMed]
64. Hughes, M.P.; Goldschmidt, L.; Eisenberg, D.S. Prevalence and species distribution of the low-complexity, amyloid-like, reversible, kinked segment structural motif in amyloid-like fibrils. *J. Biol. Chem.* **2021**, *297*, 101194. [CrossRef]
65. Murray, D.T.; Kato, M.; Lin, Y.; Thurber, K.R.; Hung, I.; McKnight, S.L.; Tycko, R. Structure of FUS protein fibrils and its relevance to self-assembly and phase separation of low-complexity domains. *Cell* **2017**, *171*, 615–627.e16. [CrossRef] [PubMed]
66. Das, R.K.; Pappu, R.V. Conformations of intrinsically disordered proteins are influenced by linear sequence distributions of oppositely charged residues. *Proc. Natl. Acad. Sci. USA* **2013**, *110*, 13392–13397. [CrossRef] [PubMed]
67. Firman, T.; Ghosh, K. Sequence charge decoration dictates coil-globule transition in intrinsically disordered proteins. *J. Chem. Phys.* **2018**, *148*, 123305. [CrossRef]
68. Enkhbayar, P.; Hikichi, K.; Osaki, M.; Kretsinger, R.H.; Matsushima, N. 310-helices in proteins are parahelices. *Proteins Struct. Funct. Bioinform.* **2006**, *64*, 691–699. [CrossRef]
69. Fiori, W.R.; Miick, S.M.; Millhauser, G.L. Increasing sequence length favors. alpha.-helix over 310-helix in alanine-based peptides: Evidence for a length-dependent structural transition. *Biochemistry* **1993**, *32*, 11957–11962. [CrossRef]
70. Doig, A.J.; Stapley, B.J.; Macarthur, M.W.; Thornton, J.M. Structures of N-termini of helices in proteins. *Protein Sci.* **1997**, *6*, 147–155. [CrossRef]
71. Pedregosa, F.; Varoquaux, G.; Gramfort, A.; Michel, V.; Thirion, B.; Grisel, O.; Blondel, M.; Prettenhofer, P.; Weiss, R.; Dubourg, V. Scikit-learn: Machine learning in Python. *JMLR* **2011**, *12*, 2825–2830.
72. Pundir, S.; Martin, M.J.; O'Donovan, C. UniProt Protein Knowledgebase. In *Protein Bioinformatics: From Protein Modifications and Networks to Proteomics*; Wu, C.H., Arighi, C.N., Ross, K.E., Eds.; Springer: New York, NY, USA, 2017; pp. 41–55.
73. Huang, D.W.; Sherman, B.T.; Tan, Q.; Kir, J.; Liu, D.; Bryant, D.; Guo, Y.; Stephens, R.; Baseler, M.W.; Lane, H.C.; et al. DAVID Bioinformatics Resources: Expanded annotation database and novel algorithms to better extract biology from large gene lists. *Nucleic Acids Res.* **2007**, *35* (Suppl. S2), W169–W175. [CrossRef] [PubMed]

that pH modulates the stability of Crkl-SH2, as shown by the change in the denaturation midpoint, the $m_{D\text{-}N}$ values were relatively robust to changes in experimental conditions, reinforcing the two-state nature of the equilibrium transition (Figure 1B). The $m_{D\text{-}N}$ value calculated from global fitting at different pH values was 3.4 ± 0.1 Kcal $mol^{-1}$ $M^{-1}$, which is consistent with a protein of 111 residues [23].

To address the mechanism of folding of Crkl-SH2, we carried out time-resolved fluorescence monitored stopped-flow experiments at different ionic strengths and pH conditions. Under all the investigated conditions, the refolding and unfolding traces could be satisfactorily fitted to a single-exponential process. Figures 2 and 3 report the logarithms of observed rate constants plotted against the Gdn-HCl concentration to generate chevron plots. Interestingly, a clear deviation from linearity, classically denoted as a "roll-over effect" [24,25], was evident in the refolding branch at a low concentration of the denaturing agent. This finding represents a classical signature for the presence of a partially folded intermediate whose accumulation transiently occurs at a specific denaturant concentration and parallels what was previously observed in the case of the folding mechanism of the N-SH2 domain from SHP2 [26]. Inspection of Figure 2 shows that, as expected, the addition of sodium-sulphate determines a stabilization of both the native and intermediate states and allows an inference of the roll-over effect with a remarkably increased level of confidence. Hence, we carried out complete pH dependence both in the presence (Figure 3A) and in the absence (Figure 3B) of this stabilizing salt. All data were fitted to a kinetic three-state model as formalized in Equation (2), and we calculated the kinetic parameters referring to the rate constants $k_{IN}$ and $k_{NI}$, the equilibrium constant $K_{IU}$ ($K_{IU} = k_{IU}/k_{UI}$) and their associated *m* values (Tables S1 and S2).

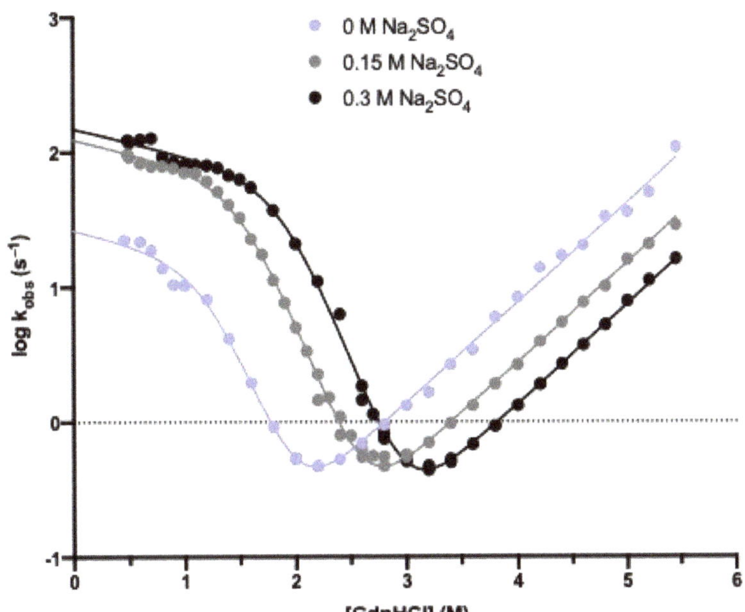


**Figure 2.** Kinetic (un)folding experiments of the Crkl SH2 domain carried out at 298 K in buffer 50 mM sodium-HEPES pH 7.2 containing different sodium-sulphate concentrations. The logarithm of the observed rate constants measured with the stopped-flow apparatus is plotted versus the concentration of guanidine-HCl. The lines are the best fit to a three-state model as formalized in Equation (2). The related kinetic parameters are listed in Table S1. For each acquisition, five traces were collected, averaged and satisfactorily fitted to a single-exponential equation.

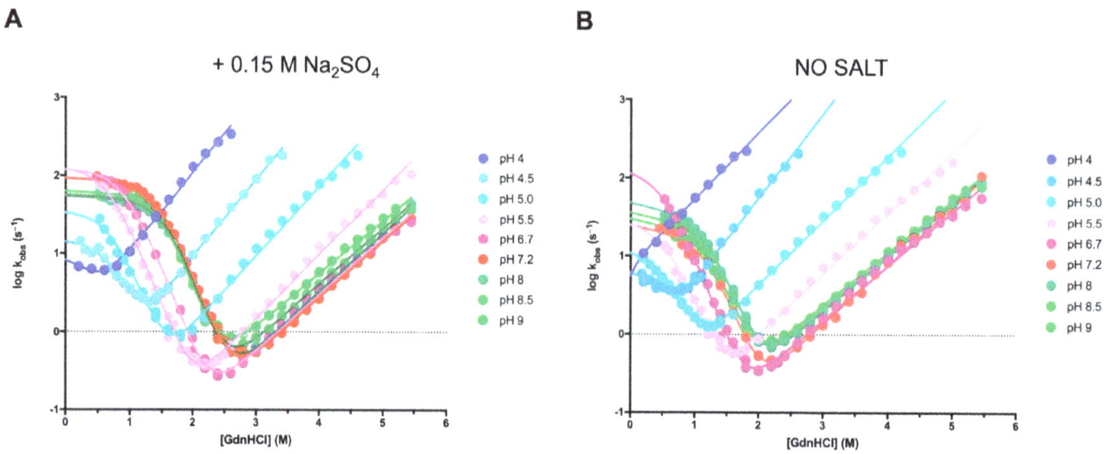


**Figure 3.** Kinetic (un)folding experiments of the Crkl SH2 domain at different pH conditions and 298 K. The logarithm of the observed rate constants measured with the stopped-flow apparatus is plotted versus the concentration of guanidine-HCl, in the presence (**A**) and absence (**B**) of 0.15 M $Na_2SO_4$. The lines are the best fit to a three-state model as formalized in Equation (2). The kinetic parameters referring to chevron plots shown in (**A**) are listed in Table S2. For each acquisition, five traces were collected, averaged and satisfactorily fitted to a single-exponential equation.

### *3.2. The Binding Reaction between the SH2 Domain of Crkl and Paxillin*

To elucidate the details of the interaction occurring between the SH2 domain of Crkl and Paxillin, we monitored binding with a peptide mimicking a specific region of Paxillin, ranging from residues 112 to 123 and carrying a dansyl fluorophore covalently attached to the N-terminus ($Pax_{112-123}$ $N_{TERM}$-Dans-GEEEHV-pY-SFPNK-$C_{TERM}$). In analogy to what was described in our previous works [26,27], the binding kinetics were followed spectroscopically with the stopped-flow apparatus by measuring the change in the FRET (Fluorescence Resonance Energy Transfer) signal upon binding. In this system, the energy is transferred from the tryptophan residue in position 14 of the SH2 domain (donor) to the dansyl group of the $Pax_{112-123}$ peptide (acceptor) when binding occurs. By following this approach, a fixed concentration of dansylated $Pax_{112-123}$ (2 μM) was rapidly mixed with increasing concentrations of the SH2 domain (from 2 to 10 μM) at 283 K. To test the contribution of electrostatic interactions in the complex formation, binding kinetic experiments were carried out at different ionic strengths (0.15, 0.3, 0.5 and 1 M sodium-chloride) and pH (from 5 to 9 pH) conditions. All the binding traces obtained by the time-resolved fluorescence monitoring were satisfactorily fitted with a single-exponential equation with flat residuals and typically displaying an R > 0.98, which allowed the extrapolation of the observed rate constants ($k_{obs}$). Then, the $k_{obs}$ values were plotted versus different concentrations of the SH2 domain (Figure 4), and data were fitted by the following linear function:

$$k_{obs} = k_{on}\,[C - SH2] + k_{off}$$

with the slope and the *y*-axis intercept of the line representing the microscopic association ($k_{on}$) and dissociation rate constants ($k_{off}$), respectively. Given the high experimental error associated with indirect measurements of the dissociation rate constants, the $k_{off}$ values were determined through displacement experiments by mixing a pre-formed Crkl-SH2 domain/dansylated $Pax_{112-123}$ complex with a high excess of non-dansylated $Pax_{112-123}$ peptide, as detailed in the Materials and Methods section. Our data displayed a clear decrease in the microscopic association rate constant as the ionic strength increased, while the value of $k_{off}$ was only marginally affected (Figure 4A, Table S3). In agreement with what has already been reported in the literature for the binding reaction of other SH2 domains [26,27], these results confirm the importance of electrostatic interactions in complex formation.

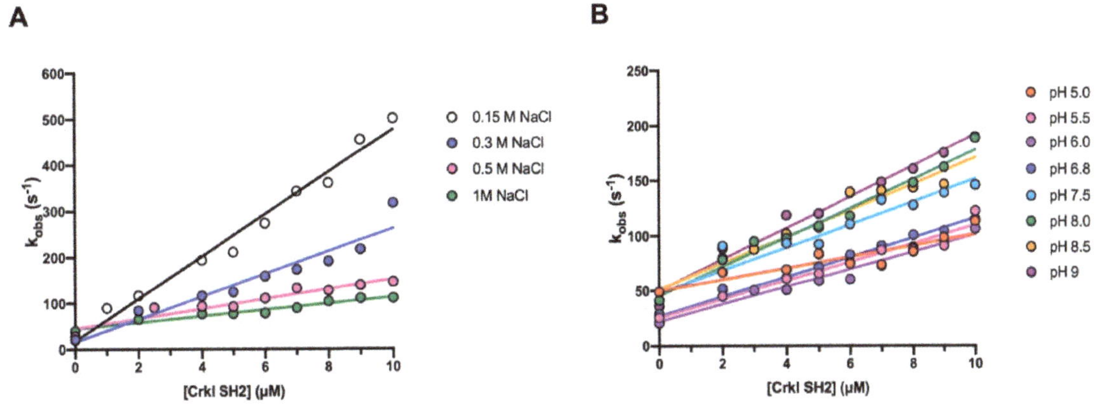


**Figure 4.** Kinetic binding experiments between the Crkl-SH2 domain and $Pax_{112–123}$ peptide. The pseudo-first-order reactions were measured by mixing 2 μM of Dans $Pax_{112–123}$ with increasing protein concentrations (ranging from 2 to 10 μM), at different ionic strengths (**A**) and pH (**B**) conditions. Lines represent the best fit to a linear equation. The related kinetic parameters are listed in Tables S3 and 1, respectively. For each acquisition, five traces were collected, averaged, and satisfactorily fitted to a single-exponential equation.

**Table 1.** Kinetics parameters obtained from pseudo-first-order binding reaction of the wild-type Crkl SH2 domain and histidine-to-alanine mutants with $Pax_{112–123}$ peptide, at different pHs and 283 K.

| WT | | | | H33A | | | |
|---|---|---|---|---|---|---|---|
| pH | $k_{on}$ (μM$^{-1}$ s$^{-1}$) | $k_{off}$ (s$^{-1}$) | $K_D$ (μM) | pH | $k_{on}$ (μM$^{-1}$ s$^{-1}$) | $k_{off}$ (s$^{-1}$) | $K_D$ (μM) |
| 5.0 | 5.2 ± 0.9 | 49.1 ± 1.7 | 9.5 ± 1.7 | 5.0 | 4.4 ± 0.3 | 33.2 ± 1.4 | 7.6 ± 0.7 |
| 5.5 | 8.4 ± 0.7 | 25.6 ± 0.4 | 3.0 ± 0.3 | 5.5 | 6.7 ± 0.4 | 22.8 ± 2.2 | 3.4 ± 0.4 |
| 6.0 | 7.9 ± 0.6 | 20.8 ± 0.3 | 2.6 ± 0.2 | 6.0 | 5.9 ± 0.3 | 19.5 ± 0.4 | 3.3 ± 0.2 |
| 6.8 | 8.9 ± 0.4 | 21.4 ± 0.2 | 2.4 ± 0.1 | 6.8 | 6.6 ± 0.4 | 22.4 ± 0.5 | 3.4 ± 0.2 |
| 7.5 | 10.4 ± 1.4 | 28.6 ± 0.5 | 2.8 ± 0.4 | 7.5 | 10.8 ± 1.9 | 24.6 ± 3.2 | 2.3 ± 0.5 |
| 8.0 | 13.2 ± 0.7 | 41.4 ± 0.9 | 3.1 ± 0.2 | 8.0 | 14.6 ± 1.1 | 29.3 ± 3.4 | 2.0 ± 0.3 |
| 8.5 | 12.0 ± 1.1 | 41.6 ± 0.9 | 3.5 ± 0.3 | 8.5 | 12.6 ± 1.4 | 26.9 ± 2.8 | 2.1 ± 0.3 |
| 9.0 | 14.1 ± 0.9 | 36.1 ± 0.6 | 2.5 ± 0.2 | 9.0 | 10.6 ± 1.0 | 44.7 ± 4.0 | 4.2 ± 0.5 |
| H80A | | | | H91A | | | |
| pH | $k_{on}$ (μM$^{-1}$ s$^{-1}$) | $k_{off}$ (s$^{-1}$) | $K_D$ (μM) | pH | $k_{on}$ (μM$^{-1}$ s$^{-1}$) | $k_{off}$ (s$^{-1}$) | $K_D$ (μM) |
| 5.0 | 4.6 ± 0.4 | 52.7 ± 3.2 | 11.5 ± 1.2 | 5.0 | * | * | * |
| 5.5 | 7.9 ± 0.9 | 20.0 ± 3.7 | 2.5 ± 0.5 | 5.5 | 5.6 ± 0.3 | 24.1 ± 0.4 | 4.3 ± 0.2 |
| 6.0 | 6.1 ± 0.3 | 19.6 ± 0.3 | 3.2 ± 0.1 | 6.0 | 6.6 ± 0.2 | 20.0 ± 0.3 | 3.0 ± 0.1 |
| 6.8 | 7.1 ± 0.4 | 21.6 ± 0.4 | 3.1 ± 0.2 | 6.8 | 7.8 ± 0.3 | 23.4 ± 0.4 | 3.0 ± 0.1 |
| 7.5 | 13.6 ± 1.1 | 27.8 ± 2.3 | 2.0 ± 0.2 | 7.5 | 10.0 ± 0.5 | 25.9 ± 0.3 | 2.6 ± 0.1 |
| 8.0 | 13.2 ± 0.7 | 30.1 ± 3.3 | 2.3 ± 0.3 | 8.0 | 13.2 ± 0.5 | 39.0 ± 0.7 | 2.9 ± 0.1 |
| 8.5 | 10.2 ± 1.2 | 32.5 ± 2.7 | 3.2 ± 0.5 | 8.5 | 13.6 ± 0.6 | 34.9 ± 0.5 | 2.6 ± 0.1 |
| 9.0 | 12.0 ± 0.9 | 34.0 ± 2.3 | 2.8 ± 0.3 | 9.0 | 12.7 ± 0.5 | 36.4 ± 0.5 | 2.9 ± 0.1 |

Note: (*) protein was not stable at this condition.

The dependence of the logarithm of $k_{on}$ and $k_{off}$ versus pH is reported in Figure 5A (wt). Fitting the curves with the Henderson–Hasselbalch equation returns a $pK_a$ of 7.1 ± 0.2, a value close to the $pK_a$ of the side chain of histidine (6.04). Given that the Crkl-SH2 domain contains four histidine residues (His33, His60, His80 and His91) (Figure 5B), these were individually replaced by alanine residues through site-directed mutagenesis with the aim of testing their role in the binding reaction. Thus, four variants of the Crkl-SH2 domain were produced (H33A, H60A, H80A and H91A) and used in kinetic binding experiments to monitor the effect of the mutation under different pH conditions. The $k_{on}$ and $k_{off}$ values obtained by the binding reactions of the H33A, H80A and H91A variants are listed in Table 1. The dependence of these values as a function of pH showed a sigmoidal behavior similar to that observed for the wild-type Crkl SH2 domain (Figure 5A). All curves were satisfactorily fitted with the same equation by sharing the $pK_a$ value of 7.1. Notably, the observed pKa is very different from the N- and C-termini of the peptide, indicating that the observed dependence is not affected by the presence of such electrostatic dipole.

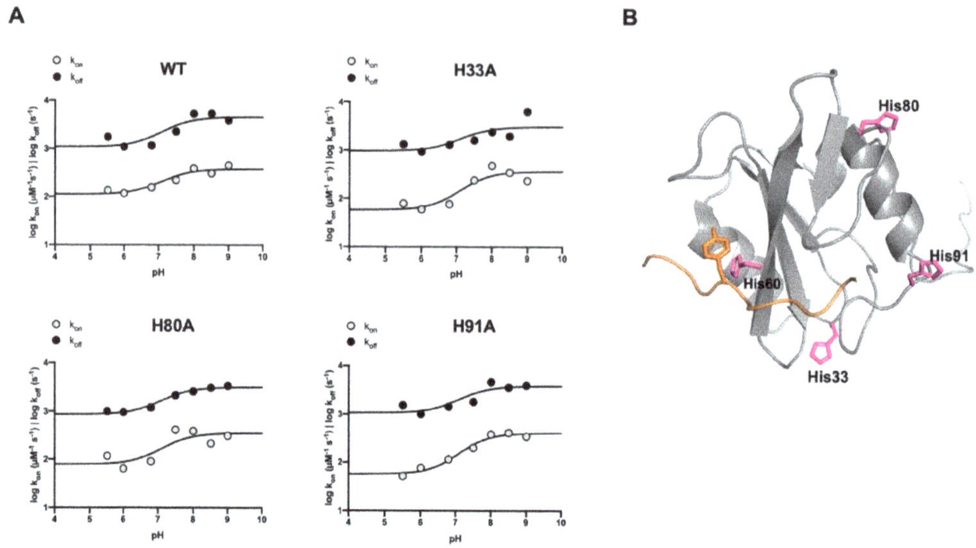


**Figure 5.** The role of histidine residues of Crkl-SH2 in the binding reaction with Paxillin. (**A**) The pH dependence of the binding reaction of wild-type and His-to-Ala SH2 domain mutants with $Pax_{112-123}$ peptide. The logarithm of microscopic association (gray circles) and dissociation (black circles) rate constants is reported as a function of pH. Curves are the best fit to the Henderson–Hasselbalch equation. The related kinetic parameters are listed in Table 1. (**B**) The three-dimensional structure of the Crkl SH2 domain (PDB code: 2EO3) was superimposed to that of the SHP2 N-SH2 domain in complex with GAB1 peptide (not shown; PDB code: 4QSY) to highlight the conserved binding pocket in the Crkl SH2 domain. In orange, a general ligand containing a phospho-tyrosine residue is shown. All the histidine residues of the Crkl SH2 domain (His 33, His 60, His 80 and His 91) are represented as sticks colored magenta.

On the contrary, for the binding reaction of the H60A mutant with dansylated $Pax_{112-123}$, no binding traces could be recorded with the stopped-flow apparatus. However, the formation of the complex between the H60A mutant and $Pax_{112-123}$ was revealed through displacement experiments (as explained in Materials and Methods) carried out under three different pH conditions, 5.5, 6.8 and 7.5. The resulting dissociation rate constant values (212 ± 11, 152 ± 10, 237 ± 21 $s^{-1}$, respectively) were approximately eight-fold higher than those obtained with the wild-type form of Crkl-SH2 under the same experimental conditions. Furthermore, in the case of the H60A variant, the dissociation rate constants appeared essentially insensitive to pH, suggesting that the pH dependence observed in the case of the wild-type Crkl-SH2 and the H33A, H80A and H91A variants may be ascribed to the protonation of His60. This finding is consistent with the proximity of His60 to the ligand, which may be observed in the complex depicted in Figure 5B.

## 4. Discussion

A powerful strategy for unveiling the folding and function of protein domains is represented by the comparison of homologous proteins sharing the same topology and showing different primary structures. From a folding perspective, it appears that the mechanism of folding of Crkl-SH2 is reminiscent to what was previously observed in the case of the N-SH2 of SHP2 [26], with both proteins displaying a pronounced roll-over effect in the refolding branch of the chevron plot. This effect, which is clearly more visible under stabilizing conditions, can nevertheless be detected under different experimental conditions, indicating that intermediate formation is rather robust to changes in protein stability. It should be noticed, however, that previous folding analysis on other SH2 domains have shown that intermediate formation may not be mandatory for the folding of this class of proteins, as exemplified in the case of the SH2 domains from Src and p85 [28,29], which both fold via a two-state mechanism. Notably, the apparent similarity between the folding mechanisms of Crkl-SH2 with N-SH2 of SHP2 does not correspond to a relevant sequence identity, with these two proteins sharing only 24.7% sequence identity. On the other hand, such value is remarkably lower than the percent sequence identity between the N-terminal and C-terminal SH2 domain of SHP2 (41.2%), whose chevron plots has appeared remarkably different among them and displayed an unfolding roll-over in the case of C-SH2 and a refolding roll-over in the case of N-SH2. This finding reinforces the notion that small changes in sequence composition may have profound effects in folding intermediate stability [30].

Because of the importance of Crkl in orchestrating several metabolic pathways [11–13], it is critical to quantitatively establish its interactions with relevant physiological partners. In this context, the results presented in this work allowed a depiction of the mechanism of interaction between Crkl-SH2 and Paxillin, as well as a comparison of them with previously characterized SH2 domains. The kinetic binding experiments indicated that the microscopic association rate constant ($k_{on}$) was substantially reduced as the ionic strength increased, whilst the microscopic dissociation rate constant ($k_{off}$) was not affected by the sodium-chloride concentrations. These observations allow the conclusion that the transition state of the binding reaction is primarily stabilized by the electrostatic recognition between the interacting partners, possibly between the phospho-tyrosine of the peptide and the charge residues of the SH2 domain, contained in its binding pocket. Furthermore, the analysis of our kinetic data obtained under different pH conditions showed that the protonation state of a histidine residue is critical in balancing both the $k_{on}$ and $k_{off}$ rate constants of the binding reaction. In this respect, mutational analysis of the domain suggests that such an effect should be ascribed to the protonation of His60, which is located at the binding site of Crkl-SH2. These findings may be important in elucidating the fine details that determine the stability of Crkl-SH2 for its ligand and, therefore, may contribute to the design of potential inhibitors in these regions.

To conclude, the results presented in this work, together with the analysis of previous results, allowed a depiction of the general features of the folding pathway of SH2 domains. Furthermore, we contributed a quantitative description of a critical interaction that is pivotal for the activation of the dysfunctional signaling pathways mediated by BCR/ABL. In this context, we pave the way for future works aimed at designing inhibitors of this aberrant interaction.

**Supplementary Materials:** The following supporting information can be downloaded at: https://www.mdpi.com/article/10.3390/biom12081014/s1. Table S1: Kinetic (un)folding parameters of the Crkl SH2 domain calculated at different $Na_2SO_4$ concentrations; Table S2: Kinetic (un)folding parameters of the Crkl SH2 domain calculated at different pH conditions in the presence of 0.15 M $Na_2SO_4$; Table S3: Kinetics parameters obtained from pseudo-first-order binding reaction between the wild-type Crkl SH2 domain and $Pax_{112–123}$ peptide, at different ionic strengths and 283 K.

**Author Contributions:** Conceptualization, C.N. and S.G.; investigation, C.N., A.T., D.S., L.P., A.D., V.P., P.P., L.M. and F.M.; writing—original draft preparation, C.N.; writing—review and editing, C.N. and S.G.; supervision, S.G.; funding acquisition, S.G. All authors have read and agreed to the published version of the manuscript.

**Funding:** Work partly supported by grants from Sapienza University of Rome (RP11715C34AEAC9B and RM1181641C2C24B9, RM11916B414C897E, RG12017297FA7223 to S.G.); by an ACIP grant (ACIP 485-21) from Institut Pasteur Paris to S.G.; by the Associazione Italiana per la Ricerca sul Cancro (Individual Grant—IG 24551 to S.G.); by the Regione Lazio (Progetti Gruppi di Ricerca LazioInnova A0375-2020-36559 to S.G.); by the Istituto Pasteur Italia (Teresa Ariaudo Research Project 2018, to A.T.); and by the European Union's Horizon 2020 Research and Innovation programme under the Marie Skłodowska Curie Grant Agreement UBIMOTIF No 860517 (to S.G.). F.M. was supported by a fellowship from the FIRC—Associazione Italiana per la Ricerca sul Cancro (Filomena Todini fellowship).

**Institutional Review Board Statement:** Not applicable.

**Informed Consent Statement:** Not applicable.

**Data Availability Statement:** Data available upon request.

**Conflicts of Interest:** The authors declare no conflict of interest.

## References

1. Mayer, B.J. Protein-protein interactions in signaling cascades. *Methods Mol. Biol.* **2006**, *332*, 79–99. [CrossRef] [PubMed]
2. Liu, B.A.; Engelmann, B.W.; Nash, P.D. The language of SH2 domain interactions defines phosphotyrosine-mediated signal transduction. *FEBS Lett.* **2012**, *586*, 2597–2605. [CrossRef] [PubMed]
3. Marasco, M.; Carlomagno, T. Specificity and regulation of phosphotyrosine signaling through SH2 domains. *J. Struct. Biol. X.* **2020**, *4*, 100026. [CrossRef] [PubMed]
4. Lappalainen, I.; Thusberg, J.; Shen, B.; Vihinen, M. Genome wide analysis of pathogenic SH2 domain mutations. *Proteins: Struct. Funct. Genet.* **2008**, *72*, 779–792. [CrossRef]
5. De Araujo, E.D.; Orlova, A.; Neubauer, H.A.; Bajusz, D.; Seo, H.S.; Dhe-Paganon, S.; Keserű, G.M.; Moriggl, R.; Gunning, P.T. Structural implications of stat3 and stat5 sh2 domain mutations. *Cancers* **2019**, *11*, 1757. [CrossRef] [PubMed]
6. Filippakopoulos, P.; Müller, S.; Knapp, S. SH2 domains: Modulators of nonreceptor tyrosine kinase activity. *Curr. Opin. Struct. Biol.* **2009**, *19*, 643–649. [CrossRef]
7. Vidal, M.; Gigoux, V.; Garbay, C. SH2 and SH3 domains as targets for anti-proliferative agents. *Crit. Rev. Oncol./Hematol.* **2001**, *40*, 175–186. [CrossRef]
8. Morlacchi, P.; Robertson, F.M.; Klostergaard, J.; McMurray, J.S. Targeting SH2 domains in breast cancer. *Future Med. Chem.* **2014**, *6*, 1909–1926. [CrossRef]
9. Bobone, S.; Pannone, L.; Biondi, B.; Solman, M.; Flex, E.; Canale, V.C.; Calligari, P.; de Faveri, C.; Gandini, T.; Quercioli, A.; et al. Targeting Oncogenic Src Homology 2 Domain-Containing Phosphatase 2 (SHP2) by Inhibiting Its Protein-Protein Interactions. *J. Med. Chem.* **2021**, *64*, 15973–15990. [CrossRef]
10. Park, T. Crk and CrkL as therapeutic targets for cancer treatment. *Cells* **2021**, *10*, 739. [CrossRef]
11. Luo, L.Y.; Hahn, W.C. Oncogenic Signaling Adaptor Proteins. *J. Genet. Genom.* **2015**, *42*, 521–529. [CrossRef] [PubMed]
12. Feller, S.M. Crk family adaptors-signalling complex formation and biological roles. *Oncogene* **2001**, *20*, 6348–6371. [CrossRef] [PubMed]
13. Sattler, M.; Salgia, R. Role of the adapter protein CRKL in signal transduction of normal hematopoietic and BCR/ABL-transformed cells. *Leukemia* **1998**, *12*, 637–644. [CrossRef]
14. Birge, R.B.; Kalodimos, C.; Inagaki, F.; Tanaka, S. Crk and CrkL adaptor proteins: Networks for physiological and pathological signaling. *Cell Commun. Signal.* **2009**, *7*, 13. [CrossRef] [PubMed]
15. Liu, B.A.; Jablonowski, K.; Raina, M.; Arcé, M.; Pawson, T.; Nash, P.D. The Human and Mouse Complement of SH2 Domain Proteins-Establishing the Boundaries of Phosphotyrosine Signaling. *Mol. Cell.* **2006**, *22*, 851–868. [CrossRef]
16. Nolte, R.T.; Eck, M.J.; Schlessinger, J.; Shoelson, S.E.; Harrison, S.C. Crystal structure of the PI 3-kinase p85 amino-terminal SH2 domain and its phosphopeptide complexes. *Nat. Struct. Biol.* **1996**, *3*, 364–374. [CrossRef]
17. Hof, P.; Pluskey, S.; Dhe-Paganon, S.; Eck, M.J.; Shoelson, S.E. Crystal Structure of the Tyrosine Phosphatase SHP-2. *Cell* **1998**, *92*, 441–450. [CrossRef]
18. Bhat, A.; Kolibaba, K.; Oda, T.; Ohno-Jones, S.; Heaney, C.; Druker, B.J. Interactions of CBL with BCR-ABL and CRKL in BCR-ABL-transformed Myeloid Cells. *J. Biol. Chem.* **1997**, *272*, 16170–16175. [CrossRef]
19. Arai, A.; Kanda, E.; Nosaka, Y.; Miyasaka, N.; Miura, O. CrkL is Recruited through Its SH2 Domain to the Erythropoietin Receptor and Plays a Role in Lyn-mediated Receptor Signaling. *J. Biol. Chem.* **2001**, *276*, 33282–33290. [CrossRef]
20. Salgia, R.; Uemura, N.; Okuda, K.; Li, J.-L.; Pisick, E.; Sattler, M.; de Jong, R.; Druker, B.; Heisterkamp, N.; Chen, L.B.; et al. CRKL Links p210 BCR/ABL with Paxillin in Chronic Myelogenous Leukemia Cells. *J. Biol. Chem.* **1995**, *270*, 29145–29150. [CrossRef]
21. Schaller, M.D. Paxillin: A focal adhesion-associated adaptor protein. *Oncogene* **2001**, *20*, 6459–6472. [CrossRef] [PubMed]
22. Singer, A.U.; Forman-Kay, J.D. pH titration studies of an SH2 domain-phosphopeptide complex: Unusual histidine and phosphate pKa values. *Protein Sci.* **1997**, *6*, 1910–1919. [CrossRef] [PubMed]
23. Myers, J.K.; Pace, C.N.; Scholtz, J.M.; Scholtz, M. Denaturant m values and heat capacity changes: Relation to changes in accessible surface areas of protein unfolding. *Protein Sci.* **1995**, *4*, 2138–2148. [CrossRef]

24. Parker, M.J.; Spencer, J.; Clarke, A.R. An Integrated Kinetic Analysis of Intermediates and Transition States in Protein Folding Reactions. *J. Mol. Biol.* **1995**, *253*, 771–786. [CrossRef]
25. Gianni, S.; Calosci, N.; Aelen, J.M.A.; Vuister, G.W.; Brunori, M.; Travaglini-Allocatelli, C. Kinetic folding mechanism of PDZ2 from PTP-BL. *Protein Eng. Des. Sel.* **2005**, *18*, 389–395. [CrossRef] [PubMed]
26. Bonetti, D.; Troilo, F.; Toto, A.; Travaglini-Allocatelli, C.; Brunori, M.; Gianni, S. Mechanism of Folding and Binding of the N-Terminal SH2 Domain from SHP2. *J. Phys. Chem. B* **2018**, *122*, 11108–11114. [CrossRef] [PubMed]
27. Nardella, C.; Malagrinò, F.; Pagano, L.; Rinaldo, S.; Gianni, S.; Toto, A. Determining folding and binding properties of the C-terminal SH2 domain of SHP2. *Protein Sci.* **2021**, *30*, 2385–2395. [CrossRef] [PubMed]
28. Wildes, D.; Anderson, L.M.; Sabogal, A.; Marqusee, S. Native state energetics of the Src SH2 domain: Evidence for a partially structured state in the denatured ensemble. *Protein Sci.* **2006**, *15*, 1769–1779. [CrossRef]
29. Visconti, L.; Malagrinò, F.; Toto, A.; Gianni, S. The kinetics of folding of the NSH2 domain from p85. *Sci. Rep.* **2019**, *9*, 4058. [CrossRef]
30. Ferguson, N.; Capaldi, A.P.; James, R.; Kleanthous, C.; Radford, S.E. Rapid Folding with and without Populated Intermediates in the Homologous Four-helix Proteins Im7 and Im9. *J. Mol. Biol.* **1999**, *286*, 1597–1608. [CrossRef]

biomolecules

MDPI

*Article*

# In-Silico Analysis of pH-Dependent Liquid-Liquid Phase Separation in Intrinsically Disordered Proteins

**Carlos Pintado-Grima †, Oriol Bárcenas † and Salvador Ventura ***

Institut de Biotecnologia i Biomedicina, Departament de Bioquímica i Biologia Molecular, Universitat Autònoma de Barcelona, Bellaterra, 08193 Barcelona, Spain; carlos.pintado@uab.cat (C.P.-G.); oriol.barcenas@autonoma.cat (O.B.)
* Correspondence: salvador.ventura@uab.cat
† These authors contributed equally to this work.

**Abstract:** Intrinsically disordered proteins (IDPs) are essential players in the assembly of biomolecular condensates during liquid–liquid phase separation (LLPS). Disordered regions (IDRs) are significantly exposed to the solvent and, therefore, highly influenced by fluctuations in the microenvironment. Extrinsic factors, such as pH, modify the solubility and disorder state of IDPs, which in turn may impact the formation of liquid condensates. However, little attention has been paid to how the solution pH influences LLPS, despite knowing that this process is context-dependent. Here, we have conducted a large-scale in-silico analysis of pH-dependent solubility and disorder in IDRs known to be involved in LLPS (LLPS-DRs). We found that LLPS-DRs present maximum solubility around physiological pH, where LLPS often occurs, and identified significant differences in solubility and disorder between proteins that can phase-separate by themselves or those that require a partner. We also analyzed the effect of mutations in the resulting solubility profiles of LLPS-DRs and discussed how, as a general trend, LLPS-DRs display physicochemical properties that permit their LLPS at physiologically relevant pHs.

**Keywords:** pH; liquid-liquid phase separation; intrinsically disordered proteins; protein solubility; protein disorder; mutations; bioinformatics

**Citation:** Pintado-Grima, C.; Bárcenas, O.; Ventura, S. In-Silico Analysis of pH-Dependent Liquid-Liquid Phase Separation in Intrinsically Disordered Proteins. *Biomolecules* **2022**, *12*, 974. https://doi.org/10.3390/biom12070974

Academic Editor: Sonia Longhi

Received: 22 June 2022
Accepted: 8 July 2022
Published: 12 July 2022



## 1. Introduction

Our view of cellular organization has been progressively changing over the last decade. Since the discovery of the first liquid droplets in *C. elegans* embryo's cells [1], numerous membrane-less organelles (MLOs) with a wide variety of biological functions have been described in different organisms [2–5]. In contrast to their classic membrane-enclosed counterparts, MLOs are dynamic supramolecular structures that can undergo reversible liquid–liquid phase separation (LLPS) in response to specific stimuli [6,7]. The reversible and tunable nature of LLPS has turned into an effective compartmentalization mechanism for the always-changing cellular milieu, allowing the selective spatiotemporal formation of biomolecular condensates [8]. MLOs are often enriched in intrinsically disordered proteins (IDPs) and/or proteins bearing unstructured regions and low-complexity domains [9–11]. These sequences play an important role in droplet formation by establishing multivalent weak intermolecular interactions [12,13]. Indeed, mutations in some of these unstructured regions might deregulate the equilibrium of LLPS and lead to the onset of neurodegenerative diseases such as amyotrophic lateral sclerosis (ALS), frontotemporal dementia (FTD) [10,14,15], or muscular dystrophies [16,17].

Intrinsically disordered regions (IDRs) are exposed to the solvent and thus influenced by the protein microenvironment, whose fluctuation can trigger conformational switches [18]. This context-dependency also applies to LLPS processes, and parameters such as ionic strength, temperature, or pH have been described as regulatory elements in these reactions [6]. Other factors apart from solvent conditions can also influence the behavior of IDPs in LLPS, including the protein concentration and the presence of other biomolecules, mainly RNA and partner proteins [19]. Therefore, the capacity of a protein to phase-separate is not binary and cannot be univocally attributed to the intrinsic properties of the sequence as it strongly depends on the specific conditions of the cellular milieu at any given time.

In recent work, we accounted for the effect of pH on the solubility [20] and disorder state [21] of IDPs using different equations that simultaneously consider the impact of the pH on sequence hydrophobicity and net charge. This allowed us to develop two novel bioinformatic tools to predict pH-dependent solubility and disorder: SolupHred [22] and DispHScan [23], respectively, which recapitulate previous and novel experimental data [24–26]. The solubility and degree of disorder of a protein at given pH influence its propensity to phase-separate, but large-scale analyses exploring these connections are scarce. To provide insights on the role of pH in LLPS, we conducted an in-silico study of solubility and disorder at different pH values for LLPS-involved disordered regions (LLPS-DRs), accounting for a total of 1600 sequences, using the SolupHred and DispHScan algorithms.

## 2. Materials and Methods

### *2.1. Dataset Generation*

LLPS regions were extracted from PhaSepDB, a manually curated database of liquid-liquid phase separation-related proteins and MLOs [27]. Afterward, the disordered nature of these regions was surveyed using the IUPRED2A server [28]. High-confidence LLPS-DRs (sequences longer than 20 amino acids with an IUPRED score $\geq 0.5$) were saved for the analyses ($n = 1600$). Two different subgroups were created according to the ability of each LLPS region to phase-separate by itself (psself) or with the help of a partner (psother). A second dataset of general disordered segments was generated for comparison purposes. IDRs longer than 20 amino acids were obtained from DisProt (release 2021_08), a manually curated database of experimentally validated IDPs [29].

### *2.2. pH-Dependent Solubility and Disorder Analyses*

The effect of pH on both LLPS-DRs' solubility and disorder was predicted by two in-house algorithms considering this key variable in their pipelines. SolupHred and DispHScan web servers recalculate protein lipophilicity and net charge as a pH function to predict IDPs' solubility and disorder in a specific pH context [22,23]. All disordered sequences were run with SolupHred and DispHScan in a pH range between 4 and 9, using a step size of 0.1 to account for small variations. The software output solubility and disorder scores, respectively. The disorder score is named the DispH score, with lower values indicating a more disordered state. Statistical significance between variables and/or datasets was assessed with Mann-Whitney-Wilcoxon two-sided test with Bonferroni correction. *p*-values were marked with asterisks to better convey statistical significance ($p > 0.05$ (ns), $p \leq 0.05$ (*), $p \leq 0.01$ (**), $p \leq 0.001$ (***), $p \leq 0.0001$ (****)). As for the statistical significance of the linear regression (whether the distribution follows a slope of 0 or not), a Wald Test with t-distribution of the test statistic was used. Statistical tests are described in the Supplementary Material of this article.

## 3. Results

### *3.1. LLPS-DRs Present Maximum Solubility around Neutral pH*

One of the main uncertainties when dealing with IDPs and LLPS is the conformational state that disordered segments present during the phase-separation process. Some studies suggest that LLPS-DRs contain aggregation-prone regions (APRs), which are needed for condensate formation, and that they act by reducing the solubility in the conditions in which LLPS occurs [30,31]. To explore this hypothesis, we analyzed the pH-dependent solubility of LLPS-DRs in the 4 to 9 pH range, comparing the obtained results with the pHs at which LLPS was described. Given that similar solubility scores may span various pH units, we assumed 10% maximum/minimum solubilities as the most representative variables for the analysis. The results revealed that solubility in LLPS-DRs tends to be maximum around neutral pH ($\mu = 6.96$, $\sigma = 1.29$, Figure 1A), whereas minimum solubilities are mainly achieved at the extremes of the analyzed pH interval (Figure 1B). When studying buffer conditions described for LLPS, a similar result was observed ($n = 181$, $\mu = 7.50$; $\sigma = 0.43$). This suggests that LLPS-DRs are soluble near physiological pH values and that LLPS occurs at pHs at which, as a group, these protein regions display significant solubility. To further support these findings, the absolute differences in maximum and minimum solubility scores were individually compared with those obtained at the pH where LLPS was described (Figure 2A). The analysis confirms that, as a trend, effective LLPS-promoting pHs resemble more those at which the correspondent proteins are maximally soluble than those with minimal solubility.

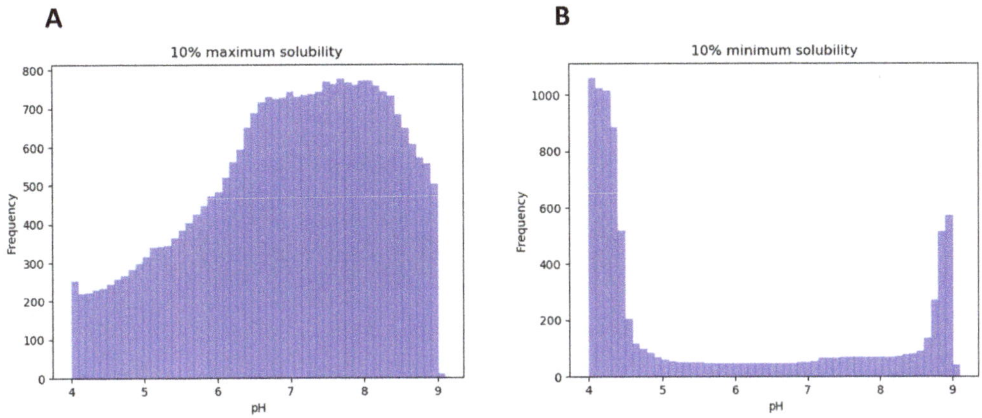


**Figure 1.** 10% maximum (**A**) and 10% minimum (**B**) solubility distribution of LLPS-DRs in the range between pH 4 and 9 using step size 0.1. Maximum solubility is mostly attained around neutral pH ($\mu = 6.95$), whereas minimum solubilities are found at the extremes of the pH interval, in more acidic or basic conditions.

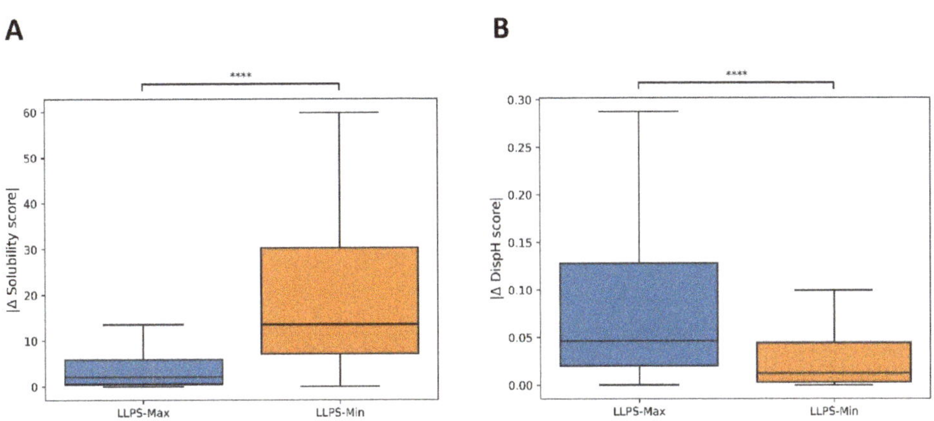


**Figure 2.** Absolute differences in LLPS-DR solubility (**A**) and disorder (**B**) scores between the predicted maximum or minimum value and the score at the pH where LLPS was described. pH conditions in LLPS buffers are closer to LLPS-DR maximum solubilities ($p = 6.080 \times 10^{-31}$) and minimum DispH scores ($p = 3.905 \times 10^{-15}$). Four asterisks (****) indicate $p \leq 0.0001$.

Overall, the results indicate that pH solution conditions favoring high solubility overlap with those at which condensation is experimentally observed, suggesting that these molecular signatures might be important for LLPS and relevant at the cellular physiological pH. This is surprising since it seems to contradict the general assumption that pro-aggregational conditions are necessary for LLPS.

### 3.2. A Link between pH-Dependent Solubility and pH-Dependent Disorder in LLPS-DRs

We performed the same kind of analysis as above to assess whether pHs at which LLPS is reported coincide with conditions that favor disorder or, on the contrary, promote compactness of LLPS-DRs (Figure 2B). The data indicate that minimum DispH scores and thus larger disorder content match better conditions at which LLPS experimentally occurs.

The relationship between solubility and conformational disorder in the LLPS-DR dataset was then analyzed. When plotting the solubility and disorder scores at neutral pH (pH = 7.0), a highly significant linear correlation was observed between these two variables ($R^2 = 0.90$, $p < 1^{-10}$) (Figure 3), suggesting that solubility and disorder are intimately ligated in this dataset. Furthermore, higher solubility scores were correlated with minimum DispH scores (maximum disorder) and vice-versa. This indicates that regions that populate a more soluble sequence space are also more prone to be disordered. This makes sense since the presence of low-soluble sequences with a high degree of disorder would render them aggregation-prone and thus harmful at the analyzed physiological pH, whereas a higher degree of compactness would protect their hydrophobic residues from exposure to the solvent, at least transiently. Conversely, highly soluble sequences would find it challenging to attain a compact conformation, and their presence in an unfolded state would not represent a significant risk of establishing aberrant hydrophobic interactions with other cellular components.

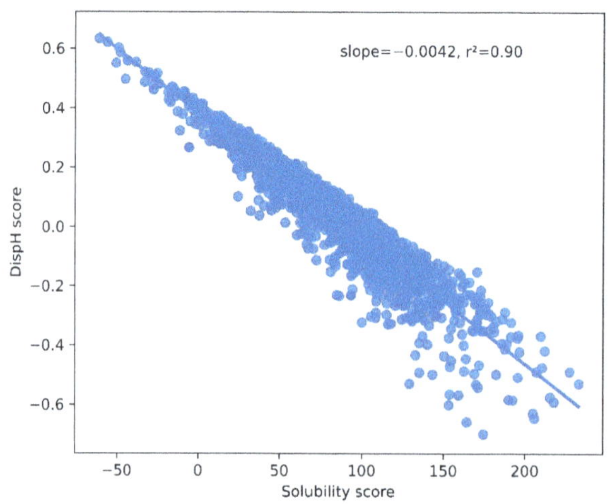


**Figure 3.** Correlation between LLPS-DR disorder and solubility scores at pH 7.0. A significant linear correlation can be observed ($R^2 = 0.90$, $p < 1^{-10}$). Higher solubility scores are associated with lower DispH scores (more disordered states).

### 3.3. Different Datasets in LLPS-DRs Present Distinct Property Distributions

Not all proteins present in MLOs can undergo independent LLPS [19]. As we said, this process is highly context-dependent [6], and many proteins require a partner, often a nucleic acid molecule, to phase-separate. Therefore, the properties of LLPS-DRs may vary according to their ability to undergo LLPS by themselves (psself) or with a partner (psother). Indeed, no differences in calculated solubility and disorder at pH 7.0 were observed between the entire LLPS-DRs dataset (without differentiating psself and psother sequences) and the IDRs present in DisProt. Conversely, when the study considered psself and psother regions separately, the significance level dramatically increased when compared with DisProt and between them. This indicates that psself and psother sequences need to be studied independently.

Psself regions showed lower solubility scores than psother and DisProt (Figure 4A), which, not surprisingly, was associated with higher DispH scores (more ordered) (Figure 4B). These differences are more apparent in the distribution plot of solubility and disorder scores for psself and psother datasets at pH 7.0 (Figure 4C,D). This indicates that psself regions tend to be less soluble and disordered than psother at pH 7.0. This allows us to reconcile our analysis with the view that aggregation propensity is a player in LLPS and to reformulate it. For psself proteins, which can separate efficiently and autonomously, the situation is not that they experiment LLPS at pHs at which their solubility is at a minimum, but instead that its intrinsic solubility at physiological pH, despite significant, is below the average solubility of disordered proteins, which facilitate homotypic self-assembly. An inverse trend applies to disorder. Still, this association only applies for psself proteins since proteins that require an interactor to be incorporated in MLOs are not only more soluble than psself ones, but also that the conjunct of IDRs in DisProt and will find difficulties to self-assemble without a scaffolding molecule. All these biophysical connections are masked when studying the LLPS dataset as a homogeneous protein group.

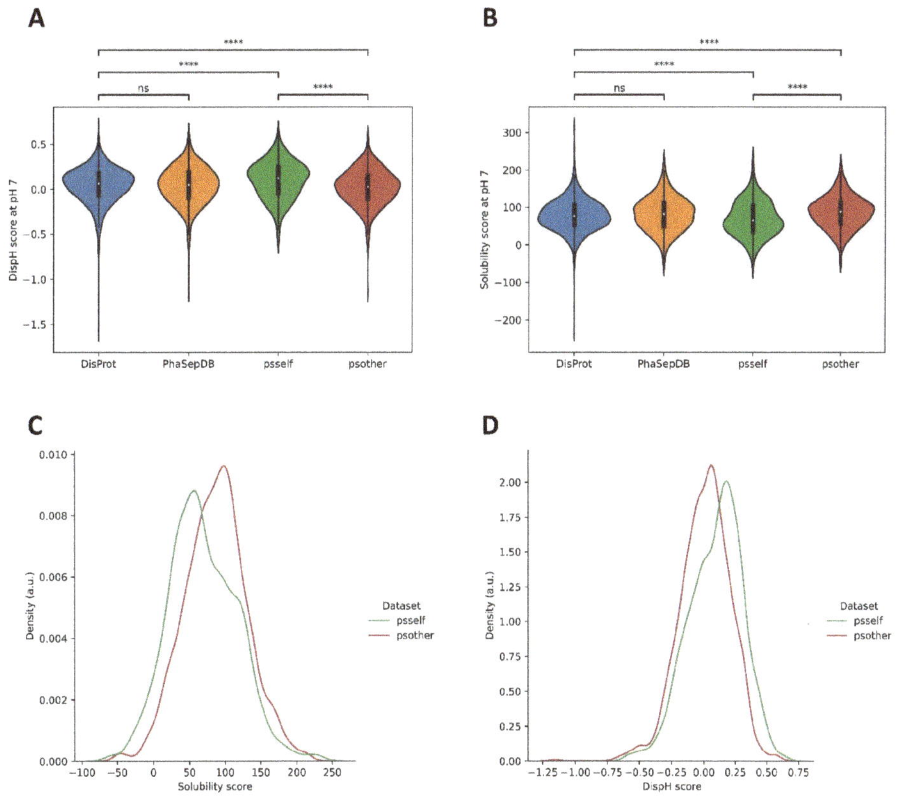


**Figure 4.** Comparison between solubility (**A**) and disorder (**B**) scores of DisProt, PhaSepDB, psself, and psother datasets at pH 7.0. Differences in disorder and solubility are non-significant between DisProt and PhaSepDB. However, when psself and psother are treated independently, major differences can be observed with DisProt and among these sub-sets. The psself dataset exhibits lower solubilities (**C**) and lower disorder (**D**) than psother sequences *. The density coordinate provides a representation of the number of observed sequences for each given score. * (**A**) Solubility: DisProt–psself ($2.321 \times 10^{-05}$); DisProt–psother ($2.000 \times 10^{-07}$); psself-psother ($2.744 \times 10^{-12}$). * (**B**) Disorder: DisProt–psself ($3.282 \times 10^{-07}$); DisProt–psother ($2.902 \times 10^{-05}$); psself-psother ($4.457 \times 10^{-13}$). Four asterisks (****) indicate $p \leq 0.0001$. "ns" indicate $p > 0.05$.

### 3.4. Psself Regions Present Lower Dispersion in Solubility and Disorder in the Physiological pH Range

When the differences between maximum and minimum solubility and disorder scores were analyzed in the entire selected pH range (from pH 4.0 to pH 9.0 with a step size of 0.1, as detailed in the Section 2), DisProt exhibited the broader dispersion of both solubility and disorder parameters, with psself being the dataset with the lowest dispersion (Figure 5A,B). This suggests that psself regions are less sensible to pH fluctuations and populate a narrower solubility-disorder space, likely compatible with cellular conditions in which LLPS can still occur even if pH deviations arise, as long as they are not very large. On the other hand, the dispersion of psother regions is significantly wider for both parameters, suggesting a lower selective pressure to keep their solubility/disorder properties restricted in a cell-compatible pH gradient, which is expected, as they are not competent for phase separation alone.

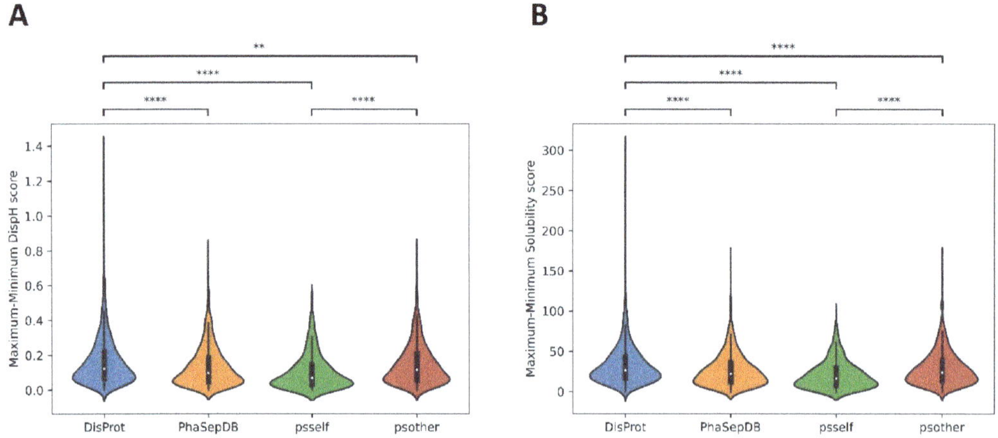


**Figure 5.** Difference between maximum and minimum solubility (**A**) and disorder (**B**) scores for DisProt, PhaSepDB, psself, and psother datasets. The DisProt dataset exhibits a wider dispersion, understood as the difference between the maximum and minimum values, in comparison with the PhaSepDB, psself, and psother datasets. In consonance with previous results, psother and psself datasets are significantly different *. * (**A**) Solubility: DisProt-PhaSepDB ($1.526 \times 10^{-19}$); DisProt–psself ($2.064 \times 10^{-31}$); DisProt–psother ($5.508 \times 10^{-06}$); psself-psother ($2.009 \times 10^{-14}$). * (**B**) Disorder: DisProt-PhaSepDB ($4.496 \times 10^{-14}$); DisProt–psself ($3.325 \times 10^{-29}$); DisProt–psother ($5.526 \times 10^{-03}$); psself-psother ($8.586 \times 10^{-17}$). Two asterisks (**) indicate $p \leq 0.01$. Four asterisks (****) indicate $p \leq 0.0001$.

### 3.5. Case Study of Independent LLPS Happening at Physiological pH

To contextualize the previous results with defined cases of LLPS, we sought the literature for IDPs whose LLPS was experimentally demonstrated to occur at physiological pH.

An example of a well-characterized LLPS-DR is the low complexity domain (LCD) of TDP-43, a protein associated with neurodegenerative disorders such as ALS or FTD [10,14,32]. Experimental studies indicate that LLPS of the TDP-43 LCD is pH-dependent [33]. LLPS was tested at three different pH values (4.0, 6.0, and 7.0) with different salt concentrations (from 0 to 300 mM NaCl). The presence of liquid droplets was observed in all salt conditions at pH 7.0 and progressively dissipated as the pH decreased, requiring higher concentrations of salt to observe LLPS. These results reveal that physiological conditions are conductive of LLPS for the TDP-43 LCD. When analyzing the resulting solubility curve predicted by SolupHred for this LLPS-DR in this pH range (Figure 6A), maximum solubility is achieved at pH 7.0. The LCDs of the hnRNPA1 [34] or the U1-70K proteins [35] have been shown to form liquid droplets at pH 7.0, where the solubility score is close to the maximum (Figure 6B,C). Importantly, dysfunctional LLPS of these proteins lead to aberrant aggregation and neurodegeneration.

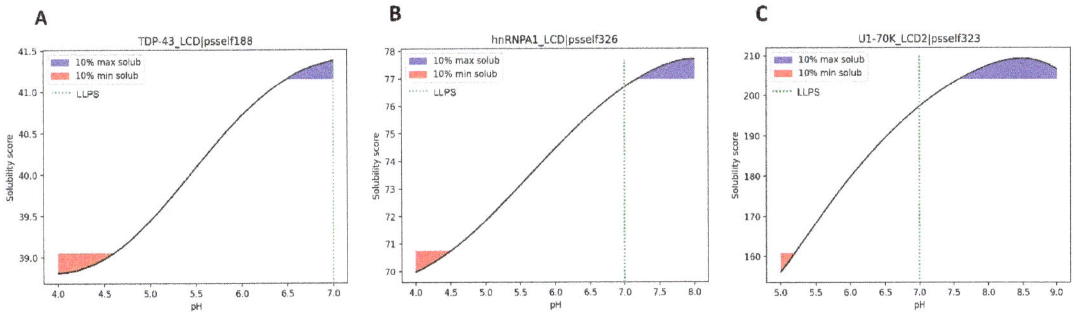


**Figure 6.** Predicted solubility curves at different pHs for the LCDs of TDP-43 (**A**), hnRNPA1 (**B**) and U1-70K (**C**). In all three cases, homotypic LLPS was observed at pH 7.0, a condition in which they display relatively high solubility scores.

Altogether, for these proteins, LLPS occurs in a high solubility range that is compatible with physiological conditions. This does not imply that LLPS necessarily occurs more efficiently at a pH in which protein solubility is strictly at its maximum, but rather that to form biomolecular condensates, proteins should be in a solubility regime that enables them to diffuse and interact homotypically and, if needed, with their partners. Conditions of very low solubility can also promote LLPS, but they are often associated with aggregation and pathogenicity.

### *3.6. Mutations in LLPS Formation and Disease*

Mutations in LLPS-DRs can hinder the capacity of proteins bearing these sequences to phase-separate. This has been studied in vitro, analyzing the formation of liquid condensates using different variants of a given sequence. Unveiling how these mutations affect the solubility pattern of these regions is vital to understanding their connection with LLPS deregulation.

It is well established that tyrosines are important residues in LLPS since their involvement in $\pi$-$\pi$ [36] or cation-$\pi$ [11,37] interactions are key for the weak multivalent contact that sustain liquid droplets. In this way, an increasing number of Tyr-to-Ser substitutions in the LCD of FUS has been linked with a reduction in the formation of droplets at pH 7.5 [9]. Therefore, we plotted the relative fluorescence intensity after 10 min, a measure of LLPS in this study, against the solubility scores obtained by SolupHred at pH 7.5 for all Y$\rightarrow$S variants. We observed that the solubility of the variants was proportional to the number of Ser residues introduced in the sequence ($R^2$ = 0.95) (Figure 7A) and thus that LLPS and solubility appear again to be correlated.

The most studied amino acid substitutions in LLPS-DRs are those leading to pathological aggregation. For example, mutations of the well-conserved Asp314 at the C-terminal LCD of hnRNPA1 to either Asn or Val divert the process of phase separation towards the formation of pathogenic amyloid fibrils, a process that is associated with ALS onset [38,39]. When studying the solubility curves of these two variants, a general decrease in maximum solubility was observed (Figure 7B), consistent with aggregation occurring inside liquid droplets, resulting in their rigidification and ultimately in the formation of stable amyloid assemblies.

Overall, tight solubility conditions are required for LLPS, and sequence modifications that alter this property, either towards more or less soluble states, impact the efficacy of the process.

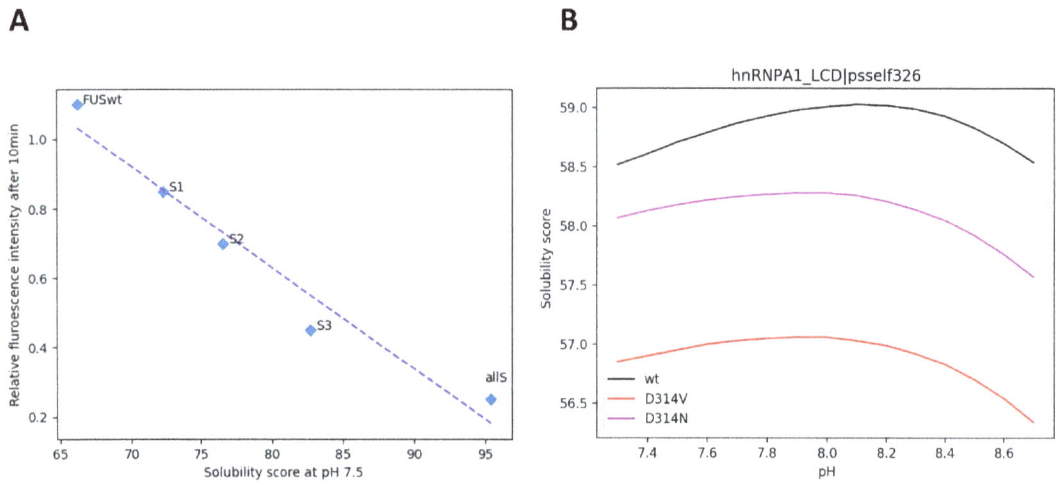


**Figure 7.** Predicted solubility scores for LLPS-impacting mutations in the LCD of FUS at pH 7.5 (**A**), and hnRNPA1 (**B**) in the 10% maximum solubility pH interval. Both solubilizing (**A**) and aggregating (**B**) mutations may alter LLPS equilibrium.

### 3.7. pH-Dependent LLPS: Optimal Condition Evaluation

Finding the optimal conditions at which LLPS may occur for a given protein is not trivial; multiple variables must be assessed when studying this phenomenon. One recent study tried to establish a generic approach to study LLPS under near-native conditions [40]. In this work, the authors induced the formation of liquid condensates by the LCDs of hnRNPA2, TDP-43, NUP98, and ERD14 proteins by a pH jump from extreme pHs (3.0 or 11.0), where the proteins did not phase separate, to physiological pH where they found that liquid droplets were formed.

The solubility curves obtained by SolupHred for these LLPS-DRs (Figure 8) in the analyzed pH regime indicate that in the case of hnRNPA2, TDP-43, and ERD14, the condensation-promoting pH precisely maps within the range in which the LCDs manifest their maximum solubility. This is not the case with NUP98, for which the solubility at physiological pH is high, but not maximum, as this value is attained at very low pHs, where, in fact, LLPS was not observed. An inspection of the sequence of this protein indicates that it is highly cationic, and the high net charge of the region at acidic pH would compromise LLPS because of electrostatic repulsion. Decreasing this effect by moving toward neutral pHs would allow phase separation to occur in conditions where the solubility is still significantly high. An important corollary of this analysis is that LLPS does not ineludibly occur in conditions where the solubility is very low. It is important to note that according to our algorithm, the pI of a given protein does not necessarily coincide with the pH conditions at which the minimum solubility score is found, or on the other way around, that the pHs more distant from the pI are not always those at which the protein would exhibit maximum solubility. This results from considering simultaneously the impact of the pH in sequence hydrophobicity and net charge, not only this last factor. Indeed, for the above-mentioned proteins, the LLPS-DR pI is a poor predictor of conditions eliciting LLPS.

The discussed results still do not allow for the standardization of a method to predict or enhance LLPS just considering the solution pH independently. However, obtaining a pH-dependent solubility profile allows for an evaluation of the physicochemical parameters that may influence the formation of liquid condensates. As a general trend, naturally occurring LLPS-DRs can do it around neutral pH, which usually matches with intervals of significant solubility. Moving away from these regions destabilizes the multivalent-weak interactions required for LLPS, increasing the repulsive net charge or over-stabilizes them, permitting LLPS, but also the evolution of liquid droplets towards the formation of pathogenic aggregates.

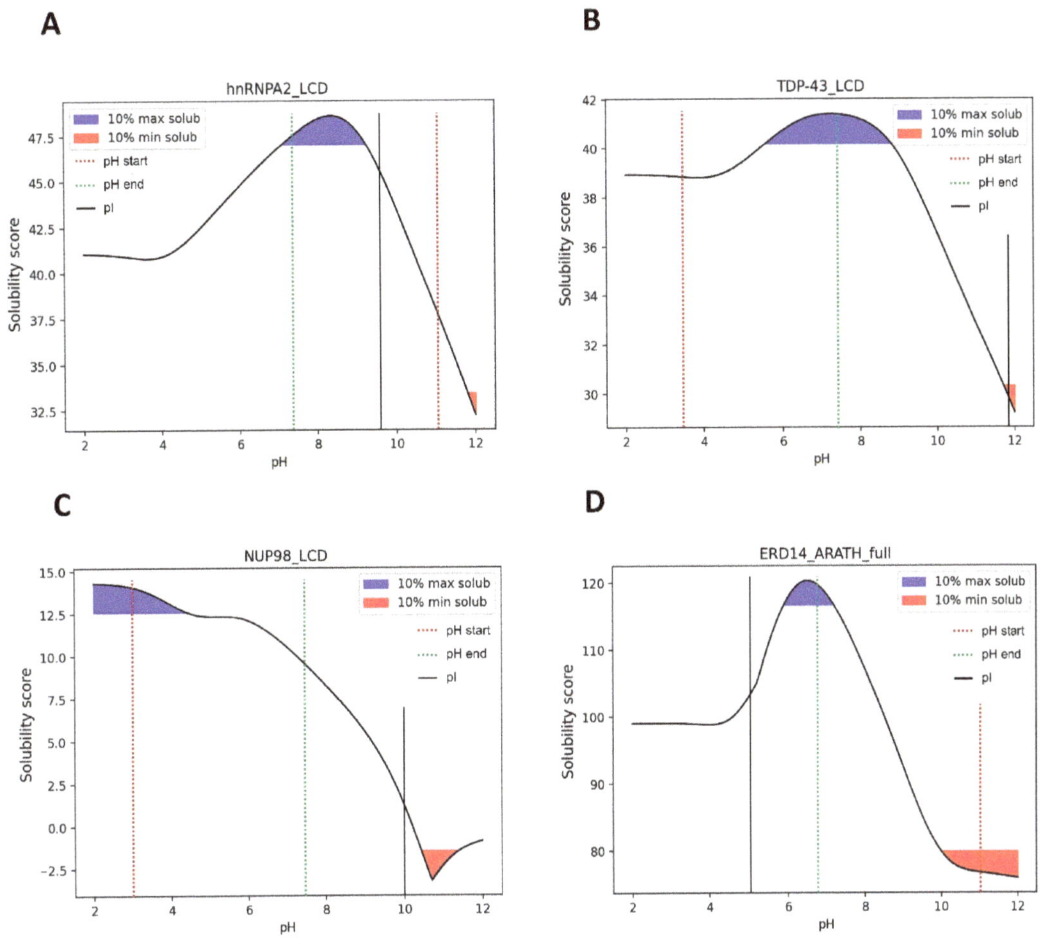


**Figure 8.** Predicted solubility curves in the pH jump interval for the LCDs of hnRNPA2 (**A**), TDP-43 (**B**), NUP98 (**C**) and ERD14 (**D**). The jump from an extreme pH (dotted red line) to near neutral pH (dotted green line) induces the formation of liquid droplets in conditions within -or close to- 10% maximum solubilities.

## 4. Discussion

The intriguing phenomenon of LLPS has been attracting the attention of biologists and biophysicists in the last years. Many different observations of MLOs associated with a wide variety of IDPs have been reported in the literature [5,12,41]. However, large-scale analyses of the properties of these regions in context to their surrounding environment have remained elusive, despite the importance of extrinsic factors in modulating this process [6]. In this work, we have conducted a bioinformatics survey to investigate the effect of pH on the solubility and disorder of LLPS-DRs, allowing for better elucidation of the role of this solvent condition in the outcome of LLPS.

Our results indicate that LLPS-DRs present maximum solubility around neutral pH. This was initially intriguing, as previous studies suggested that APRs endorsing IDRs with lower local solubilities may be necessary to drive the formation of protein condensates [30,31]. Indeed, LLPS is a highly dynamic and reversible process that likely requires weaker interactions than those provided by highly aggregating patches. Therefore, LLPS-DRs need to display a significant degree of solubility to phase-separate into liquid droplets near physiological pH conditions. Intrinsic solubility is, however, an important determinant of autonomous phase separation, since psself IDRs exhibit, as a group, less soluble sequences than psother and DisProt IDRs, a property that is accompanied by a higher propensity to populate compact states. Still, for these interaction-prone sequences, LLPS occurs at pH values where they are predicted to exhibit significant solubility, a condition that would be compatible with their functioning and preclude the aggregation of these regions. This might be biologically important since psself IDRs' solubility and disorder properties seem to have evolved to be more resistant to small pH perturbations. Our analysis provides a plausible answer to why some disordered regions phase-separate under specific pH conditions and others do not, although it is not intended to predict their behavior individually.

Mutations in psself LCDs can shift the equilibrium that sustains LLPS both to a more soluble or aggregated state, which may eventually lead to disease onset. In contrast, regions in the psother dataset need a partner to compensate for their lack of intrinsic condensation propensity. In these proteins, studying the contribution of mutations is much less straightforward, as pathogenicity may stem from poor interaction with its phase-separating partner or enhanced interaction with non-intended binders.

With this work, we aimed to start elucidating the role of pH in context-dependent phenomena such as LLPS and describe the general tendencies that arise from large-scale analysis. Given that many cellular processes and disease-associated pathways are sensible to cellular milieu fluctuations, we believe that by understanding the contribution of pH in both IDPs' solubility and disorder at a large scale, we will be a step closer to understanding the impact of the environment in such processes.

**Supplementary Materials:** The following supporting information can be downloaded at: https://www.mdpi.com/article/10.3390/biom12070974/s1, Table S1: LLPS-DRs pH data predictions.

**Author Contributions:** Conceptualization: S.V., C.P.-G. and O.B. Dataset generation and data curation: C.P.-G. Investigation and formal analysis: C.P.-G. and O.B. Writing: C.P.-G., O.B. and S.V. Funding acquisition and supervision: S.V. All authors have read and agreed to the published version of the manuscript.

**Funding:** This work was funded by European Union's Horizon 2020 research and innovation programme under GA 952334 (PhasAGE). S.V. was supported by the Spanish Ministry of Science and Innovation (PID 2019-105017RB-I00) and by ICREA Academia 2020. C.P.-G. was supported by the Secretariat of Universities and Research of the Catalan Government and the European Social Fund (2021 FI_B 00087).

**Data Availability Statement:** Data supporting the results of this study can be accessed in the Supplementary Table S1 of this article.

**Conflicts of Interest:** The authors declare no conflict of interest.

## Abbreviations

| | |
|---|---|
| IDPs | Intrinsically disordered proteins |
| IDRs | Intrinsically disordered regions |
| LCD | Low complexity domain |
| LLPS | Liquid-liquid phase separation |
| LLPS-DRs | Liquid-liquid phase separation-disordered regions |
| MLOs | Membraneless organelles |
| psself | LLPS region that can phase-separate by itself |
| psother | LLPS region that requires a partner to phase-separate |
| ALS | Amyotrophic lateral sclerosis |
| FTD | Frontotemporal dementia |
| APRs | Aggregation-prone regions |

## References


1. Brangwynne, C.P.; Eckmann, C.R.; Courson, D.S.; Rybarska, A.; Hoege, C.; Gharakhani, J.; Julicher, F.; Hyman, A.A. Germline P granules are liquid droplets that localize by controlled dissolution/condensation. *Science* **2009**, *324*, 1729–1732. [CrossRef]
2. Banani, S.F.; Lee, H.O.; Hyman, A.A.; Rosen, M.K. Biomolecular condensates: Organizers of cellular biochemistry. *Nat. Rev. Mol. Cell Biol.* **2017**, *18*, 285–298. [CrossRef]
3. Mao, Y.S.; Zhang, B.; Spector, D.L. Biogenesis and function of nuclear bodies. *Trends Genet. TIG* **2011**, *27*, 295–306. [CrossRef]
4. Decker, C.J.; Parker, R. P-bodies and stress granules: Possible roles in the control of translation and mRNA degradation. *Cold Spring Harb. Perspect. Biol.* **2012**, *4*, a012286. [CrossRef] [PubMed]
5. Alberti, S.; Hyman, A.A. Biomolecular condensates at the nexus of cellular stress, protein aggregation disease and ageing. *Nat. Rev. Mol. Cell Biol.* **2021**, *22*, 196–213. [CrossRef]
6. Alberti, S.; Gladfelter, A.; Mittag, T. Considerations and Challenges in Studying Liquid-Liquid Phase Separation and Biomolecular Condensates. *Cell* **2019**, *176*, 419–434. [CrossRef]
7. Holehouse, A.S.; Pappu, R.V. Protein polymers: Encoding phase transitions. *Nat. Mater.* **2015**, *14*, 1083–1084. [CrossRef]
8. Shin, Y.; Brangwynne, C.P. Liquid phase condensation in cell physiology and disease. *Science* **2017**, *357*, eaaf4382. [CrossRef]
9. Kato, M.; Han, T.W.; Xie, S.; Shi, K.; Du, X.; Wu, L.C.; Mirzaei, H.; Goldsmith, E.J.; Longgood, J.; Pei, J.; et al. Cell-free formation of RNA granules: Low complexity sequence domains form dynamic fibers within hydrogels. *Cell* **2012**, *149*, 753–767. [CrossRef]
10. Purice, M.D.; Taylor, J.P. Linking hnRNP Function to ALS and FTD Pathology. *Front. Neurosci.* **2018**, *12*, 326. [CrossRef]
11. Nott, T.J.; Petsalaki, E.; Farber, P.; Jervis, D.; Fussner, E.; Plochowietz, A.; Craggs, T.D.; Bazett-Jones, D.P.; Pawson, T.; Forman-Kay, J.D.; et al. Phase transition of a disordered nuage protein generates environmentally responsive membraneless organelles. *Mol. Cell* **2015**, *57*, 936–947. [CrossRef] [PubMed]
12. Ryan, V.H.; Fawzi, N.L. Physiological, Pathological, and Targetable Membraneless Organelles in Neurons. *Trends Neurosci.* **2019**, *42*, 693–708. [CrossRef] [PubMed]
13. Das, S.; Lin, Y.H.; Vernon, R.M.; Forman-Kay, J.D.; Chan, H.S. Comparative roles of charge, pi, and hydrophobic interactions in sequence-dependent phase separation of intrinsically disordered proteins. *Proc. Natl. Acad. Sci. USA* **2020**, *117*, 28795–28805. [CrossRef] [PubMed]
14. Borroni, B.; Bonvicini, C.; Alberici, A.; Buratti, E.; Agosti, C.; Archetti, S.; Papetti, A.; Stuani, C.; Di Luca, M.; Gennarelli, M.; et al. Mutation within TARDBP leads to frontotemporal dementia without motor neuron disease. *Hum. Mutat.* **2009**, *30*, E974–E983. [CrossRef] [PubMed]
15. Vance, C.; Rogelj, B.; Hortobagyi, T.; De Vos, K.J.; Nishimura, A.L.; Sreedharan, J.; Hu, X.; Smith, B.; Ruddy, D.; Wright, P.; et al. Mutations in FUS, an RNA processing protein, cause familial amyotrophic lateral sclerosis type 6. *Science* **2009**, *323*, 1208–1211. [CrossRef]
16. Batlle, C.; Ventura, S. Prion-like domain disease-causing mutations and misregulation of alternative splicing relevance in limb-girdle muscular dystrophy (LGMD) 1G. *Neural Regen. Res.* **2020**, *15*, 2239–2240. [CrossRef]
17. Batlle, C.; Yang, P.; Coughlin, M.; Messing, J.; Pesarrodona, M.; Szulc, E.; Salvatella, X.; Kim, H.J.; Taylor, J.P.; Ventura, S. hnRNPDL Phase Separation Is Regulated by Alternative Splicing and Disease-Causing Mutations Accelerate Its Aggregation. *Cell Rep.* **2020**, *30*, 1117–1128. [CrossRef]
18. Uversky, V.N. Intrinsically disordered proteins and their environment: Effects of strong denaturants, temperature, pH, counter ions, membranes, binding partners, osmolytes, and macromolecular crowding. *Protein J.* **2009**, *28*, 305–325. [CrossRef]
19. Farahi, N.; Lazar, T.; Wodak, S.J.; Tompa, P.; Pancsa, R. Integration of Data from Liquid-Liquid Phase Separation Databases Highlights Concentration and Dosage Sensitivity of LLPS Drivers. *Int. J. Mol. Sci.* **2021**, *22*, 3017. [CrossRef]
20. Santos, J.; Iglesias, V.; Santos-Suarez, J.; Mangiagalli, M.; Brocca, S.; Pallares, I.; Ventura, S. pH-Dependent Aggregation in Intrinsically Disordered Proteins Is Determined by Charge and Lipophilicity. *Cells* **2020**, *9*, 145. [CrossRef]
21. Santos, J.; Iglesias, V.; Pintado, C.; Santos-Suarez, J.; Ventura, S. DispHred: A Server to Predict pH-Dependent Order-Disorder Transitions in Intrinsically Disordered Proteins. *Int. J. Mol. Sci.* **2020**, *21*, 5814. [CrossRef]
22. Pintado, C.; Santos, J.; Iglesias, V.; Ventura, S. SolupHred: A server to predict the pH-dependent aggregation of intrinsically disordered proteins. *Bioinformatics* **2021**, *37*, 1602–1603. [CrossRef]
23. Pintado-Grima, C.; Iglesias, V.; Santos, J.; Uversky, V.N.; Ventura, S. DispHScan: A Multi-Sequence Web Tool for Predicting Protein Disorder as a Function of pH. *Biomolecules* **2021**, *11*, 1596. [CrossRef]
24. Jacoby, G.; Segal Asher, M.; Ehm, T.; Abutbul Ionita, I.; Shinar, H.; Azoulay-Ginsburg, S.; Zemach, I.; Koren, G.; Danino, D.; Kozlov, M.M.; et al. Order from Disorder with Intrinsically Disordered Peptide Amphiphiles. *J. Am. Chem. Soc.* **2021**, *143*, 11879–11888. [CrossRef]
25. Uversky, V.N.; Gillespie, J.R.; Millett, I.S.; Khodyakova, A.V.; Vasiliev, A.M.; Chernovskaya, T.V.; Vasilenko, R.N.; Kozlovskaya, G.D.; Dolgikh, D.A.; Fink, A.L.; et al. Natively unfolded human prothymosin alpha adopts partially folded collapsed conformation at acidic pH. *Biochemistry* **1999**, *38*, 15009–15016. [CrossRef]
26. Wu, K.P.; Weinstock, D.S.; Narayanan, C.; Levy, R.M.; Baum, J. Structural reorganization of alpha-synuclein at low pH observed by NMR and REMD simulations. *J. Mol. Biol.* **2009**, *391*, 784–796. [CrossRef] [PubMed]
27. You, K.; Huang, Q.; Yu, C.; Shen, B.; Sevilla, C.; Shi, M.; Hermjakob, H.; Chen, Y.; Li, T. PhaSepDB: A database of liquid-liquid phase separation related proteins. *Nucleic Acids Res.* **2020**, *48*, D354–D359. [CrossRef]

28. Meszaros, B.; Erdos, G.; Dosztanyi, Z. IUPred2A: Context-dependent prediction of protein disorder as a function of redox state and protein binding. *Nucleic Acids Res.* **2018**, *46*, W329–W337. [CrossRef]
29. Quaglia, F.; Meszaros, B.; Salladini, E.; Hatos, A.; Pancsa, R.; Chemes, L.B.; Pajkos, M.; Lazar, T.; Pena-Diaz, S.; Santos, J.; et al. DisProt in 2022: Improved quality and accessibility of protein intrinsic disorder annotation. *Nucleic Acids Res.* **2021**, *50*, D480–D487. [CrossRef]
30. Iglesias, V.; Santos, J.; Santos-Suarez, J.; Pintado-Grima, C.; Ventura, S. SGnn: A Web Server for the Prediction of Prion-Like Domains Recruitment to Stress Granules Upon Heat Stress. *Front. Mol. Biosci.* **2021**, *8*, 718301. [CrossRef]
31. Wallace, E.W.; Kear-Scott, J.L.; Pilipenko, E.V.; Schwartz, M.H.; Laskowski, P.R.; Rojek, A.E.; Katanski, C.D.; Riback, J.A.; Dion, M.F.; Franks, A.M.; et al. Reversible, Specific, Active Aggregates of Endogenous Proteins Assemble upon Heat Stress. *Cell* **2015**, *162*, 1286–1298. [CrossRef] [PubMed]
32. Neumann, M.; Sampathu, D.M.; Kwong, L.K.; Truax, A.C.; Micsenyi, M.C.; Chou, T.T.; Bruce, J.; Schuck, T.; Grossman, M.; Clark, C.M.; et al. Ubiquitinated TDP-43 in frontotemporal lobar degeneration and amyotrophic lateral sclerosis. *Science* **2006**, *314*, 130–133. [CrossRef] [PubMed]
33. Babinchak, W.M.; Haider, R.; Dumm, B.K.; Sarkar, P.; Surewicz, K.; Choi, J.K.; Surewicz, W.K. The role of liquid-liquid phase separation in aggregation of the TDP-43 low-complexity domain. *J. Biol. Chem.* **2019**, *294*, 6306–6317. [CrossRef]
34. Tsoi, P.S.; Quan, M.D.; Choi, K.J.; Dao, K.M.; Ferreon, J.C.; Ferreon, A.C.M. Electrostatic modulation of hnRNPA1 low-complexity domain liquid-liquid phase separation and aggregation. *Protein Sci. Publ. Protein Soc.* **2021**, *30*, 1408–1417. [CrossRef] [PubMed]
35. Xue, S.; Gong, R.; He, F.; Li, Y.; Wang, Y.; Tan, T.; Luo, S.Z. Low-complexity domain of U1-70K modulates phase separation and aggregation through distinctive basic-acidic motifs. *Sci. Adv.* **2019**, *5*, eaax5349. [CrossRef]
36. Vernon, R.M.; Chong, P.A.; Tsang, B.; Kim, T.H.; Bah, A.; Farber, P.; Lin, H.; Forman-Kay, J.D. Pi-Pi contacts are an overlooked protein feature relevant to phase separation. *eLife* **2018**, *7*, e31486. [CrossRef]
37. Van Lindt, J.; Lazar, T.; Pakravan, D.; Demulder, M.; Meszaros, A.; Van Den Bosch, L.; Maes, D.; Tompa, P. F/YGG-motif is an intrinsically disordered nucleic-acid binding motif. *RNA Biol.* **2022**, *19*, 622–635. [CrossRef]
38. Iglesias, V.; Conchillo-Sole, O.; Batlle, C.; Ventura, S. AMYCO: Evaluation of mutational impact on prion-like proteins aggregation propensity. *BMC Bioinform.* **2019**, *20*, 24. [CrossRef]
39. Kim, H.J.; Kim, N.C.; Wang, Y.D.; Scarborough, E.A.; Moore, J.; Diaz, Z.; MacLea, K.S.; Freibaum, B.; Li, S.; Molliex, A.; et al. Mutations in prion-like domains in hnRNPA2B1 and hnRNPA1 cause multisystem proteinopathy and ALS. *Nature* **2013**, *495*, 467–473. [CrossRef]
40. Van Lindt, J.; Bratek-Skicki, A.; Nguyen, P.N.; Pakravan, D.; Duran-Armenta, L.F.; Tantos, A.; Pancsa, R.; Van Den Bosch, L.; Maes, D.; Tompa, P. A generic approach to study the kinetics of liquid-liquid phase separation under near-native conditions. *Commun. Biol.* **2021**, *4*, 77. [CrossRef]
41. Taylor, J.P.; Brown, R.H., Jr.; Cleveland, D.W. Decoding ALS: From genes to mechanism. *Nature* **2016**, *539*, 197–206. [CrossRef] [PubMed]

biomolecules

MDPI

Article

# NMR Reveals Specific Tracts within the Intrinsically Disordered Regions of the SARS-CoV-2 Nucleocapsid Protein Involved in RNA Encountering

**Letizia Pontoriero [1,†], Marco Schiavina [1,†], Sophie M. Korn [2,†], Andreas Schlundt [2,*], Roberta Pierattelli [1,*] and Isabella C. Felli [1,*]**

1 Magnetic Resonance Center (CERM) and Department of Chemistry "Ugo Schiff", University of Florence, Via L. Sacconi 6, Sesto Fiorentino, 50019 Florence, Italy; pontoriero@cerm.unifi.it (L.P.); schiavina@cerm.unifi.it (M.S.)
2 Center for Biomolecular Magnetic Resonance (BMRZ), Institute for Molecular Biosciences, Johann Wolfgang Goethe-University, Max-von-Laue-Str. 9, 60438 Frankfurt, Germany; bochmann@bio.uni-frankfurt.de
* Correspondence: schlundt@bio.uni-frankfurt.de (A.S.); roberta.pierattelli@unifi.it (R.P.); felli@cerm.unifi.it (I.C.F.)
† These authors contributed equally to the work.

**Abstract:** The SARS-CoV-2 nucleocapsid (N) protein is crucial for the highly organized packaging and transcription of the genomic RNA. Studying atomic details of the role of its intrinsically disordered regions (IDRs) in RNA recognition is challenging due to the absence of structure and to the repetitive nature of their primary sequence. IDRs are known to act in concert with the folded domains of N and here we use NMR spectroscopy to identify the priming events of N interacting with a regulatory SARS-CoV-2 RNA element. $^{13}$C-detected NMR experiments, acquired simultaneously to $^{1}$H detected ones, provide information on the two IDRs flanking the N-terminal RNA binding domain (NTD) within the N-terminal region of the protein (NTR, 1–248). We identify specific tracts of the IDRs that most rapidly sense and engage with RNA, and thus provide an atom-resolved picture of the interplay between the folded and disordered regions of N during RNA interaction.

**Keywords:** SARS-CoV-2; COVID-19; IDP; RNA; NMR

**Citation:** Pontoriero, L.; Schiavina, M.; Korn, S.M.; Schlundt, A.; Pierattelli, R.; Felli, I.C. NMR Reveals Specific Tracts within the Intrinsically Disordered Regions of the SARS-CoV-2 Nucleocapsid Protein Involved in RNA Encountering. *Biomolecules* **2022**, *12*, 929. https://doi.org/10.3390/biom12070929

Academic Editors: Stefania Brocca, Keith Dunker, Sonia Longhi and Prakash Kulkarni

Received: 6 June 2022
Accepted: 29 June 2022
Published: 2 July 2022



## 1. Introduction

The nucleocapsid protein N of SARS-CoV-2 plays a pivotal role in the viral life cycle. The protein is organized in five different modular domains, two folded and three disordered ones, with the latter comprising almost 40% of the whole protein sequence (Supplementary Figure S1) [1,2]. It exerts various functions including packaging of genomic RNA (gRNA) inside the viral capsid [3–8] but the structural and mechanistic details of packaging remain enigmatic. The SARS-CoV-2 genome comprises a multitude of highly conserved structured *cis* regulatory RNA elements [4], which have been suggested as target sites for N in the context of packaging [9]. It is thus important to study how the disordered protein regions modulate the interaction with RNA. Recent work showed the potential of solution NMR [10–17] to describe the structural and dynamic features of different N constructs and how they interact with RNA fragments. Here we would like to explore how $^{13}$C detection can contribute to this field.

$^{13}$C-NMR emerged as a key technique to study intrinsically disordered proteins (IDPs) [18]. The large chemical shift dispersion of heteronuclei ($^{13}$C, $^{15}$N) is crucial to obtaining highly resolved spectra in the absence of a stable 3D structure. Solvent exchange often leads to the broadening of amide proton signals, in particular for exposed protein backbones, when approaching physiological pH and temperature. $^{13}$C-detected heteronuclear NMR experiments allow us to overcome this limitation. For these reasons, they constitute a valuable tool to investigate highly flexible polypeptide chains also when part of a multi-domain protein.

The contribution of the flexible regions of N to the interaction with RNA is investigated here by selecting a construct comprising the folded *N*-terminal domain NTD (44–180) and the flanking intrinsically disordered regions, IDR1 (1–43) and IDR2 (181–248). This allows us to focus on the IDRs while linked to the NTD, that is the domain deputed to bind gRNA [1]. The interaction between this N construct (1–248, referred to as N-terminal region, NTR) with RNA was studied by selecting a highly conserved *cis* element of the gRNA, namely the 5′-UTR-contained stem-loop 4 (5_SL4) [19]. This is centrally located within the 5′-UTR, has very recently been found targetable by small molecules [20] and thus represents a potential drug target to disrupt its interactions with abundant viral proteins such as N. It is described as stable [5] and is chemically versatile comprising a pentaloop, two internal loops, a bulge, and a good mix of nucleotides and types of base pairs (Supplementary Figure S1). It thus represents a bona fide example RNA for this study.

## 2. Materials and Methods

### 2.1. Protein Sample Preparation

The NTD and the NTR samples were prepared as previously described [13,21] and briefly summarized hereafter.

For the NTR construct, the gene of the N protein comprising residues 1–248 was designed based on the boundaries determined from the SARS-CoV homologue [1]. The codon-optimized gene was synthesized by Twist Bioscience and cloned into pET29b(+) vector between NdeI and XhoI restriction sites.

Uniformly $^{13}C$,$^{15}N$-labeled NTR protein was expressed in *E. coli* strain BL21 (DE3) following the Marley method [22]. The cells were grown in 1 L Luria Bertani medium at 37 °C until an optical density ($OD_{600}$) of 0.8 was reached. Then, the culture was transferred in 250 mL of labeled minimal medium supplemented with 0.25 g/L $^{15}NH_4Cl$ (Cambridge Isotope Laboratories) and 0.75 g/L $^{13}C_6$-D-glucose (Eurisotop). After 1 h of unlabeled metabolite clearance, the culture was induced with 0.2 mM isopropyl-beta-thiogalactopyranoside (IPTG) at 16 °C for 18 h. The pellet was harvested and stored at −20 °C overnight. The cell pellet was then resuspended in 25 mM 2-amino-2-(hydroxymethyl)-1,3-propanediol (TRIS), 1.0 M NaCl, 10% glycerol, and protease inhibitor cocktail (SIGMA) at pH 8.0. Cells were disrupted by sonication and the lysate was centrifuged at 30,000× *g* for 50 min at 4 °C.

The soluble fraction was dialyzed overnight against a solution of 25 mM TRIS, pH 7.2 at 4 °C. The protein solution was then loaded on a HiTrap SP FF 5 mL column and eluted in 25 CV with a 70% gradient of 25 mM TRIS and 1.0 M NaCl. Fractions containing the protein were pooled, concentrated, and loaded on a HiLoad 16/1000 Superdex 75 pg column equilibrated with 25 mM potassium phosphate, 450 mM KCl, pH 6.5. the fractions containing the protein were pooled and concentrated using centrifugal concentrators (molecular weight cut-off 10 KDa).

The gene of the single cysteine A211C mutant of the NTR protein was synthesized by Twist Bioscience and cloned into the pET29b(+) vector between NdeI and XhoI restriction sites. Uniformly $^{15}N$-labeled A211C protein was expressed and purified following the same protocol used for the NTR construct, with the addition of 5 mM dithiothreitol (DTT) in the lysis and purification buffers.

The soluble fraction was dialyzed overnight against a solution of 25 mM TRIS and 5 mM DTT, pH 7.2 at 4 °C. The protein solution was then loaded on a HiTrap SP FF 5 mL column and eluted in 25 CV with a 70% gradient of 25 mM TRIS, 1.0 M NaCl, and 5 mM DTT, pH 7.2. Fractions containing the protein were pooled and concentrated to a final concentration of 25 μM.

The sequence of the NTD (44–180) was based on SARS-CoV-2 NCBI reference genome entry NC_045512.2, identical to GenBank entry MN90894 [23]. Domain boundaries for the core NTD were defined in analogy to the available NMR structure (PDB 6YI3) [10]. An *E. coli* codon-optimized DNA construct was obtained from Eurofins Genomics and sub-cloned into the pET-21-based vector pET-Trx1a, containing an *N*-terminal $His_6$-tag, a thioredoxin-tag and a tobacco etch virus (TEV) cleavage site. After proteolytic TEV cleavage, the produced 14.9 kDa protein contains one artificial N-terminal residue (Gly0), before the start of the native protein sequence at Gly1 which corresponds to Gly44 in the full-length N protein sequence.

Uniformly $^{15}N$-labeled NTD was expressed in *E. coli* strain BL21 (DE3) in M9 minimal medium containing 1.0 g/L $^{15}NH_4Cl$ (Cambridge Isotope Laboratories) and 25 µg/mL kanamycin. Protein expression was induced at an $OD_{600}$ of 0.8 with 1 mM IPTG for 18 h at room temperature. Cell pellets were resuspended in 50 mM TRIS/HCl pH 8.0, 300 mM NaCl, 10 mM imidazole, and 100 µL protease inhibitor mix (SERVA) per 1.0 L of culture. Cells were disrupted by sonication. The supernatant was cleared by centrifugation (30 min, 9000× *g*, 4 °C). The cleared supernatant was passed over a $Ni^{2+}$-NTA gravity flow column (Sigma-Aldrich) and the $His_6$-Trx-tag was cleaved overnight at 4 °C with 0.5 mg of TEV protease per 1.0 L of culture and dialyzed into fresh buffer (50 mM TRIS/HCl pH 8.0, 300 mM NaCl, 10% glycerol). TEV protease and the cleaved tag were removed via a second $Ni^{2+}$-NTA gravity flow column, and core NTD was further purified via size exclusion on a HiLoad 16/600 SD 75 (Cytiva) in 25 mM potassium phosphate, 150 mM KCl, 2 mM Tris-(2-carboxyethyl)-phosphin (TCEP), 0.02% $NaN_3$, pH 6.5. Pure NTD protein-containing fractions were determined by SDS-PAGE, pooled and concentrated using Amicon centrifugal concentrators (molecular weight cut-off of 10 kDa).

### *2.2. RNA Production*

The 40 nucleotides (nt) SARS-CoV-2 genomic RNA element stem loop 4 (SL4) located within the 5′UTR (nt 86 to 125), extended 5′ by two guanine residues and 3′ by two cytidine residues, yielded the 44-nt sequence 5′-GG**GUG UGG CUG UCA CUC GGC UGC AUG CUU AGU GCA CUC ACGC** CC-3′ [19]. The DNA template for 5_SL4 was kindly provided in a HDV ribozyme vector by the COVID19-nmr consortium. The unlabeled RNA was produced by in-house optimized in vitro transcription and purified as described previously [5]. Final RNA samples were buffer-exchanged to 25 mM potassium phosphate, 150 mM KCl, pH 6.5, and sample quality, homogeneity and long-term stability were verified by native and denaturing PAGE as well as 1D-NMR experiments by means of the characteristic imino proton pattern.

### *2.3. Spin-Labeling Reaction for PRE Experiments*

The A211C protein solution was purified from DTT using a PD-10 desalting column and then incubated with a ten-fold excess of S-(1-oxyl-2,2,5,5,-tetramethyl-2,5,-dihydro-1H-pyrrol-3-yl) methylmethane-sulfonothiolate (MTSL) relative to the protein concentration. The reaction was performed overnight in absence of light at 4 °C while gently stirring. Then, the unreacted spin-label was eliminated using two steps of purification with a PD-10 desalting column. The protein eluted in 25 mM TRIS and 150 mM NaCl.

To reduce MTSL and obtain the diamagnetic sample, a five-fold excess of ascorbate with respect to the protein concentration was added.

### *2.4. Protein NMR Samples*

For NTR, experiments were acquired using two 500-µL-samples of 140 µM $^{13}C,^{15}N$ NTR solution in 25 mM potassium phosphate at pH 6.5, 150 mM KCl, 0.01% $NaN_3$ in $H_2O$ with 5% $D_2O$. The titration was performed in 5 mm NMR tubes. A highly concentrated batch of 5_SL4 solution in 25 mM potassium phosphate, 150 mM KCl, 0.01% $NaN_3$, pH 6.5 was prepared as previously described and added to a protein solution sample in small aliquots to reach NTR:RNA ratios of 1:0.01, 1:0.025, and 1:0.05. A second identical protein sample was used to reach NTR:RNA ratios of 1:0.1, 1:0.3, and 1:0.6.

For NTD, experiments were acquired using one 500-µL-sample of 70 µM $^{15}N$ NTD solution in 25 mM potassium phosphate at pH 6.5, 150 mM KCl, 2 mM TCEP, and 0.02% $NaN_3$ in $H_2O$ with 5% $D_2O$. A highly concentrated batch of 5_SL4 solution in 25 mM potassium phosphate, 150 mM KCl, 0.02% $NaN_3$, 2 mM TCEP, and pH 6.5 was prepared as previously described and added to a protein solution sample in small aliquots to reach NTD:RNA ratios of 1:0.1, 1:0.3, 1:1.2, and 1:2.4.

*2.5. NMR Experiments*

To follow the interaction between NTR and 5_SL4, the mr_CON//HN experiment [24] was used. To complete the available assignment [13], a 3D-(H)CBCACON experiment [25] was also acquired on a 100 μM $^{13}C$,$^{15}N$ NTR sample.

These NMR experiments were acquired on a Bruker AVANCE NEO spectrometer operating at 700.06 MHz $^{1}H$, 176.05 MHz $^{13}C$, and 70.97 MHz $^{15}N$ frequencies equipped with a cryogenically cooled probehead optimized for $^{13}C$-direct detection (TXO) at 298 K. Standard radiofrequency pulses and carrier frequencies for triple resonance experiments were used and are summarized hereafter. $^{13}C$ pulses were given at 176.7 ppm, 55.9 ppm, and 45.7 ppm for C′, $C^{\alpha}$ and $C^{ali}$ spectral regions, respectively. $^{15}N$ pulses were given at 124.0 ppm. The $^{1}H$ carrier was placed at 4.7 ppm. Q5- and Q3-shaped pulses [26] of durations of 300 and 231 μs, respectively, were used for $^{13}C$ band-selective $\pi/2$ and $\pi$ flip angle pulses except for the $\pi$ pulses that should be band selective on the $C^{\alpha}$ region (Q3, 1200 μs) and for the adiabatic $\pi$ pulse to invert both C′ and $C^{\alpha}$ (smoothed chirp 500 μs, 20% smoothing, 80 kHz sweep width, 11.3 kHz radio frequency field strength) [27]. Decoupling of $^{1}H$ and $^{15}N$ was achieved with waltz65 (100 μs) and garp4 (250 μs) decoupling sequences, respectively [26,28]. All gradients employed had a smoothed square shape.

The mr_CON//HN was acquired with an interscan delay of 1.6 s; during this delay, the HN experiment was acquired as discussed in [24]. Solvent suppression was achieved through the 3:9:19 pulse scheme [29]. For each increment of the CON experiment, acquired with 16 scans, the in-phase (IP) and antiphase (AP) components were recorded and properly combined to achieve IPAP virtual decoupling [30]. The CON spectrum was acquired with sweep widths of 5263 Hz ($^{13}C$) × 2840 Hz ($^{15}N$) and 1024 × 400 real points in the two dimensions, respectively. The HN spectrum was acquired with 32 scans, with sweep widths of 20869 Hz ($^{1}H$) × 3194 Hz ($^{15}N$) and 4096 × 400 real points in the two dimensions, respectively.

The 3D-(H)CBCACON was acquired with an interscan delay of 1 s, with 8 scans, with sweep widths of 5263 Hz ($^{13}C'$) × 2415 Hz ($^{15}N$) × 10,204 Hz ($^{13}C_{ali}$) and 1024 × 96 × 110 real points in the three dimensions, respectively.

To follow the interaction between NTD and 5_SL4 the 2D HN fingerprint spectra were acquired with the Fast-HSQC experimental variant [31] using a Bruker AVANCE III HD spectrometer operating at 700.17 MHz $^{1}H$, 176.05 MHz $^{13}C$, and 70.95 MHz $^{15}N$ frequencies equipped with a quadruple-resonance cryo-probehead optimized for $^{1}H$-direct detection (QCI) at 298 K. The $^{1}H$ carrier was placed at 4.7 ppm for non-selective hard pulses and the one for $^{15}N$ at 117 ppm. The pulse scheme includes a 60 μs delay for binomial water suppression flanking the reverse INEPT step and calculated for the $H^{N}$ central region and field strength. Decoupling of $^{15}N$ was achieved with garp4 (250 μs) [26]. The HN experiments were acquired with an interscan delay of 1 s with 32 scans with sweep widths of 11904 Hz ($^{1}H$) × 2412 Hz ($^{15}N$) and 2048 × 128 real points in the two dimensions, respectively.

For the Paramagnetic Relaxation Enhancement experiments (PRE), sensitivity improvement 2D HN HSQC [32] spectra were acquired on a Bruker AVANCE NEO spectrometer operating at 900.06 ($^{1}H$) and 91.20 ($^{15}N$) MHz equipped with a cryogenically cooled probehead (TCI). The experiments were acquired with 32 scans, with an interscan delay of 6 s, with sweep widths of 20833 Hz ($^{1}H$) × 3289 Hz ($^{15}N$) and 4096 × 400 points in the two dimensions. $^{15}N$ pulses were given at 117.0 ppm and the $^{1}H$ carrier was placed at 4.7 ppm. Decoupling of $^{15}N$ was achieved with garp (250 μs) decoupling sequences [26]. All gradients employed had a smoothed square shape.

### 2.6. Protein Visualization

The images and the surface potential of the proteins were created and calculated using Chimera 1.14 [33] by adding to the experimental NTD structure (PDB: 6YI3 [10]) an arbitrary conformer for IDR1 and IDR2 obtained through Flexible Meccano [34].

### 2.7. NMR Spectral Analysis

All the spectra were acquired and processed by using Bruker TopSpin 4.0.8 software. Calibration of the spectra was achieved using 4,4-dimethyl-4-silapentane-1-sulfonic acid (DSS) as a standard for $^{1}$H and $^{13}$C; $^{15}$N shifts were calibrated indirectly [35].

The NTR and NTD spectra were analyzed with the aid of CARA [36] and its tool NEASY [37]. All the spectra were integrated manually with NEASY taking into consideration only the well-resolved peaks. The volume of each peak, from each titration point, was divided by the volume measured in the reference spectrum acquired. The obtained ratios were plotted against the residue number. The missing values in the ratio intensity plots belong to proline residues (in the case of HN spectra), or to peaks that overlap with others, unless otherwise specified.

The Chemical Shift Perturbation (CSP) analysis was performed comparing two HN-HSQC acquired on the NTR and NTD at the same temperature and in the very same buffer (the one used for the RNA titration). The peak lists were manually inspected and only the well-resolved peaks were used to obtain the CSP values reported in the plot. The CSP values were calculated using the following equation: $CSP = \sqrt{\frac{1}{2}(\delta_H{}^2 + 0.1 \cdot \delta_N{}^2)}$, where $\delta_H$ and $\delta_N$ represent the variation in the chemical shift of the $^{1}$H and $^{15}$N nuclei, respectively.

### 2.8. Electromobility Shift Assay (EMSA)

Radioactive EMSAs were performed according to [38] with the following modifications: RNA transcripts (30 pmol) were dephosphorylated using Quick CIP (NEB) following the manufacturer's protocol and finally resuspended in $H_2O$. Subsequently, 5′ end-labelling of 15 pmol SL4 RNA with [$\gamma$-$^{32}$P]-ATP was accomplished with T4 polynucleotide kinase (NEB). Labeled RNA was separated from unincorporated [$\gamma$-$^{32}$P]-ATP by column purification (NucAway) and adjusted with binding buffer (25 mM potassium phosphate, 150 mM KCl, pH 6.5) to 0.03 pmol/µL. A master mix containing tRNA, $^{32}$P-labeled SL4 RNA, and reaction buffer was prepared and then mixed with dilutions of the NTR or NTD, respectively, to achieve the indicated protein concentrations. Binding was performed for 10 min at RT in 20-µL reaction volume in the presence of 0.6 µg tRNA from baker's yeast (Sigma), 3 nM $^{32}$P-labeled SL4 RNA, 25 mM potassium phosphate, 150 mM KCl, pH 6.5, and 1 mM $MgCl_2$. After the addition of 3 µL loading buffer (30% glycerol, bromphenol blue, xylene cyanol), the RNP complexes were resolved by PAGE (6% polyacrylamide, 5% glycerol, and 1 × TBE) at 80 V for 75 min at RT. Gels were fixed and dried and subsequently exposed to a phosphor imager screen and visualized using a GE Typhoon laser scanner under "phosphorimager" settings.

## 3. Results and Discussion

The interaction of NTR with 5_SL4 (referred to as RNA hereafter) was studied through the $^{13}$C-detected $^{13}$C-$^{15}$N CON (2D CON) experiment. Due to the very different structural and dynamic properties of the globular NTD domain and the flanking disordered regions, with the chosen setup, the NMR signals of the NTD are very weak or absent in the 2D CON. This allows to selectively pick up information about the disordered regions of NTR, yielding well-resolved NMR spectra, which reveal also information about seven proline residues (Figure 1). It thus provides highly complementary information to that available through a $^{1}$H-detected $^{1}$H-$^{15}$N HSQC (2D HN) experiment. The latter allows monitoring of most of the residues belonging to the folded domain, while those of the flexible regions suffer from extensive spectral overlap or line broadening (Figure 1). The combined use of the two NMR experiments thus provides a complete picture of NTR upon interaction with

RNA. The two experiments can also be collected simultaneously [24] without compromises in the quality of either of them. This experimental variant, referred to as mr_CON//HN, is particularly useful when dealing with multi-domain proteins constituted by globular domains and flexible regions. More than for time-saving, the approach is useful to achieve simultaneous snapshots of the protein which allow us to monitor the occurrence of the interaction from two different points of view. The two spectra obtained contain information about three different nuclei, one of them ($^{15}N$) common to the two spectra. Moreover, the 2D HN can be collected with high S/N without increasing the experimental time, just exploiting the relaxation delay of the 2D CON experiment. The NMR spectra obtained through this approach on NTR are reported in Figure 1.

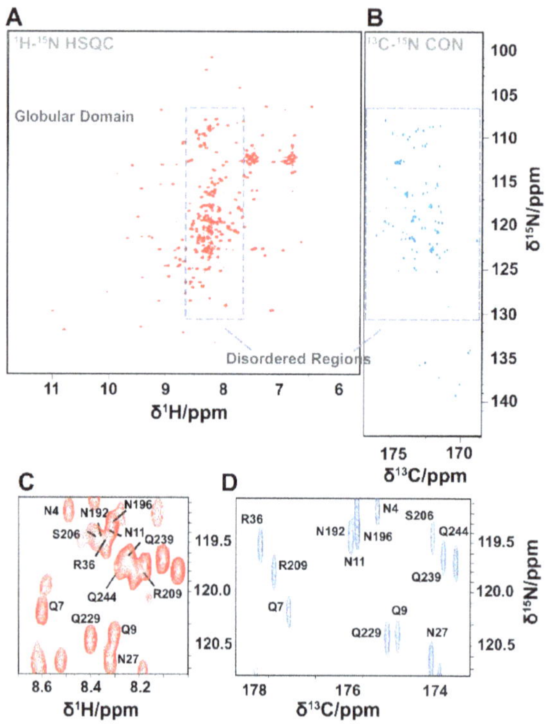


**Figure 1.** Panels A and B report the spectra obtained through the mr_CON//HN experiment. The 2D HN spectrum (**A**) shows a set of well-isolated signals deriving from the globular NTD domain as well as a number of signals, clustered in a narrow central region of the spectrum, deriving from the IDRs. The 2D CON spectrum (**B**) allows achieving the necessary resolution to investigate resonances from IDRs, including signals of proline residues. While IDR peaks fall in a very crowded region of the HN spectrum (1.1 ppm on $^{1}H$ dimension), they are well dispersed in the CON spectrum (7.2 ppm on $^{13}C$ dimension), as indicated by the two boxes. A zoom of a region of the two spectra centered at 120 ppm for $^{15}N$ is reported in panels (**C**,**D**) to stress this concept.

NMR spectroscopy reveals at the residue level the importance of the two disordered regions for the interaction with RNA. This is already evident when a sub-stoichiometric RNA concentration (0.05 equivalents) is added to NTR (Figure 2A). Inspection of the 2D HN spectra of NTR show variations in cross peak intensities, reported in Figure 2 as intensity ratios upon addition of increasing RNA equivalents, while shift changes are negligible (Supplementary Figure S3). In the very first points of the titration, a remarkable decrease in intensity is observed for the few resolved resonances of the HN signals from IDRs. In contrast, the signals that arise from the globular domain of the construct, seem to be less perturbed by the addition of a small RNA quantity. A further increase in RNA concentration leads to a measurable signal reduction of the NTD residues, with the complete disappearance of the signals upon the addition of 0.3 equivalents of RNA. In our experimental conditions, upon further addition of RNA, we observed liquid–liquid phase separation [11,39–41], not further investigated here. In contrast, the addition of RNA to the NTD (lacking the IDRs) at the same equivalent concentrations had smaller effects on line-broadening, suggesting a reduced affinity of the isolated domain (Figure 2B). This is confirmed by Electrophoretic Mobility Shift Assay (EMSA) experiments (Figure 2 and Supplementary Figure S2). The results indicate that the NTR construct has a higher affinity towards RNA compared to the NTD alone as indicated by gel shifts observed at lower concentrations. While both NTD-containing proteins show binding to RNA, the two IDRs flanking the NTD visibly increase affinity to RNA.

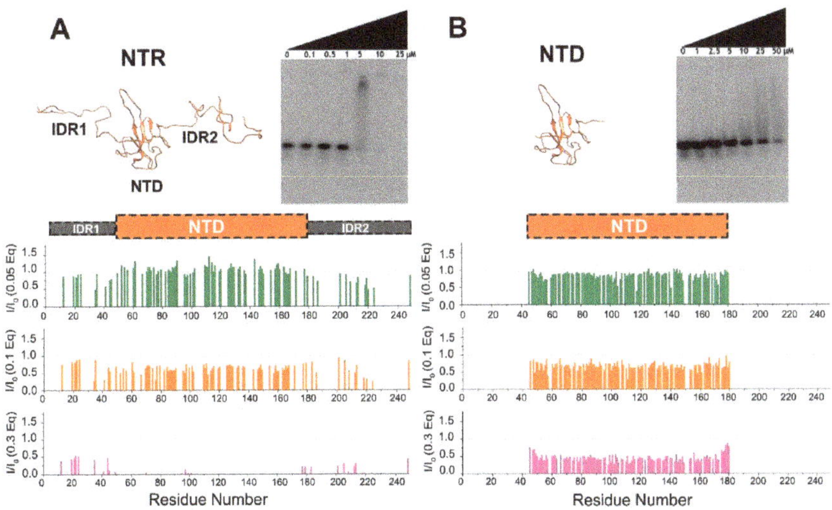


**Figure 2.** Differences in the interaction of NTR (**A**) and NTD (**B**) with 5_SL4 followed by NMR and EMSA. Upper panels show the two constructs and their different binding affinities for RNA as demonstrated by EMSA experiments. The binding of NTR to RNA occurs at a lower concentration as compared to that of NTD alone. The lower panels show plots of the HN HSQC peak intensity ratios versus residue number after the addition of increasing amounts of 5_SL4 (with equivalents as indicated) relative to protein. The structural models were obtained as described in the experimental part.

A zoom into the IDRs can be achieved through the analysis of the 2D CON spectrum. This allowed us to monitor most of the residues belonging to the highly flexible IDRs. As an example, Figure 3 shows the enlargement of selected portions of the 2D CON in diagnostic spectral regions such as that of glycine (top) and proline residues (bottom). Addition of 0.1 equivalents of RNA shows intensity changes for specific cross-peaks, suggesting the presence of preferred IDR sites for the interaction with RNA. Intensity ratios of the CON cross-peaks, obtained upon subsequent addition of RNA are reported versus the residue number in Figure 3. The most perturbed regions, indicated in the gray areas in Figure 3, comprise three different tracts (32–46, 177–203, and 216–225). These feature peculiar signatures in terms of amino acid composition as it often happens for interactions involving intrinsically disordered protein regions [42–51].

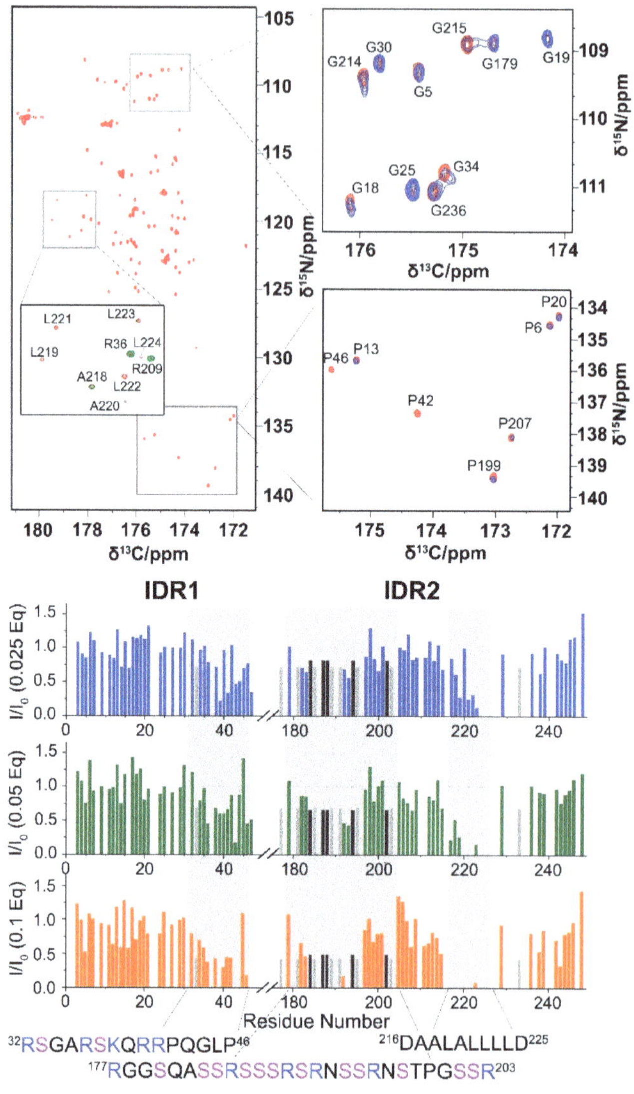


**Figure 3.** 2D CON experiments reveal differential effects of RNA-binding in specific regions of IDR1 and IDR2. The CON spectrum acquired on NTR is reported in red (**top, left**). The inset shows the superposition of the reference spectrum with NTR upon the addition of 0.01 eq of RNA (green). The enlargements of two portions of the spectra reported on the right panels (namely, the typical Gly and Pro regions) show the spectrum acquired on the NTR upon the addition of RNA (0.1 equivalents, blue) superimposed to the spectrum acquired in the absence of RNA (reference, red). The intensity ratios of CON cross-peaks are reported in the lower panel versus the residue number; spectra were acquired simultaneously to the HN spectra. Light and dark gray bars represent the intensity ratio of the envelope of signals centered at 176.6 ppm ($^{13}$C)–116.5 ppm ($^{15}$N) and 174.8 ppm ($^{13}$C)–117.9 ppm ($^{15}$N), respectively. Gray shaded areas highlight the protein regions most perturbed upon the addition of RNA.

Two of the tracts of NTR perturbed by the addition of RNA are very rich in positively charged residues: four arginine and one lysine residues in the region $^{32}$RSGARSKQRRPQG LP$^{46}$, and six arginine residues in the $^{177}$RGGSQASSRSSSRSRNSSRNSTPGSSR$^{203}$ region

("SR-rich region", Supplementary Figure S1). These segments are mapped on a conformer of NTR in Figure 4A, while Figure 4B highlights the distribution of positively charged amino acids. The two tracts extend the large patch of basic residues located in the flexible, arginine-rich loop of the NTD [52], forming an extended, yet adaptable, positively charged region. These charged residues may contribute to the interaction with the RNA backbone in a priming event driven by electrostatic interactions sensed at long-distance [53]. Notably, these two regions are likely targets of regulatory post-translational modifications, such as the phosphorylation of the serine residues within the SR-rich portion that alters the overall charge of this tract (Supplementary Figure S4) [11,54].

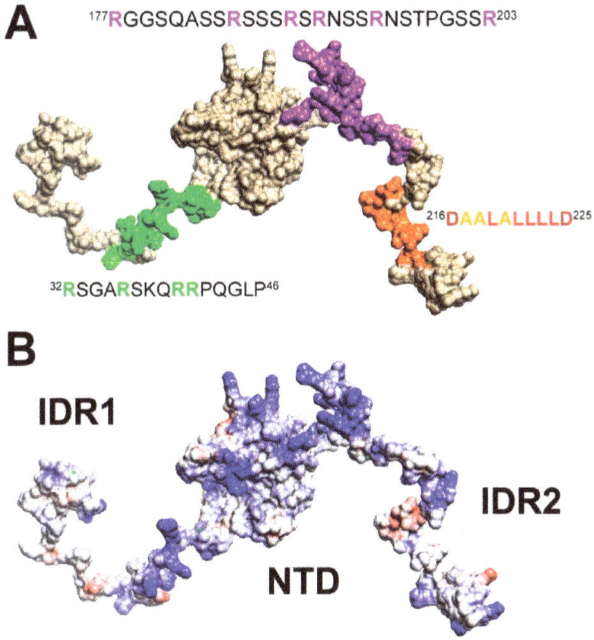


**Figure 4.** A cartoon of the NTR construct illustrating (**A**) the most perturbed regions upon the addition of RNA resulting from this study and (**B**) the large positive patch spanning both the IDRs and the globular domain. The two models were obtained as described in the experimental section.

The third region that is perturbed by the addition of RNA (216–225) has completely different properties. This region possesses a peculiar amino acid composition ($^{216}$DAALALLL LD$^{225}$, Figure 4A) and the NMR signals of the hydrophobic residues are weak, likely due to a helical propensity of this segment, which is reflected in signal broadening due to exchange with the protein-free conformation. Indeed, sequence-specific assignment of resonances in this region posed challenges to different NMR approaches before [13,14,16]. We obtained the assignment of the resonances belonging to these residues by exploiting a 3D (H)CBCACON experiment (Figure S5), thus extending the previously obtained sequence-specific assignment [13].

Differently from the two arginine-rich regions involved in the interaction with RNA (32–46 and 177–203), the 216–225 region does not present positively charged amino acids but has a highly hydrophobic nature resulting from branched-chain amino acids such as leucine [thus referred to as the poly-leucine (poly-L) region]. This hydrophobic stretch of 8 amino acids flanked by two negatively charged residues (Asp 216 and Asp 225) is likely to be engaged in transient interactions with other portions of NTR. A comparison of chemical shifts observed for the isolated NTD with those of the same nuclei in the NTR construct supports this hypothesis, and the insertion of a spin-label at position 211 indeed confirms a cross-talk between the IDR and the NTD domain (Supplementary Figure S6). Of note, the potency of the poly-L stretch to mediate protein-protein interactions has very recently been manifested in its complex with the SARS-CoV-2 nsp3 Ubl domain, while, interestingly, this interaction competes with RNA-binding of N [17]. Our data support this picture in which the poly-L region serves as an interactive hub. From our data, the observed intensity changes upon the addition of RNA in the poly-L region could derive both from direct interactions with RNA as well as from weak/fuzzy intra-molecular interactions involving different domains of NTR that are disrupted by the interaction with RNA. The latter effect might alter the dynamic properties of NTR and account for the slight increase in relative signal intensities of the globular domain observed when sub-stoichiometric amounts of RNA are added (Figure 2A). Judging by our and the previous data [17], the poly-L region might act as a regulatory motif that, within N, releases the NTD in presence of RNA and/or guides the protein to functionally relevant RNP complexes via protein-protein interactions.

Summarizing, the present results indicate that electrostatics is the main driving force for molecular recognition and the arginine-rich regions, that were found to be perturbed at the early stages of the titration, are key players to promote binding with the negatively charged RNA backbone [55,56]. Interactions between disordered protein regions with complementary charges have indeed been shown to lead to high-affinity complexes [50]. The involvement of the flexible linkers is however not limited to the arginine-rich regions but also includes the poly-L region preset in IDR2. Altogether, this suggests a complex interplay between various parts of the NTR construct.

The experimental investigation of the highly dynamic properties of N is by no means a trivial task but is of crucial importance to identifying novel approaches to interfere with SARS-CoV-2. Several insights have been recently obtained on its dynamic heterogeneity [16], on the key role of the SR-rich [11] region, on the interaction with a viral chaperone, nsp3 [17]. The interaction of NTD with different RNA fragments has been studied [10,15]. In many cases, detection of NMR signals required the use of short constructs [11,14,16,17] or changes in pH and T [16]. Increasing the complexity of the system [12] revealed very interesting insights although at the expense of residue-resolved information on the disordered regions. The proposed approach offers a tool to overcome these limitations and observe in a clean way highly flexible disordered regions within multi-domain protein constructs. As an example, the 210–248 region that comprises 56% of the IDR2 residues is challenging to observe unless smaller fragments are studied, but deletion of this region from the full-length protein has been shown to significantly alter protein function [41]. It is worth noting that this portion (219–230) shares many physicochemical properties with nucleocapsids from related coronaviruses [1–3].

## 4. Conclusions

In conclusion, $^{13}$C-detected NMR experiments such as the 2D CON allow us to access residue-resolved information on IDRs also when part of a multi-domain protein. They can be added to any high-resolution investigation performed through NMR, often based on the analysis of 2D HN NMR spectra only. The mr_CON//HN approach allows their simultaneous acquisition, providing a complete picture at residue level not only for the flexible regions but at the same time for the globular NTD domain. This complementary information is highly valuable as it reflects all components in their native context.

The NMR data, supported by EMSA data, demonstrated that the flanking disordered regions of the SARS-CoV-2 NTD initiate and enhance the binding of the protein to RNA. They revealed specific tracts of the IDRs involved in the interaction within a multi-domain, cleavage prone, structurally and dynamically complex protein as NTR is.

This represents a first step necessary to unravel the detailed molecular determinants of the N protein for specific RNA encountering and subsequent complex formation, e.g., during viral genome packaging. It paves the way for further studies with increasingly complex protein constructs, ultimately with the full-length protein, as well as with other relevant elements of the SARS-CoV-2 RNA.

**Supplementary Materials:** The following supporting information can be downloaded at: https://www.mdpi.com/article/10.3390/biom12070929/s1, Figure S1: molecular components used in this study [57,58]; Figure S2: Triplicates of EMSA gels. Figure S3: 2D HN spectra of NTR upon addiction of RNA; Figure S4: NetPhos results [59]; Figure S5: 3D-(H)CBCACON strips; Figure S6: CSP and PRE results. It also includes the additional references.

**Author Contributions:** A.S., R.P. and I.C.F. conceived the project and planned the experiments; L.P. and S.M.K. produced the protein samples; S.M.K. produced the RNA samples and performed the EMSA experiments together with A.S.; L.P. and M.S. acquired and analyzed the NMR spectra on NTR together under the guidance of I.C.F. and R.P.; S.M.K. acquired and analyzed the NMR spectra on NTD together with A.S.; A.S., R.P. and I.C.F. wrote the manuscript with the contribution from all the other authors. All authors have read and agreed to the published version of the manuscript.

**Funding:** This work was funded in part by a grant from the Italian Ministry of University and Research (FISR2020IP_02112, ID-COVID), by Fondazione CR Firenze and with the support of the Italian government program "MIUR Dipartimenti di Eccellenza 2018–2022" (58503_DIPECC). This work was also supported by the Goethe Corona Funds, by the German Research Council (DFG) within CRC902 ("Molecular Principles of RNA-based regulation") project part B18, through DFG grant number SCHL2062/2-1 to A.S., and by the Johanna Quandt Young Academy at Goethe through the financial support of A.S. (stipend number 2019/AS01).

**Institutional Review Board Statement:** Not applicable.

**Informed Consent Statement:** Not applicable.

**Data Availability Statement:** The data presented in this study are available upon request from the corresponding authors.

**Acknowledgments:** The support of the CERM/CIRMMP center of Instruct-ERIC is gratefully acknowledged. The Covid19-NMR consortium is acknowledged for providing the DNA template plasmid for the 5_SL4 RNA and for stimulating discussions. We would like to thank Volodya for his creativity and for his continuous inspiration to change the way we look at proteins, a strong motivation to improve experimental tools to characterize protein disorder.

**Conflicts of Interest:** The authors declare no conflict of interest.

## References

1. Chang, C.K.; Hou, M.H.; Chang, C.F.; Hsiao, C.D.; Huang, T.H. The SARS Coronavirus Nucleocapsid Protein—Forms and Functions. *Antivir. Res.* **2014**, *103*, 39–50. [CrossRef] [PubMed]
2. Giri, R.; Bhardwaj, T.; Shegane, M.; Gehi, B.R.; Kumar, P.; Gadhave, K.; Oldfield, C.J.; Uversky, V.N. Understanding COVID-19 via Comparative Analysis of Dark Proteomes of SARS-CoV-2, Human SARS and Bat SARS-like Coronaviruses. *Cell. Mol. Life Sci.* **2021**, *78*, 1655–1688. [CrossRef] [PubMed]
3. Chang, C.-K.; Hsu, Y.-L.; Chang, Y.-H.; Chao, F.-A.; Wu, M.-C.; Huang, Y.-S.; Hu, C.-K.; Huang, T.-H. Multiple Nucleic Acid Binding Sites and Intrinsic Disorder of Severe Acute Respiratory Syndrome Coronavirus Nucleocapsid Protein: Implications for Ribonucleocapsid Protein Packaging. *J. Virol.* **2009**, *83*, 2255–2264. [CrossRef] [PubMed]
4. Rangan, R.; Zheludev, I.N.; Hagey, R.J.; Pham, E.A.; Wayment-Steele, H.K.; Glenn, J.S.; Das, R. RNA Genome Conservation and Secondary Structure in SARS-CoV-2 and SARS-Related Viruses: A First Look. *RNA* **2020**, *26*, 937–959. [CrossRef] [PubMed]
5. Wacker, A.; Weigand, J.E.; Akabayov, S.R.; Altincekic, N.; Bains, J.K.; Banijamali, E.; Binas, O.; Castillo-Martinez, J.; Cetiner, E.; Ceylan, B.; et al. Secondary Structure Determination of Conserved SARS-CoV-2 RNA Elements by NMR Spectroscopy. *Nucleic Acids Res.* **2020**, *48*, 12415–12435. [CrossRef]
6. Cao, C.; Cai, Z.; Xiao, X.; Rao, J.; Chen, J.; Hu, N.; Yang, M.; Xing, X.; Wang, Y.; Li, M.; et al. The Architecture of the SARS-CoV-2 RNA Genome inside Virion. *Nat. Commun.* **2021**, *12*, 3917. [CrossRef]
7. de Tavares, R.C.A.; Mahadeshwar, G.; Wan, H.; Huston, N.C.; Pyle, A.M. The Global and Local Distribution of RNA Structure throughout the SARS-CoV-2 Genome. *J. Virol.* **2021**, *95*, e02190-20. [CrossRef]
8. Bai, Z.; Cao, Y.; Liu, W.; Li, J. The SARS-CoV-2 Nucleocapsid Protein and Its Role in Viral Structure, Biological Functions, and a Potential Target for Drug or Vaccine Mitigation. *Viruses* **2021**, *13*, 1115. [CrossRef]
9. Iserman, C.; Roden, C.A.; Boerneke, M.A.; Sealfon, R.S.G.; McLaughlin, G.A.; Jungreis, I.; Fritch, E.J.; Hou, Y.J.; Ekena, J.; Weidmann, C.A.; et al. Genomic RNA Elements Drive Phase Separation of the SARS-CoV-2 Nucleocapsid. *Mol. Cell* **2020**, *80*, 1078–1091.e6. [CrossRef]
10. Dinesh, D.C.; Chalupska, D.; Silhan, J.; Koutna, E.; Nencka, R.; Veverka, V.; Boura, E. Structural Basis of RNA Recognition by the SARS-CoV-2 Nucleocapsid Phosphoprotein. *PLoS Pathog.* **2020**, *16*, e1009100. [CrossRef]

11. Savastano, A.; de Opakua, A.I.; Rankovic, M.; Zweckstetter, M. Nucleocapsid Protein of SARS-CoV-2 Phase Separates into RNA-Rich Polymerase-Containing Condensates. *Nat. Commun.* **2020**, *11*, 6041. [CrossRef] [PubMed]
12. Forsythe, H.M.; Rodriguez Galvan, J.; Yu, Z.; Pinckney, S.; Reardon, P.; Cooley, R.B.; Zhu, P.; Rolland, A.D.; Prell, J.S.; Barbar, E. Multivalent Binding of the Partially Disordered SARS-CoV-2 Nucleocapsid Phosphoprotein Dimer to RNA. *Biophys. J.* **2021**, *120*, 2890–2901. [CrossRef] [PubMed]
13. Schiavina, M.; Pontoriero, L.; Uversky, V.N.; Felli, I.C.; Pierattelli, R. The Highly Flexible Disordered Regions of the SARS-CoV-2 Nucleocapsid N Protein within the 1–248 Residue Construct: Sequence-Specific Resonance Assignments through NMR. *Biomol. NMR Assign.* **2021**, *15*, 219–227. [CrossRef] [PubMed]
14. Guseva, S.; Perez, L.M.; Camacho-Zarco, A.; Bessa, L.M.; Salvi, N.; Malki, A.; Maurin, D.; Blackledge, M. $^{1}$H, $^{13}$C and $^{15}$N Backbone Chemical Shift Assignments of the N-Terminal and Central Intrinsically Disordered Domains of SARS-CoV-2 Nucleoprotein. *Biomol. NMR Assign.* **2021**, *15*, 255–260. [CrossRef]
15. Caruso, Í.P.; Sanches, K.; Da Poian, A.T.; Pinheiro, A.S.; Almeida, F.C.L. Dynamics of the SARS-CoV-2 Nucleoprotein N-Terminal Domain Triggers RNA Duplex Destabilization. *Biophys. J.* **2021**, *120*, 2814–2827. [CrossRef]
16. Redzic, J.S.; Lee, E.; Born, A.; Issaian, A.; Henen, M.A.; Nichols, P.J.; Blue, A.; Hansen, K.C.; D'Alessandro, A.; Vögeli, B.; et al. The Inherent Dynamics and Interaction Sites of the SARS-CoV-2 Nucleocapsid N-Terminal Region. *J. Mol. Biol.* **2021**, *433*, 167108. [CrossRef]
17. Bessa, L.M.; Guseva, S.; Camacho-Zarco, A.R.; Salvi, N.; Maurin, D.; Perez, L.M.; Botova, M.; Malki, A.; Nanao, M.; Jensen, M.R.; et al. The Intrinsically Disordered SARS-CoV-2 Nucleoprotein in Dynamic Complex with Its Viral Partner Nsp3a. *Sci. Adv.* **2022**, *8*, eabm4034. [CrossRef]
18. Felli, I.C.; Pierattelli, R. $^{13}$C Direct Detected NMR for Challenging Systems. *Chem. Rev.* **2022**, *122*, 9468–9496. [CrossRef]
19. Vögele, J.; Ferner, J.-P.; Altincekic, N.; Bains, J.K.; Ceylan, B.; Fürtig, B.; Grün, J.T.; Hengesbach, M.; Hohmann, K.F.; Hymon, D.; et al. $^{1}$H, $^{13}$C, $^{15}$N and $^{31}$P Chemical Shift Assignment for Stem-Loop 4 from the 5′-UTR of SARS-CoV-2. *Biomol. NMR Assign.* **2021**, *15*, 335–340. [CrossRef]
20. Sreeramulu, S.; Richter, C.; Berg, H.; Wirtz Martin, M.A.; Ceylan, B.; Matzel, T.; Adam, J.; Altincekic, N.; Azzaoui, K.; Bains, J.K.; et al. Exploring the Druggability of Conserved RNA Regulatory Elements in the SARS-CoV-2 Genome. *Angew. Chem. Int. Ed.* **2021**, *60*, 19191–19200. [CrossRef]
21. Altincekic, N.; Korn, S.M.; Qureshi, N.S.; Dujardin, M.; Ninot-Pedrosa, M.; Abele, R.; Abi Saad, M.J.; Alfano, C.; Almeida, F.C.L.; Alshamleh, I.; et al. Large-Scale Recombinant Production of the SARS-CoV-2 Proteome for High-Throughput and Structural Biology Applications. *Front. Mol. Biosci.* **2021**, *8*, 653148. [CrossRef]
22. Marley, J.; Lu, M.; Bracken, C. A Method for Efficient Isotopic Labeling of Recombinant Proteins. *J. Biomol. NMR* **2001**, *20*, 71–75. [CrossRef] [PubMed]
23. Wu, F.; Zhao, S.; Yu, B.; Chen, Y.-M.; Wang, W.; Song, Z.-G.; Hu, Y.; Tao, Z.-W.; Tian, J.-H.; Pei, Y.-Y.; et al. A New Coronavirus Associated with Human Respiratory Disease in China. *Nature* **2020**, *579*, 265–269. [CrossRef] [PubMed]
24. Schiavina, M.; Murrali, M.G.; Pontoriero, L.; Sainati, V.; Kümmerle, R.; Bermel, W.; Pierattelli, R.; Felli, I.C. Taking Simultaneous Snapshots of Intrinsically Disordered Proteins in Action. *Biophys. J.* **2019**, *117*, 46–55. [CrossRef] [PubMed]
25. Bermel, W.; Bertini, I.; Csizmok, V.; Felli, I.C.; Pierattelli, R.; Tompa, P. H-Start for Exclusively Heteronuclear NMR Spectroscopy: The Case of Intrinsically Disordered Proteins. *J. Magn. Reson.* **2009**, *198*, 275–281. [CrossRef]
26. Emsley, L.; Bodenhausen, G. Optimization of Shaped Selective Pulses for NMR Using a Quaternion Description of Their Overall Propagators. *J. Magn. Reson.* **1992**, *97*, 135–148. [CrossRef]
27. Böhlen, J.M.; Bodenhausen, G. Experimental Aspects of Chirp NMR Spectroscopy. *J. Magn. Reson. Ser. A* **1993**, *102*, 293–301. [CrossRef]
28. Geen, H.; Freeman, R. Band-Selective Radiofrequency Pulses. *J. Magn. Reson.* **1991**, *93*, 93–141. [CrossRef]
29. Piotto, M.; Saudek, V.; Sklenar, V. Gradient-Tailored Excitation for Single-Quantum NMR Spectroscopy of Aqueous Solutions. *J. Biomol. NMR* **1992**, *2*, 661–665. [CrossRef]
30. Felli, I.C.; Pierattelli, R. Spin-State-Selective Methods in Solution- and Solid-State Biomolecular $^{13}$C NMR. *Prog. Nucl. Magn. Reson. Spectrosc.* **2015**, *84–85*, 1–13. [CrossRef]
31. Mori, S.; Abeygunawardana, C.; Johnson, M.O.; Vanzijl, P.C.M. Improved Sensitivity of HSQC Spectra of Exchanging Protons at Short Interscan Delays Using a New Fast HSQC (FHSQC) Detection Scheme That Avoids Water Saturation. *J. Magn. Reson. Ser. B* **1995**, *108*, 94–98. [CrossRef] [PubMed]
32. Palmer, A.G.; Cavanagh, J.; Wright, P.E.; Rance, M. Sensitivity Improvement in Proton-Detected Two-Dimensional Heteronuclear Correlation NMR Spectroscopy. *J. Magn. Reson.* **1991**, *93*, 151–170. [CrossRef]
33. Pettersen, E.F.; Goddard, T.D.; Huang, C.C.; Couch, G.S.; Greenblatt, D.M.; Meng, E.C.; Ferrin, T.E. UCSF Chimera: A Visualization System for Exploratory Research and Analysis. *J. Comput. Chem.* **2004**, *25*, 1605–1612. [CrossRef] [PubMed]
34. Ozenne, V.; Bauer, F.; Salmon, L.; Huang, J.R.; Jensen, M.R.; Segard, S.; Bernadó, P.; Charavay, C.; Blackledge, M. Flexible-Meccano: A Tool for the Generation of Explicit Ensemble Descriptions of Intrinsically Disordered Proteins and Their Associated Experimental Observables. *Bioinformatics* **2012**, *28*, 1463–1470. [CrossRef]
35. Markley, J.L.; Bax, A.; Arata, Y.; Hilbers, C.W.; Kaptein, R.; Sykes, B.D.; Wright, P.E.; Wuethrich, K. Recommendations for the Presentation of NMR Structures of Proteins and Nucleic Acids. *Pure Appl. Chem.* **1998**, *70*, 117–142. [CrossRef]
36. Keller, R. *The Computer Aided Resonance Assignment Tutorial*; Cantina Verlag: Goldau, Switzerland, 2004; pp. 1–81.

37. Bartels, C.; Xia, T.H.; Billeter, M.; Güntert, P.; Wüthrich, K. The Program XEASY for Computer-Supported NMR Spectral Analysis of Biological Macromolecules. *J. Biomol. NMR* **1995**, *6*, 1–10. [CrossRef]
38. Ryder, S.P.; Recht, M.I.; Williamson, J.R. Quantitative Analysis of Protein-RNA Interactions by Gel Mobility Shift. *Methods Mol. Biol.* **2008**, *488*, 99–115.
39. Perdikari, T.M.; Murthy, A.C.; Ryan, V.H.; Watters, S.; Naik, M.T.; Fawzi, N.L. SARS-CoV-2 Nucleocapsid Protein Phase-separates with RNA and with Human HnRNPs. *EMBO J.* **2020**, *39*, e106478. [CrossRef]
40. Cubuk, J.; Alston, J.J.; Incicco, J.J.; Singh, S.; Stuchell-Brereton, M.D.; Ward, M.D.; Zimmerman, M.I.; Vithani, N.; Griffith, D.; Wagoner, J.A.; et al. The SARS-CoV-2 Nucleocapsid Protein Is Dynamic, Disordered, and Phase Separates with RNA. *Nat. Commun.* **2021**, *12*, 1936. [CrossRef]
41. Lu, S.; Ye, Q.; Singh, D.; Cao, Y.; Diedrich, J.K.; Yates, J.R.; Villa, E.; Cleveland, D.W.; Corbett, K.D. The SARS-CoV-2 Nucleocapsid Phosphoprotein Forms Mutually Exclusive Condensates with RNA and the Membrane-Associated M Protein. *Nat. Commun.* **2021**, *12*, 502. [CrossRef]
42. Tompa, P.; Fuxreiter, M. Fuzzy Complexes: Polymorphism and Structural Disorder in Protein-Protein Interactions. *Trends Biochem. Sci.* **2008**, *33*, 2–8. [CrossRef] [PubMed]
43. Mittag, T.; Kay, L.E.; Forman-Kay, J.D. Protein Dynamics and Conformational Disorder in Molecular Recognition. *J. Mol. Recognit.* **2009**, *23*, 105–116. [CrossRef] [PubMed]
44. Kurzbach, D.; Schwarz, T.C.; Platzer, G.; Höfler, S.; Hinderberger, D.; Konrat, R. Compensatory Adaptations of Structural Dynamics in an Intrinsically Disordered Protein Complex. *Angew. Chem. Int. Ed.* **2014**, *53*, 3840–3843. [CrossRef] [PubMed]
45. Habchi, J.; Tompa, P.; Longhi, S.; Uversky, V.N. Introducing Protein Intrinsic Disorder. *Chem. Rev.* **2014**, *114*, 6561–6588. [CrossRef]
46. Fuxreiter, M.; Tóth-Petróczy, Á.; Kraut, D.A.; Matouschek, A.; Matouschek, A.T.; Lim, R.Y.H.; Xue, B.; Kurgan, L.; Uversky, V.N. Disordered Proteinaceous Machines. *Chem. Rev.* **2014**, *114*, 6806–6843. [CrossRef]
47. Contreras-Martos, S.; Piai, A.; Kosol, S.; Varadi, M.; Bekesi, A.; Lebrun, P.; Volkov, A.N.; Gevaert, K.; Pierattelli, R.; Felli, I.C.; et al. Linking Functions: An Additional Role for an Intrinsically Disordered Linker Domain in the Transcriptional Coactivator CBP. *Sci. Rep.* **2017**, *7*, 4676. [CrossRef]
48. Arbesú, M.; Iruela, G.; Fuentes, H.; Teixeira, J.M.C.; Pons, M. Intramolecular Fuzzy Interactions Involving Intrinsically Disordered Domains. *Front. Mol. Biosci.* **2018**, *5*, 39. [CrossRef]
49. Spreitzer, E.; Usluer, S.; Madl, T. Probing Surfaces in Dynamic Protein Interactions. *J. Mol. Biol.* **2020**, *432*, 2949–2972. [CrossRef]
50. Sottini, A.; Borgia, A.; Borgia, M.B.; Bugge, K.; Nettels, D.; Chowdhury, A.; Heidarsson, P.O.; Zosel, F.; Best, R.B.; Kragelund, B.B.; et al. Polyelectrolyte Interactions Enable Rapid Association and Dissociation in High-Affinity Disordered Protein Complexes. *Nat. Commun.* **2020**, *11*, 5736. [CrossRef]
51. Murrali, M.G.; Felli, I.C.; Pierattelli, R. Adenoviral E1A Exploits Flexibility and Disorder to Target Cellular Proteins. *Biomolecules* **2020**, *10*, 1541. [CrossRef]
52. Clarkson, M.W.; Lei, M.; Eisenmesser, E.Z.; Labeikovsky, W.; Redfield, A.; Kern, D. Mesodynamics in the SARS Nucleocapsid Measured by NMR Field Cycling. *J. Biomol. NMR* **2009**, *45*, 217–225. [CrossRef] [PubMed]
53. Das, R.K.; Pappu, R.V. Conformations of Intrinsically Disordered Proteins Are Influenced by Linear Sequence Distributions of Oppositely Charged Residues. *Proc. Natl. Acad. Sci. USA* **2013**, *110*, 13392–13397. [CrossRef]
54. Carlson, C.R.; Asfaha, J.B.; Ghent, C.M.; Howard, C.J.; Hartooni, N.; Safari, M.; Frankel, A.D.; Morgan, D.O. Phosphoregulation of Phase Separation by the SARS-CoV-2 N Protein Suggests a Biophysical Basis for Its Dual Functions. *Mol. Cell* **2020**, *80*, 1092–1103.e4. [CrossRef] [PubMed]
55. Calabretta, S.; Richard, S. Emerging Roles of Disordered Sequences in RNA-Binding Proteins. *Trends Biochem. Sci.* **2015**, *40*, 662–672. [CrossRef] [PubMed]
56. Järvelin, A.I.; Noerenberg, M.; Davis, I.; Castello, A. The New (Dis)Order in RNA Regulation. *Cell Commun. Signal.* **2016**, *14*, 9. [CrossRef]
57. Popenda, M.; Szachniuk, M.; Antczak, M.; Purzycka, K.J.; Lukasiak, P.; Bartol, N.; Blazewicz, J.; Adamiak, R.W. Automated 3D Structure Composition for Large RNAs. *Nucleic Acids Res.* **2012**, *40*, e112. [CrossRef]
58. Hofacker, I.L. Vienna RNA Secondary Structure Server. *Nucleic Acids Res.* **2003**, *31*, 3429–3431. [CrossRef]
59. Blom, N.; Sicheritz-Pontén, T.; Gupta, R.; Gammeltoft, S.; Brunak, S. Prediction of Post-Translational Glycosylation and Phosphorylation of Proteins from the Amino Acid Sequence. *Proteomics* **2004**, *4*, 1633–1649. [CrossRef]

*biomolecules*

MDPI

*Article*

# Compositional Bias of Intrinsically Disordered Proteins and Regions and Their Predictions

**Bi Zhao * and Lukasz Kurgan ***

Department of Computer Science, Virginia Commonwealth University, Richmond, VA 23284, USA
* Correspondence: zhaob4@vcu.edu (B.Z.); lkurgan@vcu.edu (L.K.)

**Abstract:** Intrinsically disordered regions (IDRs) carry out many cellular functions and vary in length and placement in protein sequences. This diversity leads to variations in the underlying compositional biases, which were demonstrated for the short vs. long IDRs. We analyze compositional biases across four classes of disorder: fully disordered proteins; short IDRs; long IDRs; and binding IDRs. We identify three distinct biases: for the fully disordered proteins, the short IDRs and the long and binding IDRs combined. We also investigate compositional bias for putative disorder produced by leading disorder predictors and find that it is similar to the bias of the native disorder. Interestingly, the accuracy of disorder predictions across different methods is correlated with the correctness of the compositional bias of their predictions highlighting the importance of the compositional bias. The predictive quality is relatively low for the disorder classes with compositional bias that is the most different from the "generic" disorder bias, while being much higher for the classes with the most similar bias. We discover that different predictors perform best across different classes of disorder. This suggests that no single predictor is universally best and motivates the development of new architectures that combine models that target specific disorder classes.

**Keywords:** intrinsic disorder; intrinsically disordered proteins; intrinsic disordered regions; disorder scale; disorder propensity; amino acids; amino acid bias; predictive performance; disorder prediction

**Citation:** Zhao, B.; Kurgan, L. Compositional Bias of Intrinsically Disordered Proteins and Regions and Their Predictions. *Biomolecules* **2022**, *12*, 888. https://doi.org/10.3390/biom12070888

Academic Editors: Prakash Kulkarni, Stefania Brocca, Keith Dunker and Sonia Longhi

Received: 27 May 2022
Accepted: 10 June 2022
Published: 25 June 2022



## 1. Introduction

Intrinsically disordered regions (IDRs) are highly flexible segments in protein sequences that a lack well-defined tertiary structure and typically take form of conformational ensembles under physiological conditions [1–4]. Intrinsically disordered proteins (IDPs) include one or more IDRs. Recent bioinformatics studies have suggested that approximately a third of eukaryotic proteins have long IDRs that are composed of 30+ disordered amino acids (AAs) [5–8]. Sequences of IDRs have compositional biases, typically being enriched in charged and polar AAs and depleted in bulky hydrophobic residues [1,4,9–14]. To this end, the TOP-IDP scale was designed to quantify the intrinsic propensities of AAs for the disordered vs. structured conformations [15].

Several databases, including DisProt [16,17], PED [18,19], PDB [20,21], IDEAL [22], DIBS [23], FuzDB [24,25] and MFIB [26], provide access to the experimentally characterized IDPs and IDRs. However, they only cover a small fraction of these data, with approximately 2400 IDPs in DisProt and over 20,000 in PDB [16,27,28]. The observation that disorder is an inherent/intrinsic property of the AA sequences [1,9,10] motivates the development of accurate computational tools that predict disorder in a given protein sequence. These convenient and fast tools can be used to bridge the annotation gap and stimulate the rapid acceleration of research into IDPs and IDRs [29]. Over 100 disorder predictors have already been developed [30]. Many comprehensive studies summarize, survey and comparatively assess disorder predictors [28,30–51]. These include several community assessments, such as Critical Assessment of Structure Prediction (CASP) between CASP5 and CASP 10 [45–48,50,51], and more recently the Critical Assessment of Intrinsic Protein

Disorder (CAID) [49]. These studies describe currently available tools, identify interesting trends in the development of new methods, provide practical advice on how to identify and use the best predictors, and point to future directions.

One interesting direction is to explore the underlying diversity of intrinsic disorder [52–54]. Studies show that IDRs are instrumental for a broad spectrum of cellular functions including molecular recognition, signaling, regulation, phase separation, translation, transcription, alternative splicing, protein–protein and protein–nucleic acids interactions [53,55–70], and some of them are multifunctional [71,72]. IDRs also vary in their conformational space and they are correspondingly categorized into the native coils, native pre-molten globules and native molten globules [3,73]. Moreover, they also differ in size and placement in the sequence. Short IDRs are often located at the termini of the protein sequence while very long IDRs can span the entire length of the protein sequence [3,54,74,75]. Moreover, short IDRs were observed to have different amino acid compositions when compared to long IDRs [76,77] and correspondingly, some predictors, such as the popular IUPred [78–81], predict them separately. The diversity of sizes, locations and functions of IDRs likely results in the presence of different biases in their corresponding sequences, which cannot be captured with a single overarching TOP-IDP scale.

To this end, we investigated the compositional bias of IDRs in the context of their size and a coarsely-defined function. Moreover, using the recently released CAID results, we investigated whether the putative disorder produced by leading disorder predictors is characterized by correspondingly different AA-level biases and whether these biases influence their predictive performance. Finally, we studied whether the predictor-level biases affect their ability to accurately identify different types of disorder defined by size and function. This leads to interesting observations that may inspire the development of novel and potentially more accurate disorder predictors.

## 2. Materials and Methods

### 2.1. Data

The recent CAID experiment provides a well-annotated and large benchmark dataset that was used to assess modern disorder predictors [49]. The authors of these predictors were excluded from the process of data collection, annotation and assessment. Moreover, the underlying data were collected after these methods were trained, ensuring that the results can be reliably used to assess and compare these predictors. We obtained the experimentally annotated CAID data, including annotations of IDRs and binding IDRs from https://idpcentral.org/caid/data/1/reference/disprot-disorder.txt (accessed on 22 December 2021) and https://idpcentral.org/caid/data/1/reference/disprot-binding.txt (accessed on 22 December 2021). This dataset includes 652 protein sequences with 337,908 residues, including 838 IDRs and 54,820 disordered residues, among which there are 256 disordered binding regions and 21,389 disordered binding residues. We summarize the details in Table 1. We used these data to investigate the AA-level biases of disorder and to categorize the disorder based on the size (short, long and fully disordered) and function (binding IDRs and non-binding IDRs). We also collected predictions generated by the top 10 of 32 disorder predictors that participated in the CAID assessment from https://idpcentral.org/caid/data/1/predictions/ (accessed on 17 January 2022). These predictors include (in alphabetical order): AUCpreD [82], AUCpreD-np [82], DisoMine [83], flDPlr [84], flDPnn [84], Predisorder [85], RawMSA [86], SPOT-Disorder1 [87], SPOT-Disorder2 [88] and SPOT-Disorder-Single [89]. We excluded the ESpritz-D method that is listed in the CAID experiment since this tool was authored by the organizers of CAID and it was not officially evaluated. These data allow us to study the compositional biases of the putative disordered residues identified by these methods and to investigate the relations of these biases with the corresponding predictive performance.

**Table 1.** Summary of IDPs and IDR data in the CAID dataset.

| Protein Set | No. Proteins | No. IDRs | No. Disordered Residues | Median IDR Length | Average IDR Length |
|---|---|---|---|---|---|
| **Complete dataset** | 652 | 838 | 54,820 | 34 | 65.5 |
| **Fully disordered proteins** | 56 | 57 | 9208 | 132 | 157.6 |
| **Short IDRs** | 124 | 148 | 1810 | 12 | 12.2 |
| **Long IDRs** | 71 | 77 | 14,935 | 139 | 193.9 |
| **Disordered binding regions** | 232 | 256 | 21,389 | 54 | 83.6 |

### *2.2. Categorization of IDRs*

IDRs vary greatly in their length and function, which in our case, divides these regions into ligand binding and non-binding [54,61,62,64,65,74]. Our motivation for this coarse-grained categorization of function stems from the focus on this aspect of disorder in the recent CAID experiment [49], the high significance of the disorder-driven interactions in the context of cellular functions of disorder [61,62,64,65], and the fact that this is by far the most commonly annotated disorder function in the largest database of disorder functions annotations, DisProt [16,90].

We divided IDRs into four categories based on their length, the disordered content of the IDR-containing IDP and the annotation of binding. The disorder content is calculated as the total number of annotated disordered residues divided by the length of a given protein sequence. Using the annotations from CAID [49], which are in turn sourced from DisProt [16], IDRs are defined as the segments of at least ten consecutive disordered residues [16,91,92]. The first category are fully disordered proteins. The IDRs in this category cover at least 80% of a given IDP (disorder content $\geq$ 0.8). Approximately 10% of IDRs in our dataset belong to this category, including 57 regions and 9208 disordered residues. The second category are the short IDRs that include IDRs with $\geq$10 and <15 consecutive disordered residues that are in proteins with a disorder content < 0.3. Our dataset includes 148 short IDRs that consist of 1810 disordered residues. The third category are long IDRs that are over 70 residues in length and present in IDPs with the disorder content ranging between 0.3 and 0.8. There are 77 long IDRs with 14,935 disordered residues in our dataset. The fourth category is that of disordered binding regions. These overlap with the former three categories and their defining characteristic is that they interact with ligands. There are 256 disordered binding regions that are composed of 21,389 disordered binding residues in our dataset. While the breakdown by the region length might be seen as somehow arbitrary, we note that we did not attempt to rigorously define these categories but rather to identify large collections of IDRs that are diverse in length and cover a sufficient amount of data for performing a robust statistical analysis. We summarize these data in Table 1.

### *2.3. Computational Analysis*

Composition Profiler is a popular web-based tool that can be used to investigate the differences of amino acid compositions between collections of proteins or protein regions [93]. We applied this tool to quantify the compositional biases of AAs in various collections of IDRs and across the entire CAID dataset by comparing them with a background sample, which consists of the non-disordered residues from the CAID dataset. We note that the background is the same, allowing us to compare these scales side by side. Moreover, we computed the composition biases of the disorder predictions by comparing the putative disordered residues against the background that consists of the putative non-disordered residues generated by the top ten disorder predictors from the CAID experiment. Altogether, this analysis produced 15 scales (CAID, fully disordered; short IDRs; long IDRs; binding IDRs; plus ten predictors) that quantify the propensity of AAs for the native and predicted disorder.

We investigated the correlations between these scales to quantify their similarity. We used the Kendall rank correlation coefficients (KCCs) that measure the similarity of the orderings of given scales when the values of each scale are ranked [94]. This is motivated by the observations that the scales cover both positive and negative values (i.e., positive when residues are enriched in IDRs vs. negative when enriched in ordered regions) and that the ranges of their values differ across scales.

We also quantified the statistical significance of the differences in the predictive performance of disorder predictions. Inspired by recent works [31,32,40,95], this test aims to assess the robustness of the differences to the use of different datasets of proteins, i.e., whether a given prediction is better than another prediction across diverse datasets. First, we randomly bootstrapped 50% of proteins from the CAID dataset 100 times, and computed the corresponding 100 assessments. We compared the corresponding 100 results using the Student $t$-test if the data were normal; otherwise, we used the non-parametric Wilcoxon rank test. We tested normality using the Anderson–Darling test at the $p$-value of 0.05.

## 3. Results and Discussion

### 3.1. Compositional Biases from the TOP-IDP Scale and the CAID Data Are Consistent

We computed and investigated the AA bias (i.e., disorder scale) for the disorder in the CAID dataset. The comparison of the published TOP-IDP scale (Figure 1A) and the new scale based on the CAID dataset (Figure 1B) reveals that they are similar. The KCC of the two scales is 0.691, which means that they are highly correlated. The five-order-promoting AAs (W, F, Y, I and M) and four-disorder-promoting AAs (P, E, S and K) in TOP-IDP concur with their designation in the CAID dataset scale. The CAID scale designates the statistically disorder-promoting Q from TOP-IDP as it was not significantly different but with a slight bias towards disorder. Several other statistically significant biases in the CAID scale that include enrichment in order for L and V and enrichment in disorder for T, A, G and D are also consistent with the direction of biases in the TOP-IDP scale. The two key differences are the significant enrichment in the structured conformations for C and H in the CAID scale where these AAs have positive and not statistically significant bias toward disorder in the TOP-IDP scale. Interestingly, the TOP-IDP analyses of the bias that relies on the experimental data from DisProt ranks the AAs according to the disorder propensity as follows: P (propensity of 1.0), E (0.78), S (0.71) Q (0.66), K (0.59), A (0.45), G (0.44), D (0.41), T (0.40), R (0.39), M (0.29), N (0.28), V (0.26), H (0.26), L (0.20), F (0.12), Y (0.11), I (0.09), W (0.00) and C (0.00) [12–14]. Another study that utilizes a different source of data, primarily depending on the protein structures from PDB, finds that IDRs are depleted in W, C, F, I, Y, V, L and N; enriched in A, R, G, Q, S, P, E and K; while H, M, T and D lack a significant bias [96]. Both of these findings are in close agreement with our results, including the observation that C and H are not enriched in IDRs. The biggest outlier, cysteine (C), is considered order-promoting due to the fact that this AA forms inter- or intramolecular disulfide bonds. However, some protein domains were shown to contain disordered regions interspersed with flanking cysteines, where cysteine-induced disulfide bridges promote disorder-to-order and order-to-disorder transitions [97]. This is possibly why the TOP-IDP scale records a different bias for this AA.

### 3.2. Compositional Biases Differ between Different Categories of IDRs

We compute and investigate the disorder scales for the fully disordered proteins (Figure 1C), the short IDRs (Figure 1D), the long IDRs (Figure 1E) and the binding disordered regions (Figure 1F). Figure 1 compares these four scales with the TOP-IDP scale (Figure 1A) and the disorder in the entire CAID dataset (Figure 1B). Figure 2 gives the complete set of KCCs for all the pairs of scales. The top row in Figure 2 focuses on the correlations between the four scales and the broad collection of disorder in CAID. We find that these KCC values range from a modest level at 0.533 for the short IDRs scale to a high value at 0.828 for the binding IDRs scale. Moreover, the two scales that are highly correlated with the CAID scale, for the long IDRs (KCC = 0.797) and the binding IDRs (KCC = 0.828), are also similar to one another (KCC = 0.768). This is regardless of the fact that the binding regions are much shorter than long IDRs (Table 1). In contrast, the two scales that have modest correlations with the CAID scale, for the short IDRs (KCC = 0.533) and the fully disordered proteins (KCC = 0.596), have a similarly modest correlation with each other (KCC = 0.526). Interestingly, the correlations of the short IDRs scale with the other three targeted scales (i.e., scales for the long IDRs, binding IDRs and fully disordered) range between 0.435 and 0.526, suggesting that this scale is rather unique/dissimilar to the other three scales. This result is supported by a past study that similarly found that the AA compositions are significantly different between short IDRs (<10 residues) and long IDRs (≥30 residues) [77]. Furthermore, we find that the fully disordered scale registers relatively low KCC values between 0.526 and 0.568 when compared with the other three targeted scales. We also find that the correlations of the four scales with the TOP-IDP scale follow the same pattern as their correlations with the CAID data scale (i.e., the KCC of the binding IDRs > KCC of the long IDRs > KCC of the fully disordered IDPs > KCC of the short IDRs), except that the KCC values are lower. The lower values stem from the differences between the TOP-IDP and CAID scales that we discussed in Section 3.1. These correlation-based observations also agree with a visual inspection of the raw data in Figure 1. Scales in Figure 1E,F are relatively similar, while the scales in Figure 1C,D are different from each other and the other two scales. One of the key differences that we observe is for proline, the residue with the highest propensity for disorder in our CAID-based scale and in several other studies [12–15]. We find that proline is significantly and highly enriched in the binding and long IDRs, while being neutral for the short IDRs and fully disordered proteins. High levels of proline in the disordered binding regions concur with observations in the literature [12,98]. Moreover, proline is suggested as a modulator of secondary structures of neighboring AAs [12,99], which might explain its enrichment in the long IDRs where there is a sufficient number of residues to form residual structural elements that could be modulated and formed upon disorder-to-order transitions. Taken together, this analysis reveals three distinct types of disorder biases: one that encompasses the long and binding IDRs; the second for short IDRs; and the third for the fully disordered proteins. We also note that our results are consistent with prior studies that similarly point to substantial differences between short and long IDRs [76,77].

### 3.3. *Compositional Biases for the Putative and Native Disorder Are Highly Correlated and These Correlations Influence Predictive Performance*

We then investigate the compositional biases for the putative disorder generated by the top ten predictors evaluated in the CAID experiment. For reference, these methods secure areas under the ROC curve (AUC) values of 0.814 (flDPnn), 0.793 (flDPlr), 0.780 (RawMSA), 0.765 (DisoMine), 0.760 (SPOT-Disorder2), 0.757 (AUCpreD), 0.757 (SPOT-Disorder-Single), 0.751 (AUCpreD-np), 0.747 (Predisorder) and 0.744 (SPOT-Disorder1); and we reproduce these results from Figure 2 in the CAID article [49]. The top row in Figure 3 quantifies and compares the correlations between the CAID-based scale and the ten scales for the predicted disorder. We find that the putative disorder generated by the top ten predictors has a compositional bias that is very similar to the bias of the native disorder. The corresponding KCCs that are over 0.7 imply high correlations. This suggests that the ability of these methods to correctly predict disorder coincides with the accurate compositional bias of their predictions.

Furthermore, we find that the KCC values with the CAID-based scale range between 0.712 for SPOT-Disorder1, which is ranked 10th in CAID, and 0.850 for flDPnn, which is ranked 1st in CAID [100]. To this end, we further investigate whether these differences are correlated with the underlying predictive performance. The Pearson Correlation Coefficient (PCC) that quantifies the relation between the predictive performance measured with the AUC and corresponding KCC values of the ten predictors equals 0.703. This points to the strong effect that the level of agreement between the compositional biases of disorder predictions and the native disorder has on the performance of the best disorder predictors. This is an interesting observation since these methods utilize different training datasets, many distinctive types of inputs (e.g., protein sequences, evolutionary features, putative structural features, physicochemical properties of AAs) and various kinds of predictive models (e.g., support vector machines, decision trees, random forests, shallow and deep neural networks) [36,37,40,101]. However, the differences in their predictive performance can be largely explained by the quality of the compositional bias of the putative disorder that they generate.

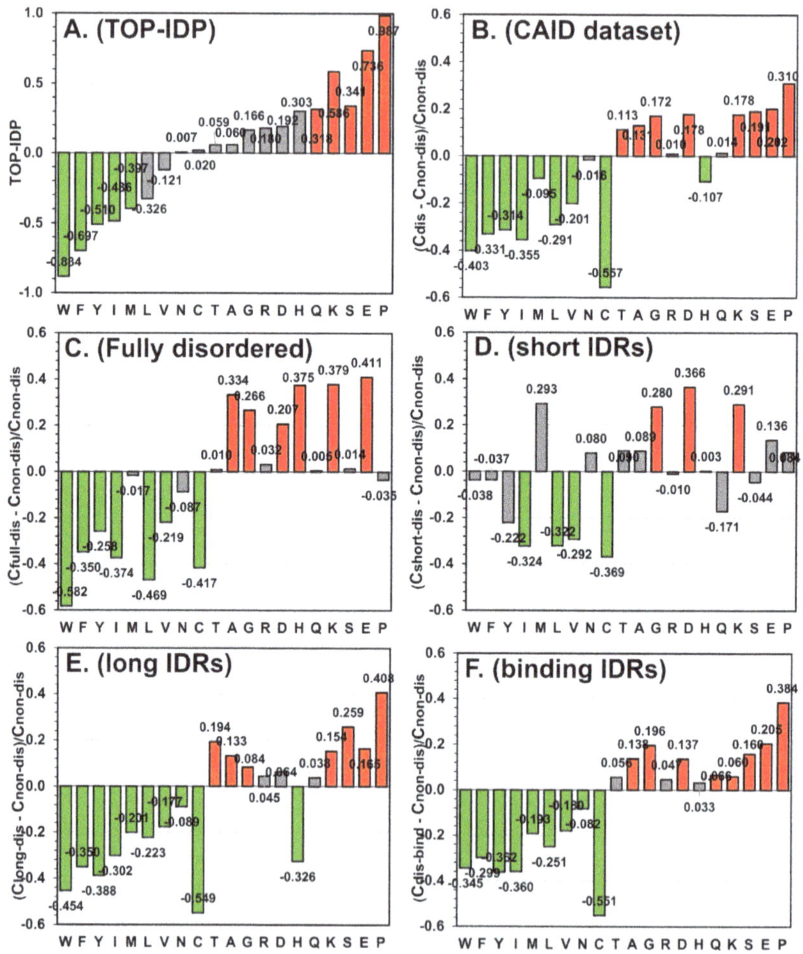


**Figure 1.** Compositional bias of intrinsic disorder measured for different collections of disordered proteins and regions. (**A**) TOP-IDP scale; (**B**) CAID dataset; (**C**) fully disordered proteins in CAID; (**D**) short IDRs in CAID; (**E**) long IDRs in CAID; and (**F**) disordered binding regions in CAID. The amino acids on the *x* axis are sorted according to the TOP-IDP scale in the way that is consistent with the original article (data for panel A was adapted from Ref. [15]), from the most order promoting to the most disorder promoting. The propensities are color-coded where green denotes statistically significant depletion; red denotes statistically significant enrichment; and gray denotes that the difference is not statistically significant at the *p*-value of 0.05. Values of the disorder propensities are shown at the top of the bars.

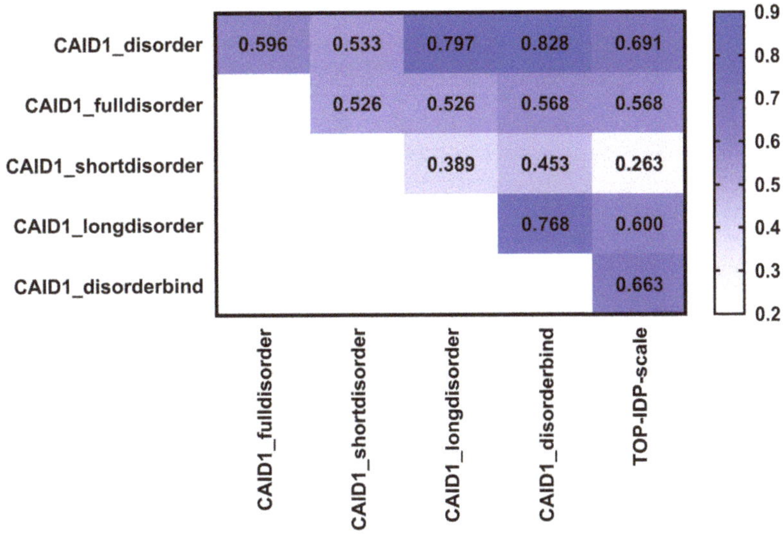


**Figure 2.** Kendall rank correlation coefficients (KCCs) between the AA biases for disorder in the overall CAID dataset, each of the four categories of IDRs (short, long, fully disordered and binding), and the TOP-IDP scale. The KCC values are color-coded from light blue for low values to dark blue for high values.

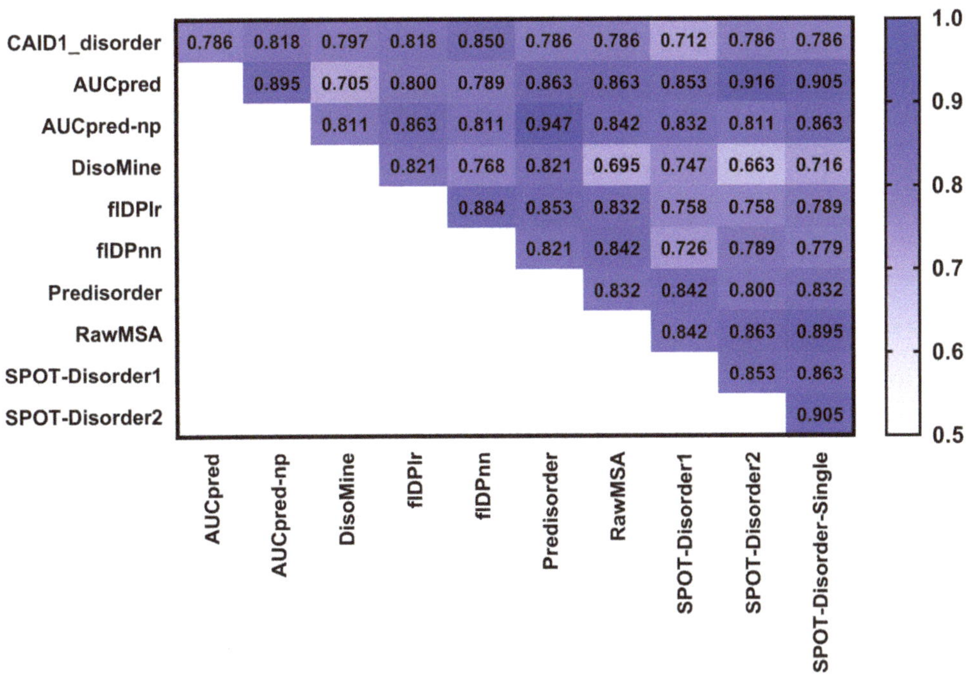


**Figure 3.** Kendall rank correlation coefficients (KCCs) between the AA biases for disorder in the overall CAID and putative disorder generated by the top ten predictors from the CAID experiment. The KCC values are color-coded from light blue for low values to dark blue for high values. Disorder predictors are sorted alphabetically.

Figure 3 also quantifies the correlations of the compositional biases of the putative disorder produced by different predictors. We find that these correlations vary widely between 0.663 (SPOT-Disorder2 with DisoMine) and 0.947 (Predisorder with AUCpred-np). This suggests that the predictions of different methods produce different biases, motivating an analysis that investigates whether their predictive performance differs across the disorder types.

### 3.4. Predictive Performance of Disorder Predictors Differs across Different Classes of IDPs

We studied the differences in the predictive performance of the top ten disorder predictors across the different types of disorder. We note that the approach in Section 2.2 catalogs IDRs in the way that some of them could belong to multiple categories, e.g., long IDRs that are binding. However, the assessment of disorder predictions must be done at the protein level, and thus we adapt the IDR-based approach to categorize IDPs. Correspondingly, we group IDPs into the following six classes: (1) fully disordered proteins (disorder content $\geq$ 0.8); (2) low disorder content proteins with short IDRs (disorder content $\leq$ 0.3 and IDRs $\geq$ 10 and <15 AAs long); (3) low disorder content proteins with binding long IDRs (disorder content $\leq$ 0.3 and binding IDRs > 15 AAs long); (4) low disorder content proteins with non-binding long IDRs (disorder content $\leq$ 0.3 and non-binding IDRs > 15 AAs long); (5) high disorder content proteins with binding IDRs (0.3 < disorder content < 0.8 and binding IDRs); and (6) high disorder content proteins with non-binding IDRs (0.3 < disorder content < 0.8 and non-binding IDRs). Table 2 provides the AUC values of the leading disorder predictors for the entire CAID dataset and each of the six classes of IDPs.

First, we analyze whether these results align with the analysis of the compositional bias from Figure 2. The lowest KCC values when compared against the CAID disorder are for the fully disordered proteins and the short IDRs (Figure 2). These two disorder types should be the hardest to predict since they have the most dissimilar bias when compared to the generic CAID disorder. Correspondingly, using Table 2, we find that the average AUC over the ten predictors for the fully disordered proteins (class 1) is 0.60, and for the proteins with short IDRs (class 2) is 0.69. In contrast, the long IDRs and binding IDRs have high values of KCC and thus they should be easier to predict based on the high similarity of their compositional bias (Figure 2). As expected, based on Table 2, the average AUC among the ten predictors for the IDPs with long IDRs (classes 3 and 4) is 0.73 and for the IDPs with binding IDRs (classes 3 and 5) is 0.71. This confirms that the compositional bias influences the predictive performance of the current methods.

Furthermore, we find that the KCC values with the CAID-based scale range between 0.712 for SPOT-Disorder1, which is ranked 10th in CAID, and 0.850 for flDPnn, which is ranked 1st in CAID [100]. To this end, we further investigate whether these differences are correlated with the underlying predictive performance. The Pearson Correlation Coefficient (PCC) that quantifies the relation between the predictive performance measured with the AUC and corresponding KCC values of the ten predictors equals 0.703. This points to the strong effect that the level of agreement between the compositional biases of disorder predictions and the native disorder has on the performance of the best disorder predictors. This is an interesting observation since these methods utilize different training datasets, many distinctive types of inputs (e.g., protein sequences, evolutionary features, putative structural features, physicochemical properties of AAs) and various kinds of predictive models (e.g., support vector machines, decision trees, random forests, shallow and deep neural networks) [36,37,40,101]. However, the differences in their predictive performance can be largely explained by the quality of the compositional bias of the putative disorder that they generate.

**Table 2.** Predictive performance measured with AUC for the top ten disorder predictors on the CAID dataset and for the six types of IDPs from the CAID dataset. The bold font identifies the methods that secure the highest AUC for a given collection of IDRs. Predictors are sorted alphabetically. We computed the results in the first row and they reproduce the original results from the CAID article [49].

| Dataset | AUCpreD | AUCpreD-np | DisoMine | flDPlr | flDPnn | Predisorder | RawMSA | SPOT-Disorder1 | SPOT-Disorder2 | SPOT-Disorder-Single |
|---|---|---|---|---|---|---|---|---|---|---|
| CAID dataset | 0.757 | 0.751 | 0.765 | 0.793 | **0.814** | 0.747 | 0.780 | 0.744 | 0.760 | 0.757 |
| Fully disordered proteins | 0.475 | 0.505 | 0.612 | 0.687 | 0.666 | 0.636 | **0.801** | 0.502 | 0.547 | 0.621 |
| Low disorder content with short IDRs | 0.715 | 0.698 | 0.654 | 0.703 | **0.736** | 0.708 | 0.651 | 0.675 | 0.687 | 0.678 |
| Low disorder content with binding long IDRs | 0.669 | 0.664 | 0.649 | 0.723 | **0.751** | 0.661 | 0.711 | 0.635 | 0.693 | 0.658 |
| Low disordered content with non-binding long IDRs | 0.801 | 0.785 | 0.747 | 0.802 | **0.816** | 0.778 | 0.806 | 0.771 | 0.779 | 0.779 |
| High disordered content with binding IDRs | 0.732 | 0.718 | 0.686 | 0.732 | 0.731 | 0.735 | **0.760** | 0.716 | 0.732 | 0.726 |
| High disordered content with non-binding IDRs | 0.824 | 0.815 | 0.799 | 0.726 | 0.737 | 0.816 | 0.811 | **0.866** | 0.808 | 0.824 |

### *3.5. Matching Disorder Predictors to Specific Classes of IDPs Substantially Improves Predictive Performance*

Using the results from Table 2, we select the best method for each IDP class and combine their predictions together, resulting in a meta-predictor. To be more specific, we normalize the scores produced by these methods using the min–max approach and use RawMSA to predict the fully disordered IDPs (class 1), flDPnn for IDPs with the low disorder content (classes 2, 3 and 4), RawMSA for the high disorder content IDPs with binding IDRs (class 5) and SPOT-Disorder1 for the high disorder content IDPs with non-binding IDRs (class 6). We quantify the predictive performance using a comprehensive collection of metrics that were utilized in the CAID assessment [49], including AUC, the area under the precision–recall curve (AUPR), F1 and the Matthews correlation coefficient (MCC). We also assessed the statistical significance of differences in the predictive performance between the meta-method and each of the top ten disorder predictors using the procedure described in Section 2.3.

Table 3 compares the predictive quality of the top ten disorder predictors and the meta-method. The AUC of the meta-method reaches 0.855 and is statistically significantly higher than the AUCs of all other predictors, including the best individual predictor, flDPnn, which secures AUC = 0.814 ($p$-value < 0.05). Similarly, the meta-method secures AUPR = 0.605, MCC = 0.474 and F1 = 0.560 when compared to the second highest AUPR = 0.479 for AUCpreD, the second highest MCC = 0. 358 and F1 = 0.462 for flDPnn; these differences are statistically significant ($p$-value < 0.05). We note large margins of improvements at approximately 0.04 for AUC and 0.13 for AUPR, which demonstrate that combining methods that best fit a given disorder class leads to substantial gains in the predictive quality. However, we emphasize that the meta-approach that we describe here is impractical since the selection of the appropriate predictor depends on prior knowledge of the disorder class.

**Table 3.** Predictive performance measured with AUC, AUPR, MCC and F1 for the top ten disorder predictors and the meta-method on the CAID dataset. The bold font identifies the highest value for a given metric. "*" means that the difference between the best-performing meta-method and a given disorder predictor is statistically significant at *p*-value of 0.05. Methods are sorted by their AUC value.

| Predictors | AUC | AUPR | MCC | F1 |
|---|---|---|---|---|
| Meta-method that selects the best predictor for each disorder class | **0.855** | **0.605** | **0.474** | **0.560** |
| flDPnn | 0.814 * | 0.475 * | 0.358 * | 0.462 * |
| flDPlr | 0.793 * | 0.422 * | 0.323 * | 0.433 * |
| RawMSA | 0.780 * | 0.414 * | 0.288 * | 0.404 * |
| DisoMine | 0.765 * | 0.388 * | 0.244 * | 0.367 * |
| SPOT-Disorder2 | 0.760 * | 0.340 * | 0.200 * | 0.351 * |
| AUCpred | 0.757 * | 0.479 * | 0.258 * | 0.399 * |
| SPOT-Disorder-Single | 0.757 * | 0.318 * | 0.221 * | 0.348 * |
| AUCpred-np | 0.751 * | 0.428 * | 0.226 * | 0.349 * |
| Predisorder | 0.747 * | 0.325 * | 0.227 * | 0.359 * |
| SPOT-Disorder1 | 0.744 * | 0.268 * | 0.143 * | 0.284 * |

## 4. Conclusions

IDRs are characterized by a sequence bias that is distinct from the sequences of structured regions. This bias at the amino acid level is captured by the TOP-IDP scale [15]. We find that this scale is largely consistent with the bias that we compute using annotations of disorder from the CAID experiment. We find that the six most disorder-promoting AAs include P, E, S, K, D and G while the most order-promoting residues are W, F, Y, I, L and C. Moreover, IDRs carry out many diverse cellular functions and differ in size and placement in the protein sequence. This diversity leads to variations in the underlying sequence biases. Prior studies demonstrate a strong amino acid composition bias of IDRs [1,4,9–14], including works that identify differences in this bias between short and long IDRs [76,77]. We analyze the compositional bias of IDRs at a finer granularity by considering four classes of disorder: fully disordered proteins, short IDRs, long IDRs and disordered binding regions. Our empirical analysis finds three distinct types of biases: one that underlies the fully disordered proteins, one that is shared by the long and binding IDRs and the third for the short IDRs.

Motivated by the large number and diversity of the sequence-based disorder predictors [30,36,37,41,42], we utilize the recently released CAID results to investigate the compositional bias of the putative disorder generated by the top performing predictors. We found that the compositional bias of the putative disorder is very similar to the bias of the native disorder. Moreover, the accuracy of the predictions across different methods is highly correlated with the level of correctness of their corresponding compositional biases. This suggests that the accurate compositional bias of the putative disorder is an important characteristic for modern disorder predictors, which to a large degree explains/determines their predictive performance.

We tie these two investigations together by quantifying and studying variations in the performance of disorder predictors across different classes of disorder. We find that an average predictive quality measured across the considered disorder predictors is relatively low for the disorder classes that have compositional bias that is the most different from the "generic" disorder bias, which include the fully disordered proteins and the short IDRs. Moreover, disorder predictions are more accurate for long IDRs and binding IDRs for which compositional bias is the most correlated with the "generic" disorder bias. This further supports the importance of compositional bias to the predictive performance of the current methods.

We also empirically find that different disorder predictors perform best across different classes of disorder. This suggests that no single predictor can claim to be universally the best. Moreover, we discover that the predictive performance of a meta-method that utilizes the best predictors for their matching disorder classes is significantly better than the performance of the best current predictors. While such a meta-method is impractical, as it requires a priori knowledge of the disorder class, this result motivates the development of new designs of disorder predictors where multiple models that target predictions of specific disorder classes are combined together. Similar methods were designed in the past where models that aim to make predictions of short and long IDRs are combined using machine learning algorithms [102–106]. These methods were rather successful in prior community assessments, with VSL2 being ranked among the most accurate methods in CASP7 [46] and MFDp ranking third in CASP10 [48]. Our study advocates further research in this vein that would consider a finer categorization of the disorder classes. Another alternative is to build a meta-model by selecting a disorder predictor based on intrinsic characteristics of the predictions (e.g., use different predictors for proteins where the putative disorder content is high vs. low or when putative binding IDRs are predicted) or the underlying protein sequence. One example of the former approach is the DISOselect tool [107]. DISOselect recommends the best-performing disorder predictor based on a tree regressor model that relies on selected sequence-derived properties, such as the estimated propensity for secondary structures, hydrophobicity and charge. However, the use of DISOselect is limited to 12 disorder predictors that exclude some of the most recent and accurate tools, for example AUCpreD, DisoMine, flDPlr, flDPnn, Predisorder, RawMSA and SPOT-Disorder2.

**Author Contributions:** Conceptualization, L.K.; methodology, L.K. and B.Z.; validation, B.Z.; investigation, B.Z.; data curation, L.K. and B.Z.; writing—original draft preparation, L.K. and B.Z.; writing—review and editing, L.K. and B.Z.; supervision, L.K.; project administration, L.K.; funding acquisition, L.K. All authors have read and agreed to the published version of the manuscript.

**Funding:** This research was funded in part by the National Science Foundation (grants 2125218 and 2146027) and by the Robert J. Mattauch Endowment funds to L.K.

**Institutional Review Board Statement:** Not applicable.

**Informed Consent Statement:** Not applicable.

**Data Availability Statement:** Not applicable.

**Acknowledgments:** We dedicate this article to Vladimir N. Uversky. We emphasize that his numerous contributions to the intrinsic disorder field made this work possible. These contributions include the TOP-IDP scale and the Composition Profiler tool that he co-developed, the numerous insightful articles on the compositional biases of IDRs that he authored and his unrelenting drive and dedication to popularize the studies of intrinsic disorder.

**Conflicts of Interest:** The authors declare no conflict of interest.

## References

1. Dunker, A.K.; Babu, M.M.; Barbar, E.; Blackledge, M.; Bondos, S.E.; Dosztanyi, Z.; Dyson, H.J.; Forman-Kay, J.; Fuxreiter, M.; Gsponer, J.; et al. What's in a name? Why these proteins are intrinsically disordered: Why these proteins are intrinsically disordered. *Intrinsically Disord. Proteins* **2013**, *1*, e24157. [CrossRef] [PubMed]
2. Uversky, V.N.; Gillespie, J.R.; Fink, A.L. Why are "natively unfolded" proteins unstructured under physiologic conditions? *Proteins-Struct. Funct. Genet.* **2000**, *41*, 415–427. [CrossRef]
3. Oldfield, C.J.; Uversky, V.N.; Dunker, A.K.; Kurgan, L. Introduction to intrinsically disordered proteins and regions. In *Intrinsically Disordered Proteins*; Salvi, N., Ed.; Academic Press: Cambridge, MA, USA, 2019; pp. 1–34.
4. Lieutaud, P.; Ferron, F.; Uversky, A.V.; Kurgan, L.; Uversky, V.N.; Longhi, S. How disordered is my protein and what is its disorder for? A guide through the "dark side" of the protein universe. *Intrinsically Disord. Proteins* **2016**, *4*, e1259708. [CrossRef] [PubMed]

5. Ward, J.J.; Sodhi, J.S.; McGuffin, L.J.; Buxton, B.F.; Jones, D.T. Prediction and functional analysis of native disorder in proteins from the three kingdoms of life. *J. Mol. Biol.* **2004**, *337*, 635–645. [CrossRef]
6. Xue, B.; Dunker, A.K.; Uversky, V.N. Orderly order in protein intrinsic disorder distribution: Disorder in 3500 proteomes from viruses and the three domains of life. *J. Biomol. Struct. Dyn.* **2012**, *30*, 137–149. [CrossRef]
7. Peng, Z.; Mizianty, M.J.; Kurgan, L. Genome-scale prediction of proteins with long intrinsically disordered regions. *Proteins* **2014**, *82*, 145–158. [CrossRef]
8. Peng, Z.; Yan, J.; Fan, X.; Mizianty, M.J.; Xue, B.; Wang, K.; Hu, G.; Uversky, V.N.; Kurgan, L. Exceptionally abundant exceptions: Comprehensive characterization of intrinsic disorder in all domains of life. *Cell Mol. Life Sci.* **2015**, *72*, 137–151. [CrossRef]
9. Romero, P.; Obradovic, Z.; Li, X.; Garner, E.C.; Brown, C.J.; Dunker, A.K. Sequence complexity of disordered protein. *Proteins* **2001**, *42*, 38–48. [CrossRef]
10. van der Lee, R.; Buljan, M.; Lang, B.; Weatheritt, R.J.; Daughdrill, G.W.; Dunker, A.K.; Fuxreiter, M.; Gough, J.; Gsponer, J.; Jones, D.T.; et al. Classification of intrinsically disordered regions and proteins. *Chem. Rev.* **2014**, *114*, 6589–6631. [CrossRef]
11. Yan, J.; Cheng, J.; Kurgan, L.; Uversky, V.N. Structural and functional analysis of "non-smelly" proteins. *Cell Mol. Life Sci.* **2020**, *77*, 2423–2440. [CrossRef]
12. Theillet, F.X.; Kalmar, L.; Tompa, P.; Han, K.H.; Selenko, P.; Dunker, A.K.; Daughdrill, G.W.; Uversky, V.N. The alphabet of intrinsic disorder: I. Act like a Pro: On the abundance and roles of proline residues in intrinsically disordered proteins. *Intrinsically Disord. Proteins* **2013**, *1*, e24360. [CrossRef] [PubMed]
13. Uversky, V.N. The alphabet of intrinsic disorder: II. Various roles of glutamic acid in ordered and intrinsically disordered proteins. *Intrinsically Disord. Proteins* **2013**, *1*, e24684. [CrossRef] [PubMed]
14. Uversky, V.N. The intrinsic disorder alphabet. III. Dual personality of serine. *Intrinsically Disord. Proteins* **2015**, *3*, e1027032. [CrossRef]
15. Campen, A.; Williams, R.M.; Brown, C.J.; Meng, J.; Uversky, V.N.; Dunker, A.K. TOP-IDP-scale: A new amino acid scale measuring propensity for intrinsic disorder. *Protein Pept. Lett.* **2008**, *15*, 956–963. [CrossRef] [PubMed]
16. Quaglia, F.; Meszaros, B.; Salladini, E.; Hatos, A.; Pancsa, R.; Chemes, L.B.; Pajkos, M.; Lazar, T.; Pena-Diaz, S.; Santos, J.; et al. DisProt in 2022: Improved quality and accessibility of protein intrinsic disorder annotation. *Nucleic Acids Res.* **2022**, *50*, D480–D487. [CrossRef] [PubMed]
17. Sickmeier, M.; Hamilton, J.A.; LeGall, T.; Vacic, V.; Cortese, M.S.; Tantos, A.; Szabo, B.; Tompa, P.; Chen, J.; Uversky, V.N.; et al. DisProt: The Database of Disordered Proteins. *Nucleic Acids Res.* **2007**, *35*, D786–D793. [CrossRef]
18. Lazar, T.; Martinez-Perez, E.; Quaglia, F.; Hatos, A.; Chemes, L.B.; Iserte, J.A.; Mendez, N.A.; Garrone, N.A.; Saldano, T.E.; Marchetti, J.; et al. PED in 2021: A major update of the protein ensemble database for intrinsically disordered proteins. *Nucleic Acids Res.* **2021**, *49*, D404–D411. [CrossRef]
19. Varadi, M.; Tompa, P. The Protein Ensemble Database. *Adv. Exp. Med. Biol.* **2015**, *870*, 335–349.
20. Le Gall, T.; Romero, P.R.; Cortese, M.S.; Uversky, V.N.; Dunker, A.K. Intrinsic disorder in the Protein Data Bank. *J. Biomol. Struct. Dyn.* **2007**, *24*, 325–342. [CrossRef]
21. Burley, S.K.; Bhikadiya, C.; Bi, C.; Bittrich, S.; Chen, L.; Crichlow, G.V.; Christie, C.H.; Dalenberg, K.; Di Costanzo, L.; Duarte, J.M.; et al. RCSB Protein Data Bank: Powerful new tools for exploring 3D structures of biological macromolecules for basic and applied research and education in fundamental biology, biomedicine, biotechnology, bioengineering and energy sciences. *Nucleic Acids Res.* **2021**, *49*, D437–D451. [CrossRef]
22. Fukuchi, S.; Amemiya, T.; Sakamoto, S.; Nobe, Y.; Hosoda, K.; Kado, Y.; Murakami, S.D.; Koike, R.; Hiroaki, H.; Ota, M. IDEAL in 2014 illustrates interaction networks composed of intrinsically disordered proteins and their binding partners. *Nucleic Acids Res.* **2014**, *42*, D320–D325. [CrossRef] [PubMed]
23. Schad, E.; Ficho, E.; Pancsa, R.; Simon, I.; Dosztanyi, Z.; Meszaros, B. DIBS: A repository of disordered binding sites mediating interactions with ordered proteins. *Bioinformatics* **2018**, *34*, 535–537. [CrossRef] [PubMed]
24. Hatos, A.; Monzon, A.M.; Tosatto, S.C.E.; Piovesan, D.; Fuxreiter, M. FuzDB: A new phase in understanding fuzzy interactions. *Nucleic Acids Res.* **2022**, *50*, D509–D517. [CrossRef] [PubMed]
25. Miskei, M.; Antal, C.; Fuxreiter, M. FuzDB: Database of fuzzy complexes, a tool to develop stochastic structure-function relationships for protein complexes and higher-order assemblies. *Nucleic Acids Res.* **2017**, *45*, D228–D235. [CrossRef] [PubMed]
26. Ficho, E.; Remenyi, I.; Simon, I.; Meszaros, B. MFIB: A repository of protein complexes with mutual folding induced by binding. *Bioinformatics* **2017**, *33*, 3682–3684. [CrossRef]
27. Zhou, J.; Oldfield, C.J.; Yan, W.; Shen, B.; Dunker, A.K. Identification of Intrinsic Disorder in Complexes from the Protein Data Bank. *ACS Omega* **2020**, *5*, 17883–17891. [CrossRef]
28. Walsh, I.; Giollo, M.; Di Domenico, T.; Ferrari, C.; Zimmermann, O.; Tosatto, S.C. Comprehensive large-scale assessment of intrinsic protein disorder. *Bioinformatics* **2015**, *31*, 201–208. [CrossRef]
29. Kurgan, L.; Radivojac, P.; Sussman, J.L.; Dunker, A.K. On the Importance of Computational Biology and Bioinformatics to the Origins and Rapid Progression of the Intrinsically Disordered Proteins Field. In *Pacific Symposium on Biocomputing*; World Scientific: Singapore, 2020; pp. 149–158.
30. Zhao, B.; Kurgan, L. Surveying over 100 predictors of intrinsic disorder in proteins. *Expert Rev. Proteom.* **2021**, *18*, 1019–1029. [CrossRef]

31. Katuwawala, A.; Oldfield, C.J.; Kurgan, L. Accuracy of protein-level disorder predictions. *Brief. Bioinform.* **2020**, *21*, 1509–1522. [CrossRef]
32. Katuwawala, A.; Kurgan, L. Comparative Assessment of Intrinsic Disorder Predictions with a Focus on Protein and Nucleic Acid-Binding Proteins. *Biomolecules* **2020**, *10*, 1636. [CrossRef]
33. Necci, M.; Piovesan, D.; Dosztanyi, Z.; Tompa, P.; Tosatto, S.C.E. A comprehensive assessment of long intrinsic protein disorder from the DisProt database. *Bioinformatics* **2018**, *34*, 445–452. [CrossRef] [PubMed]
34. Peng, Z.L.; Kurgan, L. Comprehensive comparative assessment of in-silico predictors of disordered regions. *Curr. Protein Pept. Sci.* **2012**, *13*, 6–18. [CrossRef]
35. Deng, X.; Eickholt, J.; Cheng, J. A comprehensive overview of computational protein disorder prediction methods. *Mol. Biosyst.* **2012**, *8*, 114–121. [CrossRef]
36. Liu, Y.; Wang, X.; Liu, B. A comprehensive review and comparison of existing computational methods for intrinsically disordered protein and region prediction. *Brief. Bioinform.* **2019**, *20*, 330–346. [CrossRef] [PubMed]
37. Meng, F.; Uversky, V.N.; Kurgan, L. Comprehensive review of methods for prediction of intrinsic disorder and its molecular functions. *Cell Mol. Life Sci.* **2017**, *74*, 3069–3090. [CrossRef] [PubMed]
38. Varadi, M.; Vranken, W.; Guharoy, M.; Tompa, P. Computational approaches for inferring the functions of intrinsically disordered proteins. *Front. Mol. Biosci.* **2015**, *2*, 45. [CrossRef] [PubMed]
39. Li, J.; Feng, Y.; Wang, X.; Li, J.; Liu, W.; Rong, L.; Bao, J. An Overview of Predictors for Intrinsically Disordered Proteins over 2010–2014. *Int. J. Mol. Sci.* **2015**, *16*, 23446–23462. [CrossRef]
40. Zhao, B.; Kurgan, L. Deep learning in prediction of intrinsic disorder in proteins. *Comput. Struct. Biotechnol. J.* **2022**, *20*, 1286–1294. [CrossRef]
41. Kurgan, L. Resources for computational prediction of intrinsic disorder in proteins. *Methods* **2022**, *204*, 132–141. [CrossRef]
42. Meng, F.; Uversky, V.; Kurgan, L. Computational Prediction of Intrinsic Disorder in Proteins. *Curr. Protoc. Protein Sci.* **2017**, *88*, 2–16. [CrossRef]
43. Dosztanyi, Z.; Meszaros, B.; Simon, I. Bioinformatical approaches to characterize intrinsically disordered/unstructured proteins. *Brief. Bioinform.* **2010**, *11*, 225–243. [CrossRef] [PubMed]
44. He, B.; Wang, K.; Liu, Y.; Xue, B.; Uversky, V.N.; Dunker, A.K. Predicting intrinsic disorder in proteins: An overview. *Cell Res.* **2009**, *19*, 929–949. [CrossRef] [PubMed]
45. Jin, Y.; Dunbrack, R.L., Jr. Assessment of disorder predictions in CASP6. *Proteins* **2005**, *61* (Suppl. 7), 167–175. [CrossRef] [PubMed]
46. Bordoli, L.; Kiefer, F.; Schwede, T. Assessment of disorder predictions in CASP7. *Proteins* **2007**, *69* (Suppl. 8), 129–136. [CrossRef] [PubMed]
47. Noivirt-Brik, O.; Prilusky, J.; Sussman, J.L. Assessment of disorder predictions in CASP8. *Proteins* **2009**, *77* (Suppl. 9), 210–216. [CrossRef]
48. Monastyrskyy, B.; Kryshtafovych, A.; Moult, J.; Tramontano, A.; Fidelis, K. Assessment of protein disorder region predictions in CASP10. *Proteins* **2014**, *82* (Suppl. 2), 127–137. [CrossRef]
49. Necci, M.; Piovesan, D.; Predictors, C.; DisProt, C.; Tosatto, S.C.E. Critical assessment of protein intrinsic disorder prediction. *Nat. Methods* **2021**, *18*, 472–481. [CrossRef]
50. Melamud, E.; Moult, J. Evaluation of disorder predictions in CASP5. *Proteins* **2003**, *53* (Suppl. 6), 561–565. [CrossRef]
51. Monastyrskyy, B.; Fidelis, K.; Moult, J.; Tramontano, A.; Kryshtafovych, A. Evaluation of disorder predictions in CASP9. *Proteins* **2011**, *79* (Suppl. 10), 107–118. [CrossRef]
52. Necci, M.; Piovesan, D.; Tosatto, S.C. Large-scale analysis of intrinsic disorder flavors and associated functions in the protein sequence universe. *Protein Sci.* **2016**, *25*, 2164–2174. [CrossRef]
53. Deiana, A.; Forcelloni, S.; Porrello, A.; Giansanti, A. Intrinsically disordered proteins and structured proteins with intrinsically disordered regions have different functional roles in the cell. *PLoS ONE* **2019**, *14*, e0217889. [CrossRef]
54. Howell, M.; Green, R.; Killeen, A.; Wedderburn, L.; Picascio, V.; Rabionet, A.; Peng, Z.L.; Larina, M.; Xue, B.; Kurgan, L.; et al. Not That Rigid Midgets and Not So Flexible Giants: On the Abundance and Roles of Intrinsic Disorder in Short and Long Proteins. *J. Biol. Syst.* **2012**, *20*, 471–511. [CrossRef]
55. Uversky, V.N.; Oldfield, C.J.; Dunker, A.K. Showing your ID: Intrinsic disorder as an ID for recognition, regulation and cell signaling. *J. Mol. Recognit.* **2005**, *18*, 343–384. [CrossRef] [PubMed]
56. Babu, M.M. The contribution of intrinsically disordered regions to protein function, cellular complexity, and human disease. *Biochem. Soc. Trans.* **2016**, *44*, 1185–1200. [CrossRef] [PubMed]
57. Hahn, S. Phase Separation, Protein Disorder, and Enhancer Function. *Cell* **2018**, *175*, 1723–1725. [CrossRef]
58. Peng, Z.; Xue, B.; Kurgan, L.; Uversky, V.N. Resilience of death: Intrinsic disorder in proteins involved in the programmed cell death. *Cell Death Differ.* **2013**, *20*, 1257–1267. [CrossRef]
59. Zhou, J.; Zhao, S.; Dunker, A.K. Intrinsically Disordered Proteins Link Alternative Splicing and Post-translational Modifications to Complex Cell Signaling and Regulation. *J. Mol. Biol.* **2018**, *430*, 2342–2359. [CrossRef]
60. Ahmed, S.S.; Rifat, Z.T.; Lohia, R.; Campbell, A.J.; Dunker, A.K.; Rahman, M.S.; Iqbal, S. Characterization of intrinsically disordered regions in proteins informed by human genetic diversity. *PLoS Comput. Biol.* **2022**, *18*, e1009911. [CrossRef]

61. Hu, G.; Wu, Z.; Uversky, V.N.; Kurgan, L. Functional Analysis of Human Hub Proteins and Their Interactors Involved in the Intrinsic Disorder-Enriched Interactions. *Int. J. Mol. Sci.* **2017**, *18*, 2761. [CrossRef]
62. Zhao, B.; Katuwawala, A.; Oldfield, C.J.; Hu, G.; Wu, Z.; Uversky, V.N.; Kurgan, L. Intrinsic Disorder in Human RNA-Binding Proteins. *J. Mol. Biol.* **2021**, *433*, 167229. [CrossRef]
63. Peng, Z.; Mizianty, M.J.; Xue, B.; Kurgan, L.; Uversky, V.N. More than just tails: Intrinsic disorder in histone proteins. *Mol. Biosyst.* **2012**, *8*, 1886–1901. [CrossRef] [PubMed]
64. Wang, C.; Uversky, V.N.; Kurgan, L. Disordered nucleiome: Abundance of intrinsic disorder in the DNA- and RNA-binding proteins in 1121 species from Eukaryota, Bacteria and Archaea. *Proteomics* **2016**, *16*, 1486–1498. [CrossRef] [PubMed]
65. Wu, Z.; Hu, G.; Yang, J.; Peng, Z.; Uversky, V.N.; Kurgan, L. In various protein complexes, disordered protomers have large per-residue surface areas and area of protein-, DNA- and RNA-binding interfaces. *FEBS Lett.* **2015**, *589*, 2561–2569. [CrossRef] [PubMed]
66. Peng, Z.; Oldfield, C.J.; Xue, B.; Mizianty, M.J.; Dunker, A.K.; Kurgan, L.; Uversky, V.N. A creature with a hundred waggly tails: Intrinsically disordered proteins in the ribosome. *Cell Mol. Life Sci.* **2014**, *71*, 1477–1504. [CrossRef] [PubMed]
67. Buljan, M.; Chalancon, G.; Dunker, A.K.; Bateman, A.; Balaji, S.; Fuxreiter, M.; Babu, M.M. Alternative splicing of intrinsically disordered regions and rewiring of protein interactions. *Curr. Opin. Struct. Biol.* **2013**, *23*, 443–450. [CrossRef] [PubMed]
68. Meng, F.; Na, I.; Kurgan, L.; Uversky, V.N. Compartmentalization and Functionality of Nuclear Disorder: Intrinsic Disorder and Protein-Protein Interactions in Intra-Nuclear Compartments. *Int. J. Mol. Sci.* **2015**, *17*, 24. [CrossRef]
69. Yan, J.; Dunker, A.K.; Uversky, V.N.; Kurgan, L. Molecular recognition features (MoRFs) in three domains of life. *Mol. Biosyst.* **2016**, *12*, 697–710. [CrossRef]
70. Zhao, B.; Katuwawala, A.; Uversky, V.N.; Kurgan, L. IDPology of the living cell: Intrinsic disorder in the subcellular compartments of the human cell. *Cell Mol. Life Sci.* **2020**, *78*, 2371–2385. [CrossRef]
71. Meng, F.; Kurgan, L. High-throughput prediction of disordered moonlighting regions in protein sequences. *Proteins* **2018**, *86*, 1097–1110. [CrossRef]
72. Sluchanko, N.N.; Bustos, D.M. Intrinsic disorder associated with 14-3-3 proteins and their partners. *Prog. Mol. Biol. Transl. Sci.* **2019**, *166*, 19–61.
73. Uversky, V.N. Unusual biophysics of intrinsically disordered proteins. *Biochim. Biophys. Acta* **2013**, *1834*, 932–951. [CrossRef] [PubMed]
74. Uversky, V.N. The most important thing is the tail: Multitudinous functionalities of intrinsically disordered protein termini. *FEBS Lett.* **2013**, *587*, 1891–1901. [CrossRef] [PubMed]
75. Nielsen, J.T.; Mulder, F.A.A. There is Diversity in Disorder-"In all Chaos there is a Cosmos, in all Disorder a Secret Order". *Front. Mol. Biosci.* **2016**, *3*, 4. [CrossRef] [PubMed]
76. Romero, P.; Obradovic, Z.; Kissinger, C.; Villafranca, J.E.; Dunker, A.K. Identifying disordered regions in proteins from amino acid sequence. In Proceedings of the 1997 Ieee International Conference on Neural Networks, Houston, TX, USA, 12–12 June 1997; Volume 1–4, pp. 90–95.
77. Radivojac, P.; Obradovic, Z.; Smith, D.K.; Zhu, G.; Vucetic, S.; Brown, C.J.; Lawson, J.D.; Dunker, A.K. Protein flexibility and intrinsic disorder. *Protein Sci.* **2004**, *13*, 71–80. [CrossRef]
78. Dosztanyi, Z. Prediction of protein disorder based on IUPred. *Protein Sci.* **2018**, *27*, 331–340. [CrossRef]
79. Erdos, G.; Pajkos, M.; Dosztanyi, Z. IUPred3: Prediction of protein disorder enhanced with unambiguous experimental annotation and visualization of evolutionary conservation. *Nucleic Acids Res.* **2021**, *49*, W297–W303. [CrossRef]
80. Meszaros, B.; Erdos, G.; Dosztanyi, Z. IUPred2A: Context-dependent prediction of protein disorder as a function of redox state and protein binding. *Nucleic Acids Res.* **2018**, *46*, W329–W337. [CrossRef]
81. Dosztanyi, Z.; Csizmok, V.; Tompa, P.; Simon, I. IUPred: Web server for the prediction of intrinsically unstructured regions of proteins based on estimated energy content. *Bioinformatics* **2005**, *21*, 3433–3434. [CrossRef]
82. Wang, S.; Ma, J.; Xu, J. AUCpreD: Proteome-level protein disorder prediction by AUC-maximized deep convolutional neural fields. *Bioinformatics* **2016**, *32*, i672–i679. [CrossRef]
83. Orlando, G.; Raimondi, D.; Codice, F.; Tabaro, F.; Vranken, W. Prediction of disordered regions in proteins with recurrent neural networks and protein dynamics. *J. Mol. Biol.* **2022**, *434*, 167579. [CrossRef]
84. Hu, G.; Katuwawala, A.; Wang, K.; Wu, Z.; Ghadermarzi, S.; Gao, J.; Kurgan, L. flDPnn: Accurate intrinsic disorder prediction with putative propensities of disorder functions. *Nat. Commun.* **2021**, *12*, 4438. [CrossRef] [PubMed]
85. Deng, X.; Eickholt, J.; Cheng, J. PreDisorder: Ab initio sequence-based prediction of protein disordered regions. *BMC Bioinform.* **2009**, *10*, 436. [CrossRef] [PubMed]
86. Mirabello, C.; Wallner, B. rawMSA: End-to-end Deep Learning using raw Multiple Sequence Alignments. *PLoS ONE* **2019**, *14*, e0220182. [CrossRef] [PubMed]
87. Hanson, J.; Yang, Y.; Paliwal, K.; Zhou, Y. Improving protein disorder prediction by deep bidirectional long short-term memory recurrent neural networks. *Bioinformatics* **2017**, *33*, 685–692. [CrossRef]
88. Hanson, J.; Paliwal, K.K.; Litfin, T.; Zhou, Y. SPOT-Disorder2: Improved Protein Intrinsic Disorder Prediction by Ensembled Deep Learning. *Genom. Proteom. Bioinform.* **2019**, *17*, 645–656. [CrossRef]
89. Hanson, J.; Paliwal, K.; Zhou, Y. Accurate Single-Sequence Prediction of Protein Intrinsic Disorder by an Ensemble of Deep Recurrent and Convolutional Architectures. *J. Chem. Inf. Model.* **2018**, *58*, 2369–2376. [CrossRef]

90. Katuwawala, A.; Ghadermarzi, S.; Kurgan, L. Computational prediction of functions of intrinsically disordered regions. *Prog. Mol. Biol. Transl. Sci.* **2019**, *166*, 341–369.
91. Hatos, A.; Hajdu-Soltesz, B.; Monzon, A.M.; Palopoli, N.; Alvarez, L.; Aykac-Fas, B.; Bassot, C.; Benitez, G.I.; Bevilacqua, M.; Chasapi, A.; et al. DisProt: Intrinsic protein disorder annotation in 2020. *Nucleic Acids Res.* **2020**, *48*, D269–D276. [CrossRef]
92. Piovesan, D.; Tosatto, S.C.E. Mobi 2.0: An improved method to define intrinsic disorder, mobility and linear binding regions in protein structures. *Bioinformatics* **2018**, *34*, 122–123. [CrossRef]
93. Vacic, V.; Uversky, V.N.; Dunker, A.K.; Lonardi, S. Composition Profiler: A tool for discovery and visualization of amino acid composition differences. *BMC Bioinform.* **2007**, *8*, 211. [CrossRef]
94. Kendall, M.G. A new measure of rank correlation. *Biometrika* **1938**, *30*, 81–93. [CrossRef]
95. Wang, K.; Hu, G.; Wu, Z.; Su, H.; Yang, J.; Kurgan, L. Comprehensive Survey and Comparative Assessment of RNA-Binding Residue Predictions with Analysis by RNA Type. *Int. J. Mol. Sci.* **2020**, *21*, 6879. [CrossRef] [PubMed]
96. Lise, S.; Jones, D.T. Sequence patterns associated with disordered regions in proteins. *Proteins* **2005**, *58*, 144–150. [CrossRef] [PubMed]
97. Bhopatkar, A.A.; Uversky, V.N.; Rangachari, V. Disorder and cysteines in proteins: A design for orchestration of conformational see-saw and modulatory functions. *Prog. Mol. Biol. Transl. Sci.* **2020**, *174*, 331–373. [PubMed]
98. Kini, R.M.; Evans, H.J. A hypothetical structural role for proline residues in the flanking segments of protein-protein interaction sites. *Biochem. Biophys. Res. Commun.* **1995**, *212*, 1115–1124. [CrossRef]
99. Richardson, J.S.; Richardson, D.C. Amino-Acid Preferences for Specific Locations at the Ends of Alpha-Helices. *Science* **1988**, *240*, 1648–1652. [CrossRef]
100. Lang, B.; Babu, M.M. A community effort to bring structure to disorder. *Nat. Methods* **2021**, *18*, 454–455. [CrossRef]
101. Fan, X.; Kurgan, L. Accurate prediction of disorder in protein chains with a comprehensive and empirically designed consensus. *J. Biomol. Struct. Dyn.* **2014**, *32*, 448–464. [CrossRef]
102. Mizianty, M.J.; Peng, Z.L.; Kurgan, L. MFDp2: Accurate predictor of disorder in proteins by fusion of disorder probabilities, content and profiles. *Intrinsically Disord. Proteins* **2013**, *1*, e24428. [CrossRef]
103. Mizianty, M.J.; Uversky, V.; Kurgan, L. Prediction of intrinsic disorder in proteins using MFDp2. *Methods Mol. Biol.* **2014**, *1137*, 147–162.
104. Mizianty, M.J.; Stach, W.; Chen, K.; Kedarisetti, K.D.; Disfani, F.M.; Kurgan, L. Improved sequence-based prediction of disordered regions with multilayer fusion of multiple information sources. *Bioinformatics* **2010**, *26*, i489–i496. [CrossRef] [PubMed]
105. Obradovic, Z.; Peng, K.; Vucetic, S.; Radivojac, P.; Dunker, A.K. Exploiting heterogeneous sequence properties improves prediction of protein disorder. *Proteins* **2005**, *61* (Suppl. 7), 176–182. [CrossRef] [PubMed]
106. Peng, K.; Radivojac, P.; Vucetic, S.; Dunker, A.K.; Obradovic, Z. Length-dependent prediction of protein intrinsic disorder. *BMC Bioinform.* **2006**, *7*, 208. [CrossRef] [PubMed]
107. Katuwawala, A.; Oldfield, C.J.; Kurgan, L. DISOselect: Disorder predictor selection at the protein level. *Protein Sci.* **2020**, *29*, 184–200. [CrossRef]

*biomolecules*

MDPI

*Article*

# Identification of Intrinsically Disordered Proteins and Regions in a Non-Model Insect Species *Ostrinia nubilalis* (Hbn.)

**Miloš Avramov [1], Éva Schád [2], Ágnes Révész [3], Lilla Turiák [3], Iva Uzelac [1], Ágnes Tantos [2], László Drahos [3] and Željko D. Popović [1,*]**

[1] Department of Biology and Ecology, Faculty of Sciences, University of Novi Sad, 21000 Novi Sad, Serbia; milos.avramov@dbe.uns.ac.rs (M.A.); iva.uzelac@dbe.uns.ac.rs (I.U.)

[2] Institute of Enzymology, Research Centre for Natural Sciences, 1117 Budapest, Hungary; schad.eva@gmail.com (É.S.); tantos.agnes@ttk.hu (Á.T.)

[3] Institute of Organic Chemistry, Research Centre for Natural Sciences, 1117 Budapest, Hungary; revesz.agnes@ttk.hu (Á.R.); turiak.lilla@ttk.hu (L.T.); drahos.laszlo@ttk.hu (L.D.)

* Correspondence: zeljko.popovic@dbe.uns.ac.rs

**Abstract:** Research in previous decades has shown that intrinsically disordered proteins (IDPs) and regions in proteins (IDRs) are as ubiquitous as highly ordered proteins. Despite this, research on IDPs and IDRs still has many gaps left to fill. Here, we present an approach that combines wet lab methods with bioinformatics tools to identify and analyze intrinsically disordered proteins in a non-model insect species that is cold-hardy. Due to their known resilience to the effects of extreme temperatures, these proteins likely play important roles in this insect's adaptive mechanisms to sub-zero temperatures. The approach involves IDP enrichment by sample heating and double-digestion of proteins, followed by peptide and protein identification. Next, proteins are bioinformatically analyzed for disorder content, presence of long disordered regions, amino acid composition, and processes they are involved in. Finally, IDP detection is validated with an in-house 2D PAGE. In total, 608 unique proteins were identified, with 39 being mostly disordered, 100 partially disordered, 95 nearly ordered, and 374 ordered. One-third contain at least one long disordered segment. Functional information was available for only 90 proteins with intrinsic disorders out of 312 characterized proteins. Around half of the 90 proteins are cytoskeletal elements or involved in translational processes.

**Keywords:** intrinsically disordered proteins (IDPs); intrinsically disordered protein regions (IDRs); LC–MS/MS; IUPred analysis; *Ostrinia nubilalis*; cold hardiness

**Citation:** Avramov, M.; Schád, É.; Révész, Á.; Turiák, L.; Uzelac, I.; Tantos, Á.; Drahos, L.; Popović, Ž.D. Identification of Intrinsically Disordered Proteins and Regions in a Non-Model Insect Species *Ostrinia nubilalis* (Hbn.). *Biomolecules* **2022**, *12*, 592. https://doi.org/10.3390/biom12040592

Academic Editors: Prakash Kulkarni, Stefania Brocca, Keith Dunker and Sonia Longhi

Received: 7 March 2022
Accepted: 11 April 2022
Published: 18 April 2022



## 1. Introduction

Intrinsically disordered proteins exist and function without a well-defined three-dimensional structure, occupying a conformational space through a series of fluctuating structural states, often described as conformational ensembles [1].

Intrinsic disorder can span the full length of a polypeptide chain or be localized in specific regions of globular, ordered proteins as intrinsically disordered protein regions or IDRs [2]. A protein's amino acid sequence contains the code for disorder [3], as certain residues were found to be particularly abundant in intrinsically disordered proteins (IDPs) and IDRs, such as Ala, Glu, Ser, Gln, Lys, and Pro [4].

The biological importance of intrinsic disorder is well described: many proteins that are involved in signaling and regulatory pathways, as well as different intermolecular interactions, possess segments that are unstructured. This allows them to recognize a wide array of binding partners, often undergoing disorder-to-order transition upon interaction with their targets [5,6]. Disorder is present in all domains of life, even in viruses. It is established that the total disorder content, measured in the amount of disordered residues present in the entire proteome, is higher in eukaryotes than in prokaryotes and viruses [7–9]. Despite the abundance of IDPs/IDRs in proteomes, as well as their key roles in regulatory processes, experimental characterization of these proteins is still sparse. The fact that IDPs and ID-containing proteins are difficult to crystalize makes them unsuitable for structure-resolving methods such as X-ray crystallography, while high protein disorder can prohibit structure determination using cryo-EM. Additionally, cryo-EM has a rather strict and high size restriction, so its usefulness in the structure determination of individual proteins is rather limited [10,11]. It has also been reported that nearly a half of eukaryotic proteins are considered parts of the "dark proteome", meaning there is a lack of information on their folded structure which prevents them from being used in homology modeling [12], further complicating the annotation of IDPs and IDRs [13,14]. While experimental characterization has its share of obstacles, different computational methods have been developed that can be used to predict disorder in protein sequences as well as to facilitate functional annotation of such proteins *in silico* [15–18].

In this article, a combined approach to identifying intrinsically disordered proteins is presented. Wet lab methods are employed to generate protein samples and prepare them for bioinformatic analysis. A particular experimental setup provides biological context for the subsequent computational analysis of identified proteins. For example, intrinsically disordered proteins are known to possess cold stability and resistance to cold treatment [19], so they likely play important protective roles in the mechanisms behind acclimation and adaptation to cold winter temperatures. In order to study this aspect of IDPs, a pipeline for the identification of such proteins in non-cold-acclimated and cold-acclimated diapausing larvae of a moth insect species was developed. The insect in question, the European corn borer (ECB, *Ostrinia nubilalis*, Hbn.), is a Eurasian species of moth that was introduced to North Africa and North America in the early 1900s by accident [20]. Larvae of the ECB are notorious as pests that feed on more than 200 economically important species of grains, fruits, and vegetables [21,22]. The larvae used in this study were acclimated to temperatures below 0 °C, triggering specific molecular, biochemical, and behavioral adaptations in this species that should also lead to changes in the content of its proteome. This way, protein identification and characterization can be examined in the context of the cold hardiness molecular ecophysiology in this species. Thus, the broader aim of this study is to explore how intrinsic disorder is implicated in the adaptation of this species to cold weather conditions. In the months preceding winter, the fifth instar larvae of the ECB enter diapause, a form of life cycle arrest observed mostly in insects of temperate and polar zones, in order to prepare for the coming cold temperatures and food scarcity [23]. Diapause consists of three major phases—pre-diapause, diapause, and post-diapause, allowing an organism to gradually adapt to the changes in the environment and ensuring its survival [24]. An organism undergoes a wide array of molecular and biochemical adaptations during these different diapause phases, such as depression of metabolism [25,26], alteration of metabolic enzyme activity [27], synthesis of specific cryoprotectants [28,29], and changes in the lipid composition of storage molecules and membranes [30–32], as well as changes in the expression of genes and proteins involved in stress protection and the regulation of cell cycle and programmed cell death [33–38]. The listed adaptations allow the diapausing larvae of the ECB to develop cold hardiness and successfully overwinter [23,39].

An added value of this approach is its robustness; hence, it can be adjusted and used in research with other types of biological materials and experimental setups.

## 2. Materials and Methods

### 2.1. Experimental Design

The generalized experimental workflow is presented in Figure 1. Diapausing (winter) fifth instar larvae of the ECB were collected from the fields of the Maize Research Institute in Zemun Polje (44°87′ N, 20°33′ E), Serbia, during the winter season of 2018/2019. The collected larvae were first acclimated at 15 °C for two weeks, after which one group was frozen in liquid nitrogen and stored at −80 °C until analysis. The remainder of the larvae were placed in insect homes made out of waxed cardboard and gradually chilled by lowering the temperature by 1 °C each day, with additional acclimation for two weeks when specific checkpoint temperatures were reached (5, −3, and −16 °C). After the final acclimation, larvae were frozen in liquid nitrogen and stored at −80 °C until analysis. In total, two experimental groups were formed—one non-cold-acclimated (NCA) diapausing group and one cold-acclimated (CA) diapausing group (Figure 1). The groups consisted of 5 biological replicates each.

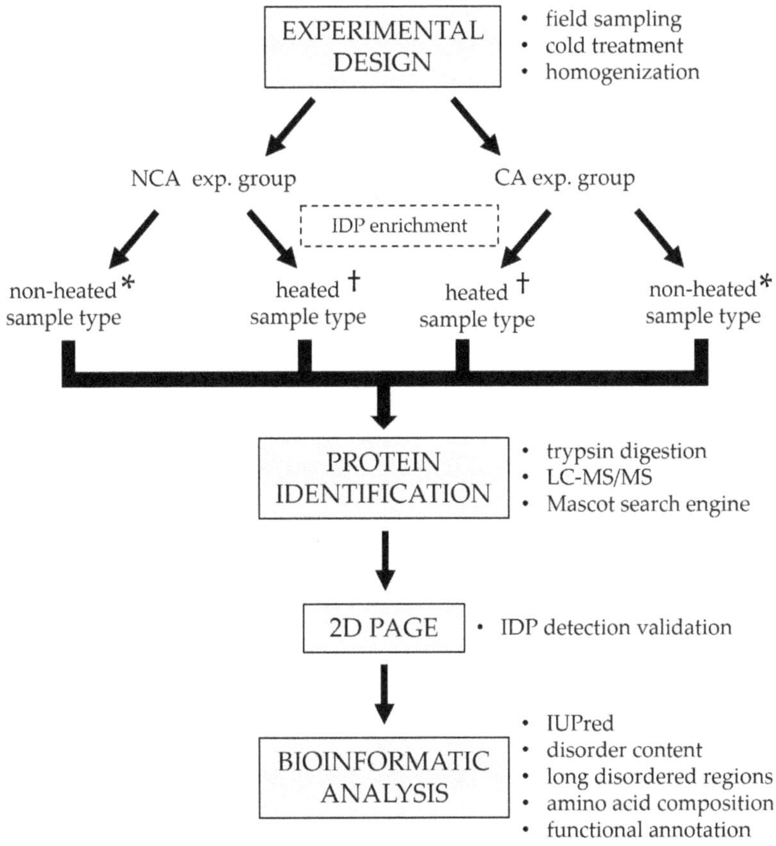


**Figure 1.** Generalized workflow of the experiment. Detailed explanations are given in the relevant subsections. Aliquots from samples with the same symbol in the superscript (* or †) were pooled before being run on 2D PAGE. NCA—non-cold-acclimated diapausing group; CA—cold-acclimated diapausing group.

### 2.2. Sample Preparation

Whole-body larvae (5 per sample) were homogenized in ice-cold 50 mM K-phosphate buffer pH 7.5 with 1 mM DTE to make a 20% *w*/*v* homogenate. The homogenates were then additionally lysed with sonication for 2 min (24 sonic pulses lasting 5 s each, with 10 s pauses in-between). Sonication was followed by centrifugation for 10 min at 7000 rpm, 4 °C to remove insoluble debris and lipids. Supernatants were divided into two microtubes per sample. One microtube from each sample was placed in a water bath at 100 °C for 5 min in order to remove globular proteins and enrich the content of IDPs (heated sample type, Figure 1). The other microtube was left untreated for comparison (non-heated sample type, Figure 1). After the heat treatment, all samples were centrifuged for 10 min at 12,000 rpm, 4 °C to further purify them and the supernatants were transferred to clean microtubes. Total proteins were assayed on a 250 μL microplate using the commercial Quick Start™ Bradford Protein Assay kit (Bio-Rad, Hercules, CA, USA, cat. no. 5000203), according to the manufacturer's protocol. After the assay, a 5× concentrated protease inhibitor cocktail (Roche cOmplete™ ULTRA tablets, Merck KGaA, Darmstadt, Germany, cat. no. 5892970001) was added to the samples to a final concentration of 1×.

*2.3. Protein Identification*

Protein identification was done using shotgun LC–MS/MS and the Mascot search engine [40]. Aliquots containing up to 20 µg of total protein were taken from every whole-body homogenate, and proteins were double-digested in-solution using Trypsin/Lys-C mixture, followed by Trypsin digestion (Figure 1). First, the samples were prepared for digestion in Microcon-10 kDa centrifugal filters according to the following steps:

1. Rinse the filters with 200 µL of LC-MS grade $H_2O$ by centrifuging at 13,500× *g*, 4 °C for 10 min, with ~30 µL of water remaining in the filter after the rinse; discard the elute from the outer vial;
2. Add a solution of up to 20 µg of protein to the filter and fill up to 200 µL with 200 mM $NH_5CO_3$, centrifuge at 13,500× *g*, 4 °C for 10 min; discard the eluate from the outer vial;
3. Add 200 µL of 200 mM $NH_4CO_3$, centrifuge at 13,500× *g*, 4 °C for 10 min; discard the eluate from the outer vial;
4. Add 200 µL of 50 mM $NH_4CO_3$, centrifuge at 13,500× *g*, 4 °C for 10 min; discard the eluate from the outer vial;
5. Place the filter upside-down in a new outer vial and centrifuge at 1000× *g* for 2 min to transfer the protein solution from the filter to the outer vial, then pipette the solution into a 0.5 mL Lo-Bind Eppendorf tube.

After the preparation, the protein samples were digested according to the following protocol:

1. Add 1.5 µL of LC-MS MeOH to the protein sample for a final MeOH concentration of 5%;
2. Add 5 µL of 0.5% Rapigest and 2 µL of 200 mM DTT to the protein sample and incubate at 60 °C for 30 min;
3. Cool the sample to room temperature, add 5 µL of 200 mM $NH_5CO_3$ and 2.5 µL of 200 mM iodoacetamide;
4. Incubate in the dark for 30 min at room temperature;
5. Add 1 µL of stock Trypsin/Lys-C Mix (20 µg of the mixture in 80 µL of LC–MS grade $H_2O$) and incubate at 37 °C for 1 h;
6. Add trypsin in a 1:25 trypsin:protein ratio at 37 °C for 1 h;
7. Stop the digestion by adding 1.5 µL of formic acid for a final concentration of at least 2% *v*/*v*;
8. Completely dry the samples in a vacuum dryer at 50 °C.

After the digestion, the samples were desalted and cleaned up using Pierce C 18 Spin Columns placed in 2 mL Lo-Bind Eppendorf tubes according to the following steps:

1. Add 200 µL of 50% MeOH to the column and centrifuge at 1500 rpm for 1 min. Repeat the step once more and then discard the eluate;
2. Add 200 µL of 0.5% TFA, 5% ACN solution to the column and centrifuge at 1500 rpm for 1 min. Repeat the step once more and then discard the eluate;
3. Add 200 µL of 0.1% TFA to the column and centrifuge at 1500 rpm for 1 min. Repeat the step once more and then discard the eluate;
4. Dissolve the dried sample in 50 µL of 0.1% TFA and apply it on the column. Centrifuge at 1500 rpm for 1 min;
5. Collect the eluate and apply it on the column again. Centrifuge at 1500 rpm for 1 min;
6. Add 100 µL of 0.1% TFA to the column and centrifuge at 1500 rpm for 1 min. Repeat the step once more;
7. Place the column in a new 2 mL Lo-Bind Eppendorf tube;

8. Add 50 μL of 0.1% TFA, 70% ACN solution to the column and centrifuge at 1500 rpm for 1 min to elute the sample. Repeat the step once more;
9. Completely dry the sample in a vacuum dryer at 50 °C and store it in a freezer until analysis.

Tryptic digests were subjected to nano-LC–MS/MS analysis using a Dionex Ultimate 3000 RSLC nanoLC system (Sunnyvale, CA, USA) coupled to Bruker Maxis II ETD Q-TOF instrument (Bremen, Germany) via a CaptiveSpray nanoBooster ionization source. Peptides were separated online using Acquity M-Class BEH130 C18 analytical column (1.7 μm, 130 Å, 75 μm × 250 mm Waters, Milford, MA, USA) following trapping on an Acclaim PepMap 100 C-18 trap column (5 μm, 100 Å, 100 μm × 20 mm, Thermo Fisher Scientific, Waltham, MA, USA). The temperature was set at 48 °C, and a flow rate of 300 nl/min was applied. The gradient method was from 4% B to 50% B in 90 min; solvent A was 0.1% formic acid in water, while solvent B was 0.1% formic acid in acetonitrile.

Sample ionization was achieved in the positive electrospray ionization mode. Data-dependent analysis was performed using a fixed cycle time of 2.5 s. MS spectra were acquired over a mass range of 150–2200 $m/z$ at 3 Hz, while CID was performed at 16 Hz for abundant precursors and at 4 Hz for ones of low abundance.

Data were evaluated with ProteinScape 3.0 software (Bruker Daltonic GmbH, Bremen, Germany) using the Mascot search engine version 2.5.1 (Matrix Science, London, UK). MS/MS spectra were searched against *O. nubilalis*, *O. furnacalis* (a species closely related to the ECB), as well as all lepidopteran protein sequences available in the NCBI database, due to the limited availability of *O. nubilalis* protein sequences. The following parameters were applied: trypsin as enzyme, 7 ppm peptide mass tolerance, 0.05 Da fragment mass tolerance, and 2 missed cleavages. Carbamidomethylation on cysteines was set as a fixed modification, with deamidation (NQ) and oxidation (M) as variable modifications.

### 2.4. 2D PAGE

To validate the results of IDP detection, proteins from both untreated and heat-treated samples were separated using a modified in-house 2D PAGE method [41]. Aliquots from the treated and untreated samples were taken and pooled in two mixtures, respectively (Figure 1). In the first dimension, the mixtures were run on a discontinuous native PAGE (12.5% separating gel, 20 μg of total proteins per well) for 50 min at 180 V. After the run, individual lanes were cut out as strips and placed in 1.5 M Tris-HCl pH 8.8 containing 8 M urea for 45 min to solubilize the proteins. The separating gel (12.5%) for the second dimension was prepared by adding 8 M urea to the standard native gel solution. After casting the gel, a strip with solubilized proteins was placed on top of it instead of a stacking gel, making sure not to introduce any bubbles between the separating gel and the strip. The second dimension was run for 30 min at 400 V. After the run, the gels were stained using the Pierce™ Silver Stain Kit (Thermo Scientific, Waltham, MA, USA, cat. no. 24612) to visualize the protein spots.

### 2.5. Bioinformatic Analysis

After protein identification, FASTA sequences for all identified proteins were downloaded from the NCBI database to be used for the prediction of structural disorder (Figure 1). The structural disorder of proteins was determined with the IUPred long disorder predictor (https://iupred2a.elte.hu/, accessed on 24 March 2022) [42], which is based on estimating the total pair-wise inter-residue interaction energy gained upon folding of a polypeptide chain. An amino acid is considered to be disordered if its IUPred score is at least 0.5. Mean disorder was computed as the average of residue scores, which range from 0.0 to 1.0. Overall disorder rate (percental disorder, ranging from 0 to 100%) represents the fraction of disordered amino acids in a polypeptide chain. Proteins are considered globular if their overall disorder rate is below 10%; nearly ordered if the rate is between 10% and 30%; partially disordered if the rate is between 30% and 70%; (mostly) disordered if the rate is above 70%. All proteins were further analyzed for the presence of long intrinsically disordered regions (long IDRs)—sequences of at least 20 consecutive disordered amino acids. Additionally, the amino acid composition of the proteins was analyzed to determine the absolute number of individual amino acids that make up each polypeptide, as well as their ratios.

Lastly, functional characterization was performed on the identified sequences (Figure 1). Functional information on the identified proteins was collected from various databases such as UniProt (www.uniprot.org/, accessed on 18 January 2022) [43], Pfam (http://pfam.xfam.org/, accessed on 18 January 2022) [44], Interpro (https://www.ebi.ac.uk/interpro/, accessed on 18 January 2022) [45], and GeneOntology (http://geneontology.org/, accessed on 18 January 2022) [46,47]. Data on their molecular functions, cellular localization, the biological processes they are involved in, and the domains they contained was collected. All of the analyses were performed using homemade PERL scripts run locally.

## 3. Results

Protein identification was performed directly from the individual homogenates, as described in the Methods section. In total, 820 proteins were identified—506 in the non-cold-acclimated (NCA) group and 314 in the cold-acclimated (CA) group. Accounting for shared entries between the two experimental groups, our investigation yielded a total of 608 unique proteins (Table S1), with almost 50% of hits (290) being linked to polypeptides that have only been predicted from nucleotide sequences. Out of that total, 294 were present only in the NCA experimental group, with 102 only in the CA experimental group; the remaining 212 proteins were found in both (Figure 2A).

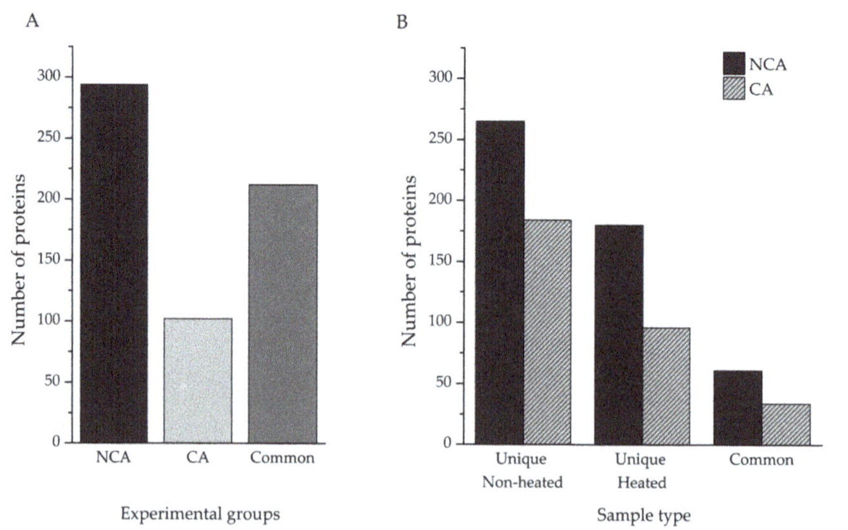


**Figure 2.** (**A**) Total unique and common proteins isolated from different experimental groups. (**B**) Effect of sample heating on the number of identified proteins. Unique Non-heated—proteins found only in the non-heated samples; Unique Heated—proteins found only in heated samples; Common—proteins that were found in both heated and non-heated samples; NCA—non-cold-acclimated diapausing group; CA—cold-acclimated diapausing group.

Identification of proteins from complex mixtures, such as larval extracts used in this study, can be challenging for a number of reasons. The signal of less abundant proteins can be masked by the ones that are overrepresented in the samples, and proteins embedded in large, multi-subunit complexes may remain invisible to LC–MS/MS. Heat treatment of such samples can enable the detection of those proteins, with the added advantage of enriching proteins with significant disorder content. A comparison of total identified proteins was made between the heat-treated and untreated samples of both experimental groups (Figure 2B). Heating the samples resulted in the identification of an additional 180 unique proteins, compared to the non-heated sample in the NCA group, while 265 heat-sensitive proteins were eliminated. The two sample types had 61 proteins in common. Within the CA group, 96 proteins were found only in the heated sample, 184 in the non-heated samples, and 34 proteins were shared between the two sample types.

To assess the disorder content of the identified proteins, a reliable disorder predictor, IUPred, was used to determine the disorder tendencies of the hits (Table S1). Percental disorder was calculated for both the non-cold-acclimated and cold-acclimated diapausing groups (506 and 314 proteins, respectively). Proteins with an average percental disorder of 70% or higher were considered as mostly disordered (MDPs), with partially disordered proteins (PDPs) if the average percental disorder was between 30% and 70%, nearly ordered (NOPs) for values between 10% and 30%, and ordered (OPs) if the percental value was no higher than 10%. In the NCA group, MDPs accounted for 31 of all identified proteins; 81 were PDPs, 75 were NOPs, and the remaining 319 were OPs. In the CA group, 16 proteins were MDPs, 51 were PDPs, and 45 were NOPs; there were 198 OPs (Figure 3A). The heat treatment had a remarkable effect on the distribution of proteins, with various degrees of intrinsic disorder in both experimental groups. The heat-treated samples contained more partially and mostly disordered proteins compared to the non-heated samples, while still retaining a significant portion of heat-resistant ordered proteins (Figure 3B).

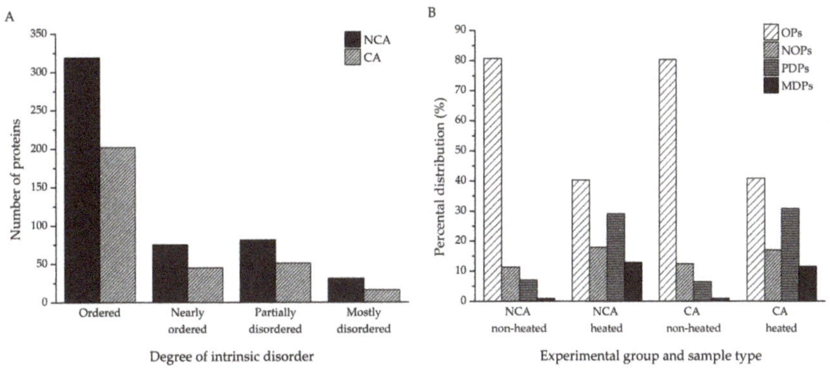


**Figure 3.** (**A**) Total number of proteins with varying degrees of intrinsic disorder—ordered (OPs, 10% at most), nearly ordered (NOPs, 10–30%), partially disordered (PDPs, 30–70%), mostly disordered (MDPs, at least 70%). (**B**) Effect of heat treatment on the percental distribution of proteins with varying degrees of intrinsic disorder in the two sample types of both experimental groups. NCA—non-cold-acclimated diapausing group; CA—cold-acclimated diapausing group.

Since percental disorder in itself is not necessarily informative of function, it is better to identify long intrinsically disordered regions (long IDRs), which have a higher potential to possess biological relevance. A region is considered a long IDR if it contains at least 20 consecutive disordered amino acids. Further analysis was performed to determine whether the proteins identified in this study contain long IDRs (Table S1). The results show that the proteins from both the NCA (Figure 4A) and CA (Figure 4B) experimental groups follow mostly similar patterns when it comes to the distribution of long IDRs in their sequences. Ordered proteins are devoid of IDRs, as are most of the NOPs. The majority of PDPs contain either 1 or 2 such regions. Certain muscle proteins, on the other hand, such as myosin heavy chain, are particularly enriched in long disordered regions. Depending on the isoform, they possess between 6 and as many as 14 such segments in their sequence. When it comes to MDPs, most of them possess one disordered segment, followed by proteins containing two, four, and three long IDRs, respectively.

The amino acid composition of every identified protein was analyzed, and the ratios of individual amino acids that make up the polypeptides were determined (Figure 5). Depending on the degree of structural disorder, the proteins in this dataset are comprised of varying amounts of disorder- and order-promoting amino acids. On average, mostly disordered proteins are composed of 68.7% disorder-promoting and 31.3% order-promoting amino acids. Partially disordered proteins have a similar composition to MDPs (65.2%/34.8%), while nearly ordered and ordered proteins trend towards a more balanced distribution of disorder- and order-promoting amino acids—60.5%/39.5% and 56.1%/43.9%, respectively.

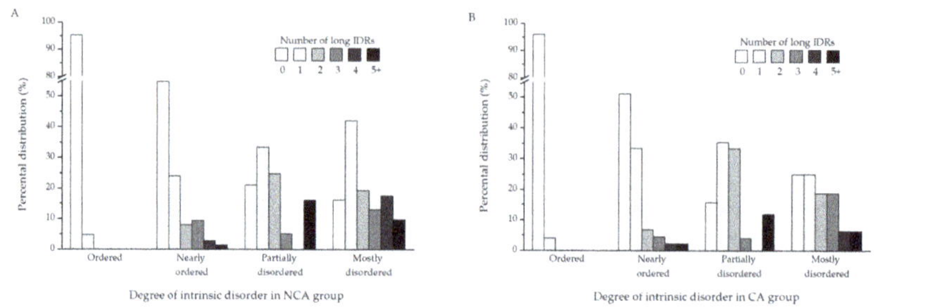


**Figure 4.** Distribution of long intrinsically disordered regions in proteins with various degrees of intrinsic disorder in the (**A**) NCA (non-cold-acclimated diapausing) and (**B**) CA (cold-acclimated diapausing) experimental groups. The different numbers of long IDRs in proteins are color-coded.

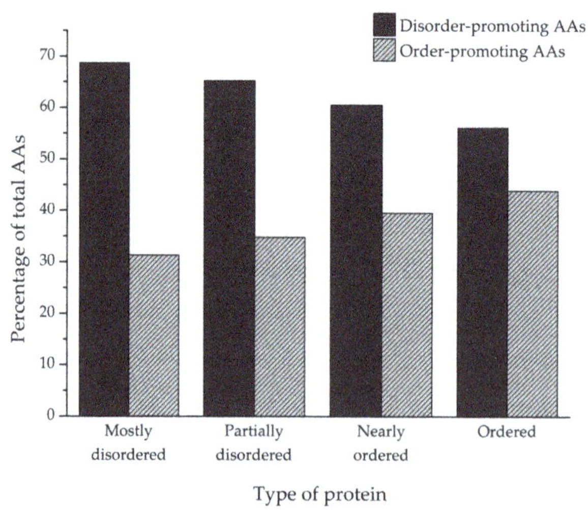


**Figure 5.** Ratios of disorder-promoting (P, E, S, K, Q, H, D, R, G, A) and order-promoting amino acids (T, C, N, V, L, M, I, Y, F, W) in mostly disordered (MDPs), partially disordered (PDPs), nearly ordered (NOPs), and ordered proteins (OPs).

The MDPs and PDPs identified in this study are particularly enriched in disorder-promoting amino acids (Figure 6A), such as glutamate (12.07% and 11.77% of total amino acids in a sequence on average, respectively), lysine (9.63% and 8.54%), and glutamine (6.69% and 5.34%), compared to nearly ordered (7.69% E, 8.39% K, and 4.28% Q) and ordered proteins (6.54% E, 7.2% K, and 3.48% Q). The only exception is glycine, which is more prevalent in nearly ordered (7.41%) and ordered proteins (7.46%) than in MDPs and PDPs (5.95% and 5.85%, respectively). Additionally, MDPs contain almost double the amount of proline as the other protein groups. When it comes to order-promoting amino acids (Figure 6B), MDPs are almost depleted in cysteine (0.39%), tyrosine (1.86%), phenylalanine (1.88%), isoleucine (3.67%), and valine (5.39%) in comparison to nearly ordered (0.98% C, 2.49% Y, 3.48% F, 5.23% I, and 7.26% V) and ordered proteins (1.88% C, 3.3% Y, 4.04% F, 5.78% I, and 7.43% V). Leucine is the standout order-promoting amino acid that partially disordered proteins contain in amounts similar to NOPs and OPs (7.88% compared to 7.54% and 8.58%, respectively), unlike MDPs (5.53%). The distribution of the remaining amino acids is more or less similar between the three groups of proteins.

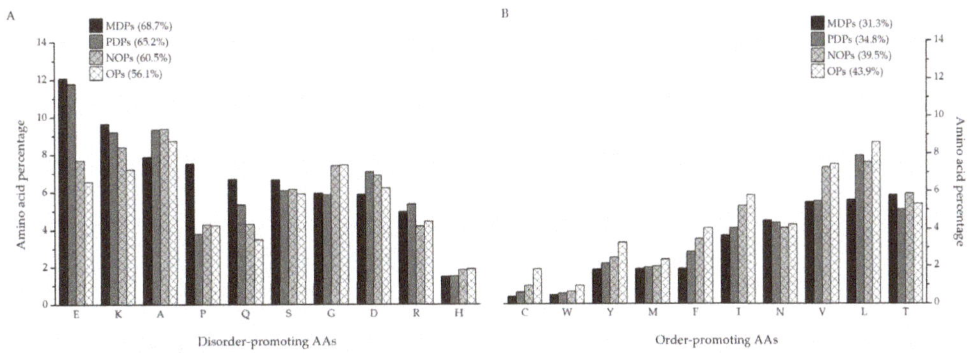


**Figure 6.** Ratios of individual disorder (**A**) and order-promoting amino acids (**B**) in mostly disordered (MDPs), partially disordered (PDPs), nearly ordered (NOPs), and ordered proteins (OPs), ordered by abundance in MDPs.

*In silico* structure predictions should ideally be supported by detailed *in vitro* structural studies, which are difficult to carry out on a proteomic scale. In order to validate the computational analyses and identification of intrinsically disordered proteins, a specific two-dimensional electrophoretic assay [41] was performed. The assay can provide experimental information on the large-scale structural state of a protein solution. It is based on the heat-stability and resistance to chemical denaturation of IDPs, resulting in a pattern where disordered proteins align in a diagonal line in the second dimension. For this step, two 2D PAGE were performed, one for each sample type (untreated and heat-treated). In order to ensure that as many unique proteins were covered by the 2D PAGE, aliquots from both experimental groups (NCA and CA) were pooled and run as a singular sample on the respective gels. As seen in Figure 7A, a large proportion of the proteins from the heat-treated sample are aligned on the diagonal line, signifying that they are mostly or fully disordered. The heat-stable globular proteins are generally found above the diagonal line in this setup, as indicated by the arrows in Figure 7A. Additionally, it can be seen that proteins from the non-heated sample, which is rich in globular polypeptides, mostly stayed in the first gel and did not transfer into the second gel. In fact, the proteins did not migrate far during the separation in the first dimension (Figure 7B). This is likely due to the abundance of high molecular weight arylphorins, common storage proteins in the hemolymph of the ECB [48], which prevented other proteins from separating. Heating of the samples and removal of globular proteins resolved this issue, which allowed the proteins to separate in the first dimension and transfer into the second gel.

To gain an insight into the biological importance of IDPs in the cold adaptation process of the ECB, we performed a bioinformatic functional analysis of the identified disordered proteins using the data from online knowledgebases Uniprot, Pfam, Interpro, and Gene Ontology. Our results have revealed that only 312 of the proteins have a Uniprot entry and at least one data point from the other listed knowledgebases. Out of that number, 90 proteins are either mostly or partially disordered or contain at least 1 long IDR (Figure 8, Total unique).

The largest functional group (24 unique hits) comprises the structural components of the cytoskeleton or proteins associated with it, such as actin filament organization proteins or regulators of muscle contraction. The second-largest group comprises proteins functioning as molecular chaperones (21 unique hits), followed by proteins involved in translational processes (10 unique hits). The rest of the proteins cover a wide range of biological processes and molecular functions, such as protein and amino acid metabolism

(9 unique hits), binding of nucleic acids (6 unique hits), binding of chitin and cuticle formation (4 unique hits), and others (Figure 8, Total unique). Additionally, heat treatment of the samples increased the number of proteins that could be identified and functionally analyzed. Proteins that are involved in chitin-binding and cuticle formation were found only in the heated samples, as were the majority of nucleic-acid-binding proteins (5 total hits compared to 1 total hit). More proteins belonging to the Cytoskeleton category were also present in the heated samples (19 total hits) compared to the non-heated ones (10 total hits). More proteins that act as molecular chaperones, on the other hand, were present in the non-heated samples (16 total hits) than in the heated samples (6 total hits) (Figure 8, Non-heated, Heated).

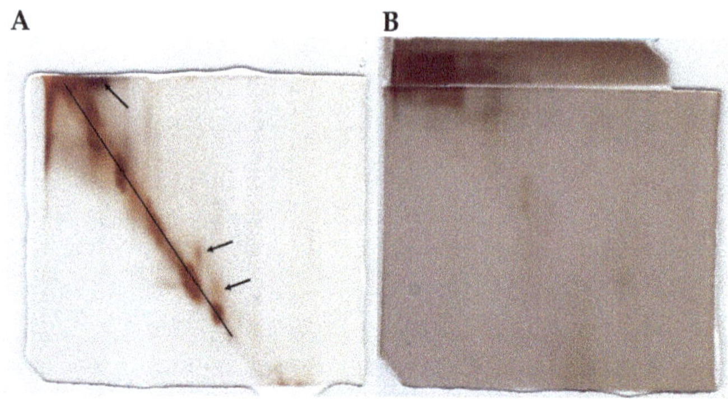


**Figure 7.** Results of in-house 2D PAGE for detecting intrinsically disordered proteins. (**A**) Proteins from heat-treated samples have successfully entered the second dimension. The black line represents the diagonal along which IDPs are located. Arrows denote ordered proteins that stay above the diagonal. (**B**) Proteins from non-heated samples are mostly locked in the gel from the first dimension (strip overlaying the larger gel).

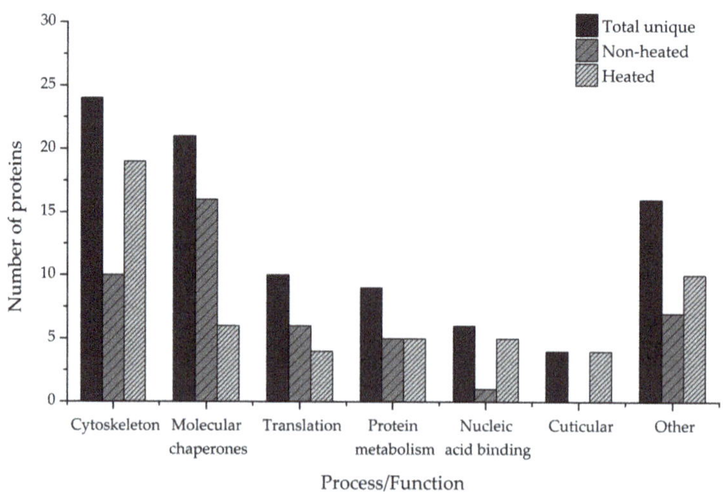


**Figure 8.** Biological processes and molecular functions of intrinsically disordered proteins and proteins containing long IDRs. The category Other encompasses processes and functions that make up less than 4% of total hits each. Total unique—all uniquely identified proteins; Non-heated—all proteins identified in the non-heated sample types; Heated—all proteins identified in the heated sample types.

## 4. Discussion

Proteomic identification of intrinsically disordered proteins is still a field where improvements are needed. Here, we applied an effort where enrichment based on heat treatment and bioinformatics analysis were combined to identify as many IDPs as possible that are involved in the cold adaptation of a non-model insect species *O. nubilalis*.

Almost three times as many unique proteins were identified in the NCA group than in the CA group (294 and 102, respectively, Figure 2A). The discrepancy is likely due to the experimental design, as cold acclimation in this species leads to a general depression of metabolic rate [25,26] and redirection of metabolic pathways that are of low priority towards the synthesis of low-molecular-weight cryoprotective compounds [27,29,49], among other things. When it comes to the effects of sample heating on the total number of proteins identified, as a means of IDP enrichment, fewer proteins were identified in the heat-treated samples, as expected. However, heat treatment enabled the identification of many proteins that were masked from LC–MS/MS analysis in the non-treated samples (Figure 2B). Additionally, it is important to highlight that nearly 50% of the proteins identified in this study have previously only been predicted from nucleotide sequences and were thus experimentally validated.

Structural disorder was predicted for the identified proteins using the IUPred algorithm. Depending on the amount of intrinsic disorder, the proteins were divided into four categories—mostly disordered proteins (MDPs, at least 70% percental disorder), partially disordered proteins (PDPs, 30–70% percental disorder), nearly ordered proteins (NOPs, 10–30% percental disorder), and ordered proteins (OPs, 10% percental disorder at most). According to our results, 30% of all identified proteins are either partially or mostly disordered or belong to the group of nearly ordered proteins that contain long IDRs (Figures 3A and 4), in accordance with previous meta-studies on the prevalence of protein intrinsic disorder in eukaryotic proteomes [7–9,50]. The majority of these proteins were revealed after IDP enrichment by sample heating, as only a few of them were identified specifically in the non-heated samples (Figure 3B). Heating the samples led to the removal of globular and heat-sensitive disordered proteins, while the majority of disordered proteins remained unaffected due to their stability in denaturing conditions, such as high temperatures [51–53]. As such, the heat treatment had a remarkable effect on the distribution of proteins with various degrees of intrinsic disorder in both NCA and CA experimental groups. The heat-treated samples contained more partially and mostly disordered proteins compared to the non-treated samples while still retaining a significant portion of heat-resistant ordered proteins. These findings further underscore the importance of sample boiling, without which we would have not only identified fewer novel proteins, but also missed out on the disordered proteins. The remaining 374 proteins (61% of total) scored 10% percental disorder at most and were labeled as ordered globular proteins. The reason that most of these proteins have percental values above 0%, which would indicate a complete absence of intrinsic disorder, and are still considered ordered is that fully structured proteins are quite rare. In fact, only around 7% of proteins deposited in the Protein Data Bank (PDB) do not contain any disordered residues at all [54].

Adding on to the fact that even globular proteins can contain unstructured segments, all proteins were analyzed for the presence of such long intrinsically disordered regions (long IDRs). These are segments of at least 20 consecutive disorder-promoting amino acids. They can be associated with sites of post-translational modification, act as flexible linkers to facilitate domain movements, or function as regions of molecular recognition and binding [55,56]. Proteins that scored below 70% (PDPs, NOPs, and OPs, 569 in total) were analyzed for the presence of IDRs first. Around 75% of these proteins (427 in total) did not contain any long disordered regions. Most of them were the ones labeled as ordered proteins (355 out of 374 OPs) and nearly ordered proteins (52 out of 95 NOPs). However, a fifth of partially disordered proteins did not contain any long IDRs either (20 out of 100 PDPs). This discrepancy between the measured percental disorder and the presence/absence of long IDRs in these groups of proteins is probably indicative of the localization of disorder in their structures. Partially disordered proteins without long IDRs likely contain disordered residues in contiguous segments that do not reach the 20 amino acids threshold of long IDRs, yet the segments are long enough to be reflected on the measured IUPred score of at least 10%. When it comes to ordered proteins, the small amount of detected disorder that they do possess is likely contained within the singular long IDR that some of these proteins contain. In total, 142 proteins that scored below 70% percental disorder contain at least one long IDR, with most of them (55%) containing exactly one. More than a fifth of them contain two LDRs, while the remaining proteins contain at least 3 and up to 14 long disordered regions. Next, mostly disordered proteins were subjected to the same analysis. However, this was done in order to further characterize the nature of their structural disorder. The majority of these proteins (38%) contain one long disordered region in their sequences, but two and three such segments are also frequent. A small number of these proteins contain four or more long IDRs. Interestingly, it can be seen that around 20% of the identified proteins contain none, despite their high disorder content (Figure 4A,B).

An interesting approach is to also analyze the amino acid composition of the identified proteins. Considering that intrinsic disorder is encoded in the primary sequence of proteins [3], this type of analysis would provide valuable information on the distribution of individual amino acids in the varying degrees of structural disorder. Depending on their intrinsic properties, amino acids can promote either order or disorder in protein structure in varying degrees. In that sense, amino acids can be designated as order- (W, F, Y, I, M, L, V, N, C, T) or disorder-promoting (A, G, R, D, H, Q, K, S, E, P), with tryptophan being the most order-promoting and proline the most disorder-promoting amino acid [57]. On average, more than two-thirds of amino acids in the primary sequence of MDPs and PDPs identified in this study are disorder-promoting (Figure 5). In particular, these proteins are enriched in polar charged amino acids such as glutamic acid (E, 12.07% and 11.77%, respectively), lysine (K, 9.63% and 9.19%), and polar non-charged glutamine (Q, 6.69% and 5.34%) in comparison to NOPs and OPs that have noticeably lower amounts of these amino acids in their sequences (Figure 6A). Another amino acid that MDPs are conspicuously enriched in compared to the other groups of proteins is proline (7.52%, Figure 6A), which is known to disrupt the formation of secondary structures in protein [58,59]. On the other hand, it can be noticed that PDPs contain a considerable amount of leucine (L, 7.88%, Figure 6B), which is considered an order-promoting amino acid. Compositional analysis has shown that disorder-containing muscle proteins, such as tropomyosins, are particularly rich in leucine (10% or higher content), indicating the importance of this amino acid for the function of muscle proteins. The aforementioned tropomyosins, for example, are involved in the regulation of muscle contraction and contain leucine zippers in their C-termini [60], similar to many nucleic-acid-binding proteins such as transcription factors. Functional analysis shows that the identified proteins are involved in a wide array of biological processes and, as such, fulfill many different functions. The majority of the proteins (24) are in some way connected with the organization and operation of the cytoskeletal network. Either they are structural constituents of the cytoskeleton or they regulate the contraction of muscle fibers. Next, a faction of the disorder-containing proteins (21) was found to act as molecular chaperones and assist in the proper folding of nascent proteins or handling of misfolded polypeptides. The last major group of proteins (10) is the proteins that are involved in translational processes, such as translation elongation, or are structural components of ribosomes. The remaining proteins cover a myriad of functions and processes, such as the metabolism of carbohydrates, lipids, and proteins, formation of insect cuticle, binding of nucleic acids, synthesis of amino acids, and oxidoreductive processes in connection with the electron transport chain (Figure 8). Sample heating also led to the identification of novel functions/processes, such as proteins involved in the formation of insect cuticle. These proteins were found exclusively in the heated samples. Likewise, the fraction of proteins involved in the cytoskeletal network was enriched after sample heating, as were the proteins involved in the binding of nucleic acids (Figure 8, Heated). On the other hand, heat treatment led to the removal of many molecular chaperones from the affected samples, so this group of proteins was more prominent in the non-heated samples (Figure 8, Non-heated). These results lend even more credence to the importance of heat treating the samples when it comes to the identification and functional analysis of intrinsically disordered proteins.

The findings in this study, first and foremost, demonstrate the need for further, more thorough research into the identification and characterization of intrinsically disordered proteins and intrinsically disordered protein regions. Secondly, by combining wet and dry lab methods, as proposed here, valuable information on the pervasiveness and function of IDPs/IDRs can be uncovered. Even a simple experimental design, such as the acclimation of insect larvae to low temperatures, caused an evident differentiation in the proteome content between the two experimental groups, both in quantity and quality. This is a reflection of the different metabolic states these experimental groups are in, as well as the various changes that have occurred at the molecular and biochemical levels. As such, the roles and functions of the identified proteins can be inferred from this differentiation, even if actual functional data is missing from the relevant databases. Additionally, this gives direction on where next to take the research and what proteins to focus on.

**Supplementary Materials:** The following supporting information can be downloaded at: https://www.mdpi.com/article/10.3390/biom12040592/s1, Table S1: Identified proteins, their IDs, percental disorder content and number of long IDRs present in their sequences.

**Author Contributions:** Conceptualization, M.A., Ž.D.P. and Á.T.; methodology, Ž.D.P. and Á.T.; validation, M.A., É.S. and Á.R.; formal analysis, É.S., Á.R. and L.T.; investigation, M.A.; resources, Á.T., L.D. and I.U.; data curation, É.S. and M.A.; writing—original draft preparation, M.A.; writing—review and editing, M.A., Ž.D.P., Á.T., É.S., Á.R. and I.U.; visualization, M.A.; supervision, Á.T. and Ž.D.P.; project administration, Á.T.; funding acquisition, Á.T., L.D. and Ž.D.P. All authors have read and agreed to the published version of the manuscript.

**Funding:** This research was funded by: the Ministry of Education, Science and Technological Development of the Republic of Serbia, grant no. 451-03-68/2022-14/200125; the National Research, Development and Innovation Office (Budapest, Hungary), grant nos. NKFIH PD-132135 and K-131762; the International Visegrad Fund (Bratislava, Slovakia), grant no. 51700031.

**Institutional Review Board Statement:** Not applicable.

**Informed Consent Statement:** Not applicable.

**Data Availability Statement:** The data presented in this study are available from the corresponding author upon reasonable request.

**Conflicts of Interest:** The authors declare no conflict of interest. The funders had no role in the design of the study; in the collection, analyses, or interpretation of data; in the writing of the manuscript, or in the decision to publish the results.

## References

1. Chen, J. Towards the physical basis of how intrinsic disorder mediates protein function. *Arch. Biochem. Biophys.* **2012**, *524*, 123–131. [CrossRef] [PubMed]
2. Tompa, P. Intrinsically unstructured proteins. *Trends Biochem. Sci.* **2002**, *27*, 527–533. [CrossRef]
3. Uversky, V.N.; Gillespie, J.R.; Fink, A.L. Why Are "Natively Unfolded" Proteins Unstructured Under Physiologic Conditions? *Proteins* **2000**, *41*, 415–427. [CrossRef]
4. Uversky, V.N.; Dunker, A.K. Understanding Protein Non-Folding. *Biochim. Biophys. Acta* **2010**, *1804*, 1231–1264. [CrossRef] [PubMed]
5. Dyson, H.J.; Wright, P.E. Intrinsically unstructured proteins and their functions. *Nat. Rev. Mol. Cell. Biol.* **2005**, *6*, 197–208. [CrossRef]
6. Wright, P.E.; Dyson, H.J. Intrinsically disordered proteins in cellular signalling and regulation. *Nat. Rev. Mol. Cell. Biol.* **2015**, *16*, 18–29. [CrossRef]
7. Pancsa, R.; Tompa, P. Structural Disorder in Eukaryotes. *PLoS ONE* **2012**, *7*, e34687. [CrossRef]
8. Xue, B.; Blocquel, D.; Habchi, J.; Uversky, A.V.; Kurgan, L.; Uversky, V.N.; Longhi, S. Structural Disorder in Viral Proteins. *Chem. Rev.* **2014**, *114*, 6880–6911. [CrossRef]

9. Peng, Z.; Yan, J.; Fan, X.; Mizianty, M.J.; Xue, B.; Wang, K.; Hu, G.; Uversky, V.N.; Kurgan, L. Exceptionally abudant exceptions: Comprehensive characterization of intrinsic disorder in all domains of life. *Cell. Mol. Life Sci.* **2015**, *72*, 137–151. [CrossRef]
10. Piovesan, D.; Tabaro, F.; Mičetić, I.; Necci, M.; Quaglia, F.; Oldfield, C.J.; Aspromonte, M.C.; Davey, N.E.; Davidović, R.; Dosztányi, Z.; et al. Disprot 7.0: A major update of the database of disordered proteins. *Nucleic Acids Res.* **2017**, *45*, D219–D227. [CrossRef]
11. Bari, K.J.; Prakashchand, D.D. Fundamental Challenges and Outlook in Simulating Liquid–Liquid Phase Separation of Intrinsically Disordered Proteins. *J. Phys. Chem. Lett.* **2021**, *12*, 1644–1656. [CrossRef] [PubMed]
12. Perdigão, N.; Heinrich, J.; Stolte, C.; Sabir, K.S.; Buckley, M.J.; Tabor, B.; Signal, B.; Gloss, B.S.; Hammang, C.J.; Rost, B.; et al. Unexpected Features of the dark proteome. *Proc. Natl. Acad. Sci. USA* **2015**, *112*, 15898–15903. [CrossRef] [PubMed]
13. Bhowmick, A.; Brookes, D.H.; Yost, S.R.; Dyson, H.J.; Forman-Kay, J.D.; Gunter, D.; Head-Gordon, M.; Hura, G.L.; Pande, V.S.; Wemmer, D.E.; et al. Finding Our Way in the Dark Proteome. *J. Am. Chem. Soc.* **2016**, *138*, 9730–9742. [CrossRef] [PubMed]
14. Kulkarni, P.; Uversky, V.N. Intrinsically Disordered Proteins: The Dark Horse of the Dark Proteome. *Proteomics* **2018**, *18*, e1800061. [CrossRef] [PubMed]
15. He, B.; Wang, K.; Liu, Y.; Xue, B.; Uversky, V.N.; Dunker, A.K. Predicting intrinsic disorder in proteins: An overview. *Cell Res.* **2009**, *19*, 929–949. [CrossRef]
16. Dosztányi, Z.; Mészáros, B.; Simon, I. Bioinformatical approaches to characterize intrinsically disordered/unstructured proteins. *Brief. Bioinform.* **2010**, *11*, 225–243. [CrossRef]
17. Piovesan, D.; Tosatto, S.C.E. INGA 2.0: Improving protein function prediction for the dark proteome. *Nucleic Acids Res.* **2019**, *47*, W373–W378. [CrossRef]
18. Necci, M.; Piovesan, D.; Predictors, C.; DisProt, C.; Tosatto, S.C.E. Critical assessment of protein intrinsic disorder prediction. *Nat. Methods.* **2021**, *18*, 472–481. [CrossRef]
19. Tantos, A.; Friedrich, P.; Tompa, P. Cold stability of intrinsically disordered proteins. *FEBS Lett.* **2009**, *583*, 465–469. [CrossRef]
20. Krumm, J.T.; Hunt, T.E.; Skoda, S.R.; Hein, G.L.; Lee, D.J.; Clark, P.L.; Foster, J.E. Genetic variability of the European corn borer, *Ostrinia nubilalis*, suggests gene flow between populations in the Midwestern United States. *J. Insect Sci.* **2008**, *8*, 72. [CrossRef]
21. Molinari, F.; Demaria, D.; Vittone, G. *Ostrinia nubilalis* Hübner (Lepidoptera, Pyralidae) as a threat for apple. *Int. Organ. Biol. Integr. Control Noxious Anim. Plants WPRS Bull.* **2010**, *54*, 247–250.
22. Robinson, G.S.; Ackery, P.R.; Kitching, I.J.; Beccaloni, G.W.; Hernández, L.M. HOSTS—A Database of the World's Lepidopteran Hostplants. 2010. Available online: https://www.nhm.ac.uk/our-science/data/hostplants/ (accessed on 15 January 2022).
23. Beck, S.D.; Hanec, W. Diapause in the European corn borer, *Pyrausta nubilalis* (Hübn.). *J. Insect Physiol.* **1960**, *4*, 304–318. [CrossRef]
24. Koštál, V. Eco-physiological phases of insect diapause. *J. Insect Physiol.* **2006**, *52*, 113–127. [CrossRef] [PubMed]
25. Hahn, D.A.; Denlinger, D.L. Energetics of insect diapause. *Annu. Rev. Entomol.* **2011**, *56*, 103–121. [CrossRef]
26. Popović, Ž.D.; Maier, V.; Avramov, M.; Uzelac, I.; Gošić-Dondo, S.; Blagojević, D.; Koštál, V. Acclimations to Cold and Warm Conditions Differently Affect the Energy Metabolism of Diapausing Larvae of the European Corn Borer *Ostrinia nubilalis* (Hbn.). *Front. Physiol.* **2021**, *12*, 2126. [CrossRef]
27. Uzelac, I.; Avramov, M.; Čelić, T.; Vukašinović, E.; Gošić-Dondo, S.; Purać, J.; Kojić, D.; Blagojević, D.; Popović, Ž.D. Effect of Cold Acclimation on Selected Metabolic Enzymes During Diapause in The European Corn Borer *Ostrinia nubilalis* (Hbn.). *Sci. Rep.* **2020**, *10*, 9085. [CrossRef]
28. Koštál, V.; Zahradníčková, H.; Šimek, P.; Zelený, J. Multiple component system of sugars and polyols in the overwintering spruce bark beetle, *Ips typographus*. *J. Insect Physiol.* **2007**, *53*, 580–586. [CrossRef]
29. Kojić, D.; Popović, Ž.D.; Orčić, D.; Purać, J.; Orčić, S.; Vukašinović, E.L.; Nikolić, T.V.; Blagojević, D.P. The influence of low temperature and diapause phase on sugar and polyol content in the European corn borer *Ostrinia nubilalis* (Hbn). *J. Insect Physiol.* **2018**, *109*, 107–113. [CrossRef]
30. Vukašinović, E.L.; Pond, D.W.; Worland, M.R.; Kojić, D.; Purać, J.; Blagojević, D.P.; Grubor-Lajšić, G. Diapause induces changes in the composition and biophysical properties of lipids in larvae of the European corn borer, *Ostrinia nubilalis* (Lepidoptera: Crambidae). *Comp. Biochem. Physiol. Part B Biochem. Mol. Biol.* **2013**, *165*, 219–225. [CrossRef]
31. Vukašinović, E.L.; Pond, D.W.; Worland, M.R.; Kojić, D.; Purać, J.; Popović, Ž.D.; Grubor-Lajšić, G. Diapause induces remodeling of the fatty acid composition of membrane and storage lipids in overwintering larvae of *Ostrinia nubilalis*, Hubn. (Lepidoptera: Crambidae). *Comp. Biochem. Physiol. Part B Biochem. Mol. Biol.* **2015**, *184*, 36–43. [CrossRef]
32. Vukašinović, E.L.; Pond, D.W.; Grubor-Lajšić, G.; Worland, M.R.; Kojić, D.; Purać, J.; Popović, Ž.D.; Blagojević, D.P. Temperature adaptation of lipids in diapausing *Ostrinia nubilalis*: An experimental study to distinguish environmental versus endogenous controls. *J. Comp. Physiol. Part B* **2018**, *188*, 27–36. [CrossRef] [PubMed]
33. Denlinger, D.L. Regulation of diapause. *Annu. Rev. Entomol.* **2002**, *478*, 93–122. [CrossRef] [PubMed]
34. Rinehart, J.P.; Li, A.; Yocum, G.D.; Robich, R.M.; Hayward, S.A.L.; Denlinger, D.L. Up-regulation of heat shock proteins is essential for cold survival during insect diapause. *Proc. Natl. Acad. Sci. USA* **2007**, *104*, 11130–11137. [CrossRef] [PubMed]
35. MacRae, T.H. Gene expression, metabolic regulation and stress tolerance during diapause. *Cell. Mol. Life Sci.* **2010**, *67*, 2405–2424. [CrossRef] [PubMed]
36. Popović, Ž.D.; Subotić, A.; Nikolić, T.V.; Radojičić, R.; Blagojević, D.P.; Grubor-Lajšić, G.; Koštál, V. Expression of stress-related genes in diapause of European corn borer (*Ostrinia nubilalis*, Hbn.). *Comp. Biochem. Physiol. B Biochem. Mol. Biol.* **2015**, *186*, 1–7. [CrossRef]

37. Purać, J.; Kojić, D.; Petri, E.; Popović, Ž.D.; Grubor-Lajšić, G.; Blagojević, D.P. Cold Adaptation Responses in Insects and Other Arthropods: An "Omics" Approach. In *Short Views on Insect Genomics and Proteomics*; Raman, C., Goldsmith, M., Agunbiade, T., Eds.; Springer International Publishing: Cham, Switzerland, 2016; Volume 4, pp. 89–112. [CrossRef]
38. Koštál, V.; Štětina, T.; Poupardin, R.; Korbelová, J.; Bruce, A.W. Conceptual framework of the eco-physiological phases of insect diapause development justified by transcriptomic profiling. *Proc. Natl. Acad. Sci. USA* **2017**, *114*, 8532–8537. [CrossRef]
39. Beck, S.D. Photoperiodic induction of diapause in an insect. *Biol. Bull.* **1962**, *122*, 1–12. [CrossRef]
40. Perkins, D.N.; Pappin, D.J.; Creasy, D.M.; Cottrell, J.S. Probability-based protein identification by searching sequence databases using mass spectrometry data. *Electrophoresis* **1999**, *20*, 3551–3567. [CrossRef]
41. Tantos, A.; Tompa, P. Identification of Intrinsically Disordered Proteins by a Special 2D Electrophoresis. In *Intrinsically Disordered Protein Analysis*; Uversky, V.N., Dunker, A.K., Eds.; Springer: New York, NY, USA, 2012; Volume 2, pp. 215–222. [CrossRef]
42. Dosztányi, Z.; Csizmok, V.; Tompa, P.; Simon, I. IUPred: Web server for the prediction of intrinsically unstructured regions of proteins based on estimated energy content. *Bioinformatics* **2005**, *21*, 3433–3434. [CrossRef]
43. Consortium, T.U. UniProt: The universal protein knowledgebase in 2021. *Nucleic Acids Res.* **2021**, *49*, D480–D489. [CrossRef]
44. Mistry, J.; Chuguransky, S.; Williams, L.; Qureshi, M.; Salazar, G.A.; Sonnhammer, E.L.; Tosatto, S.C.; Paladin, L.; Raj, S.; Richardson, L.J.; et al. Pfam: The protein families database in 2021. *Nucleic Acids Res.* **2021**, *49*, D412–D419. [CrossRef] [PubMed]
45. Blum, M.; Chang, H.Y.; Chuguransky, S.; Grego, T.; Kandasaamy, S.; Mitchell, A.; Nuka, G.; Paysan-Lafosse, T.; Qureshi, M.; Raj, S.; et al. The InterPro protein families and domains and database: 20 years on. *Nucleic Acids Res.* **2021**, *49*, D344–D354. [CrossRef] [PubMed]
46. Ashburner, M.; Ball, C.A.; Blake, J.A.; Botstein, D.; Butler, H.; Cherry, J.M.; Davis, A.P.; Dolinski, K.; Dwight, S.S.; Eppig, J.T.; et al. Gene Ontology: Tool for the unification of biology. *Nat. Genet.* **2000**, *25*, 25–29. [CrossRef] [PubMed]
47. Consortium, T.G.O. The Gene Ontology resource: Enriching a GOld mine. *Nucleic Acids Res.* **2021**, *49*, D325–D334. [CrossRef] [PubMed]
48. Taški, K.J.; Stanić, B.Đ.; Grubor-Lajšić, G.N. The presence of an arylphorin-type storage protein at different stages of *Ostrinia nubilalis* (Lepidoptera: Pyrilidae). *Matica Srpska J. Nat. Sci.* **2004**, *106*, 5–13. [CrossRef]
49. Grubor-Lajšić, G.; Block, W.; Palanački, V.; Glumac, S. Cold hardiness parameters of overwintering diapause larvae of *Ostrinia nubilalis* in Vojvodina, Yugoslavia. *CryoLetters* **1991**, *12*, 177–182.
50. Bondos, S.E.; Dunker, A.K.; Uversky, V.N. On the roles of intrinsically disordered proteins and regions in cell communication and signaling. *Cell Commun. Signal.* **2021**, *19*, 88. [CrossRef]
51. Cortese, M.S.; Baird, J.P.; Uversky, V.N.; Dunker, A.K. Uncovering the Unfoldome: Enriching Cell Extracts for Unstructured Proteins by Acid Treatment. *J. Proteome Res.* **2005**, *4*, 1610–1618. [CrossRef]
52. Galea, C.A.; Pagala, V.R.; Obenauer, J.C.; Park, C.G.; Slaugher, C.A.; Kriwacki, R.W. Proteomic Studies of the Intrinsically Unstructured Mammalian Proteome. *J. Proteome Res.* **2006**, *5*, 2839–2848. [CrossRef]
53. Zhang, Y.; Launay, H.; Schramm, A.; Lebrun, R.; Gontero, B. Exploring intrinsically disordered proteins in *Chlamydomonas reinhardtii*. *Sci. Rep.* **2018**, *8*, 6805. [CrossRef]
54. Le Gall, T.; Romero, P.R.; Cortese, M.S.; Uversky, V.N.; Dunker, A.K. Intrinsic Disorder in the Protein Data Bank. *J. Biomol. Struct. Dyn.* **2007**, *24*, 325–342. [CrossRef] [PubMed]
55. Lobley, A.; Swindells, M.B.; Orengo, C.A.; Jones, D.T. Inferring Function Using Patterns of Native Disorder in Proteins. *PLoS Comput. Biol.* **2007**, *3*, e162. [CrossRef] [PubMed]
56. Davey, N.E. The functional importance of structure in unstructured protein regions. *Curr. Opin. Struct. Biol.* **2019**, *56*, 155–163. [CrossRef] [PubMed]
57. Campen, A.; Williams, R.M.; Brown, C.J.; Meng, J.; Uversky, V.N.; Dunker, A.K. TOP-IDP-Scale: A New Amino Acid Scale Measuring Propensity for Intrinsic Disorder. *Protein Pept. Lett.* **2008**, *15*, 956–963. [CrossRef] [PubMed]
58. Imai, K.; Mitaku, S. Mechanisms of secondary structure breakers in soluble proteins. *Biophysics* **2005**, *1*, 55–65. [CrossRef] [PubMed]
59. Morgan, A.A.; Rubenstein, E. Proline: The Distribution, Frequency, Positioning, and Common Functional Roles of Proline and Polyproline Sequences in the Human Proteome. *PLoS ONE* **2013**, *8*, e53785. [CrossRef] [PubMed]
60. Brown, J.H.; Zhou, Z.; Reshetnikova, L.; Robinson, H.; Yammani, R.D.; Tobacman, L.S.; Cohen, C. Structure of the mid-region of tropomyosin: Bending and binding sites for actin. *Proc. Natl. Acad. Sci. USA* **2005**, *102*, 18878–18883. [CrossRef]

biomolecules

MDPI

Article

# Distribution of Charged Residues Affects the Average Size and Shape of Intrinsically Disordered Proteins

**Greta Bianchi [1], Marco Mangiagalli [1], Alberto Barbiroli [2], Sonia Longhi [3], Rita Grandori [1], Carlo Santambrogio [1,*] and Stefania Brocca [1,*]**

1 Department of Biotechnology and Biosciences, University of Milano-Bicocca, 20126 Milan, Italy; greta.bianchi@unimib.it (G.B.); marco.mangiagalli@unimib.it (M.M.); rita.grandori@unimib.it (R.G.)
2 Departement of Food, Environmental and Nutritional Sciences, University of Milan, 20133 Milan, Italy; alberto.barbiroli@unimi.it
3 Laboratory Architecture et Fonction des Macromolécules Biologiques (AFMB), UMR 7257, Centre National de la Recherche Scientifique (CNRS), Aix Marseille University, 13288 Marseille, France; sonia.longhi@univ-amu.fr
* Correspondence: carlo.santambrogio@unimib.it (C.S.); stefania.brocca@unimib.it (S.B.); Tel.: +39-02-6448-3363 (C.S.); +39-02-6448-3518 (S.B.)

**Abstract:** Intrinsically disordered proteins (IDPs) are ensembles of interconverting conformers whose conformational properties are governed by several physico-chemical factors, including their amino acid composition and the arrangement of oppositely charged residues within the primary structure. In this work, we investigate the effects of charge patterning on the average compactness and shape of three model IDPs with different proline content. We model IDP ensemble conformations as ellipsoids, whose size and shape are calculated by combining data from size-exclusion chromatography and native mass spectrometry. For each model IDP, we analyzed the wild-type protein and two synthetic variants with permuted positions of charged residues, where positive and negative amino acids are either evenly distributed or segregated. We found that charge clustering induces remodeling of the conformational ensemble, promoting compaction and/or increasing spherical shape. Our data illustrate that the average shape and volume of the ensembles depend on the charge distribution. The potential effect of other factors, such as chain length, number of proline residues, and secondary structure content, is also discussed. This methodological approach is a straightforward way to model IDP average conformation and decipher the salient sequence attributes influencing IDP structural properties.

**Keywords:** charge clustering; polyelectrolytes; average shape of conformational ensembles; charged-residue patterning; hydrodynamic radius; solvent-accessible surface area; proline content; conformational compactness; ellipsoid model

**Citation:** Bianchi, G.; Mangiagalli, M.; Barbiroli, A.; Longhi, S.; Grandori, R.; Santambrogio, C.; Brocca, S. Distribution of Charged Residues Affects the Average Size and Shape of Intrinsically Disordered Proteins. *Biomolecules* **2022**, *12*, 561. https://doi.org/10.3390/biom12040561

Academic Editor: Supriyo Bhattacharya

Received: 5 March 2022
Accepted: 7 April 2022
Published: 9 April 2022



## 1. Introduction

Intrinsically disordered proteins (IDPs) and regions have a biased sequence composition compared to folded counterparts, being enriched in disorder-promoting and charged amino acids and depleted in order promoting ones [1–3]. The high number of charged residues (Asp, Glu, Arg, Lys) has enabled modeling IDPs as either polyelectrolytes or polyampholytes, depending on the presence of same- or opposite-sign charges, respectively. The charge state of polyampholytes is often described by the total fraction of charged residues (*FCR*), obtained as the sum of the fractions of positive ($f_+$) and negative residues ($f_-$), and by the net charge per residue (*NCPR*), calculated as the difference between $f_+$ and $f_-$ [4]. In addition to these coarse-grain parameters, the linear distribution of positive and negative charges, described by $\kappa$ or sequence charge decoration parameters [5,6], is also an important feature in determining protein compactness. More in detail, computational and experimental data show that charge segregation promotes protein compaction [7–10].

IDPs consist of fluctuating and interconverting conformations that constitute "conformational ensembles". Size-exclusion chromatography (SEC), which enables molecule separation based on their hydrodynamic radius ($R_h$), is one of the most popular and easy to apply techniques to study the compaction of proteins, including IDPs. Experimentally, $R_h$ can be determined from the chromatographic elution volume, using a calibration curve obtained with proteins of known $R_h$, or known molecular mass belonging to the same structural class [11,12]. Achieving a more quantitative description of IDP ensembles requires methods capable of dealing with heterogeneous molecular systems, such as nuclear magnetic resonance (NMR), small-angle X-ray scattering (SAXS), mass spectrometry (MS) combined with labeling techniques, high-speed atomic force microscopy, and Förster resonance energy transfer and non-denaturing mass spectrometry (native MS) to cite a few [13–17]. Native MS has been extensively employed to characterize the properties of heterogeneous conformational ensembles, enabling the detection of even poorly populated states [18–21]. Indeed, gentle ionization conditions, such as those obtained by nano-electrospray ionization (nanoESI), preserve non-covalent interactions under the vanishing-solvent conditions of the electrospray, leading to protein ionization and transfer to the gas phase. The final protein net charge is mainly dictated by structural compactness under controlled conditions. Thus, charge state distributions (CSDs) in nanoESI spectra reveal the main components of conformational ensembles [17,20,22]. Unfolded/disordered proteins achieve higher charge states than their globular counterparts. For both folded and unfolded chains, the average charge state correlates with the solvent-accessible surface area (*SASA*), reflecting chain compactness [17,23–25].

In spite of the seminal and breaking-through studies by Pappu and co-workers that illuminated the relationships between charge distribution and conformational properties of IDPs [7,8,10], a full understanding of how the sequence of IDPs encodes their conformation is still lacking, thereby preventing, for instance, the *ex nihilo* design of IDPs with a precise set of desired conformational properties. With the goal of shedding light on these still unsolved issues, here we have studied the effect of charge segregation on three model IDPs that exhibit similar content in overall charged residues, net charge, and hydropathy, but different content of proline residues and secondary structure, and slightly different size. Charged residues within these model IDPs were permutated to obtain different $\kappa$-variants (Figure 1), and the three sets of proteins were characterized by SEC and ESI-MS. Experimentally derived $R_h$ and *SASA* values were used to obtain coarse-grained structural information on these IDP ensembles using a recently published model, originally developed for globular proteins, that approximates the geometry of a protein to an ellipsoid [26].

Results show how the changes in average volume and shape triggered by the distribution of charged residues are variously affected by the frequency of proline residues. In addition, we discussed the potential role of other factors such as secondary structure content and amino acid chain length.

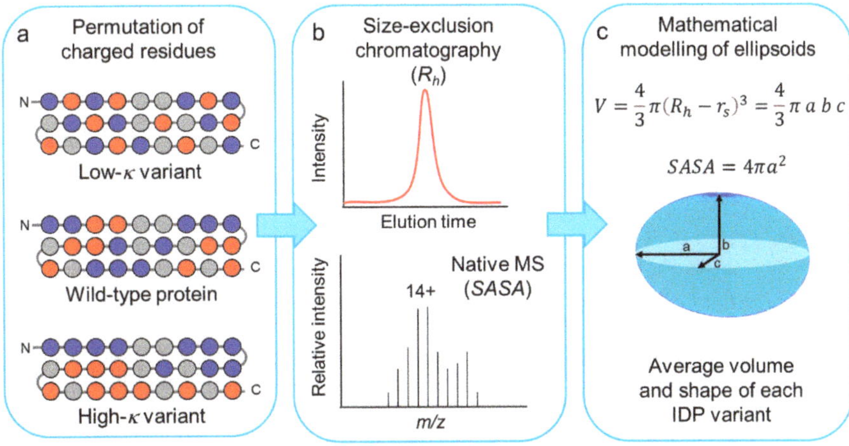


**Figure 1.** Scheme of the experimental plan used in this work. (**a**) Scheme of the primary structures of a protein set, derived from a generic wild-type IDP by distributing more evenly the oppositely charged residues (low-κ variant) or by clustering them in two blocks at the N- and C-moieties (high-κ variant). Only charged residues were permutated, preserving the original location in the sequence of non-charged residues (see also Figure S1). Blue and red spheres indicate positively and negatively charged residues, respectively. Gray spheres indicate all the other amino acid residues. (**b**) The conformational ensemble of each model IDP was investigated by size-exclusion chromatography (SEC) and native mass spectrometry (MS), to derive experimental values of $R_h$ and *SASA*. (**c**) $R_h$ and *SASA* values were combined to calculate the volume and depict the average shape from the ensemble of each model IDP.

## 2. Materials and Methods

### 2.1. Gene Design and Cloning

The model molecules employed in this study are IDPs derived from the measles virus N protein, $N_{TAIL}$ [27], from the Hendra virus P protein, PNT4 [28], and from the human medium neurofilament protein, NFM (UniProtKB ID: P07197) [29,30]. The region used in this work (residues 790–916) belongs to the KE-rich tail of NFM, which is predicted to be intrinsically disordered. The rules followed for NFM gene design are those used for PNT4 and $N_{TAIL}$ [8]. Briefly, we conceived low-κ and high-κ variants sharing with wild type (wt) the same number of charged residues and the same position of non-charged residues and differing just in the distribution of positively (Lys, Arg) and negatively (Glu, Asp) charged residues along the sequence. In high-κ sequences, positively and negatively charged residues are clustered in the N- and C-terminal regions, respectively. On the contrary, in low-κ sequences, positively and negatively charged residues are more evenly distributed than in the wt sequence. Synthetic genes encoding for NFM were optimized for expression in *Escherichia coli* (Genscript, Piscataway, NJ, USA) and cloned into the pET-21a vector (EMD, Millipore, Billerica, MA, USA) between the *Nde*I and *Xho*I sites (Jena Biosciences, Jena, Germany). Each synthetic gene encodes a protein with an N-terminal hexa-histidine (6xHis) tag, while a stop codon has been inserted immediately before the *Xho*I restriction site, thereby excluding from the coding region the vector-encoded 6xHis tag. The amino acid sequences are shown in Figure S1. *Escherichia coli* DH5α™ strain (Invitrogen, Waltham, MA, USA) was used for plasmid DNA propagation.

### 2.2. Production and Purification of κ Variants

The *E. coli* strain BL21 (DE3) (EMD Millipore, Billerica, MA, USA) was used for protein heterologous production. Cultures were grown in ZYM-5052 medium [31], and recombinant IDPs were extracted and purified as described by Tedeschi and co-authors [9]. Briefly, recombinant proteins were purified from the soluble fraction of the bacterial lysate by gravity-flow, immobilized-metal affinity chromatography using a nickel-nitrilotriacetic acid agarose resin (ABT, Torrejon de Ardoz, Madrid, Spain). The fractions exhibiting the highest concentration were pooled, and buffers were exchanged for phosphate-buffered saline (PBS, 150 mM NaCl, 50 mM sodium phosphate, pH 7.0) or ultrapure ammonium acetate buffer (ammonium acetate 50 mM, pH 6.95, Merck KGaA, Darmstadt, Germany) by gel filtration on PD-10 columns (GE Healthcare, Little Chalfont, UK). Protein concentration was determined by Bradford protein assay (Bio-Rad, Hercules, CA, USA), using bovine serum albumin as a standard.

### 2.3. Bioinformatics Analysis

Sequence analysis of model proteins was performed using CIDER [32] and IUPred [33] web servers. IUPred provides a score that characterizes the disordered tendency of each position along the sequence. The score ranges from 0 to 1, with predicted scores above 0.5 indicating disorder. CIDER was used with default parameters to compute $\kappa$ values and local sequence properties such as *NCPR*, *FCR*, and the mean hydrophobicity in the 0–9 scaled Kyte-Doolittle hydropathy score.

### 2.4. Far-UV Circular Dichroism (CD) Spectroscopy

Far-UV CD analyses were carried out in PBS using a Jasco J-815 spectropolarimeter (Jasco Europe, Lecco, Italy) in a 1-mm path-length quartz cuvette. Measurements were performed at variable wavelengths (190–260 nm) with a scanning velocity of 20 nm/min and a data pitch of 0.2 nm. All spectra were corrected for buffer contribution, averaged from three independent acquisitions, and smoothed by the Means-Movement algorithm implemented in the Spectra Manager package (Jasco Europe, Lecco, Italy). Experiments were performed in triplicate. Mean ellipticity values per residue ([θ]) were calculated as described by Tedeschi and co-authors [9]. The deconvolution of CD spectra to assess secondary structure content was performed using the BestSel program [34].

### 2.5. Analytical SEC

Recombinant IDPs produced in this work were analyzed by SEC within the day they were purified. Chromatographic separations were carried out on a Superose 12 10/300 GL column (GE Healthcare, Milan, Italy), in mobile phase PBS, at a flow rate 0.5 mL/min. The chromatographic system was composed of a Waters Delta 600 pump, a 600 Controller, and a 2487 Dual λ Absorbance Detector; all managed through the Empower Pro Software (Waters Corporation, Milford, MA, USA). Chromatograms were recorded at 220 nm. The calibration curve was built using the following standards: Apo-ferritin (horse spleen, 443 kDa, $R_h$ 6.1 nm), Alcohol dehydrogenase (yeast, 150 kDa, $R_h$ 4.6 nm), BSA (bovine serum, 66 kDa, $R_h$ 3.5 nm), Ovalbumin (chicken egg, 43 kDa, $R_h$ 2.8 nm), Carbonic anhydrase (bovine erythrocytes, 29 kDa, $R_h$ 2.1 nm), Cytochrome C (horse heart, 12.4 kDa, $R_h$ 1.7 nm [35].

Firstly, for each standard protein the distribution coefficient ($K_d$) was calculated:

$$K_d = \frac{V_e - V_0}{V_t - V_0} \tag{1}$$

where $V_e$ is the elution volume, $V_0$ the void volume, and $V_t$ the total volume. Uracil (0.112 kDa) and Blue dextran (2000 kDa) were used for $V_t$ and $V_0$ determination.

Finally, the calibration curve Log($R_h$) vs. $K_d$ was built and the interpolated linear equation used to calculate IDPs hydrodynamic radii from their $K_d$ values. IDPs were run at least in triplicate.

The theoretical radius ($R_t$) was calculated according to the empirical Equation (2) [36].

$$R_t = (1.24\ P_{pro} + 0.904)\ (0.00759\ |Q| + 0.963)\ S_{his*} \tag{2}$$

where $P_{pro}$ is the number of proline residues, $|Q|$ the absolute net charge and the $S_{his*}$ is 0.901 or 1 depending on whether a 6xHistag is present or absent, respectively.

$R_h$ values were used to calculate the compaction index (*CI*), which provides a simple and continuous descriptor useful for comparing conformational properties of IDPs of different lengths [23,37]. The *CI* derived from the experimental value of $R_h$ ($CI_R$) was calculated by applying the following equation [37]:

$$CI_R = \frac{R^D - R_h}{R^D - R^{NF}} \tag{3}$$

where $R_h$ is the experimental value, $R^D$ and $R^{NF}$ are the theoretical values of a chemically denatured or a folded protein, calculated on the basis of power-law Equations (4) and (5), which describe their dependence on the number of residues, $N$ [11].

$$R^{NF} = 4.92 \cdot N^{0.285} \tag{4}$$

$$R^D = 2.49 \cdot N^{0.509} \tag{5}$$

### 2.6. Native MS Analyses

Protein solutions in 50 mM ammonium acetate, pH 7.0, were brought to a concentration of 10 μM, and samples under non-denaturing conditions were directly injected at room temperature into an Orbitrap Fusion mass spectrometer (Thermo Fisher Scientific, Waltham, MA, USA) equipped with a nano-electrospray ion source. Metal-coated borosilicate capillaries with medium-length emitter tips of 1 μm internal diameter were used to infuse the sample. To assess the effect of electrostatic interactions, protein samples were also analyzed at higher ionic strength (200 mM ammonium acetate pH 7.0) and low pH (no buffer, 1% formic acid, pH 2.5). The following instrumental setting was applied: ion spray voltage, 1.1–1.2 kV; ion-transfer tube temperature, 275 °K; AGC target, $4 \times 10^5$; maximum injection time, 100 ms. Spectra were averaged over 1-min acquisition. Multi-Gaussian fitting of MS spectra was performed employing the program OriginPro 2020 (OriginLab Corporation, Northampton, MO, USA), and *CI* of single conformers ($CI^i_{SASA}$) and ensembles ($\overline{CI}_{SASA}$) were calculated as follows [16]:

$$CI^i_{SASA} = \frac{A^c - A^0}{A^c - A^f} \tag{6}$$

$$\overline{CI}_{SASA} = \sum_{i=1}^{n} w_i.\, CI^i_{SASA} \tag{7}$$

where $A^c$ and $A^f$ are the solvent-accessible surface areas derived by native MS for reference, random coil (*c*) and folded (*f*) proteins of the same size of the protein under study, $A^0$ is the solvent-accessible surface areas derived by native MS for the conformer (exploiting the charge state—*SASA* relationship), $w_i$ is the relative amount of the conformer with compaction index $CI^i_{SASA}$.

Statistical significance of experimental differences was estimated by performing a Welch's *t*-test on three independent datasets.

### 2.7. Application of Ellipsoid Model

The ellipsoid model assumes that the average conformation of a given protein can be represented by an ellipsoid with semi-axes *a*, *b*, and *c* ($a \geq b \geq c$) [26]. The experimental ellipsoid volume depicting the conformation of the IDP averaged over the ensemble can be estimated by the volume of a sphere given by the following formula:

$$V = \frac{4}{3}\pi(R_h - r_s)^3 \tag{8}$$

where $r_s$ represents the hydration shell (generally assumed to be 5 Å) [38,39], and $R_h$ the hydrodynamic radius obtained by SEC experiments. The geometrical volume of an ellipsoid is expressed as:

$$V = \frac{4}{3}\pi abc, \quad (9)$$

To calculate *a*, the quadratic relationship with *SASA* given by the model of Wu and co-authors [26] can be exploited:

$$SASA = 4\pi a^2, \quad (10)$$

Then, *b* and *c* values can be approximated by weighted averages between the extreme conditions of prolate ($a > b = c$) and oblate ($a = b > c$) ellipsoids, according to the equations published by Wu and co-authors [26].

Thomsen's approximation was employed to calculate the ellipsoid surface area (maximal discrepancy to real surface ~1%).

The ellipsoid *flattening* was described through the values of $f_b$ and $f_c$, calculated according to the formulas:

$$f = \frac{(a-b)}{a};\ f = \frac{(a-c)}{a} \quad (11)$$

Both indices report the eccentricity of axial elliptic sections of the ellipsoid, and span in the range [0;1), where 0 corresponds to a circular section.

## 3. Results

### 3.1. Design of Model IDPs by Permutation of Charged Residues

The model IDPs used in this work are the viral proteins PNT4 and $N_{TAIL}$ and a C-terminal IDR from the human NFM. These IDPs are similar in length, theoretical hydrodynamic radius ($R_t$), charge density, and charge segregation, as witnessed by their $\kappa$ value (Table 1). Values of $\kappa$ vary between 0 and 1, with 0 indicating evenly mixed positive and negative residues, and 1 referring to the complete segregation of oppositely charged residues along the linear sequence [4]. In our model proteins, the number of positive and negative charges is well balanced, producing a rather low *NCPR* (mean absolute value 0.038 ± 0.017), and opposite charges are evenly distributed along the sequence, thereby resulting in rather low $\kappa$ values (mean value 0.167 ± 0.041). The three proteins differ in the fraction of proline residues, which is 0.7%, 5.2%, and 11.4% for NFM, $N_{TAIL}$, and PNT4, respectively. Among disorder-promoting residues, proline residues are also recognized to disfavor α-helical and β-structures [40], and to promote extended conformations by conferring rigidity to the backbone [36]. For each model IDP, a "high-$\kappa$" and a "low-$\kappa$" variants were designed by permuting charged residues while keeping the position of all other residues unchanged. Table 1 summarizes, for each model protein and its variants, the $\kappa$ parameter, *NCPR*, and *FCR* values calculated using the CIDER webserver [28].

**Table 1.** Features of the three model proteins and their derived $\kappa$ variants. Sequence features were computed using CIDER [28]; the theoretical radius $R_t$ was calculated according to Marsh and Forman-Kay [35].

| Protein | Number of Residues | Number of Prolines | Mean Hydropathy | *FCR* | *NCPR* | $R_t$ (nm) | $\kappa$ | Variant |
|---|---|---|---|---|---|---|---|---|
| $N_{TAIL}$ | 134 | 7 | 3.35 | 0.299 | −0.045 | 2.64 | 0.078<br>0.153<br>0.431 | Low $\kappa$<br>wt<br>High $\kappa$ |
| NFM | 136 | 1 | 3.40 | 0.390 | −0.051 | 2.54 | 0.037<br>0.134<br>0.516 | Low $\kappa$<br>wt<br>High $\kappa$ |
| PNT4 | 114 | 13 | 3.26 | 0.298 | 0.018 | 2.54 | 0.044<br>0.213<br>0.421 | Low $\kappa$<br>wt<br>High $\kappa$ |

In the high-$\kappa$ variants, positive and negative charged residues are clustered in two distinct blocks at the N- and C-terminal moieties of the sequence, while in low-$\kappa$ variants, these residues are evenly alternated along the sequence, as highlighted by their *NCPR* profiles (Figure 2a–c, upper panels, and Figure S1). The degree of disorder predicted by IUPred [33] is conserved within each set of model proteins derived by permutation from the respective wt sequence (Figure 2a–c, lower panels). The three sets of proteins were recombinantly produced and purified by immobilized-metal affinity chromatography and experimentally assessed by CD analysis in the far-UV (Figure S2). The CD spectra of wt IDPs display the typical trait of structural disorder with a negative peak at ~200 nm (black line in Figure S2). Worthy to note, all the spectra of wt IDPs present a small shoulder at ~220 nm, which indicates the presence of some elements of helical secondary structure. Despite the common high level of disorder predicted by IUPred, deconvolution of CD spectra indicates that in all the three model IDPs the α-helical content tends to increase along with the values of $\kappa$ (Figure S2, inset).

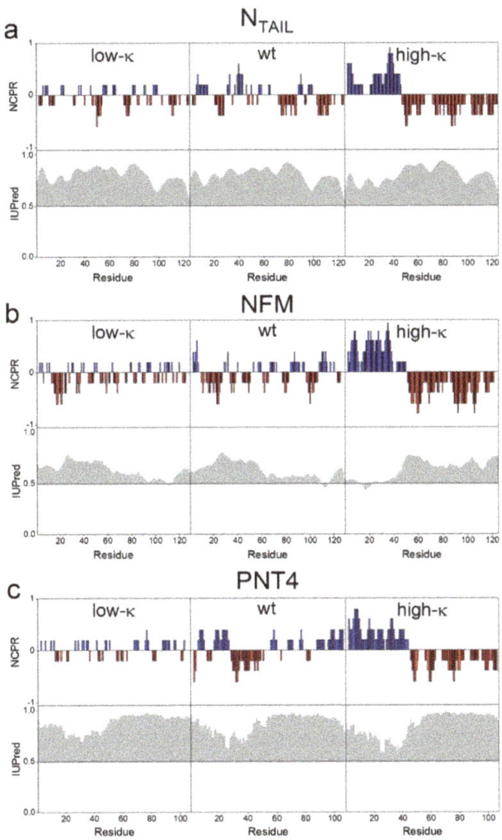


**Figure 2.** Comparative bioinformatic analyses of $N_{TAIL}$ (**a**), NFM (**b**) and PNT4 (**c**). Upper panels: The *FCR*, fraction of charged residues, was calculated by CIDER [32]. Each model protein contains charged residues at high density, with red and blue bars indicating negative and positive charges, respectively. The increase in $\kappa$ value is reflected in the progressively more "blocky" distribution of charged residues. Lower panel: each protein is predicted to be predominantly disordered by IUPred [33]. The discrepancy from the disorder threshold value (0.5) in the IUPred score is shaded in gray. The IUPred and CIDER outputs were generated using the default options of the respective web server.

### 3.2. Impact of Charge Clustering on the $R_h$ of the Model IDPs

Size-exclusion chromatography was employed to estimate the $R_h$ values of the three sets of model IDPs (Table 2). Experimental $R_h$ values of wt $N_{TAIL}$ and wt PNT4 (2.71 ± 0.09 and 2.34 ± 0.11 nm, respectively) are close to the theoretical ones (Table 1) and similar to the previously determined ones [9]. The $R_h$ of wt NFM (3.31 ± 0.12) is determined here for the first time. We observed that $R_h$ decreases as $\kappa$ increases for $N_{TAIL}$ and NFM, but not for PNT4 (Table 2). To compare the compaction properties of IDPs with different chain lengths, $R_h$ data were used for the calculation of the $R_h$-based *CI* ($CI_R$, defined in Equation (2)). The value of *CI* ranges from 0 to 1, corresponding to minimal and maximal compaction, respectively [37]. Analysis of the $CI_R$ confirms that $N_{TAIL}$ and NFM significantly respond to charge segregation, while PNT4 average compactness is not affected by the $\kappa$ value (Figure 3a).

**Table 2.** Hydrodynamic radii ($R_h$) and average solvent accessible surface area (*SASA*) of the three model proteins and their derived $\kappa$ variants. Mean values and standard deviations from three independent measurements are reported. Volume, surface area and flattening indices of the ellipsoids were derived from the model proposed by Wu and co-authors [23].

| Protein | Variant | $R_h$ (nm) | *SASA* (nm$^2$) | Volume (nm$^3$) | $f_b$ * | $f_c$ * |
|---|---|---|---|---|---|---|
| $N_{TAIL}$ | Low $\kappa$ | 2.78 ± 0.03 | 113.2 ± 2.1 | 49.5 ± 2.2 | 0.26 ± 0.02 | 0.41 ± 0.16 |
| | wt | 2.73 ± 0.03 | 105.6 ± 1.4 | 46.3 ± 1.8 | 0.25 ± 0.01 | 0.39 ± 0.16 |
| | High $\kappa$ | 2.58 ± 0.05 | 89.3 ± 1.0 | 37.5 ± 2.3 | 0.24 ± 0.02 | 0.38 ± 0.16 |
| NFM | Low $\kappa$ | 3.37 ± 0.05 | 136.2 ± 2.0 | 99.1 ± 5.0 | 0.14 ± 0.02 | 0.23 ± 0.12 |
| | wt | 3.31 ± 0.04 | 124.5 ± 2.4 | 93.2 ± 4.8 | 0.12 ± 0.02 | 0.19 ± 0.10 |
| | High $\kappa$ | 3.05 ± 0.10 | 81.5 ± 0.4 | 69.7 ± 8.0 | -0.03 ± 0.08 | 0.02 ± 0.09 |
| PNT4 | Low $\kappa$ | 2.39 ± 0.04 | 106.8 ± 3.0 | 28.1 ± 2.1 | 0.42 ± 0.03 | 0.53 ± 0.13 |
| | wt | 2.36 ± 0.05 | 106.8 ± 2.5 | 26.8 ± 2.1 | 0.44 ± 0.03 | 0.54 ± 0.12 |
| | High $\kappa$ | 2.43 ± 0.05 | 69.0 ± 1.0 | 29.9 ± 2.4 | 0.19 ± 0.03 | 0.31 ± 0.15 |

* flattening indices relative to *b* (1-*b*/*a*) and *c* (1-*c*/*a*) axis.

### 3.3. Impact of Charge Clustering on the Conformational Ensemble of the Model IDPs

Native MS was employed to assess the conformational properties of the three sets of IDPs. In this approach, the CSDs resulting from the nanoESI process reflect the overall compactness and relative amounts of the main conformers in the original solution [17,18,22]. Native-MS spectra obtained under non-denaturing conditions for the three variants of $N_{TAIL}$ (Figure 4a), NFM, and PNT4 (Figure S3) display multimodal CSDs, highlighting the heterogeneous conformational ensemble typical of IDPs. Multi-Gaussian deconvolution of the MS spectra of the wt IDPs (Figure 4b–d, central row) indicates that these variants exist in three main conformational components. For each component, the *SASA* and the corresponding *CI* ($CI^i_{SASA}$ defined in Equation (7)) were calculated as recently described [17]. The components were classified as "extended" ($CI^i_{SASA} < 0.25$), "intermediate" ($0.25 < CI^i_{SASA} < 0.75$), and "compact" ($CI^i_{SASA} > 0.75$) (Figure S4). In all the model IDPs, the three main conformational components observed in the wt IDPs also characterize the ensemble of low-$\kappa$ variants, but not that of high-$\kappa$ variants, which includes only the "intermediate" and "compact" components (Figure 4). These data indicate that charge clustering induces a loss of heterogeneity of conformational components, in favor of more compact states, in agreement with the increase in secondary structure observed by CD spectroscopy on our model proteins and also with results obtained on p27 by ion-mobility MS [8]. To gain a more comprehensive view of charge clustering effects on IDP conformation, we calculated the *CI* based on the average *SASA* ($\overline{CI}_{SASA}$), which weights the $CI^i_{SASA}$ (Figure S4) by the relative abundance (Figure S5) of the corresponding conformational component. The analysis of $\overline{CI}_{SASA}$ indicates that the protein compactness increases with $\kappa$ (Figure 3b). These results are in good agreement with those obtained by SEC, confirming the general trend of protein compaction at increasing $\kappa$ values and the peculiar behavior of PNT4.

In this latter case, the $\overline{CI}_{SASA}$ does not vary for low-$\kappa$ and wt variants, and it strongly increases just for high-$\kappa$ variants (Figure 3). Overall, the largest differences between MS and SEC results are obtained for the high-$\kappa$ variants. To rule out possible technical artifacts, control MS experiments were carried out, exposing high-$\kappa$ variants to acidic pH (formic acid 1%, pH 2.5) or higher ionic strength (ammonium acetate 200 mM). Indeed, electrostatic interactions are expected to be attenuated by the extensive protonation of all ionizable groups under very low pH conditions or by the charge shielding by salt ions. The resulting spectra show an increased amount of the components at high charge states, indicating that protein compaction is actually driven by in-solution electrostatic interactions (Figure S6).

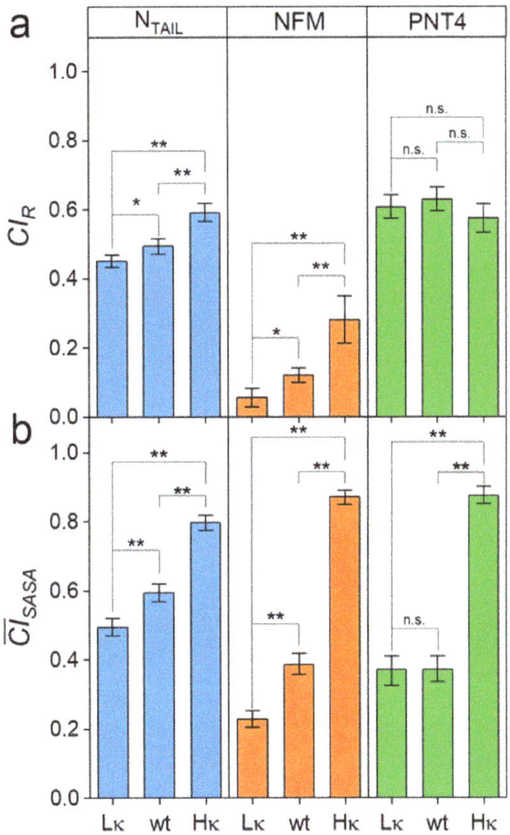


**Figure 3.** Compactness of the model IDPs. (**a**) *CI* derived from the $R_h$ ($CI_R$); (**b**) *CI* derived from the average *SASA* of the conformational ensemble ($\overline{CI}_{SASA}$) of $N_{TAIL}$, NFM and PNT4 variants (L$\kappa$: low-$\kappa$; wt: wild type; H$\kappa$: high-$\kappa$). Mean values of three independent measurements are shown with error bars indicating standard deviations. Statistical analyses were carried out using Welch's *t*-test, n.s.: not significant $p > 0.05$, *: $p < 0.05$, **: $p < 0.01$.

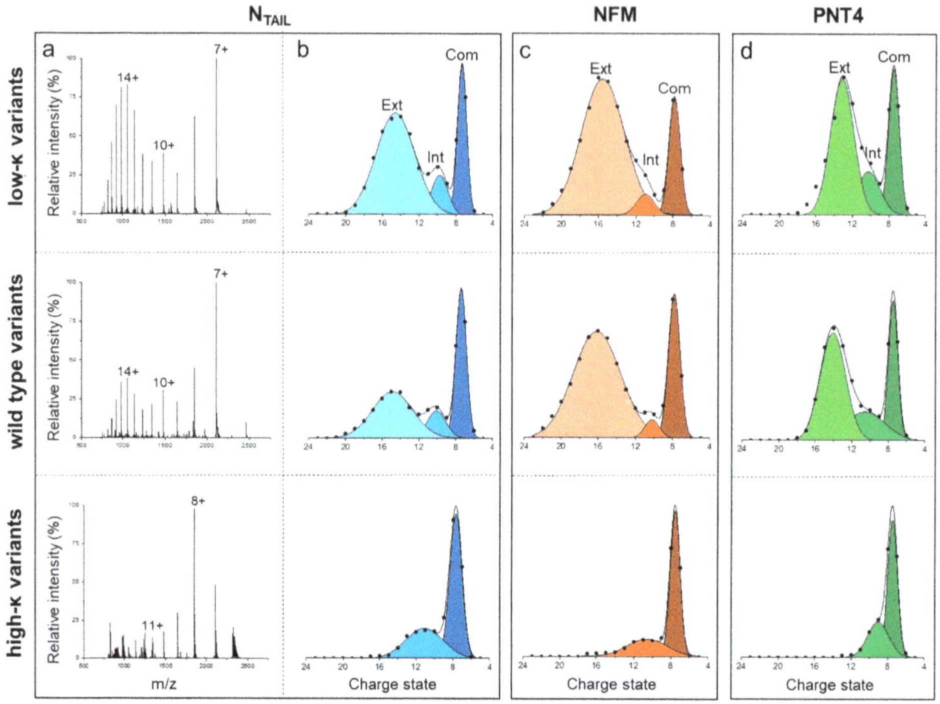


**Figure 4.** Native-MS analyses. (**a**) NanoESI-MS spectra of $N_{TAIL}$ variants acquired under non-denaturing conditions (50 mM ammonium acetate pH 7.0). The most intense signal of each peak-envelope is labeled by the corresponding charge state. (**b–d**) Multi-Gaussian deconvolution of the MS spectra obtained for $N_{TAIL}$ (**b**), NFM (**c**) and PNT4 (**d**), in the low-κ (upper row), wt (central row) and high-κ (bottom row) variants. Extended (Ext), intermediate (Int) and compact (Com) species are colored with different shades and labeled in the upper panels. MS spectra of NFM and PNT4 variants are reported in Figure S3.

### 3.4. Average Shape of the Model IDPs

The geometric, ensemble-averaged shape of each protein under investigation was predicted by combining the results for $R_h$ and *SASA*, as reported by Wu and co-authors [26]. The model was originally applied to approximate the shape of globular proteins to an ellipsoid, whose elongation (prolate-shaped) and/or flattening (oblate-shaped) describe the protein conformational transitions. The volume of the ellipsoid can be estimated from the experimentally derived $R_{h,}$ through Equation (8). By collating Equations (8) and (9), one obtains:

$$V = \frac{4}{3}\pi abc = \frac{4}{3}\pi(R_h - r_s)^3 \quad (12)$$

The average length of the *a*-axis was calculated through Equation (10), while the length of the *b* and *c* axes were obtained as described by Wu and co-authors [26].

The application of this model to the nine IDPs under investigation resulted in the values shown in Table 2 and Table S1 and represented in Figure 5, in which ellipsoid volumes and shapes are related to κ values. Comparing wt variants, NFM has the largest volume, followed by $N_{TAIL}$ and PNT4. Considering the effects induced by charge clustering, and therefore moving from the lowest towards the highest κ values, a clear linear and negative correlation can be observed in the case of NFM and $N_{TAIL}$ (overall reduction in volume of ~30% and ~25%, respectively) (Figure 5, Table 2). On the other hand, a neglectable effect was observed in the case of PNT4, for which the volume remains almost constant among the three variants, reflecting little variation of their $R_h$ value.

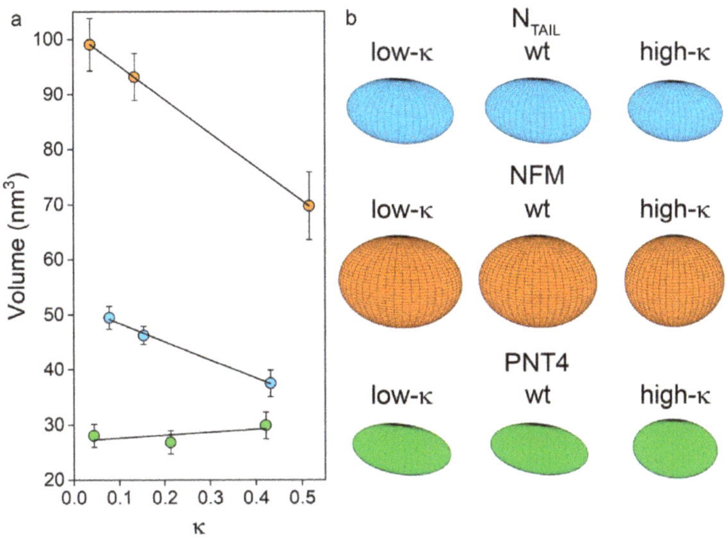


**Figure 5.** Relationship between ellipsoid volume and $\kappa$ values. (**a**) Regression of ellipsoid volume and $\kappa$ for $N_{TAIL}$ (light blue), NFM (orange) and PNT4 (green). The equation of trend lines are: y = −33.4 x + 51.7, $R^2$ = 0.987 for $N_{TAIL}$, y = −61.2 x + 101.4, $R^2$ = 0.998 for NFM and y = 4.3 x + 27.2, $R^2$ = 0.710 for PNT4. Mean values of three independent measurements are represented, with error bars indicating standard deviations. (**b**) Geometry of the model proteins as obtained by applying the ellipsoid model.

The shape of an ellipsoid depends on the length ratio of the *a*, *b*, and *c* axes, which in turn was derived from the experimental data of *SASA* (Table 2, Figure 3). The shape of an ellipsoid can be described by flattening indices (i.e., $f_b$ and $f_c$), which report the eccentricity of axial elliptic sections. These indices vary in the range [0; 1), where 0 corresponds to circle sections, while elliptic sections of increasing eccentricity are obtained as the index approaches 1 (Table 2). Comparing wt variants, NFM has the most spherical conformation, followed by $N_{TAIL}$ and PNT4 (which has the most prolate ensemble). As the $\kappa$ value increases, the spheroid reshaping reflects the trends observed by native MS and reported in terms of $\overline{CI}_{SASA}$ with $N_{TAIL}$ experiencing the smallest changes, and NFM and PNT4 the greatest ones (Table 2, Figure 3b). Indeed, on the basis of the flattening indices, the oblateness of $N_{TAIL}$ is not significantly affected by $\kappa$, while NFM and PNT4 tend to approach a spherical shape.

## 4. Discussion

Computational and experimental works have already shown that charge clustering causes an overall increase in protein conformational compactness [7–10]. However, few data are available in terms of quantitative description of various conformational components within a heterogeneous ensemble. Our work highlights that the conformational ensembles of IDPs can be experimentally dissected by native MS to capture components of different *SASA* and abundance. Our results show that charge segregation triggers a loss of heterogeneity of conformational components, in favor of more compact and intermediate states. At the same time, we used SEC to monitor the average $R_h$ and observed an overall shrinkage resulting from charge clusterization.

To integrate the two kinds of information resulting from MS and SEC, and to obtain coarse-grained information on the shape of IDP ensembles, we applied a recently published model, which approximates the shape of globular proteins to ellipsoids [26]. The applicability of this "ellipsoid model" to IDPs, herein explored for the first time, is supported by three observations: (i) the relationship between CSD and *SASA* was proved to be independent of the folded or disordered nature of the proteins [20,23,25]; (ii) the ellipsoid model was successfully applied to depict the conformational changes induced by denaturation [26]. The broad molecular mass range of globular proteins for which the model was shown to hold true (i.e., ~9 kDa to ~70 kDa) [26] argues for the applicability of this model to the three model IDPs herein investigated whose mass falls within this range.

This model substantially helped us in translating and rationalizing the conformational effects induced by charge clustering into the shrinkage and loss of oblateness of each IDP ensemble, while providing evidence of singular, protein-specific compaction behaviors. The observation that each ellipsoid undergoes volume and shape changes in a protein-specific manner argues for a multifactorial response to charge segregation. Although referring to a small set of proteins, and hence likely not directly generalizable to all IDPs, our data suggest that proline content, chain length, and secondary structure content are potentially all involved in the response to charge segregation.

Proline content appears to play a relevant role in modulating the average conformational properties of the ensemble. Indeed, the abundance of proline residues (PNT4 > $N_{TAIL}$ > NFM) promotes the ellipsoid oblateness in wt variants and counteracts the volume shrinking induced by $\kappa$. This is in line with the observations that proline disfavors $\alpha$- and $\beta$ structures [36,41] because of the conformational constraints imposed by its pyrrolidine ring [42] and the higher stiffness conferred by the preference towards the trans conformation of the Xaa-Pro peptide bonds [36]. Our data show that an increase from 0.7 to 5.2%, and then to 11% in proline content causes a significant reduction in the compaction response associated with charge clustering. Remarkably, the mean frequency of proline residues is $4.57 \pm 0.05$ and $8.11 \pm 0.63$ in databases of structured (i.e., PDB Select 25 [43]) and disordered proteins (i.e., DisProt [44,45]), respectively. In this scenario, proline residues would strongly hinder compaction driven by electrostatic interactions and reduce IDP propensity for induced folding. This indirectly supports the hypothesis that a high proline content is a compositional trait typical of "unfoldable IDPs", in contrast to IDPs prone to undergo induced folding, which instead exhibit, at least locally, compositional features nearly overlapping with those of folded proteins [2,36,46]. This hypothesis is corroborated by the analyses of large protein datasets [46].

Polypeptide length may also affect the ellipsoid oblateness in wt variants and counteract $\kappa$-induced volume shrinking. Indeed, PNT4 (the shortest protein under investigation) responds to increasing $\kappa$ with small volume changes and pronounced shape remodeling (from highly prolate ellipsoid to a more spherical geometry in the high-$\kappa$ variant), whereas NFM (the longest chain herein studied) shows the greatest volume excursion among variants. Unfortunately, it is difficult to disentangle the contribution of chain length and proline content to charge clustering responsiveness: indeed, the attempt at rationalizing our experimental data and at dissecting the effect of protein length is hampered by the fact that PNT4 has the highest fraction of proline residue and NFM the lowest among our model proteins, thus making the effect of size and proline content overlapping.

Finally, the role of secondary structure content appears controversial. For each of the three proteins, charge clustering triggers an increase in the $\alpha$-helical content. This could be related to the loss of heterogeneity among conformational components in favor of more compact and intermediate states observed by MS experiments. However, $\alpha$-helical content does not correlate with compaction in terms of $CI_R$ and volume shrinkage (e.g., PNT4). This behavior seems to be consistent with previous studies indicating that the propensity of IDPs for compactness, unlike that of globular proteins, is not correlated with $\alpha$-helical content [36,47]. Unfortunately, the paucity of data concerning the effects of charge segregation on IDP secondary structure makes it difficult to detail trends and deserves more extensive and systematic study.

Overall, our experimental data, complemented by the ellipsoid model, indicate that the extent of compaction and shape remodeling triggered by charge separation is modulated by multiple parameters that can concur, either individually or collectively, to counteract the expected response. Among the possible sequence features affecting IDP conformational responsiveness to charge clustering, the Lys/Arg and Asp/Glu ratio, recently reported by

Zeng and co-authors [48], is a plausible factor that deserves further investigation. Many additional ones are probably at play and still remain elusive, thereby preventing our ability to fully rationalize and model the conformational behavior of IDPs.

## 5. Conclusions

In summary, the effect of charge segregation on the conformational properties of IDP ensembles was studied by applying a mathematical model that integrates experimental data from two orthogonal techniques, i.e., SEC and native MS. This original approach was proved to be more informative compared to the single techniques, delineating a distinct and protein-specific compaction behavior in terms of the size and shape of each conformational ensemble. The structural information afforded by this approach relies on techniques that are more accessible compared to more elaborate techniques, such as ion mobility, NMR, or SAXS, usually applied for the study of IDP ensembles. Potentially transposable on a larger scale, i.e., by using available experimental datasets of $SASA$ and $R_h$, this approach could also serve as an asset to a more systematic study of the individual factors influencing the compaction behavior of IDPs triggered by charge segregation.

Although we do not pretend to extend our findings to all IDPs, our work identified proline content, protein size, and intrinsic content in ordered secondary structure as factors governing IDP responsiveness. We hope that the present study will stimulate and foster future studies aimed at a systematic analysis of the elements that contribute to the conformational behavior of IDPs in response to charge clustering. In addition to unraveling the physicochemical rules underlying the response to charge segregation, these elements may account for sequence-specific and biologically relevant properties of proteins, such as the propensity to undergo induced folding or to exhibit partner-mediated conformational polymorphism. The next challenge will be to decipher the hierarchy of elements governing IDP conformation and how they can be modeled to better predict IDP behavior.

**Supplementary Materials:** The following supporting information can be downloaded at: https://www.mdpi.com/article/10.3390/biom12040561/s1: Table S1: Size of semi-axes of ellipsoids; Figure S1: Amino acid sequences of model IDPs; Figure S2: Far-UV CD spectra of model IDPs; Figure S3: Native MS analyses of model IDPs; Figure S4: Compactness of conformational components of model IDPs studied by native MS; Figure S5: Relative abundance of conformational components of model IDPs studied by native MS; Figure S6: NanoESI-MS spectra of model IDPs under conditions affecting electrostatic interactions.

**Author Contributions:** Conceptualization, G.B., M.M., R.G., C.S. and S.B.; Formal analysis, C.S.; Investigation, G.B., A.B. and C.S.; Methodology, A.B., S.L., R.G., C.S. and S.B.; Supervision, C.S. and S.B; Visualization, M.M., C.S. and S.B.; Writing—original draft, S.B.; Writing—review and editing, G.B., M.M., A.B., S.L., R.G. and S.B. All authors have read and agreed to the published version of the manuscript.

**Funding:** This research received no external funding. This work was partly supported by Fondo di Ateneo of the University Milano-Bicocca (FAR 2020-ATE-0031, SB; FAR 2020-ATE-0319, CS), by the CNRS (S.L.). G.B. acknowledges support by the Italian national PhD program. M.M. benefits from an Assegno di Ricerca of the University Milano-Bicocca.

**Institutional Review Board Statement:** Not applicable.

**Informed Consent Statement:** Not applicable.

**Data Availability Statement:** Not applicable.

**Acknowledgments:** The authors thank Giulia Tedeschi for her contribution in the experimental set up.

**Conflicts of Interest:** The authors declare no conflict of interest.

## References


1. Uversky, V.N. Intrinsically Disordered Proteins and Their "Mysterious" (Meta) Physics. *Front. Phys.* **2019**, *7*, 10. [CrossRef]
2. Theillet, F.-X.; Kalmar, L.; Tompa, P.; Han, K.-H.; Selenko, P.; Dunker, A.K.; Daughdrill, G.W.; Uversky, V.N. The Alphabet of Intrinsic Disorder: I. Act like a Pro: On the Abundance and Roles of Proline Residues in Intrinsically Disordered Proteins. *Intrinsically Disord. Proteins* **2013**, *1*, e24360. [CrossRef] [PubMed]
3. Ruff, K.M. Predicting Conformational Properties of Intrinsically Disordered Proteins from Sequence. In *Intrinsically Disordered Proteins*; Kragelund, B.B., Skriver, K., Eds.; Methods in Molecular Biology; Springer US: New York, NY, USA, 2020; Volume 2141, pp. 347–389. ISSN 978-1-07-160523-3.
4. Mao, A.H.; Crick, S.L.; Vitalis, A.; Chicoine, C.L.; Pappu, R.V. Net Charge per Residue Modulates Conformational Ensembles of Intrinsically Disordered Proteins. *Proc. Natl. Acad. Sci. USA* **2010**, *107*, 8183–8188. [CrossRef] [PubMed]
5. Das, R.K.; Pappu, R.V. Conformations of Intrinsically Disordered Proteins Are Influenced by Linear Sequence Distributions of Oppositely Charged Residues. *Proc. Natl. Acad. Sci. USA* **2013**, *110*, 13392–13397. [CrossRef]
6. Sawle, L.; Ghosh, K. A Theoretical Method to Compute Sequence Dependent Configurational Properties in Charged Polymers and Proteins. *J. Chem. Phys.* **2015**, *143*, 085101. [CrossRef] [PubMed]
7. Das, R.K.; Huang, Y.; Phillips, A.H.; Kriwacki, R.W.; Pappu, R.V. Cryptic Sequence Features within the Disordered Protein p27 $^{Kip1}$ Regulate Cell Cycle Signaling. *Proc. Natl. Acad. Sci. USA* **2016**, *113*, 5616–5621. [CrossRef]
8. Beveridge, R.; Migas, L.G.; Das, R.K.; Pappu, R.V.; Kriwacki, R.W.; Barran, P.E. Ion Mobility Mass Spectrometry Uncovers the Impact of the Patterning of Oppositely Charged Residues on the Conformational Distributions of Intrinsically Disordered Proteins. *J. Am. Chem. Soc.* **2019**, *141*, 4908–4918. [CrossRef]
9. Tedeschi, G.; Salladini, E.; Santambrogio, C.; Grandori, R.; Longhi, S.; Brocca, S. Conformational Response to Charge Clustering in Synthetic Intrinsically Disordered Proteins. *Biochim. Et Biophys. Acta Gen. Subj.* **2018**, *1862*, 2204–2214. [CrossRef]
10. Sherry, K.P.; Das, R.K.; Pappu, R.V.; Barrick, D. Control of Transcriptional Activity by Design of Charge Patterning in the Intrinsically Disordered RAM Region of the Notch Receptor. *Proc. Natl. Acad. Sci. USA* **2017**, *114*, E9243–E9252. [CrossRef]
11. Uversky, V.N. What Does It Mean to Be Natively Unfolded?: Natively Unfolded Proteins. *Eur. J. Biochem.* **2002**, *269*, 2–12. [CrossRef]
12. Schramm, A.; Bignon, C.; Brocca, S.; Grandori, R.; Santambrogio, C.; Longhi, S. An Arsenal of Methods for the Experimental Characterization of Intrinsically Disordered Proteins—How to Choose and Combine Them? *Arch. Biochem. Biophys.* **2019**, *676*, 108055. [CrossRef] [PubMed]
13. Kodera, N.; Noshiro, D.; Dora, S.K.; Mori, T.; Habchi, J.; Blocquel, D.; Gruet, A.; Dosnon, M.; Salladini, E.; Bignon, C.; et al. Structural and Dynamics Analysis of Intrinsically Disordered Proteins by High-Speed Atomic Force Microscopy. *Nat. Nanotechnol.* **2021**, *16*, 181–189. [CrossRef] [PubMed]
14. Gomes, G.-N.W.; Krzeminski, M.; Namini, A.; Martin, E.W.; Mittag, T.; Head-Gordon, T.; Forman-Kay, J.D.; Gradinaru, C.C. Conformational Ensembles of an Intrinsically Disordered Protein Consistent with NMR, SAXS, and Single-Molecule FRET. *J. Am. Chem. Soc.* **2020**, *142*, 15697–15710. [CrossRef] [PubMed]
15. Müller-Späth, S.; Soranno, A.; Hirschfeld, V.; Hofmann, H.; Rüegger, S.; Reymond, L.; Nettels, D.; Schuler, B. Charge Interactions Can Dominate the Dimensions of Intrinsically Disordered Proteins. *Proc. Natl. Acad. Sci. USA* **2010**, *107*, 14609–14614. [CrossRef]
16. Corti, R.; Marrano, C.A.; Salerno, D.; Brocca, S.; Natalello, A.; Santambrogio, C.; Legname, G.; Mantegazza, F.; Grandori, R.; Cassina, V. Depicting Conformational Ensembles of α-Synuclein by Single Molecule Force Spectroscopy and Native Mass Spectroscopy. *Int. J. Mol. Sci.* **2019**, *20*, 5181. [CrossRef]
17. Santambrogio, C.; Natalello, A.; Brocca, S.; Ponzini, E.; Grandori, R. Conformational Characterization and Classification of Intrinsically Disordered Proteins by Native Mass Spectrometry and Charge-State Distribution Analysis. *Proteomics* **2019**, *19*, 1800060. [CrossRef]
18. Li, J.; Santambrogio, C.; Brocca, S.; Rossetti, G.; Carloni, P.; Grandori, R. Conformational Effects in Protein Electrospray-Ionization Mass Spectrometry: Native Protein Esi-Ms. *Mass Spec. Rev.* **2016**, *35*, 111–122. [CrossRef]
19. Konijnenberg, A.; van Dyck, J.F.; Kailing, L.L.; Sobott, F. Extending Native Mass Spectrometry Approaches to Integral Membrane Proteins. *Biol. Chem.* **2015**, *396*, 991–1002. [CrossRef]
20. Kaltashov, I.A.; Bobst, C.E.; Abzalimov, R.R. Mass Spectrometry-Based Methods to Study Protein Architecture and Dynamics: MS-Based Methods to Study Protein Architecture and Dynamics. *Protein Sci.* **2013**, *22*, 530–544. [CrossRef]
21. Mehmood, S.; Allison, T.M.; Robinson, C.V. Mass Spectrometry of Protein Complexes: From Origins to Applications. *Annu. Rev. Phys. Chem.* **2015**, *66*, 453–474. [CrossRef]
22. Natalello, A.; Santambrogio, C.; Grandori, R. Are Charge-State Distributions a Reliable Tool Describing Molecular Ensembles of Intrinsically Disordered Proteins by Native MS? *J. Am. Soc. Mass Spectrom.* **2017**, *28*, 21–28. [CrossRef] [PubMed]
23. Testa, L.; Brocca, S.; Grandori, R. Charge-Surface Correlation in Electrospray Ionization of Folded and Unfolded Proteins. *Anal. Chem.* **2011**, *83*, 6459–6463. [CrossRef] [PubMed]
24. Hall, Z.; Robinson, C.V. Do Charge State Signatures Guarantee Protein Conformations? *J. Am. Soc. Mass Spectrom.* **2012**, *23*, 1161–1168. [CrossRef] [PubMed]
25. Kaltashov, I.A.; Mohimen, A. Estimates of Protein Surface Areas in Solution by Electrospray Ionization Mass Spectrometry. *Anal. Chem.* **2005**, *77*, 5370–5379. [CrossRef] [PubMed]

26. Wu, H.; Zhang, R.; Zhang, W.; Hong, J.; Xiang, Y.; Xu, W. Rapid 3-Dimensional Shape Determination of Globular Proteins by Mobility Capillary Electrophoresis and Native Mass Spectrometry. *Chem. Sci.* **2020**, *11*, 4758–4765. [CrossRef]
27. Longhi, S.; Receveur-Bréchot, V.; Karlin, D.; Johansson, K.; Darbon, H.; Bhella, D.; Yeo, R.; Finet, S.; Canard, B. The C-Terminal Domain of the Measles Virus Nucleoprotein Is Intrinsically Disordered and Folds upon Binding to the C-Terminal Moiety of the Phosphoprotein. *J. Biol. Chem.* **2003**, *278*, 18638–18648. [CrossRef]
28. Habchi, J.; Mamelli, L.; Darbon, H.; Longhi, S. Structural Disorder within Henipavirus Nucleoprotein and Phosphoprotein: From Predictions to Experimental Assessment. *PLoS ONE* **2010**, *5*, e11684. [CrossRef]
29. Yuan, A.; Rao, M.V.; Veeranna; Nixon, R.A. Neurofilaments and Neurofilament Proteins in Health and Disease. *Cold Spring Harb. Perspect. Biol.* **2017**, *9*, a018309. [CrossRef]
30. Herrmann, H.; Aebi, U. Intermediate Filaments: Structure and Assembly. *Cold Spring Harb. Perspect. Biol.* **2016**, *8*, a018242. [CrossRef]
31. Studier, F.W. Protein Production by Auto-Induction in High-Density Shaking Cultures. *Protein Expr. Purif.* **2005**, *41*, 207–234. [CrossRef]
32. Holehouse, A.S.; Das, R.K.; Ahad, J.N.; Richardson, M.O.G.; Pappu, R.V. CIDER: Resources to Analyze Sequence-Ensemble Relationships of Intrinsically Disordered Proteins. *Biophys. J.* **2017**, *112*, 16–21. [CrossRef] [PubMed]
33. Dosztanyi, Z.; Csizmok, V.; Tompa, P.; Simon, I. IUPred: Web Server for the Prediction of Intrinsically Unstructured Regions of Proteins Based on Estimated Energy Content. *Bioinformatics* **2005**, *21*, 3433–3434. [CrossRef] [PubMed]
34. Micsonai, A.; Wien, F.; Bulyáki, É.; Kun, J.; Moussong, É.; Lee, Y.-H.; Goto, Y.; Réfrégiers, M.; Kardos, J. BeStSel: A Web Server for Accurate Protein Secondary Structure Prediction and Fold Recognition from the Circular Dichroism Spectra. *Nucleic Acids Res.* **2018**, *46*, W315–W322. [CrossRef] [PubMed]
35. Irvine, G.B. Determination of Molecular Size by Size-Exclusion Chromatography (Gel Filtration). *Curr. Protoc. Cell Biol.* **2000**, *6*, 5.5.1–5.5.16. [CrossRef]
36. Marsh, J.A.; Forman-Kay, J.D. Sequence Determinants of Compaction in Intrinsically Disordered Proteins. *Biophys. J.* **2010**, *98*, 2383–2390. [CrossRef]
37. Brocca, S.; Testa, L.; Sobott, F.; Šamalikova, M.; Natalello, A.; Papaleo, E.; Lotti, M.; De Gioia, L.; Doglia, S.M.; Alberghina, L.; et al. Compaction Properties of an Intrinsically Disordered Protein: Sic1 and Its Kinase-Inhibitor Domain. *Biophys. J.* **2011**, *100*, 2243–2252. [CrossRef]
38. Sinha, S.K.; Chakraborty, S.; Bandyopadhyay, S. Thickness of the Hydration Layer of a Protein from Molecular Dynamics Simulation. *J. Phys. Chem. B* **2008**, *112*, 8203–8209. [CrossRef]
39. Pal, S.; Bandyopadhyay, S. Effects of Protein Conformational Flexibilities and Electrostatic Interactions on the Low-Frequency Vibrational Spectrum of Hydration Water. *J. Phys. Chem. B* **2013**, *117*, 5848–5856. [CrossRef]
40. Rath, A.; Davidson, A.R.; Deber, C.M. The Structure of ?Unstructured? Regions in Peptides and Proteins: Role of the Polyproline II Helix in Protein Folding and Recognition. *Biopolymers* **2005**, *80*, 179–185. [CrossRef]
41. Dunker, A.K.; Lawson, J.D.; Brown, C.J.; Williams, R.M.; Romero, P.; Oh, J.S.; Oldfield, C.J.; Campen, A.M.; Ratliff, C.M.; Hipps, K.W.; et al. Intrinsically Disordered Protein. *J. Mol. Graph. Model.* **2001**, *19*, 26–59. [CrossRef]
42. Adzhubei, A.A.; Sternberg, M.J.E. Left-Handed Polyproline II Helices Commonly Occur in Globular Proteins. *J. Mol. Biol.* **1993**, *229*, 472–493. [CrossRef] [PubMed]
43. Berman, H.M. The Protein Data Bank. *Nucleic Acids Res.* **2000**, *28*, 235–242. [CrossRef] [PubMed]
44. Sickmeier, M.; Hamilton, J.A.; LeGall, T.; Vacic, V.; Cortese, M.S.; Tantos, A.; Szabo, B.; Tompa, P.; Chen, J.; Uversky, V.N.; et al. DisProt: The Database of Disordered Proteins. *Nucleic Acids Res.* **2007**, *35*, D786–D793. [CrossRef] [PubMed]
45. Piovesan, D.; Tabaro, F.; Mičetić, I.; Necci, M.; Quaglia, F.; Oldfield, C.J.; Aspromonte, M.C.; Davey, N.E.; Davidović, R.; Dosztányi, Z.; et al. DisProt 7.0: A Major Update of the Database of Disordered Proteins. *Nucleic Acids Res* **2017**, *45*, D219–D227. [CrossRef] [PubMed]
46. Vymětal, J.; Vondrášek, J.; Hlouchová, K. Sequence Versus Composition: What Prescribes IDP Biophysical Properties? *Entropy* **2019**, *21*, 654. [CrossRef]
47. Blocquel, D.; Habchi, J.; Gruet, A.; Blangy, S.; Longhi, S. Compaction and Binding Properties of the Intrinsically Disordered C-Terminal Domain of Henipavirus Nucleoprotein as Unveiled by Deletion Studies. *Mol. BioSyst.* **2012**, *8*, 392–410. [CrossRef]
48. Zeng, X.; Ruff, K.M.; Pappu, R.V. Competing Interactions Give Rise to Two-State Behavior and Switch-like Transitions in Charge-Rich Intrinsically Disordered Proteins. *bioRxiv* **2022**. [CrossRef]

biomolecules

MDPI

Review

# Topological Considerations in Biomolecular Condensation

Debapriya Das and Ashok A. Deniz *

Department of Integrative Structural and Computational Biology, The Scripps Research Institute, 10550 N. Torrey Pines Rd., La Jolla, CA 92037, USA
* Correspondence: deniz@scripps.edu

**Abstract:** Biomolecular condensation and phase separation are increasingly understood to play crucial roles in cellular compartmentalization and spatiotemporal regulation of cell machinery implicated in function and pathology. A key aspect of current research is to gain insight into the underlying physical mechanisms of these processes. Accordingly, concepts of soft matter and polymer physics, the thermodynamics of mixing, and material science have been utilized for understanding condensation mechanisms of multivalent macromolecules resulting in viscoelastic mesoscopic supramolecular assemblies. Here, we focus on two topological concepts that have recently been providing key mechanistic understanding in the field. First, we will discuss how percolation provides a network-topology-related framework that offers an interesting paradigm to understand the complex networking of dense 'connected' condensate structures and, therefore, their phase behavior. Second, we will discuss the idea of entanglement as another topological concept that has deep roots in polymer physics and important implications for biomolecular condensates. We will first review some historical developments and fundamentals of these concepts, then we will discuss current advancements and recent examples. Our discussion ends with a few open questions and the challenges to address them, hinting at unveiling fresh possibilities for the modification of existing knowledge as well as the development of new concepts relevant to condensate science.

**Keywords:** intrinsically disordered proteins; polymer physics; percolation; entanglement; RNA; topology; polymer rheology; biomolecular condensates

**Citation:** Das, D.; Deniz, A.A. Topological Considerations in Biomolecular Condensation. *Biomolecules* **2023**, *13*, 151. https://doi.org/10.3390/biom13010151

Academic Editors: Prakash Kulkarni, Stefania Brocca, Keith Dunker and Sonia Longhi

Received: 1 December 2022
Revised: 7 January 2023
Accepted: 9 January 2023
Published: 11 January 2023



## 1. Introduction

Biomolecular condensation via phase separation (PS) of proteins and nucleic acids is believed to play a pivotal role in cellular compartmentalization and spatiotemporal regulation of cellular biochemistry, which are associated with an array of essential biological functions and debilitating neurodegenerative dysfunctions [1–11]. Intracellular biomolecular phase-separated structures, also known as membrane-less organelles, are viscoelastic dynamic mesoscopic supramolecular assemblies with various cluster size distributions within the biological milieu [5,12–15]. A key driving force of PS is the multivalency of different interaction domains comprising the condensates [5,12,16], resulting in the formation of dense noncovalent networks/crosslinks within the system. The traditional mean-field Flory–Huggins theory of homopolymer in solution illustrates the thermodynamics and the physical understanding of the phase transition [17–22]. The derived Flory–Huggins interaction parameter $(\chi)$ allows us to quantify the intricate balance between chain–chain, chain–solvent and solvent–solvent interactions, thereby dictating the phase separation propensity of the system, depending on the solvent quality. Although the mean-field model has been widely utilized to understand the characteristics of phase behavior, it may not offer a good approximation to understand the complex nature of phase transitions involving larger macromolecules such as proteins and RNA, which are believed to be major drivers of intracellular bimolecular condensation [5,23–25]. The phase separation of these molecules possessing multiple intrinsically disordered regions (IDRs) and/or folded domains involves a mosaic of sequence-dependent, structurally, and conformationally heterogeneous dynamic multivalent interactions. In order to gain insight into the mechanistic understanding of these processes, the theory of linear or branched associative polymers has been proposed using stickers-and-spacers network architecture [5,16,26–31]. In this framework, depending on the system-specific sticker–sticker interaction, the physical crosslinking amongst them leads to two types of transitions: (1) phase separation or density transition above a critical protein concentration ($C_{sat}$), forming a polymer-rich dense phase ($C_{dense}$) cohabiting with a polymer-deficient dispersed phase ($C_{dil}$); (2) percolation, which is a topology-related geometric transition that depends on the connectivity probability and leads to the formation of system-spanning clusters [23,28,29,31,32] (Figure 1). Phase separation and percolation may be coupled or decoupled depending on the system and other specific parameters. Percolation theories, which were developed in early work in connection with graph and network theories, can be employed to better understand the intricate details of biomolecular condensation [23,32–37]. Another captivating topological concept is polymer entanglement, which has recently been implicated in biomolecular condensates and their rheological properties [38–40] (Figure 1). In this review, we discuss the underlying physical origins of percolation and polymer entanglement and their relevance in the context of PS and gelation of associative biopolymers from both a historical and scientific standpoint. We note that the application of these exciting concepts to biomolecular condensates is at a relatively early stage, and we discuss current limitations and future directions in the final section of this article. We envision that the application and invocation of topology-network-related theories may shed light on different aspects of biomolecular condensation within the cellular milieu.

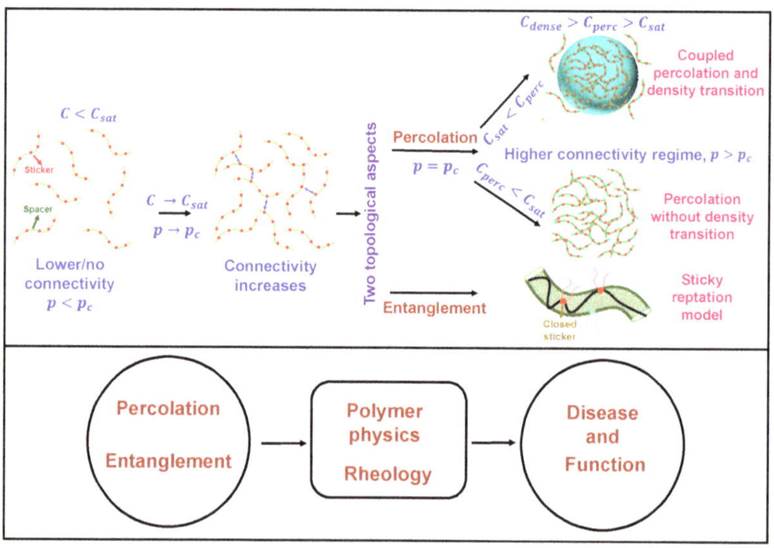


**Figure 1.** General overview of two topological concepts discussed here, namely percolation and entanglement in the context of biomolecular condensation. In the upper panel (on left side) the red circles describe the stickers and the green circles constitute the spacer. The blue two-sided arrows describe sticker-sticker interactions. On right side, the droplet has been shown by a light blue sphere with color coded intersticker interactions. In the 'sticky reptation model' the polymer chain confined in the tube (olive) is shown by a black strand. The red circles with strand describe the 'closed stickers' and cyan circles describes the next available polymer (sticker) chain for reptation depending upon the chain diffusion pattern. This concept has been discussed in detail in later sections of this review.

## 2. Percolation Physics

### 2.1. Percolation Physics: A Historical and Scientific Overview

***Theoretical foundation and different models.*** For a simple visualization, we can consider percolation as a simplified probabilistic model for a porous rock in which the interior of the rock is depicted to be a random maze through which fluid can flow. In this context, an important question to ask is which part of the rock will become wet after being submerged in the fluid. Mathematically, the porous material can be depicted by a random graph with vertices and edges, as first described by Broadbent and Hammersley in the 1950s. They first introduced the term 'percolation' in the context of their novel mathematical problems concerning the flow of a liquid through a random maze, hence the name 'percolation' [35,36]. A percolation model is defined as a collection of points with a spatial distribution in which certain pairs are shown to be connected. Depending on the model, the nature of connectedness is random, which suggests that each of these connected structures has a certain statistical probability of occurring. We focus here on the bond percolation model, which can be most intuitively mapped with biomolecular condensates (Figure 2), and note the existence of other models such as site, continuous, and hybrid percolation models. Before we proceed further, to motivate the following discussion of percolation theory through a more concrete link between protein/RNA condensates/networks and lattice percolation models, we point to Figure 2. Here, the reader can see a simple conceptual mapping of a reversibly crosslinked condensate-forming macromolecular system as commonly depicted in the field (Figure 2A; e.g., of disordered proteins or RNA) and bond percolation on a 2-D square lattice (Figure 2C), via an intermediate map of Figure 2B. In a related point, we also would like to emphasize that although fluid flow was used to conceptually introduce percolation models and its historical background, the relevant concept for biomolecular condensates is percolation through bonds, as depicted in Figure 2A,B.

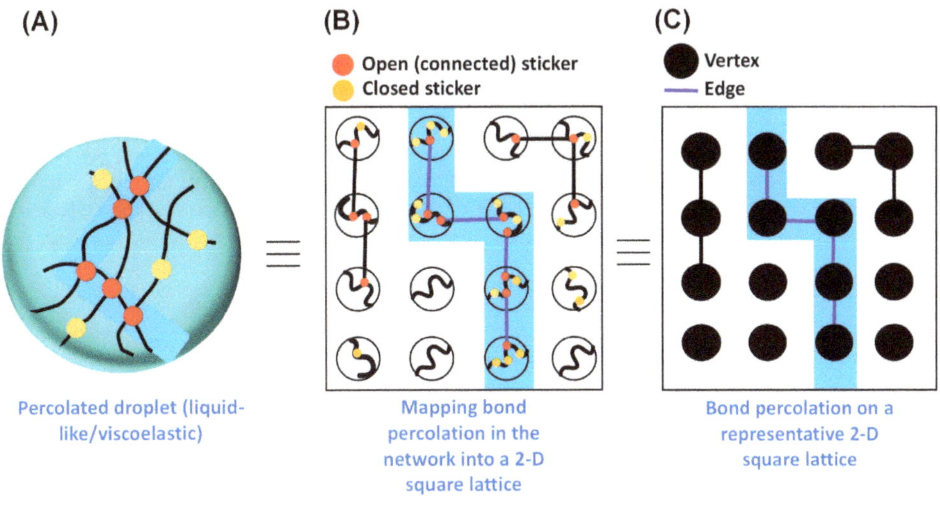


**Figure 2.** Mapping the bond percolation of a representative 2-D square lattice onto the physical crosslinks in the percolated network within the phase separating biomacromolecules. (**A**) Schematic representation of physical crosslinking formed by sticker–sticker interactions within the percolated droplet. Polymer chains are shown in black. Red circles define the open (connected) stickers, and yellow circles define the closed stickers. The percolating cluster (open path) is shown by the light blue shade. (**B**) Conceptualization of bond percolation in the context of percolated droplet. (**C**) Bond percolation on a representative 2-D square lattice, as proposed by Broadbent and Hammersley in the context of an arbitrary linear graph with vertices and edges. The color code remains the same as (**A**).

Broadbent and Hammersley originally introduced the bond percolation model in the context of graph theory [36]. According to this model, in an arbitrary linear graph, a certain pair of vertices or points forming an edge in the graph are connected with probability $p$ independent of the connectivity of other pairs, considering no edge formation between the pairs without linkages (Figure 2C) [35,40,41]. Figure 2C shows a general model of bond percolation on a two-dimensional square lattice in which the points of the model represent the lattice sites and each closest neighboring pair is linked with probability $p$. The points possess fixed locations, and the linkage (bonds) formation can occur randomly, and the properties of the model are determined by the topology of the network. Therefore, in the case of a square lattice, in the bond percolation model, lattice edges are the relevant entities, and the substance (fluid) seeps through the adjacent bonds. This idea may directly be mapped onto the concept of passage of liquid through the open path (physical crosslinks) formed by the sticker–sticker interaction, as depicted by Figure 2A,B. In a broader sense, if we recall the concept of percolation through a porous rock, the open edges allow the fluid to pass through, with the closed edges blocking the percolation.

In percolation theory, the phrase 'percolation threshold', denoted as $p_c$ defines the (connectivity) probability that 'marks the birth' of an infinitely connected cluster. In other words, it measures how likely a particular point is to be a part of an infinite cluster [35,40,42]. In the context of fluids, this is the probability that a fluid introduced at the point will percolate away through the 'open paths' within the system perpetually (Figure 3A–C). The cluster size increases as the number of linkages increases, and at a given critical density of linkages, it crosses the percolation threshold, and the extent of cluster size increment may become infinite, at which point the system is considered to be percolating. When $p < p_c$, the system lacks infinitely connected components, whereas above $p_c$, the system will possess at least one such cluster (Figure 3A–C). Therefore, $p_c$ marks the critical transition point from a low (local) to a dense (global) connectivity regime.

Next, we will briefly discuss the analytical treatment of the percolation problem in one dimension as a simple example.

***A simple example–percolation problem in one dimension*** Consider a one-dimensional lattice with an infinite number of equally spaced nodes [41] (Figure 3D). The probability of bonds between adjacent sites is denoted as $p$ (open) giving rise to a $(1-p)$ probability of no bond. The question is as follows: what is the critical value of the percolation threshold $p_c$ or the bond probability at which an infinite cluster arrives for the first time?

Let us denote $\prod(p, L)$ as the probability of percolation at $p$ for a lattice of linear size $L$. Therefore, in line with our previous discussion, two scenarios can emerge, which are as follows,

$$\lim_{L\to\infty} \prod(p, L) = \begin{cases} 0 \; for \; p < p_c \\ 1 \; for \; p \geq p_c \end{cases}$$

In the case of the 1-D finite lattice of size $L$, all nodes are occupied with probability $\prod(p, L) = p^{L-1}$, as the events of occupation are independent of each other, and it gives rise to the following,

$$\lim_{L\to\infty} \prod(p, L) = \lim_{L\to\infty} p^{L-1} = \begin{cases} 0 \; for \; p < 1 \\ 1 \; for \; p = 1 \end{cases}$$

which implies $p_c = 1$.

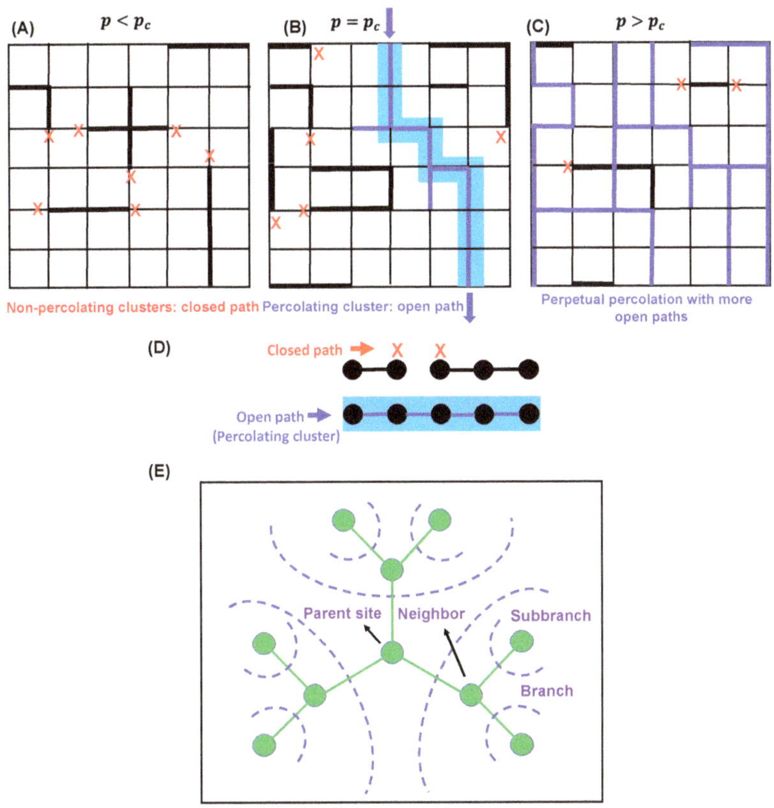


**Figure 3.** Three lattices with different percolation thresholds ($p_c$) conditions: (**A**) $p < p_c$, (**B**) $p = p_c$, and (**C**). Percolation occurs when $p = p_c$. In (**B**), the bond connectivity is shown by blue lines, and the percolating cluster is shown in light blue shade. Red crosses describe closed paths. In (**C**), the bond connectivity is shown by blue lines, red crosses describe closed paths, and the clusters are not shown for simplicity. (**D**) Bond percolation in one dimension. The vertices are shown by black circles, and the edges (bond connectivity) are shown by black lines. Red cross describes the closed paths, and blue line describes the open paths for percolation. Percolating cluster is shown by light blue shade. (**E**) Typical sketch of a Bethe lattice with coordination number 3. Parent site, branches, and subbranches are shown by olive circles. The blue dashed lines depict possible directions of branching.

This solution is in line with the idea that for a 1-D lattice, percolating cluster formation can occur with all adjacent sites forming bonds only when $p_c = 1$, because even a single 'no bond' situation would block a cluster to percolate through the lattice [41]. The one-dimensional percolation problem demonstrates several traits present in higher-dimensional systems, and it furnishes a clear starting point to understand the fields of scaling concepts, phase transition, renormalization group theories, and so forth [35,41]. Therefore, the mathematical treatment of a simple one-dimensional problem may aid in delving deeper into the more complex percolation problems in higher dimensions.

***Percolation: a topology-driven phenomenon*** A percolation process describes the transition from an initial structure comprising a set of isolated objects to a system with an inter-connected structure as a function of increasing density. In any geometric structure or field, the presence of points of two opposing edges or planes belonging to the same connected component indicates that the system or structure has the potential to undergo percolation. As the connectivity increases, the intrusion of the fluid approaches completion. Therefore, it is a topology-driven phenomenon, as the addition of more 'connectivity' to the structure modifies the underlying topology [42–47]. We note that the percolation threshold depends on different parameters of the model, including the lattice type in different dimensions. For instance, for the bond percolation model, $p_c = 1$ for a one-dimensional lattice, as noted above, whereas $p_c = 0.5$ for a two-dimensional square lattice [40,44,45]. In general, the percolation threshold decreases as the coordination number of the lattice increases in each dimension. Increasing functionality would progressively introduce more complexity in the percolating behavior of the system [40]. Next, we will discuss the relevance of percolation transitions in the context of gelation and phase separation.

*2.2. Percolation Approach: Gelation and Phase Separation*

***General concepts of gelation*** To put it simply, a polymerization process is initiated starting with a liquid containing monomers with higher reaction functionality $(f)$. This eventually results in a transition from liquid to solid (gel). This idea was first described by the classical model of gelation developed by Flory and Stockmayer [20,48]. This is a mean-field approach based on several assumptions, such as not considering the possibility of intramolecular linkage formation (cyclization) and treating all unreacted functional groups as equally active at any stage of the reaction. According to their theory, gelation behavior is observed in systems with higher functionality and with a possibility of unrestricted growth capability resulting in the formation of indefinitely large three-dimensional molecules. Flory's theory furnishes a general 'critical' value $\alpha_c$ for the formation of this infinitely large network, which is as follows,

$$\alpha_c = \frac{1}{f-1} \quad (1)$$

where $f$ is the functionality of the branch units and $\alpha$ is the probability of the chain branching as opposed to chain termination, depending on various parameters such as the ratios of the reactants and the reaction capability of the functional groups. Approximately, we can consider the branching probability $\alpha$ to be equivalent (not necessarily equal) to the extent of the reaction, related to $p$ [49]. Concisely, when the degree of branching and crosslinking events exceeds a critical value, three-dimensional polymerization causes gelation due to network formation to an indefinite extent. Following that, we will direct our efforts toward understanding gelation in light of the percolation approach.

***Gelation: a bond percolation transition*** Flory–Stockmayer theory is the cornerstone of percolation models undergoing a transition from a state of local connectedness to one in which the connections extend indefinitely. From this perspective, gelation can be described as the connectivity transition from sol to gel that can be modeled by bond percolation theory, such that all sites of the lattice are occupied by monomers [5,16,29,33,40]. The extent of the networking increases as a function of crosslinking from 0 to 1. When the system reaches the percolation threshold or the gel point, it undergoes a connectivity transition [48]. In this case, we must consider two scenarios: (1) when the system is slightly below $p_c$, it is a polydisperse mixture of branched polymers; (2) when the system is slightly above $p_c$, the network is not fully developed, and only one structure seeps (percolates) through the entire system, as discussed in depth by Rubinstein and Colby [40]. The sol fraction $(P_{sol})$ is the fraction of monomers that are part of the finite-size polymers, and the gel fraction $\left(P_{gel}\right)$ is the fraction of all the monomers that belong to the gel network. From these ideas, we can depict the following conditions as shown by Equation (2a–c) [40].

$$P_{sol} + P_{gel} = 1 \quad (2a)$$

$$P_{sol} = 1,\ P_{gel} = 0,\ p \leq p_c \quad (2b)$$

$$P_{sol} < 1,\ P_{gel} > 0,\ p > p_c \quad (2c)$$

As previously discussed, percolation effects are dependent on the lattice type and functionality; in this context, it is worthwhile to discuss the mean-field gelation model, which corresponds to bond percolation on a Bethe lattice (Figure 3E) [40,41]. The simplest random bond percolation model on a Bethe lattice directly considers the functionality of the monomers by adopting this functionality for the lattice which, unlike a simple cubic lattice model, assumes the absence of any intermolecular crosslinking and is convenient for analytical treatment of the model. Consistent with Flory's equation and the analytical treatment of the one-dimensional percolation problem, the critical occupation probability or the gel point for the bond percolation model of an 'infinite-dimensional' Bethe lattice is given by the following equation [40],

$$p_c = \frac{1}{f-1} \tag{3}$$

where each site possesses $f$ number of neighboring sites; therefore, each branch gives rise to $f-1$ subbranches. Here, below the gel point, only finite-size branched clusters exist, and above the gel point, in addition to that, at least one infinite polymer exists. Figure 3E shows a sketch of a typical Bethe lattice with functionality (coordination number) 3, with a large number of independent branching probabilities $(p)$ starting from the parent site. Interestingly, a distinct feature of percolation on a Bethe lattice is the presence of a significant number of infinite polymers in the system just above the gel point, as opposed to the regular lattice in which only one infinite polymer exists above the gel point [40,41]. Next, we will shed light on understanding the interplay between percolation and sol–gel transitions in the context of biomolecular phase condensation.

***Interplay between percolation and biomolecular condensation*** Macromolecular systems such as proteins can be considered in the framework of sticker–spacer-based associative polymer models, founded on an equilibrium theory originally developed by Semenov, Dobrynin, and Rubinstein in the context of reversible network formations in solutions of polymers with many associating groups, namely stickers (which are generally the functional monomeric units, charged moiety, or hydrophobic group) per chain [5,16,27–29,31]. As opposed to the mean-field assumption, this model takes into account the specific pairwise attractive interaction between stickers. The spacers are considered to be noninteracting and, thus, behave as ideal chains that are interspersed between stickers, without much influence in the formation of physical crosslinks but contribute toward the excluded volume effects implicated in the overall association of the polymers [5,16,50]. The reversible intersticker interactions give rise to two physical events: (1) intermolecular clustering and gelation transition, and (2) phase separation as a function of increasing intersticker interaction potential. The phase behavior of associative polymers is theoretically based on the classical gelation theory proposed by Flory and Stockmayer and the theory of polymer solutions developed by Flory [21,48,49]. Because of the reversible nature of the crosslink formation, a specific polymer chain can reversibly be a part of a sol phase (finite cluster) or gel phase (infinite cluster along with finite clusters), as opposed to chemical gelation in which the bonds are not reversible. Associative polymer models with sticker–spacer paradigms offer a useful platform for elucidating the physical attributes of biomolecular condensation.

During biomolecular condensation, a percolation transition occurs when protein and/or nucleic acid molecules (such as RNA) are topologically connected into a system such that the connectivity percolates throughout the system, giving rise to a droplet spanning network matrix (Figure 2) [51]. The critical concentration $(C_{perc})$ for connectivity transition or the percolation threshold depends on the types and valence behavior of the stickers, sticker–sticker interaction potential, and spacer-mediated solvation effects [5,33]. When $C_{perc} < C_{sat}$, a percolation transition can occur without phase separation, forming an 'infinite polymer' or gel depending on the degree of reversible crosslink formation. Interestingly, when $C_{sat} < C_{perc} < C_{dense}$, the polymer solution should be able to undergo phase separation coupled with percolation (PSCP), leading to the development of a droplet-spanning percolated matrix [16,24,51]. As theorized by Semenov and Rubinstein [28] and also discussed by Choi et al. [5], for a system comprising associative polymers in a solvent with $n$ number of self-interacting stickers, the percolation threshold of the system is given by the following equation,

$$C_{perc} = \frac{1}{\lambda n^2} \quad (4)$$

Here, the stickers are considered to be phantom chains. $n$ = apparent valence of stickers, and $\lambda = v_b e^{-\left(\frac{\varepsilon}{k_B T}\right)}$, where $v_b$ = intersticker crosslinking volume, $\varepsilon$ = effective interaction energy between the stickers ($\varepsilon \leq 0$), $k_B$ = Boltzmann constant, and $T$ = temperature of the system.

Graph-based Monte Carlo simulations carried out by Choi et al. described the concept of phase-separation-aided bond percolation (PSBP) using the sticker–spacer framework of associative polymers [33]. The mean-field model ignores the effect of growing network connectivity and forming clusters below $p_c$. These clusters form as a result of pairwise sticker interactions between different polymers involved in physical crosslinking, as well as the bond cooperativity effect, which deals with the effective intersticker interaction influenced by the previously generated interaction and, thus, can alter the percolation behavior of the system. Overall, these concepts are in line with the classical gelation model which was pictured by Flory and Stockmayer and the Flory–Huggins model of polymer solutions, with the correction for the mean-field approach. In light of this, associative polymer theories with the inclusion of the percolation approach can describe sticker–spacer-based macromolecular phase separation-assisted bond percolation (PSBP).

### 2.3. Current Implementation and Biological Implications

The concept of percolation effects has recently been applied to the area of biomolecular condensates and assemblies, which has helped us to delve deeper into the mechanistic characteristics of PS, liquid–solid transition (gelation), percolation effects on phase transitions, nano- and mesoscale cluster formation, and so forth. In 2012, Li et al. reported on the phase behavior of systems in which phase transitions are fueled by multivalent interactions between poly-SH3 and proline-rich (poly-PRM) molecules [24]. They showed that these interactions are driven by the unique association ability of the SH3$_n$-PRM$_n$ molecules, implying a valence-specific percolation threshold for phase separation to occur. Further evidence indicated that macroscopic phase separation is thermodynamically coupled to a sol–gel transition within the droplet state, which is an example of PSCP that eventually leads to the formation of gel, as previously discussed. They noted that, as well as being generic features of multivalent macromolecular biological systems, these phenomena with sharp phase transitions could impact the cellular signaling pathways or contribute to the structural and functional ability of cellular components [24]. By adopting this synthetic system, SH3$_n$-PRM$_n$, Harmon et al. performed Monte Carlo simulations using a coarse-grained lattice model with different valencies to demonstrate the effect of intrinsically disordered linkers, namely Flory random coil linkers (FRC) and self-avoiding random coil linkers (SARC), and their effective solvation volume on gelation with and without phase separation [16,50]. Their studies revealed that at bulk concentrations below the Flory–Stockmayer limit, gelation with phase separation results in positive global cooperativity and leads to the generation of a percolated network. On the other hand, gelation without phase separation is preferred in systems with zero or negative global cooperativity, and the transition takes place at or above the Flory–Stockmayer limit. The authors speculated that cell-signaling regulation is primarily modulated by gelation-driven phase separation of multivalent proteins, with specific interaction motifs or linear domains leading to the formation of percolated networks based on the theory of associative polymers. A few years ago, Franzmann et al. investigated the pH-regulated PS of Sup35 and subsequent solidification into a porous mesh-like polymer network or crosslinked protein gel driven by the intrinsically disordered prion domain [52]. This phenomenon is consistent with the idea of gelation driven by phase separation, but the complex mechanism underlying the formation of crosslinked meshwork remains elusive. We can speculate that the percolation transition might play an important role in the conversion from liquid-like droplets to reversible permeable gel or gel-like condensates. This intracellular phase separation coupled with gelation offers a beneficial way for cells to respond to sudden environmental stress.

Interestingly, recent work by Kar et al. has shown that subsaturated solutions of FET family proteins contain a variety of nanoscale clusters, even though micron-scale phase separation is not seen in solutions below an effective $C_{sat}$ [32]. In general, a subsaturated solution is expected to contain mostly dispersed monomers, along with very few small clusters at a time, and the phase separation is governed by the unique Flory interaction parameter $\chi$ [21]. Thus, interestingly, their results do not reconcile with this conventional notion. The authors discuss how the results are instead consistent with the presence of multiple relevant energy scales in the system, including one that relates to percolation clustering. The generation of smaller networks below the gel point can be understood from the viewpoint of percolation theory, in which below the gel point, the connectivity is low, thereby forming percolation clusters (termed pre-percolation clusters in the work of Kar et al.). Above the gel point, percolation commences, and the size distribution of the clusters increases as a function of increasing connectivity. The authors also report the results of simulations that also are consistent with a model involving percolation clustering. Notably, Li et al. had previously reported the presence of mesoscale percolation clusters below $C_{sat}$ during PRM-$SH3_5$ titrations characterized by dynamic light scattering (DLS) and small-angle X-ray scattering (SAXS) and connected the observations to percolation [24].

Recently, Cho et al. demonstrated that many RNA-binding proteins form clusters (potentially similar to percolation clusters, as speculated by Kar et al.) under biologically relevant unstressed conditions, which could eventually drive the onset of phase separation under stressed conditions [53]. Recent work by Zhao et al. featured the generation of supramolecular clusters in the subsaturated solution of SARS-CoV-2 N-protein prior to the formation of phase-separated droplets [54]. Seim et al. demonstrated the intricate interplay between homotypic and heterotypic interactions, which drives the phase separation coupled to percolation in their fungal protein Whi3 and RNA system [51]. Interestingly they also observed the presence of heterogeneous distribution of percolation clusters in the sol (dilute) phase cohabiting with the dense phase, which is the embodiment of PSCP. Previously, Vorontsova et al. showed the presence of mesoscopic clusters with low occurrence in the subsaturated solutions of lysozyme [55]. In that case, the protein-rich clusters of a definite size, independent of the protein concentration variation, indicate microphase separation as opposed to percolation-type clustering and were suggested to be the precursors to the formation of protein aggregates, amyloid fibrils, and crystals [55–58]. Another example is work by Frey et al., which showed that phenylalanine-glycine (FG) repeats of nuclear pore proteins undergo sol–gel transition via noncovalent reversible crosslinking, which is critical for viability in yeast [59].

In computational work, Ranganathan et al. demonstrated that for a multivalent sticker–spacer protein complex, there is a dynamic interplay between two competing processes: (1) protein–protein interactions limited by diffusion and (2) loss of available valency within the smaller clusters engendering kinetically trapped metastable multi-droplet states [60]. They observed a slowdown of the dynamics of the condensed phase in the regime favoring large clusters, which may result in functional loss. This is an interesting phenomenon in which the metastability of the dynamic cluster controls the progress (kinetics) of the phase transition reaction, and percolation behavior might play an important role in the increasing network connectivity event. Overall, percolation is a networking transition governed by specific multivalent interactions which may (PSCP) or may not result in phase separation. The PSCP paradigm is pertinent to defining the phase behavior of multivalent biomolecules with the sticker–spacer framework and engenders sequence-, chemistry-, and topology-specific clusters, which results in network fluids, as opposed to with pure LLPS [6]. In the case of percolation without phase separation, a system-spanning percolated network is formed. All these physical states may be functionally relevant on the mesoscale, depending on the structural and dynamical properties of the condensate-/system-spanning physical crosslink engendered from the sticker–sticker network. Nevertheless, all of these intriguing observations point directly to the possibility that percolation coupled or decoupled with phase separation may play a vital role in biology and may fill the gap between micro- and macroscopic phase separation in cellular biochemistry.

## 3. Entanglement Effects

Another intriguing topological concept, entanglement, emerged in the polymer physics field more than half a century ago, providing a new understanding of the physical properties of polymer melts and polymer motion in gels. The basic idea is that because polymer chains cannot cross through each other (without breaking bonds), under the above conditions, any polymer chain can be viewed as existing within a set of obstacles made up of all the other polymer chains surrounding it. A theory for this situation was developed by de Gennes for polymer motion in an environment of fixed obstacles such as crosslinked gels [39,61] and is illustrated in Figure 4A. Lateral motions of the polymer chain are, therefore, difficult because they are constrained by this crosslinked or entangled matrix of obstacles (in the original paper, the obstacles do not move). Thus, the polymer moves by a reptation motion, along the polymer 'longitudinal' directions. The model by de Gennes provided several predictions, including that the translational diffusion constant of the chain would scale as $M^{-2}$ (very small for larger polymers; M is the polymer molecular weight; compare to a predicted $M^{-0.33}$ scaling for a spherical particle following the Stokes–Einstein equation). Edwards and Doi developed the related tube model (Figure 4B), in which the dynamics of the polymer chain are restricted within a tube formed by the (mean field of the) surrounding entangled chains, as discussed above, and similarly resulting in a reptation motion [62–64]. The early papers by Edwards et al. also made interesting predictions, including that the viscosity of entangled polymer solutions should follow a $M^3$ scaling law (M is the polymer molecular weight), which rapidly increases for longer polymers [63]. Later single-molecule studies directly visualized this type of reptation motion in concentrated solutions of DNA [65] and actin filaments [66]. These models and various subsequent extensions and theoretical advancements that included incorporation of sticker interactions provide a mechanistic basis for the understanding of many rheological and microscopic dynamical properties of such polymer systems [38,67,68]. Recent work has begun to discuss the relevance of these concepts for biomolecular condensates.

Several reports have noted the potential for entanglement effects to constrain the dynamics of long polymeric components of biomolecular condensates. One interesting example has been discussed in regard to the dynamics of the nucleolus by Riback et al. [69] Here, measurements of rRNA dynamics were used to infer an entangled network, with rRNA production at transcriptional sites and their cleavage/processing resulting in vectorial motion and facilitation of release of pre-ribosomal particles at the nucleolus periphery. Another interesting example has been discussed by Nguyen et al. in the context of nucleotide expansion repeat sequences, which are linked to diseases such as Huntington's disease and ALS [70]. Here, coarse-grained computational simulations revealed that these sequences form dense networks in condensates, with expanded molecular conformations (predicted by Flory [49]) and with reptation-like slow dynamics. Another example has been discussed for the case of TIS granules. These granules consist of mesh-like condensates that have common surface area with the endoplasmic reticulum and are important for the trafficking of membrane proteins. Using in vivo and in vitro experiments, Ma et al. showed that a minimal model of RNA-binding protein and mRNAs with disordered regions can recapitulate the formation of such irregular structures, presumably with entanglement effects contributing to the overall morphology and dynamics [71].

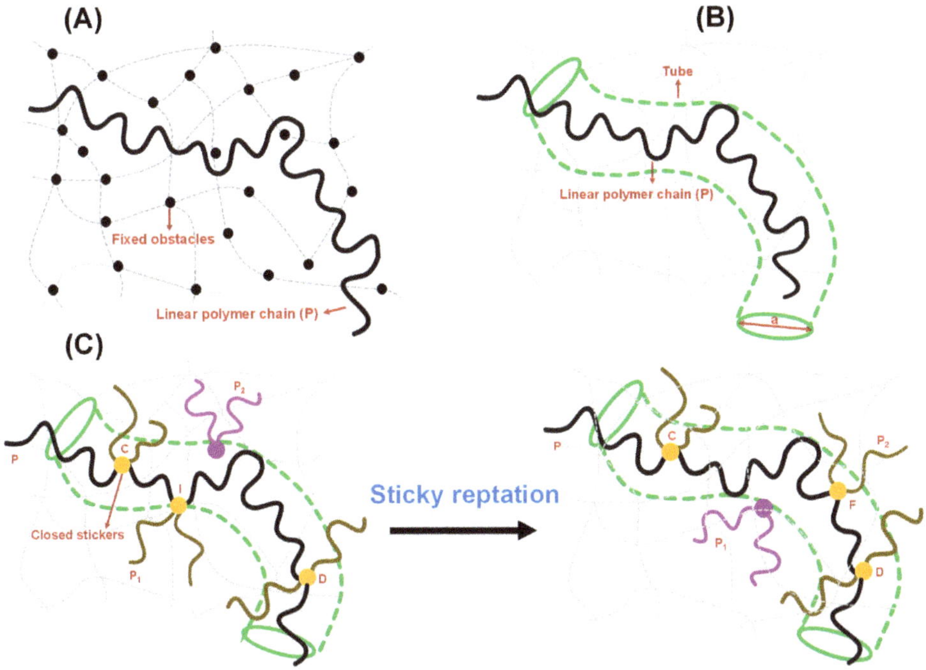


**Figure 4.** Different models of polymer entanglement. (**A**) Conceptualization of reptation of a polymer chain (P) in the presence of fixed obstacles, as theorized by de Gennes. The chain can freely move between the fixed obstacles but is not allowed to cross any of them. The fixed obstacles are shown by black circles, and the linear polymer chain is shown by a black strand. The grey dashed lines and curves describe the polymer network. (**B**) Conceptualization of reptation of an infinitely long polymer chain based on the tube model proposed by Edwards and Doi. The polymer chain, shown by black strand, is confined in the tube (olive) of a certain diameter 'a' and allowed to move along the contour of the tube. The grey dashed lines and curves describe the polymer network. (**C**) Sticky reptation model of polymer entanglement as proposed by Leibler, Rubinstein, and Colby in the context of associative polymers possessing several 'associating' groups (stickers). Initial stage (on left): The linear chain P (black strand) has a crosslink I with chain $P_1$ (dark gold strand). $P_2$ (purple) represents the next available chain for crosslink formation. Final stage (on right): a new crosslink F is formed with another chain $P_2$ (dark gold). In general, the chain that belongs to the crosslink is shown by dark gold strands with yellow circle representing the 'closed stickers', otherwise it is shown by purple. During this period, the center of mass of section CD of chain P is moved in a random manner. Details are explained in the main text.

A variation of such entanglement effects is the case in which intermolecular interactions are topologically enforced by the formation of interlinked closed geometries such as rings or loops, envisioned in the form of an 'Olympic gel' of interlinked rings by de Gennes [39]. An interesting biological example of such a condensate is represented in the thousands of interlinked DNA rings in the kinetoplast DNA of Leishmania tarentolae, for which dissociation can only be achieved by a bond-breakage process. In a related study, Michieletto et al. have shown how biochemical reactions that alter the topology of entangled fluids can result in complex patterns of time-dependent rheological properties of these soft materials [72,73].

Given the prevalence of long RNA/protein modules and the potential for transient looped structures in biomolecular condensates, it is likely that entanglement effects play substantial roles in many condensates and their biological functions.

## 4. Polymer Rheology and Biomolecular Condensation

A growing body of evidence suggests that biomolecular condensation occurs via thermodynamically reversible PS, resulting in droplets with liquid-like properties. Recently, several studies have shown that many protein and nucleic acid droplets exhibit viscoelastic behavior, which is a characteristic of non-Newtonian fluid as opposed to a fluid such as water, a Newtonian fluid [72,74–76]. Recently, Michieletto and Marenda discussed several possible reasons behind this complex non-trivial behavior of condensates, such as the aging of the fluid as an outcome of a local increase in protein concentration driven by liquid–liquid phase separation, bridging-induced phase separation (BIPS) and the effects of percolating network formation based on the sticker–spacer framework of associative polymer models, which may/may not lead to the formation of physical gel depending on the connectivity pattern and other parameters of the system [72]. The viscoelastic behavior of condensates and gels is believed to be biologically relevant and implicated in disease and functions [15,77,78].

To understand the viscoelasticity of condensates, it is crucial to discuss the theory of reversible networks that sets the cornerstone of the modern view of condensate rheology. Reversible polymer networks are viscoelastic, showing intermediate features between Newtonian fluids and Hookean solids. They show enhanced viscoelastic behavior compared to polymers that lack associative groups or stickers [38]. The rheological properties of a physically reversible network are attributed to two criteria [79–82]: (1) the extra macroscopic relaxation process due to the making and breaking of temporary junctions, sticker-based crosslinks; and (2) the microscopic lifetime of junction points/sticker–sticker crosslinks implicated in the slower rate of crosslink formation and destruction compared to thermal motion of the polymer chain/strand. The Rouse model and reptation theories pictured by de Gennes took a mean-field approach to describe the relaxation process of polymer chains [38,61,67,83–85]. According to them, if the relaxation timescale of chains of similar size is the same, the dynamical properties of a single chain can be explained by considering the neighboring chains as a frictional environment, whereas the reptation model considers the neighboring chains as a tube-like confinement [62–64,86]. Reptation dynamics is a snake-like diffusion of a chain along the length of the tube, and the relaxation time of the entangled polymer melt/gel is the time it takes to reptate out of the tube [40,61]. The simple reptation model is not valid near the gel point due to the presence of precursor chains of sol of different sizes and topologies and the unique dynamical feature of the gel matrix governed by the sticker–sticker interactions [79]. The scaling law developed by Rubinstein and Semenov appears to be more suitable for quantifying the change in linear viscoelasticity in connection with the degree of gelation [27,80]. When the gel network is fully formed without leaving any sol chain in the system, the mean-field approach of reptation model is sufficient to describe the viscoelasticity of the system. Leibler, Rubinstein, and Colby demonstrated a sticky reptation model for the dynamics of entangled networks possessing several temporary crosslinks [38]. Figure 4C depicts a fundamental process of chain diffusion in a reversible gel governed by sticker–sticker interactions, as proposed by Leibler, Rubinstein, and Colby. According to this model, a closed sticker that belongs to the crosslink I (yellow circle) between the chains P (black) and $P_1$ (dark gold) is allowed to move distances of the order of the confining tube diameter. Therefore, crosslink I does not allow the diffusion of unentangled loops of the chain P between closed stickers C (yellow circle) and D (yellow circle). CI and DI, which are the parts of the chain P between the closed stickers, undergo Rouse-like motions with almost fixed ends, meaning their center of mass changes around their average positions [38]. When the crosslink I opens, the free sticker moves. If it is assumed that the equilibration time of the strand CD is shorter than the lifetime of the open sticker, and the sticker C and D remains closed within this timescale, the sticker would either recombine with chain $P_1$ at the crosslink I, resulting in zero net displacements, or it would associate with a different chain $P_2$ (purple), resulting in the formation of a new crosslink F (yellow circle) (Figure 4C) [38]. During the process from breaking crosslink I to making the crosslink F, the center of mass of the section of chain

P moves to a new average position with the assumption that the stickers remain closed during the equilibration of the strand. Such displacement motion along the tube results in a reptation, such as the diffusion of the linear chain P. Overall, they suggested that if the relaxation timescale is shorter than the lifetime of the crosslink, the network exhibits elastic behavior, whereas the chain diffusion along the confining tube is governed by the sequential destruction of only a few crosslinks on a longer timescale.

Keeping the essence of the polymer rheology theories in mind, it is suggestive that the droplet-spanning percolated network with precise dynamical properties, the degree of crosslinking, relaxation due to dissociation and reassociation of stickers, and the microscopic lifetime of sticker-mediated crosslinks may contribute to the viscoelasticity and other material properties of the condensates. Crosstalk among soft matter physics, rheology, polymer physics, and fluid mechanics is necessary to elucidate the physical underpinning of condensate viscoelasticity.

## 5. Concluding Remarks and Future Directions

In recent years, the concept of network theory, in which the 'links' represent the interaction between the elements has gained significant importance in analyzing and predicting the behavior of complex biomolecular systems [87]. For example, protein–protein interactions networks are substantially implicated in cellular structure and function [88]. The revolution of network theory prompted the idea of the application of topology-based models to characterize a multitude of principal biological phenomena including biomolecular condensation, protein folding, gelation, and connectivity arrangements at molecular, cellular, and tissue scales in response to stress and ailments. Furthermore, there is increasing application of concepts from polymer physics to biological systems and materials. Along these lines, understanding the physical bases of percolation and entanglement in these systems is expected to be important for better definition of their links to biomolecular condensation. As discussed earlier, the application of percolation/entanglement concepts in this area is at relatively early stages and has current limitations as well. Correspondingly, substantial future work is needed (and expected) towards testing the applicability, generality and implications of these ideas.

Along these lines, the discoveries and concepts reviewed here may lay the groundwork for addressing a plethora of unexplored areas. These include more direct tests of percolation and entanglement approaches through the use of single-molecule or advanced rheology measurements [74–76] combined with molecular/cell biology and computational tools. It will also be important to carry out systematic studies to test the applicability and limitations of percolation and entanglement concepts, both in model and complex protein and RNA systems in vitro and in vivo. Additionally, more detailed studies of the distributions and dynamic properties of percolation clusters are needed, which can then allow more in-depth analysis using analytical theory and computational results. Here again, advanced single-molecule/particle and imaging methods will likely be particularly useful. The dynamical behavior of these clusters and, eventually, their conversion into the larger system-spanning droplets will be of great importance for investigation. As it accounts for the formation of connected clusters (or lattice animals, which essentially stands for a set of distinct connected clusters, also called animals, and which could be considered to be the equivalent of connected percolation clusters), the percolation approach could be useful in elucidating the formation of microgels, the first stage of the gelation process, with the spherical cross-linked microscopic network containing only finite clusters [89]. Other lines of future study include a more detailed mechanistic understanding of the physical underpinnings of sol–gel transitions coupled/decoupled with phase separation with the incorporation of different models, the interplay of protein/RNA conformational properties and complex/dynamic substructure in multicomponent and active-matter systems, better mapping of different types of sticker–spacer architecture in terms of percolation, and links to function [1,5,23,32,72,90–92] and the interplay/relevance of other mechanisms of cluster formation. Another captivating area to investigate is the condition where

$C_{perc} < C_{sat}$, which essentially means that the system exceeds the connectivity threshold, thereby switching from dispersed monomers/clusters (sol) to the system spanning percolated matrix or physical gel without phase separation [23]. Physical gels are characterized by reversible noncovalent crosslinking. In the context of associative polymers with the sticker–spacer framework, if the bulk concentration of the interaction motifs is above the gel point but below $C_{sat}$, a connectivity transition occurs without droplet formation. We surmise that depending on the connectivity feature of the structures, the system will either form a physical gel or a distribution of network clusters with percolating behavior, but further investigation is necessary to elucidate the physical underpinnings of this shift in biological contexts. Gelation without phase separation is biologically relevant in many aspects [93–96]. Studies by Halfmann demonstrated that the phase transition of low-complexity sequence proteins to an amorphous solid or glass, a process based on the principle of vitrification, can be viewed as a phenomenon of gelation without phase separation [94]. The bacterial cytosol also exhibits sol-to-gel conversion akin to glass transition and impacts the mobility and fluidity of the cytoplasmic component in a size-dependent fashion [95,96]. Another interesting example is a report of analytical theory developed to understand the biology of actin networks, showing that actin-binding proteins that modulate connectivity can result in complex percolation-related behavior that can alter rheology and function [97]. Entanglement concepts are also being applied to explain several phenomena related to the diffusion and rheology of biomolecular condensates. Recent work by Nguyen et al. demonstrated that the mobility of RNA inside the highly viscous and dense droplets follows the reptation model of polymer entanglement [70]. Recently, Tom et al. demonstrated that at a relatively low concentration of $Mg^{2+}$ induces short polyA-RNA sequences to form droplets that appear as internally arrested species [98]. They discovered that RNA chains exhibit slow translational dynamics, potentially with contributions from the entanglement effect within the densely packed RNA–RNA networking in the droplet state.

Because biomolecular condensation involves large, complex networking connectivity and intricate interactions between the interacting modules, it is a challenging task to quantify or decouple all these mechanistic aspects. We believe that the amalgamation of different models, techniques, and theories, along with the existing knowledge of percolation models, polymer entanglement, and phase-transition physics will further illuminate the inner workings of condensate science and their functions in biology.

**Author Contributions:** Conceptualization, D.D. and A.A.D.; Writing—original draft preparation, review and editing, D.D. and A.A.D. All authors have read and agreed to the published version of the manuscript.

**Funding:** The authors gratefully acknowledge support from the US NIH/NIGMS (Grant R35 GM130375 to AAD).

**Institutional Review Board Statement:** Not applicable.

**Informed Consent Statement:** Not applicable.

**Data Availability Statement:** Not applicable.

**Acknowledgments:** The authors thank current and past lab members and Rohit Pappu for useful discussions.

**Conflicts of Interest:** The authors declare no conflict of interest.

## References

1. Alberti, S.; Hyman, A.A. Biomolecular condensates at the nexus of cellular stress, protein aggregation disease and ageing. *Nat. Rev. Mol. Cell Biol.* **2021**, *22*, 196–213. [CrossRef]
2. Banani, S.F.; Lee, H.O.; Hyman, A.A.; Rosen, M.K. Biomolecular condensates: Organizers of cellular biochemistry. *Nat. Rev. Mol. Cell Biol.* **2017**, *18*, 285–298. [CrossRef] [PubMed]
3. Hyman, A.A.; Weber, C.A.; Julicher, F. Liquid-liquid phase separation in biology. *Annu. Rev. Cell Dev. Biol.* **2014**, *30*, 39–58. [CrossRef]

4. Tom, J.K.A.; Deniz, A.A. Complex dynamics of multicomponent biological coacervates. *Curr. Opin. Colloid Interface Sci.* **2021**, *56*, 101488. [CrossRef]
5. Choi, J.M.; Holehouse, A.S.; Pappu, R.V. Physical Principles Underlying the Complex Biology of Intracellular Phase Transitions. *Annu. Rev. Biophys.* **2020**, *49*, 107–133. [CrossRef] [PubMed]
6. Mittag, T.; Pappu, R.V. A conceptual framework for understanding phase separation and addressing open questions and challenges. *Mol. Cell* **2022**, *82*, 2201–2214. [CrossRef]
7. Nesterov, S.V.; Ilyinsky, N.S.; Uversky, V.N. Liquid-liquid phase separation as a common organizing principle of intracellular space and biomembranes providing dynamic adaptive responses. *Biochim. Biophys. Acta Mol. Cell Res.* **2021**, *1868*, 119102. [CrossRef]
8. Roden, C.; Gladfelter, A.S. RNA contributions to the form and function of biomolecular condensates. *Nat. Rev. Mol. Cell Biol.* **2021**, *22*, 183–195. [CrossRef]
9. Boeynaems, S.; Alberti, S.; Fawzi, N.L.; Mittag, T.; Polymenidou, M.; Rousseau, F.; Schymkowitz, J.; Shorter, J.; Wolozin, B.; Van Den Bosch, L.; et al. Protein Phase Separation: A New Phase in Cell Biology. *Trends Cell Biol.* **2018**, *28*, 420–435. [CrossRef]
10. Kilgore, H.R.; Young, R.A. Learning the chemical grammar of biomolecular condensates. *Nat. Chem. Biol.* **2022**, *18*, 1298–1306. [CrossRef]
11. Dignon, G.L.; Best, R.B.; Mittal, J. Biomolecular Phase Separation: From Molecular Driving Forces to Macroscopic Properties. *Annu. Rev. Phys. Chem.* **2020**, *71*, 53–75. [CrossRef] [PubMed]
12. Alberti, S.; Gladfelter, A.; Mittag, T. Considerations and Challenges in Studying Liquid-Liquid Phase Separation and Biomolecular Condensates. *Cell* **2019**, *176*, 419–434. [CrossRef] [PubMed]
13. Banerjee, P.R.; Milin, A.N.; Moosa, M.M.; Onuchic, P.L.; Deniz, A.A. Reentrant Phase Transition Drives Dynamic Substructure Formation in Ribonucleoprotein Droplets. *Angew. Chem. Int. Ed. Engl.* **2017**, *56*, 11354–11359. [CrossRef] [PubMed]
14. Toretsky, J.A.; Wright, P.E. Assemblages: Functional units formed by cellular phase separation. *J. Cell Biol.* **2014**, *206*, 579–588. [CrossRef]
15. Shin, Y.; Brangwynne, C.P. Liquid phase condensation in cell physiology and disease. *Science* **2017**, *357*, eaaf4382. [CrossRef]
16. Harmon, T.S.; Holehouse, A.S.; Rosen, M.K.; Pappu, R.V. Intrinsically disordered linkers determine the interplay between phase separation and gelation in multivalent proteins. *Elife* **2017**, *6*, e30294. [CrossRef]
17. Posey, A.E.; Holehouse, A.S.; Pappu, R.V. Phase Separation of Intrinsically Disordered Proteins. In *Methods in Enzymology*; Academic Press: Cambridge, MA, USA, 2018; Volume 611, pp. 1–30.
18. Brangwynne, C.P.; Tompa, P.; Pappu, R.V. Polymer Physics of Intracellular Phase Transitions. *Nat. Phys.* **2015**, *11*, 899–904. [CrossRef]
19. Martin, E.W.; Mittag, T. Relationship of Sequence and Phase Separation in Protein Low-Complexity Regions. *Biochemistry* **2018**, *57*, 2478–2487. [CrossRef]
20. Flory, P.J. Molecular Size Distribution in Three Dimensional Polymers. I. Gelation1. *J. Am. Chem. Soc.* **1941**, *63*, 3083–3090. [CrossRef]
21. Flory, P.J. Thermodynamics of High Polymer Solutions. *J. Chem. Phys.* **1942**, *10*, 51–61. [CrossRef]
22. Flory, P.J. Constitution of Three-dimensional Polymers and the Theory of Gelation. *J. Phys. Chem.* **1942**, *46*, 132–140. [CrossRef]
23. Deniz, A.A. Percolation physics and density transition frameworks converge in biomolecular condensation. *Proc. Natl. Acad. Sci. USA* **2022**, *119*, e2210177119. [CrossRef] [PubMed]
24. Li, P.; Banjade, S.; Cheng, H.C.; Kim, S.; Chen, B.; Guo, L.; Llaguno, M.; Hollingsworth, J.V.; King, D.S.; Banani, S.F.; et al. Phase transitions in the assembly of multivalent signalling proteins. *Nature* **2012**, *483*, 336–340. [CrossRef] [PubMed]
25. Musacchio, A. On the Role of Phase Separation in the Biogenesis of Membraneless Compartments. *EMBO J.* **2022**, *41*, e109952. [CrossRef] [PubMed]
26. Tanaka, F. *Polymer Physics: Applications to Molecular Asssociation and Thermoreversible Gelation*; Cambridge University Press: Cambridge, UK, 2011.
27. Rubinstein, M.; Semenov, A.N. Thermoreversible gelation in solutions of associating polymers. 2. Linear dynamics. *Macromolecules* **1998**, *31*, 1386–1397. [CrossRef]
28. Semenov, A.N.; Rubinstein, M. Thermoreversible gelation in solutions of associative polymers. 1. Statics. *Macromolecules* **1998**, *31*, 1373–1385. [CrossRef]
29. Choi, J.M.; Dar, F.; Pappu, R.V. LASSI: A lattice model for simulating phase transitions of multivalent proteins. *PLoS Comput. Biol.* **2019**, *15*, e1007028. [CrossRef]
30. Tanaka, F. Phase formation of associating polymers: Gelation, phase separation and microphase formation. *Adv. Colloid Interface Sci.* **1996**, *63*, 23–40. [CrossRef]
31. Rubinstein, M.; Dobrynin, A.V. Solutions of associative polymers. *Trends Polym. Sci.* **1997**, *5*, 181–186.
32. Kar, M.; Dar, F.; Welsh, T.J.; Vogel, L.T.; Kuhnemuth, R.; Majumdar, A.; Krainer, G.; Franzmann, T.M.; Alberti, S.; Seidel, C.A.M.; et al. Phase-separating RNA-binding proteins form heterogeneous distributions of clusters in subsaturated solutions. *Proc. Natl. Acad. Sci. USA* **2022**, *119*, e2202222119. [CrossRef]
33. Choi, J.M.; Hyman, A.A.; Pappu, R.V. Generalized models for bond percolation transitions of associative polymers. *Phys. Rev. E* **2020**, *102*, 042403. [CrossRef]
34. Steif, J.E. *A Mini Course on Percolation Theory*; University of Jyväskylä: Jyväskylän yliopisto, Finland, 2011.

35. Essam, J.W. Percolation Theory. *Rep. Prog. Phys.* **1980**, *43*, 833. [CrossRef]
36. Broadbent, S.R.; Hammersley, J.M. Percolation processes. *Math. Proc. Cambridge Philos. Soc.* **1957**, *53*, 629–641. [CrossRef]
37. Stauffer, D.; Aharony, A. *Introduction to Percolation Theory*; Taylor & Francis: Abingdon, UK, 1994.
38. Leibler, L.; Rubinstein, M.; Colby, R.H. Dynamics of reversible networks. *Macromolecules* **2002**, *24*, 4701–4707. [CrossRef]
39. De Gennes, P.G. *Scaling Concept in Polymer Physics*; Cornell University Press: Ithaca, NY, USA, 1979.
40. Rubinstein, M.; Colby, R.H. *Polymer Physics*; Oxford University Press: Oxford, UK; New York, NY, USA, 2003.
41. Christensen, K. *Percolation Theory*; MIT: Cambridge, MA, USA, 2002.
42. Jouannot-Chesney, P.; Jernot, J.-P.; Lantuéjoul, C. Percolation Transition and Topology. *Image Anal. Stereol.* **2017**, *36*, 95–103. [CrossRef]
43. Sanderson, D.J.; Nixon, C.W. Topology, connectivity and percolation in fracture networks. *J. Struct. Geol.* **2018**, *115*, 167–177. [CrossRef]
44. Stauffer, D. Scaling theory of percolation clusters. *Phys. Rep.* **1979**, *54*, 1–74. [CrossRef]
45. Sykes, M.F.; Essam, J.W. Exact Critical Percolation Probabilities for Site and Bond Problems in Two Dimensions. *J. Math. Phys.* **1964**, *5*, 1117. [CrossRef]
46. Bobrowski, O.; Skraba, P. Homological percolation and the Euler characteristic. *Phys. Rev. E* **2020**, *101*, 032304. [CrossRef]
47. Neher, R.A.; Mecke, K.; Wagner, H. Topological estimation of percolation thresholds. *Stat. Mech.* **2008**, *2008*, P01011. [CrossRef]
48. Stockmayer, W.H. Theory of Molecular Size Distribution and Gel Formation in Branched-Chain Polymers. *J. Chem. Phys.* **1943**, *11*, 45–55. [CrossRef]
49. Flory, P.J. Introductory lecture. *Faraday Discuss. Chem. Soc.* **1974**, *57*, 7–18. [CrossRef]
50. Harmon, T.S.; Holehouse, A.S.; Pappu, R.V. Differential solvation of intrinsically disordered linkers drives the formation of spatially organized droplets in ternary systems of linear multivalent proteins. *New J. Phys.* **2018**, *20*, 045002. [CrossRef]
51. Seim, I.; Posey, A.E.; Snead, W.T.; Stormo, B.M.; Klotsa, D.; Pappu, R.V.; Gladfelter, A.S. Dilute phase oligomerization can oppose phase separation and modulate material properties of a ribonucleoprotein condensate. *Proc. Natl. Acad. Sci. USA* **2022**, *119*, e2120799119. [CrossRef]
52. Franzmann, T.M.; Jahnel, M.; Pozniakovsky, A.; Mahamid, J.; Holehouse, A.S.; Nuske, E.; Richter, D.; Baumeister, W.; Grill, S.W.; Pappu, R.V.; et al. Phase separation of a yeast prion protein promotes cellular fitness. *Science* **2018**, *359*, eaao5654. [CrossRef] [PubMed]
53. Cho, N.H.; Cheveralls, K.C.; Brunner, A.D.; Kim, K.; Michaelis, A.C.; Raghavan, P.; Kobayashi, H.; Savy, L.; Li, J.Y.; Canaj, H.; et al. OpenCell: Endogenous tagging for the cartography of human cellular organization. *Science* **2022**, *375*, eabi6983. [CrossRef]
54. Zhao, H.; Wu, D.; Nguyen, A.; Li, Y.; Adao, R.C.; Valkov, E.; Patterson, G.H.; Piszczek, G.; Schuck, P. Energetic and structural features of SARS-CoV-2 N-protein co-assemblies with nucleic acids. *iScience* **2021**, *24*, 102523. [CrossRef] [PubMed]
55. Vorontsova, M.A.; Chan, H.Y.; Lubchenko, V.; Vekilov, P.G. Lack of Dependence of the Sizes of the Mesoscopic Protein Clusters on Electrostatics. *Biophys. J.* **2015**, *109*, 1959–1968. [CrossRef]
56. Maes, D.; Vorontsova, M.A.; Potenza, M.A.; Sanvito, T.; Sleutel, M.; Giglio, M.; Vekilov, P.G. Do protein crystals nucleate within dense liquid clusters? *Acta Crystallogr. Sect. F Struct. Biol. Commun.* **2015**, *71*, 815–822. [CrossRef]
57. Giege, R. A historical perspective on protein crystallization from 1840 to the present day. *FEBS J.* **2013**, *280*, 6456–6497. [CrossRef]
58. Sleutel, M.; Van Driessche, A.E. Role of clusters in nonclassical nucleation and growth of protein crystals. *Proc. Natl. Acad. Sci. USA* **2014**, *111*, E546–E553. [CrossRef]
59. Frey, S.; Richter, R.P.; Görlich, D. FG-rich repeats of nuclear pore proteins form a three-dimensional meshwork with hydrogel-like properties. *Science* **2006**, *314*, 815–817. [CrossRef]
60. Ranganathan, S.; Shakhnovich, E.I. Dynamic metastable long-living droplets formed by sticker-spacer proteins. *Elife* **2020**, *9*, e56159. [CrossRef]
61. de Gennes, P.G. Reptation of a Polymer Chain in the Presence of Fixed Obstacles. *J. Chem. Phys.* **1971**, *55*, 572–579. [CrossRef]
62. Doi, M.; Edwards, S.F. Dynamics of concentrated polymer systems. Part 1. Brownian motion in the equilibrium state. *J. Chem. Soc. Faraday Trans.* **1978**, *74*, 1789–1801. [CrossRef]
63. Edwards, S.F.; Grant, J.W.V. The effect of entanglements on the viscosity of a polymer melt. *J. Phys. A Math. Nucl. Gen.* **1973**, *6*, 1186. [CrossRef]
64. Doi, M.; Edwards, S.F. *The Theory of Polymer Dynamics*, 3rd ed.; Clarendon Press: Oxford, UK, 1986.
65. Perkins, T.T.; Quake, S.R.; Smith, D.E.; Chu, S. Relaxation of a single DNA molecule observed by optical microscopy. *Science* **1994**, *264*, 822–826. [CrossRef]
66. Käs, J.; Strey, H.; Sackmann, E. Direct imaging of reptation for semiflexible actin filaments. *Nature* **1994**, *368*, 226–229. [CrossRef]
67. Golde, T.; Glaser, M.; Tutmarc, C.; Elbalasy, I.; Huster, C.; Busteros, G.; Smith, D.M.; Herrmann, H.; Kas, J.A.; Schnauss, J. The role of stickiness in the rheology of semiflexible polymers. *Soft Matter* **2019**, *15*, 4865–4872. [CrossRef]
68. Muthukumar, M. Screening effect on viscoelasticity near the gel point. *Macromolecules* **1989**, *22*, 4656–4658. [CrossRef]
69. Riback, J.A.; Eeftens, J.M.; Lee, D.S.; Quinodoz, S.A.; Beckers, L.; Becker, L.A.; Brangwynne, C.P. Viscoelastic RNA entanglement and advective flow underlie nucleolar form and function. *BioRxiv* **2022**. [CrossRef]
70. Nguyen, H.T.; Hori, N.; Thirumalai, D. Condensates in RNA repeat sequences are heterogeneously organized and exhibit reptation dynamics. *Nat. Chem.* **2022**, *14*, 775–785. [CrossRef] [PubMed]

71. Ma, W.; Zheng, G.; Xie, W.; Mayr, C. In vivo reconstitution finds multivalent RNA-RNA interactions as drivers of mesh-like condensates. *Elife* **2021**, *10*, e64252. [CrossRef] [PubMed]
72. Michieletto, D.; Marenda, M. Rheology and Viscoelasticity of Proteins and Nucleic Acids Condensates. *J. Am. Chem. Soc. Au* **2022**, *2*, 1506–1521. [CrossRef] [PubMed]
73. Michieletto, D.; Neill, P.; Weir, S.; Evans, D.; Crist, N.; Martinez, V.A.; Robertson-Anderson, R.M. Topological digestion drives time-varying rheology of entangled DNA fluids. *Nat. Commun.* **2022**, *13*, 4389. [CrossRef]
74. Jawerth, L.; Fischer-Friedrich, E.; Saha, S.; Wang, J.; Franzmann, T.; Zhang, X.; Sachweh, J.; Ruer, M.; Ijavi, M.; Saha, S.; et al. Protein condensates as aging Maxwell fluids. *Science* **2020**, *370*, 1317–1323. [CrossRef] [PubMed]
75. Alshareedah, I.; Moosa, M.M.; Pham, M.; Potoyan, D.A.; Banerjee, P.R. Programmable viscoelasticity in protein-RNA condensates with disordered sticker-spacer polypeptides. *Nat. Commun.* **2021**, *12*, 6620. [CrossRef]
76. Ghosh, A.; Kota, D.; Zhou, H.X. Shear relaxation governs fusion dynamics of biomolecular condensates. *Nat. Commun.* **2021**, *12*, 5995. [CrossRef] [PubMed]
77. Michieletto, D.; Gilbert, N. Role of nuclear RNA in regulating chromatin structure and transcription. *Curr. Opin. Cell Biol.* **2019**, *58*, 120–125. [CrossRef]
78. Mathieu, C.; Pappu, R.V.; Taylor, J.P. Beyond aggregation: Pathological phase transitions in neurodegenerative disease. *Science* **2020**, *370*, 56–60. [CrossRef] [PubMed]
79. Wu, S.; Chen, Q. Advances and New Opportunities in the Rheology of Physically and Chemically Reversible Polymers. *Macromolecules* **2021**, *55*, 697–714. [CrossRef]
80. Lodge, A.S.; Armstrong, R.C.; Wagner, M.H.; Winter, H.H. Constitutive Equations from Gaussian Molecular Network Theories in Polymer Rheology. *Pure Appl. Chem.* **1980**, *54*, 1349–1359.
81. Tanaka, F.; Edwards, S.F. Viscoelastic properties of physically cross-linked networks. Transient network theory. *Macromolecules* **1992**, *25*, 1516–1523. [CrossRef]
82. Stukalin, E.B.; Cai, L.H.; Kumar, N.A.; Leibler, L.; Rubinstein, M. Self-Healing of Unentangled Polymer Networks with Reversible Bonds. *Macromolecules* **2013**, *46*, 7525–7541. [CrossRef]
83. Baxandall, L.G. Dynamics of reversibly cross-linked chains. *Macromolecules* **1989**, *22*, 1982–1988. [CrossRef]
84. Jiang, N.; Zhang, H.; Tang, P.; Yang, Y. Linear Viscoelasticity of Associative Polymers: Sticky Rouse Model and the Role of Bridges. *Macromolecules* **2020**, *53*, 3438–3451. [CrossRef]
85. Shao, J.; Jiang, N.; Zhang, H.; Yang, Y.; Tang, P. Sticky Rouse Model and Molecular Dynamics Simulation for Dual Polymer Networks. *Macromolecules* **2022**, *55*, 535–549. [CrossRef]
86. Edwards, S.F.; Vilgis, T.A. The effect of entanglements in rubber elasticity. *Polymer* **1986**, *27*, 483–492. [CrossRef]
87. Antal, M.A.; Bode, C.; Csermely, P. Perturbation waves in proteins and protein networks: Applications of percolation and game theories in signaling and drug design. *Curr. Protein Pept. Sci.* **2009**, *10*, 161–172. [CrossRef]
88. Deb, D.; Vishveshwara, S.; Vishveshwara, S. Understanding protein structure from a percolation perspective. *Biophys. J.* **2009**, *97*, 1787–1794. [CrossRef]
89. Vilgis, T.A. *Polymer Networks in Comprehensive Polymer Science and Supplements*; Allen, G., Bevington, J.C., Eds.; Pergamon Press: Pergamon, Amsterdam, 1989; pp. 227–279.
90. Boland, C.S.; Khan, U.; Ryan, G.; Barwich, S.; Charifou, R.; Harvey, A.; Backes, C.; Li, Z.; Ferreira, M.S.; Möbius, M.E.; et al. Sensitive electromechanical sensors using viscoelastic graphene-polymer nanocomposites. *Science* **2016**, *354*, 1257–1260. [CrossRef]
91. Levon, K.; Margolina, A.; Patashinsky, A.Z. Multiple percolation in conducting polymer blends. *Macromolecules* **1993**, *26*, 4061–4063. [CrossRef]
92. Farag, M.; Cohen, S.R.; Borcherds, W.M.; Bremer, A.; Mittag, T.; Pappu, R.V. Condensates formed by prion-like low-complexity domains have small-world network structures and interfaces defined by expanded conformations. *Nat. Commun.* **2022**, *13*, 7722. [CrossRef] [PubMed]
93. Cai, J.; Townsend, J.P.; Dodson, T.C.; Heiney, P.A.; Sweeney, A.M. Eye patches: Protein assembly of index-gradient squid lenses. *Science* **2017**, *357*, 564–569. [CrossRef] [PubMed]
94. Halfmann, R.A. glass menagerie of low complexity sequences. *Curr. Opin. Struct. Biol.* **2016**, *38*, 18–25. [CrossRef] [PubMed]
95. Parry, B.R.; Surovtsev, I.V.; Cabeen, M.T.; O'Hern, C.S.; Dufresne, E.R.; Jacobs-Wagner, C. The bacterial cytoplasm has glass-like properties and is fluidized by metabolic activity. *Cell* **2014**, *156*, 183–194. [CrossRef] [PubMed]
96. Munder, M.C.; Midtvedt, D.; Franzmann, T.; Nüske, E.; Otto, O.; Herbig, M.; Ulbricht, E.; Müller, P.; Taubenberger, A.; Maharana, S.; et al. A pH-driven transition of the cytoplasm from a fluid- to a solid-like state promotes entry into dormancy. *eLife* **2016**, *5*, e09347. [CrossRef]
97. Bueno, C.; Liman, J.; Schafer, N.P.; Cheung, M.S.; Wolynes, P.G. A generalized Flory-Stockmayer kinetic theory of connectivity percolation and rigidity percolation of cytoskeletal networks. *PLoS Comput. Biol.* **2022**, *18*, e1010105. [CrossRef]
98. Tom, J.K.A.; Onuchic, P.L.; Deniz, A.A. Short PolyA RNA Homopolymers Undergo $Mg^{2+}$-Mediated Kinetically Arrested Condensation. *J. Phys. Chem. B* **2022**, *126*, 9715–9725. [CrossRef]

biomolecules MDPI

*Review*

# Illuminating Intrinsically Disordered Proteins with Integrative Structural Biology

**Rachel Evans [1], Sravani Ramisetty [2], Prakash Kulkarni [2,3] and Keith Weninger [1,*]**

1 Department of Physics, North Carolina State University, Raleigh, NC 27695, USA
2 Department of Medical Oncology and Therapeutics Research, City of Hope National Medical Center, Duarte, CA 91010, USA
3 Department of Systems Biology, City of Hope National Medical Center, Duarte, CA 91010, USA
* Correspondence: krwening@ncsu.edu

**Abstract:** Intense study of intrinsically disordered proteins (IDPs) did not begin in earnest until the late 1990s when a few groups, working independently, convinced the community that these 'weird' proteins could have important functions. Over the past two decades, it has become clear that IDPs play critical roles in a multitude of biological phenomena with prominent examples including coordination in signaling hubs, enabling gene regulation, and regulating ion channels, just to name a few. One contributing factor that delayed appreciation of IDP functional significance is the experimental difficulty in characterizing their dynamic conformations. The combined application of multiple methods, termed integrative structural biology, has emerged as an essential approach to understanding IDP phenomena. Here, we review some of the recent applications of the integrative structural biology philosophy to study IDPs.

**Keywords:** intrinsically disordered proteins; integrative structural biology; unfolded; unstructured; flexible; protein function

**Citation:** Evans, R.; Ramisetty, S.; Kulkarni, P.; Weninger, K. Illuminating Intrinsically Disordered Proteins with Integrative Structural Biology. *Biomolecules* **2023**, *13*, 124. https://doi.org/10.3390/biom13010124

Academic Editor: Barbara Zambelli

Received: 1 December 2022
Revised: 1 January 2023
Accepted: 4 January 2023
Published: 7 January 2023



## 1. Introduction

Intrinsically disordered proteins (IDPs) are presently a prime focus of the protein biochemistry research enterprise, but that was not always the case. Although IDPs represent around a third of all proteins in eukaryotes [1–4], they were not a fashionable topic for researchers until the late 1990s. The existence of natively unfolded or disordered segments within otherwise folded proteins was well known from the early days of X-ray crystallography where parts of proteins that were not part of structure solutions were presumed dynamic and flexible. Those regions were often excised to aid crystallization. A few examples of focused IDP study appear through the literature in the 1960s and 1970s [5]. The discrepancy of some protein mobilities in size-exclusion chromatography compared to well-folded standards was an early observation interpreted as being due to flexible, disordered proteins [6]. In the 1970s, NMR studies could reveal disordered conformations, for example in glucagon [7]. From the 1960s to the 1980s, components of ribosomes [5] and histones [8] were also considered to have flexibility or disorder.

Despite these studies discussing properties of IDPs, the idea that biological functions could derive directly from the disordered properties was generally not considered. Gradually, appreciation for functional impacts of flexible linkers between domains or disorder-to-order transitions accrued [9]. One notable example of ahead of its time thinking was Paul Sigler's musings where he synthesized several results about transcription factors in 1988 [10], resulting in a proposal of a key functional role for the disordered domains. Perhaps, it was not more widely adopted in part following his naming of the functional domain as an 'acid blob' or 'negative noodle', alluding to the role of the overall charge of the disordered domain in this proposed function. The failure of the broader field to seriously consider that functions could directly result from the nature of the disordered chain in IDPs has been suggested by several authors [11–13] to be a result of the then dominance of the lock-and-key view of enzyme/substrate functional interactions, reinforced by the stunning successes of X-ray crystallography to provide snapshots of carefully folded proteins stable 'active sites'. This bias, combined with the experimental difficulties in characterizing these disordered conformational ensembles, prevented earlier appreciation of the functional roles we now know IDPs have. Indeed, even today, IDPs are thought to comprise a significant fraction of the 'dark' proteome, the proteome that is genetically expected to exist but not yet observed and characterized [14].

In the 1990s, the proliferation of gene-based techniques to interrogate protein function, genomic sequencing, and bioinformatics advances, along with technical improvements in NMR, led to increased appreciation of the functional importance of IDPs. Uversky was among the first few scientists to discuss significant wide-spread functions for IDPs in ways that were highly influential and brought recognition of this potential to the wider field [15–18].

At the present time, we appreciate many important functions for IDPs in physiology. A few examples are coordinating signaling networks (hubs), contributing to gene regulation, and modulating ion channel function [19–25]. The mechanisms that underlie these functions are equally diverse: multi-valency, fuzzy complexes, hubs, switches, and non-genetic switches based on ensemble switching [25–29]. In addition to physiological functions, disease associated pathology involving IDPs are recognized with prominent examples including amyloid-β, Tau, and α-synuclein [30,31]. Recently, consideration of physiochemical properties of IDPs, such as possible LLPS phenomena, have further expanded the sorts of functions that are contemplated for IDPs [32,33].

Some credit for establishing clear 'structure-function' paradigms arising from the disordered properties of IDPs must go to the practice of combining several distinct characterization approaches to draw conclusions, an approach termed integrative structural biology [34–36]. Integrative structural biology seeks to combine multiple characterization approaches with different sensitivities to provide a more complete understanding of biomolecular conformational ensembles and dynamics. The tendencies for IDPs to rapidly fluctuate while sampling wide ranges of conformation space rather than remaining in each state makes them well suited for applications of multiple experimental probes to reveal different aspects of their behaviors. Such integrative structural biology approaches are becoming more common. Impressively, Uversky anticipated the utility of multiple characterization methods to enhance the understanding of IDP functions. He amusingly illustrated the necessity of using integrative structural biology approaches for IDP studies with a parable about confusion when examining an elephant without the proper global perspective [37]. Here, we review some of the latest successes in combining methods through the integrative structural biology approach to characterize IDP conformations and address their myriad functions.

## 2. Summary of Methods

From a general experimental perspective, confirming that conclusions are consistent with multiple different experimental methods inspires increased confidence. For example, some methods require modification of molecules with extrinsic labels (fluorescence or EPR for example). Consistency with other measurement methods that do not use the modifications or use different modifications can confirm that such modifications do not affect the results in detrimental ways. In the integrative structural biology approach applied to IDPs, using different methods also has greater benefits because different methods have sensitivities to distinct length or time scales and even different concentration ranges (Figure 1). IDPs have behaviors that span broad ranges in these properties, making the use of multiple methods almost essential. For one example, liquid–liquid phase separation (LLPS) phenomena occur for a number of IDPs where concentrations are a key controlling factor [38]. Before discussing applications of integrative structural biology approaches to

IDPs (see Figure 1), we first briefly describe some of the key individual methods used to characterize IDPs.

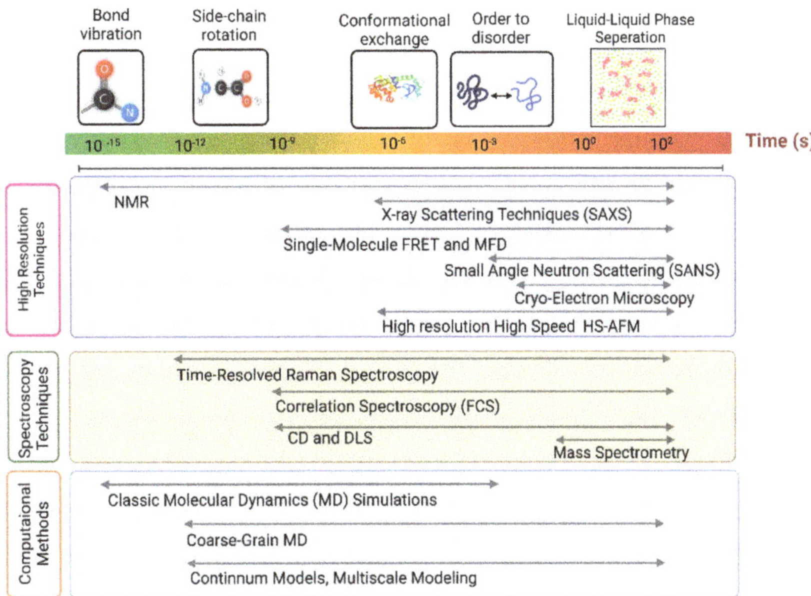


**Figure 1.** Comparison of sensitivities of methods commonly applied in IDP studies. Limits on temporal resolutions are not intended to be precise in this figure.

### 2.1. Nuclear Magnetic Resonance

Nuclear magnetic resonance (NMR) has been used to study ordered proteins with atomic resolution since the 1950s [39,40] and applied to IDPs for several decades [37,41]. NMR relies on the local environment of each nucleus to produce a unique chemical shift signal which provides information on the conformation and close surroundings [37,42,43]. IDPs do not have a stable local environment so NMR alone lacks the ability to characterize disordered regions. However, NMR methods such as paramagnetic relaxation enhancement (PRE), secondary chemical shift (SCS), residual dipolar couplings (RDCs), and others can be used to characterize the conformational dynamics of IDP structures. Although NMR is a powerful technique, it is important to note that it is not without its limitations. Long IDPs must be divided into smaller sequences, experiments are often conducted at low temperatures which can decrease some kinetic activity, generally need high concentrations, and tags should be removed before conducting experiments [44]. NMR provides averages of ensembles but is limited in full conformational distribution determination for IDPs. A similar method, electron paramagnetic resonance (EPR), requires attachment of extrinsic spin labels and can be used at low temperatures to probe individual states and collect information on distance distributions [43,45].

### 2.2. Scattering Methods

Small angle X-ray scattering (SAXS) and small angle neutron scattering (SANS) are other methods commonly used to study IDPs which provide information on a global scale compared to the local scale NMR offers. A SAXS or SANS scattering profile can differentiate between globular and disordered proteins and determine a protein's size and overall shape [46,47], although these interpretations are low-resolution, require high protein concentrations, and are dependent on model selection [48,49]. These methods are commonly used to characterize IDPs [47]. An IDP will react to changes in its environments that allow the protein to bind or unbind to other molecules present in the cell. By changing the experimental conditions to mimic these signals (pH, temperature, additives, etc.), the behavior of an IDP changes on a global scale, which SAXS and SNAS are well equipped to measure.

### 2.3. Label-Based Approaches

Fluorescence correlation spectroscopy (FCS) is an optical technique commonly used to study the diffusion of fluorescently labeled molecular ensembles by measuring the time correlation of fluorescent fluctuations in the detected signal [50,51]. FCS is minimally invasive and does not require high protein concentrations [52,53]. As a solitary method, it is a powerful tool for studying the interactions between an IDP and its associated molecules, such as the Alzheimer's related protein Tau and tubulin dimers [54]. Although FCS alone cannot reveal information about secondary protein structures, the conformational dynamics of a protein can be determined when it is combined with the results from other fluorescent methods [50]. Electron paramagnetic resonance (EPR) spectroscopy detects unpaired electrons and is commonly coupled with site-directed spin labeling (SDSL) to study a protein's folding and unfolding events, interaction sites, and side chain mobility [45,55–57]. Although traditional labeling of a protein can change its conformational properties, the most common spin labels introduced via site-directed mutagenesis onto cysteine residues are relatively small, which decreases the risk deviating from wild-type behaviors [56,58]. SDSL-EPR spectroscopy is a sensitive and practical way to study the disorder-to-order transitions an IDP undergoes during binding events in near-native conditions [58,59]. This method is also capable of revealing IDPs or regions of an IDP that remain unstructured upon binding and complex formation [56]. Another EPR technique commonly used in the study of IDPs is Double Electron Electron Resonance (DEER), also called Pulsed Electron Double Resonance (PELDOR), which is well suited to determine spin site distances [45,60,61]. Because DEER requires spin-labeling, the distance measurements possess an inherent uncertainty due to potential (unintended) impacts on molecular conformation from the presence of the labels [45]. Multiparameter fluorescence detection (MFD) is an approach which collects fluorescent information such as intensity, lifetime, anisotropy, excitation and fluorescence spectra, and fluorescence quantum yield [62–67]. MFD is useful for improving the resolution of ensemble fluorescence experiments to reveal differences between similar sub-populations [65,67,68].

### 2.4. Single-Molecule Approaches

NMR, EPR (DEER) and SAXS are powerful methods that can be used to collect data about IDPs; however, the information provided is limited to the characteristics of an ensemble. Instead of averaging the properties of an ensemble, single molecule techniques can resolve dynamics and conformations of individual molecules [69–71]. Single molecule fluorescence (or Forster) resonance energy transfer (smFRET) is an optical spectroscopy approach to measuring the distance between two fluorophores of choice, but the fluorophore and position of labeling must be carefully considered to minimize the possibility of changing the dynamics of the protein. This is particularly useful in the study of IDPs because of the irregular folding dynamics of each protein as well as protein–protein interactions and protein aggregation [72–75]. smFRET has been applied to studies of many IDPs including the human proteins histone H1 and its partner nuclear protein prothymosin-alpha (ProTa), SNARE complexes such as syntaxin and SNAP-25, and Prostate-associated Gene 4 (PAGE4) [51,72,76–87].

### 2.5. Atomic Force Microscopy

Electron microscopy (EM) was among some of the first methods used to obtain structural data for proteins; however, it has had limited use for studying IDPs [34]. An exciting new technique being applied to study IDPs is high-speed AFM [88,89]. High-speed atomic force microscopy (HS-AFM) is a method particularly suited for studying protein functions in near-native conditions with no labeling necessary. HS-AFM has the capability to observe

IDPs transitioning between states of order and disorder and partial folding under certain conditions with a broader range of applicable length scales than FRET [90]. Interactions with surfaces that might shift energy landscapes and thus conformational ensembles is a concern, but the practice of this method is advancing rapidly.

### 2.6. Cryo-EM and X-ray Crystallography

Cryo-electron microscopy (cryo-EM) is a rapidly advancing technique that is gaining popularity as a method for protein structural analysis [91]. Before the development of commercially available direct electron detectors and data analysis software for cryo-EM, X-ray crystallography was the method of choice to investigate protein structure [92]. However, X-ray crystallography does not provide insights on the properties of a disordered region due to its atomic flexibility, resulting in non-coherent X-rays. Instead, the lack of structural data, or missing electron density, is used to determine where disordered regions are located [37]. Unlike X-ray crystallography, cryo-EM does not require sample crystallization; instead, the proteins are frozen in a thin layer of solution [91,93,94]. Cryo-EM works well for proteins with large molecular weight and can survey multiple conformational states. However, similar to X-ray crystallography, cryo-EM only works with a moderate level of heterogeneity and regions of disorder are represented with poor resolution [92,93,95]. Time resolved measurements with both X-ray and cryo-EM methods for folded proteins are developing [96–100].

### 2.7. Solvent Accessibility Methods

Because solvent accessibility is associated with protein folding and stability, it can be a useful parameter when classifying and modeling an IDP [101].

#### 2.7.1. Hydrogen-Deuterium-Exchange

Hydrogen-deuterium-exchange (HDex or HDX) measures differences in deuterium uptake that are reflected in the solvent accessibility of the protein under native conditions in solution [102]. Information gathered from HDX is useful for studying folding intermediates as well as protein dynamics as the protein performs its function [102–104]. IDPs can be difficult to study using HDX because of their flexibility, heterogeneity in solution, and fast deuteration times [102]. Lowering the pH of the solution decreases the exchange rate and provides reasonable experimental time windows for the study of IDPs using HDX. To avoid the affect that lowering the pH can have on a protein's structure and dynamics, pulse labeling HDX has been used to study IDPs [103–105].

#### 2.7.2. Crosslinking Mass Spectrometry

Crosslinking mass spectrometry (XL-MS) uses a "bottom-up" approach that supplies information on interaction sites rather than the "top-down" approach of native MS which informs overall protein structure [106]. XL-MS can be used to study the interaction sites between proteins or within a single protein. A cross-linking reaction, which can be performed in the protein's native environment, covalently links nearby amino acids that react with the crosslinker of choice [104]. Another advantage of XL-MS is the low protein concentration required to perform experiments. Two residues often targeted in XL-MS are lysine and arginine which are frequently abundant in disordered regions or disordered proteins, causing XL-MS to gain popularity as a method for IDP studies [106–109]. However, studying dynamic proteins such as IDPs with XL-MS can be challenging because the results often reflect only a fraction of the conformations or residue distances of the ensemble [104].

#### 2.7.3. Proteolysis

Proteolysis, the enzymatic digestion of a protein into amino acids, disproportionally affects unstructured sequence regions [110]. IDPs are digested more quickly than ordered proteins due to their flexibility and the accessibility of protease susceptible sequences [13,111,112].

The rates of digestion are quantified via SDS-PAGE or liquid chromatography mass spectroscopy (Figure 2), which can then be used to loosely determine degree of disorder [113].

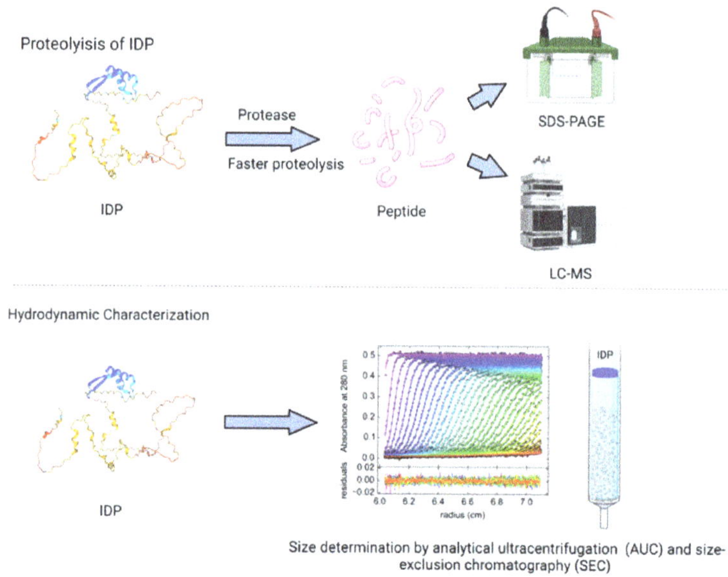


**Figure 2.** IDPs undergo faster proteolysis with enzymatic digestion compared to the structured proteins. The digested peptides are further analyzed using SDS-PAGE and LS-MS techniques. Techniques such as analytical ultracentrifugation and size-exclusion chromatography help to characterize the hydrodynamic size of IDPs.

### 2.8. *Spectroscopies*

Spectroscopy is an invaluable tool for probing and studying characteristics of IDPs.

#### 2.8.1. Circular Dichroism

Circular dichroism, the difference between the absorption coefficient of left- and right-handed circularly polarized light, is measured via circular dichroism (CD) spectroscopy [114]. CD spectroscopy is a powerful technique used to investigate secondary structure elements [115,116]. IDPs possess dynamic secondary structures that can be well assessed and characterized in an average sense by CD spectroscopy analysis [114,117]. Structural dynamics of an IDP can be reduced or promoted by altering their physical or chemical environment, which can then be quantified using CD spectroscopy. The two spectral regions used to study CD in proteins are the near-UV (250–300 nm) which correspond to the aromatic side chains and the far-UV (175–250 nm) that inform about secondary structures. Because an IDP moves through secondary structure as it changes conformations, far-UV CD spectroscopy is particularly useful for reporting the presence of alpha helices and beta sheets [118]. Time-resolved approaches using synchrotron light sources can provide information on dynamic processes in proteins down to nanosecond timescales [119], which eventually may prove useful for IDPs.

#### 2.8.2. Fourier Transform Infrared Spectroscopy

Fourier transform infrared spectroscopy (FTIR) is another spectroscopic method used to study the secondary structure of proteins [120]. FTIR relies on the absorption of infrared light at the frequency of the sample's molecular vibrational modes. The vibrational modes of a polypeptide chain, a repeated sequence of peptide bonds inherent to proteins, can produce up to nine bands measured by FTIR, the two most studied being the amide I and amide II bands [121]. Specifically, the amide I band is used to observe secondary structure formation. FTIR is also commonly used to study the aggregation of IDPs, such as the Parkinson's disease associated IDP α-synuclein [120,122,123].

2.8.3. Raman Spectroscopy

Raman spectroscopy obtains its name from its use of Raman scattering, or the inelastic scattering of light, due to a system's molecular vibrations [121,124]. Comparable to FTIR spectroscopy, the measured change in energy from the incident light can be correlated to the protein's vibrational modes and secondary structure [125,126]. Raman spectroscopy can be performed at dilute concentrations which is advantageous in the study of IDPs due to their aggregation tendences at high concentrations [127]. The conformational changes of an IDP are also well characterized by Raman spectral analysis. Raman optical activity (ROA) is another Raman scattering technique that measures the change in vibrational spectra due to left- and right-circularly polarized light and can add information about secondary and tertiary structures [124,128].

2.8.4. Mass Spectrometry

Native mass spectrometry (MS) is a technique used in structural biology for studying the structure and stoichiometry of proteins through their mass to charge (m/z) ratio. MS has the capability to inform on multiple conformational states present in a heterogenous mixture and is often combined with other methods, such as ion mobility MS (IM-MS), which can separate the proteins by size and charge [104,129,130]. Time resolved MS has been successfully used to measure dynamic processes in proteins [131].

*2.9. Hydrodynamic Characterizations*

The hydrodynamic properties of a protein are necessary for conformation classification and can be determined with methods such as dynamic light scattering (DLS), FCS (see Section 2.3), size-exclusion chromatography (SEC, also known as gel filtration), and analytical ultracentrifugation (AUC) [46]. DLS, SEC, and AUC are complementary methods for studying the hydrodynamic (Stokes) radius, $R_H$ [132]. DLS is a simple and noninvasive technique that can be used to obtain information on a protein's hydrodynamic dimensions [133,134]. DLS measures the scattering of light caused by Brownian motion and has been applied to the study of IDPs with high aggregation tendencies [135,136]. SEC uses porous beads to separate molecules based on hydrodynamic dimensions [46]. AUC uses centrifugal force generated in a centrifuge to separate molecules based on their hydrodynamic properties (Figure 2). AUC can experimentally determine the sedimentation coefficient, s, which is inversely related to the Stokes radius [132]. Various combinations of these techniques, as well as molecular simulations, have been used to calculate and confirm the hydrodynamic characteristics of IDPs [78,137].

*2.10. Computational Methods*

All atom molecular dynamics simulation (MD simulation) is a computational method used to predict the behavior of proteins, especially when combined with parameters from data acquired via X-ray crystallography, SAXS, NMR, or other techniques [36]. MD has been increasingly applied to characterize conformational ensembles of IDPs [138–142]. MD simulation is a highly valuable tool for data analysis and structural modeling but is not without its limitations. Force fields that are used for MD simulations of structured proteins fail to succeed when applied to IDPs and the inhomogeneous conformational landscape occupied by any single IDP also presents modeling challenges [143]. MD simulations are a key tool in integrative structural biology due to their ability to combine information from many methods and create a unified model of a protein's structure and conformational changes [144,145].

Until recently, a protein's tertiary structure was unpredictable based on its amino acid sequence. The residue sequence in disordered regions varies in composition when compared to ordered proteins [146,147]. Several disordered regions of proteins have been predicted by a group of algorithms within the PONDR (predictor of natural disordered regions) family [148,149]. Using more than one predictor and averaging the results provide a more robust disorder profile than a single algorithm [144]. The associative memory, water-mediated, structure and energy model (AWSEM) is a course-grained force field model that has been used to predict protein structure, folding, and aggregation [144,150,151]. AWSEM's optimized force fields have correctly predicted protein structures dependent solely on sequence [150,152,153]. AlphaFold, a machine learning model created by DeepMind, has made significant strides in the field of structural biology after successfully predicting the three-dimensional structure of proteins based on sequencing data [154]. Regions of low confidence in AlphaFold's predictions correlate to disordered regions and confirm previous estimates that more than 30% of protein regions are disordered [154].

## 3. The Integrative Structural Biology Approach to IDPs and Examples

IDPs by nature fluctuate on many timescales among wide ranges of conformations. Their conformational ensembles can be altered by accessory proteins or post-translational modification. Thus, using an integrated, multiscale approach (integrative structural biology) rather than a single isolated technique is more prudent for accurately characterizing the dynamics and fluctuating conformational landscapes inherent to IDPs. Using a battery of methods with different sensitivities, complemented by advanced computational simulations, is essential to characterize the full range of the conformation space. Studying an IDP is like putting together a jigsaw puzzle where each method provides a limited number of pieces. Method one may gather all the blue pieces together, while method two helps arrange the edges. The full picture comes together only when the information from one method can be placed into its larger context with complementary methods. Therefore, instead of relying on the limited data provided by a single experimental method, integrative structural biology is an approach that combines the data from various methods to form models and a more complete understanding of these proteins [34–36,61,80,140,155–163].

As the application of integrative approaches to study IDPs is increasing at a rapid pace, here we will highlight only a few of the many successes. Our goal is to illustrate some different combinations of methods or cases where unexpected functions are uncovered. We will not discuss the important related topic of using combinations of methods to characterize dynamic assemblies of folded domains connected by disordered linkers [34,159,164–166].

### 3.1. Ubiquitin

To examine the robustness of an integrative structural biology approach, the ubiquitin protein in its denatured state was observed by combining results from multiple methods [156]. Ubiquitin is a regulatory protein involved in cell regulation with a tertiary structure that is denatured as urea concentration increases. Data collected from smFRET, NMR, and SAXS had good agreement for the distance distributions for unfolded ubiquitin. Local structure and dynamics were derived from NMR restraints while the overall shape was provided by SAXS measurements. Intramolecular distances and distributions within subpopulations as well as dynamic properties of the protein's conformational changes were uncovered by smFRET. In this study, combining the results of smFRET, NMR, and SAXS provided a complete picture of the conformational ensemble of this unfolded protein.

### 3.2. Nucleoporins

Phenylalanine-glycine-rich nucleoporins have also been studied using an integrative structural biology approach [167]. A combination of SAXS, smFRET measurements was compared with MD simulations that used different models for solvent interactions. The ultimate agreement of experiment and simulations in this work highlights successful approaches to improve theoretical force fields used to model IDPs.

### 3.3. Aggregation-Prone Synaptic Proteins

The aggregation of some IDPs is associated with the pathology of diseases, such as α-Synuclein (αS) with Parkinson's disease or amyloid-β (Aβ) and Tau with Alzheimer's disease. αS has previously been investigated using X-ray diffraction [168] and NMR [157], but more recent studies [158] have used smFRET combined with MD simulations and NMR measurements to provide information on its structure and dynamics. Good agreement was found with other methods, and the conditions found to promote aggregation pointed toward possible therapeutic approaches to target αS.

Similarly, Aβ has been investigated [163]. Fluorophores were used to label both the N- and C- termini and FRET was observed in both free-diffusion and immobilized modes. Again, results aligned with previously reported data while adding information on possible reasons for aggregation of monomeric Aβ.

In the mid-1990s, before the wide acceptance of IDPs, studies of the Tau protein showed that its overall shape and conformation suggested it was similar to a denatured protein with no tertiary structure [169]. Since then, integrated structural biology has enhanced our understanding of these IDPs which are implicated in neurodegenerative disorders. There is evidence to support that both the aggregation of Tau and increased Tau-tubulin binding influence the pathology of disease. smFRET data combined with Monte Carlo simulations provide possible Tau conformations on binding to tubulin dimers [169].

### 3.4. Sic1

In *Saccharomyces cerevisiae*, Sic1 is a disordered protein involved in cell cycle regulation and DNA replication initiation [170–172]. Sic1 forms a complex with a subunit of ubiquitin ligase, Cdc4, after the phosphorylation of at least six of the nine Cdc4 phosphodegron (CPD) sites on Sic1, seven of which are located on the 90 residue N-terminal (Figure 3A) [170,172]. Phosphorylation followed by ubiquitination results in the degradation of Sic1, which allows DNA replication to begin [170,171,173,174]. An integrative structural biology approach to Sic1 characterization that used NMR, SAXS, and smFRET (Figure 3B,C) focused on the seven CPD sites on the disordered N-terminal [170]. Phosphorylated Sic1 (pSic1) has different binding properties than Sic1, but neither phosphorylation nor Cdc4 binding creates a disorder-to-order transition of Sic1. SAXS and smFRET of both Sic1 and pSic1 were constrained by including NMR-PRE data and indicated a subtle conformational change in Sic1 after phosphorylation. Analysis of SAXS and smFRET data showed that these methods were individually capable of accurately measuring the root-mean-squared radius of gyration $R_g$ and the root-mean-squared end-to-end distance $R_{e\text{-}e}$, respectively. SAXS data alone show little change in conformational properties between Sic1 and pSic1; however, SAXS+PRE restrained ensembles show an expansion of $R_{e\text{-}e}$ which is consistent with the change in distance observed by smFRET.

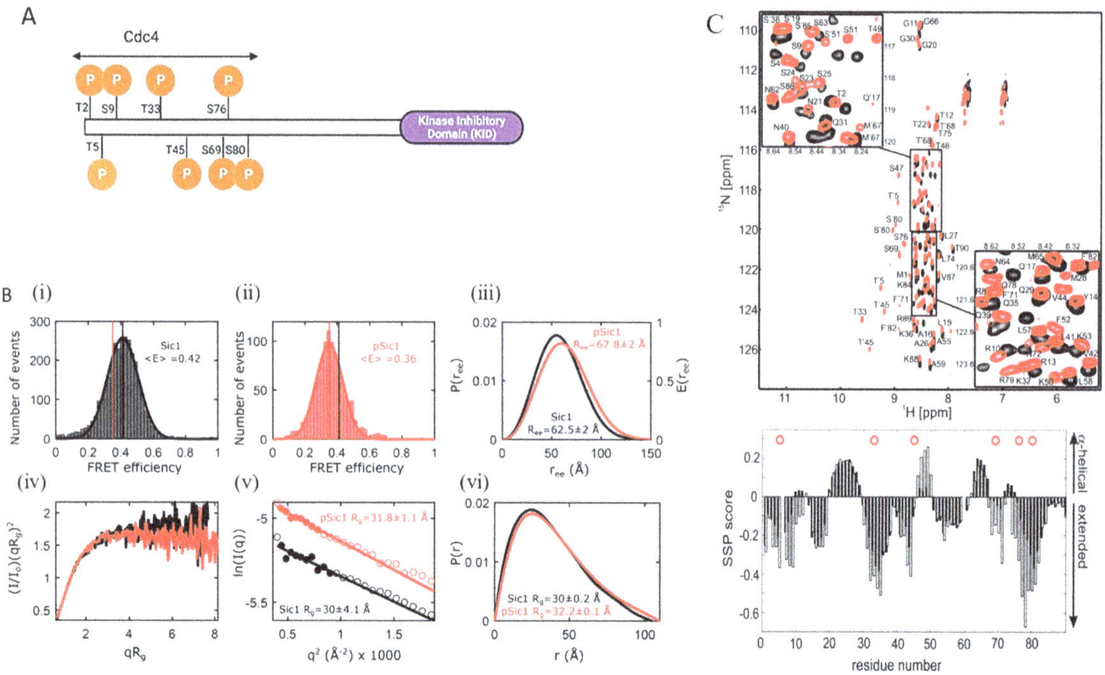

**Figure 3.** An integrative approach to elucidate IDP structure. (**A**) Schematic representation of full-length intrinsically disordered Sic1 protein and CPD phosphorylation sites in the N-terminal domain. The minimum functional KID fragment, the last 70 residues, is indicated in the purple box. (**B**) top three panels (**i–iii**) show smFRET efficiency histograms of Sic1 and pSic1 and end to end probability distributions. iv–vi show SAXS data for Sic1 and pSic1 and deduced $R_g$, which was estimated to be approximately 30 Å for Sic1 and 32 Å for pSic1 [170]. (B) is adapted with permission from [170]. Copyright 2020, American Chemical Society. (**C**) Upper panel displays $^1$HN-$^{15}$N correlation spectra of Sic1 (black) and pSic1 (red). The lower panel shows experimentally determined secondary structural propensity (SSP) values (described in [175]) for Sic1 (black bars) and pSic1 (open bars). Note that the helical vs. extended interpretations are marked on the right axis. Red circles indicate the locations of the phosphorylation sites [175]. (**C**) is reproduced with permission from [175]. Copyright 2008, National Academy of Sciences, USA.

### 3.5. N-WASP

An integrative approach allowed characterization of the conformational ensemble of the disordered domain of the neural Aldrich syndrome protein (N-WASP) [176], which regulates actin assembly pathways [177]. MD modeling generated conformational ensembles of the protein, which were validated by NMR and SAXS measurements. Using the SAXS and NMR data to benchmark simulations and guide selection of optimal force fields allowed the MD simulations to reveal both the global and local details of the conformational ensemble of this disordered protein. The simulations provided information about the transient underlying secondary structure within the ensembles. The use of experimentally derived restraints to guide computational modeling [178–180] or, more generally, cross validating simulation with experiments is a powerful tool to apply to IDP studies because it provides insights into both global and local structural features of the conformational ensembles.

### 3.6. SNAP-25

SNAP-25 is a SNARE protein that is a key player in neurotransmitter release. SNAP-25 is an intrinsically disordered protein that undergoes a disorder-to-order transition upon binding its partners syntaxin and synaptobrevin where it folds into colinear alpha helices to form the SNARE complex. SNARE complex formation is associated with membrane fusion of synaptic vesicles to the pre-synaptic terminal to release neurotransmitters. Integrated structural biology investigations of SNAP-25 combining smFRET, AUC, DLS, CD (circular dichroism) and SEC characterized the conformational ensemble in the isolated disordered state as consistent with a simple, semi-flexible polymer model with no underlying structure [78]. Interestingly, smFRET measurements of SNAP-25, in a binary complex with syntaxin (lacking synaptobrevin) that is on the pathway to full SNARE complex, found the transient tendency to switch between the folded alpha helix and a disordered conformation [87]. Returning to the isolated protein using additional methods of single molecule MFD, double electron-electron resonance (DEER), and MD, transient helix–coil transitions in short regions of SNAP-25 that occur in sub-millisecond timescales were observed despite a disordered fluctuating ensemble being the dominant conformational feature [161]. It was suggested that these transient alpha helix forming tendencies could play a role in priming SNAP-25 to zip into the SNARE complex rapidly upon binding with the requisite partners, assisting in the speed of neurotransmitter release. This example illustrates the value of the integrative structural biology approach for addressing measurements at the many length scales and timescales required to characterize IDP conformational ensembles, especially those with switching tendencies [82].

### *3.7. p27*

p27 is a member of the Kip family of cyclin-dependent kinase inhibitors that plays an important role in controlling the cell cycle in eukaryotes [181]. Binding of the disordered C-terminal domain of p27 with cyclin-dependent kinase (Cdk2) and cyclins results in a disordered-to-ordered transition that has regulatory impact on the cell cycle. By integrating results from single-molecule multiparameter fluorescence spectroscopy, stopped-flow experiments, and molecular dynamics simulation, the multi-step process of assembling this fuzzy complex involving the disordered domain of p27 was mapped out [182,183].

The unbound p27 was found to be compact at the scale of a random coil by an integrated approach but rapidly fluctuating with dynamics covering orders of magnitudes of time scales from nanoseconds to milliseconds [184,185]. The interaction with its binding partners induced a multi-step process where p27 switches among conformational ensembles until favorable conformation is encountered to advance the binding process. In the end, p27 binds its partners in a more extended conformation than in isolation but remains dynamic without a fixed structure in a fuzzy complex. Elucidation of this pathway suggests that the assembly of the complex starts with a first recognition step involving conformational selections among rapidly fluctuating states, followed by a period waiting for a switch between distinct conformational sub-ensembles permitting progression to a later step where an induced fit phenomena completes the assembly. The complexity of the binding sequence is suggested to offer multiple opportunities for regulation of the assembly by other cellular signals.

### *3.8. PAGE4*

Prostate-associated gene 4 (PAGE4) is an IDP that is expressed only in the prostate and only during early developmental stages and in the cancerous state [184]. An integrated structural biology approach identified phosphorylation-induced changes in the conformational ensemble of this IDP that were connected to impact cellular signaling pathway important to cancer progression. Combining experimental results from NMR, PRE, SAXS and smFRET studies with MD simulations revealed distinct changes in the conformational ensemble upon phosphorylation by different kinases [79,80,83,142,144,185,186]. In particular, phosphorylation by homeodomain-interacting protein kinase 1 (HIPK1) leads to a more compact ensemble, whereas phosphorylation by CDC-Like Kinase 2 (CLK2) expands the ensemble. The change in the conformational state was connected to signaling in prostate cancer by its ability to regulate interactions with the transcription factor c-Jun. HIPK1 treatment resulted in increased c-Jun dependent transcription activity in cell models of prostate cancer while CLK2 phosphorylation caused the opposite [79,80]. Given that CLK2 and PAGE4 are expressed only in androgen-dependent prostate cancer cells whereas HIPK1 is expressed in all prostate cancer cells (both androgen-dependent and -independent), these phosphorylation states that result in the expanded and contracted conformational ensembles were correlated with androgen sensitivity in prostate cancer [83,144,185–189]. Modeling these changing transcription factor interactions in a cellular androgen control pathway suggested that the PAGE4 phosphorylation state could oscillate in time, which could result in temporal oscillations of androgen sensitivity in prostate cancer [187–189]. This model suggests direct connections between changes in the conformational ensemble of an IDP and cell phenotypes in a cancer model. Such a complete picture would not have been obtained without the use of the integrated structural biology approach.

## 4. Summary and Conclusions

It took more than two decades for IDPs to be recognized as legitimate biological entities [190] with important functions in myriad biological functions from prebiotic evolution, multicellularity, and cell fate determination to phenotypic plasticity, adaptive evolution, and disease pathology. Several of Uversky's contributions to the IDP field have shed new light on these important components of the proteome including remarkable conceptual advances from a dynamical systems perspective [191,192]. Therefore, this Special Issue of Biomolecules dedicated to VladimirUversky on his 60th birthday, is a celebration of his many contributions over the past three decades.

**Author Contributions:** Conceptualization, R.E., P.K. and K.W.; writing—original draft preparation, R.E.; writing—review and editing, R.E., S.R., P.K. and K.W.; graphics—S.R. All authors have read and agreed to the published version of the manuscript.

**Funding:** K.W. research is funded by the National Institute of General Medical Sciences (GM132263).

**Institutional Review Board Statement:** Not applicable.

**Informed Consent Statement:** Not applicable.

**Data Availability Statement:** Not applicable.

**Conflicts of Interest:** The authors declare no conflict of interest.

## References

1. Peng, Z.; Yan, J.; Fan, X.; Mizianty, M.J.; Xue, B.; Wang, K.; Hu, G.; Uversky, V.N.; Kurgan, L. Exceptionally Abundant Exceptions: Comprehensive Characterization of Intrinsic Disorder in All Domains of Life. *Cell. Mol. Life Sci.* **2015**, *72*, 137–151. [CrossRef] [PubMed]
2. Xue, B.; Dunker, A.K.; Uversky, V.N. Orderly Order in Protein Intrinsic Disorder Distribution: Disorder in 3500 Proteomes from Viruses and the Three Domains of Life. *J. Biomol. Struct. Dyn.* **2012**, *30*, 137–149. [CrossRef]
3. Ward, J.J.; Sodhi, J.S.; McGuffin, L.J.; Buxton, B.F.; Jones, D.T. Prediction and Functional Analysis of Native Disorder in Proteins from the Three Kingdoms of Life. *J. Mol. Biol.* **2004**, *337*, 635–645. [CrossRef] [PubMed]
4. Dunker, A.K.; Obradovic, Z.; Romero, P.; Garner, E.C.; Brown, C.J. Intrinsic Protein Disorder in Complete Genomes. *Genome Inform. Ser. Workshop Genome Inform.* **2000**, *11*, 161–171. [PubMed]
5. DeForte, S.; Uversky, V.N. Intrinsically Disordered Proteins in PubMed: What Can the Tip of the Iceberg Tell Us about What Lies Below? *RSC Adv.* **2016**, *6*, 11513–11521. [CrossRef]
6. Chao, L.-P.; Roboz Einstein, E. Estimation of the Molecular Weight of Flexible Disordered Proteins by Exclusion Chromatography. *J. Chromatogr. A* **1969**, *42*, 485–492. [CrossRef]
7. Boesch, C.; Bundi, A.; Oppliger, M.; Wuthrich, K. 1H Nuclear-Magnetic-Resonance Studies of the Molecular Conformation of Monomeric Glucagon in Aqueous Solution. *Eur. J. Biochem.* **1978**, *91*, 209–214. [CrossRef]
8. Boublik, M.; Bradbury, E.M.; Crane-Robinson, C.; Johns, E.W. An Investigation of the Conformational Changes of Histone F2b by High Resolution Nuclear Magnetic Resonance. *Eur. J. Biochem.* **1970**, *17*, 151–159. [CrossRef]
9. Huber, R.; Bennett, W.S. Functional Significance of Flexibility in Proteins. *Biopolymers* **1983**, *22*, 261–279. [CrossRef]
10. Sigler, P.B. Acid Blobs and Negative Noodles. *Nature* **1988**, *333*, 210–212. [CrossRef]
11. Nishikawa, K. Natively Unfolded Proteins: An Overview. *Biophysics* **2009**, *5*, 53–58. [CrossRef] [PubMed]

12. Uversky, V.N.; Dunker, A.K. Understanding Protein Non-Folding. *Biochim. Biophys. Acta (BBA)—Proteins Proteom.* **2010**, *1804*, 1231–1264. [CrossRef] [PubMed]
13. Dyson, H.J.; Wright, P.E. Intrinsically Unstructured Proteins and Their Functions. *Nat. Rev. Mol. Cell Biol.* **2005**, *6*, 197–208. [CrossRef] [PubMed]
14. Kulkarni, P.; Uversky, V.N. Intrinsically Disordered Proteins: The Dark Horse of the Dark Proteome. *Proteomics* **2018**, *18*, 1800061. [CrossRef] [PubMed]
15. Wright, P.E.; Dyson, H.J. Intrinsically Unstructured Proteins: Re-Assessing the Protein Structure-Function Paradigm. *J. Mol. Biol.* **1999**, *293*, 321–331. [CrossRef]
16. Uversky, V.N.; Gillespie, J.R.; Fink, A.L. Why Are "Natively Unfolded" Proteins Unstructured under Physiologic Conditions? *Proteins Struct. Funct. Genet.* **2000**, *41*, 415–427. [CrossRef]
17. Tompa, P. Intrinsically Unstructured Proteins. *Trends Biochem. Sci.* **2002**, *27*, 527–533. [CrossRef]
18. Dunker, A.K.; Lawson, J.D.; Brown, C.J.; Williams, R.M.; Romero, P.; Oh, J.S.; Oldfield, C.J.; Campen, A.M.; Ratliff, C.M.; Hipps, K.W.; et al. Intrinsically Disordered Protein. *J. Mol. Graph. Model.* **2001**, *19*, 26–59. [CrossRef]
19. Ferrie, J.J.; Karr, J.P.; Tjian, R.; Darzacq, X. "Structure"-Function Relationships in Eukaryotic Transcription Factors: The Role of Intrinsically Disordered Regions in Gene Regulation. *Mol. Cell* **2022**, *82*, 3970–3984. [CrossRef]
20. Trnka, M.J.; Pellarin, R.; Robinson, P.J. Role of Integrative Structural Biology in Understanding Transcriptional Initiation. *Methods* **2019**, *159–160*, 4–22. [CrossRef]
21. Choi, U.B.; Kazi, R.; Stenzoski, N.; Wollmuth, L.P.; Uversky, V.N.; Bowen, M.E. Modulating the Intrinsic Disorder in the Cytoplasmic Domain Alters the Biological Activity of the N-Methyl-D-Aspartatesensitive Glutamate Receptor. *J. Biol. Chem.* **2013**, *288*, 22506–22515. [CrossRef]
22. Hu, G.; Wu, Z.; Uversky, V.; Kurgan, L. Functional Analysis of Human Hub Proteins and Their Interactors Involved in the Intrinsic Disorder-Enriched Interactions. *Int. J. Mol. Sci.* **2017**, *18*, 2761. [CrossRef] [PubMed]
23. Deiana, A.; Forcelloni, S.; Porrello, A.; Giansanti, A. Intrinsically Disordered Proteins and Structured Proteins with Intrinsically Disordered Regions Have Different Functional Roles in the Cell. *PLoS ONE* **2019**, *14*, e0217889. [CrossRef] [PubMed]
24. Wright, P.E.; Dyson, H.J. Intrinsically Disordered Proteins in Cellular Signalling and Regulation. *Nat. Rev. Mol. Cell Biol.* **2015**, *16*, 18–29. [CrossRef] [PubMed]
25. Kulkarni, P.; Leite, V.B.P.; Roy, S.; Bhattacharyya, S.; Mohanty, A.; Achuthan, S.; Singh, D.; Appadurai, R.; Rangarajan, G.; Weninger, K.; et al. Intrinsically Disordered Proteins: Ensembles at the Limits of Anfinsen's Dogma. *Biophys. Rev.* **2022**, *3*, 011306. [CrossRef]
26. Dunker, A.K.; Cortese, M.S.; Romero, P.; Iakoucheva, L.M.; Uversky, V.N. Flexible Nets. The Roles of Intrinsic Disorder in Protein Interaction Networks. *FEBS J.* **2005**, *272*, 5129–5148. [CrossRef]
27. Csermely, P.; Palotai, R.; Nussinov, R. Induced Fit, Conformational Selection and Independent Dynamic Segments: An Extended View of Binding Events. *Nat. Preced.* **2010**, *35*, 539–546. [CrossRef]
28. Berlow, R.B.; Dyson, H.J.; Wright, P.E. Expanding the Paradigm: Intrinsically Disordered Proteins and Allosteric Regulation. *J. Mol. Biol.* **2018**, *430*, 2309–2320. [CrossRef]
29. Fung, H.Y.J.; Birol, M.; Rhoades, E. IDPs in Macromolecular Complexes: The Roles of Multivalent Interactions in Diverse Assemblies. *Curr. Opin. Struct. Biol.* **2018**, *49*, 36–43. [CrossRef]
30. Coskuner-Weber, O.; Mirzanli, O.; Uversky, V.N. Intrinsically Disordered Proteins and Proteins with Intrinsically Disordered Regions in Neurodegenerative Diseases. *Biophys. Rev.* **2022**, *14*, 679–707. [CrossRef]
31. Martinelli, A.; Lopes, F.; John, E.; Carlini, C.; Ligabue-Braun, R. Modulation of Disordered Proteins with a Focus on Neurodegenerative Diseases and Other Pathologies. *Int. J. Mol. Sci.* **2019**, *20*, 1322. [CrossRef] [PubMed]
32. Dignon, G.L.; Best, R.B.; Mittal, J. Biomolecular Phase Separation: From Molecular Driving Forces to Macroscopic Properties. *Annu. Rev. Phys. Chem.* **2020**, *71*, 53–75. [CrossRef] [PubMed]
33. Brangwynne, C.P.; Tompa, P.; Pappu, R.v. Polymer Physics of Intracellular Phase Transitions. *Nat. Phys.* **2015**, *11*, 899–904. [CrossRef]
34. Rout, M.P.; Sali, A. Principles for Integrative Structural Biology Studies. *Cell* **2019**, *177*, 1384–1403. [CrossRef] [PubMed]
35. Ward, A.B.; Sali, A.; Wilson, I.A. Integrative Structural Biology. *Science* **2013**, *339*, 913–915. [CrossRef] [PubMed]
36. Masrati, G.; Landau, M.; Ben-Tal, N.; Lupas, A.; Kosloff, M.; Kosinski, J. Integrative Structural Biology in the Era of Accurate Structure Prediction: The Era of Accurate Structure Prediction. *J. Mol. Biol.* **2021**, *433*, 167127. [CrossRef]
37. Cohen, I.R.; Lajtha, A.; Lambris, J.D.; Paoletti, R. *Intrinsically Disordered Proteins Studied by NMR Spectroscopy*; Felli, I.C., Pierattelli, R., Eds.; Advances in Experimental Medicine and Biology; Springer International Publishing: Cham, Switzerland, 2015; Volume 870, ISBN 978-3-319-20163-4.
38. Musacchio, A. On the Role of Phase Separation in the Biogenesis of Membraneless Compartments. *EMBO J.* **2022**, *41*, 1–20. [CrossRef] [PubMed]
39. Saunders, M.; Wishnia, A.; Kirkwood, J.G. The Nuclear Magnetic Resonance Spectrum of Ribonuclease. *J. Am. Chem. Soc.* **1957**, *79*, 3289–3290. [CrossRef]
40. Kowalsky, A. Nuclear Magnetic Resonance Studies of Proteins. *J. Biol. Chem.* **1962**, *237*, 1807–1819. [CrossRef]
41. Dyson, H.J.; Wright, P.E. NMR Illuminates Intrinsic Disorder. *Curr. Opin. Struct. Biol.* **2021**, *70*, 44–52. [CrossRef]

42. Mureddu, L.; Vuister, G.W. Simple High-Resolution NMR Spectroscopy as a Tool in Molecular Biology. *FEBS J.* **2019**, *286*, 2035–2042. [CrossRef] [PubMed]
43. Konrat, R. NMR Contributions to Structural Dynamics Studies of Intrinsically Disordered Proteins. *J. Magn. Reson.* **2014**, *241*, 74–85. [CrossRef]
44. Prestel, A.; Bugge, K.; Staby, L.; Hendus-Altenburger, R.; Kragelund, B.B. *Characterization of Dynamic IDP Complexes by NMR Spectroscopy*, 1st ed.; Elsevier Inc.: Amsterdam, The Netherlands, 2018; Volume 611, ISBN 9780128156490.
45. Drescher, M. *EPR in Protein Science*; Springer: Berlin/Heidelberg, Germany, 2011; pp. 91–119.
46. Fontana, A.; de Laureto, P.P.; Spolaore, B.; Frare, E. *Intrinsically Disordered Protein Analysis*; Uversky, V.N., Dunker, A.K., Eds.; Methods in Molecular Biology; Springer: New York, NY, USA, 2012; Volume 896, ISBN 978-1-4614-3703-1.
47. Bernadó, P.; Svergun, D.I. Structural Analysis of Intrinsically Disordered Proteins by Small-Angle X-ray Scattering. *Mol. BioSyst.* **2012**, *8*, 151–167. [CrossRef] [PubMed]
48. Pauw, B.R. Corrigendum: Everything SAXS: Small-Angle Scattering Pattern Collection and Correction (2013 J. Phys.: Condens. Matter 25 383201). *J. Phys. Condens. Matter* **2014**, *26*, 239501. [CrossRef] [PubMed]
49. Fuertes, G.; Banterle, N.; Ruff, K.M.; Chowdhury, A.; Mercadante, D.; Koehler, C.; Kachala, M.; Estrada Girona, G.; Milles, S.; Mishra, A.; et al. Decoupling of Size and Shape Fluctuations in Heteropolymeric Sequences Reconciles Discrepancies in SAXS vs. FRET Measurements. *Proc. Natl. Acad. Sci. USA* **2017**, *114*, E6342–E6351. [CrossRef] [PubMed]
50. Luitz, M.P.; Barth, A.; Crevenna, A.H.; Bomblies, R.; Lamb, D.C.; Zacharias, M. Covalent Dye Attachment Influences the Dynamics and Conformational Properties of Flexible Peptides. *PLoS ONE* **2017**, *12*, e0177139. [CrossRef] [PubMed]
51. Nasir, I.; Onuchic, P.L.; Labra, S.R.; Deniz, A.A. Single-Molecule Fluorescence Studies of Intrinsically Disordered Proteins and Liquid Phase Separation. *Biochim. Biophys. Acta (BBA)—Proteins Proteom.* **2019**, *1867*, 980–987. [CrossRef]
52. Yu, L.; Lei, Y.; Ma, Y.; Liu, M.; Zheng, J.; Dan, D.; Gao, P. A Comprehensive Review of Fluorescence Correlation Spectroscopy. *Front. Phys.* **2021**, *9*, 644450. [CrossRef]
53. Haustein, E.; Schwille, P. Fluorescence Correlation Spectroscopy: Novel Variations of an Established Technique. *Annu. Rev. Biophys. Biomol. Struct.* **2007**, *36*, 151–169. [CrossRef]
54. Li, X.-H.; Culver, J.A.; Rhoades, E. Tau Binds to Multiple Tubulin Dimers with Helical Structure. *J. Am. Chem. Soc.* **2015**, *137*, 9218–9221. [CrossRef]
55. Drescher, M.; Jeschke, G. (Eds.) *EPR Spectroscopy*; Topics in Current Chemistry; Springer: Berlin/Heidelberg, Germany, 2012; Volume 321, ISBN 978-3-642-28346-8.
56. Le Breton, N.; Martinho, M.; Mileo, E.; Etienne, E.; Gerbaud, G.; Guigliarelli, B.; Belle, V. Exploring Intrinsically Disordered Proteins Using Site-Directed Spin Labeling Electron Paramagnetic Resonance Spectroscopy. *Front. Mol. Biosci.* **2015**, *2*, 21. [CrossRef] [PubMed]
57. Lorenzi, M.; Sylvi, L.; Gerbaud, G.; Mileo, E.; Halgand, F.; Walburger, A.; Vezin, H.; Belle, V.; Guigliarelli, B.; Magalon, A. Conformational Selection Underlies Recognition of a Molybdoenzyme by Its Dedicated Chaperone. *PLoS ONE* **2012**, *7*, e49523. [CrossRef] [PubMed]
58. Klare, J.P.; Steinhoff, H.-J. Spin Labeling EPR. *Photosynth. Res.* **2009**, *102*, 377–390. [CrossRef] [PubMed]
59. Longhi, S.; Belle, V.; Fournel, A.; Guigliarelli, B.; Carrière, F. Probing Structural Transitions in Both Structured and Disordered Proteins Using Site-Directed Spin-Labeling EPR Spectroscopy. *J. Pept. Sci.* **2011**, *17*, 315–328. [CrossRef]
60. Van Doorslaer, S.; Murphy, D.M. EPR Spectroscopy in Catalysis. In *EPR Spectroscopy*; Springer: Berlin, Germany, 2011; pp. 1–39.
61. Peter, M.F.; Gebhardt, C.; Mächtel, R.; Muñoz, G.G.M.; Glaenzer, J.; Narducci, A.; Thomas, G.H.; Cordes, T.; Hagelueken, G. Cross-Validation of Distance Measurements in Proteins by PELDOR/DEER and Single-Molecule FRET. *Nat. Commun.* **2022**, *13*, 4396. [CrossRef]
62. Widengren, J.; Kudryavtsev, V.; Antonik, M.; Berger, S.; Gerken, M.; Seidel, C.A.M. Single-Molecule Detection and Identification of Multiple Species by Multiparameter Fluorescence Detection. *Anal. Chem.* **2006**, *78*, 2039–2050. [CrossRef]
63. Ma, J.; Yanez-Orozco, I.S.; Rezaei Adariani, S.; Dolino, D.; Jayaraman, V.; Sanabria, H. High Precision FRET at Single-Molecule Level for Biomolecule Structure Determination. *J. Vis. Exp.* **2017**, *123*, e55623. [CrossRef]
64. Margittai, M.; Widengren, J.; Schweinberger, E.; Schröder, G.F.; Felekyan, S.; Haustein, E.; König, M.; Fasshauer, D.; Grubmüller, H.; Jahn, R.; et al. Single-Molecule Fluorescence Resonance Energy Transfer Reveals a Dynamic Equilibrium between Closed and Open Conformations of Syntaxin 1. *Proc. Natl. Acad. Sci. USA* **2003**, *100*, 15516–15521. [CrossRef]
65. Rothwell, P.J.; Berger, S.; Kensch, O.; Felekyan, S.; Antonik, M.; Wöhrl, B.M.; Restle, T.; Goody, R.S.; Seidel, C.A.M. Multiparameter Single-Molecule Fluorescence Spectroscopy Reveals Heterogeneity of HIV-1 Reverse Transcriptase:Primer/Template Complexes. *Proc. Natl. Acad. Sci. USA* **2003**, *100*, 1655–1660. [CrossRef]
66. Eggeling, C.; Berger, S.; Brand, L.; Fries, J.R.; Schaffer, J.; Volkmer, A.; Seidel, C.A.M. Data Registration and Selective Single-Molecule Analysis Using Multi-Parameter Fluorescence Detection. *J. Biotechnol.* **2001**, *86*, 163–180. [CrossRef]
67. Hamilton, G.; Sanabria, H. Multiparameter Fluorescence Spectroscopy of Single Molecules. In *Spectroscopy and Dynamics of Single Molecules*; Elsevier: Amsterdam, The Netherlands, 2019; pp. 269–333.
68. Sisamakis, E.; Valeri, A.; Kalinin, S.; Rothwell, P.J.; Seidel, C.A.M. Accurate Single-Molecule FRET Studies Using Multiparameter Fluorescence Detection. In *Methods in Enzymology*; Academic Press: New York, NY, USA, 2010; pp. 455–514.
69. LeBlanc, S.; Kulkarni, P.; Weninger, K. Single Molecule FRET: A Powerful Tool to Study Intrinsically Disordered Proteins. *Biomolecules* **2018**, *8*, 140. [CrossRef] [PubMed]

70. Holmstrom, E.D.; Holla, A.; Zheng, W.; Nettels, D.; Best, R.B.; Schuler, B. Accurate Transfer Efficiencies, Distance Distributions, and Ensembles of Unfolded and Intrinsically Disordered Proteins from Single-Molecule FRET. In *Methods in Enzymology*; Academic Press: New York, NY, USA, 2018; Volume 611, pp. 287–325, ISBN 9780128156490.
71. Barth, A.; Opanasyuk, O.; Peulen, T.O.; Felekyan, S.; Kalinin, S.; Sanabria, H.; Seidel, C.A.M. Unraveling Multi-State Molecular Dynamics in Single-Molecule FRET Experiments. I. Theory of FRET-Lines. *J. Chem. Phys.* **2022**, *156*, 5–8. [CrossRef] [PubMed]
72. Metskas, L.A.; Rhoades, E. Single-Molecule FRET of Intrinsically Disordered Proteins. *Annu. Rev. Phys. Chem.* **2020**, *71*, 391–414. [CrossRef]
73. Hofmann, H. Understanding Disordered and Unfolded Proteins Using Single-Molecule FRET and Polymer Theory. *Methods Appl. Fluoresc.* **2016**, *4*, 042003. [CrossRef] [PubMed]
74. Schuler, B.; Soranno, A.; Hofmann, H.; Nettels, D. Single-Molecule FRET Spectroscopy and the Polymer Physics of Unfolded and Intrinsically Disordered Proteins. *Annu. Rev. Biophys.* **2016**, *45*, 207–231. [CrossRef]
75. Tan, P.S.; Lemke, E.A. Probing Differential Binding Mechanisms of Phenylalanine-Glycine-Rich Nucleoporins by Single-Molecule FRET. *Methods Enzymol.* **2018**, *611*, 327–346. [CrossRef] [PubMed]
76. Borgia, A.; Borgia, M.B.; Bugge, K.; Kissling, V.M.; Heidarsson, P.O.; Fernandes, C.B.; Sottini, A.; Soranno, A.; Buholzer, K.J.; Nettels, D.; et al. Extreme Disorder in an Ultrahigh-Affinity Protein Complex. *Nature* **2018**, *555*, 61–66. [CrossRef]
77. Sakon, J.J.; Weninger, K.R. Detecting the Conformation of Individual Proteins in Live Cells. *Nat. Methods* **2010**, *7*, 203–205. [CrossRef]
78. Choi, U.B.; McCann, J.J.; Weninger, K.R.; Bowen, M.E. Beyond the Random Coil: Stochastic Conformational Switching in Intrinsically Disordered Proteins. *Structure* **2011**, *19*, 566–576. [CrossRef]
79. Mooney, S.M.; Qiu, R.; Kim, J.J.; Sacho, E.J.; Rajagopalan, K.; Johng, D.; Shiraishi, T.; Kulkarni, P.; Weninger, K.R. Cancer/Testis Antigen PAGE4, a Regulator of c-Jun Transactivation, Is Phosphorylated by Homeodomain-Interacting Protein Kinase 1, a Component of the Stress-Response Pathway. *Biochemistry* **2014**, *53*, 1670–1679. [CrossRef]
80. He, Y.; Chen, Y.; Mooney, S.M.; Rajagopalan, K.; Bhargava, A.; Sacho, E.; Weninger, K.; Bryan, P.N.; Kulkarni, P.; Orban, J. Phosphorylation-Induced Conformational Ensemble Switching in an Intrinsically Disordered Cancer/Testis Antigen. *J. Biol. Chem.* **2015**, *290*, 25090–25102. [CrossRef] [PubMed]
81. Gomes, G.-N.; Gradinaru, C.C. Insights into the Conformations and Dynamics of Intrinsically Disordered Proteins Using Single-Molecule Fluorescence. *Biochim. Biophys. Acta (BBA)—Proteins Proteom.* **2017**, *1865*, 1696–1706. [CrossRef] [PubMed]
82. Choi, U.B.; Sanabria, H.; Smirnova, T.; Bowen, M.E.; Weninger, K.R. Spontaneous Switching among Conformational Ensembles in Intrinsically Disordered Proteins. *Biomolecules* **2019**, *9*, 114. [CrossRef] [PubMed]
83. Rajagopalan, K.; Qiu, R.; Mooney, S.M.; Rao, S.; Shiraishi, T.; Sacho, E.; Huang, H.; Shapiro, E.; Weninger, K.R.; Kulkarni, P. The Stress-Response Protein Prostate-Associated Gene 4, Interacts with c-Jun and Potentiates Its Transactivation. *Biochim. Biophys. Acta (BBA)—Mol. Basis Dis.* **2014**, *1842*, 154–163. [CrossRef] [PubMed]
84. Hofmann, H.; Soranno, A.; Borgia, A.; Gast, K.; Nettels, D.; Schuler, B. Polymer Scaling Laws of Unfolded and Intrinsically Disordered Proteins Quantified with Single-Molecule Spectroscopy. *Proc. Natl. Acad. Sci. USA* **2012**, *109*, 16155–16160. [CrossRef] [PubMed]
85. Soranno, A.; Buchli, B.; Nettels, D.; Cheng, R.R.; Müller-Späth, S.; Pfeil, S.H.; Hoffmann, A.; Lipman, E.A.; Makarov, D.E.; Schuler, B. Quantifying Internal Friction in Unfolded and Intrinsically Disordered Proteins with Single-Molecule Spectroscopy. *Proc. Natl. Acad. Sci. USA* **2012**, *109*, 17800–17806. [CrossRef]
86. Brucale, M.; Schuler, B.; Samorì, B. Single-Molecule Studies of Intrinsically Disordered Proteins. *Chem. Rev.* **2014**, *114*, 3281–3317. [CrossRef]
87. Weninger, K.; Bowen, M.E.; Choi, U.B.; Chu, S.; Brunger, A.T. Accessory Proteins Stabilize the Acceptor Complex for Synaptobrevin, the 1:1 Syntaxin/SNAP-25 Complex. *Structure* **2008**, *16*, 308–320. [CrossRef]
88. Miyagi, A.; Tsunaka, Y.; Uchihashi, T.; Mayanagi, K.; Hirose, S.; Morikawa, K.; Ando, T. Visualization of Intrinsically Disordered Regions of Proteins by High-Speed Atomic Force Microscopy. *ChemPhysChem* **2008**, *9*, 1859–1866. [CrossRef]
89. Kodera, N.; Ando, T. Guide to Studying Intrinsically Disordered Proteins by High-Speed Atomic Force Microscopy. *Methods* **2022**, *207*, 44–56. [CrossRef]
90. Kodera, N.; Ando, T. Visualization of Intrinsically Disordered Proteins by High-Speed Atomic Force Microscopy. *Curr. Opin. Struct. Biol.* **2022**, *72*, 260–266. [CrossRef] [PubMed]
91. Nwanochie, E.; Uversky, V.N. Structure Determination by Single-Particle Cryo-Electron Microscopy: Only the Sky (and Intrinsic Disorder) Is the Limit. *Int. J. Mol. Sci.* **2019**, *20*, 4186. [CrossRef] [PubMed]
92. Benjin, X.; Ling, L. Developments, Applications, and Prospects of Cryo-electron Microscopy. *Protein Sci.* **2020**, *29*, 872–882. [CrossRef] [PubMed]
93. Abriata, L.A.; Dal Peraro, M. Will Cryo-Electron Microscopy Shift the Current Paradigm in Protein Structure Prediction? *J. Chem. Inf. Model.* **2020**, *60*, 2443–2447. [CrossRef] [PubMed]
94. Musselman, C.A.; Kutateladze, T.G. Characterization of Functional Disordered Regions within Chromatin-Associated Proteins. *iScience* **2021**, *24*, 102070. [CrossRef]
95. Bonomi, M.; Vendruscolo, M. Determination of Protein Structural Ensembles Using Cryo-Electron Microscopy. *Curr. Opin. Struct. Biol.* **2019**, *56*, 37–45. [CrossRef]
96. Schmidt, M. Macromolecular Movies, Storybooks Written by Nature. *Biophys. Rev.* **2021**, *13*, 1191–1197. [CrossRef]

97. Brändén, G.; Neutze, R. Advances and Challenges in Time-Resolved Macromolecular Crystallography. *Science* **2021**, *373*, eaba0954. [CrossRef]
98. Malla, T.N.; Schmidt, M. Transient State Measurements on Proteins by Time-Resolved Crystallography. *Curr. Opin. Struct. Biol.* **2022**, *74*, 102376. [CrossRef]
99. Frank, J. Time-Resolved Cryo-Electron Microscopy: Recent Progress. *J. Struct. Biol.* **2017**, *200*, 303–306. [CrossRef]
100. Dandey, V.P.; Budell, W.C.; Wei, H.; Bobe, D.; Maruthi, K.; Kopylov, M.; Eng, E.T.; Kahn, P.A.; Hinshaw, J.E.; Kundu, N.; et al. Time-Resolved Cryo-EM Using Spotiton. *Nat. Methods* **2020**, *17*, 897–900. [CrossRef] [PubMed]
101. Ali, S.; Hassan, M.; Islam, A.; Ahmad, F. A Review of Methods Available to Estimate Solvent-Accessible Surface Areas of Soluble Proteins in the Folded and Unfolded States. *Curr. Protein Pept. Sci.* **2014**, *15*, 456–476. [CrossRef] [PubMed]
102. Hodge, E.A.; Benhaim, M.A.; Lee, K.K. Bridging Protein Structure, Dynamics, and Function Using Hydrogen/Deuterium-exchange Mass Spectrometry. *Protein Sci.* **2020**, *29*, 843–855. [CrossRef] [PubMed]
103. Zhang, Y.; Rempel, D.L.; Zhang, J.; Sharma, A.K.; Mirica, L.M.; Gross, M.L. Pulsed Hydrogen–Deuterium Exchange Mass Spectrometry Probes Conformational Changes in Amyloid Beta (A$\beta$) Peptide Aggregation. *Proc. Natl. Acad. Sci. USA* **2013**, *110*, 14604–14609. [CrossRef]
104. Beveridge, R.; Calabrese, A.N. Structural Proteomics Methods to Interrogate the Conformations and Dynamics of Intrinsically Disordered Proteins. *Front. Chem.* **2021**, *9*, 603639. [CrossRef]
105. Illes-Toth, E.; Rempel, D.L.; Gross, M.L. Pulsed Hydrogen–Deuterium Exchange Illuminates the Aggregation Kinetics of $\alpha$-Synuclein, the Causative Agent for Parkinson's Disease. *ACS Chem. Neurosci.* **2018**, *9*, 1469–1476. [CrossRef]
106. Piersimoni, L.; Kastritis, P.L.; Arlt, C.; Sinz, A. Cross-Linking Mass Spectrometry for Investigating Protein Conformations and Protein–Protein Interactions–A Method for All Seasons. *Chem. Rev.* **2022**, *122*, 7500–7531. [CrossRef]
107. Ubbiali, D.; Fratini, M.; Piersimoni, L.; Ihling, C.H.; Kipping, M.; Heilmann, I.; Iacobucci, C.; Sinz, A. Direct Observation of "Elongated" Conformational States in A-Synuclein upon Liquid-Liquid Phase Separation. *Angew. Chem. Int. Ed.* **2022**, *134*, e202205726. [CrossRef]
108. Chen, D.; Drombosky, K.W.; Hou, Z.; Sari, L.; Kashmer, O.M.; Ryder, B.D.; Perez, V.A.; Woodard, D.R.; Lin, M.M.; Diamond, M.I.; et al. Tau Local Structure Shields an Amyloid-Forming Motif and Controls Aggregation Propensity. *Nat. Commun* **2019**, *10*, 2493. [CrossRef]
109. Niemeyer, M.; Moreno Castillo, E.; Ihling, C.H.; Iacobucci, C.; Wilde, V.; Hellmuth, A.; Hoehenwarter, W.; Samodelov, S.L.; Zurbriggen, M.D.; Kastritis, P.L.; et al. Flexibility of Intrinsically Disordered Degrons in AUX/IAA Proteins Reinforces Auxin Co-Receptor Assemblies. *Nat. Commun* **2020**, *11*, 2277. [CrossRef]
110. Suskiewicz, M.J.; Sussman, J.L.; Silman, I.; Shaul, Y. Context-Dependent Resistance to Proteolysis of Intrinsically Disordered Proteins. *Protein Sci.* **2011**, *20*, 1285–1297. [CrossRef] [PubMed]
111. Johnson, D.E.; Xue, B.; Sickmeier, M.D.; Meng, J.; Cortese, M.S.; Oldfield, C.J.; le Gall, T.; Dunker, A.K.; Uversky, V.N. High-Throughput Characterization of Intrinsic Disorder in Proteins from the Protein Structure Initiative. *J. Struct. Biol.* **2012**, *180*, 201–215. [CrossRef]
112. Baker, E.S.; Luckner, S.R.; Krause, K.L.; Lambden, P.R.; Clarke, I.N.; Ward, V.K. Inherent Structural Disorder and Dimerisation of Murine Norovirus NS1-2 Protein. *PLoS ONE* **2012**, *7*, e30534. [CrossRef] [PubMed]
113. Hamdi, K.; Salladini, E.; O'Brien, D.P.; Brier, S.; Chenal, A.; Yacoubi, I.; Longhi, S. Structural Disorder and Induced Folding within Two Cereal, ABA Stress and Ripening (ASR) Proteins. *Sci. Rep.* **2017**, *7*, 15544. [CrossRef] [PubMed]
114. Chemes, L.B.; Alonso, L.G.; Noval, M.G.; de Prat-Gay, G. Circular Dichroism Techniques for the Analysis of Intrinsically Disordered Proteins and Domains. In *Intrinsically Disordered Protein Analysis*; Humana Press: Totowa, NJ, USA, 2012; pp. 387–404.
115. Micsonai, A.; Moussong, É.; Murvai, N.; Tantos, Á.; Tőke, O.; Réfrégiers, M.; Wien, F.; Kardos, J. Disordered–Ordered Protein Binary Classification by Circular Dichroism Spectroscopy. *Front. Mol. Biosci.* **2022**, *9*, 863141. [CrossRef] [PubMed]
116. Ezerski, J.C.; Zhang, P.; Jennings, N.C.; Waxham, M.N.; Cheung, M.S. Molecular Dynamics Ensemble Refinement of Intrinsically Disordered Peptides According to Deconvoluted Spectra from Circular Dichroism. *Biophys. J.* **2020**, *118*, 1665–1678. [CrossRef]
117. Uversky, V.N. Natively Unfolded Proteins: A Point Where Biology Waits for Physics. *Protein Sci.* **2002**, *11*, 739–756. [CrossRef]
118. Na, J.-H.; Lee, W.-K.; Yu, Y. How Do We Study the Dynamic Structure of Unstructured Proteins: A Case Study on Nopp140 as an Example of a Large, Intrinsically Disordered Protein. *Int. J. Mol. Sci.* **2018**, *19*, 381. [CrossRef]
119. Auvray, F.; Dennetiere, D.; Giuliani, A.; Jamme, F.; Wien, F.; Nay, B.; Zirah, S.; Polack, F.; Menneglier, C.; Lagarde, B.; et al. Time Resolved Transient Circular Dichroism Spectroscopy Using Synchrotron Natural Polarization. *Struct. Dyn.* **2019**, *6*, 054307. [CrossRef]
120. Natalello, A.; Ami, D.; Doglia, S.M. Fourier transform infrared spectroscopy of intrinsically disordered proteins: Measurement procedures and data analyses. In *Intrinsically Disordered Protein Analysis*; Humana Press: Totowa, NJ, USA, 2012; pp. 229–244.
121. Uversky, V.N. Biophysical Methods to Investigate Intrinsically Disordered Proteins: Avoiding an "Elephant and Blind Men" Situation. *Adv. Exp. Med. Biol.* **2015**, *870*, 215–260.
122. Sethi, A.; Anunciado, D.; Tian, J.; Vu, D.M.; Gnanakaran, S. Deducing Conformational Variability of Intrinsically Disordered Proteins from Infrared Spectroscopy with Bayesian Statistics. *Chem. Phys.* **2013**, *422*, 143–155. [CrossRef] [PubMed]
123. Takekiyo, T.; Yamada, N.; Nakazawa, C.T.; Amo, T.; Asano, A.; Yoshimura, Y. Formation of A-synuclein Aggregates in Aqueous Ethylammonium Nitrate Solutions. *Biopolymers* **2020**, *111*, e23352. [CrossRef] [PubMed]

124. Zhu, F.; Isaacs, N.W.; Hecht, L.; Barron, L.D. Raman Optical Activity: A Tool for Protein Structure Analysis. *Structure* **2005**, *13*, 1409–1419. [CrossRef] [PubMed]
125. Sane, S.U.; Cramer, S.M.; Przybycien, T.M. A Holistic Approach to Protein Secondary Structure Characterization Using Amide I Band Raman Spectroscopy. *Anal. Biochem.* **1999**, *269*, 255–272. [CrossRef] [PubMed]
126. Berjot, M.; Marx, J.; Alix, A.J.P. Determination of the Secondary Structure of Proteins from the Raman Amide I Band: The Reference Intensity Profiles Method. *J. Raman Spectrosc.* **1987**, *18*, 289–300. [CrossRef]
127. Maiti, N.C.; Apetri, M.M.; Zagorski, M.G.; Carey, P.R.; Anderson, V.E. Raman Spectroscopic Characterization of Secondary Structure in Natively Unfolded Proteins: α-Synuclein. *J. Am. Chem. Soc.* **2004**, *126*, 2399–2408. [CrossRef]
128. Syme, C.D.; Blanch, E.W.; Holt, C.; Jakes, R.; Goedert, M.; Hecht, L.; Barron, L.D. A Raman Optical Activity Study of Rheomorphism in Caseins, Synucleins and Tau: New Insight into the Structure and Behaviour of Natively Unfolded Proteins. *Eur. J. Biochem.* **2002**, *269*, 148–156. [CrossRef]
129. Stuchfield, D.; France, A.P.; Migas, L.G.; Thalhammer, A.; Bremer, A.; Bellina, B.; Barran, P.E. *The Use of Mass Spectrometry to Examine IDPs: Unique Insights and Caveats*, 1st ed.; Elsevier Inc.: Amsterdam, The Netherlands, 2018; Volume 611, ISBN 9780128156490.
130. Santambrogio, C.; Natalello, A.; Brocca, S.; Ponzini, E.; Grandori, R. Conformational Characterization and Classification of Intrinsically Disordered Proteins by Native Mass Spectrometry and Charge-State Distribution Analysis. *Proteomics* **2019**, *19*, e1800060. [CrossRef]
131. Lento, C.; Wilson, D.J. Subsecond Time-Resolved Mass Spectrometry in Dynamic Structural Biology. *Chem. Rev.* **2022**, *122*, 7624–7646. [CrossRef]
132. Salvay, A.G.; Communie, G.; Ebel, C. *Sedimentation Velocity Analytical Ultracentrifugation for Intrinsically Disordered Proteins*; John Wiely & Sons: New York, NY, USA, 2012; pp. 91–105.
133. Gast, K.; Fiedler, C. Dynamic and Static Light Scattering of Intrinsically Disordered Proteins. In *Intrinsically Disordered Protein Analysis*; Springer: New York, NY, USA, 2012; pp. 137–161.
134. Al-Ghobashy, M.A.; Mostafa, M.M.; Abed, H.S.; Fathalla, F.A.; Salem, M.Y. Correlation between Dynamic Light Scattering and Size Exclusion High Performance Liquid Chromatography for Monitoring the Effect of PH on Stability of Biopharmaceuticals. *J. Chromatogr. B* **2017**, *1060*, 1–9. [CrossRef]
135. Leite, J.P.; Gimeno, A.; Taboada, P.; Jiménez-Barbero, J.J.; Gales, L. Dissection of the Key Steps of Amyloid-β Peptide 1–40 Fibrillogenesis. *Int. J. Biol. Macromol.* **2020**, *164*, 2240–2246. [CrossRef] [PubMed]
136. Hochmair, J.; Exner, C.; Betzel, C.; Mandelkow, E.; Wegmann, S. Light Microscopy and Dynamic Light Scattering to Study Liquid-Liquid Phase Separation of Tau Proteins In Vitro. In *Protein Aggregation*; Humana: New York, NY, USA, 2023; pp. 225–243.
137. Tomasso, M.E.; Tarver, M.J.; Devarajan, D.; Whitten, S.T. Hydrodynamic Radii of Intrinsically Disordered Proteins Determined from Experimental Polyproline II Propensities. *PLoS Comput. Biol.* **2016**, *12*, e1004686. [CrossRef] [PubMed]
138. Wang, W. Recent Advances in Atomic Molecular Dynamics Simulation of Intrinsically Disordered Proteins. *Phys. Chem. Chem. Phys.* **2021**, *23*, 777–784. [CrossRef]
139. Dokholyan, N.V. Experimentally-Driven Protein Structure Modeling. *J. Proteom.* **2020**, *220*, 103777. [CrossRef]
140. Hsu, C.C.; Buehler, M.J.; Tarakanova, A. The Order-Disorder Continuum: Linking Predictions of Protein Structure and Disorder through Molecular Simulation. *Sci. Rep.* **2020**, *10*, 2068. [CrossRef] [PubMed]
141. Best, R.B. Computational and Theoretical Advances in Studies of Intrinsically Disordered Proteins. *Curr. Opin. Struct. Biol.* **2017**, *42*, 147–154. [CrossRef]
142. Lin, X.; Roy, S.; Jolly, M.K.; Bocci, F.; Schafer, N.P.; Tsai, M.-Y.; Chen, Y.; He, Y.; Grishaev, A.; Weninger, K.; et al. PAGE4 and Conformational Switching: Insights from Molecular Dynamics Simulations and Implications for Prostate Cancer. *J. Mol. Biol.* **2018**, *430*, 2422–2438. [CrossRef]
143. Kasahara, K.; Terazawa, H.; Takahashi, T.; Higo, J. Studies on Molecular Dynamics of Intrinsically Disordered Proteins and Their Fuzzy Complexes: A Mini-Review. *Comput. Struct. Biotechnol. J.* **2019**, *17*, 712–720. [CrossRef]
144. Lin, X.; Kulkarni, P.; Bocci, F.; Schafer, N.P.; Roy, S.; Tsai, M.Y.; He, Y.; Chen, Y.; Rajagopalan, K.; Mooney, S.M.; et al. Structural and Dynamical Order of a Disordered Protein: Molecular Insights into Conformational Switching of Page4 at the Systems Level. *Biomolecules* **2019**, *9*, 77. [CrossRef]
145. Blackledge, M.; Ferrage, F.; Kadeřávek, P.; Salvi, N.; Zapletal, V.; Jaseňáková, Z.; Zachrdla, M.; Padrta, P.; Narasimhan, S.; Marquardsen, T.; et al. Convergent Views on Disordered Protein Dynamics from NMR and Computational Approaches. *Biophys. J.* **2022**, *121*, 3785–3794. [CrossRef]
146. Romero, P.; Obradovic, Z.; Li, X.; Garner, E.C.; Brown, C.J.; Dunker, A.K. Sequence Complexity of Disordered Protein. *Proteins Struct. Funct. Genet.* **2001**, *42*, 38–48. [CrossRef] [PubMed]
147. Obradovic, Z.; Peng, K.; Vucetic, S.; Radivojac, P.; Dunker, A.K. Exploiting Heterogeneous Sequence Properties Improves Prediction of Protein Disorder. *Proteins Struct. Funct. Bioinform.* **2005**, *61*, 176–182. [CrossRef]
148. Xue, B.; Dunbrack, R.L.; Williams, R.W.; Dunker, A.K.; Uversky, V.N. PONDR-FIT: A Meta-Predictor of Intrinsically Disordered Amino Acids. *Biochim. Biophys. Acta (BBA)—Proteins Proteom.* **2010**, *1804*, 996–1010. [CrossRef] [PubMed]
149. Peng, K.; Radivojac, P.; Vucetic, S.; Dunker, A.K.; Obradovic, Z. Length-Dependent Prediction of Protein Intrinsic Disorder. *BMC Bioinform.* **2006**, *7*, 208. [CrossRef] [PubMed]
150. Sirovetz, B.J.; Schafer, N.P.; Wolynes, P.G. Protein Structure Prediction: Making AWSEM AWSEM-ER by Adding Evolutionary Restraints. *Proteins Struct. Funct. Bioinform.* **2017**, *85*, 2127–2142. [CrossRef]

151. Kmiecik, S.; Gront, D.; Kolinski, M.; Wieteska, L.; Dawid, A.E.; Kolinski, A. Coarse-Grained Protein Models and Their Applications. *Chem. Rev.* **2016**, *116*, 7898–7936. [CrossRef]
152. Chen, M.; Lin, X.; Zheng, W.; Onuchic, J.N.; Wolynes, P.G. Protein Folding and Structure Prediction from the Ground Up: The Atomistic Associative Memory, Water Mediated, Structure and Energy Model. *J. Phys. Chem. B* **2016**, *120*, 8557–8565. [CrossRef]
153. Chen, M.; Lin, X.; Lu, W.; Onuchic, J.N.; Wolynes, P.G. Protein Folding and Structure Prediction from the Ground Up II: AAWSEM for α/β Proteins. *J. Phys. Chem. B* **2017**, *121*, 3473–3482. [CrossRef]
154. Ruff, K.M.; Pappu, R.v. AlphaFold and Implications for Intrinsically Disordered Proteins. *J. Mol. Biol.* **2021**, *433*, 167208. [CrossRef]
155. Ehm, T.; Shinar, H.; Meir, S.; Sekhon, A.; Sethi, V.; Morgan, I.L.; Rahamim, G.; Saleh, O.A.; Beck, R. Intrinsically Disordered Proteins at the Nano-Scale. *Nano Futures* **2021**, *5*, 1–15. [CrossRef]
156. Aznauryan, M.; Delgado, L.; Soranno, A.; Nettels, D.; Huang, J.; Labhardt, A.M.; Grzesiek, S.; Schuler, B. Comprehensive Structural and Dynamical View of an Unfolded Protein from the Combination of Single-Molecule FRET, NMR, and SAXS. *Proc. Natl. Acad. Sci. USA* **2016**, *113*, E5389–E5398. [CrossRef] [PubMed]
157. Dedmon, M.M.; Lindorff-Larsen, K.; Christodoulou, J.; Vendruscolo, M.; Dobson, C.M. Mapping Long-Range Interactions in α-Synuclein Using Spin-Label NMR and Ensemble Molecular Dynamics Simulations. *J. Am. Chem. Soc.* **2005**, *127*, 476–477. [CrossRef] [PubMed]
158. Ferrie, J.J.; Haney, C.M.; Yoon, J.; Pan, B.; Lin, Y.-C.; Fakhraai, Z.; Rhoades, E.; Nath, A.; Petersson, E.J. Using a FRET Library with Multiple Probe Pairs To Drive Monte Carlo Simulations of α-Synuclein. *Biophys. J.* **2018**, *114*, 53–64. [CrossRef]
159. Hamilton, G.L.; Saikia, N.; Basak, S.; Welcome, F.S.; Wu, F.; Kubiak, J.; Zhang, C.; Hao, Y.; Seidel, C.A.; Ding, F.; et al. Fuzzy Supertertiary Interactions within PSD-95 Enable Ligand Binding. *eLife* **2022**, *11*, e77242. [CrossRef]
160. Thomasen, F.E.; Lindorff-Larsen, K. Conformational Ensembles of Intrinsically Disordered Proteins and Flexible Multidomain Proteins. *Biochem. Soc. Trans.* **2022**, *50*, 541–554. [CrossRef] [PubMed]
161. Saikia, N.; Yanez-Orozco, I.S.; Qiu, R.; Hao, P.; Milikisiyants, S.; Ou, E.; Hamilton, G.L.; Weninger, K.R.; Smirnova, T.I.; Sanabria, H.; et al. Integrative Structural Dynamics Probing of the Conformational Heterogeneity in Synaptosomal-Associated Protein 25. *Cell Rep. Phys. Sci.* **2021**, *2*, 100616. [CrossRef] [PubMed]
162. Choi, U.B.; Xiao, S.; Wollmuth, L.P.; Bowen, M.E. Effect of Src Kinase Phosphorylation on Disordered C-Terminal Domain of N-Methyl-D-Aspartic Acid (NMDA) Receptor Subunit GluN2B Protein. *J. Biol. Chem.* **2011**, *286*, 29904–29912. [CrossRef]
163. Meng, F.; Bellaiche, M.M.J.; Kim, J.-Y.; Zerze, G.H.; Best, R.B.; Chung, H.S. Highly Disordered Amyloid-β Monomer Probed by Single-Molecule FRET and MD Simulation. *Biophys. J.* **2018**, *114*, 870–884. [CrossRef]
164. Brunger, A.T.; Strop, P.; Vrljic, M.; Chu, S.; Weninger, K.R. Three-Dimensional Molecular Modeling with Single Molecule FRET. *J. Struct. Biol.* **2011**, *173*, 497–505. [CrossRef]
165. Choi, U.B.; Strop, P.; Vrljic, M.; Chu, S.; Brunger, A.T.; Weninger, K.R. Single-Molecule FRET–Derived Model of the Synaptotagmin 1–SNARE Fusion Complex. *Nat. Struct. Mol. Biol.* **2010**, *17*, 318–324. [CrossRef]
166. Lerner, E.; Barth, A.; Hendrix, J.; Ambrose, B.; Birkedal, V.; Blanchard, S.C.; Börner, R.; Sung Chung, H.; Cordes, T.; Craggs, T.D.; et al. FRET-Based Dynamic Structural Biology: Challenges, Perspectives and an Appeal for Open-Science Practices. *eLife* **2021**, *10*, e60416. [CrossRef] [PubMed]
167. Mercadante, D.; Milles, S.; Fuertes, G.; Svergun, D.I.; Lemke, E.A.; Gräter, F. Kirkwood–Buff Approach Rescues Overcollapse of a Disordered Protein in Canonical Protein Force Fields. *J. Phys. Chem. B* **2015**, *119*, 7975–7984. [CrossRef] [PubMed]
168. Araki, K.; Yagi, N.; Nakatani, R.; Sekiguchi, H.; So, M.; Yagi, H.; Ohta, N.; Nagai, Y.; Goto, Y.; Mochizuki, H. A Small-Angle X-ray Scattering Study of Alpha-Synuclein from Human Red Blood Cells. *Sci. Rep.* **2016**, *6*, 30473. [CrossRef]
169. Schweers, O.; Schönbrunn-Hanebeck, E.; Marx, A.; Mandelkow, E. Structural Studies of Tau Protein and Alzheimer Paired Helical Filaments Show No Evidence for Beta-Structure. *J. Biol. Chem.* **1994**, *269*, 24290–24297. [CrossRef]
170. Gomes, G.-N.W.; Krzeminski, M.; Namini, A.; Martin, E.W.; Mittag, T.; Head-Gordon, T.; Forman-Kay, J.D.; Gradinaru, C.C. Conformational Ensembles of an Intrinsically Disordered Protein Consistent with NMR, SAXS, and Single-Molecule FRET. *J. Am. Chem. Soc.* **2020**, *142*, 15697–15710. [CrossRef]
171. Liu, B.; Chia, D.; Csizmok, V.; Farber, P.; Forman-Kay, J.D.; Gradinaru, C.C. The Effect of Intrachain Electrostatic Repulsion on Conformational Disorder and Dynamics of the Sic1 Protein. *J. Phys. Chem. B* **2014**, *118*, 4088–4097. [CrossRef] [PubMed]
172. Gomes, G.-N.W.; Namini, A.; Gradinaru, C.C. Integrative Conformational Ensembles of Sic1 Using Different Initial Pools and Optimization Methods. *Front. Mol. Biosci.* **2022**, *9*, 910956. [CrossRef]
173. Sala, D.; Cosentino, U.; Ranaudo, A.; Greco, C.; Moro, G. Dynamical Behavior and Conformational Selection Mechanism of the Intrinsically Disordered Sic1 Kinase-Inhibitor Domain. *Life* **2020**, *10*, 110. [CrossRef]
174. Nash, P.; Tang, X.; Orlicky, S.; Chen, Q.; Gertler, F.B.; Mendenhall, M.D.; Sicheri, F.; Pawson, T.; Tyers, M. Multisite Phosphorylation of a CDK Inhibitor Sets a Threshold for the Onset of DNA Replication. *Nature* **2001**, *414*, 514–521. [CrossRef]
175. Mittag, T.; Orlicky, S.; Choy, W.; Tang, X.; Lin, H.; Sicheri, F.; Kay, L.E.; Tyers, M.; Forman-Kay, J.D. Dynamic equilibrium engagement of a polyvalent ligand with a single-site receptor. *Proc. Natl. Acad. Sci. USA* **2008**, *105*, 17772–17777. [CrossRef]
176. Chan-Yao-Chong, M.; Deville, C.; Pinet, L.; van Heijenoort, C.; Durand, D.; Ha-Duong, T. Structural Characterization of N-WASP Domain V Using MD Simulations with NMR and SAXS Data. *Biophys. J.* **2019**, *116*, 1216–1227. [CrossRef] [PubMed]
177. Hansen, M.D.H.; Kwiatkowski, A.V. Control of Actin Dynamics by Allosteric Regulation of Actin Binding Proteins. *Int. Rev. Cell Mol. Biol.* **2013**, *303*, 1–25.

178. Robustelli, P.; Piana, S.; Shaw, D.E. Developing a Molecular Dynamics Force Field for Both Folded and Disordered Protein States. *Proc. Natl. Acad. Sci. USA* **2018**, *115*, E4758–E4766. [CrossRef] [PubMed]
179. Chan-Yao-Chong, M.; Durand, D.; Ha-Duong, T. Molecular Dynamics Simulations Combined with Nuclear Magnetic Resonance and/or Small-Angle X-ray Scattering Data for Characterizing Intrinsically Disordered Protein Conformational Ensembles. *J. Chem. Inf. Model.* **2019**, *59*, 1743–1758. [CrossRef] [PubMed]
180. Rauscher, S.; Gapsys, V.; Gajda, M.J.; Zweckstetter, M.; de Groot, B.L.; Grubmüller, H. Structural Ensembles of Intrinsically Disordered Proteins Depend Strongly on Force Field: A Comparison to Experiment. *J. Chem. Theory Comput.* **2015**, *11*, 5513–5524. [CrossRef] [PubMed]
181. Ou, L.; Waddell, M.B.; Kriwacki, R.W. Mechanism of Cell Cycle Entry Mediated by the Intrinsically Disordered Protein P27 Kip1. *ACS Chem. Biol.* **2012**, *7*, 678–682. [CrossRef] [PubMed]
182. Tsytlonok, M.; Hemmen, K.; Hamilton, G.; Kolimi, N.; Felekyan, S.; Seidel, C.A.M.; Tompa, P.; Sanabria, H. Specific Conformational Dynamics and Expansion Underpin a Multi-Step Mechanism for Specific Binding of P27 with Cdk2/Cyclin A. *J. Mol. Biol.* **2020**, *432*, 2998–3017. [CrossRef] [PubMed]
183. Tsytlonok, M.; Sanabria, H.; Wang, Y.; Felekyan, S.; Hemmen, K.; Phillips, A.H.; Yun, M.-K.; Waddell, M.B.; Park, C.-G.; Vaithiyalingam, S.; et al. Dynamic Anticipation by Cdk2/Cyclin A-Bound P27 Mediates Signal Integration in Cell Cycle Regulation. *Nat. Commun.* **2019**, *10*, 1676. [CrossRef]
184. Das, R.K.; Huang, Y.; Phillips, A.H.; Kriwacki, R.W.; Pappu, R.v. Cryptic Sequence Features within the Disordered Protein P27 Kip1 Regulate Cell Cycle Signaling. *Proc. Natl. Acad. Sci. USA* **2016**, *113*, 5616–5621. [CrossRef]
185. Zeng, Y.; He, Y.; Yang, F.; Mooney, S.M.; Getzenberg, R.H.; Orban, J.; Kulkarni, P. The Cancer/Testis Antigen Prostate-Associated Gene 4 (PAGE4) Is a Highly Intrinsically Disordered Protein. *J. Biol. Chem.* **2011**, *286*, 13985–13994. [CrossRef]
186. Kulkarni, P.; Jolly, M.K.; Jia, D.; Mooney, S.M.; Bhargava, A.; Kagohara, L.T.; Chen, Y.; Hao, P.; He, Y.; Veltri, R.W.; et al. Phosphorylation-Induced Conformational Dynamics in an Intrinsically Disordered Protein and Potential Role in Phenotypic Heterogeneity. *Proc. Natl. Acad. Sci. USA* **2017**, *114*, E2644–E2653. [CrossRef] [PubMed]
187. Kulkarni, P.; Dunker, A.; Weninger, K.; Orban, J. Prostate-Associated Gene 4 (PAGE4), an Intrinsically Disordered Cancer/Testis Antigen, Is a Novel Therapeutic Target for Prostate Cancer. *Asian J. Androl.* **2016**, *18*, 695. [CrossRef] [PubMed]
188. Salgia, R.; Jolly, M.; Dorff, T.; Lau, C.; Weninger, K.; Orban, J.; Kulkarni, P. Prostate-Associated Gene 4 (PAGE4): Leveraging the Conformational Dynamics of a Dancing Protein Cloud as a Therapeutic Target. *J. Clin. Med.* **2018**, *7*, 156. [CrossRef]
189. Jolly, M.K.; Kulkarni, P.; Weninger, K.; Orban, J.; Levine, H. Phenotypic Plasticity, Bet-Hedging, and Androgen Independence in Prostate Cancer: Role of Non-Genetic Heterogeneity. *Front. Oncol.* **2018**, *8*, 50. [CrossRef] [PubMed]
190. Uversky, V.N.; Kulkarni, P. Intrinsically Disordered Proteins: Chronology of a Discovery. *Biophys. Chem.* **2021**, *279*, 106694. [CrossRef]
191. Uversky, V.N. Unusual Biophysics of Intrinsically Disordered Proteins. *Biochim. Biophys. Acta* **2013**, *1834*, 932–951. [CrossRef]
192. Kulkarni, P.; Bhattacharya, S.; Achuthan, S.; Behal, A.; Jolly, M.K.; Kotnala, S.; Mohanty, A.; Rangarajan, G.; Salgia, R.; Uversky, V. Intrinsically Disordered Proteins: Critical Components of the Wetware. *Chem. Rev.* **2022**, *122*, 6614–6633. [CrossRef]

*biomolecules*

*Review*

# Reorganization of Cell Compartmentalization Induced by Stress

**Anna S. Fefilova †, Iuliia A. Antifeeva †, Anastasia A. Gavrilova, Konstantin K. Turoverov *, Irina M. Kuznetsova and Alexander V. Fonin**

Laboratory of Structural Dynamics, Stability and Folding of Proteins, Institute of Cytology of RAS, 194064 St. Petersburg, Russia
* Correspondence: kkt@incras.ru
† These authors contributed equally to this work.

**Abstract:** The discovery of intrinsically disordered proteins (IDPs) that do not have an ordered structure and nevertheless perform essential functions has opened a new era in the understanding of cellular compartmentalization. It threw the bridge from the mostly mechanistic model of the organization of the living matter to the idea of highly dynamic and functional "soft matter". This paradigm is based on the notion of the major role of liquid-liquid phase separation (LLPS) of biopolymers in the spatial-temporal organization of intracellular space. The LLPS leads to the formation of self-assembled membrane-less organelles (MLOs). MLOs are multicomponent and multifunctional biological condensates, highly dynamic in structure and composition, that allow them to fine-tune the regulation of various intracellular processes. IDPs play a central role in the assembly and functioning of MLOs. The LLPS importance for the regulation of chemical reactions inside the cell is clearly illustrated by the reorganization of the intracellular space during stress response. As a reaction to various types of stresses, stress-induced MLOs appear in the cell, enabling the preservation of the genetic and protein material during unfavourable conditions. In addition, stress causes structural, functional, and compositional changes in the MLOs permanently present inside the cells. In this review, we describe the assembly of stress-induced MLOs and the stress-induced modification of existing MLOs in eukaryotes, yeasts, and prokaryotes in response to various stress factors.

**Keywords:** membrane-less organelles; intrinsically disordered proteins; liquid-liquid phase separation; stress

**Citation:** Fefilova, A.S.; Antifeeva, I.A.; Gavrilova, A.A.; Turoverov, K.K.; Kuznetsova, I.M.; Fonin, A.V. Reorganization of Cell Compartmentalization Induced by Stress. *Biomolecules* **2022**, *12*, 1441. https://doi.org/10.3390/biom12101441

Academic Editors: Prakash Kulkarni, Stefania Brocca, Keith Dunker and Sonia Longhi

Received: 10 August 2022
Accepted: 1 October 2022
Published: 8 October 2022



## 1. Introduction

Any organism and, accordingly, its cells are constantly subject to environmental changes that are often stressful. In fact, it is hard to imagine real life conditions lacking occasional stressful impact. Constant temperature, pressure, humidity, etc., is a privilege of the laboratory environment. Throughout its existence each cell and the whole organism must constantly overcome different negative conditions. A failure to adjust to external pressure by the cellular systems leads to various diseases and pathological states at the organismal level.

Cellular stress may be triggered by both physical and biological factors, such as changes in pH, temperature, osmotic pressure, UV radiation, cell cycle disorders, changes in the metabolites and nutrients availability, DNA damage, cellular aging, and various diseases [1]. It should be noted that these factors are interrelated. Thus, a change in the cytoplasmic pH in eukaryotic cells can be caused by osmotic and thermal shock, as well as by the processes of aging and carcinogenesis [2].

The adaptive response of a cell to stress is the activation of various signaling pathways that are specifically determined by the type and severity of injury [1]. For eukaryotes, the most typical pathways are the heat shock response (HSR), unfolded protein responses of the mitochondria ($UPR^{MT}$), the unfolded protein responses of the endoplasmic reticulum ($UPR^{EM}$), and integrated "general" stress response, which is activated by a wide range of physiological conditions, such as amino acid deficiency, viral infection, and endoplasmic reticulum stress [3]. For many serious diseases, such as cancer, viral infection, and neurodegeneration, the association between the disease onset and the disruption of cellular stress response has been proven [4]. For example, inactivation of p53 in cancer, "hijacking" of cellular stress responses by viruses to increase the rate of replication by increasing the number of chaperones, and mutation of key signal transducers such as ATF6 in UPR in neurodegenerative diseases [4].

Regardless of the type of cellular response, stress conditions cause global arrest of the gene expression and protein synthesis, inhibition of most of the "normal" signaling pathways, activation of autophagy, accumulation of a large number of unfolded, partially unfolded, misfolded proteins and RNA that have not been translated [4]. Revolutionary changes in the ideas about the organization of the intracellular space that occurred in the mid-2010s made it possible to form a unified view on the molecular mechanisms underlying cell physiology [5]. First, it became obvious that adaptive, fast, and reversible reprogramming of regulatory pathways in response to a stimulus is achieved with the help of the formation/disassembly of liquid-droplet compartments and, secondly, the concentration of proteins via phase separation is necessary for this mechanism [6]. Intrinsically disordered proteins play a central role in these processes. The structure of disordered proteins presents an ensemble of different conformers, which simultaneously co-exist in solution, and dynamically transits between different conformational states separated by low energy barriers. Due to conformational heterogeneity and the presence of low complexity domains in IDPs sequences, these proteins are capable of spontaneous phase separation in highly concentrated solutions and are the main drivers of MLOs formation [7,8]. Additionally, the promiscuity and plasticity of binding allow IDPs to interact with multiple partners in networks of protein interactions and provide important functional advantages in molecular recognition through transient protein–protein interactions [9]. Short interaction-prone segments within the IDP, called molecular recognition tags, are potential binding sites that can undergo a disorder-to-order transition when binding to their partners [9]. The polyvalence of IDP depends on the cooperation of many separate, weak, non-covalent interactions that combine to give a highly specific end state [10].

The transformability and pliability of MLOs, provided by unique properties of IDPs composing them, greatly benefit cellular systems ensuring quick and timely response to life-threatening challenges. A clear illustration of that is fast reorganization of cellular compartmentalization under stress conditions. [6]. Initially, the majority of studies devoted to stress-induced MLOs focused on cytoplasmic compartments, especially stress-granules. However, an increasing number of reports have been published demonstrating a multiple nuclear MLOs sensitive to stress and potentially involved in stress-response mechanisms. Some MLOs are stress-induced and form de novo in response to stress, whereas others exist and function in unstressed cells and during stress-response undergo adaptive structural and functional changes ((Table 1, Figure 1) [6,11]. Stress-induced MLOs have been found across eukarya and bacteria life domains advocating early evolutionary development of this cell survival strategy (Tables 1–3). In this review, we attempted to classify and give general description of the MLOs formed anew or reorganized during stress-response in prokaryotic and eukaryotic cells and summarize the available data from a unified point of view.

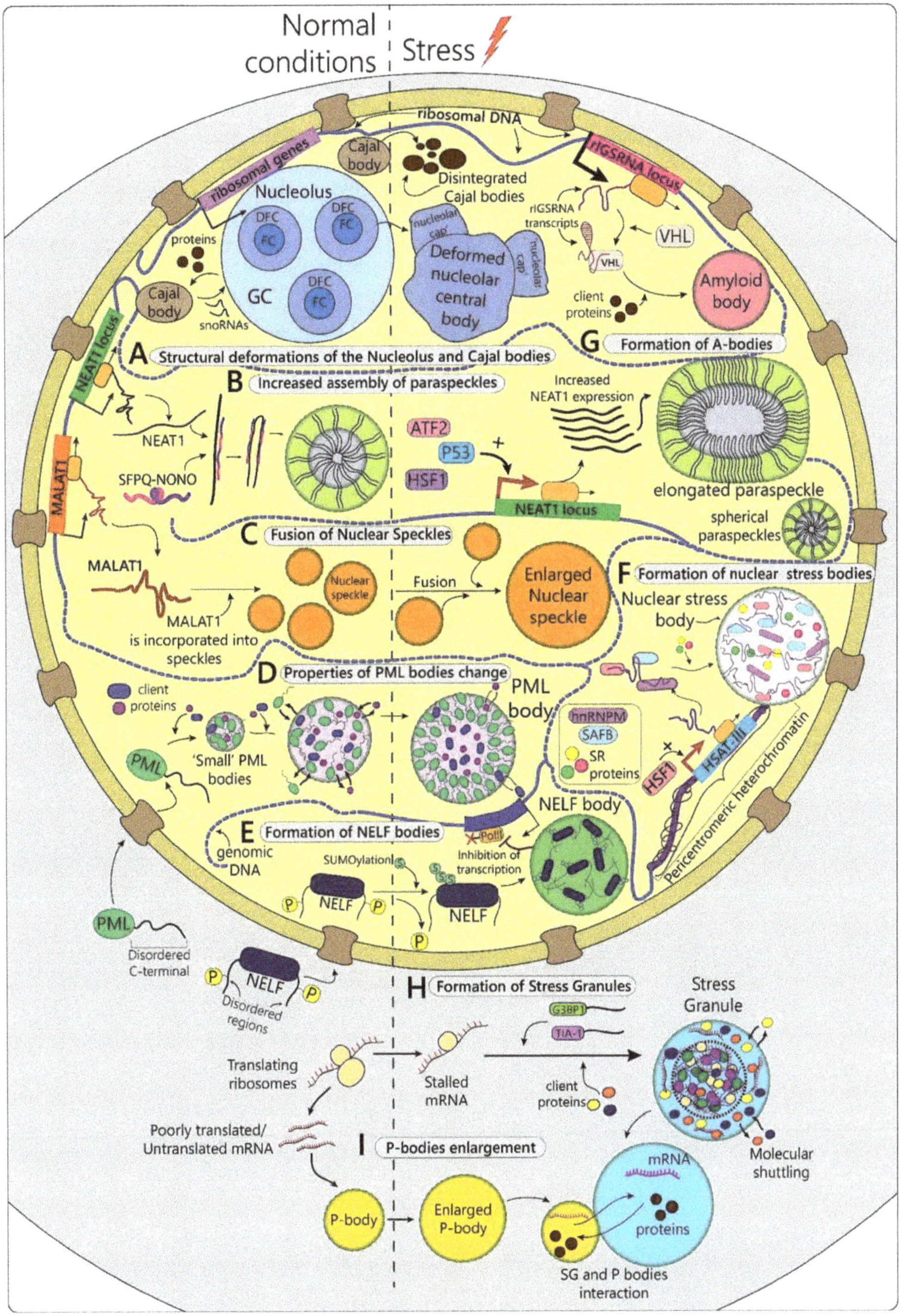


**Figure 1.** Illustration of biomolecular compartments formed or rearranged in response to stress in eukaryotic cells (**A–D**) Nuclear MLOs that undergo structural and functional changes in response to

stress. (**A**) Nucleolus and Cajal bodies are structurally deformed in response to stressful stimuli. Under some types of stress, nucleolus loses its tripartite structure while nucleolar central bodies surrounded by nucleolar caps appear. Cajal bodies are reduced in size and/or disintegrate. (**B**) Under normal conditions, paraspeckles assemble due to NEAT1 lncRNA and SFPQ-NONO heterodimer interactions. In response to stress paraspeckles increase in size and numbers as a result of enhanced NEAT1 transcription. NEAT1 transcription is activated by various stress-sensitive transcription factors, such as HSF1, p53, ATF2. Burst in the amount of NEAT1 transcripts leads to the formation of more spherical paraspeckles as well as the assembly of so-called elongated paraspeckles that were suggested to be a result of block copolymer micellization. (**C**) Nuclear speckles, that incorporate MALAT1 lncRNA, during stress increase in size but decrease in number, which is suggested to be a result of their fusion. (**D**) PML bodies that are formed by multiple isoforms of PML protein, upon stress significantly change properties. For example, under $H_2O_2$-induced oxidative stress PML bodies increase in size while the mobility of their components reduces. (**E–G**) Stress-induced nuclear MLOs. (**E**) NELF bodies form anew in response to stress after removal of inhibitory phosphorylation tag from the NELF protein and its subsequent SUMOylation. These modifications allow NELF to phase separate and form NELF bodies at the active transcription sites. NELF bodies inhibit RNA Pol II activity downregulating gene expression. (**F**) Nuclear stress bodies (nSB) form with the onset of stress after HSF1 factor activates transcription of HSatIII lncRNA from pericentromeric heterochromatin regions. HSatIII transcripts interaction with HSF1 and other protein results in condensation and assembly of nSBs. (**G**) A-bodies form in a nucleolus vicinity or within it as a result of rIGSRNA transcription. rIGSRNA is transcribed from the intergenic regions of the ribosomal DNA during stress. Increased local concentration of nascent rIGSRNA sequester VHL and other amyloidogenic proteins that together drive assembly and solidification of A-body. (**H**,**I**) Cytoplasmic MLOs involved in stress regulation. (**H**) Stress granules are stress-induced cytoplasmic MLOs that require accumulation of stalled initiation complexes for assembly. SG proteins, such as G3BP1 and TIA-1, are recruited by the repressed mRNA, a process that promotes their phase separation. Formed stress granules have two organizational layers—the low-dynamic core and highly dynamic external shell. The shell actively exchanges the mRNA and protein content with the surrounding cytoplasm (**I**) P-bodies in unstressed cells sequester poorly translated and repressed mRNAs for degradation. During stress, P-bodies enlarge in size and are able to approach stress granules and perform mutual content exchange via direct interaction.

**Table 1.** Examples of LLPS (or suggested to be LLPS) compartments formed or rearranged in response to stress in eukaryotic cells.

| Stress-Linked Organelle | Main Components | Organism | Structural Changes in Response to Stress | Function |
|---|---|---|---|---|
| **Nuclear membrane-less organelles** | | | | |
| Nucleolus | Fibrillarin, nucleophosmin, rRNA, snoRNPs, Nop58, etc. | Eukarya | Release of ribosomal proteins, change in the nucleolar proteome. Nucleolar segregation upon DNA damage or rRNA transcription. Nucleolar fragmentation upon inhibition of RNA Pol II transcription or protein kinases. Nucleolar and FC enlargement upon viral infection | Ribosome biogenesis |

**Table 1.** *Cont.*

| | Stress-Linked Organelle | Main Components | Organism | Structural Changes in Response to Stress | Function |
|---|---|---|---|---|---|
| MLOs subject to change and rearrangement in response to stress | Cajal bodies | Coilin, SMN1, snRNA, snoRNA, scaRNAs, etc. | Animals and plants | CBs decrease in number and size in response to starvation. CBs undergo disruption and formation of coilin nucleoplasmic microfoci upon UV-C irradiation, osmotic stress, and heat shock. Fusion of transformed CBs with the nucleolus upon GRV infection in plants | Maturation of snoRNA, snRNA, histone mRNA |
| | Paraspeckles | lncRNA Neat1, NONO, SFPQ, FUS, etc. | Mammals | Increase in paraspeckles numbers upon different types of stress: hypoxia, temperature, sulforaphane treatment, softening of the cellular substrate, etc. | Storage of RNAs and proteins involved in the transcription regulation and pre-mRNA processing. |
| | Nuclear speckles | snRNP, SR proteins, lncRNA MALAT1, etc. | Mammals and plants | Enlargement and rounding probably via fusions and reincorporation of splicing factors for temporal storage during stress. | Splicing regulation and storage of proteins |
| | PML-bodies | PML, SUMO-1, Sp100, etc | Mammals Absent in flies, plants and yeasts | Enlargement and decrease in the content mobility upon oxidative stress induced by H2O2. Degradation or cytoplasmic relocalization of the PML isoforms upon oxidative stress induced by As2O3. Decrease in the number and size of PML bodies upon heat stress, heavy metal addition, and expression of adenovirus E1A. | Regulation of the p53-dependent signaling, DNA damage response, DNA repair, telomere homeostasis |
| Transient assembly in response to stress | NELF bodies | NELF | Human cells | Stress-induced assembly at PolII-active transcription sites driven by NELF protein dephosphorylation and SUMOylation. | Inhibition of RNA Pol II transcription |
| | Nuclear stress-bodies | HSF1, HSatIII lncRNA, SAFB, hnRNPM | Primates | Stress-induced formation at sites of HSatIII transcription activated by HSF1 transcription factor. | Protein storage and regulation of mRNA splicing |
| | A-bodies | rIGSRNA, VHL | Mammals, fungi, insects, plants | Assembly and solidification upon the onset of stress at the sites of rIGSRNA transcription. | Temporal storage of amyloidogenic proteins |

**Table 1.** *Cont.*

| | Stress-Linked Organelle | Main Components | Organism | Structural Changes in Response to Stress | Function |
|---|---|---|---|---|---|
| | | | **Cytoplasmic membrane-less organelles** | | |
| Assembly | Stress-granules | G3BP1, TIA-1, FUS, hnRNPA1, untranslated mRNA, etc | Eukaryotic cells | Reversible assembly in response to stress as a result of accumulation of translationally repressed mRNA in the cytoplasm. | mRNA storage and triage, regulation of translation |
| Rearrangement | P-bodies | DDX6, EDC-4, LSM-4, EIF4E-T, poorly translated and untranslated mRNA | Eukaryotic cells | Increase in the number and size of P-bodies under stress conditions. | mRNA translation, processing and degradation |

## 2. Eukaryotes

In eukaryotic cells, phase-separated biopolymers undergo significant structural alterations that affect the regulation of stress-specific signaling pathways (Table 1, Figure 1). Some of the stress-responsive MLOs function in the unstressed cells (nucleolus, Cajal bodies, P-bodies, etc.) and upon stress, they undergo significant alterations of properties and potentially performed roles, while other condensates are only present in cells that experience stress or recover from it (cytoplasmic stress granules, A-bodies, etc.) and, thus, are specifically required to combat stress (Figure 1). Moreover, inhibition/activation of the corresponding stress receptors is often accompanied by the formation of biomolecular condensates on the surface of cell organelles [12]. Additionally, reprograming of gene expression programs in stressed cells is associated with the formation of super-enhancers, complexes necessary for activating the transcription of the corresponding genes, as a result of phase separation [13]. Therefore, phase separation is widely used by eukaryotic cells to promote survival during unfavorable conditions.

### *2.1. Nuclear MLOs*

#### 2.1.1. Nucleolus

The nucleolus is a dynamic subnuclear structure which has primarily been known for its role in ribosome biosynthesis but has recently gained attention for its novel role in sensing and coordinating cellular stress response. The numerous protein, DNA, and RNA components are spatially organized in three distinct sub-nucleolar compartments, corresponding to the steps of the ribosome biogenesis (Figure 2A): (1) pre-rRNA transcription from rDNA occurs in the fibrillar center (FC) or at the border between the FC and dense fibrillar component (DFC), surrounding the FC; (2) rRNA processing occurs in DFC; (3) pre-ribosome subunit assembly takes place within the granular component (GC), encapsulating FC and DFC. FCs are enriched in components of the RNA Pol I machinery, such as UBF. The DFC component is enriched in pre-rRNA processing factors, such as snoRNPs, fibrillarin, and Nop58. GC is an accumulation of dense particles with a mean diameter of 10–20 nm, which correspond to the most mature precursors of ribosome subunits. The GC is enriched with the protein nucleophosmin (NPM1) [14], which is also involved in ribosome biogenesis [15,16].

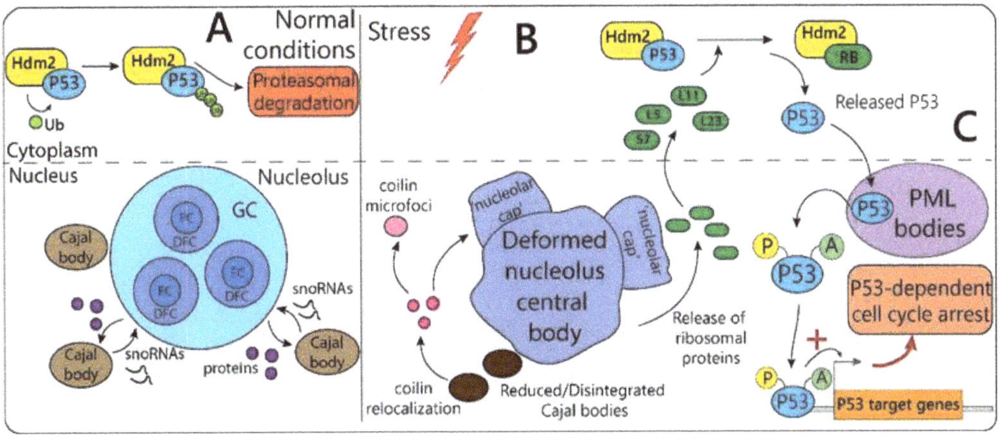


**Figure 2.** Functional and structural changes in nucleolus and Cajal bodies in response to stress. (**A**) Under normal environmental conditions, nucleolus and CBs are formed in the nucleus via mechanisms of phase separation. Nucleolus is a multiphase compartment composed of three internal layers: fibrillar center (FC), dense fibrillar component (DFC), and granular component (GC). Close proximity allows for content (snoRNAs and proteins) exchange between CBs and nucleolus. P53 is inhibited by direct binding of Hdm2 E3 ubiquitin ligase, ubiquitination by it and degradation by proteasome, and also by export to the cytoplasm. (**B**) The onset of stress results in structural deformations of the nucleolus and CBs. Coilin relocates from disintegrated CBs into 'nucleolar caps' or special microfoci. Released from the nucleolus ribosomal proteins displace p53 from p53-Hdm2 tandem via mechanism of competitive binding. (**C**) Activation of p53 in response to stress. Released p53 is relocated to the nucleus, first to PML bodies, where post-translational modification necessary for p53 activation takes place. Then, active p53 binds to promoters of its target genes, initiating cell cycle arrest.

Nucleolus morphology, structural integrity, and composition are heavily affected by different stressful stimuli (Table 1, Figure 1A). Two types of nucleolus stress-induced structural deformations have been described: segregation and fragmentation [16]. Nucleoli segregate in response to DNA damage (e.g., UV light [17]) or inhibition of rRNA transcription (e.g., RNA Pol I or topoisomerase II impairment [18]). This process involves condensation with subsequent separation of the FC and GC, accompanied by the formation of 'nucleolar caps' around the so-called central body (nucleolus deformed residue) (Figure 1A) [16,19]. On the other hand, inhibition of RNA Pol II or protein kinases leads to the unravelling of the FC, the process called nucleolar fragmentation [20,21].

One of the most prominent mechanisms of nucleolus-dependent regulation of stress response is associated with stabilization and activation of "genome guardian" tumor suppressor p53 (Figure 2B) [22]. Under normal conditions, the p53 function is blocked by inhibitory binding of E3 ubiquitin ligase Hdm2 (also called Mdm2 in mice), which interacts with the p53 transcription activation domain, preventing it from inducing its target genes. Moreover, Hdm2 shuttles p53 from the nucleus to the cytoplasm, a process facilitated by the export of ribosomal subunits [23], where ubiquitinylated p53 can be degraded by the proteasome. In either case of nucleolar segregation or fragmentation triggered by stress, aberrant expression and re-localization of many ribosomal proteins (RBs) are observed. These alterations in ribosome biogenesis initialize p53-dependent cell cycle arrest via several different mechanisms: (1) p53 release from the complex with Hdm2 (Figure 2B); (2) enhancement of the p53 translational profile; and (3) inhibition of co-ribosomal export of p53-Hdm2. The first mechanism is underliedby the competition between p53 and released from the nucleolus RBs for Hdm2 binding, leading to the recession of p53 proteasomal degradation. For example, under ribosomal stress, liberated ribosomal proteins (such as L5, L11, L23, and S7) directly interact with Hdm2 blocking its association with p53 (Figure 2B) [16,24]. Then, the elevation of active p53 levels under stress conditions is also facilitated by its increased translation. For instance, under genotoxic stress, the released L26 from the 60S ribosomal subunit ribosomal protein binds to the 5′ untranslated region of p53 mRNA and upregulates its translation [25]. Under normal conditions, the association of L26 with p53 mRNA is additionally repressed by Mdm2-induced polyubiquitylation and proteasomal degradation of L26. However, under genotoxic stress, this process is inhibited [25]. Finally, the last described pathway involves inhibition of p53/Hdm2 co-export with ribosomal subunits from the nucleolus to the cytoplasm where p53 proteasomal degradation occurs [23].

It is known that many different viruses target proteins to the nucleolus and recruit nucleolar proteins to facilitate virus replication. It obviously affects the morphology and composition of the nucleolus. For example, the coronavirus infection increases nucleolar size and, in particular, the enlargement of FC, as well as alternates the nucleolar proteome (e.g., localization of nucleocapsid (N) protein of coronavirus to the DFC of nucleolus, an increase in the amount of nucleolin within nucleolus) [26]. Viral infections may also induce the nucleolar accumulation of chaperones such as Hsp70. The Hsc70s (heat shock cognate proteins 70) are located to the nucleolus during the recovery period after stress [27]. It has been shown that under cellular starvation in the serum-free medium, the level of nucleophosmin in the nucleoli was diminished while its amount in the nucleoplasm increased. When the normal serum content has been restored, the nucleophosmin relocated back to the nucleolus [28]. Additionally, a wide range of anticancer agents induced the nucleoplasm translocation of nucleophosmin [15].

#### 2.1.2. Cajal Bodies

Cajal bodies (CBs) are nuclear MLOs that have been functionally linked to the nucleolus. CBs are often observed in a close spatial proximity to the nucleolus (and in some cases even within it) [29]. They share a certain degree of compositional overlap (for example, proteins fibrillarin, nucleolin, Nopp140, NAP57); moreover, the constant flux of proteins and various RNA species between these two nuclear entities has been revealed (Figures 1A and 2A) [30]. CBs are involved in the maturation of small nucleolar RNA (snoRNAs) which are necessary for rRNA post-transcriptional modifications. In this way, CBs facilitate the nucleolus in rRNA biogenesis. Given that functional and spatial interconnection of nucleolus and CBs and the nucleolar role in the stress response, it should not be a surprise that CBs are also responsive to stress [16]. Besides snoRNAs, CBs are also centers for small nuclear RNA (snRNA) and histone mRNA processing. CBs assemble at snRNAs transcriptional loci and sometimes at sites of active histone mRNA transcription [29]. Additionally, a distinct type of small non-coding RNAs, called scaRNAs (small Cajal body-specific RNAs), is specifically localized to CBs. scaRNAs guide RNA modifications on snRNAs [29].

CBs are conserved MLOs found in plant and animal cells. Additionally, structures compositionally and functionally similar to CBs have been reported in other organisms. The major CBs scaffold protein coilin is widely used as a molecular marker of CBs. However, in some organisms (e.g., Drosophila, C. elegans, yeast), coilin or its obvious homologues are absent which impedes the CBs homologues identification. In budding yeast, the analogue of CBs named "nucleolar body" is found within the nucleolus. These bodies are enriched with the same components as mammalian CBs such as precursor forms of U3 snoRNAs and TGS-1 (conserved methyltransferase catalyzing the formation of the 5′ terminal tri-methyl-CAP structure in sno- and snRNAs) [31].

Coilin-deficient animals (flies, mice) and plants (Arabidopsis) lack CBs; however, they still remain viable [32–34]. On the other hand, coilin gene disruption (and therefore CBs loss) is semi-lethal for zebrafish and murine embryos (especially late in the gestation period when embryos rapidly grow) [32]. Additionally, coilin knockout mice display reduced litter size and litter number, compared to wildtype controls, and mutant males have smaller testes, which could reduce or delay sperm production and mutant females might produce fewer mature oocytes [32]. Embryonic fibroblasts derived from these animals lack typical CBs but contain residual bodies containing a subset of typical CB components [35]. For zebrafish embryos, functional CBs are absolutely required for completion of the developmental process and concomitant cell survival [36]. Depletion of coilin in zebrafish embryos leads to splicing defects that could be partially restored by injection of fully assembled snRNPs [36]. Thus, according to the collected data, CBs are not essential for the developed organism under normal conditions. At the same time, these cellular structures are highly conserved and withstood a great evolutionary pressure and, therefore, bear a significant natural selection benefit. This allows us to suggest that CBs' key role may lie in the maintenance of the cellular homeostasis in abnormal or quickly changing conditions, as well as for highly specific parts of the life cycle, such as embryogenesis. For instance, the suppression of coilin gene expression can confer salt tolerance on *N. benthamiana* plants, confirming the role of CBs in the plant cells response to osmotic shock [37].

Typically CBs disintegrate in response to various types of stress with its core proteins being relocated (e.g., coilin [38,39]) or undergo proteasomal degradation (e.g., FLASH protein [40]) (Table 1, Figures 1A and 2B). It has been shown that cellular starvation decreases the number and size of CBs [41]. The UV-C irradiation, osmotic stress, and heat shock reversibly disrupt CBs, with the formation of the coilin-containing nucleoplasmic microfoci (Figure 2B) [38,39]. The chilling stress of soybean root meristem cells reduces the number of CBs with the subsequent recovery of their amount after the stress. However, this reduction may be caused by the hindering of CBs formation or by their fusion [42]. The CBs disassembly may be caused by the alteration of intermolecular interactions associated with the stress-induced posttranslational CB proteins modifications (e.g., SUMOylation of CB proteins upon stress [43]) and/or by the degradation of CB components via the proteasomal pathway. Thus, the involvement of proteasome activator subunit PA28g in the UV-C-induced coilin nuclear redistribution was clearly demonstrated [38]. It has also been shown that coilin is not degraded during stress, as its cellular levels remain constant, but rather it changes its localization. Thus, the inhibition of RNA polymerase II transcription by 5,6-dichloro-1-b-D-ribobenzimidazole causes the transition of coilin into the cap-like structures associated with the nucleolus (Figure 2B) [16].

It has been shown that different viral infections lead to diverse CBs responses. For example, HSV-1 infection induces the relocation of some CBs proteins (coilin, SMN, and fibrillarin) to the damaged centrosomes [44]. Adenoviruses induce the redistribution of the coilin and some other CB components in the periphery of viral replication centers to participate in the processing of virial transcripts [45]. In plants, groundnut rosette virus (GRV) induces the fusion of the transformed CBs containing viral ORF3 protein with the nucleolus [46]. However, the data regarding the functional importance of these structural changes have been rather contradictory. For example, it was reported that knockdown of coilin in Nicotiana plants may increase the accumulation of the barley stripe mosaic virus and tomato golden mosaic virus promoting the virus spread. On the other hand, the same study using the same knockdown system reported a decline in virus accumulation in the case of the turnip vein clearing virus and the potato virus Y, also linked to downregulation of symptoms progression [19,37].

Overall, the available data suggest that CBs are highly sensitive to various types of stress. However, the question remains whether the observed structural alterations are a consequence of cell response to stress or a part of its regulation and if the latter, then the exact mechanisms are awaiting clarification.

#### 2.1.3. Paraspeckles

Another example of MLOs that respond to stress with structural and functional changes is paraspeckles (Table 1, Figures 1B and 3A). Paraspeckles are nuclear condensates which assembly is driven by the long noncoding RNA NEAT1 (Nuclear Paraspeckle Assembly Transcript 1). NEAT1 is a single-exon transcript that is alternatively spliced in human cells to produce short 3.7-kb (NEAT1_1) and long 22.7-kb (NEAT1_2) isoforms. Long NEAT1_2 is essential for paraspeckle formation. Its knockdown with antisense oligonucleotides resulted in a complete disintegration of paraspeckles in both human and murine cells [47]. The paraspeckles most probably assemble co-transcriptionally at the nascent NEAT1 RNA, however, they may migrate throughout the nucleoplasm upon maturation. The process of paraspeckle formation starts with the expression of NEAT1 transcripts followed by binding of the members of DBHS (Drosophila Behavior Human Splicing) family—proteins SFPQ and NONO—which together form SFPQ-NONO functionally active heterodimers (Figures 1B and 3A). The initial binding of SFPQ and NONO to NEAT1 is essential for NEAT1 stability. Additionally, SFPQ-NONO represent paraspeckle structural scaffold themselves as RNA-protein interaction leads to oligomerization of SFPQ-NONO heterodimers into longer chains of polymers along NEAT1_2 transcripts increasing the system multivalency (Figure 3A). The knockdown of either SFPQ or NONO completely oblates paraspeckle formation [48]. At the final step of assembly, the SFPQ-NONO-NEAT1 system attracts the additional proteins such as FUS and the phase separates, forming the mature paraspeckle [49]. The paraspeckles are composed of a core part containing the middle hydrophobic part of NEAT1_2 transcripts and the shell part containing the 5′ and 3′ hydrophilic ends of NEAT1_2 [49,50] (Figures 1B and 3A). The core and the shell also have different protein compositions. Interestingly, paraspeckles can become elongated, forming cylindrical shapes over time (Figures 1B and 3A).

The paraspeckle assembly has been tightly linked to cellular adaptation to changing external conditions. Working as a storage hub for RNAs and proteins involved in the transcription regulation and pre-mRNA processing, paraspeckles modulate various cellular pathways, such as circadian cycling and response to various stressors (mitochondrial stress, hypoxia, heat shock, viral infection, etc.). During normal conditions, when cells are unstressed, paraspeckles are still ubiquitously observed in cellulo but not in vivo. In mice raised in stable laboratory conditions, paraspeckles are rarely found within tissues, and usually appear in terminally differentiated cells such as at the tips of crypts in the large intestine or corpus luteum [48,51]. It has been shown that NEAT1 knockout (KO) mice, which lack paraspeckles, are viable and fertile, however, nearly half of the naturally mated female mice stochastically failed to become pregnant probably due to the dramatic decrease in serum progesterone level due to corpus luteum impairment in the KO animals [51]. In cell culture, paraspeckles were reported in all the cell types except embryonic cells; however, their differentiation was shown to be accompanied by the paraspeckles formation [48]. Altogether these data indicate that paraspeckles, while not vital MLOs, aid cells in adjusting to specific, not yet clearly identified, changes in environmental conditions as well as in the internal cellular state.

Most cells can reversibly multiply the number of paraspeckles upon different types of stress (Figure 1B). The increase in the paraspeckles level has been shown under hypoxia conditions [52], temperature elevation [53], sulforaphane treatment [53], as well as for softening of the cellular substrate [54]. The number of nuclear paraspeckles correlates with the amount of the expressed NEAT1_2, while corresponding protein levels remain unchanged. Therefore, the stress-dependent accumulation of paraspeckles is triggered by enhanced transcription of NEAT1, activated by various stress-responsive transcription factors, such as HIF-2$\alpha$ during hypoxia [52], HSF1 during heat shock [53], p53 in replication stress [55], or ATF2 during mitochondrial stress [56], each of which binds to the corresponding element located in the NEAT1 promoter (Figure 1B).

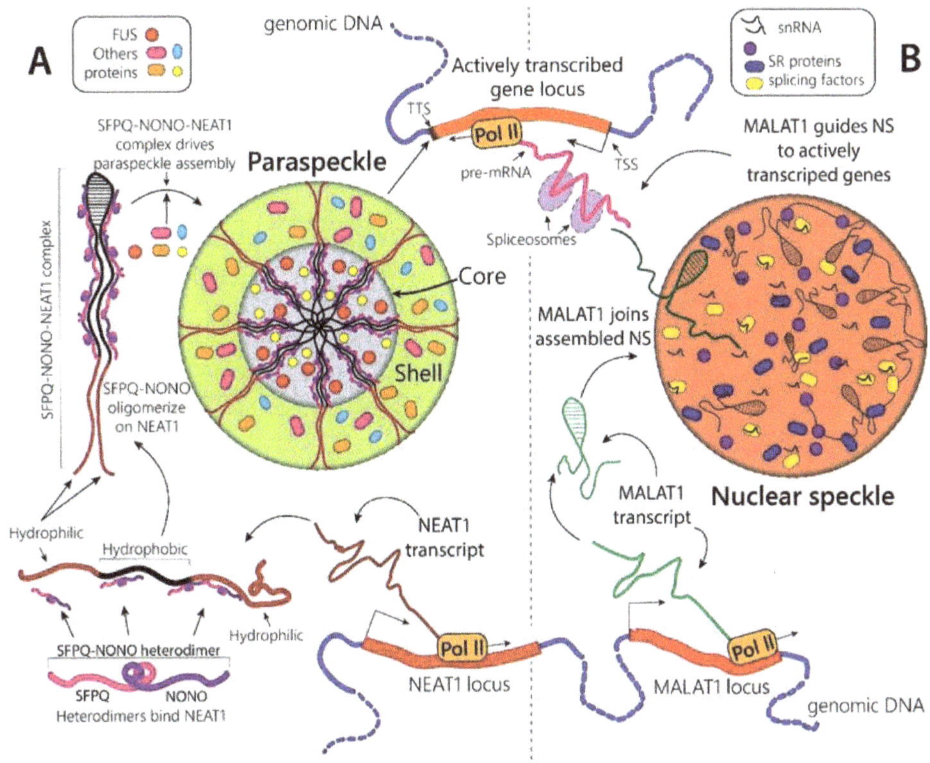


**Figure 3.** Assembly and cooperation of nuclear speckles and paraspeckles. (**A**) Paraspeckles assemble at sites of NEAT1 lncRNA expression. Nascent NEAT1 transcripts sequester SFPQ-NONO heterodimers that oligomerize on the synthesized NEAT1, stabilizing it and forming SFPQ-NONO-NEAT1 complexes. These tripartite complexes assemble into the condensate that attracts multiple client proteins, forming mature paraspeckle. The hydrophobic and hydrophilic regions of NEAT1 fluctuate towards center and edges of paraspeckle, respectively, forming the core and the shell. (**A**) MALAT1 lncRNA is incorporated into nuclear speckles but is not required for their formation. NS contains various splicing factors, and it was found that MALAT1 accumulates at the sites of active transcription, potentially guiding NS to the spliceosomes. (**A**,**B**) NEAT1 and MALAT1 are expressed from the adjacent genomic sites. NS and paraspeckles colocalize at the actively transcribed gene loci. Paraspeckles are more enriched at transcriptional start sites (TSS) and transcriptional termination sites (TTS). NS primarily localized across gene bodies.

It has also been shown that viral infections predominantly increase the paraspeckles number. The elevated amount of paraspeckles enhances the sequestration of the SFPQ protein which is a suppressor of several anti-viral genes (e.g., RIG-I and IL-8). Such sequestration causes the de-repression of the respective genes with the following production of the gene products. This mechanism has been observed for Hepatitis D, Influenza, polyI:C infection, and Hantavirus [57–59]. Moreover, paraspeckles are involved in the nuclear retention of the viral mRNA, for example, REV-dependent HIV-1 transcripts [60]. Paraspeckles also play an essential role in the antibacterial immune response. For example, upregulation of NEAT1 is observed in response to salmonella infection [61].

The paraspeckles are observed in two distinct shapes: spherical shape, typical for other MLOs, and unusual elongated shape (Figure 1B) [49]. Some stress events trigger the formation of spherical paraspeckles (temperature [52], hypoxia [53]), while others the formation of elongated paraspeckles. Thus, mitochondrial stress caused by depletion of mitochondrial proteins leads to the generation of elongated paraspeckles [56]. The shift from sphere to cylinder-like shape has been associated with alterations in post-transcriptional processing of the NEAT1-favoring production of long NEAT1_2 over short NEAT1_1 [49,56]. This is in accordance with the suggested block copolymer micellization model of paraspeckles elongation in which cylindrical micelles depend on the NEAT1_2 level and are stabilized above its certain threshold [50] (Figure 1B). The dynamic of the micellization process is distinct from that of the liquid-liquid phase separation and was suggested to facilitate the regulation of paraspeckle size [50]. Thus, paraspeckles are the only currently known MLOs that may assemble not as a result of the LLPS process alone. However, if it is indeed the case, then it is reasonable to expect a discovery of analogical assembly mechanisms for other biological condensates.

#### 2.1.4. Nuclear Speckles

Nuclear speckles (NS) are nuclear MLOs involved in splicing regulation. NS are also sometimes called 'interchromatin granule clusters' as they are located in the interchromatin regions of the nucleoplasm of mammalian cells. NSs contain pre-mRNA splicing factors, including snRNPs and SR proteins [62]. Additionally, long non-coding RNA MALAT1, a single-exon transcript over 7 kb in length, is enriched in NS through its specific interactions with NS-retained proteins (Table 1, Figures 1C and 3B). MALAT1 was found to regulate the SR splicing factors distribution to NS via direct interaction and modulation of their phosphorylation state [63]. SR proteins cycle between phosphorylated and dephosphorylated states, which is essential for pre-mRNA processing. MALAT1 depletion results in both dephosphorylation of SR proteins and differential changes in alternative splicing events in several mRNAs, mostly exon inclusions [63]. However, MALAT1 is dispensable for NS formation or cellular viability and MALAT1-deficient mice did not demonstrate abnormalities in alternative splicing patterns [64].

It was demonstrated that MALAT1 localizes to actively expressed genomic loci, most likely via its proteins partners targeting long non-coding RNA to newly synthesized pre-mRNA transcript (Figure 3B) [65,66]. Additionally, with the help of various genome mapping methods, it was uncovered that NS are associated with chromosome regions characterized by high levels of active RNA polymerase II transcription [67,68]. These discoveries have led to the suggestion that MALAT1 acts as a molecular leash delivering splicing machinery contained in the NS at the sites of active gene transcription [66,69] (Figure 3B). Moreover, it has been experimentally shown that association of Hsp70 genes and four genes flanking the Hsp70 locus with nuclear speckles causes a several-fold boost in expression of these genes following heat shock [68]. Authors suggested that this NS-dependent upregulation is a result of the decreased exosomal degradation of the nascent transcript combined with increased transcriptional rate [68]. Based on these results, a so-called "gene expression amplification" model was proposed. According to this model, nuclear speckles act as gene expression hubs capable of increasing the net production of transcripts of genes positioned in the NS vicinity [68].

Interestingly, the core RNA of paraspeckles NEAT1 is positioned in the genomic environment of MALAT1 and two RNAs are transcribed from the adjacent regions in the genome [69] (Figure 1B,C and Figure 3A,B). Despite that, these RNAs partition to different MLOs and never colocalize to the same condensate. On the other hand, the nuclear speckles and paraspeckles has been found to localize together at hundreds of active gene loci, however, primarily bound to distinct parts of the genes: NEAT1 was found at transcriptional start sites (TSS) and transcriptional termination sites (TTS), whereas MALAT1 primarily localized across gene bodies [69] (Figure 3A,B). This might indicate a cooperation between these two biological condensates in the upregulation of gene expression. It is not yet clear if this mutually functional complementarity is maintained during the stress condition.

NS accumulates various splicing factors as well as components of the splicing machinery. The alternative mRNA splicing is significantly impacted by stressful stimuli via changes in localization, interactions, expression, and chemical modifications of splicing factors and spliceosome components [70]. Additionally, changes in alternative splicing patterns are used by cells to regulate gene expression in order to combat stress [70,71]. For a significant part of the transcriptome, splicing is downregulated in response to heat shock with the exception to genes involved in stress response [71]. NS are inevitably involved in these regulatory pathways; however, specific mechanistic details remain unknown.

Similar morphological changes have been observed in NS across various stress conditions. Typically, enlargement and rounding of NS condensates accompanied by the reduction in their total number is reported (Figure 1C). This has been shown for transcription arrest caused by heat shock (45 °C for 15 min) [72], treatment with transcription inhibitors, such as actinomycin D [73,74], genotoxic stress induced by Etoposide [75], heavy metal stress [74], and osmotic stress [76]. This aberrant morphology was attributed to two processes: 1) proteins migrating back to NS for storage upon stress [75]; 2) NS particles fusion [74] (Figure 1C). Moreover, increased NS mobility, characterized by long-range directional migration across interchromatin space was demonstrated for several different types of stress [74]. Interestingly, this motion terminates with condensates coalescence, suggesting that NS mergence is not a stochastic, but rather a controlled process [74].

#### 2.1.5. PML-Bodies

PML bodies are nuclear polyfunctional compartments that are involved in the regulation of transcription, stress response, differentiation, and transition of cells to the senescent state and are present in cells under normal conditions [77]. The major protein of these compartments is the promyelocytic leukemia (PML) protein (Table 1, Figure 1D). The main components of PML bodies in human cells are the six nuclear isoforms of the protein of promyelocytic leukemia (PML) formed via alternative splicing, therefore, they differ in size and amino acid sequence of their C-terminal domains [78–80].

Analysis of morphology and dynamics of PML bodies showed the existence of at least two populations of PML bodies in U2OS and HeLa cells with a diameter of about 0.6 μm and 1.2 μm. In the population of "small" PML bodies, all bodies are spherical and all PML isoforms dynamically exchange with nucleoplasm. It has been suggested that such bodies act as liquid "seeds" of functionally active PML bodies, forming due to weak intermolecular interactions and providing the necessary concentration of PML isoforms for the formation of intermolecular disulfide bonds between PML monomers (Figure 1D). The "large" mature bodies have a toroidal morphology and scaffold with low mobility formed predominantly by PML-V and PML-VI [81,82].

PML bodies are one of the key regulators of the p53-dependent stress response (Figure 2C) [83]. In response to stress, p53 undergoes a number of post-translational modifications necessary for the activation of this protein and the subsequent induction of the expression of cyclin-dependent kinase inhibitor genes, which, in turn, contribute to the inhibition of proliferative gene expression and cell cycle arrest [84]. Nuclear PML bodies are one of the main platforms that provide the post-translational modifications of this protein necessary for the activation of the p53-dependent signaling pathway (Figure 2C) [77]. PML bodies promote activation of p53 target genes which are oxidative stress-induced, for example, Trp53inp1 or Sesn2 are part of the p53 anti-oxidant response [83]. According to recent data, the PML-IV isoform makes a decisive contribution to p53 activation, forming PML-IV-CBP-p53 complexes in PML bodies [85]. Under hypoxic conditions, PML bodies suppress the AKT-mTOR signaling pathway by inhibiting PP2 phosphatase within these organelles [86]. PML bodies also promote activation of the DNA damage response via the ATM/ATR-p53-p21 pathway [78].

According to the previously existing model of PML bodies formation [87,88], oxidative stress should induce solidification of PML bodies due to the disulfide-mediated multimerization of PML monomers and enhancement of intermolecular electrostatic interactions by K487 deacetylation and K490 SUMOylation [89,90]. However, using the FRAP method to characterize liquid properties of condensates, it has been shown for PML-/- HeLa cells as well as for wild-type cells, that oxidative stress induced by $H_2O_2$ alters the dynamicity of the main proteins of canonical PML bodies, as well as the PML bodies associated with alternative telomere lengthening (APBs) (complete immobilization of PML-V and decrease mobility of PML-I and PML-II between nucleoplasm and these organelles) while its localization and morphology are still practically unchanged [81,82]. Peroxide treatment of U2OS cells causes a slight increase in the size of APBs. The exchange rate of PML-III, PML-IV, and PML-VI between PML bodies and nucleoplasm remained unchanged upon oxidative stress. At the same time, hydrogen peroxide treatment completely immobilized the PML-V isoform within PML bodies and reduced the diffusion of PML-I and PML-II isoforms. The arrested PML-V diffusion may be caused by or promote the disulfide bonds formation between these isoforms, which is forced by a strong tendency for its $\alpha$-helical motif to form hyper stable oligomers and a low diffusion rate of this isoform under normal condition [81]. The dynamic of the C-terminal domain of PML-II and PML-V as well as their mutants with K490R substitution, disrupting the PML SUMOylation, in normal and oxidative stress conditions caused by $H_2O_2$ treatment has also been studied. A slight decrease in the exchange rate and a decrease in the proportion of the mobile fraction of the wild-type PML-II C-terminal domain have been observed. For the mutant form of the PML-II C-terminal domain with the K490R substitution, a slight increase in the exchange rate has been revealed. Additionally, oxidative stress caused a significant decrease in the diffusion rate of the C-terminal domain of PML-V and its mutant form with the K490R substitution between the bodies and the nucleoplasm. The dynamics of the exchange of the mutant form of the C-terminal domain of PMLV, K490R, under the conditions of the acute oxidative stress, slows down significantly more than that of the wild-type domain [82]. The induction of oxidative stress by $As_2O_3$ resulted in degradation of most of the PML isoforms, leaving the SUMO at the core of the nuclear bodies. PML-I, PML-II, and PML-VI isoforms dissociated to cytoplasm upon arsenic treatment [79]. The exposure of cells to other types of stress such as heat stress, heavy metal addition, and expression of adenovirus E1A demonstrated the decrease in the number and size of PML bodies and the formation of smaller PML-containing structures called ‘microstructures’. Such microstructures are formed from parental PML bodies as a result of fission or budding from its surface. They are mobile and able to fuse with each other as they move through the nucleoplasm. The over-expression of SUMO-1 prevents the formation of microstructures [27].

During nuclear dissociation during mitosis, PML bodies are not disassembled, but are transformed into the so-called mitotic accumulation of PML proteins (MAPPs), which can be visualized using confocal fluorescence microscopy [91]. In the early G1 phase of the cell cycle, MAPPs, in turn, are transformed into the so-called cytoplasmic assemblies of PML and nucleoporins (CyPN) [92]. CyPNs are large gel-like structures prepared for nuclear import containing KPBN1 importin and at least 20 FG-porins are involved in the formation of a selective barrier inside nuclear pores. Like MAPPs, these structures can be easily visualized using confocal fluorescence microscopy. During the PML translocation into the core, CyPN is disassembled.

#### 2.1.6. NELF-Bodies

Unlike MLOs discussed previously, that can be observed in the nucleus of the unstressed cells, there is also a group of biological condensates only present in cells that experience stress or recover from it. Recently, a novel stress-induced condensate formed by a negative elongation factor (NELF) has been described (Table 1, Figure 1E) [93]. In order to successfully resist stress, cells need to quickly reprogram a multitude of regulatory pathways and shut down processes that are not essential for immediate survival. So, one of the first steps of stress-response is downregulation of transcription and translation. NELF is a negative regulator of transcription that directly inhibits RNA polymerase (Pol) II activity at the elongation step via binding [93,94]. It has been found that NELF is able to undergo LLPS in vitro and upon stress forms nuclear condensates in cellulo, that potentially stabilize its interaction with chromatin and enhance inhibitory potential. NELF protein contains intrinsically disordered regions, so-called "tentacles", that are essential for its phase separation (Figure 1E) [93]. Under normal conditions, NELF is present in the cell, but its activity is blocked by CDK9-dependent phosphorylation. Upon stress induction, NELF is quickly dephosphorylated and SUMOylated that promotes its condensation and formation of NELF bodies at transcriptional loci of many housekeeping genes [93]. NELF bodies block Pol II enzymatic activity, promoting global downregulation of transcription and aiding cell survival mechanism (Figure 1E) [93].

#### 2.1.7. Nuclear Stress-Bodies

Nuclear stress bodies (nSBs) form de novo in the cell nucleus in response to stress. nSB assembly starts with the activation of expression of a so-called human highly repetitive satellite 3 long non-coding RNA (HSatIII) [95] (Table 1, Figure 1F). HSatIII contains multiple tandem repeats of nucleotide sequences and is transcribed from the pericentromeric heterochromatin. Its synthesis is triggered by binding of the HSF1 (heat shock factor 1) transcription factor. Increased local concentration of the nascent HSatIII transcripts attracts HSF1 and other proteins providing the platform for nSBs nucleation (Figure 1F) [96]. On average, several 1-2 μm nSBs assembled in one cell, all of them form and remain in a vicinity of HSatIII loci located on several chromosomes. The main protein components of nSBs are the heat shock regulators, HSF1 and HSF2; hnRNP proteins, SAFB and hnRNPM; and other mRNA splicing factors (Figure 1F) [95,96]. nSBs possess properties of a phase-separated condensates. However, they are subject to hardening in the conditions of a prolonged stress that leads to reduced cellular viability [97]. Concentrations of HSF1 and SAFB mark two successive phases in nSBs evolution. HSF1 is predominant during the eruption of the stress response with gradual decline of its levels during a stress recovery period, whereas SAFB is incorporated into nSBs with the delay and peaks after the stress termination [98]. The biological significance of nSBs is not entirely clear. There is evidence that nSBs may be involved in the regulation of mRNA splicing. For example, an increased import of SR proteins into nSBs was detected in response to stress [6]. Additionally, nSBs components positively impact cell survival, thus HSF1 and SAFB knockdowns promoted apoptosis [6]. Overall, further studies are necessary to shed the light onto nSBs biogenesis and the role in the regulation of cell survival during stress.

#### 2.1.8. A-Bodies

Like nuclear stress bodies and NELF bodies, amyloid bodies (A-bodies) assemble transiently in the cell nucleus in response to stress. A-bodies are droplet-like foci containing hundreds of proteins in the amyloidogenic state (Table 1, Figure 1G) [99]. Although solid-like MLOs are often considered pathological, A-bodies formation is reversible and useful for temporal storage of molecules. A-bodies are formed in several stages. At first, the stress (heat, acidosis, etc.) induces synthesis of non-coding RNA molecules called rIGSRNA (ribosomal intergenic spacer RNA). rIGSRNA transcripts are expressed from intron regions of ribosomal DNA consisting of numerous dinucleotide repeats (Figure 1G) [99]. Then, a local increase in the concentration of rIGSRNA molecules, represented by sequences with a low degree of complexity, causes the formation of 'seeds' for bimolecular condensates, to which amyloidogenic proteins containing ACM (amyloid-converting motif) are recruited, in particular E3 ubiquitin ligase VHL (Von Hippel–Lindau tumor suppressor). Different types of stress induce transcription from various areas of rIGS and production of distinct rIGSRNA isoforms. In turn, different rIGSRNAs sequester different proteins subsets [100]. Electrostatic interactions between negatively charged low-complexity RNA and disordered positively charged regions of ACM-containing proteins rich in arginine and histidine residues promote condensation [101]. At this stage A-bodies display properties of liquid-like dynamic condensates, such as fusion and content mobility. Then, a high local accumulation of hydrophobic fibrillation propensity domains of ACM creates conditions for the transformation of bimolecular condensates into a gel-like state and then into aggregates of amyloid fibrils [102]. Mature A-bodies completely immobilize stored proteins. The breakdown of A-bodies is initiated after the termination of stress and is carried out in an

Hsp70/90-dependent manner [102]. The released proteins do not undergo degradation but change the topology of the polypeptide chain to the native conformation and return to functional state. Thus, the key function attributed to A-bodies is protein storage and the isolation of potentially toxic amyloid fibrils preventing their interaction with the rest of the cellular proteome.

### *2.2. Cytoplasmic MLOs*

#### 2.2.1. Stress-Granules

Stress granules (SG) are cytoplasmic membrane-less organelles that transiently assemble in eukaryotic cells in response to various types of endogenous (for instance, impaired proteostasis, genotoxic stress, etc.) and exogenous stresses: temperature, oxidative stress, UV irradiation, nutrient deprivation, hypoxia, viral infection, and many others [103,104]. They have been found to be the key regulators of cellular stress response, reducing detrimental consequences of stress-induced damage. At least partially, this is achieved via incorporation into SGs translationally stalled mRNAs, RNA-binding proteins, and translation initiation factors and, thus, temporal isolation of these molecules from the rest of the cellular milieu (Table 1, Figures 1H and 4). Translation is one of the most energy-consuming cellular mechanisms, therefore, it is one of the first to be inhibited in response to stress in order to save cellular resources for the stress response. mRNA and translation factors are recruited into SGs upon stress and re-enter normal translation process after normalization of conditions and release from granules, making SGs temporal 'storage' capsules for indispensable molecules. Importantly, mRNA coding for stress response factors, such as heat shock proteins mRNA, are not incorporated into stress granules, allowing cells to activate expression of proteins essential for survival [105]. The cytoprotective benefits of such mechanism include global energy savings on mRNA and protein degradation and post-stress resynthesis, as well as hindrance of the toxic aggregation of partially unfolded protein biopolymers in the cytoplasm.

The highly dynamic nature of SGs allows them to quickly modulate cellular translation and proteostasis during unfavorable conditions thus promoting cell survival [106]. SGs have vast therapeutic potential as their deregulation has been linked to progression of multiple neurodegenerative disorders [107], oncogenesis and resistance to treatment of cancer cells [103,108], and viral replication inside the host [109] and other pathologies.

SGs formation takes several steps, the first of which starts when cellular stress leads to translation arrest via various pathways, including phosphorylation of translation initiation factors eIF2 and eIF4. Abrupted translation causes dissociation of polyribosome accumulation of free mRNA in the cytoplasm, which is then able to interact with RNA-binding motifs of SG scaffold proteins (G3BP1/2, TIA-1, and others), driving their liquid-liquid phase transition and nucleation of initial SG condensates (Figures 1H and 4A,B) [110,111]. Further maturation of SGs relies on higher-order heterotypic interactions between scaffold proteins leading to 'hardening' of the central 'core' of SG, around which client proteins form a more dynamic layer (Figure 4B) [111]. Functional activity of SGs depends on the composition of the dynamic outer phase.

The liquid droplet properties of SG ensure constant trafficking of molecules between the granule and the surrounding cytoplasm, allowing for a timely response to the onset and termination of stress. Upon restoration of normal conditions, SGs are quickly disassembled, and released mRNA is re-recruited by translation machinery. Decay of SGs is facilitated by chaperones inhibiting the mRNA–SG core proteins interaction [112,113]. Upon termination of stress, SGs can also be degraded via an autophagosomal mechanism [114] while violation of this process can lead to the formation of cytotoxic amyloid fibrils [114,115].

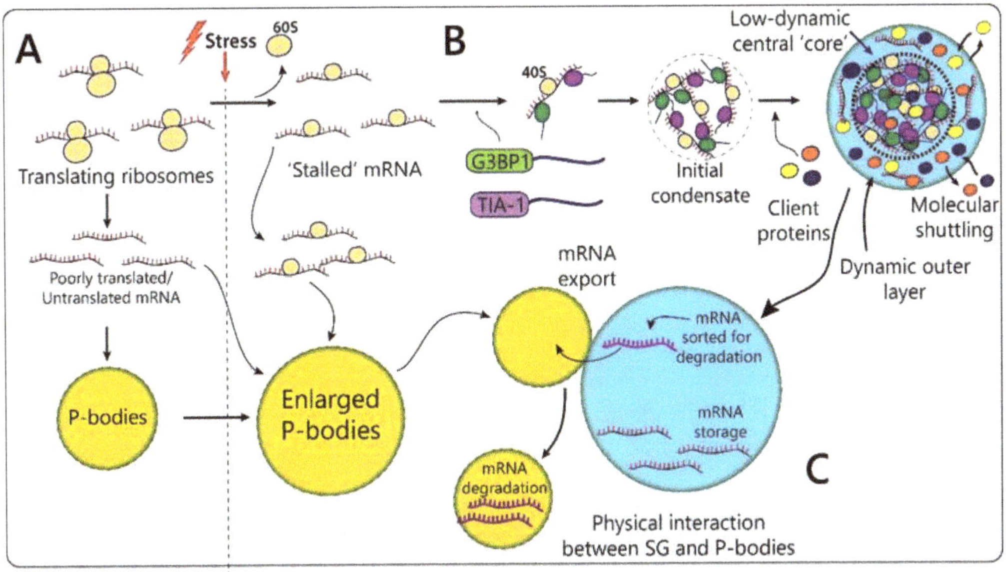


**Figure 4.** Stress granules and P-bodies interplay during stress response. (**A**) Under normal conditions, SGs are absent from the cytoplasm, mRNA translation is normal, and poorly translated or repressed mRNAs are sequestered into P-bodies. (**B**) Upon stress treatment, translation is inhibited and 'stalled' translation initiation complexes are recruited to P-bodies, causing their enlargement, or interact with SG proteins, such as G3BP1 and TIA-1. This interaction leads to phase separation and formation of initial pre-mature SG that then attracts more proteins. Mature SG has a low-dynamic central core and dynamic outer layer. (**C**) Stress conditions promote physical association between SG and P-bodies. It was suggested that within SG, the mRNA molecule undergoes sorting and the mRNA destined for decay are exported directly into P-bodies via temporal fusion between SG and P-bodies.

Several neurodegenerative diseases, including amyotrophic lateral sclerosis (ALS), Alzheimer's disease (AD), and frontotemporal dementia (FTD), are associated with abnormal SG biogenesis. Transformation of SG into toxic aggregates of amyloid fibrils is promoted by incorporation of the disease-associated mutant forms of proteins, such as TIA-1, TIAR, FUS (RNA-binding protein fused in sarcoma), hnRNPA1 (heterogeneous nuclear ribonucleoprotein A1), TDP-43 (transactive response DNA binding protein 43 kDa), and PABP1 (polyadenylate-binding protein 1) [116,117].

RNA is the major component of SG as 78–95% of SG composition are RNA molecules [118]. Despite extensive high throughput analysis of SG transcriptome, the mechanisms that drive the enrichment of certain RNA transcripts into SGs but not the others remain unknown. Previous studies have shown that all cellular mRNAs are represented in SGs to some extent, however, the magnitude of their concentration relative to cytoplasm differs drastically, suggesting that yet unknown factors promote preferential recruitment of certain RNAs to SG [104,118]. One parameter that was found to positively correlate with SG recruitment was the length of the transcript [104]. However, other studies demonstrated that mRNAs of the same length show different levels of SGs incorporation. These data suggest that individual mRNA molecules carry specific information that significantly affects their enrichment into these organelles. This could be attributed to primary nucleotide sequence, secondary structure, modifications of RNA nucleotides and the last especially has been the research focus in recent years. A curious contradiction can be found between two papers published recently [119,120]. Both studies performed a comparative analysis of mRNA partitioning into SG between wildtype and METTL3 methyltransferase knockout mouse embryonic stem cells (mESC). METTL3 is the key writer enzyme of m6A RNA modification, the most common type of chemical modification found in mRNA. One of the reports showed an association between m6A modifications and average mRNA enrichment into SG [119]. However, a report published by Khong et al. found no evidence that METTL3 depletion affects mRNA composition of SGs, leading to the conclusion that m6A edits play little or no role in this process [120]. This contradiction can be explained by a deeper assessment of the properties of the METTL3 knockouts used by the two groups. The study that suggested positive correlation between m6A modifications and mRNA SG enrichment used knockout cells with a complete loss of m6A upon induced deletion of METTL3. On the other hand, the other work was performed on METTL3 knockout mES cell line that only had partial loss (~60%) of m6A levels. Moreover, this knockout cell line has been found to have an activated expression of a shortened partially functional METTL3 isoform [121]. Altogether, these data suggests that m6A chemical markers, and potentially other types of RNA modifications, are important for SG transcriptome regulation, however, even residual amounts of m6A may be able to fulfill the functional needs.

#### 2.2.2. P-Bodies

Along with stress granules, the most important cytoplasmic MLOs involved in the regulation of the stress response are Processing-bodies (P-bodies) (Table 1, Figures 1I and 4) [122]. Unlike temporary stress-induced stress granules, P-bodies are constantly present in most of the cell types, and they enlarge and multiply during stress (Figure 4C) [123]. These dynamic compartments are mainly composed of poorly translated mRNA molecules, proteins that contribute to translation inhibition or to different aspects of mRNA degradation, such as 3′-deadenylation, 5′-decapping, 5′-3′ exonuclease activity, and nonsense-mediated decay [124,125]. Additionally, during stress P-bodies, similarly to SG, we incorporate repressed translation initiation complexes, a process that contributes to their enlargement. The marker proteins of these organelles are DDX6, AGO1/3, DCP2, XRN4, EDC3, EIF4E-T, LSM1-7, SMG7, HNRNPM, and CPEB1 [126,127]. A critical role in the formation of P-bodies is played by the phase transitions of helicase scaffold proteins DDX6, EDC-4, LSM-4, and EIF4E-T upon interaction with untranslated mRNA. Inhibition of these proteins causes disassembly of P-bodies [122]. However, the details of the assembly mechanism of mature P-bodies in unstressed cells with low levels of untranslated mRNA remain elusive. It has been established that P-bodies contain hundreds of mRNA types and, probably, in the absence of stress, they serve as a depot for adaptive switching of protein synthesis programs with minimal energy consumption during the cell life cycle [127].

There is strong evidence for functional interplay and cooperation between stress granules and P-bodies (Figure 1H,I and Figure 4C). Two MLOs have been found to share some of the protein and mRNA content, while also having molecules uniquely attributed to one or another. The proteomic analysis showed that the protein composition of stress granules and P-bodies overlaps by 10–25% [128]. Moreover, the composition of stress-induced P-bodies resembles stress granules to an even greater extent [11]. Both SG and P-bodies contain components of the RNA-induced silencing complex (RISC), microRNAs, and argonaute proteins that are needed for RNA interference-induced silencing of mRNA. Additionally, both organelles include RNA-editing enzymes with antiviral activity, such as APOBEC3G (apolipoprotein B mRNA-editing enzyme catalytic subunit 3G) [129]. The presence of so many various catalytically active complexes suggests that these MLOs are the centers for post-transcriptional regulation of gene expression.

SG and P-bodies also were found to directly interact by coming into close spatial proximity that is promoted by regulated molecular tethering (Figure 4C) [125,130]. It has been shown that oxidative stress induced by arsenite promoted the convergence of P-bodies and stress granules and subsequent content exchange in HeLa cells [125]. Two components of mRNA decay machinery TTP and BRF1 were found to promote the SG's and P-bodies' physical association [130]. The authors of the study suggested a model of coordinated regulation by SG and P-bodies of mRNA biogenesis during stress. According to this model, firstly, mRNA accumulates to SG for sorting, processing, and storage. Then, mRNA molecules destined for degradation are directly transported into P-bodies via TTP/BRF1 fusions for decay (Figure 4C) [130].

It was also found that, like stress granules, P-bodies can have a multiphase structure in Drosophila oocytes [131]. This type of P-bodies is characterized by two immiscible regions, one containing *gurken* mRNA and the other *bicoid* mRNA. Additionally, in *Drosophila* oocytes, it has been shown that P-bodies and associated U-bodies (MLOs responsible for the assembly and storage of uridine-rich small nuclear ribonucleoproteins that are essential for pre-mRNA splicing) enlarge during starvation [132]. In contrast to mammalian and drosophila cells, yeast P-bodies appear only under stress conditions [133]. Besides that, P-bodies in yeast may be formed under nutrient stress caused by glucose starvation. The resulting bodies are enriched in mRNAs encoding specific mitochondrial oxidative phosphorylation factors such as ATP11, ILM1, MRPL38, and AIM2. At the same time, P-bodies induced by osmotic stresses were depleted by ATP11 [134].

In aging somatic cells of *C. elegans* under stressful conditions, P-bodies regulate proteostasis by recruiting the IFE-2 isoform of the transcription initiation factor eIF4E into these organelles, which contributes to the blocking of protein biosynthesis and increases the lifespan of cells [135].

### 2.3. MLOs Associated with Membrane-Bound Organelles

Dysfunction of cellular homeostasis under stress conditions causes activation of the stress response due to inhibition/activation of specific signaling receptors. As a rule, in eukaryotic cells, these processes occur on the surface of the membranes of cell organelles. The efficient occurrence of this type of reactions often requires the formation of biomolecular condensates on the membrane surface [136]. In this case, the concentrations of proteins required for phase separation are an order of magnitude lower than in the solution. Serine-threonine kinase Target of Rapamycin Complex 1 (TORC1), which is a megadalton complex of four proteins, under normal conditions regulates the synthesis of various biomolecules and inhibits autophagy. The arrest of this receptor activity is accompanied by the formation of TOROID (TORC1 Organized in Inhibited Domain) clusters on the surface of lysosomal membranes. TORC1 reactivation is accompanied by TOROID disassembly [12]. In Drosophila S2 cells, in response to nutrient deficiency, so-called Sec-bodies are formed due to the interaction of the intrinsically disordered protein Sec16 and subunits of the COPII complex [137]. This enables inhibition of protein transport and prevention of damage to vesicle border proteins. Calcium ions are a universal second messenger of various cellular processes that determine cell metabolism [138]. In this regard, under stress conditions, there is a change in the regulation of $Ca^{2+}$-dependent signaling pathways [139]. Catabolic processes observed during the activation of various types of stress responses are regulated by the transport of calcium ions from the ER to mitochondria [139]. An increase in the concentration of calcium ions in mitochondria causes an increase in the production of reactive oxygen species by mitochondria and arrest of the cell cycle, and inhibition of the transport of calcium ions from the ER to mitochondria causes cell death [139]. The so-called MAMs (mitochondria associated membranes) are the platform for calcium transport from the ER to mitochondria [140]. These structures provide the necessary machinery and distance between the ER and the outer mitochondrial membrane for efficient transport of calcium ions. One of the key players in the MAM machinery required for the transport of calcium ions is the family of 1,4,5 triphosphate inositol receptors (IP3R) localized in the ER membrane [140]. In response to an external stimulus, these receptors are activated, which form a complex with the VDAC1 channel localized on the outer mitochondrial membrane and the Grp-75 chaperone, which makes it possible to ensure and coordinate the transport of calcium ions [139]. One of the regulators of IP3R activity, and, accordingly, calcium transport from the ER to mitochondria, is the PML protein [141]. This predominantly nuclear tumor suppressor exists in several isoforms, some of which are capable of cytoplasmic localization [142]. The localization of PML in MAMs is mediated by the cytoplasmic p53 fraction, usually in response to stress [143]. PML is able to form microdomains in MAMs, including IP3R, AKT kinase, and PP2 phosphatase, which ensure phosphorylation and correct operation of IP3R [141]. At the same time, the localization of PML to MAM in the cells of primary mouse fibroblasts contributes to a decrease in autophagy [143]. As is known, MAMs play one of the central roles in the initiation of autophagy, the abundance of MAM is significantly reduced during natural and pathological aging [144]. The key $UPR^{MT}$ stress response receptor, IPE1, also forms clusters in MAMs in response to stress, thereby inhibiting ER-associated mRNA and, accordingly, suppressing the synthesis of new proteins [145]. Endoplasmic reticulum membranes play an extremely important role in the regulation of P-bodies biogenesis. It has been shown that the interaction of endoplasmic reticulum membranes with P-bodies regulates their composition and functional activity [146]. In addition, ER membranes are a platform for the fusion of stress granules and P-bodies.

### 2.4. Yeast MLOs

Yeast cells contain both MLOs that have clear analogues in other eukaryotes, as well as a number of unique yeast-specific condensates (Table 2).

**Table 2.** Examples of LLPS (or suggested to be LLPS) compartments formed or rearranged in response to in yeast cells.

| MLO-Type | Main Components | Stress Factors | Structural Changes in Response to Stress | Main Functions |
|---|---|---|---|---|
| Stress granules | mRNA, Pub1, Pbp1, eIF4GII | Impaired proteostasis, genotoxic stress, temperature, UV irradiation, nutrient deprivation, hypoxia, viral infection, etc. | Assembly of gel-like structures in the cytoplasm. | Storage of capped and polyadenylated mRNAs and their protection from degradation in P-bodies. Regulation of TORC1 signaling |
| P-bodies | mRNA, Dcp2p and Pat1p [147] | Nutrient deprivation, oxidative and osmotic stress | Assembly of liquid droplets in the cytoplasm. Yeast P-bodies mRNA and proteins composition depends on the type of stress. | Translation repression and mRNA turnover: 3′-deadenylation, 5′-decapping, 5′-3′ exonuclease activity, nonsense-mediated decay |
| eIF2B bodies | eIF2B | Glucose deprivation | Formation of eIF2B bodies as a result of eIF2B accumulation in the cytoplasm [147]. | Involved in inhibition of translation initiation |
| Proteasome storage granules | Proteasome 19S and 20S subunits [147] | Glucose deprivation | Relocalization of proteasome subunits and formation of proteasome storage granules in the cytoplasm. | Storage of proteasome subunits |

Yeast stress granules, in contrast to mammals and Drosophila, exhibit the gel-like properties [110]. Their formation occurs in several stages and is coordinated. In the first step, RNA and RNA-binding proteins interact to form large ribonucleoprotein complexes (RNP complexes). Further, RNP complexes fuse into larger compartments through additional RNA-mediated interactions and, above all, through the binding of prion-like domains. As a result, a solid core is formed, surrounded by a liquid shell [137,148]. At present, the molecular mechanism of SG assembly in fission yeast is not completely clear. It is known that their formation does not depend on the phosphorylation of eIF2$\alpha$, and glucose starvation-induced yeast SGs lack 40S ribosomal subunits and eIF3, which is a characteristic component and is required for mammalian SG assembly [149]. Yeast has fewer eIF3 subunits than mammals, whereas mammalian eIF4G has an eIF3-binding domain not found in yeast. Therefore, the assembly of yeast SGs is independent of the eIF4G/eIF3 interaction. Multicellular animals have several eIF2$\alpha$ kinases, whereas budding yeasts have only one that also affects the assembly mechanism of SG [150]. It is known that stress granules in the yeast *Schizosaccharomyces pombe* and *Saccharomyces cerevisiae* contain orthologues of proteins found in mammalian SG. In particular, Nxt3, Ubp3, Pub1, PbP1 proteins, orthologues of G3BP, USP10, TIA-1, and Ataxin-2, respectively, were identified. Under heat stress, like their human orthologues, Nxt3 and Ubp3 interact with the RNA-binding protein Pabp and are involved in the formation of stress granules. However, unlike G3BP1 and USP10, neither deletion nor overexpression of nxt3(+) or ubp3(+) affect SG assembly in yeast. Similar results were observed in mutants defective in ataxia-2 and TIA-like proteins, which are important components of SG [151].

Yeast P-bodies are formed independently of stress granules, however, they still contribute to their occurrence. Additionally, when translation is inhibited by glucose deprivation, P-bodies are formed first, then Pab1 accumulates in association with P-bodies, and stress granules appear last [152]. Yeast P-bodies contain the proteins Dcp1p, Dcp2p, Edc3p, Dhh1p, Pat1p, Lsm1p, Xrn1p, Ccr4p, and Pop2p. Studies of yeast P-bodies show that there are clear dependencies in the assembly of specific components. For example, recruitment of Dcp1p to P-bodies is mediated by Dcp2p. The second clear relationship is that Pat1p is required to recruit the Lsm1-7p complex [153]. However, in yeast, deletion of any of the genes encoding P-body components does not compromise their integrity, indicating that they are redundant and cooperative [154]. It has been established that in yeast cells, P-bodies are visible only upon induction of stress [155] and have the properties of liquid droplets since they are soluble by 1,6-hexanediol. P-bodies are heterogeneous in mRNA and proteins depending on the type of stress. Study [155] identified RNAs in yeast P-bodies induced by 10 min glucose fasting or osmotic stress using high concentrations of CaCl2 and NaCl. A total of 1544 glucose starvation mRNAs were present in P-bodies, and 35% of them were stress specific [155]. An analysis of RNA length showed that P-bodies induced by glucose starvation contained shorter RNAs compared to the total pool of activated mRNAs under the corresponding stress conditions, whereas P-bodies induced by osmotic stress contained longer RNAs. This indicates that, at least in yeast, transcript length may be important for P-body recruitment.

Nutrient stress induces the formation of cytoplasmic aggregates called eIF2B bodies. These MLOs contain subunits of the eIF2B and eIF2 protein complexes and are induced during stress caused by glucose deprivation [147]. One of the major control points in translation initiation involves the activation of eukaryotic initiation factor 2 (eIF2) by eIF2B. eIF2, in its active GTP-bound form, interacts with methionyl-tRNA to form a ternary complex (TC). In yeast, this TC can be associated with initiation factors eIF1, eIF3, and eIF5 to form a multifactorial complex (MFC). The MFC recruits the 40S ribosomal subunit to the mRNA to enable further translation. eIF2B is required for converting eIF2 into a translationally active form. Thus, the eIF2B-dependent response is a highly regulated step in the translation initiation pathway. As a result of stress, phosphorylation of eIF2$\alpha$ by Gcn2p kinase occurs, which leads to a decrease in the cellular pool of active eIF2-GTP and, consequently, to a decrease in the rate of translation initiation. As a result, eIF2B accumulates in the cytoplasm and combines into eIF2B bodies [147]. Yeast eIF2B bodies occur in less than 10% of cells under normal conditions in the logarithmic growth phase but are rapidly induced by stress caused by glucose deprivation. It is important to note that the emergence of eIF2B bodies does not depend on the formation of stress granules. eIF2B bodies are dynamic structures that form faster than stress granules but disassemble more slowly depending on the presence of glucose.

The 26S proteasome is responsible for the proteolysis of a large number of proteins, including important cell cycle regulators. The 26S proteasome cleaves polyubiquitylated substrates in an ATP-dependent manner and can also degrade specific non-ubiquitylated target proteins. In growing and dividing yeast, proteasomes are assembled both in the nucleus and in the cytoplasm. During the transition of the cell to a state of rest or starvation for glucose, the proteasome subunits form large cytoplasmic proteasome storage granules. These granules function as a kind of "reservoir" that stores proteasome subunits until glucose appears in the medium [156].

Acidification of yeast intracellular milieu induces the formation of reversible fibril-like structures [157]. It was shown for IDPR proteins Cdc-19 kinase [158] and glutamine synthetase Gln1 [159]. Regulation of amyloid formation of ATP-producing Cdc-19 yeast kinase are considered as a possible indirect mechanism of SG disassembly. In stress conditions, Cdc-19 fibrillation blocks ATP production. According to [160], after stress glycolytic metabolite fructose-1,6-bisphosphate initiates recruitment of chaperones to Cdc-19 fibrils and promotes solubilization of this kinase. In turn, this causes the synthesis of ATP, a metabolite necessary for the disassembly of stress granules.

## 3. Prokaryotes

Bacteria in nature demonstrate remarkable stress resistance. This is an essential property for survival as the majority of prokaryotic organisms inhabit areas with rapidly and unpredictably changing environmental conditions, such as temperature, pH, salt, oxidation, nutrition, water, and chemical elements availability [161]. In addition to that, prokaryotes invading multicellular organisms must overcome host-defense systems. For example, *E. coli* bacterium upon infection faces bile salts, gastric acid with pH ranging from 2.5 to 4.5, and gastrointestinal tract organic acids [161]. Development of proper adaptation mechanisms and quick responses to various stressors were certainly a great evolutionary requirement that pushed bacteria to evolve complex regulatory networks able to quickly sense dangerous changes in their surroundings and rapidly respond with differential expression of a plethora of regulatory genes. Several major bacterial stress responses have been described, including the general stress response regulated in *E. coli* by sigma38 (rpoS) protein [162], envelope stress response modulated in *E.coli* by sigma(E) factor [163], the heat shock response [164] regulated in *E.coli* by sigma factor-32 ($\sigma$32), and the cold shock response regulated by cold-shock proteins [165]. Unlike eukaryotic cells, prokaryotes lack any membrane organelles and the formation of LLPS-driven condensates is a very potent mechanism to substitute for the absence of membrane organelles and spatiotemporally organize thousands of stress factors in a bacterial cytoplasm [166–168]. This hypothesis found confirmation in multiple works reporting LLPS-driven cellular moieties in prokaryotes. Just a couple of examples are RNA polymerase clusters (RNAP) in *E. coli* [169], the ParABS protein system responsible for the segregation of bacterial plasmids and chromosomes during proliferation [170], PopZ microdomains, and SpmX condensates in *Caulobacter crescentus* [171,172] and many others.

LLPS condensates are highly responsive to environmental changes making them perfect tools to navigate stress response mechanisms. Similar to eukaryotes, prokaryotic cells were found to both assemble temporal specialized stress-induced membrane-less organelles (such as BR bodies) and rearrange existing condensates in order to combat stress (for instance, SSB and Dps condensates) (Table 3).

One instance of rearrangement of pre-existing structures in response to stress can be SSB condensates. Single-stranded DNA-binding proteins (SSB) play vital role in cellular metabolism and survival by binding single-stranded DNA, forming DNA-protein filaments, and preventing potential harmful interactions during DNA replication and DNA damage response (Figure 5A). SSB proteins were found to be present in much larger numbers that are necessary to protect the replication fork during normal DNA duplication [176]. In *E. coli* excess, the SSB protein is stored in a form of structures resembling droplets bound to bacterial membrane, which are rapidly (within 5 min) disassembled upon DNA damage releasing SSB into the bacterial cytoplasm (Figure 5B) [176]. Another study demonstrated that SSB from *E. coli* forms LLPS condensates at physiological conditions in vitro via its intrinsically disordered linker and these biological condensates are quickly disintegrated upon presence of ssDNA [177]. In addition to SSB protein itself, SSB condensates also sequester multiple DNA damage response factors binding to SSB mainly via its C-terminal peptide and are released from the droplets upon DNA damage stress combined with SSB [177]. This mechanism ensures a quick reaction to a highly dangerous condition of single-stranded DNA accumulation with release of necessary DNA reparation machinery 'ready-to-go' and active, preserving the precious time and resources for the time-consuming process of protein synthesis and post-translational modification (Figure 5B).

**Table 3.** Examples of LLPS (or suggested to be LLPS) compartments formed or rearranged in response to in prokaryotic cells.

| Stress-Linked Organelle | Scaffolding Component | Organism | Structural Changes in Response to Stress | Function |
|---|---|---|---|---|
| SSB condensates | Single-stranded DNA-binding protein (SSB) | *Escherichia coli* | Disassembled in response to stress that causes DNA damage and accumulation of ssDNA. | Serve as storage capsules for SSB protein and other DNA repairing enzymes. |
| Dps condensates | Dps (DNA-binding protein from starved cells) | *Escherichia coli* | Transform into denser structures in response to stress. | Compact nucleoid during stress conditions, while preserving transcription of genes. |
| BR bodies (containing RNase E) | RNase E endonuclease | *Caulobacter crescentus, Sinorhizobium meli-loti, Agrobacterium tumefacienes, Escherichia coli, and Cyanobacteria* | Assembled in bacterial cytoplasm in response to stress. | Isolation of untranslated mRNA during stress. Centers for mRNA decay and degradation. |
| BR bodies (containing RNase Y) [173] | RNase Y endonuclease | *Bacillus subtilis* | Assembled in bacterial cytoplasm in response to stress. | Isolation of untranslated mRNA during stress. Centers for mRNA decay and degradation. |
| BR bodies (containing RNase J) [174] | RNase J endonuclease | *Helicobacter pylori* | Assembled in bacterial cytoplasm in response to stress. | Isolation of untranslated mRNA during stress. Centers for mRNA decay and degradation. |
| Granular bodies | IbpA heat shock protein | *Acholeplasma laidlawii* | Assembled in response to stress. | Regulation of heat shock response. |
| PolyP granules [175] | polyphosphate (polyP) | *Pseudomonas aeruginosa* | Assembled under nitrogen starvation. | Regulation of bacterial cell cycle exit during starvation survival response. |

Another DNA-binding protein Dps (DNA-binding protein from starved cells) carrying a DNA-protecting function in *E. coli* also undergoes significant structural rearrangements in response to stress. During the stationary phase in *E. coli* bacteria, Dps massively but transiently binds to nucleoid compacting it (Figure 5A). However, upon serious stress, such as starvation, heat shock, and oxidative stress, Dps heavily covers the nucleoid that leads to formation of condensates, which were proposed to be liquid-liquid phase separated organelles (Figure 5B) [178]. This process is probably driven by the intrinsically disordered N-terminal region of Dps, which has been demonstrated to be essential for Dps DNA-binding activity [179]. Interestingly, the Dps-formed condensates remain permeable for RNA polymerase enzyme, whereas other DNA-binding proteins are excluded, enabling active gene transcription while preventing destruction of the genome.

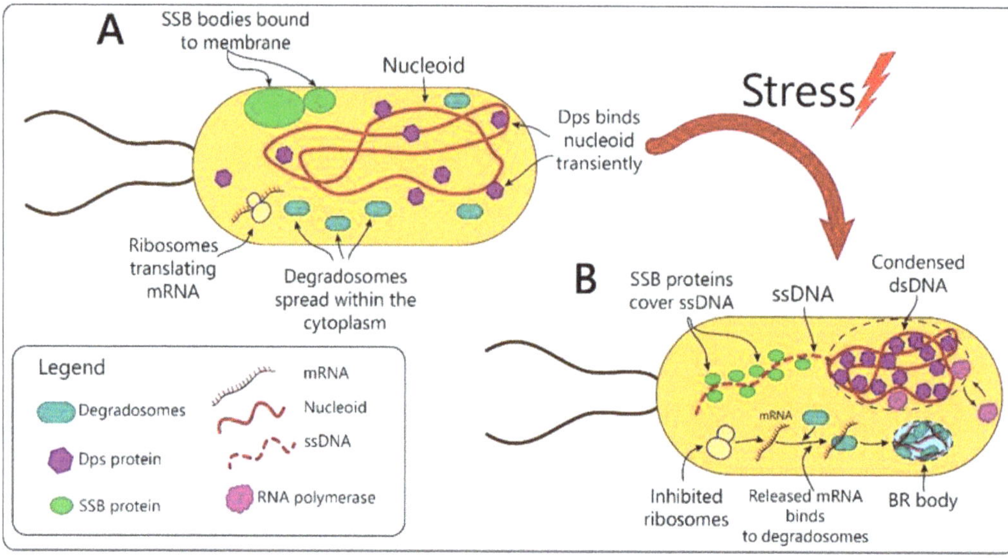


**Figure 5.** Illustration of changes happening in SSB condensates, Dps condensates, and BR bodies formed by RNaseE degradosomes during stress response in prokaryotes. (**A**) Illustration of unstressed bacterial cell. SSB proteins are sequestered into droplet-like SSB condensates bound to membrane. Dps protein before stress transiently binds to bacterial genome but unable to reach sufficient concentrations to trigger condensate assembly. RNase E tetramers bound to protein partners (degradosomes) are spread throughout the cytoplasm. RNA transcription and translation proceed normally. (**B**) Illustration of cell upon the onset of stress. SSB condensates are disintegrated upon accumulation of single-stranded DNA (ssDNA) and free SSB complexes bind to ssDNA protecting it. Dps complexes heavily cover nucleoid driving the formation of phase separated organelle, which remains permeable for RNA polymerase enzyme. Stress leads to inhibition of translation and release of mRNA from polyribosomes. Untranslated mRNA binds to degradosomes complexes leading to phase separation and assembly of BR bodies.

A certain degree of analogy could be drawn between cytoplasmic stress granules (SG) in eukaryotes and bacterial RNP bodies (BR-bodies). BR bodies are formed as a result of liquid-liquid phase separation of RNaseE endonuclease tetramers called degradosomes (Figure 5A) [180]. The intrinsically disordered C-terminal domain of RNase E facilitates its LLPS transition while multiple protein-partner and RNA binding domains recruit other proteins required for mRNA processing (RNA chaperons, DEAD-box helicases, etc.) as well as RNA molecules [180]. Many microorganisms were found to contain BR bodies, for example, *Caulobacter crescentus, Sinorhizobium meli-loti, Agrobacterium tumefacienes*, *E. coli*, and *Cyanobacteria* [166], all of these bacteria species encode RNase E protein. Additionally, in Bacillus subtilis [173] and Helicobacter pylori [174], entities similar to BR bodies were found, but they were formed by different types of endonucleases.

Similarly to SGs, BR bodies form as a result of the accumulation of free mRNA in the cytoplasm, which is released from the polyribosomes as a result of stress-induced inhibition of translation. Degradosomes interact with untranslated mRNA, a process that drives assembly of BR-bodies (Figure 5B) [180,181]. Although the complete set of BR bodies functional properties is yet to be uncovered, BR bodies are known to modulate mRNA decay and degradation in *E. coli* and *C. crescentus* bacteria [182,183]. Additionally, these condensates demonstrate selective permeability against highly structured RNA molecules, such as rRNA and tRNA, preventing their incorporation into the organelles and, therefore, isolating from the mRNA molecules [182].

Another example of membrane-less organelle formed transiently and only during the stress response is the so-called granular body found in mycoplasma *Acholeplasma laidlawii* [184,185]. *A. laidlawii* granular bodies form in response to heat shock and contain a small heat shock protein, IbpA, which has a subset of interacting partners [186] and assembles into globular-type oligomers and fibrils [187]. *A. laidlawii* belongs to Mollicutes, a class of microorganisms that possess the smallest known genome sizes among autonomously replicating organisms [188] and, thus, developed highly evolutionary optimized gene regulatory networks and metabolic pathways. Having biomolecular condensate-like structure formation as a first-line response towards unfavorable conditions suggests the universal biological significance of membrane-less organelles for cellular survival during stress.

## 4. Factors Regulating Reorganization of MLOs in Stress Response

The main advantage of MLOs that allows such structures to regulate signaling pathways, compared to "classical" organelles, is a fast and reversible response to external stimuli [7]. This is due to the fact that the condensates formed as a result of LLPS of IDPs and other conformationally heterogeneous polymers are metastable structures. Accordingly, a slight change in external conditions can cause a change in the state of such a system and lead to a change in the physical properties of the condensate. The scaffold proteins of the vast majority of MLOs are IDPs and proteins containing IDPRs [5]. The transition of these proteins to the liquid-drop state may be due to a change in the network of their inter- and intramolecular interactions [7]. This can be caused either by a change in the physical characteristics of the environment or by a change in conditionally "biological" factors: post-translational modifications, changes in the concentration of scaffold proteins, and interactions with partners which mostly does not require de novo protein synthesis (i.e., transcription, translation).

### 4.1. "Physical" Factors

Stress conditions are accompanied by changes in the intracellular space of temperature, pH, ionic strength of the solution, osmotic pressure, concentration of metabolites, and reactive oxygen species [1]. Often, a change in one of the physical parameters of the system entails a change in several more. So, heat shock in yeast cells and drosophila causes acidification of the cytoplasmic space [2]. Osmotic shock causes a change in the concentration of salts in the intracellular space, as a result of which the operation of ion channels changes, which in turn can cause a change in cytoplasmic pH [189]. In the cells of a number of bacteria, osmotic stress causes a decrease in the pH of the cytosol. The lack of nutrients in yeast cells causes a decrease in pH in the cytoplasmic space from 7.4 to 6.0 [190]. Cytosol acidification in mammalian and yeast cells is also associated with impaired ion transport under conditions of metabolite deficiency [2]. During aging and related neurodegenerative diseases, deregulation of the transport of calcium ions from the ER to mitochondria is usually observed [138]. Mitochondria are the key organelles involved in the production of energy and metabolites necessary for the cell, therefore, the dysfunction of these organelles is critical for the cell, causing saturation of the cytoplasm with $H^+$ ions, which leads to acidification of the intracellular space. The pH in the cytoplasm of tumor cells is also significantly shifted to a more acidic region compared to the characteristic values of healthy cells [2]. As is known, electrostatic inter/intramolecular interactions are one of the main driving forces contributing to the phase separation of IDPs. Accordingly, a change in pH contributes to a change in the network of such interactions, primarily due to a change in the charge of the side groups of amino acid residues [191]. For example, the phase transitions of most of the proteins that make up the stress granules, including the scaffold proteins G3BP1, Pub1, DDX, are pH dependent [192–194]. At the same time, G3BP1 can form condensates in the cell in response to pH acidification. Hypoxic conditions associated with pH acidification cause the formation in the nucleoplasm of a special type of A-bodies, the protein composition of which only corresponds to the composition of A-bodies resulting from heat shock by only 20% [101]. Even a small change in temperature can have a significant effect on the interaction of the phase separation of IDPs, changing the network of interactions of "protein–solvent" [195,196]. Depending on the balance between protein–protein interactions, protein-solvent interactions, and protein conformational entropy, the separation of such systems into phases can occur in different temperature ranges, for example, when lower-critical solution temperature (LCST) and upper-critical solution temperature (UCST) are reached [197]. The same picture can be observed when the salt composition of the solution changes, for example, in osmotic stress conditions [198].

### 4.2. "Biological" Factors

Changes in physical environmental factors have a nonspecific effect on all proteins potentially predisposed to LLPS and do not allow for fine regulation of the properties of MLOs. In this regard, transitions of this type play a significant role only at the initial stages of the formation of MLOs. Nonspecific interactions of scaffold proteins of MLOs with mRNA, lncRNA, and rIGSRNA also play a significant role in initiating the formation of MLOs, but not during their maturation [199]. The main factor regulating maturation, attachment of client proteins, and properties of MLOs are post-translational modifications (PTM) of intrinsically disordered proteins [200]. PTMs allow us to specifically change the conditions necessary for the phase separation of a particular protein, depending on the cellular context [201]. Phosphorylation, acetylation, methylation, SUMOylation, and poly-ADP-ribosylation of a number of scaffold proteins of stress granules, including G3BP1, has a significant effect on the correct assembly and functioning of these organelles [202–205]. On the other hand, phosphorylation of FUS proteins, TDP-43, can reduce the critical concentrations of these proteins required for their phase separation, which makes it possible to weaken the incorporation of these proteins into stress granules, in turn, inhibiting the degradation of SGs [206,207]. O-linked N-acetylglucosaminylation of the hNRNPA1 protein performs the same function [208]. A change in the profile of SUMOylation and acetylation of PML isoforms in response to stress causes a change in the composition of PML bodies and their physical properties [89,209]. The additional evidence of the PTM role in regulation of MLOs assembly/disassembly process may be the reduction in Huntingtin aggregation in the cytosol and chromatin-associated Huntingtin aggregates in the nucleus by SUMO-targeted ubiquitin ligase, Slx5 [210].

Except for PTM, chaperones and autophagosomal degradation play an important role in regulating the properties of MLOs under stress [211]. Chaperone activity ensures correct disassembly of stress-induced MLOs and prevents their degradation into insoluble toxic aggregates. Thus, the HSPB8/BAG3/HSP70 complex prevents the hardening of stress granules [212].

## 5. Discussion

Stress causes the formation of new intracellular environment. The adaptive reaction of cells to the new intracellular environment proceeds at the following levels of cellular organization: genomic, transcriptional, and translational. The genome reorganization is an extreme and mostly irreversible response of the cells to the stress action which is usually observed under conditions of chronic stress [213]. The genome reorganization often leads to pathological cellular transformations.

Cell survival under "physiological stress" (i.e., under the conditions when cells are principally able to return to pre-stress state) is carried out by rearranging its translational and transcriptional profiles. Such cellular program is primarily aimed to change its expression profile and preserve the necessary biomacromolecules. Membrane-less organelles are essential in these processes. The reorganization of cell compartmentalization in response to stress is a fast reversible and adaptive process, primarily aimed at preventing damage to the genetic and protein material of the cell in an aggressive environment. Apparently, this stage of the stress response is a "fire" reaction of the cell to stress, allowing it to survive until the switching of cellular expression programs occurs. The rapid and reversible formation of biomolecular condensates under stress conditions in eukaryotic and prokaryotic cells makes it possible to temporarily exclude the key "complex" biopolymers that provide cell homeostasis from the intracellular space. The synthesis of these molecules (mRNA, rRNA, transcription elongation and initiation factors, and other proteins) is extremely energy consuming. The simultaneous synthesis of these molecules under conditions of nutritional deficiency after stress action can cause cell death.

However, the function of membrane-less organelles under stress conditions is not limited to the protein and genetic preservation. Membrane-less organelles are primarily biomolecular reactors that ensure the occurrence of various biochemical reactions. A number of key enzymes involved in the metabolism of carbohydrates, amino acids, fatty acids, nucleotides in animal cells, yeast, and bacteria are included in the composition of condensates and are activated in response to stress [12]. Therefore, the reorganization of biomolecular condensates under stress conditions is directly related to the activation and regulation of stress signaling pathways.

Reorganization of biomolecular condensates under stress conditions is a systemic response. Formation and alteration of the properties of nuclear membrane-less organelles correlate with changes in the cellular expression profile. A systematic analysis was shown that perturbations of at least 128 genes cause nucleolar enlargement with subsequent formation of stress granules, an increase in the number of Cajal bodies, and splicing speckles in mammalian cells [214]. In addition, this work established a correlation between an increase in the nucleolus and a decrease in P-bodies, wherein no relationship was found between changes in gene expression and the formation of cytoplasmic stress granules [215].

One of the possible regulators of the reorganization of intracellular condensates in response to stress is a change in the localization of their components. Thus, stress conditions cause translocation into the nucleus of the A-bodies scaffold protein VHL [216]. Additionally, transcription inhibition leads to the structural alterations of the nucleolus resulting in the formation of nucleolar caps containing coilin, PML [16], and PATL1 [217]—scaffold proteins of Cajal bodies, PML and P-bodies under normal conditions.

Apparently, RNA turnover can play the same role. The biogenesis of mRNA, rRNA, rDNA, and lncRNA is one of the main regulators of gene expression [201,218–220]. These molecules are the key components of stress-induced organelles formed in the cytoplasm, nucleoplasm, and nucleolus under stress conditions. Regulation of these molecules intracellular composition is carried out by recruiting them into stress granules, nuclear stress-bodies, A-bodies, paraspeckles, nuclear speckles, and other membrane-less organelles. It has recently been shown that the regulation of rRNA processing in the nucleolus is a key step in the Ribosome Biogenesis Stress Response pathway [221], which makes it possible to indirectly regulate the structure of the nucleolus, as well as the level of mRNA in the cytoplasm and the formation of stress granules. Under conditions of severe stress, fragmentation of the nucleolus is observed, which in turn causes the release of ribosomal proteins into the cytoplasm, accompanied by inhibition of Hdm2, accumulation of p53, and subsequent induction of apoptosis (Figure 2C) [222].

The formation of new and reorganization of already existing compartments under stress conditions is associated with a change in their material properties. The gelation of MLOs and even the formation of functional amyloid fibrils by them in response to stress is observed in cells of various kingdoms of life. This is due to the need to limit the dynamics of the exchange of the contents of membrane-less organelles in an aggressive intracellular environment. Intrinsically disordered proteins are key actors of these processes. First of all, this is because the material properties of proteins strongly depend on the properties of the environment [196].

Stress-induced reorganization of intracellular milieu is a conservative process and occurs in a similar way in bacterial, yeast, plant, and animal cells. Biomolecular condensates formed in the cells of these organisms are usually regulated by proteins with similar functions. The accumulated data will allow us to state that this form of reorganization of biopolymers is a universal mechanism of stress response.

## 6. Conclusions

The analysis of literature data presented in this work showed that stress-induced rearrangement of liquid-drop cell compartments is a systemic process that regulates cells stress response at the translational and transcriptional levels. In response to unfavorable conditions, both a stress-responsive reorganization of the existing biomolecular condensates and de novo formation of new membrane-less organelles occur in the intracellular environment of eukaryotes and prokaryotes. The phase separation of biopolymers underlying these changes (reorganization) provides a fast, adequate, adaptive, and controlled cell response to any kind of stress. The cell response to stress illustrates the role of biomolecular condensates formed via LLPS for cell physiology

**Author Contributions:** Writing—original draft preparation, A.V.F., I.A.A., A.S.F. and A.A.G.; writing—review and editing, A.V.F., I.A.A., A.S.F., K.K.T. and I.M.K. All authors have read and agreed to the published version of the manuscript.

**Funding:** This research was funded by Russian Science Foundation, grant number 22-15-00429 (K.K.T.).

**Institutional Review Board Statement:** Not applicable.

**Informed Consent Statement:** Not applicable.

**Data Availability Statement:** Not applicable.

**Conflicts of Interest:** The authors declare no conflict of interests.

## References


1. Sekine, Y.; Houston, R.; Sekine, S. Cellular metabolic stress responses via organelles. *Exp. Cell Res.* **2021**, *400*, 112515. [CrossRef] [PubMed]
2. Jin, X.; Zhou, M.; Chen, S.; Li, D.; Cao, X.; Liu, B. Effects of pH alterations on stress- and aging-induced protein phase separation. *Cell. Mol. Life Sci.* **2022**, *79*, 380. [CrossRef]
3. Dutta, N.; Garcia, G.; Higuchi-Sanabria, R. Hijacking Cellular Stress Responses to Promote Lifespan. *Front. Aging* **2022**, *3*, 860404. [CrossRef] [PubMed]
4. Fulda, S.; Gorman, A.M.; Hori, O.; Samali, A. Cellular stress responses: Cell survival and cell death. *Int. J. Cell Biol.* **2010**, *2010*, 214074. [CrossRef]
5. Antifeeva, I.A.; Fonin, A.V.; Fefilova, A.S.; Stepanenko, O.V.; Povarova, O.I.; Silonov, S.A.; Kuznetsova, I.M.; Uversky, V.N.; Turoverov, K.K. Liquid–liquid phase separation as an organizing principle of intracellular space: Overview of the evolution of the cell compartmentalization concept. *Cell. Mol. Life Sci.* **2022**, *79*, 251. [CrossRef]
6. Fefilova, A.S.; Fonin, A.V.; Vishnyakov, I.E.; Kuznetsova, I.M.; Turoverov, K.K. Stress-Induced Membraneless Organelles in Eukaryotes and Prokaryotes: Bird's-Eye View. *Int J. Mol. Sci.* **2022**, *23*, 5010. [CrossRef]
7. Turoverov, K.K.; Kuznetsova, I.M.; Fonin, A.V.; Darling, A.L.; Zaslavsky, B.Y.; Uversky, V.N. Stochasticity of Biological Soft Matter: Emerging Concepts in Intrinsically Disordered Proteins and Biological Phase Separation. *Trends Biochem. Sci.* **2019**, *44*, 716–728. [CrossRef] [PubMed]
8. Fonin, A.V.; Darling, A.L.; Kuznetsova, I.M.; Turoverov, K.K.; Uversky, V.N. Intrinsically disordered proteins in crowded milieu: When chaos prevails within the cellular gumbo. *Cell. Mol. Life Sci.* **2018**, *75*, 3907–3929. [CrossRef]
9. Uversky, V.N. Functional roles of transiently and intrinsically disordered regions within proteins. *FEBS J.* **2015**, *282*, 1182–1189. [CrossRef] [PubMed]
10. Uversky, V.N. Intrinsically Disordered Proteins and Their "Mysterious" (Meta)Physics. *Front. Phys.* **2019**, *7*, 10. [CrossRef]
11. van Leeuwen, W.; Rabouille, C. Cellular stress leads to the formation of membraneless stress assemblies in eukaryotic cells. *Traffic* **2019**, *20*, 623–638. [CrossRef]
12. Prouteau, M.; Loewith, R. Regulation of Cellular Metabolism through Phase Separation of Enzymes. *Biomolecules* **2018**, *8*, 160. [CrossRef]
13. Hnisz, D.; Shrinivas, K.; Young, R.A.; Chakraborty, A.K.; Sharp, P.A. A Phase Separation Model for Transcriptional Control. *Cell* **2017**, *169*, 13–23. [CrossRef] [PubMed]
14. Sirri, V.; Urcuqui-Inchima, S.; Roussel, P.; Hernandez-Verdun, D. Nucleolus: The fascinating nuclear body. *Histochem. Cell Biol.* **2008**, *129*, 13–31. [CrossRef] [PubMed]
15. Yang, K.; Yang, J.; Yi, J. Nucleolar Stress: Hallmarks, sensing mechanism and diseases. *Cell Stress* **2018**, *2*, 125–140. [CrossRef] [PubMed]
16. Boulon, S.; Westman, B.J.; Hutten, S.; Boisvert, F.M.; Lamond, A.I. The nucleolus under stress. *Mol. Cell* **2010**, *40*, 216–227. [CrossRef] [PubMed]
17. Al-Baker, E.A.; Oshin, M.; Hutchison, C.J.; Kill, I.R. Analysis of UV-induced damage and repair in young and senescent human dermal fibroblasts using the comet assay. *Mech. Ageing Dev.* **2005**, *126*, 664–672. [CrossRef]

18. Govoni, M.; Farabegoli, F.; Pession, A.; Novello, F. Inhibition of topoisomerase II activity and its effect on nucleolar structure and function. *Exp. Cell Res.* **1994**, *211*, 36–41. [CrossRef]
19. Shav-Tal, Y.; Blechman, J.; Darzacq, X.; Montagna, C.; Dye, B.T.; Patton, J.G.; Singer, R.H.; Zipori, D. Dynamic sorting of nuclear components into distinct nucleolar caps during transcriptional inhibition. *Mol. Biol. Cell* **2005**, *16*, 2395–2413. [CrossRef]
20. David-Pfeuty, T.r.s. Potent inhibitors of cyclin-dependent kinase 2 induce nuclear accumulation of wild-type p53 and nucleolar fragmentation in human untransformed and tumor-derived cells. *Oncogene* **1999**, *18*, 7409–7422. [CrossRef]
21. Haaf, T.; Ward, D.C. Inhibition of RNA polymerase II transcription causes chromatin decondensation, loss of nucleolar structure, and dispersion of chromosomal domains. *Exp. Cell Res.* **1996**, *224*, 163–173. [CrossRef]
22. Rubbi, C.P.; Milner, J. Disruption of the nucleolus mediates stabilization of p53 in response to DNA damage and other stresses. *EMBO J.* **2003**, *22*, 6068–6077. [CrossRef] [PubMed]
23. Zhang, Y.; Lu, H. Signaling to p53: Ribosomal proteins find their way. *Cancer Cell* **2009**, *16*, 369–377. [CrossRef] [PubMed]
24. Lindström, M.S. Emerging functions of ribosomal proteins in gene-specific transcription and translation. *Biochem. Biophys. Res. Commun.* **2009**, *379*, 167–170. [CrossRef] [PubMed]
25. Ofir-Rosenfeld, Y.; Boggs, K.; Michael, D.; Kastan, M.B.; Oren, M. Mdm2 regulates p53 mRNA translation through inhibitory interactions with ribosomal protein L26. *Mol. Cell* **2008**, *32*, 180–189. [CrossRef]
26. Dove, B.K.; You, J.H.; Reed, M.L.; Emmett, S.R.; Brooks, G.; Hiscox, J.A. Changes in nucleolar morphology and proteins during infection with the coronavirus infectious bronchitis virus. *Cell. Microbiol.* **2006**, *8*, 1147–1157. [CrossRef] [PubMed]
27. Eskiw, C.H.; Dellaire, G.; Mymryk, J.S.; Bazett-Jones, D.P. Size, position and dynamic behavior of PML nuclear bodies following cell stress as a paradigm for supramolecular trafficking and assembly. *J. Cell Sci.* **2003**, *116*, 4455–4466. [CrossRef]
28. Chan, P.K.; Aldrich, M.; Busch, H. Alterations in immunolocalization of the phosphoprotein B23 in HeLa cells during serum starvation. *Exp. Cell Res.* **1985**, *161*, 101–110. [CrossRef]
29. Trinkle-Mulcahy, L.; Sleeman, J.E. The Cajal body and the nucleolus: "In a relationship" or "It's complicated"? *RNA Biol.* **2017**, *14*, 739–751. [CrossRef]
30. Matera, A.G. Nuclear bodies: Multifaceted subdomains of the interchromatin space. *Trends Cell Biol.* **1999**, *9*, 302–309. [CrossRef]
31. Cioce, M.; Lamond, A.I. Cajal bodies: A long history of discovery. *Annu. Rev. Cell Dev. Biol.* **2005**, *21*, 105–131. [CrossRef]
32. Walker, M.P.; Tian, L.; Matera, A.G. Reduced viability, fertility and fecundity in mice lacking the cajal body marker protein, coilin. *PLoS ONE* **2009**, *4*, e6171. [CrossRef]
33. Deryusheva, S.; Gall, J.G. Small Cajal body-specific RNAs of Drosophila function in the absence of Cajal bodies. *Mol. Biol. Cell* **2009**, *20*, 5250–5259. [CrossRef]
34. Collier, S.; Pendle, A.; Boudonck, K.; van Rij, T.; Dolan, L.; Shaw, P. A distant coilin homologue is required for the formation of cajal bodies in Arabidopsis. *Mol. Biol. Cell* **2006**, *17*, 2942–2951. [CrossRef]
35. Tucker, K.E.; Berciano, M.T.; Jacobs, E.Y.; LePage, D.F.; Shpargel, K.B.; Rossire, J.J.; Chan, E.K.; Lafarga, M.; Conlon, R.A.; Matera, A.G. Residual Cajal bodies in coilin knockout mice fail to recruit Sm snRNPs and SMN, the spinal muscular atrophy gene product. *J. Cell Biol.* **2001**, *154*, 293–307. [CrossRef]
36. Strzelecka, M.; Trowitzsch, S.; Weber, G.; Lührmann, R.; Oates, A.C.; Neugebauer, K.M. Coilin-dependent snRNP assembly is essential for zebrafish embryogenesis. *Nat. Struct. Mol. Biol.* **2010**, *17*, 403–409. [CrossRef]
37. Love, A.J.; Yu, C.; Petukhova, N.V.; Kalinina, N.O.; Chen, J.; Taliansky, M.E. Cajal bodies and their role in plant stress and disease responses. *RNA Biol.* **2017**, *14*, 779–790. [CrossRef]
38. Cioce, M.; Boulon, S.; Matera, A.G.; Lamond, A.I. UV-induced fragmentation of Cajal bodies. *J. Cell Biol.* **2006**, *175*, 401–413. [CrossRef]
39. Handwerger, K.E.; Wu, Z.; Murphy, C.; Gall, J.G. Heat shock induces mini-Cajal bodies in the Xenopus germinal vesicle. *J. Cell Sci.* **2002**, *115*, 2011–2020. [CrossRef]
40. Bongiorno-Borbone, L.; De Cola, A.; Barcaroli, D.; Knight, R.A.; Di Ilio, C.; Melino, G.; De Laurenzi, V. FLASH degradation in response to UV-C results in histone locus bodies disruption and cell-cycle arrest. *Oncogene* **2010**, *29*, 802–810. [CrossRef]
41. Andrade, L.E.; Tan, E.M.; Chan, E.K. Immunocytochemical analysis of the coiled body in the cell cycle and during cell proliferation. *Proc. Natl. Acad. Sci. USA* **1993**, *90*, 1947–1951. [CrossRef]
42. Stępiński, D. Cajal body dynamics in soybean root meristem cells under chilling stress and recovery. *Environ. Exp. Bot.* **2020**, *180*, 104241. [CrossRef]
43. Navascues, J.; Bengoechea, R.; Tapia, O.; Casafont, I.; Berciano, M.T.; Lafarga, M. SUMO-1 transiently localizes to Cajal bodies in mammalian neurons. *J. Struct. Biol.* **2008**, *163*, 137–146. [CrossRef]
44. Morency, E.; Sabra, M.; Catez, F.; Texier, P.; Lomonte, P. A novel cell response triggered by interphase centromere structural instability. *J. Cell Biol.* **2007**, *177*, 757–768. [CrossRef]
45. James, N.J.; Howell, G.J.; Walker, J.H.; Blair, G.E. The role of Cajal bodies in the expression of late phase adenovirus proteins. *Virology* **2010**, *399*, 299–311. [CrossRef]
46. Kim, S.H.; Ryabov, E.V.; Kalinina, N.O.; Rakitina, D.V.; Gillespie, T.; MacFarlane, S.; Haupt, S.; Brown, J.W.S.; Taliansky, M. Cajal bodies and the nucleolus are required for a plant virus systemic infection. *EMBO J.* **2007**, *26*, 2169–2179. [CrossRef]
47. Sasaki, Y.T.F.; Ideue, T.; Sano, M.; Mituyama, T.; Hirose, T. MEN$\varepsilon/\beta$ noncoding RNAs are essential for structural integrity of nuclear paraspeckles. *Proc. Natl. Acad. Sci. USA* **2009**, *106*, 2525–2530. [CrossRef]
48. Nakagawa, S.; Naganuma, T.; Shioi, G.; Hirose, T. Paraspeckles are subpopulation-specific nuclear bodies that are not essential in mice. *J. Cell Biol.* **2011**, *193*, 31–39. [CrossRef]

49. McCluggage, F.; Fox, A.H. Paraspeckle nuclear condensates: Global sensors of cell stress? *BioEssays News Rev. Mol. Cell. Dev. Biol.* **2021**, *43*, e2000245. [CrossRef]
50. Yamazaki, T.; Yamamoto, T.; Yoshino, H.; Souquere, S.; Nakagawa, S.; Pierron, G.; Hirose, T. Paraspeckles are constructed as block copolymer micelles. *Embo J.* **2021**, *40*, e107270. [CrossRef]
51. Nakagawa, S.; Shimada, M.; Yanaka, K.; Mito, M.; Arai, T.; Takahashi, E.; Fujita, Y.; Fujimori, T.; Standaert, L.; Marine, J.C.; et al. The lncRNA Neat1 is required for corpus luteum formation and the establishment of pregnancy in a subpopulation of mice. *Development* **2014**, *141*, 4618–4627. [CrossRef] [PubMed]
52. Choudhry, H.; Albukhari, A.; Morotti, M.; Haider, S.; Moralli, D.; Smythies, J.; Schödel, J.; Green, C.M.; Camps, C.; Buffa, F.; et al. Tumor hypoxia induces nuclear paraspeckle formation through HIF-2α dependent transcriptional activation of NEAT1 leading to cancer cell survival. *Oncogene* **2015**, *34*, 4482–4490. [CrossRef] [PubMed]
53. Lellahi, S.M.; Rosenlund, I.A.; Hedberg, A.; Kiær, L.T.; Mikkola, I.; Knutsen, E.; Perander, M. The long noncoding RNA NEAT1 and nuclear paraspeckles are up-regulated by the transcription factor HSF1 in the heat shock response. *J. Biol. Chem.* **2018**, *293*, 18965–18976. [CrossRef] [PubMed]
54. Todorovski, V.; Fox, A.H.; Choi, Y.S. Matrix stiffness-sensitive long noncoding RNA NEAT1 seeded paraspeckles in cancer cells. *Mol. Biol. Cell* **2020**, *31*, 1654–1662. [CrossRef]
55. Adriaens, C.; Standaert, L.; Barra, J.; Latil, M.; Verfaillie, A.; Kalev, P.; Boeckx, B.; Wijnhoven, P.W.G.; Radaelli, E.; Vermi, W.; et al. p53 induces formation of NEAT1 lncRNA-containing paraspeckles that modulate replication stress response and chemosensitivity. *Nat. Med.* **2016**, *22*, 861–868. [CrossRef]
56. Wang, Y.; Hu, S.B.; Wang, M.R.; Yao, R.W.; Wu, D.; Yang, L.; Chen, L.L. Genome-wide screening of NEAT1 regulators reveals cross-regulation between paraspeckles and mitochondria. *Nat. Cell Biol.* **2018**, *20*, 1145–1158. [CrossRef]
57. Imamura, K.; Imamachi, N.; Akizuki, G.; Kumakura, M.; Kawaguchi, A.; Nagata, K.; Kato, A.; Kawaguchi, Y.; Sato, H.; Yoneda, M.; et al. Long noncoding RNA NEAT1-dependent SFPQ relocation from promoter region to paraspeckle mediates IL8 expression upon immune stimuli. *Mol. Cell* **2014**, *53*, 393–406. [CrossRef]
58. Beeharry, Y.; Goodrum, G.; Imperiale, C.J.; Pelchat, M. The Hepatitis Delta Virus accumulation requires paraspeckle components and affects NEAT1 level and PSP1 localization. *Sci. Rep.* **2018**, *8*, 6031. [CrossRef]
59. Ma, H.; Han, P.; Ye, W.; Chen, H.; Zheng, X.; Cheng, L.; Zhang, L.; Yu, L.; Wu, X.; Xu, Z.; et al. The Long Noncoding RNA NEAT1 Exerts Antihantaviral Effects by Acting as Positive Feedback for RIG-I Signaling. *J. Virol.* **2017**, *91*, e02250–e16. [CrossRef]
60. Zhang, Q.; Chen, C.Y.; Yedavalli, V.S.; Jeang, K.T. NEAT1 long noncoding RNA and paraspeckle bodies modulate HIV-1 posttranscriptional expression. *Mbio* **2013**, *4*, e00596-00512. [CrossRef]
61. Imamura, K.; Takaya, A.; Ishida, Y.I.; Fukuoka, Y.; Taya, T.; Nakaki, R.; Kakeda, M.; Imamachi, N.; Sato, A.; Yamada, T.; et al. Diminished nuclear RNA decay upon Salmonella infection upregulates antibacterial noncoding RNAs. *Embo J.* **2018**, *37*, e97723. [CrossRef] [PubMed]
62. Spector, D.L.; Lamond, A.I. Nuclear speckles. *Cold Spring Harb. Perspect. Biol.* **2011**, *3*, a000646. [CrossRef] [PubMed]
63. Tripathi, V.; Ellis, J.D.; Shen, Z.; Song, D.Y.; Pan, Q.; Watt, A.T.; Freier, S.M.; Bennett, C.F.; Sharma, A.; Bubulya, P.A.; et al. The nuclear-retained noncoding RNA MALAT1 regulates alternative splicing by modulating SR splicing factor phosphorylation. *Mol. Cell* **2010**, *39*, 925–938. [CrossRef] [PubMed]
64. Zhang, B.; Arun, G.; Mao, Y.S.; Lazar, Z.; Hung, G.; Bhattacharjee, G.; Xiao, X.; Booth, C.J.; Wu, J.; Zhang, C.; et al. The lncRNA Malat1 Is Dispensable for Mouse Development but Its Transcription Plays a cis-Regulatory Role in the Adult. *Cell Rep.* **2012**, *2*, 111–123. [CrossRef]
65. Engreitz, J.M.; Sirokman, K.; McDonel, P.; Shishkin, A.A.; Surka, C.; Russell, P.; Grossman, S.R.; Chow, A.Y.; Guttman, M.; Lander, E.S. RNA-RNA Interactions Enable Specific Targeting of Noncoding RNAs to Nascent Pre-mRNAs and Chromatin Sites. *Cell* **2014**, *159*, 188–199. [CrossRef]
66. West, J.A.; Davis, C.P.; Sunwoo, H.; Simon, M.D.; Sadreyev, R.I.; Wang, P.I.; Tolstorukov, M.Y.; Kingston, R.E. The long noncoding RNAs NEAT1 and MALAT1 bind active chromatin sites. *Mol. Cell* **2014**, *55*, 791–802. [CrossRef]
67. Quinodoz, S.A.; Ollikainen, N.; Tabak, B.; Palla, A.; Schmidt, J.M.; Detmar, E.; Lai, M.M.; Shishkin, A.A.; Bhat, P.; Takei, Y.; et al. Higher-Order Inter-chromosomal Hubs Shape 3D Genome Organization in the Nucleus. *Cell* **2018**, *174*, 744–757.e724. [CrossRef]
68. Kim, J.; Venkata, N.C.; Hernandez Gonzalez, G.A.; Khanna, N.; Belmont, A.S. Gene expression amplification by nuclear speckle association. *J. Cell Biol.* **2019**, *219*, e201904046. [CrossRef]
69. Kopp, F.; Mendell, J.T. Functional Classification and Experimental Dissection of Long Noncoding RNAs. *Cell* **2018**, *172*, 393–407. [CrossRef]
70. Dutertre, M.; Sanchez, G.; Barbier, J.; Corcos, L.; Auboeuf, D. The emerging role of pre-messenger RNA splicing in stress responses: Sending alternative messages and silent messengers. *RNA Biol.* **2011**, *8*, 740–747. [CrossRef]
71. Biamonti, G.; Caceres, J.F. Cellular stress and RNA splicing. *Trends Biochem. Sci.* **2009**, *34*, 146–153. [CrossRef]
72. Spector, D.L.; Fu, X.D.; Maniatis, T. Associations between distinct pre-mRNA splicing components and the cell nucleus. *Embo J.* **1991**, *10*, 3467–3481. [CrossRef] [PubMed]
73. Melcák, I.; Cermanová, S.; Jirsová, K.; Koberna, K.; Malínský, J.; Raska, I. Nuclear pre-mRNA compartmentalization: Trafficking of released transcripts to splicing factor reservoirs. *Mol. Biol. Cell* **2000**, *11*, 497–510. [CrossRef]
74. Kim, J.; Han, K.Y.; Khanna, N.; Ha, T.; Belmont, A.S. Nuclear speckle fusion via long-range directional motion regulates speckle morphology after transcriptional inhibition. *J. Cell Sci.* **2019**, *132*, jcs226563. [CrossRef] [PubMed]

75. Raina, K.; Rao, B.J. Mammalian nuclear speckles exhibit stable association with chromatin: A biochemical study. *Nucleus* **2022**, *13*, 58–73. [CrossRef] [PubMed]
76. Spector, D.L.; O'Keefe, R.T.; Jiménez-García, L.F. Dynamics of transcription and pre-mRNA splicing within the mammalian cell nucleus. *Cold Spring Harb. Symp. Quant. Biol.* **1993**, *58*, 799–805. [CrossRef] [PubMed]
77. Corpet, A.; Kleijwegt, C.; Roubille, S.; Juillard, F.; Jacquet, K.; Texier, P.; Lomonte, P. PML nuclear bodies and chromatin dynamics: Catch me if you can! *Nucleic Acids Res.* **2020**, *48*, 11890–11912. [CrossRef] [PubMed]
78. Guan, D.; Kao, H.Y. The function, regulation and therapeutic implications of the tumor suppressor protein, PML. *Cell Biosci.* **2015**, *5*, 60. [CrossRef]
79. Hands, K.J.; Cuchet-Lourenco, D.; Everett, R.D.; Hay, R.T. PML isoforms in response to arsenic: High-resolution analysis of PML body structure and degradation. *J. Cell Sci.* **2014**, *127*, 365–375. [CrossRef] [PubMed]
80. Nisole, S.; Maroui, M.A.; Mascle, X.H.; Aubry, M.; Chelbi-Alix, M.K. Differential Roles of PML Isoforms. *Front. Oncol.* **2013**, *3*, 125. [CrossRef] [PubMed]
81. Fonin, A.V.; Silonov, S.A.; Shpironok, O.G.; Antifeeva, I.A.; Petukhov, A.V.; Romanovich, A.E.; Kuznetsova, I.M.; Uversky, V.N.; Turoverov, K.K. The Role of Non-Specific Interactions in Canonical and ALT-Associated PML-Bodies Formation and Dynamics. *Int. J. Mol. Sci.* **2021**, *22*, 5821. [CrossRef]
82. Fonin, A.V.; Silonov, S.A.; Fefilova, A.S.; Stepanenko, O.V.; Gavrilova, A.A.; Petukhov, A.V.; Romanovich, A.E.; Modina, A.L.; Zueva, T.S.; Nedelyaev, E.M.; et al. New Evidence of the Importance of Weak Interactions in the Formation of PML-Bodies. *Int. J. Mol. Sci.* **2022**, *23*, 1613. [CrossRef]
83. Niwa-Kawakita, M.; Wu, H.C.; Thé, H.; Lallemand-Breitenbach, V. PML nuclear bodies, membrane-less domains acting as ROS sensors? *Semin. Cell Dev. Biol.* **2018**, *80*, 29–34. [CrossRef]
84. Rufini, A.; Tucci, P.; Celardo, I.; Melino, G. Senescence and aging: The critical roles of p53. *Oncogene* **2013**, *32*, 5129–5143. [CrossRef] [PubMed]
85. Matt, S.; Hofmann, T.G. Crosstalk between p53 modifiers at PML bodies. *Mol. Cell. Oncol.* **2018**, *5*, e1074335. [CrossRef]
86. Trotman, L.C.; Alimonti, A.; Scaglioni, P.P.; Koutcher, J.A.; Cordon-Cardo, C.; Pandolfi, P.P. Identification of a tumour suppressor network opposing nuclear Akt function. *Nature* **2006**, *441*, 523–527. [CrossRef] [PubMed]
87. Sahin, U.; Ferhi, O.; Jeanne, M.; Benhenda, S.; Berthier, C.; Jollivet, F.; Niwa-Kawakita, M.; Faklaris, O.; Setterblad, N.; de Thé, H.; et al. Oxidative stress–induced assembly of PML nuclear bodies controls sumoylation of partner proteins. *J. Cell Biol.* **2014**, *204*, 931–945. [CrossRef]
88. Sahin, U.; de Thé, H.; Lallemand-Breitenbach, V. PML nuclear bodies: Assembly and oxidative stress-sensitive sumoylation. *Nucleus* **2014**, *5*, 499–507. [CrossRef] [PubMed]
89. Guan, D.; Lim, J.H.; Peng, L.; Liu, Y.; Lam, M.; Seto, E.; Kao, H.Y. Deacetylation of the tumor suppressor protein PML regulates hydrogen peroxide-induced cell death. *Cell Death Dis.* **2014**, *5*, e1340. [CrossRef]
90. Saito, M.; Hess, D.; Eglinger, J.; Fritsch, A.W.; Kreysing, M.; Weinert, B.T.; Choudhary, C.; Matthias, P. Acetylation of intrinsically disordered regions regulates phase separation. *Nat. Chem. Biol.* **2019**, *15*, 51–61. [CrossRef]
91. Lång, A.; Eriksson, J.; Schink, K.O.; Lång, E.; Blicher, P.; Połeć, A.; Brech, A.; Dalhus, B.; Bøe, S.O. Visualization of PML nuclear import complexes reveals FG-repeat nucleoporins at cargo retrieval sites. *Nucleus* **2017**, *8*, 404–420. [CrossRef] [PubMed]
92. Lång, A.; Lång, E.; Bøe, S.O. PML Bodies in Mitosis. *Cells* **2019**, *8*, 893. [CrossRef]
93. Rawat, P.; Boehning, M.; Hummel, B.; Aprile-Garcia, F.; Pandit, A.S.; Eisenhardt, N.; Khavaran, A.; Niskanen, E.; Vos, S.M.; Palvimo, J.J.; et al. Stress-induced nuclear condensation of NELF drives transcriptional downregulation. *Mol. Cell* **2021**, *81*, 1013–1026 e1011. [CrossRef] [PubMed]
94. Ngian, Z.K.; Lin, W.Q.; Ong, C.T. NELF-A controls Drosophila healthspan by regulating heat-shock protein-mediated cellular protection and heterochromatin maintenance. *Aging Cell* **2021**, *20*, e13348. [CrossRef] [PubMed]
95. Goenka, A.; Sengupta, S.; Pandey, R.; Parihar, R.; Mohanta, G.C.; Mukerji, M.; Ganesh, S. Human satellite-III non-coding RNAs modulate heat-shock-induced transcriptional repression. *J. Cell Sci.* **2016**, *129*, 3541–3552. [CrossRef] [PubMed]
96. Jolly, C.; Metz, A.; Govin, J.r.m.; Vigneron, M.; Turner, B.M.; Khochbin, S.; Vourc'h, C. Stress-induced transcription of satellite III repeats. *J. Cell Biol.* **2003**, *164*, 25–33. [CrossRef]
97. Gaglia, G.; Rashid, R.; Yapp, C.; Joshi, G.N.; Li, C.G.; Lindquist, S.L.; Sarosiek, K.A.; Whitesell, L.; Sorger, P.K.; Santagata, S. HSF1 phase transition mediates stress adaptation and cell fate decisions. *Nat. Cell Biol.* **2020**, *22*, 151–158. [CrossRef] [PubMed]
98. Aly, M.K.; Ninomiya, K.; Adachi, S.; Natsume, T.; Hirose, T. Two distinct nuclear stress bodies containing different sets of RNA-binding proteins are formed with HSATIII architectural noncoding RNAs upon thermal stress exposure. *Biochem. Biophys. Res. Commun.* **2019**, *516*, 419–423. [CrossRef] [PubMed]
99. Audas, T.E.; Audas, D.E.; Jacob, M.D.; Ho, J.J.; Khacho, M.; Wang, M.; Perera, J.K.; Gardiner, C.; Bennett, C.A.; Head, T.; et al. Adaptation to Stressors by Systemic Protein Amyloidogenesis. *Dev. Cell* **2016**, *39*, 155–168. [CrossRef] [PubMed]
100. Wang, M.; Tao, X.; Jacob, M.D.; Bennett, C.A.; Ho, J.J.D.; Gonzalgo, M.L.; Audas, T.E.; Lee, S. Stress-Induced Low Complexity RNA Activates Physiological Amyloidogenesis. *Cell Rep.* **2018**, *24*, 1713–1721 e1714. [CrossRef] [PubMed]
101. Marijan, D.; Tse, R.; Elliott, K.; Chandhok, S.; Luo, M.; Lacroix, E.; Audas, T.E. Stress-specific aggregation of proteins in the amyloid bodies. *FEBS Lett.* **2019**, *593*, 3162–3172. [CrossRef] [PubMed]
102. Wang, M.; Bokros, M.; Theodoridis, P.R.; Lee, S. Nucleolar Sequestration: Remodeling Nucleoli Into Amyloid Bodies. *Front. Genet.* **2019**, *10*, 1179. [CrossRef]

103. Mahboubi, H.; Stochaj, U. Cytoplasmic stress granules: Dynamic modulators of cell signaling and disease. *Biochim. Et Biophys. Acta. Mol. Basis Dis.* **2017**, *1863*, 884–895. [CrossRef] [PubMed]
104. Khong, A.; Matheny, T.; Jain, S.; Mitchell, S.F.; Wheeler, J.R.; Parker, R. The Stress Granule Transcriptome Reveals Principles of mRNA Accumulation in Stress Granules. *Mol. Cell* **2017**, *68*, 808–820.e805. [CrossRef] [PubMed]
105. Verma, A.; Sumi, S.; Seervi, M. Heat shock proteins-driven stress granule dynamics: Yet another avenue for cell survival. *Apoptosis Int. J. Program. Cell Death* **2021**, *26*, 371–384. [CrossRef] [PubMed]
106. Ivanov, P.; Kedersha, N.; Anderson, P. Stress Granules and Processing Bodies in Translational Control. *Cold Spring Harb Perspect. Biol.* **2019**, *11*, a032813. [CrossRef]
107. Wolozin, B.; Ivanov, P. Stress granules and neurodegeneration. *Nat. Rev. Neurosci.* **2019**, *20*, 649–666. [CrossRef]
108. Asadi, M.R.; Rahmanpour, D.; Moslehian, M.S.; Sabaie, H.; Hassani, M.; Ghafouri-Fard, S.; Taheri, M.; Rezazadeh, M. Stress Granules Involved in Formation, Progression and Metastasis of Cancer: A Scoping Review. *Front. Cell Dev. Biol.* **2021**, *9*, 745394. [CrossRef]
109. Poblete-Durán, N.; Prades-Pérez, Y.; Vera-Otarola, J.; Soto-Rifo, R.; Valiente-Echeverría, F. Who Regulates Whom? An Overview of RNA Granules and Viral Infections. *Viruses* **2016**, *8*, 180. [CrossRef]
110. Protter, D.S.W.; Parker, R. Principles and Properties of Stress Granules. *Trends Cell Biol.* **2016**, *26*, 668–679. [CrossRef]
111. Wheeler, J.R.; Matheny, T.; Jain, S.; Abrisch, R.; Parker, R. Distinct stages in stress granule assembly and disassembly. *eLife* **2016**, *5*, e18413. [CrossRef] [PubMed]
112. Moujaber, O.; Mahboubi, H.; Kodiha, M.; Bouttier, M.; Bednarz, K.; Bakshi, R.; White, J.; Larose, L.; Colmegna, I.; Stochaj, U. Dissecting the molecular mechanisms that impair stress granule formation in aging cells. *Biochim. Et Biophys. Acta. Mol. Cell Res.* **2017**, *1864*, 475–486. [CrossRef] [PubMed]
113. Omer, A.; Patel, D.; Moran, J.L.; Lian, X.J.; Di Marco, S.; Gallouzi, I.E. Autophagy and heat-shock response impair stress granule assembly during cellular senescence. *Mech. Ageing Dev.* **2020**, *192*, 111382. [CrossRef] [PubMed]
114. Maharjan, N.; Künzli, C.; Buthey, K.; Saxena, S. C9ORF72 Regulates Stress Granule Formation and Its Deficiency Impairs Stress Granule Assembly, Hypersensitizing Cells to Stress. *Mol. Neurobiol.* **2017**, *54*, 3062–3077. [CrossRef]
115. Zhao, Y.G.; Codogno, P.; Zhang, H. Machinery, regulation and pathophysiological implications of autophagosome maturation. *Nat. Rev. Mol. Cell Biol.* **2021**, *22*, 733–750. [CrossRef]
116. Cao, X.; Jin, X.; Liu, B. The involvement of stress granules in aging and aging-associated diseases. *Aging Cell* **2020**, *19*, e13136. [CrossRef]
117. Elbaum-Garfinkle, S. Matter over mind: Liquid phase separation and neurodegeneration. *J. Biol. Chem.* **2019**, *294*, 7160–7168. [CrossRef]
118. Namkoong, S.; Ho, A.; Woo, Y.M.; Kwak, H.; Lee, J.H. Systematic Characterization of Stress-Induced RNA Granulation. *Mol. Cell* **2018**, *70*, 175–187.e178. [CrossRef]
119. Ries, R.J.; Pickering, B.F.; Poh, H.X.; Namkoong, S.; Jaffrey, S.R. $m^6A$ governs length-dependent enrichment of mRNAs in stress granules. *bioRxiv* **2022**. [CrossRef]
120. Khong, A.; Matheny, T.; Huynh, T.N.; Babl, V.; Parker, R. Limited effects of m6A modification on mRNA partitioning into stress granules. *Nat. Commun.* **2022**, *13*, 3735. [CrossRef]
121. Poh, H.X.; Mirza, A.H.; Pickering, B.F.; Jaffrey, S.R. Understanding the source of METTL3-independent $m^6A$ in mRNA. *bioRxiv* **2021**. [CrossRef]
122. Riggs, C.L.; Kedersha, N.; Ivanov, P.; Anderson, P. Mammalian stress granules and P bodies at a glance. *J. Cell Sci.* **2020**, *133*, jcs242487. [CrossRef] [PubMed]
123. Kedersha, N.; Anderson, P. Mammalian stress granules and processing bodies. *Methods Enzymol.* **2007**, *431*, 61–81. [CrossRef]
124. Luo, Y.; Na, Z.; Slavoff, S.A. P-Bodies: Composition, Properties, and Functions. *Biochemistry* **2018**, *57*, 2424–2431. [CrossRef]
125. Souquere, S.; Mollet, S.; Kress, M.; Dautry, F.; Pierron, G.; Weil, D. Unravelling the ultrastructure of stress granules and associated P-bodies in human cells. *J. Cell Sci.* **2009**, *122*, 3619–3626. [CrossRef] [PubMed]
126. Ayache, J.; Bénard, M.; Ernoult-Lange, M.; Minshall, N.; Standart, N.; Kress, M.; Weil, D. P-body assembly requires DDX6 repression complexes rather than decay or Ataxin2/2L complexes. *Mol. Biol Cell* **2015**, *26*, 2579–2595. [CrossRef] [PubMed]
127. Hubstenberger, A.; Courel, M.; Bénard, M.; Souquere, S.; Ernoult-Lange, M.; Chouaib, R.; Yi, Z.; Morlot, J.B.; Munier, A.; Fradet, M.; et al. P-Body Purification Reveals the Condensation of Repressed mRNA Regulons. *Mol. Cell* **2017**, *68*, 144–157 e145. [CrossRef]
128. Youn, J.Y.; Dyakov, B.J.A.; Zhang, J.; Knight, J.D.R.; Vernon, R.M.; Forman-Kay, J.D.; Gingras, A.C. Properties of Stress Granule and P-Body Proteomes. *Mol. Cell* **2019**, *76*, 286–294. [CrossRef] [PubMed]
129. Gallois-Montbrun, S.; Kramer, B.; Swanson, C.M.; Byers, H.; Lynham, S.; Ward, M.; Malim, M.H. Antiviral protein APOBEC3G localizes to ribonucleoprotein complexes found in P bodies and stress granules. *J. Virol.* **2007**, *81*, 2165–2178. [CrossRef]
130. Kedersha, N.; Stoecklin, G.; Ayodele, M.; Yacono, P.; Lykke-Andersen, J.; Fritzler, M.J.; Scheuner, D.; Kaufman, R.J.; Golan, D.E.; Anderson, P. Stress granules and processing bodies are dynamically linked sites of mRNP remodeling. *J. Cell Biol.* **2005**, *169*, 871–884. [CrossRef]
131. Weil, T.T.; Parton, R.M.; Herpers, B.; Soetaert, J.; Veenendaal, T.; Xanthakis, D.; Dobbie, I.M.; Halstead, J.M.; Hayashi, R.; Rabouille, C.; et al. Drosophila patterning is established by differential association of mRNAs with P bodies. *Nat. Cell Biol.* **2012**, *14*, 1305–1313. [CrossRef] [PubMed]
132. Buckingham, M.; Liu, J.L. U bodies respond to nutrient stress in Drosophila. *Exp. Cell Res.* **2011**, *317*, 2835–2844. [CrossRef] [PubMed]

133. Buchan, J.R.; Muhlrad, D.; Parker, R. P bodies promote stress granule assembly in Saccharomyces cerevisiae. *J. Cell Biol.* **2008**, *183*, 441–455. [CrossRef] [PubMed]
134. Wang, C.; Schmich, F.; Srivatsa, S.; Weidner, J.; Beerenwinkel, N.; Spang, A. Context-dependent deposition and regulation of mRNAs in P-bodies. *eLife* **2018**, *7*, e29815. [CrossRef] [PubMed]
135. Rieckher, M.; Markaki, M.; Princz, A.; Schumacher, B.; Tavernarakis, N. Maintenance of Proteostasis by P Body-Mediated Regulation of eIF4E Availability during Aging in Caenorhabditis elegans. *Cell Rep.* **2018**, *25*, 199–211 e196. [CrossRef] [PubMed]
136. Ditlev, J.A. Membrane-associated phase separation: Organization and function emerge from a two-dimensional milieu. *J. Mol. Cell Biol.* **2021**, *13*, 319–324. [CrossRef]
137. Zacharogianni, M.; Aguilera-Gomez, A.; Veenendaal, T.; Smout, J.; Rabouille, C. A stress assembly that confers cell viability by preserving ERES components during amino-acid starvation. *eLife* **2014**, *3*, e04132. [CrossRef]
138. Müller, M.; Ahumada-Castro, U.; Sanhueza, M.; Gonzalez-Billault, C.; Court, F.A.; Cárdenas, C. Mitochondria and Calcium Regulation as Basis of Neurodegeneration Associated With Aging. *Front. Neurosci.* **2018**, *12*, 470. [CrossRef]
139. Loncke, J.; Kaasik, A.; Bezprozvanny, I.; Parys, J.B.; Kerkhofs, M.; Bultynck, G. Balancing ER-Mitochondrial Ca(2+) Fluxes in Health and Disease. *Trends Cell Biol.* **2021**, *31*, 598–612. [CrossRef]
140. Barazzuol, L.; Giamogante, F.; Calì, T. Mitochondria Associated Membranes (MAMs): Architecture and physiopathological role. *Cell Calcium* **2021**, *94*, 102343. [CrossRef]
141. Giorgi, C.; Ito, K.; Lin, H.K.; Santangelo, C.; Wieckowski, M.R.; Lebiedzinska, M.; Bononi, A.; Bonora, M.; Duszynski, J.; Bernardi, R.; et al. PML regulates apoptosis at endoplasmic reticulum by modulating calcium release. *Science* **2010**, *330*, 1247–1251. [CrossRef] [PubMed]
142. Carracedo, A.; Ito, K.; Pandolfi, P.P. The nuclear bodies inside out: PML conquers the cytoplasm. *Curr. Opin. Cell Biol.* **2011**, *23*, 360–366. [CrossRef] [PubMed]
143. Missiroli, S.; Bonora, M.; Patergnani, S.; Poletti, F.; Perrone, M.; Gafà, R.; Magri, E.; Raimondi, A.; Lanza, G.; Tacchetti, C.; et al. PML at Mitochondria-Associated Membranes Is Critical for the Repression of Autophagy and Cancer Development. *Cell Rep.* **2016**, *16*, 2415–2427. [CrossRef] [PubMed]
144. Yang, M.; Li, C.; Yang, S.; Xiao, Y.; Xiong, X.; Chen, W.; Zhao, H.; Zhang, Q.; Han, Y.; Sun, L. Mitochondria-Associated ER Membranes—The Origin Site of Autophagy. *Front. Cell Dev. Biol.* **2020**, *8*, 595. [CrossRef] [PubMed]
145. Belyy, V.; Tran, N.H.; Walter, P. Quantitative microscopy reveals dynamics and fate of clustered IRE1α. *Proc. Natl. Acad. Sci. USA* **2020**, *117*, 1533–1542. [CrossRef]
146. Lee, J.E.; Cathey, P.I.; Wu, H.; Parker, R.; Voeltz, G.K. Endoplasmic reticulum contact sites regulate the dynamics of membraneless organelles. *Science* **2020**, *367*, eaay7108. [CrossRef] [PubMed]
147. Moon, S.L.; Parker, R. Analysis of eIF2B bodies and their relationships with stress granules and P-bodies. *Sci Rep.* **2018**, *8*, 12264. [CrossRef]
148. Mollet, S.; Cougot, N.; Wilczynska, A.; Dautry, F.; Kress, M.; Bertrand, E.; Weil, D. Translationally repressed mRNA transiently cycles through stress granules during stress. *Mol. Biol. Cell* **2008**, *19*, 4469–4479. [CrossRef]
149. Ohn, T.; Kedersha, N.; Hickman, T.; Tisdale, S.; Anderson, P. A functional RNAi screen links O-GlcNAc modification of ribosomal proteins to stress granule and processing body assembly. *Nat. Cell Biol.* **2008**, *10*, 1224–1231. [CrossRef] [PubMed]
150. Mitchell, S.F.; Walker, S.E.; Rajagopal, V.; Aitken, C.E.; Lorsch, J.R. Recruiting knotty partners: The roles of translation initiation factors in mRNA recruitment to the eukaryotic ribosome. In *Ribosomes: Structure, Function, and Dynamics*; Rodnina, M.V., Wintermeyer, W., Green, R., Eds.; Springer Vienna: Vienna, Austria, 2011; pp. 155–169.
151. Wang, C.Y.; Wen, W.L.; Nilsson, D.; Sunnerhagen, P.; Chang, T.H.; Wang, S.W. Analysis of stress granule assembly in Schizosaccharomyces pombe. *RNA (New York N.Y.)* **2012**, *18*, 694–703. [CrossRef]
152. Ernoult-Lange, M.; Baconnais, S.; Harper, M.; Minshall, N.; Souquere, S.; Boudier, T.; Bénard, M.; Andrey, P.; Pierron, G.; Kress, M.; et al. Multiple binding of repressed mRNAs by the P-body protein Rck/p54. *RNA (New York N.Y.)* **2012**, *18*, 1702–1715. [CrossRef]
153. Teixeira, D.; Parker, R. Analysis of P-body assembly in Saccharomyces cerevisiae. *Mol. Biol. Cell* **2007**, *18*, 2274–2287. [CrossRef] [PubMed]
154. Leung, A.K.; Sharp, P.A. Quantifying Argonaute proteins in and out of GW/P-bodies: Implications in microRNA activities. *Adv. Exp. Med. Biol.* **2013**, *768*, 165–182. [CrossRef]
155. Qi, M.Y.; Wang, Z.Z.; Zhang, Z.; Shao, Q.; Zeng, A.; Li, X.Q.; Li, W.Q.; Wang, C.; Tian, F.J.; Li, Q.; et al. AU-rich-element-dependent translation repression requires the cooperation of tristetraprolin and RCK/P54. *Mol. Cell. Biol.* **2012**, *32*, 913–928. [CrossRef]
156. Laporte, D.; Salin, B.; Daignan-Fornier, B.; Sagot, I. Reversible cytoplasmic localization of the proteasome in quiescent yeast cells. *J. Cell Biol.* **2008**, *181*, 737–745. [CrossRef] [PubMed]
157. Munder, M.C.; Midtvedt, D.; Franzmann, T.; Nüske, E.; Otto, O.; Herbig, M.; Ulbricht, E.; Müller, P.; Taubenberger, A.; Maharana, S.; et al. A pH-driven transition of the cytoplasm from a fluid- to a solid-like state promotes entry into dormancy. *eLife* **2016**, *5*, e09347. [CrossRef] [PubMed]
158. Saad, S.; Cereghetti, G.; Feng, Y.; Picotti, P.; Peter, M.; Dechant, R. Reversible protein aggregation is a protective mechanism to ensure cell cycle restart after stress. *Nat. Cell Biol.* **2017**, *19*, 1202–1213. [CrossRef] [PubMed]

159. Petrovska, I.; Nüske, E.; Munder, M.C.; Kulasegaran, G.; Malinovska, L.; Kroschwald, S.; Richter, D.; Fahmy, K.; Gibson, K.; Verbavatz, J.M.; et al. Filament formation by metabolic enzymes is a specific adaptation to an advanced state of cellular starvation. *eLife* **2014**, *3*. [CrossRef] [PubMed]
160. Cereghetti, G.; Wilson-Zbinden, C.; Kissling, V.M.; Diether, M.; Arm, A.; Yoo, H.; Piazza, I.; Saad, S.; Picotti, P.; Drummond, D.A.; et al. Reversible amyloids of pyruvate kinase couple cell metabolism and stress granule disassembly. *Nat. Cell Biol.* **2021**, *23*, 1085–1094, e02409. [CrossRef] [PubMed]
161. Boor, K.J. Bacterial stress responses: What doesn't kill them can make then stronger. *PLoS Biol.* **2006**, *4*, e23. [CrossRef] [PubMed]
162. Huo, Y.X.; Rosenthal, A.Z.; Gralla, J.D. General stress response signalling: Unwrapping transcription complexes by DNA relaxation via the sigma38 C-terminal domain. *Mol. Microbiol.* **2008**, *70*, 369–378. [CrossRef] [PubMed]
163. Alba, B.M.; Gross, C.A. Regulation of the Escherichia coli sigma-dependent envelope stress response. *Mol. Microbiol.* **2004**, *52*, 613–619. [CrossRef]
164. Maleki, F.; Khosravi, A.; Nasser, A.; Taghinejad, H.; Azizian, M. Bacterial Heat Shock Protein Activity. *J. Clin. Diagn. Res. JCDR* **2016**, *10*, BE01–BE03. [CrossRef]
165. Keto-Timonen, R.; Hietala, N.; Palonen, E.; Hakakorpi, A.; Lindström, M.; Korkeala, H. Cold Shock Proteins: A Minireview with Special Emphasis on Csp-family of Enteropathogenic Yersinia. *Front. Microbiol.* **2016**, *7*, 1151. [CrossRef]
166. Nandana, V.; Schrader, J.M. Roles of liquid-liquid phase separation in bacterial RNA metabolism. *Curr. Opin. Microbiol.* **2021**, *61*, 91–98. [CrossRef] [PubMed]
167. Azaldegui, C.A.; Vecchiarelli, A.G.; Biteen, J.S. The emergence of phase separation as an organizing principle in bacteria. *Biophys. J.* **2021**, *120*, 1123–1138. [CrossRef] [PubMed]
168. Gao, Z.; Zhang, W.; Chang, R.; Zhang, S.; Yang, G.; Zhao, G. Liquid-Liquid Phase Separation: Unraveling the Enigma of Biomolecular Condensates in Microbial Cells. *Front. Microbiol.* **2021**, *12*, 751880. [CrossRef] [PubMed]
169. Ladouceur, A.M.; Parmar, B.S.; Biedzinski, S.; Wall, J.; Tope, S.G.; Cohn, D.; Kim, A.; Soubry, N.; Reyes-Lamothe, R.; Weber, S.C. Clusters of bacterial RNA polymerase are biomolecular condensates that assemble through liquid-liquid phase separation. *Proc. Natl. Acad. Sci. USA* **2020**, *117*, 18540–18549. [CrossRef] [PubMed]
170. Babl, L.; Giacomelli, G.; Ramm, B.; Gelmroth, A.K.; Bramkamp, M.; Schwille, P. CTP-controlled liquid-liquid phase separation of ParB. *J. Mol. Biol.* **2022**, *434*, 167401. [CrossRef] [PubMed]
171. Lasker, K.; von Diezmann, L.; Zhou, X.; Ahrens, D.G.; Mann, T.H.; Moerner, W.E.; Shapiro, L. Selective sequestration of signalling proteins in a membraneless organelle reinforces the spatial regulation of asymmetry in Caulobacter crescentus. *Nat. Microbiol.* **2020**, *5*, 418–429. [CrossRef]
172. Saurabh, S.; Chong, T.N.; Bayas, C.; Dahlberg, P.D.; Cartwright, H.N.; Moerner, W.E.; Shapiro, L. ATP-responsive biomolecular condensates tune bacterial kinase signaling. *Sci. Adv.* **2022**, *8*, eabm6570. [CrossRef] [PubMed]
173. Hamouche, L.; Billaudeau, C.; Rocca, A.; Chastanet, A.; Ngo, S.; Laalami, S.; Putzer, H. Dynamic Membrane Localization of RNase Y in Bacillus subtilis. *mBio* **2020**, *11*, e03337–e19. [CrossRef] [PubMed]
174. Tejada-Arranz, A.; Galtier, E.; El Mortaji, L.; Turlin, E.; Ershov, D.; De Reuse, H. The RNase J-Based RNA Degradosome Is Compartmentalized in the Gastric Pathogen Helicobacter pylori. *mBio* **2020**, *11*, e01173–e20. [CrossRef] [PubMed]
175. Racki, L.R.; Tocheva, E.I.; Dieterle, M.G.; Sullivan, M.C.; Jensen, G.J.; Newman, D.K. Polyphosphate granule biogenesis is temporally and functionally tied to cell cycle exit during starvation in Pseudomonas aeruginosa. *Proc. Natl. Acad. Sci. USA* **2017**, *114*, E2440–E2449. [CrossRef] [PubMed]
176. Zhao, T.; Liu, Y.; Wang, Z.; He, R.; Xiang Zhang, J.; Xu, F.; Lei, M.; Deci, M.B.; Nguyen, J.; Bianco, P.R. Super-resolution imaging reveals changes in Escherichia coli SSB localization in response to DNA damage. *Genes Cells Devoted Mol. Cell. Mech.* **2019**, *24*, 814–826. [CrossRef] [PubMed]
177. Harami, G.M.; Kovács, Z.J.; Pancsa, R.; Pálinkás, J.; Baráth, V.; Tárnok, K.; Málnási-Csizmadia, A.; Kovács, M. Phase separation by ssDNA binding protein controlled via protein-protein and protein-DNA interactions. *Proc. Natl. Acad. Sci. USA* **2020**, *117*, 26206–26217. [CrossRef] [PubMed]
178. Janissen, R.; Arens, M.M.A.; Vtyurina, N.N.; Rivai, Z.; Sunday, N.D.; Eslami-Mossallam, B.; Gritsenko, A.A.; Laan, L.; de Ridder, D.; Artsimovitch, I.; et al. Global DNA Compaction in Stationary-Phase Bacteria Does Not Affect Transcription. *Cell* **2018**, *174*, 1188–1199 e1114. [CrossRef] [PubMed]
179. Ceci, P.; Cellai, S.; Falvo, E.; Rivetti, C.; Rossi, G.L.; Chiancone, E. DNA condensation and self-aggregation of Escherichia coli Dps are coupled phenomena related to the properties of the N-terminus. *Nucleic Acids Res.* **2004**, *32*, 5935–5944. [CrossRef]
180. Al-Husini, N.; Tomares, D.T.; Bitar, O.; Childers, W.S.; Schrader, J.M. α-Proteobacterial RNA Degradosomes Assemble Liquid-Liquid Phase-Separated RNP Bodies. *Mol. Cell* **2018**, *71*, 1027–1039 e1014. [CrossRef] [PubMed]
181. Strahl, H.; Turlan, C.; Khalid, S.; Bond, P.J.; Kebalo, J.M.; Peyron, P.; Poljak, L.; Bouvier, M.; Hamoen, L.; Luisi, B.F.; et al. Membrane recognition and dynamics of the RNA degradosome. *PLoS Genet.* **2015**, *11*, e1004961. [CrossRef] [PubMed]
182. Al-Husini, N.; Tomares, D.T.; Pfaffenberger, Z.J.; Muthunayake, N.S.; Samad, M.A.; Zuo, T.; Bitar, O.; Aretakis, J.R.; Bharmal, M.M.; Gega, A.; et al. BR-Bodies Provide Selectively Permeable Condensates that Stimulate mRNA Decay and Prevent Release of Decay Intermediates. *Mol. Cell* **2020**, *78*, 670–682 e678. [CrossRef] [PubMed]
183. Lopez, P.J.; Marchand, I.; Joyce, S.A.; Dreyfus, M. The C-terminal half of RNase E, which organizes the Escherichia coli degradosome, participates in mRNA degradation but not rRNA processing in vivo. *Mol. Microbiol.* **1999**, *33*, 188–199. [CrossRef] [PubMed]

184. Vishnyakov, I.E.; Levitskii, S.A.; Manuvera, V.A.; Lazarev, V.N.; Ayala, J.A.; Ivanov, V.A.; Snigirevskaya, E.S.; Komissarchik, Y.Y.; Borchsenius, S.N. The identification and characterization of IbpA, a novel α-crystallin-type heat shock protein from mycoplasma. *Cell Stress Chaperones* **2012**, *17*, 171–180. [CrossRef] [PubMed]
185. Vishnyakov, I.E.; Levitskii, S.A.; Borchsenius, S.N. The effect of heat shock on phytopathogenic mycoplasma Acholeplasma laidlawii PG-8A. *Cell Tissue Biol.* **2015**, *9*, 149–157. [CrossRef]
186. Vishnyakov, I.E.; Bogachev, M.I.; Salafutdinov, I.; Borchsenius, S.N.; Kayumov, A.R. The Temperature-Dependent Selectivity of Potential Interaction Partners for the Small Heat Shock Protein IbpA from Acholeplasma laidlawii. *BioNanoScience* **2016**, *6*, 437–442. [CrossRef]
187. Strózecka, J.; Chrusciel, E.; Górna, E.; Szymanska, A.; Ziętkiewicz, S.; Liberek, K. Importance of N- and C-terminal regions of IbpA, Escherichia coli small heat shock protein, for chaperone function and oligomerization. *J. Biol. Chem.* **2012**, *287*, 2843–2853. [CrossRef]
188. Lazarev, V.N.; Levitskii, S.A.; Basovskii, Y.I.; Chukin, M.M.; Akopian, T.A.; Vereshchagin, V.V.; Kostrjukova, E.S.; Kovaleva, G.Y.; Kazanov, M.D.; Malko, D.B.; et al. Complete genome and proteome of Acholeplasma laidlawii. *J. Bacteriol.* **2011**, *193*, 4943–4953. [CrossRef]
189. Bright, C.M.; Ellis, D. Intracellular pH changes induced by hypoxia and anoxia in isolated sheep heart Purkinje fibres. *Exp. Physiol.* **1992**, *77*, 165–175. [CrossRef]
190. Dechant, R.; Binda, M.; Lee, S.S.; Pelet, S.; Winderickx, J.; Peter, M. Cytosolic pH is a second messenger for glucose and regulates the PKA pathway through V-ATPase. *EMBO J.* **2010**, *29*, 2515–2526. [CrossRef]
191. Adame-Arana, O.; Weber, C.A.; Zaburdaev, V.; Prost, J.; Jülicher, F. Liquid Phase Separation Controlled by pH. *Biophys. J.* **2020**, *119*, 1590–1605. [CrossRef]
192. Riback, J.A.; Katanski, C.D.; Kear-Scott, J.L.; Pilipenko, E.V.; Rojek, A.E.; Sosnick, T.R.; Drummond, D.A. Stress-Triggered Phase Separation Is an Adaptive, Evolutionarily Tuned Response. *Cell* **2017**, *168*, 1028–1040 e1019. [CrossRef] [PubMed]
193. Hondele, M.; Sachdev, R.; Heinrich, S.; Wang, J.; Vallotton, P.; Fontoura, B.M.A.; Weis, K. DEAD-box ATPases are global regulators of phase-separated organelles. *Nature* **2019**, *573*, 144–148. [CrossRef] [PubMed]
194. Guillén-Boixet, J.; Kopach, A.; Holehouse, A.S.; Wittmann, S.; Jahnel, M.; Schlüßler, R.; Kim, K.; Trussina, I.; Wang, J.; Mateju, D.; et al. RNA-Induced Conformational Switching and Clustering of G3BP Drive Stress Granule Assembly by Condensation. *Cell* **2020**, *181*, 346–361 e317. [CrossRef] [PubMed]
195. Li, S.; Yoshizawa, T.; Yamazaki, R.; Fujiwara, A.; Kameda, T.; Kitahara, R. Pressure and Temperature Phase Diagram for Liquid-Liquid Phase Separation of the RNA-Binding Protein Fused in Sarcoma. *J. Phys. Chemistry. B* **2021**, *125*, 6821–6829. [CrossRef]
196. Franzmann, T.M.; Alberti, S. Protein Phase Separation as a Stress Survival Strategy. *Cold Spring Harb Perspect. Biol.* **2019**, *11*, a034058. [CrossRef]
197. Dignon, G.L.; Zheng, W.; Kim, Y.C.; Mittal, J. Temperature-Controlled Liquid-Liquid Phase Separation of Disordered Proteins. *ACS Cent. Sci.* **2019**, *5*, 821–830. [CrossRef]
198. Jalihal, A.P.; Pitchiaya, S.; Xiao, L.; Bawa, P.; Jiang, X.; Bedi, K.; Parolia, A.; Cieslik, M.; Ljungman, M.; Chinnaiyan, A.M.; et al. Multivalent Proteins Rapidly and Reversibly Phase-Separate upon Osmotic Cell Volume Change. *Mol. Cell* **2020**, *79*, 978–990.e975. [CrossRef]
199. Onoguchi-Mizutani, R.; Akimitsu, N. Long noncoding RNA and phase separation in cellular stress response. *J. Biochem.* **2022**, *171*, 269–276. [CrossRef]
200. Snead, W.T.; Gladfelter, A.S. The Control Centers of Biomolecular Phase Separation: How Membrane Surfaces, PTMs, and Active Processes Regulate Condensation. *Mol. Cell* **2019**, *76*, 295–305. [CrossRef]
201. Drino, A.; Schaefer, M.R. RNAs, Phase Separation, and Membrane-Less Organelles: Are Post-Transcriptional Modifications Modulating Organelle Dynamics? *BioEssays News Rev. Mol. Cell. Dev. Biol.* **2018**, *40*, e1800085. [CrossRef]
202. Yang, P.; Mathieu, C.; Kolaitis, R.M.; Zhang, P.; Messing, J.; Yurtsever, U.; Yang, Z.; Wu, J.; Li, Y.; Pan, Q.; et al. G3BP1 Is a Tunable Switch that Triggers Phase Separation to Assemble Stress Granules. *Cell* **2020**, *181*, 325–345 e328. [CrossRef] [PubMed]
203. Huang, C.; Chen, Y.; Dai, H.; Zhang, H.; Xie, M.; Zhang, H.; Chen, F.; Kang, X.; Bai, X.; Chen, Z. UBAP2L arginine methylation by PRMT1 modulates stress granule assembly. *Cell Death Differ.* **2020**, *27*, 227–241. [CrossRef]
204. Duan, Y.; Du, A.; Gu, J.; Duan, G.; Wang, C.; Gui, X.; Ma, Z.; Qian, B.; Deng, X.; Zhang, K.; et al. PARylation regulates stress granule dynamics, phase separation, and neurotoxicity of disease-related RNA-binding proteins. *Cell Res.* **2019**, *29*, 233–247. [CrossRef] [PubMed]
205. Jongjitwimol, J.; Baldock, R.A.; Morley, S.J.; Watts, F.Z. Sumoylation of eIF4A2 affects stress granule formation. *J. Cell Sci.* **2016**, *129*, 2407–2415. [CrossRef]
206. Carey, J.L.; Guo, L. Liquid-Liquid Phase Separation of TDP-43 and FUS in Physiology and Pathology of Neurodegenerative Diseases. *Front. Mol. Biosci.* **2022**, *9*, 826719. [CrossRef] [PubMed]
207. Hofweber, M.; Dormann, D. Friend or foe-Post-translational modifications as regulators of phase separation and RNP granule dynamics. *J. Biol. Chem.* **2019**, *294*, 7137–7150. [CrossRef]
208. Roth, S.; Khalaila, I. The effect of O-GlcNAcylation on hnRNP A1 translocation and interaction with transportin1. *Exp. Cell Res.* **2017**, *350*, 210–217. [CrossRef]
209. Cheng, X.; Kao, H.Y. Post-translational modifications of PML: Consequences and implications. *Front. Oncol.* **2012**, *2*, 210. [CrossRef]

210. Ohkuni, K.; Pasupala, N.; Peek, J.; Holloway, G.L.; Sclar, G.D.; Levy-Myers, R.; Baker, R.E.; Basrai, M.A.; Kerscher, O. SUMO-Targeted Ubiquitin Ligases (STUbLs) Reduce the Toxicity and Abnormal Transcriptional Activity Associated With a Mutant, Aggregation-Prone Fragment of Huntingtin. *Front. Genet.* **2018**, *9*, 379. [CrossRef]
211. Alberti, S.; Carra, S. Quality Control of Membraneless Organelles. *J. Mol. Biol.* **2018**, *430*, 4711–4729. [CrossRef]
212. Mateju, D.; Franzmann, T.M.; Patel, A.; Kopach, A.; Boczek, E.E.; Maharana, S.; Lee, H.O.; Carra, S.; Hyman, A.A.; Alberti, S. An aberrant phase transition of stress granules triggered by misfolded protein and prevented by chaperone function. *EMBO J.* **2017**, *36*, 1669–1687. [CrossRef] [PubMed]
213. Horne, S.D.; Chowdhury, S.K.; Heng, H.H. Stress, genomic adaptation, and the evolutionary trade-off. *Front. Genet.* **2014**, *5*, 92. [CrossRef] [PubMed]
214. Berchtold, D.; Battich, N.; Pelkmans, L. A Systems-Level Study Reveals Regulators of Membrane-less Organelles in Human Cells. *Mol. Cell* **2018**, *72*, 1035–1049 e1035. [CrossRef] [PubMed]
215. Singh, A.; Kandi, A.R.; Jayaprakashappa, D.; Thuery, G.; Purohit, D.J.; Huelsmeier, J.; Singh, R.; Pothapragada, S.S.; Ramaswami, M.; Bakthavachalu, B. The transcriptional response to oxidative stress is independent of stress-granule formation. *Mol. Biol. Cell* **2022**, *33*, ar25. [CrossRef] [PubMed]
216. Lee, S.; Neumann, M.; Stearman, R.; Stauber, R.; Pause, A.; Pavlakis, G.N.; Klausner, R.D. Transcription-dependent nuclear-cytoplasmic trafficking is required for the function of the von Hippel-Lindau tumor suppressor protein. *Mol. Cell. Biol.* **1999**, *19*, 1486–1497. [CrossRef] [PubMed]
217. Marnef, A.; Weil, D.; Standart, N. RNA-related nuclear functions of human Pat1b, the P-body mRNA decay factor. *Mol. Biol. Cell* **2012**, *23*, 213–224. [CrossRef]
218. Borbolis, F.; Syntichaki, P. Cytoplasmic mRNA turnover and ageing. *Mech. Ageing Dev.* **2015**, *152*, 32–42. [CrossRef]
219. Campos-Melo, D.; Hawley, Z.C.E.; Droppelmann, C.A.; Strong, M.J. The Integral Role of RNA in Stress Granule Formation and Function. *Front. Cell Dev. Biol.* **2021**, *9*, 621779. [CrossRef]
220. Sharp, P.A.; Chakraborty, A.K.; Henninger, J.E.; Young, R.A. RNA in formation and regulation of transcriptional condensates. *RNA (New York N.Y.)* **2022**, *28*, 52–57. [CrossRef]
221. Szaflarski, W.; Leśniczak-Staszak, M.; Sowiński, M.; Ojha, S.; Aulas, A.; Dave, D.; Malla, S.; Anderson, P.; Ivanov, P.; Lyons, S.M. Early rRNA processing is a stress-dependent regulatory event whose inhibition maintains nucleolar integrity. *Nucleic Acids Res.* **2022**, *50*, 1033–1051. [CrossRef]
222. Pfister, A.S. Emerging Role of the Nucleolar Stress Response in Autophagy. *Front. Cell. Neurosci.* **2019**, *13*, 156. [CrossRef]

biomolecules

MDPI

*Review*

# The Role of Intrinsically Disordered Proteins in Liquid–Liquid Phase Separation during Calcium Carbonate Biomineralization

**Aneta Tarczewska, Klaudia Bielak, Anna Zoglowek, Katarzyna Sołtys, Piotr Dobryszycki, Andrzej Ożyhar and Mirosława Różycka ***

Department of Biochemistry, Molecular Biology and Biotechnology, Faculty of Chemistry, Wroclaw University of Science and Technology, Wybrzeże Wyspiańskiego 27, 50-370 Wroclaw, Poland
* Correspondence: miroslawa.rozycka@pwr.edu.pl

**Abstract:** Some animal organs contain mineralized tissues. These so-called hard tissues are mostly deposits of calcium salts, usually in the form of calcium phosphate or calcium carbonate. Examples of this include fish otoliths and mammalian otoconia, which are found in the inner ear, and they are an essential part of the sensory system that maintains body balance. The composition of ear stones is quite well known, but the role of individual components in the nucleation and growth of these biominerals is enigmatic. It is sure that intrinsically disordered proteins (IDPs) play an important role in this aspect. They have an impact on the shape and size of otoliths. It seems probable that IDPs, with their inherent ability to phase separate, also play a role in nucleation processes. This review discusses the major theories on the mechanisms of biomineral nucleation with a focus on the importance of protein-driven liquid–liquid phase separation (LLPS). It also presents the current understanding of the role of IDPs in the formation of calcium carbonate biominerals and predicts their potential ability to drive LLPS.

**Keywords:** liquid–liquid phase separation; intrinsically disordered proteins; biomineralization; calcium carbonate; otoliths; nucleation pathways

**Citation:** Tarczewska, A.; Bielak, K.; Zoglowek, A.; Sołtys, K.; Dobryszycki, P.; Ożyhar, A.; Różycka, M. The Role of Intrinsically Disordered Proteins in Liquid–Liquid Phase Separation during Calcium Carbonate Biomineralization. *Biomolecules* **2022**, *12*, 1266. https://doi.org/10.3390/biom12091266

Academic Editors: Prakash Kulkarni, Stefania Brocca, Keith Dunker and Sonia Longhi

Received: 5 August 2022
Accepted: 7 September 2022
Published: 9 September 2022



## 1. Introduction

Liquid–liquid phase separation (LLPS) is one of the key mechanisms affecting how macromolecular assemblies, including membrane-less organelles (MLOs), are formed and regulated [1]. This reversible, thermodynamically-driven process relies on the separation of a homogeneous solution into two distinct liquid phases with different concentrations of solutes. LLPS occurs as a two-phase system based on the concentration of molecules and the physio-chemical parameters of the microenvironment [2]. Phase-separated condensates, especially MLOs, are multicomponent assemblies of proteins and other macromolecules, e.g., nucleic acids [2–4]. The interactions between the components are weak, transient, and multivalent [5,6]. The proteins that are found to reside in a condensate may play diverse roles (e.g., scaffold, co-scaffold, clients, and regulators) in maintaining the condensate integrity, composition, and biochemical properties. Scaffolds can self-associate and drive LLPS, so they are primarily responsible for condensate formation. Clients, on the other hand, are low-valency molecules that are recruited to the condensate through their interactions with scaffold biomolecules [7]. Their content may be adjusted to the changing conditions within and outside the condensate [8].

LLPS as a physical process has been known for decades in polymer science, but it has been rediscovered in eukaryotic cells [9]. At present, it appears that it is a universal phenomenon that plays an important role in the interior organization of eukaryotic cells, in the formation of MLOs in prokaryotes [10,11], and during viral life cycles [12]. Notably, some extracellular protein interactions facilitate LLPS [13]. Biomineralization is the process by which organisms produce minerals under biological control. The control of biomineralization is aimed at creating specific minerals composed of inorganic and organic fractions.

A very common inorganic component of biominerals is calcium carbonate in the form of various, usually non-calcite polymorphs [14,15]. Although the mechanisms of biomineral nucleation processes have been studied for years, their principles are still enigmatic. Calcite, the most stable polymorph of calcium carbonate, serves as a model for the primal and accepted-for-years theory of crystal growth, known as a classical theory. Some recent studies, however, showed that the formation of calcium carbonate frequently does not follow the classical model [16,17]. Since 2000, when Gower et al. launched the polymer-induced liquid precursor (PILP) concept of biomineral precursors, it has been widely accepted that the early events of biomineral formation may follow diverse alternative pathways [18]. Moreover, further experimental results concerning calcium carbonate mineralization presented prenucleation clusters as a key precursor phase in mineral formation [19]. Currently, a concept that involves the formation of dense liquid precursors of amorphous calcium carbonate (ACC) via LLPS has become a popular topic of investigation [20,21].

As indicated above, the molecular and biochemical mechanisms involved in the biomineralization pathway remain puzzling. Additionally, the significance of intrinsically disordered proteins (IDPs), which are an abundant organic component of hard tissues [22], in the formation of liquid precursors of biominerals remains to be solved. Research on the interactions between proteins and divalent cations is essential for understanding the resulting liquid precursors. In the available literature, there are only a few examples describing such interactions. However, they may help to understand the functional and pathological phase behaviours in the biomineralization process.

In this paper, the role of proteins, mainly IDPs, in the formation of calcium carbonate biominerals is reviewed. We focus on different mechanisms of mineral formation and discuss the potential role of LLPS in the nucleation processes. Since the discovery of protein-driven phase separation [9,23], the concept of LLPS has been deeply integrated into the life sciences, especially into research focusing on cell biology. Phase separation has also been considered in studies concentrating on mineralization processes, yet often the term, phase separation (and LLPS in particular), is masked by coalescence, whereas liquid condensates/droplets are often referred to as coacervates. In this review, we gathered the knowledge that is scattered and hidden under non-uniform terms. We also discuss the potential importance of IDPs in LLPS during calcium carbonate biomineralization.

## 2. Role of Proteins and Divalent Ions in LLPS

First, it was suspected that strong stereospecific interactions between protein components played a major role in the formation of MLOs. However, subsequent studies have shed light on the importance of very weak interactions, e.g., electrostatic, hydrophobic, and $\pi$-$\pi$ interactions [5]. Currently, we know that both strong and weak interactions, which occur simultaneously, contribute to the entire complex interaction network and facilitate condensate formation. Additionally, the interaction of proteins with solvent plays a critical role in the regulation of phase transitions; thus, an important feature affecting protein-induced LLPS is solubility [24]. Most proteins that undergo LLPS have poor solubility in water. Placing such proteins in the structure forming during phase separation is more energetically beneficial than allowing them to come into contact with water [25]. This is particularly important in the case of proteins containing low sequence complexity and a richness in residues that tend to aggregate [26]. Since the physio-chemical properties of the solvent strongly impact protein solubility, LLPS occurs as a function of parameters such as osmolality, ionic strength, pH, or temperature [2,27–29].

Another key factor underlying LLPS is multivalence, i.e., the availability of many different binding sites in the molecule. Multivalent proteins can form heterologous electrostatic interactions with different, oppositely-charged proteins or homologous interactions between their repetitive domains [30]. Multivalent proteins have a critical phase separation threshold that is often related to the number of domains it contains and the availability of ligands [25]. Multivalence is especially characteristic for IDPs; therefore, this class of proteins is often involved in promoting phase separation. IDPs do not fold into unique, three-dimensional globular structures under physiological conditions. Changes in the cellular environment and conformational properties allow IDPs to take on numerous conformations induced by the attachment of ligands, binding to the membrane surface, or various types of post-translational modifications [31]. NMR analyses of IDPs after LLPS show that disordered regions of proteins retain conformational flexibility in the condensed phase [27]. Interestingly, IDPs can also form complexes with other macromolecules or metal ions and consequently undergo, at least in fragments, disordered-to-ordered transitions [32]. IDPs with an inherent propensity for LLPS affect various cellular functions, such as signaling, cell division, intracellular transport, cell cycle control, and regulation of transcription and translation. Unfortunately, in some cases, structural features of IDPs can promote the formation of abnormal conformations prone to aggregation, which in turn causes severe diseases associated with protein misfolding, such as Alzheimer's, Parkinson's, or Huntington's disease [33]. Interestingly, not all fibrous structures cause disease. Amyloid aggregation of a large number of IDPs is associated with the biogenesis of functional amyloids, which positively influence various biological functions, e.g., melanin pigment formation, bacterial biofilm formation, or biominerals [34].

Recently, it was shown that divalent cations also have the ability to tune protein phase behaviour. However, it remains a largely unexplored area. The first report describing LLPS of proteins in the presence of divalent cations comes from 2020. Singh et al. showed that LLPS of tau protein is modulated by zinc ions, which strongly enhance the propensity for tau to undergo LLPS by lowering the critical concentration of protein [35]. Surprisingly, none of the other divalent metal ions tested (manganese(II), iron(II), cobalt(II), nickel(II), and copper(II)) were found to promote the phase separation of tau. However, the mechanism by which zinc ions promote LLPS of tau is not known. Singh et al. suggested that local folding of tau, resulting from zinc binding, could cause an increased density of positive and negative charges within particular regions. This, in turn, would lead to stronger attractive intermolecular interactions, facilitating LLPS [35]. Another proposed theory is that zinc ions promote LLPS of tau by facilitating the formation of transient intermolecular cross-links between protein molecules [35]. However, these suggestions need to be further studied.

Divalent cations can modulate phase transitions both directly and indirectly through interactions with other proteins. EF-hand domain protein 2 (EFhd2) is a conserved calcium-binding protein. It is expressed in various tissues but predominantly in the central nervous system [36,37]. EFhd2 has been found to be associated with tau aggregates in the mouse model, JNPL3, and as a tau-associated protein in Alzheimer's diseased brains [36,38]. Recent studies have shown that EFhd2 modulates the phase transition of tau and directly alters tau liquid phase behaviour to form solid-like structures in vitro, and this phenomenon is controlled by calcium ions [39]. Notably, both EFhd2 and tau, in the absence of calcium ions, lead to the formation of solid-like structures containing both. On the other hand, in the presence of calcium ions, EFhd2 and tau phase separate together into liquid droplets [39].

Divalent cations can also modulate the LLPS of transcription factors. It was shown that zinc and copper(II) ions induce LLPS of the F region of the *Aedes aegypti* ecdysteroid receptor [40]. Since this region seems to affect the dimerization of nuclear receptors, the interactions with other proteins, and the stabilization of ligand binding, LLPS of the ecdysteroid receptor might contribute to the regulation of transcriptional activation.

Protein interactions driving LLPS may vary depending on the nature of the amino acid sequence [41]. Proteins are polyelectrolytes that can have both positive and negative charges [42]. Well-described examples of LLPS (often referred to as coacervation) are those based on interactions between polycationic proteins and polyanionic RNA molecules [43]. Less is known concerning the ability of polyanionic proteins to undergo LLPS in a similar charge-dependent manner. Mayfield et al. identified a previously unknown mechanism of calcium-dependent LLPS occurring within the endoplasmic/sarcoplasmic reticulum (ER/SR) that explains efficient calcium ion buffering and storage [44]. It was shown that calcium ions modulate LLPS of the polyanionic protein and major calcium-binding protein of the SR of skeletal muscle, calsequestrin-1 (CASQ1). The process was reversible and occurred within cells. CASQ1 is an IDP that influences its capacity for LLPS. It was also shown that the LLPS of CASQ1 is regulated via phosphorylation by the secretory pathway kinase Fam20C, which phosphorylates structurally-conserved regions of CASQ1 [44]. Thus, the phosphorylated protein (pCASQ1) more readily entered the LLPS state. Mayfield et al. [44] suggested that this likely arises from the increased disorder and conformational flexibility of pCASQ1. Additionally, they hypothesized that calcium-dependent LLPS of polyanionic IDPs is a widespread and evolutionarily-conserved phenomenon that might represent a major mechanism underlying calcium ion handling and signaling [44].

## 3. Intrinsically Disordered Proteins in Biomineralization

Although biominerals are built on a scaffold of collagenous proteins, IDPs are the most prominent regulatory effector of the process. IDPs, with no defined secondary structure, act as specific regulators of biomineral formation by influencing their nucleation and orienting their growth and controlling polymorph selection (Figure 1). It is estimated that proteins involved in biomineralization processes in humans and other organisms (e.g., fish, mollusks, and diatoms), as well as in the formation of eggshells, have a high average disorder level of 53% [22,45].

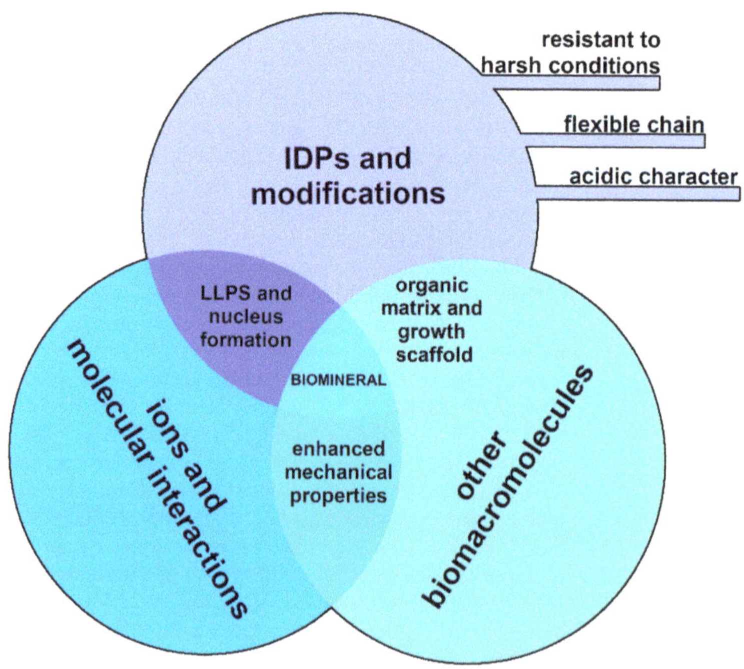


**Figure 1.** A conceptual diagram of the biomineral formation constituents and the relationships between them.

The basis for the biomineralization activity of IDPs lies in their ability to interact with their inorganic mineral counterparts. Proteins involved in biomineralization can be incorporated into biominerals at the inter- and intracrystalline levels [46]. The protein–mineral interphase, in the case of IDPs, is mainly governed by electrostatic interactions resulting from the characteristic composition of the protein primary sequence and post-translational modifications [47]. IDPs show an increased frequency of acidic amino acids, such as Asp, Glu, and commonly phosphorylated Ser, compared to the average in the SwissProt database [48]. The first stages of calcium carbonate biomineral nuclei formation involve the uptake of calcium ions from the environment by negatively-charged amino acid side residues. In this way, IDPs form a specific link between the organic matrix and the inorganic part of the biominerals. This linkage is further enhanced by the frequent phosphorylation of biomineralization-related IDPs [47].

The composite character of all biominerals is their undeniable advantage. The small presence (up to 5% *w/w*) of organic matrices often containing hundreds of types of proteins affects their mechanical properties [49,50]. In nacre, the improvement of mechanical properties, such as high resilience, ductility, and energy dissipation, is attributed to the presence of a thin layer of organic components that counterbalances crack propagation [48,51]. An in vitro study on the effect of disordered Otolith Matrix Macromolecule 64 (OMM-64) on the biomineralization of calcium carbonate also confirmed that IDPs enhanced the mechanical properties of crystals. Compared to non-bionic calcium carbonate, biominerals with embedded OMM-64 protein showed improved properties, such as greater flexibility, as determined by atomic force microscopy (AFM) [52]. The ability to modulate crystal growth, morphology, or mechanical properties is associated with intra- and intercrystallite interactions at the IDP-mineral level. These interactions are mediated by electrostatic interactions between inorganic ions and side residues of acidic amino groups or functional groups [47]. The increased participation of acidic amino acids in the IDP sequence is likely to mediate protein incorporation into the crystal structure, as in the case of Asp, which was shown to replace bicarbonate ions with its carboxyl group in biominerals [53].

In addition to the characteristic features of the amino acid residue chain, biomineralizing IDPs also exhibits a number of post-translational modifications, including extensive phosphorylation and glycosylation. One well-studied glycosylated IDP involved in the sea urchin biomineralization process is spicule matrix protein (SM30) and its various isoforms [54,55]. Extensive phosphorylation on the other hand was confirmed for zebrafish Starmaker (Stm) protein in a proteomic study, as residues 25 Ser and 3 Thr were identified as phosphorylated. Notably, this protein was the only one in this study to have this type of modification [50].

The attachment of phosphate groups to biomineralization-related IDPs influences their molecular properties, changing their isoelectric point and making them even more acidic [56]. This modification increases the ability to bind calcium ions more effectively. The comparative studies on Stm protein mineralization activity with and without phosphorylation indicates that phosphorylated Stm reduces the size and influences the morphology of the crystals that are greater than the non-modified version of Stm [57].

In in vitro studies, casein kinase 2 is typically used for phosphorylation [57], nonetheless the proteomic studies identified Fam20C kinase as an important biomineralization player [49,58]. Fam20C was confirmed to phosphorylate small integrin-binding ligand and N-linked glycoproteins (SIBLINGs) in bone, and the mutations of Fam20C result in bone dysfunction [58]. The same kinase was identified in fish otoliths [49] and the shells of *Pinctada fucata* oysters, where its expression is enhanced in the shell repair process [59].

The modifications of biomineralization-related proteins, both phosphorylation and glycosylation, alter the electrostatic interactions with inorganic counterparts of the biominerals. However, to define in one sentence the role of glycans in biomineralization-related proteins is beyond the means of research and nature, given the possible variations and heterogeneity of this modification [55]. In the case of recombinant aragonite protein 24 (AP24) nacre protein, the glycosylated AP24 influences the directions of crystal growth and inhibits the nucleation process, while non-glycosylated proteins stabilize the mineral phase more efficiently. At the same time, the lack of modification does not influence the hydrogels formation activity [60]. In the case of SpSM30B/C, the influence on biomineralization remains similar in both glycosylated and non-glycosylated variants; nonetheless, the non-glycosylated SpSM30B/C does not form aggregates effectively, contrary to the glycosylated version [61].

## 4. The Role of Proteins in Formation of Calcium Carbonate

To illustrate the importance of individual IDPs in the biomineralization of calcium carbonate, it is worth reviewing the functions of specific proteins among different taxa. One of the most widely-understood proteins from molluscan shells is aspein. This protein consists of four regions, including a signal peptide (cleaved before the protein is released to the extracellular matrix), an uncharged N-terminal region, and two domains enriched in Asp, i.e., a DA and poly-D domain. The Asp-rich domains are crucial for the regulation of biomineral formation [62]. Aspein promotes calcite formation over aragonite. It was shown that the protein binds to calcium ions with higher affinity than it binds to magnesium ions, thus increasing the local concentration of calcium ions and inducing calcite formation [62,63]. Additionally, the Asp-rich domains are not conserved between species, suggesting that sequence specificity is not a determinant of aspein's biomineralization activity [64].

In crustaceans, such as crayfish, one of the widely described-proteins is acidic calcification-associated peptide-1 (CAP-1), which is involved in the mineralization of the exoskeleton. CAP-1 has a dual structure: the N-terminal part is most likely responsible for maintaining the conformation of the protein, while the C-terminal part is associated with calcium carbonate mineralization [65]. The C-terminal part possesses calcium ion binding ability, which in turn also increases their local concentration. Moreover, the protein exhibits carbonate growth inhibitory properties, which are further enhanced by the presence of phosphoserine in the C-terminal part [66].

In stony corals, coral acid-rich proteins (CARPs) are a group of proteins involved in calcium carbonate precipitation. To date, the characterized proteins of this group are highly acidic and possess calcium-binding domains [61]. Their N-terminal region, which is extremely rich in acidic amino acid residues, is directly involved in calcium carbonate formation. Recombinant N-terminal fragments of CARP-1 and -2, as well as full-length CARP-3 and -4, are able to spontaneously precipitate calcium carbonate in an experiment with artificial seawater [67]. In addition, the ability to control crystal polymorphs was confirmed for the CARP-3 protein [68].

The aforementioned SM30 is a sea urchin spicule protein that exists in multiple isoforms. It is an acidic, glycosylated protein with a C-type lectin domain at the N-terminus and a disordered C-terminal region [69,70]. Studies on the B/C isoform hybrid of the SM30 protein from *Strongylocentrotus purpuratus* show that SM30 is capable of aggregating and forming a hydrogel that controls the biomineralization process by initially stabilizing ACC, thereby forming single crystals of calcite and promoting directed crystal growth [61].

Nonetheless, in our laboratory, we have extensively studied proteins related to the biomineralization of human otoconia and fish otoliths. These calcium carbonate biominerals located in semicircular canals of the inner ear are responsible for the sense of gravity, balance, the perception of linear and angular acceleration, pressure changes, and sound vibrations [71]. Otoconia, as in *Homo sapiens*, are similar to sand suspended in the gel mass of utricles and saccules, while fish otoliths resemble stones placed in three otolithic end organs: saccules, lagenas, and utricles. Otoconia remain inert during their lifespan, while otoliths grow diurnally, accumulating new layers of calcium carbonate and organic matrix [72–74]. The mineralization process takes place in an acellular medium, the endolymph, which is a fluid rich in structural materials such as calcium, proteins, and other macromolecules [75]. The organic matrix of human otoconia and zebrafish otoliths constitutes up to 5% of the biomineral mass [76,77]. Although the percentage content of the organic matrix is quite low, proteins, such as OMM-64, Stm, Starmaker-like (Stm-l), otolin 1, and otolith matrix protein-1 (OMP-1; the orthologue of mammalian otoconin 90 (Oc90)-Otoc1), are required for normal otolith growth [78–82].

Stm from zebrafish was the first protein found to be capable of controlling the process of calcium carbonate biomineralization. Stm controls the size, shape, and polymorph of the mineral component of the otolith [83]. Stm acts as an inhibitor of crystal growth; the decrease in crystal size depends on the protein concentration. The ability of Stm to act as a crystal nucleation factor and inhibitor of crystal growth is directly related to its degree of phosphorylation, which adds a negative charge and increases the binding affinity to calcium ions [57].

Stm-l, an IDP from medaka (*Oryzias latipes*), is able to adopt a more ordered and rigid structure under the influence of the environment and has a negative effect on the size of precipitating crystals. However, the higher number of crystals formed in the presence of the protein suggested that Stm-l could also act as a crystal nucleator [81]. According to Różycka et al., in vaterite crystals, the occurrence of Stm-l is probably limited to its nucleation site, whereas in calcite, the distribution of the protein occurs throughout almost the entire crystal. The time-dependent mineralization tests allow visualization of the sequential deposition of Stm-l in forming calcite. The protein acted as a nucleator of crystal growth through the condensation and formation of intermediate phases at the early stages of the process. Then, Stm-l regulated crystal growth by adhering to step edges on calcite, which resulted in ellipsoidal to spherical shapes of crystals and a reduction in crystal/crystallite sizes [84].

Similar to Stm-l, OMM-64, a highly-acidic IDP rich in Asp and Glu residues, can undergo transitions to more ordered states [80]. Additionally, the presence of calcium ions resulted in protein compaction. In vitro biomineralization experiments showed that OMM-64 plays a biological role similar to that of Stm and Stm-l and controls both the size and the shape of calcium carbonate crystals. As shown with two-photon fluorescence experiments, the enhanced density of the protein in the central part of the crystals suggested the participation of OMM-64 in the nucleation of calcium carbonate crystals. The nucleation of crystals can be initiated by the adsorption of calcium ions exposed to OMM-64 acidic residues and their local concentration, accompanied by the collapse of the protein molecule. Hyperphosphorylation of OMM-64 strengthens the inhibitory effect of the protein in the biomineralization process [52,80].

One of the major components of the otolith matrix is otolin-1, a 48 kDa collagen-like protein [78,85–88]. This secreted otoconin is present both in the organic matrix of human otoconia and zebrafish otoliths. In zebrafish, otolin-1 can be found in the otolith itself and on the boundaries of the structure as a link between the otolith and the sensory epithelia [85]. Otolin-1 is composed of four domains: a 23-amino acid signal peptide, a non-collagenous N-terminal domain, a central collagen-like domain, and a globular C-terminal C1q domain liable for protein molecule trimerization [89]. The presence of calcium ions influences the secondary and tertiary structure of recombinant otolin-1, especially the thermal stability [90]. Recombinant human and zebrafish otolin-1 forms high-order oligomers [90]. The oligomerization of zebrafish protein is dependent on its concentration and the presence of calcium ions, whereas human protein exhibits the same oligomeric stage regardless of these factors. Despite the high sequence similarity (45.51% identity and 56.58% similarity), these two homologues show differences that may be reflected in the nature of otoliths and otoconia [90]. Otolin-1 could be the crucial element of the organic matrix of otoconia and otoliths, serving as a high-order oligomeric scaffolding protein stabilized by calcium ions [90].

The process of otoconia and otolith growth involves a series of temporally- and spatially-specific events that are tightly coordinated by numerous proteins [73]. Otolin-1 and OMM-64 extracted from rainbow trout (*Oncorhynchus mykiss*) otoliths led to the formation of aragonite crystals in an in vitro biomineralization assay, while otolin-1 and OMM-64 separately induced small calcite and vaterite crystals, respectively [79]. Similarly, recombinant murine otolin-1 influenced the size and shape of the obtained crystals, but the effect was enhanced by Oc90 [91]. Interestingly, it has been shown that another IDP, dentin matrix protein 1 (DMP1), which is an extracellular matrix protein essential for the biomineralization of calcium phosphate in bone and dentin [92], was present in the inner ear, specifically in otoconia [93]. Later studies indicated that the 57K fragment of DMP1 [94] formed oligomers in the presence of calcium ions and affected the morphology of calcium carbonate crystals in vitro. These studies suggest that DMP1 shows a previously unknown regulatory function for the biomineralization of otoconia [95].

## 5. Calcium Carbonate Nucleation Process

The key features of biominerals, such as lattice orientation, particle size, and size distribution, are determined by the conditions prevailing during the first phase of crystal growth—nucleation. To this day, the process of nucleation of crystals from a solution remains poorly characterized mainly due to the difficulty of measurements and their interpretation at the low (atomic) level of matter organization, but also because in nature, nucleation involves a number of unknown or hardly characterizable factors. These factors include surface wettability or inhomogeneity, which affect the nucleation barrier and the nucleation rate. Another difficulty stems from the interdisciplinary nature of this phenomenon, which has caused diversity in the terminology used by researchers from different fields [20,96,97].

In general, the mechanisms aiming to explain the calcium carbonate nucleation kinetics from solutions can be divided into two distinct groups, i.e., classical nucleation theory (CNT) and alternative multistep (non-classical) pathways (Figure 2). The classical nucleation mechanism first introduced in 1878 by Gibbs [98] and developed during the past century [99–101] provides a fairly simple explanation of how crystals nucleate in homogeneous and heterogeneous pathways. This process is limited by the energy barrier resulting from the cost of generating a phase interface and, with it, the interfacial tension between the nucleus (also known as a cluster) and its surroundings. In homogenous nucleation, the process of nuclei formation is driven by the stochastic fluctuation of monomer association in the supersaturated solution, while in heterogeneous nucleation (most common in nature), the process is accelerated due to the presence of foreign molecules (including proteins), which can act as heterogeneous nuclei and reduce the free energy barrier [102]. Briefly, CNT assumes the presence of an unstable pre-critical cluster that grows by successive and reversible attachment (and detachment) of monomers to its surface that are components of the final crystal. As the cluster grows, the Gibbs free energy of the system increases, but only to the maximum value for the critical size of the nucleus and the formation of the metastable cluster. After exceeding the critical size, a stable solid form of the post-critical nucleus is formed, and the free energy is released during crystal growth. Smaller nuclei are thermodynamically unstable and dissolve again [19,96,103].

Recently, the development of advanced experimental and bioinformatics analyses has provided evidence that the formation of calcium carbonate frequently does not follow the CNT [16,17,104]. The alternative mechanisms are based on the observation of the formation of stable or metastable precursors, most likely created by the collisions and coalescence of their constituent components. These results clearly conflict with the nucleation picture presented in the CNT (Figure 2A). The presence of such individuals indicates the appearance of additional minima in the graph illustrating the Gibbs free energy of the calcium carbonate precipitation reaction (Figure 2B). This means that multistep nucleation pathways comprise more than one barrier that must be overcome, along with a number of local minima corresponding to the formation of precursors with different sizes and probably different structural arrangements [19,103,105]. In the next steps of the process, either the formation of the crystalline phase within the post-critical nucleus and subsequent crystal growth or the formation of a stable ACC may take place [16,106,107].

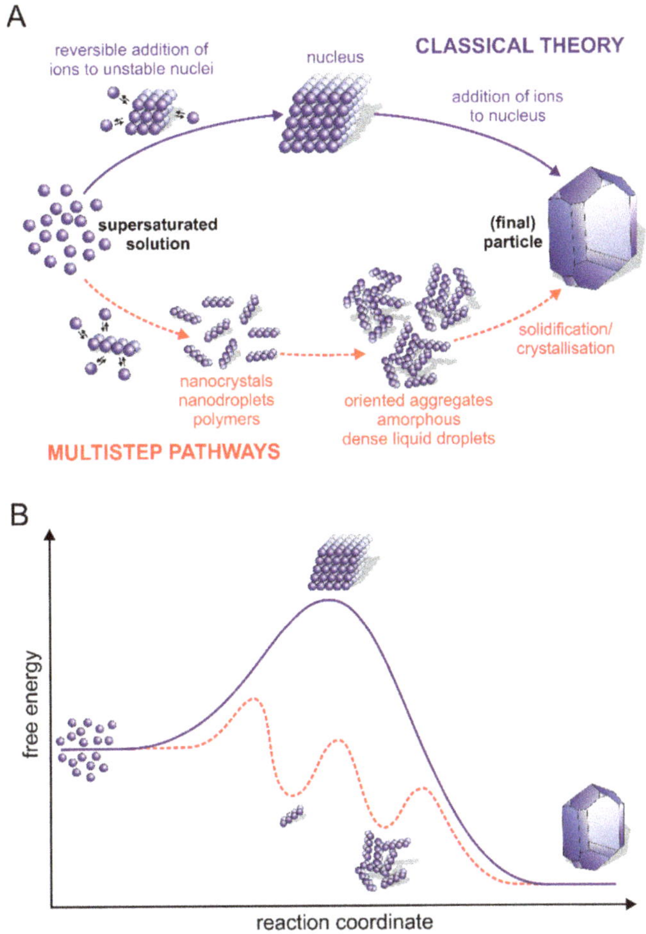


**Figure 2.** Classical and multistep nucleation pathways. (**A**) Schematic presentation of events occurring during the formation of crystals from the bulk liquid to crystalline pathways according to classical theory (blue) and multistep pathways (red). (**B**) Comparison of free energy along the pathway of crystal nucleation following the classical nucleation mechanism (blue) and multistep pathways (red) [20,108].

The concept of PILP, one of the non-classical mechanisms of calcium carbonate nucleation, was proposed in 2000 by Gower et al. [18]. They observed the formation of droplets of fluidic ACC precursor in the presence of negatively-charged polyelectrolytes during the calcium carbonate precipitation process with an in vitro model system. In the PILP pathway, nucleation is a multistep process where the polymer associates with calcium and bicarbonate ions to form an intermediate liquid phase prior to solid nucleation [109]. The liquid-like character of the early-stage amorphous precursor was evidenced by the coalescence of the droplets, which grow from tens of nanometres to a couple of microns, and by in situ AFM [18,110,111].

Based on the results of potentiometric titration and analytical ultracentrifugation (AUC) for undersaturated, saturated, and supersaturated solutions, Gebauer et al. proposed another concept that refers to prenucleation clusters (PNCs)—non-classical theory [1]. They observed that during the titration, the amount of free calcium ions was always less than the total calcium ions added, suggesting the formation of long-lived calcium carbonate clusters called PNCs. Once a critical point is reached, nucleation occurs, and the free calcium ions are consumed by the growing particles. Before nucleation, small cluster species with a hydrodynamic diameter of ~2 nm (corresponding to approximately 70 calcium and bicarbonate ions) were mostly detected in AUC, but the second-largest cluster species (hydrodynamic diameter of 4 to 6 nm) was also present, suggesting further nucleation via cluster aggregation. The proposed hypothesis was also confirmed by the fact that smaller clusters could not be detected in the post-nucleation phase [17,19]. The presence of PNCs has also been corroborated in solutions saturated with respect to calcite by cryo-TEM experiments; however, in contrast to the observations of Gebauer et al. [19], the obtained results showed that the prenucleation clusters persisted even after nucleation [112].

The amorphous precursor strategy refers to the approach by which organisms make use of the flexibility of ACC to control the kinetics of biomineral formation and the spatial distribution of the final calcium carbonate polymorphs. Despite the numerous examples of the transformation of synthetic and biogenic metastable ACC into a crystalline phase, the factors involved in the polymorph selection mechanisms as well as the effects of ACC precursors on the structural characteristics of the final products are still puzzling [113,114]. It was shown that stable hydrated ACC becomes dehydrated during transformation into the crystalline phase, which suggests that there might exist specific mechanisms involved in the stabilization, destabilization, and transformation of ACC involving some proteins and other ions [115]. It has been proposed that the formation of ACC in the precursor phase causes the lowering of interfacial free energy during the formation of crystalline phases [116].

Interestingly, recent studies increasingly use LLPS to explain the behaviour of calcium carbonate-containing solutions in the context of the mineralization mechanism [20]. Due to problems in determining the thermodynamics of the transient clusters that form during the nucleation process, molecular dynamics simulations were used to investigate the initial stability of the clusters with respect to the composition of the solution and formation pathway. Initially, high ion concentrations were used for the bioinformatics simulations to increase the frequency of ion association and to facilitate obtaining a cross-section of thermodynamic changes over time. Then, the clusters formed were transferred to a lower concentrated environment to demonstrate their stability under different conditions. It was found that the earliest formed clusters adopted chain, ring, and low-density branched structures. At high concentrations, growth occurred at the diffusion limit, with barriers opposing ion attachment with ambient thermal energy [117,118]. Low-density configurations were observed for small clusters. However, such configurations were quickly replaced by more condensed states with ion additives [104].

The dynamic nature of the clusters was quantitatively defined by the ion diffusivity components. The dependence of the diffusion coefficient of calcium ions within the clusters at different growth stages on the two solid phases of calcium carbonate, calcite and ACC, in several solvents has been described [119]. The ion diffusion properties were analysed in different solvents in the bulk ACC and calcite, indicating that the clusters are droplets of the dense, ion-rich hydrated calcium carbonate liquid phase. The ion diffusivity decreased with the increasing density of the liquid phase, but the rate of diffusion gradually decreased and approached a constant value characteristic of the depleted liquid phase. The lack of an energy barrier for cluster formation is characteristic of solutions that have passed their stability limit and undergo spontaneous phase separation via the spinoid pathway. The availability of the spinodal region, at low concentrations, is important for the mineralization process (Figure 3). Thermodynamically, this means that there is a line of liquid–liquid coexistence between the dense and depleted solution phases. Both liquids are in metastable equilibrium with respect to the calcium carbonate solid phases over a wide range of dissolution conditions (Figure 3). The solubility of all polymorphs is represented by a single solubility line (SL) that bounds the dashed unsaturated solution field. This representation highlights that all calcium carbonate solid phases (calcite, aragonite, vaterite, and presumably ACC) show the same general retrograde solubility behaviour. The nucleation of the solid phases progresses towards a high concentration on the dark blue line side of the liquid–liquid coexistence (L-L). The field between the binodal and spinodal lines bounded by the L-L line indicates the conditions under which nucleation of a dense liquid phase is possible. In the region bounded by the spinoid line, the solution is unstable, and liquid–liquid separation occurs spontaneously (Figure 3) [104].

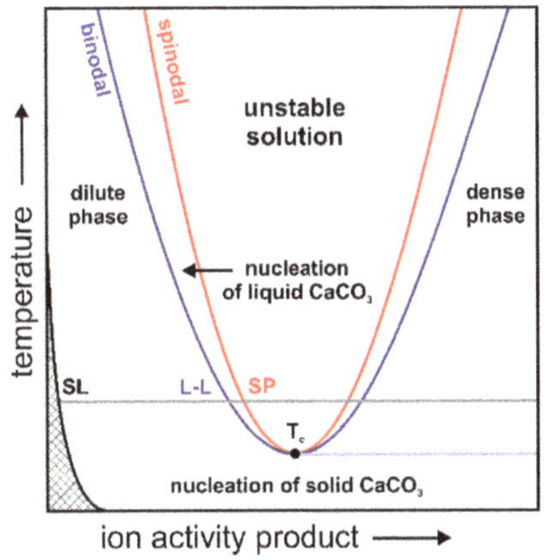


**Figure 3.** Schematic representation of the calcium carbonate phase relationships. The grey line represents an example solubility line (SL) of different polymorphs at constant temperature, the grey, checkered field corresponds to the unsaturated solution field, the L-L line represents the binodal (blue) and SP spinodal (red) lines, and $T_c$ is the critical temperature point [120].

The existence of a dense liquid phase makes it possible to apply both the classical model of ion crystallization and a model related to phase separation via clusters. Increasing the product of ion activity and the liquid–liquid coexistence line encounter makes homogeneous nucleation of a dense liquid phase possible. The formation of a dense liquid in a short time is more likely than direct crystallization because the excess of free energy at the solution–liquid interface is greatly reduced compared to the solution–crystal interface, thereby resulting in a lower thermodynamic barrier for liquid–liquid separation than for crystallization [104].

## 6. Liquid–Liquid Phase Separation in the Formation of Hard Tissue

Recently, research attention has been focused on assembly processes, including the formation of mineralized components of the body via LLPS. Faatz et al. were one of the first to show that spherical particles of ACC can be formed by LLPS [121]. Additionally, Wolf et al. demonstrated that ACC can grow in the absence of organic polymers [122]. In other studies, inorganic salts were shown to undergo LLPS at high temperatures [123,124]. In experiments in the presence of a poly-Asp additive, the data indicate that the polypeptide stabilizes a condensed phase of liquid-like droplets of calcium carbonate during PILP formation [109]. However, since liquid precursors can be detected in samples without any polymer additives, PILP may be considered a polymer-stabilized rather than polymer-induced state [109,122]. Therefore, an important question arises regarding the role of IDPs during the formation of biominerals: are IDPs inducers or modulators of the process? The available literature does not include many examples discussing their role in the formation of liquid-phase condensates and organic calcium carbonate components of the body. Recent studies on a nacre-like, aragonite-forming protein, Pif80, from *Pinctada funcata*, indicate that it has the ability to drive LLPS. Pif80 is a functional fragment of a Pif protein originating from proteolytic cleavage [125]. The protein has a high content of acidic residues; moreover, it exhibits a high degree of intrinsic disorder [126]. Bahn et al. showed that recombinant Pif80 (rPif80) underwent LLPS in the presence of calcium ions, thereby forming a dense protein-rich phase [127]. Pif80-containing liquid condensates occurred in solutions containing either chloride or bicarbonate ions as counter-ions, and the process was only mildly influenced by pH. These behaviours support the idea that electrostatic interaction is the major driving force for LLPS of Pif80. The mineralization experiments performed in conditions that allow for LLPS revealed the scenario in which Pif80 condensates and PILP-like calcium carbonate granules can coexist. Importantly, the PILP-like calcium carbonate granules contained an amorphous phase of the salt. Based on that, the authors suggested that the Pif80 condensates worked as stabilizing agents of PILP-like ACC inhibiting the growth of calcite. It is worth emphasizing that Pif80 was the first matrix protein for which the ability to undergo LLPS was shown [127].

As already mentioned, proteins capable of undergoing LLPS can be classified into four types: scaffolds (drivers), co-scaffolds (co-drivers), clients, and regulators [3,128]. This classification might reflect some functions of molecules (other than proteins) in biomineralization. Scaffolds are essential constituents of each condensate and are responsible for its integrity. In biomineralization, the role of scaffolds is assigned to collagen and chitin. On the other hand, co-scaffolds are components that need another co-scaffold to phase separate [129]. It is known that acidic IDPs in the organic matrix regulate the stability and polymorph selection of calcium carbonate at the molecular level [130]. They can induce nucleation, adsorb specifically onto some crystal faces, and/or intercalate in a controlled manner into the crystal lattice [84,131]. Some of them (e.g., rPif80) need an inorganic fraction (e.g., calcium ions) for LLPS [127]. Clients are dispensable components and reside in condensates only under certain conditions. This role might be assigned to carbonic anhydrases or calcium-binding proteins, for example [132]. The last type consists of molecules called regulators, which promote LLPS but are not located in the condensates (e.g., modifying enzymes) [128]. Many proteins involved in biomineralization are extensively post-translationally modified (e.g., phosphorylated, glycosylated, and proteolytically cleaved). For modifying enzymes, the regulatory role should be assigned [133,134].

## 7. Can LLPS Impact the Formation of Otoliths?

In our laboratory, we have been studying the molecular properties of proteins involved in the mineralization of fish otoliths and human otoconia for years. Similar to Pif80, proteins found in fish otoliths are negatively-charged polyampholytes (Figure 4) that can bind calcium ions. At present, there are no data in the literature indicating the ability of otolithic proteins to drive LLPS. Therefore, we performed in silico analyses to test the probability that LLPS is induced by OMM-64 protein from *Oncorhynchus mykiss;* Stm from *Danio rerio;* the Stm orthologue, Stm-l, from *Oryzias latipes;* and otolin-1 from *Danio rerio* (Figure 5).

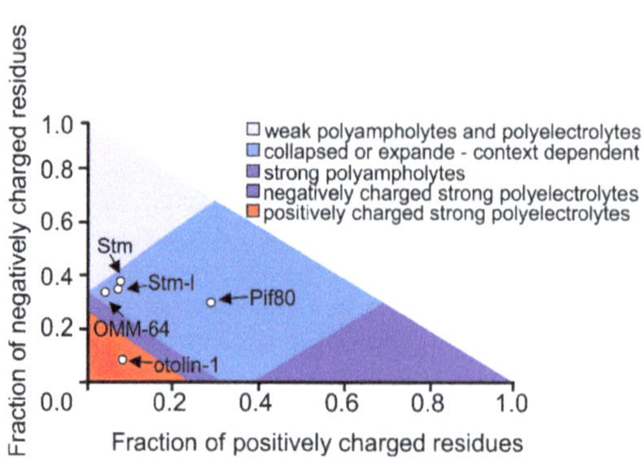


**Figure 4.** Phase diagram of proteins involved in the formation of fish otoliths. For the CIDER analysis [135], the following sequences were used: OMM-64 A0A060XQP6, Stm A2VD23, Stm-l A0A3B3H599, otolin-1 A5PN28, and Pif80 C7G0B5 (544-1007 aa residues [121]). [Accessed on 11 July 2022].

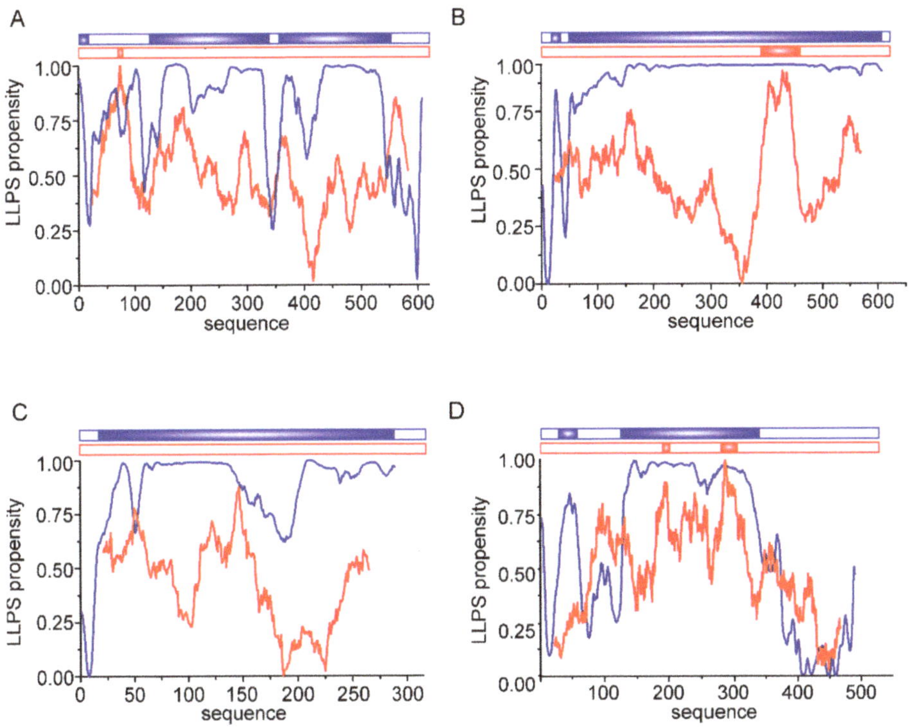


**Figure 5.** In silico analysis of otolithic proteins. The graph demonstrates the normalized results of in silico analysis of representative otolithic proteins: (**A**) OMM-64, (**B**) Stm (**C**), Stm-l, and (**D**) otolin-1 performed using FuzDrop (blue) and PScore (red). The protein regions that obtained positive scores are indicated as a scheme at the top of each panel. [Accessed on 12 May 2022].

At present, there are several computational tools for predicting the LLPS propensity of a given protein [136]. Each tool has a different approach and takes into consideration various characteristics, thereby enabling interactions that facilitate the formation of liquid condensates. The integrity of liquid condensates is maintained by weak and transient interactions between complementary binding partner regions. The possible interaction modes include charge–charge, cation-$\pi$, dipole–dipole, or stacking $\pi$-$\pi$ matches of chemical groups [5]. For the in silico analysis of otolithic proteins, two predictors, FuzzDrop [137–139] and PScore [140], were chosen. The FuzzDrop algorithm is based on a model in which interactions within condensates are maintained by multivalent interactions between disordered regions, and a high score is given to regions that are unlikely to become ordered upon binding. This determines the local sequence bias with respect to composition, hydrophobicity, and structural disorder by examining a large number of possible sequence contexts [139]. On the other hand, the PScore approach gives the prediction of LLPS propensity based on the calculated likelihood of IDRs forming long-range planar $\pi$–$\pi$ contacts [140].

According to the FuzzDrop predictor, the analysed otolith proteins involved in the formation of calcium carbonate biominerals showed a very high probability of driving LLPS. The calculated propensity for LLPS (pLLPS) equals 0.9991 for OMM-64, 0.9953 for Stm, 0.9946 for Stm-l, and 0.9969 for otolin-1 on a 0–1 scale. Such high values of pLLPS indicate that they may spontaneously undergo LLPS. Moreover, they may function as condensate drivers (pLLPS $\geq$ 0.6) [139]. The residue-based analysis revealed that except for otolin-1, almost all of the studied protein sequences had a high probability for LLPS. OMM-64, Stm, and Stm-l may be considered disordered polyelectrolytes [80,81,141], whereas collagen-like otolin-1 contains two externally-ordered domains. The central fragment of the protein possesses collagenous structures [85], while on its C-termini, a calcium ion-dependent C1q domain is present [89]. The C1q domains show no probable ability to drive LLPS, in contrast to the central region. To our knowledge, at present, there are no reports on the ability of collagenous proteins to drive LLPS. All four analysed proteins also contain regions with a potential tendency for context-dependent interactions. Such regions are able to adopt various binding modes, depending on their binding partner [142]. These regions may also contribute to the formation of an interaction network between proteins determining otolith morphology and material properties. The analysed proteins also contain some aggregation hotspots (not shown). In particular, OMM-64 contains several regions with a tendency to aggregate. In otoliths, it accumulates in ring-like structures. Tohse et al. found OMM-64 in high-molecular-weight aggregates (HMWAs) of the otolith matrix [79]. It is possible that the selected regions may be involved in the interaction leading to aggregation via the liquid-to-solid transition.

Otolith proteins were also analysed by the PScore predictor. This program predicts the likelihood of IDRs driving LLPS based on the propensity for long-range planar pi–pi contacts. This type of interaction is typically linked to the interactions between aromatic rings. Analysed herein, proteins are depleted in aromatic residues, but the rationale for using this predictor was the fact that orbitals of bonded sp2-hybridized atoms are present in other chemical groups of proteins, including amide groups, carboxyl, and guanidinium [140]. Therefore, proteins containing small residues allowing an exposition of the protein backbone are quite likely to form these contacts. Examples exist in which LLPS-driving regions are those that contain repeats enriched with small residues such as Gly-Pro or Gly-Arg residues [5,140]. The otolithic proteins analysed herein are rich in Gly and Pro residues, but as presented in Figure 4, OMM-64, Stm, and otolin-1 contain only short fragments, which obtained positive results in the PScore analysis. The Stm-l protein has no such region.

To summarize, according to our in silico analysis, otolithic proteins may have the potential to drive LLPS. Considering what we have recently learned about the importance of spontaneous LLPS in biological systems, it is likely that otolithic IDPs also drive the formation of dense condensates. At present, however, the significance of that potential ability remains to be solved.

## 8. Conclusions

Biomineralization leads to the formation of stiff components of the body that function as structures where inorganic salts form crystals, which are incorporated into the complex organic mesh. The presence of an organic matrix in biominerals influences more than just its material properties. Organic compounds, among which IDPs play a major role, may induce nucleation, function as regulators of the gross volume of the biomineral, and determine the pattern of growth of the mineral phase. Although studied for years, the mechanisms by which organic components play a role in nucleation and growth in the formation of mineral bodily components remain under debate. Undoubtedly, a better understanding of this process holds promise for a variety of fields, including drug and cell-therapy engineering, cancer/tumor target engineering, bone tissue engineering, and other advanced biomedical engineering [143]. Organic compounds that could influence the shape, size, and properties of biominerals could be used to induce the formation of biominerals with improved and strictly desired properties. The presence of organic molecules can also affect the incorporation of contaminating metal by substituting calcium ions in calcite. The application of this approach is, for example, promising for the remediation of toxic or radioactive metals in environments where calcite is stable over the long term [144]. Moreover, since calcium carbonate is abundant in the oceans, as many organisms use it to produce protective shell structures or skeletal elements, a better understanding of biomineralization pathways may be important for environmental and climate changing studies [145].

LLPS seems to be a widespread mechanism for the supramolecular organization of molecules. It often facilitates the assembly of proteins, both intra- and extracellularly. It is a thermodynamically driven process that guarantees the harmony of intracellular processes and likely extracellular processes as well, including the formation of mineralized components of the body. Notably, LLPS has only recently been appreciated in biomineralization studies. At present, it appears that only the tip of the iceberg has been discovered in that regard, and more fascinating discoveries will come.

**Author Contributions:** Conceptualization, A.T., K.S., P.D. and M.R.; writing—original draft preparation, A.T., K.B., A.Z., K.S., P.D. and M.R.; writing—review and editing, A.T., K.B., A.Z., K.S., P.D., A.O. and M.R.; visualization, A.T., K.B. and M.R.; formal analysis, A.T. All authors have read and agreed to the published version of the manuscript.

**Funding:** This work was supported by the Polish National Science Center (UMO-2020/39/B/ST10/01253) and, in part, by a statutory activity subsidy from the Polish Ministry of Education and Science for the Faculty of Chemistry of Wroclaw University of Science and Technology.

**Acknowledgments:** Taking this opportunity, we wish Vladimir all the best! Fruitful experiments and even more discoveries that are fascinating.

**Conflicts of Interest:** The authors declare no conflict of interest. The funders had no role in the design of the study; in the collection, analyses, or interpretation of data; in the writing of the manuscript; or in the decision to publish the results.

## References

1. Gomes, E.; Shorter, J. The Molecular Language of Membraneless Organelles. *J. Biol. Chem.* **2019**, *294*, 7115–7127. [CrossRef] [PubMed]
2. Hyman, A.A.; Weber, C.A.; Jülicher, F. Liquid-Liquid Phase Separation in Biology. *Annu. Rev. Cell Dev. Biol.* **2014**, *30*, 39–58. [CrossRef] [PubMed]
3. Banani, S.F.; Rice, A.M.; Peeples, W.B.; Lin, Y.; Jain, S.; Parker, R.; Rosen, M.K. Compositional Control of Phase-Separated Cellular Bodies. *Cell* **2016**, *166*, 651–663. [CrossRef]
4. Shin, Y.; Brangwynne, C.P. Liquid Phase Condensation in Cell Physiology and Disease. *Science* **2017**, *357*, eaaf4382. [CrossRef] [PubMed]
5. Brangwynne, C.P.; Tompa, P.; Pappu, R.V. Polymer Physics of Intracellular Phase Transitions. *Nat. Phys.* **2015**, *11*, 899–904. [CrossRef]
6. Posey, A.E.; Holehouse, A.S.; Pappu, R.V. Phase Separation of Intrinsically Disordered Proteins. In *Methods in Enzymology*; Academic Press: Cambridge, MA, USA, 2018; Volume 611, pp. 1–30. ISBN 9780128156490. [CrossRef]
7. Banani, S.F.; Lee, H.O.; Hyman, A.A.; Rosen, M.K. Biomolecular Condensates: Organizers of Cellular Biochemistry. *Nat. Rev. Mol. Cell Biol.* **2017**, *18*, 285–298. [CrossRef]
8. Ditlev, J.A.; Case, L.B.; Rosen, M.K. Who's In and Who's Out—Compositional Control of Biomolecular Condensates. *J. Mol. Biol.* **2018**, *430*, 4666–4684. [CrossRef]
9. Brangwynne, C.P.; Eckmann, C.R.; Courson, D.S.; Rybarska, A.; Hoege, C.; Gharakhani, J.; Jülicher, F.; Hyman, A.A. Germline P Granules Are Liquid Droplets That Localize by Controlled Dissolution/Condensation. *Science* **2009**, *324*, 1729–1732. [CrossRef]
10. Azaldegui, C.A.; Vecchiarelli, A.G.; Biteen, J.S. The Emergence of Phase Separation as an Organizing Principle in Bacteria. *Biophys. J.* **2021**, *120*, 1123–1138. [CrossRef]
11. Sołtys, K.; Tarczewska, A.; Bystranowska, D.; Sozańska, N. Getting Closer to Decrypting the Phase Transitions of Bacterial Biomolecules. *Biomolecules* **2022**, *12*, 907. [CrossRef] [PubMed]
12. Brocca, S.; Grandori, R.; Longhi, S.; Uversky, V. Liquid-Liquid Phase Separation by Intrinsically Disordered Protein Regions of Viruses: Roles in Viral Life Cycle and Control of Virus-Host Interactions. *Int. J. Mol. Sci.* **2020**, *21*, 9045. [CrossRef] [PubMed]
13. Chiu, Y.P.; Sun, Y.C.; Qiu, D.C.; Lin, Y.H.; Chen, Y.Q.; Kuo, J.C.; Huang, J.R. Liquid-Liquid Phase Separation and Extracellular Multivalent Interactions in the Tale of Galectin-3. *Nat. Commun.* **2020**, *11*, 1229. [CrossRef] [PubMed]
14. Boskey, A.L. Biomineralization: An Overview. *Connect. Tissue Res.* **2003**, *44* (Suppl. 1), 5–9. [CrossRef] [PubMed]
15. Hołubowicz, R.; Porębska, A.; Poznar, M.; Różycka, M.; Dobryszycki, P. Biomineralization–Precision of Shape, Structure and Properties Controlled by Proteins. *Postepy Biochem.* **2015**, *61*, 364–380.
16. Demichelis, R.; Raiteri, P.; Gale, J.D.; Quigley, D.; Gebauer, D. Stable Prenucleation Mineral Clusters Are Liquid-like Ionic Polymers. *Nat. Commun.* **2011**, *2*, 590. [CrossRef]
17. Gebauer, D.; Cölfen, H. Prenucleation Clusters and Non-Classical Nucleation. *Nano Today* **2011**, *6*, 564–584. [CrossRef]
18. Gower, L.B.; Odom, D.J. Deposition of Calcium Carbonate Films by a Polymer-Induced Liquid-Precursor (PILP) Process. *J. Cryst. Growth* **2000**, *210*, 719–734. [CrossRef]
19. Gebauer, D.; Volkel, A.; Colfen, H. Stable Prenucleation Calcium Carbonate Clusters. *Science* **2008**, *332*, 1819–1822. [CrossRef]

20. Qin, D.; He, Z.; Li, P.; Zhang, S. Liquid-Liquid Phase Separation in Nucleation Process of Biomineralization. *Front. Chem.* **2022**, *10*, 834503. [CrossRef]
21. Avaro, J.T.; Wolf, S.L.P.; Hauser, K.; Gebauer, D. Stable Prenucleation Calcium Carbonate Clusters Define Liquid–Liquid Phase Separation. *Angew. Chem. Int. Ed.* **2020**, *59*, 6155–6159. [CrossRef] [PubMed]
22. Boskey, A.L.; Villarreal-Ramirez, E. Intrinsically Disordered Proteins and Biomineralization. *Matrix Biol.* **2016**, *52–54*, 43–59. [CrossRef] [PubMed]
23. Brangwynne, C.P.; Mitchison, T.J.; Hyman, A.A. Active Liquid-like Behavior of Nucleoli Determines Their Size and Shape in Xenopus Laevis Oocytes. *Proc. Natl. Acad. Sci. USA* **2011**, *108*, 4334–4339. [CrossRef]
24. Zaslavsky, B.Y.; Uversky, V.N. In Aqua Veritas: The Indispensable yet Mostly Ignored Role of Water in Phase Separation and Membrane-Less Organelles. *Biochemistry* **2018**, *57*, 2437–2451. [CrossRef]
25. Alberti, S. Phase Separation in Biology. *Curr. Biol.* **2017**, *27*, 1097–1102. [CrossRef]
26. Martin, E.W.; Mittag, T. Relationship of Sequence and Phase Separation in Protein Low-Complexity Regions. *Biochemistry* **2018**, *57*, 2478–2487. [CrossRef]
27. Mitrea, D.M.; Kriwacki, R.W. Phase Separation in Biology; Functional Organization of a Higher Order. *Cell Commun. Signal.* **2016**, *14*, 1. [CrossRef]
28. Wang, B.; Zhang, L.; Dai, T.; Qin, Z.; Lu, H.; Zhang, L.; Zhou, F. Liquid–Liquid Phase Separation in Human Health and Diseases. *Signal Transduct. Target. Ther.* **2021**, *6*, 290. [CrossRef]
29. Franzmann, T.M.; Alberti, S.; Morimoto, R.I.; Hartl, F.U.; Kelly, J.W. Protein Phase Separation as a Stress Survival Strategy. *Cold Spring Harb. Perspect. Biol.* **2019**, *11*, a034058. [CrossRef]
30. Uversky, V.N. Intrinsically Disordered Proteins in Overcrowded Milieu: Membrane-Less Organelles, Phase Separation, and Intrinsic Disorder. *Curr. Opin. Struct. Biol.* **2017**, *44*, 18–30. [CrossRef]
31. Chen, J.; Kriwacki, R.W. Intrinsically Disordered Proteins: Structure, Function and Therapeutics. *J. Mol. Biol.* **2018**, *430*, 2275–2277. [CrossRef] [PubMed]
32. Bhattarai, A.; Emerson, I.A. Dynamic Conformational Flexibility and Molecular Interactions of Intrinsically Disordered Proteins. *J. Biosci.* **2020**, *45*, 29. [CrossRef] [PubMed]
33. Alberti, S.; Dormann, D. Liquid–Liquid Phase Separation in Disease. *Annu. Rev. Genet.* **2019**, *53*, 171–194. [CrossRef] [PubMed]
34. Mukhopadhyay, S. The Dynamism of Intrinsically Disordered Proteins: Binding-Induced Folding, Amyloid Formation, and Phase Separation. *J. Phys. Chem. B* **2020**, *124*, 11541–11560. [CrossRef]
35. Singh, V.; Xu, L.; Boyko, S.; Surewicz, K.; Surewicz, W.K. Zinc Promotes Liquid-Liquid Phase Separation of Tau Protein. *J. Biol. Chem.* **2020**, *295*, 5850–5856. [CrossRef]
36. Vega, I.E.; Traverso, E.E.; Ferrer-Acosta, Y.; Matos, E.; Colon, M.; Gonzalez, J.; Dickson, D.; Hutton, M.; Lewis, J.; Yen, S.H. A Novel Calcium-Binding Protein Is Associated with Tau Proteins in Tauopathy. *J. Neurochem.* **2008**, *106*, 96–106. [CrossRef] [PubMed]
37. Ferrer-Acosta, Y.; Rodríguez Cruz, E.N.; Vaquer, A.d.C.; Vega, I.E. Functional and Structural Analysis of the Conserved EFhd2 Protein. *Protein Pept. Lett.* **2013**, *20*, 573–583. [CrossRef]
38. Ferrer-Acosta, Y.; Rodríguez-Cruz, E.N.; Orange, F.; De Jesús-Cortés, H.; Madera, B.; Vaquer-Alicea, J.; Ballester, J.; Guinel, M.J.-F.; Bloom, G.S.; Vega, I.E. EFhd2 Is a Novel Amyloid Protein Associated with Pathological Tau in Alzheimer's Disease. *J. Neurochem.* **2013**, *125*, 921–931. [CrossRef]
39. Vega, I.E.; Umstead, A.; Kanaan, N.M. EFhd2 Affects Tau Liquid-Liquid Phase Separation. *Front. Neurosci.* **2019**, *13*, 845. [CrossRef]
40. Więch, A.; Tarczewska, A.; Ożyhar, A.; Orłowski, M. Metal Ions Induce Liquid Condensate Formation by the F Domain of Aedes Aegypti Ecdysteroid Receptor. New Perspectives of Nuclear Receptor Studies. *Cells* **2021**, *10*, 571. [CrossRef]
41. Wang, J.; Choi, J.-M.; Holehouse, A.S.; Lee, H.O.; Zhang, X.; Jahnel, M.; Maharana, S.; Lemaitre, R.; Pozniakovsky, A.; Drechsel, D.; et al. A Molecular Grammar Governing the Driving Forces for Phase Separation of Prion-like RNA Binding Proteins. *Cell* **2018**, *174*, 688–699. [CrossRef] [PubMed]
42. Alberti, S.; Gladfelter, A.; Mittag, T. Considerations and Challenges in Studying Liquid-Liquid Phase Separation and Biomolecular Condensates. *Cell* **2019**, *176*, 419–434. [CrossRef] [PubMed]
43. Dutagaci, B.; Nawrocki, G.; Goodluck, J.; Ashkarran, A.A.; Hoogstraten, C.G.; Lapidus, L.J.; Feig, M. Charge-Driven Condensation of RNA and Proteins Suggests Broad Role of Phase Separation in Cytoplasmic Environments. *Elife* **2021**, *10*, e64004. [CrossRef] [PubMed]
44. Mayfield, J.E.; Pollak, A.J.; Worby, C.A.; Xu, J.C.; Tandon, V.; Newton, A.C.; Dixon, J.E. $Ca^{2+}$-Dependent Liquid-Liquid Phase Separation Underlies Intracellular Ca2+ Stores. *bioRxiv* **2021**. [CrossRef]
45. Wojtas, M.; Dobryszycki, P.; Ożyhar, A. *Intrinsically Disordered Proteins in Biomineralization, Advanced Topics in Biomineralization*; IntechOpen: London, UK, 2012. [CrossRef]
46. Press, A.I.N.; Marin, F.; Pokroy, B.; Luquet, G.; Layrolle, P.; De Groot, K. Protein Mapping of Calcium Carbonate Biominerals by Immunogold. *Biomaterials* **2007**, *28*, 2368–2377. [CrossRef]
47. Gilbert, P.U.P.A.; Abrecht, M.; Frazer, B.H. The Organic-Mineral Interface in Biominerals. *Rev. Mineral. Geochem.* **2005**, *59*, 157–185. [CrossRef]
48. Kalmar, L.; Homola, D.; Varga, G.; Tompa, P. Structural Disorder in Proteins Brings Order to Crystal Growth in Biomineralization. *Bone* **2012**, *51*, 528–534. [CrossRef]

49. Thomas, O.R.B.B.; Swearer, S.E.; Kapp, E.A.; Peng, P.; Tonkin-Hill, G.Q.; Papenfuss, A.; Roberts, A.; Bernard, P.; Roberts, B.R. The Inner Ear Proteome of Fish. *FEBS J.* **2019**, *286*, 66–81. [CrossRef]
50. Kalka, M.; Markiewicz, N.; Ptak, M.; Sone, E.D.; Ożyhar, A.; Dobryszycki, P.; Wojtas, M. In Vivo and in Vitro Analysis of Starmaker Activity in Zebrafish Otolith Biomineralization. *FASEB J.* **2019**, *33*, 6877–6886. [CrossRef]
51. George, M. Rigid Biological Systems as Models for Synthetic Composites. *Science* **2005**, *310*, 1144–1147. [CrossRef]
52. Poznar, M.; Stolarski, J.; Sikora, A.; Mazur, M.; Olesiak-Bańska, J.; Brach, K.; Ożyhar, A.; Dobryszycki, P. Fish Otolith Matrix Macromolecule-64 (OMM-64) and Its Role in Calcium Carbonate Biomineralization. *Cryst. Growth Des.* **2020**, *20*, 5808–5819. [CrossRef]
53. Kim, Y.-Y.; Carloni, J.D.; Demarchi, B.; Sparks, D.; Reid, D.G.; Kunitake, M.E.; Tang, C.C.; Duer, M.J.; Freeman, C.L.; Pokroy, B.; et al. Tuning Hardness in Calcite by Incorporation of Amino Acids. *Nat. Mater.* **2016**, *15*, 903–910. [CrossRef] [PubMed]
54. Jain, G.; Pendola, M.; Koutsoumpeli, E.; Johnson, S.; Evans, J.S. Glycosylation Fosters Interactions between Model Sea Urchin Spicule Matrix Proteins. Implications for Embryonic Spiculogenesis and Biomineralization. *Biochemistry* **2018**, *57*, 3032–3035. [CrossRef] [PubMed]
55. Evans, J.S. Glycosylation: A "Last Word" in the Protein-Mediated Biomineralization Process. *Crystals* **2020**, *10*, 818. [CrossRef]
56. Alvares, K. The Role of Acidic Phosphoproteins in Biomineralization. *Connect. Tissue Res.* **2014**, *55*, 34–40. [CrossRef]
57. Wojtas, M.; Wo, M.; Andrzej, O.; Dobryszycki, P.; Wołcyrz, M.; Ożyhar, A.; Dobryszycki, P. Phosphorylation of Intrinsically Disordered Starmaker Protein Increases Its Ability To Control the Formation of Calcium Carbonate Crystals. *Cryst. Growth Des.* **2012**, *12*, 158–168. [CrossRef]
58. Tagliabracci, V.S.; Engel, J.L.; Wen, J.; Wiley, S.E.; Worby, C.A.; Kinch, L.N.; Xiao, J.; Grishin, N.V.; Dixon, J.E. Secreted Kinase Phosphorylates Extracellular Proteins That Regulate Biomineralization. *Science* **2012**, *336*, 1150–1153. [CrossRef]
59. Du, J.; Liu, C.; Xu, G.; Xie, J.; Xie, L.; Zhang, R. Fam20C Participates in the Shell Formation in the Pearl Oyster, Pinctada Fucata. *Sci. Rep.* **2018**, *8*, 3563. [CrossRef]
60. Chang, E.P.; Perovic, I.; Rao, A.; Cölfen, H.; Evans, J.S. Insect Cell Glycosylation and Its Impact on the Functionality of a Recombinant Intracrystalline Nacre Protein, AP24. *Biochemistry* **2016**, *55*, 1024–1035. [CrossRef]
61. Jain, G.; Pendola, M.; Rao, A.; Cölfen, H.; Evans, J.S. A Model Sea Urchin Spicule Matrix Protein Self-Associates To Form Mineral-Modifying Protein Hydrogels. *Biochemistry* **2016**, *55*, 4410–4421. [CrossRef]
62. Takeuchi, T.; Sarashina, I.; Iijima, M.; Endo, K. In Vitro Regulation of $CaCO_3$ Crystal Polymorphism by the Highly Acidic Molluscan Shell Protein Aspein. *FEBS Lett.* **2008**, *582*, 591–596. [CrossRef] [PubMed]
63. Tsukamoto, D.; Sarashina, I.; Endo, K. Structure and Expression of an Unusually Acidic Matrix Protein of Pearl Oyster Shells. *Biochem. Biophys. Res. Commun.* **2004**, *320*, 1175–1180. [CrossRef] [PubMed]
64. Isowa, Y.; Sarashina, I.; Setiamarga, D.H.; Endo, K. A Comparative Study of the Shell Matrix Protein Aspein in Pterioid Bivalves. *J. Mol. Evol.* **2012**, *75*, 11–18. [CrossRef]
65. Inoue, H.; Ohira, T.; Nagasawa, H. Significance of the N- and C-Terminal Regions of CAP-1, a Cuticle Calcification-Associated Peptide from the Exoskeleton of the Crayfish, for Calcification. *Peptides* **2007**, *28*, 566–573. [CrossRef] [PubMed]
66. Arakaki, A.; Shimizu, K.; Oda, M.; Sakamoto, T.; Nishimura, T.; Kato, T. Biomineralization-Inspired Synthesis of Functional Organic/Inorganic Hybrid Materials: Organic Molecular Control of Self-Organization of Hybrids. *Org. Biomol. Chem.* **2015**, *13*, 974–989. [CrossRef]
67. Mass, T.; Drake, J.L.; Haramaty, L.; Kim, J.D.; Zelzion, E.; Bhattacharya, D.; Falkowski, P.G. Cloning and Characterization of Four Novel Coral Acid-Rich Proteins That Precipitate Carbonates In Vitro. *Curr. Biol.* **2013**, *23*, 1126–1131. [CrossRef]
68. Laipnik, R.; Bissi, V.; Sun, C.Y.; Falini, G.; Gilbert, P.U.P.A.; Mass, T. Coral Acid Rich Protein Selects Vaterite Polymorph in Vitro. *J. Struct. Biol.* **2020**, *209*, 107431. [CrossRef]
69. Wilt, F.H. Biomineralization of the Spicules of Sea Urchin Embryos. *Zoolog. Sci.* **2002**, *19*, 253–261. [CrossRef]
70. Illies, M.R.; Peeler, M.T.; Dechtiaruk, A.M.; Ettensohn, C.A. Identification and Developmental Expression of New Biomineralization Proteins in the Sea Urchin Strongylocentrotus Purpuratus. *Dev. Genes Evol.* **2002**, *212*, 419–431. [CrossRef] [PubMed]
71. Popper, A.N.; Ramcharitar, J.; Campana, S.E. Why Otoliths? Insights from Inner Ear Physiology and Fisheries Biology. *Mar. Freshw. Res.* **2005**, *56*, 497–504. [CrossRef]
72. Ross, M.D.; Pote, K.G. Some Properties of Otoconia. *Philos. Trans. R. Soc. Lond. B Biol. Sci.* **1984**, *304*, 445–452. [CrossRef] [PubMed]
73. Lundberg, Y.W.; Xu, Y.; Thiessen, K.D.; Kramer, K.L. Mechanisms of Otoconia and Otolith Development. *Dev. Dyn.* **2015**, *244*, 239–253. [CrossRef]
74. Thomas, O.R.B.; Swearer, S.E. Otolith Biochemistry—A Review. *Rev. Fish. Sci. Aquac.* **2019**, *27*, 458–489. [CrossRef]
75. Payan, P.; De Pontual, H.; Bœuf, G.; Mayer-gostan, N. Endolymph Chemistry and Otolith Growth in Fish. *Comptes Rendus Palevol* **2004**, *3*, 535–547. [CrossRef]
76. Walther, L.E.; Wenzel, A.; Buder, J.; Bloching, M.B.; Kniep, R.; Blödow, A. Detection of Human Utricular Otoconia Degeneration in Vital Specimen and Implications for Benign Paroxysmal Positional Vertigo. *Eur. Arch. Oto-Rhino-Laryngol.* **2014**, *271*, 3133–3138. [CrossRef] [PubMed]
77. Kang, Y.-J.J.; Stevenson, A.K.; Yau, P.M.; Kollmar, R. Sparc Protein Is Required for Normal Growth of Zebrafish Otoliths. *J. Assoc. Res. Otolaryngol.* **2008**, *9*, 436–451. [CrossRef]

78. Murayama, E.; Herbomel, P.; Kawakami, A.; Takeda, H.; Nagasawa, H. Otolith Matrix Proteins OMP-1 and Otolin-1 Are Necessary for Normal Otolith Growth and Their Correct Anchoring onto the Sensory Maculae. *Mech. Dev.* **2005**, *122*, 791–803. [CrossRef]
79. Tohse, H.; Takagi, Y.; Nagasawa, H. Identification of a Novel Matrix Protein Contained in a Protein Aggregate Associated with Collagen in Fish Otoliths. *FEBS J.* **2008**, *275*, 2512–2523. [CrossRef]
80. Poznar, M.; Hołubowicz, R.; Wojtas, M.; Gapiński, J.; Banachowicz, E.; Patkowski, A.; Ożyhar, A.; Dobryszycki, P. Structural Properties of the Intrinsically Disordered, Multiple Calcium Ion-Binding Otolith Matrix Macromolecule-64 (OMM-64). *Biochim. Biophys. Acta-Proteins Proteom.* **2017**, *1865*, 1358–1371. [CrossRef]
81. Różycka, M.; Wojtas, M.; Jakób, M.; Stigloher, C.; Grzeszkowiak, M.; Mazur, M.; Ożyhar, A. Intrinsically Disordered and Pliable Starmaker-Like Protein from Medaka (Oryzias Latipes) Controls the Formation of Calcium Carbonate Crystals. *PLoS ONE* **2014**, *9*, e114308. [CrossRef]
82. Petko, J.A.; Millimaki, B.B.; Canfield, V.A.; Riley, B.B.; Levenson, R. Otoc1: A Novel Otoconin-90 Ortholog Required for Otolith Mineralization in Zebrafish. *Dev. Neurobiol.* **2008**, *68*, 209–222. [CrossRef] [PubMed]
83. Sollner, C.; Burghammer, M.; Busch-Nentwich, E.; Berger, J.; Schwarz, H.; Riekel, C.; Nicolson, T.; Söllner, C.; Burghammer, M.; Busch-Nentwich, E.; et al. Control of Crystal Size and Lattice Formation by Starmaker in Otolith Biomineralization. *Science* **2003**, *302*, 282–286. [CrossRef] [PubMed]
84. Różycka, M.; Coronado, I.; Brach, K.; Olesiak-Bańska, J.; Samoć, M.; Zarębski, M.; Dobrucki, J.; Ptak, M.; Weber, E.; Polishchuk, I.; et al. Lattice Shrinkage by Incorporation of Recombinant Starmaker-Like Protein within Bioinspired Calcium Carbonate Crystals. *Chemistry* **2019**, *25*, 12740–12750. [CrossRef]
85. Murayama, E.; Takagi, Y.; Ohira, T.; Davis, J.G.; Greene, M.I.; Nagasawa, H. Fish Otolith Contains a Unique Structural Protein, Otolin-1. *Eur. J. Biochem.* **2002**, *269*, 688–696. [CrossRef] [PubMed]
86. Murayama, E.; Takagi, Y.; Nagasawa, H. Immunohistochemical Localization of Two Otolith Matrix Proteins in the Otolith and Inner Ear of the Rainbow Trout, Oncorhynchus Mykiss: Comparative Aspects between the Adult Inner Ear and Embryonic Otocysts. *Histochem. Cell Biol.* **2004**, *121*, 155–166. [CrossRef]
87. Andrade, L.R.; Lins, U.; Farina, M.; Kachar, B.; Thalmann, R. Immunogold TEM of Otoconin 90 and Otolin-Relevance to Mineralization of Otoconia, and Pathogenesis of Benign Positional Vertigo. *Hear. Res.* **2012**, *292*, 14–25. [CrossRef] [PubMed]
88. Davis, J.G.; Oberholtzer, J.C.; Burns, F.R.; Greene, M.I. Molecular Cloning and Characterization of an Inner Ear-Specific Structural Protein. *Science* **1995**, *267*, 1031–1034. [CrossRef]
89. Dobryszycki, P.; Hołubowicz, R.; Wojtas, M.; Taube, M.; Kozak, M.; Ożyhar, A.; Dobryszycki, P. Effect of Calcium Ions on Structure and Stability of the C1q-like Domain of Otolin-1 from Human and Zebrafish. *FEBS J.* **2017**, *284*, 4278–4297. [CrossRef]
90. Bielak, K.; Hołubowicz, R.; Zoglowek, A.; Żak, A.; Kędzierski, P.; Ożyhar, A.; Dobryszycki, P. N′-Terminal- and $Ca^{2+}$-Induced Stabilization of High-Order Oligomers of Full-Length Danio Rerio and Homo Sapiens Otolin-1. *Int. J. Biol. Macromol.* **2022**, *209*, 1032–1047. [CrossRef]
91. Moreland, K.T.; Hong, M.; Lu, W.; Rowley, C.W.; Ornitz, D.M.; De Yoreo, J.J.; Thalmann, R. In Vitro Calcite Crystal Morphology Is Modulated by Otoconial Proteins Otolin-1 and Otoconin-90. *PLoS ONE* **2014**, *9*, e95333. [CrossRef]
92. Silvent, J.; Sire, J.-Y.; Delgado, S. The Dentin Matrix Acidic Phosphoprotein 1 (DMP1) in the Light of Mammalian Evolution. *J. Mol. Evol.* **2013**, *76*, 59–70. [CrossRef]
93. Xu, Y.; Zhang, H.; Yang, H.; Zhao, X.; Lovas, S.; Lundberg, Y.W. Expression, Functional, and Structural Analysis of Proteins Critical for Otoconia Development. *Dev. Dyn.* **2010**, *239*, 2659–2673. [CrossRef] [PubMed]
94. Qin, C.; Brunn, J.C.; Cook, R.G.; Orkiszewski, R.S.; Malone, J.P.; Veis, A.; Butler, W.T. Evidence for the Proteolytic Processing of Dentin Matrix Protein 1: Identification and characterization of processed fragments and cleavage sites. *J. Biol. Chem.* **2003**, *278*, 34700–34708. [CrossRef] [PubMed]
95. Porębska, A.; Różycka, M.; Hołubowicz, R.; Szewczuk, Z.; Ożyhar, A.; Dobryszycki, P. Functional Derivatives of Human Dentin Matrix Protein 1 Modulate Morphology of Calcium Carbonate Crystals. *FASEB J.* **2020**, *34*, 6147–6165. [CrossRef] [PubMed]
96. De Yoreo, J.J.; Vekilov, P.G.; De Yoreo, J.J.; Vekilov, P.G. Principles of Crystal Nucleation and Growth. *Rev. Mineral. Geochem.* **2003**, *54*, 57–93. [CrossRef]
97. Debenedetti, P.G. *Metastable Liquids*; Princeton University Press: Princeton, NJ, USA, 1996; Volume 1, ISBN 9780691085951.
98. Gibbs, J.W. On the Equilibrium of Heterogeneous Substances. *Am. J. Sci.* **1878**, *s3–16*, 441–458. [CrossRef]
99. Becker, R.; Döring, W. Kinetische Behandlung Der Keimbildung in Übersättigten Dämpfen. *Ann. Phys.* **1935**, *416*, 719–752. [CrossRef]
100. Farkas, L. Keimbildungsgeschwindigkeit in Übersättigten Dämpfen. *Z. Phys. Chem.* **1927**, *125U*, 236–242. [CrossRef]
101. Volmer, M.; Weber, A. Keimbildung in Übersättigten Gebilden. *Z. Phys. Chem.* **1926**, *119U*, 277–301. [CrossRef]
102. Kashchiev, D.; van Rosmalen, G.M. Review: Nucleation in Solutions Revisited. *Cryst. Res. Technol.* **2003**, *38*, 555–574. [CrossRef]
103. Gebauer, D.; Raiteri, P.; Gale, J.D.; Cölfen, H. On classical and non-classical views on nucleation. *Am. J. Sci.* **2018**, *318*, 969–988. [CrossRef]
104. Wallace, A.F.; Hedges, L.O.; Fernandez-Martinez, A.; Raiteri, P.; Gale, J.D.; Glenn, W.; Whitelam, S.; Banfield, J.F.; De Yoreo, J.J. Microscopic Evidence for Liquid-Liquid Separation in Supersaturated $CaCO_3$ Solutions. *Science* **2013**, *341*, 885–889. [CrossRef] [PubMed]
105. Koishi, A. *Carbonate Mineral Nucleation Pathways*; Université Grenoble Alpes: Grenoble, France, 2017.

106. Evans, J.S. "Liquid-like" Biomineralization Protein Assemblies: A Key to the Regulation of Non-Classical Nucleation. *CrystEngComm* **2013**, *15*, 8388–8394. [CrossRef]
107. Meldrum, F.C.; Sear, R.P. Materials Science. Now You See Them. *Science* **2008**, *322*, 1802–1803. [CrossRef] [PubMed]
108. Gebauer, D. How Can Additives Control the Early Stages of Mineralisation? *Minerals* **2018**, *8*, 179. [CrossRef]
109. Bewernitz, M.A.; Gebauer, D.; Long, J.; Cölfen, H.; Gower, L.B. A Metastable Liquid Precursor Phase of Calcium Carbonate and Its Interactions with Polyaspartate. *Faraday Discuss.* **2012**, *159*, 291. [CrossRef]
110. Gower, L.B. Biomimetic Model Systems for Investigating the Amorphous Precursor Pathway and Its Role in Biomineralization. *Chem. Rev.* **2008**, *108*, 4551–4627. [CrossRef]
111. Wolf, S.L.P.; Caballero, L.; Melo, F.; Cölfen, H. Gel-Like Calcium Carbonate Precursors Observed by in Situ AFM. *Langmuir* **2017**, *33*, 158–163. [CrossRef]
112. Pouget, E.M.; Bomans, P.H.; Goos, J.A.; Frederik, P.M.; de With, G.; Sommerdijk, N.A. The Initial Stages of Template-Controlled $CaCO_3$ Formation Revealed by Cryo-TEM. *Science* **2009**, *323*, 1455–1458. [CrossRef]
113. Addadi, L.; Raz, S.; Weiner, S. Taking Advantage of Disorder: Amorphous Calcium Carbonate and Its Roles in Biomineralization. *Adv. Mater.* **2003**, *15*, 959–970. [CrossRef]
114. Weiner, S.; Mahamid, J.; Politi, Y.; Ma, Y.; Addadi, L. Overview of the Amorphous Precursor Phase Strategy in Biomineralization. *Front. Mater. Sci. China* **2009**, *3*, 104–108. [CrossRef]
115. Radha, A.V.; Forbes, T.Z.; Killian, C.E.; Gilbert, P.U.; Navrotsky, A. Transformation and Crystallization Energetics of Synthetic and Biogenic Amorphous Calcium Carbonate. *Proc. Natl. Acad. Sci. USA* **2010**, *107*, 16438–16443. [CrossRef] [PubMed]
116. Alexandra, N. Energetic Clues to Pathways to Biomineralization: Precursors, Clusters, and Nanoparticles. *Proc. Natl. Acad. Sci. USA* **2004**, *101*, 12096–12101. [CrossRef]
117. Quigley, D.; Freeman, C.L.; Harding, J.H.; Rodger, P.M. Sampling the Structure of Calcium Carbonate Nanoparticles with Metadynamics. *J. Chem. Phys.* **2011**, *134*, 44703. [CrossRef]
118. Tribello, G.A.; Bruneval, F.; Liew, C.; Parrinello, M. A Molecular Dynamics Study of the Early Stages of Calcium Carbonate Growth. *J. Phys. Chem. B* **2009**, *113*, 11680–11687. [CrossRef]
119. Wang, J.; Hou, T. Application of Molecular Dynamics Simulations in Molecular Property Prediction II: Diffusion Coefficient. *J. Comput. Chem.* **2011**, *32*, 3505–3519. [CrossRef]
120. Gebauer, D.; Kellermeier, M.; Gale, J.D.; Bergström, L.; Cölfen, H. Pre-Nucleation Clusters as Solute Precursors in Crystallisation. *Chem. Soc. Rev.* **2014**, *43*, 2348–2371. [CrossRef]
121. Faatz, M.; Gröhn, F.; Wegner, G. Amorphous Calcium Carbonate: Synthesis and Potential Intermediate in Biomineralization. *Adv. Mater.* **2004**, *16*, 996–1000. [CrossRef]
122. Wolf, S.E.; Leiterer, J.; Kappl, M.; Emmerling, F.; Tremel, W. Early Homogenous Amorphous Precursor Stages of Calcium Carbonate and Subsequent Crystal Growth in Levitated Droplets. *J. Am. Chem. Soc.* **2008**, *130*, 12342–12347. [CrossRef]
123. Wang, X.; Chou, I.M.; Hu, W.; Burruss, R.C. In Situ Observations of Liquid-Liquid Phase Separation in Aqueous MgSO4 Solutions: Geological and Geochemical Implications. *Geochim. Cosmochim. Acta* **2013**, *103*, 1–10. [CrossRef]
124. Wang, X.; Wan, Y.; Hu, W.; Chou, I.M.; Cao, J.; Wang, X.; Wang, M.; Li, Z. In Situ Observations of Liquid–Liquid Phase Separation in Aqueous $ZnSO_4$ Solutions at Temperatures up to 400 °C: Implications for $Zn^{2+}$–$SO_4{}^{2-}$ Association and Evolution of Submarine Hydrothermal Fluids. *Geochim. Cosmochim. Acta* **2016**, *181*, 126–143. [CrossRef]
125. Suzuki, M.; Saruwatari, K.; Kogure, T.; Yamamoto, Y.; Nishimura, T.; Kato, T.; Nagasawa, H. An Acidic Matrix Protein, Pif, Is a Key Macromolecule for Nacre Formation. *Science* **2009**, *325*, 1388–1390. [CrossRef] [PubMed]
126. Evans, J.S. Aragonite-Associated Biomineralization Proteins Are Disordered and Contain Interactive Motifs. *Bioinformatics* **2012**, *28*, 3182–3185. [CrossRef] [PubMed]
127. Bahn, S.Y.; Jo, B.H.; Choi, Y.S.; Cha, H.J. Control of Nacre Biomineralization by Pif80 in Pearl Oyster. *Sci. Adv.* **2017**, *3*, e1700765. [CrossRef]
128. Farahi, N.; Lazar, T.; Wodak, S.J.; Tompa, P.; Pancsa, R. Integration of Data from Liquid-Liquid Phase Separation Databases Highlights Concentration and Dosage Sensitivity of LLPS Drivers. *Int. J. Mol. Sci.* **2021**, *22*, 3017. [CrossRef]
129. Lin, Y.; Protter, D.S.W.; Rosen, M.K.; Parker, R. Formation and Maturation of Phase-Separated Liquid Droplets by RNA-Binding Proteins. *Mol. Cell* **2015**, *60*, 208–219. [CrossRef]
130. Belcher, A.M.; Wu, X.H.; Christensen, R.J.; Hansma, P.K.; Stucky, G.D.; Morse, D.E. Control of crystal phase switching and orientation by soluble mollusc-shell proteins. *Nature* **1996**, *381*, 56–58. [CrossRef]
131. Marin, F.; Luquet, G. Unusually Acidic Proteins in Biomineralization. In *Handbook of Biomineralization*; Bäuerlein, E., Behrens, P., Epple, M., Eds.; Wiley-VCH Verlag GmbH & Co. KGaA: Weinheim, Germany, 2007; Volume 1, pp. 273–290. [CrossRef]
132. Mummadisetti, M.P.; Drake, J.L.; Falkowski, P.G. The Spatial Network of Skeletal Proteins in a Stony Coral. *J. R. Soc. Interface* **2021**, *18*, 20200859. [CrossRef]
133. George, A.; Veis, A. Phosphorylated Proteins and Control over Apatite Nucleation, Crystal Growth, and Inhibition. *Chem. Rev.* **2008**, *108*, 4670–4693. [CrossRef]
134. Drickamer, K.; Taylor, M.E. Evolving Views of Protein Glycosylation. *Trends Biochem. Sci.* **1998**, *23*, 321–324. [CrossRef]
135. Holehouse, A.S.; Das, R.K.; Ahad, J.N.; Richardson, M.O.G.; Pappu, R.V. CIDER: Resources to Analyze Sequence-Ensemble Relationships of Intrinsically Disordered Proteins. *Biophys. J.* **2017**, *112*, 16–21. [CrossRef] [PubMed]

136. Pancsa, R.; Vranken, W.; Mészáros, B. Computational Resources for Identifying and Describing Proteins Driving Liquid–Liquid Phase Separation. *Brief. Bioinform.* **2021**, *22*, bbaa408. [CrossRef] [PubMed]
137. Vendruscolo, M.; Fuxreiter, M. Sequence Determinants of the Aggregation of Proteins Within Condensates Generated by Liquid-Liquid Phase Separation. *J. Mol. Biol.* **2022**, *434*, 167201. [CrossRef]
138. Hardenberg, M.; Horvath, A.; Ambrus, V.; Fuxreiter, M.; Vendruscolo, M. Widespread Occurrence of the Droplet State of Proteins in the Human Proteome. *Proc. Natl. Acad. Sci. USA* **2020**, *117*, 33254–33262. [CrossRef]
139. Hatos, A.; Tosatto, S.C.E.; Vendruscolo, M.; Fuxreiter, M. FuzDrop on AlphaFold: Visualizing the Sequence-Dependent Propensity of Liquid–Liquid Phase Separation and Aggregation of Proteins. *Nucleic Acids Res.* **2022**, *50*, 337–344. [CrossRef]
140. Vernon, R.M.C.; Chong, P.A.; Tsang, B.; Kim, T.H.; Bah, A.; Farber, P.; Lin, H.; Forman-Kay, J.D. Pi-Pi Contacts Are an Overlooked Protein Feature Relevant to Phase Separation. *Elife* **2018**, *7*, e31486. [CrossRef]
141. Kapłon, T.M.; Michnik, A.; Drzazga, Z.; Richter, K.; Kochman, M.; Ozyhar, A. The Rod-Shaped Conformation of Starmaker. *Biochim. Biophys. Acta* **2009**, *1794*, 1616–1624. [CrossRef]
142. Horvath, A.; Miskei, M.; Ambrus, V.; Vendruscolo, M.; Fuxreiter, M. Sequence-Based Prediction of Protein Binding Mode Landscapes. *PLoS Comput. Biol.* **2020**, *16*, e1007864. [CrossRef]
143. Chen, Y.; Feng, Y.; Deveaux, J.G.; Masoud, M.A.; Chandra, F.S.; Chen, H.; Zhang, D.; Feng, L. Biomineralization Forming Process and Bio-Inspired Nanomaterials for Biomedical Application: A Review. *Minerals* **2019**, *9*, 68. [CrossRef]
144. Fujita, Y.; Redden, G.D.; Ingram, J.C.; Cortez, M.M.; Ferris, F.G.; Smith, R.W. Strontium Incorporation into Calcite Generated by Bacterial Ureolysis. *Geochim. Cosmochim. Acta* **2004**, *68*, 3261–3270. [CrossRef]
145. Clark, M.S. Molecular Mechanisms of Biomineralization in Marine Invertebrates. *J. Exp. Biol.* **2020**, *223*, jeb206961. [CrossRef] [PubMed]

*biomolecules*

MDPI

*Review*

# A Trajectory of Discovery: Metabolic Regulation by the Conditionally Disordered Chloroplast Protein, CP12

**Cassy Gérard, Frédéric Carrière, Véronique Receveur-Bréchot, Hélène Launay * and Brigitte Gontero ***

Aix Marseille Univ, CNRS, BIP, UMR 7281, IMM, FR3479, 31 Chemin J. Aiguier, CEDEX 9, 13 402 Marseille, France; cgerard@imm.cnrs.fr (C.G.); carriere@imm.cnrs.fr (F.C.); veronique.brechot@imm.cnrs.fr (V.R.-B.)

* Correspondence: helene.launay@univ-amu.fr (H.L.); bmeunier@imm.cnrs.fr (B.G.)

**Abstract:** The chloroplast protein CP12, which is widespread in photosynthetic organisms, belongs to the intrinsically disordered proteins family. This small protein (80 amino acid residues long) presents a bias in its composition; it is enriched in charged amino acids, has a small number of hydrophobic residues, and has a high proportion of disorder-promoting residues. More precisely, CP12 is a conditionally disordered proteins (CDP) dependent upon the redox state of its four cysteine residues. During the day, reducing conditions prevail in the chloroplast, and CP12 is fully disordered. Under oxidizing conditions (night), its cysteine residues form two disulfide bridges that confer some stability to some structural elements. Like many CDPs, CP12 plays key roles, and its redox-dependent conditional disorder is important for the main function of CP12: the dark/light regulation of the Calvin-Benson-Bassham (CBB) cycle responsible for $CO_2$ assimilation. Oxidized CP12 binds to glyceraldehyde-3-phosphate dehydrogenase and phosphoribulokinase and thereby inhibits their activity. However, recent studies reveal that CP12 may have other functions beyond the CBB cycle regulation. In this review, we report the discovery of this protein, its features as a disordered protein, and the many functions this small protein can have.

**Keywords:** Calvin-Benson-Bassham cycle; conditionally disordered protein; history of modern science; metabolism regulation; moonlighting protein; protein-protein interaction

**Citation:** Gérard, C.; Carrière, F.; Receveur-Bréchot, V.; Launay, H.; Gontero, B. A Trajectory of Discovery: Metabolic Regulation by the Conditionally Disordered Chloroplast Protein, CP12. *Biomolecules* **2022**, *12*, 1047. https://doi.org/10.3390/biom12081047

Academic Editors: Prakash Kulkarni, Stefania Brocca, Keith Dunker and Sonia Longhi

Received: 27 June 2022
Accepted: 26 July 2022
Published: 28 July 2022



## 1. Introduction

As V. Uversky mentioned, the discovery of the natural abundance and functional importance of intrinsically disordered proteins (IDPs) has changed protein science [1]. It is now widely accepted that the protein structure-function paradigm that dominated scientific minds for more than 100 years does not hold true for all proteins, and IDPs or proteins containing disordered regions (IDR)s are widespread in all areas of life. IDPs and IDRs differ from structured globular proteins and domains in many respects, such as their amino acid composition, complexity of sequence, hydrophobicity, charge, flexibility, and rate of amino acid substitutions over evolutionary time [2]. They play significant roles in many biological processes, such as control of the cell cycle, transcriptional activation, and signaling, and they frequently interact with many partners or function as central hubs in protein interaction networks. However, in the plant kingdom only a few IDPs have been studied through a recent analysis of 12 plant genomes, which revealed that the occurrence of disorder in plants is similar to that in many other eukaryotes [3]. In plants, most of the information on IDPs comes from *Arabidopsis thaliana*, and among them, to cite but a few, the late embryogenesis abundant (LEA) proteins that are important IDPs are mainly associated with environmental stress [4]. In the algae research field, a recent experimental study reported that 682 proteins from a chlorophyte, *Chlamydomonas reinhardtii*, were heat-resistant, and 299 were predicted to be disordered by four different disorder predictors [5]. However, only a few algal proteins that are fully or partially disordered have been studied [6–9]. Among them, only the essential pyrenoid component 1 (EPYC1) [10–14], and, above all, the chloroplast protein of 12 kDa (CP12), as evidenced below, have been experimentally studied in depth.

## 2. Discovery of a Small Protein, CP12, in Photosynthetic Organisms

In 1996, cDNA clones were reported from expression libraries for a nuclear-encoded chloroplast protein in three higher plants: pea, spinach, and tobacco, which was named CP12 [15]. This was the first report on CP12, and at this stage, not much was known about this protein. The authors found that this protein consists of about 75 amino acid residues, and it had an abnormal electrophoretic mobility in sodium dodecyl sulfate-polyacrylamide gel electrophoresis (SDS-PAGE) experiments. In addition, they noticed that the CP12 proteins from the three higher plants had a high content of charged amino acid residues, and their conclusion at this time was that this protein was highly hydrophilic and very likely a good candidate for a soluble stroma-located protein within the chloroplast. In all of the three species, two conserved cysteine residues are present at the N- and at the C-terminus that are separated by eight amino acid residues with a central proline residue. The secondary structure prediction suggested that CP12 has a central helix and is organized into two domains, with each containing two cysteine residues that could form disulfide bonds (Figure 1). In this pioneer work, they also showed that CP12 could interact with an enzyme from the Calvin-Benson-Bassham (CBB) cycle, the glyceraldehyde -3-phosphate dehydrogenase (GAPDH, EC 1.2.1.13) [15].

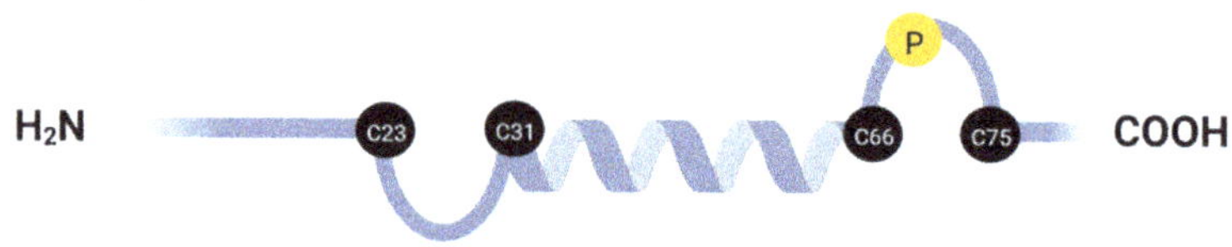


**Figure 1.** Schematic representation of the predicted secondary structure of mature CP12. Cysteine residues proposed to form peptide loops and that can form two disulfide bridges (C23–C31 and C66–C75) when CP12 is oxidized are indicated with black circles. These two loops are separated by an alpha helix. The proline residue conserved in CP12 from Plantae is shown with a yellow circle. Numbering is from the *C. reinhardtii* mature CP12 sequence. This figure was created with BioRender (https://biorender.com/ (accessed on 25 July 2022)) and adapted from Wedel et al. [16].

Later, in a work performed on spinach leaves, the same group showed that CP12 not only interacts with GAPDH but also with another enzyme of the CBB cycle, the phosphoribulokinase (PRK, EC 2.7.1.19) [16]. They showed that CP12 could form a 600 kDa complex with the two enzymes mentioned above. This ternary complex was not affected by the presence of NAD(H), but the presence of NADP or NADPH, as well as the high concentration of a reducing agent, dithiothreitol (DTT), led to its dissociation. The authors proposed a model in which the ternary complex exists under dark and dissociates under light. Their results and those of other groups in the literature suggested that the enzymes within the complex were inactive and became active upon dissociation [17–19]. Since the CBB cycle does not operate in the dark and become active in the light, the association–dissociation of this complex could be a means to regulate the CBB cycle upon dark-light transitions and reciprocally.

At the same time, Avilan et al. found a complex made up of PRK and GAPDH in the green alga, *C. reinhardtii* [20]. In the green algae, there is a unique form of GAPDH, the homotetrameric $A_4$, while in higher plants there is also an $A_2B_2$ where the B-type subunit has a C-terminal extremity that presents homology to the C-terminus of CP12. These authors deeply analyzed the characteristics and the kinetics of the enzymes within the complex and those of the enzymes that dissociated from it [20–23]. They purified this complex to homogeneity, and, later, the presence of CP12 in the PRK/GAPDH complex reported previously was revealed by MALDI-ToF mass spectrometry [24]. N-terminal sequencing by Edman degradation of the ternary complex allowed one to determine the first amino acid residues of each protein and revealed that the sequence of the mature

CP12 starts at SGQPA [25]. Therefore, the complex isolated by Avilan et al. had the same compositions as that found by Wedel et al. Indeed, in 1998, Wedel et al. showed that CP12 was present not only in higher plants but also in *C. reinhardtii*, as well as in many other species (mosses, cyanobacteria). The presence of CP12 in this complex agreed with the cryo-electron microscopy performed on this purified complex, which suggested that other components besides PRK and GAPDH were present [26]. The activities of the enzymes involved in this complex were regulated in vitro by metabolites such as NADP(H) [16,27,28]. All together, these results provided new ideas for the regulation of photosynthesis and were further investigated by many groups.

In *C. reinhardtii*, it was shown that not only the regulatory properties of GAPDH but also its kinetics parameters were affected by CP12. Native GAPDH and recombinant algal GAPDH displayed Michaelis-Menten kinetics with NADH and NADPH as cofactors, with a marked preference for NADPH. Both forms displayed positive cooperativity towards the substrate, 1,3-bisphosphoglycerate (BPGA), but interestingly, these kinetic analyses showed that the native GAPDH had a two-fold lower catalytic constant for the reduction of BPGA, as well as a two-fold lower pseudo-affinity ($K_{0.5}$) for BPGA compared to recombinant GAPDH [24]. These results were surprising, but using mass spectrometry the authors showed that the native GAPDH was still associated with CP12. At the same time, as only a partial sequence of the *C. reinhardtii* CP12 was obtained by PCR, the same authors cloned the entire cDNA of this algal protein and subsequently expressed the protein in *Escherichia coli* [25].

If some results suggested that the PRK/GAPDH/CP12 allowed for the regulation of these enzymes, the presence of this small protein raised a question about its role in the formation of the complex. The role of CP12 in the assembly pathway of the algal PRK/GAPDH/CP12 complex was thus investigated as no complex could be reconstituted in vitro with the native PRK and the recombinant GAPDH devoid of CP12.

In darkened spinach leaves, Scheibe's group also showed that GAPDH can exist under two inactive aggregated states, one that corresponded to a hexadecameric $A_8B_8$ form and another one that corresponded to the PRK/GAPDH/CP12 complex. Only the dissociation of these edifices with reducing treatment mimicking light resulted in the activity of the released enzymes [29]. The role of CP12 was, by then, far from being understood. Of interest, in the literature, many high oligomerization states of either GAPDH or PRK, in spinach but also in *Phaesolus vulgaris*, have been reported [30]. In spinach, the oligomeric enzymes had latent activity that only appeared upon dissociation [31,32]. In the 20th century, the existence of supramolecular complexes was not recognized in living cells and their existence was seen as artefactual. Therefore, the data were differentially interpreted, but it is very likely that the high molecular mass of GAPDH and PRK with latent activity in fact corresponded to supramolecular complexes.

In the green algae as in the higher plants, it was later shown that the four cysteine residues could form two disulfide bridges, one bridging the N-terminal cysteine pair (residues 23 and 31 in *C. reinhardtii*) and one bridging the C-terminal pair (residues 66 and 75 in *C. reinhardtii*). It was shown using surface plasmon resonance that CP12 under its oxidized state, with two disulfide bridges, was able to bind sequentially to GAPDH with a high affinity ($K_D$ equal to 0.44 nM), and then this subcomplex was able to bind to PRK ($K_D$ equal to 60 nM). The affinity of CP12 for GAPDH was higher than the one found (μM range) for the Arabidopsis complex [33]. The entity composed of one tetrameric GAPDH, one dimeric PRK, and CP12 (the stoichiometry of which was yet unknown) was defined as a unit. This entity was then able to dimerize to provide the native complex. CP12 therefore acted as a linker in the assembly of the ternary PRK/GAPDH/CP12 complex [25]. Later, native mass spectrometry revealed that two monomeric CP12 molecules were bound to one GAPDH tetramer [34]. Consequently, the stoichiometry inside the ternary complex is two tetrameric GAPDH, two dimeric PRK, and four monomeric CP12. Studies using mutagenesis and limited proteolysis have allowed the residues involved in the interaction between CP12 and GAPDH from *C. reinhardtii* to be mapped and to show that this interaction involves the S-loop arginine residues of GAPDH and the C-terminus of CP12 [35].

## 3. CP12, a Flexible Protein

The first observation of CP12 as an IDP was its abnormal behavior under SDS-PAGE. The protein migrates as a 15 kDa under oxidized form and 25 kDa under its reduced form for *C. reinhardtii*, while the expected theoretical molecular mass of the monomer is 8.5 kDa (Figure 2A,B). Moreover, using size-exclusion chromatography, the elution volume of *C. reinhardtii* CP12 released from the PRK/GAPDH/CP12 complex corresponded to an apparent molecular mass of 35 ± 4 kDa using a column calibrated with globular proteins (Figure 2C). This could correspond to a tetrameric globular form that has never been proven or to an elongated form. These enigmatic behaviors of CP12 were only understood after the concept of IDP was claimed [2,36–38]. The size exclusion elution volume mentioned above correlates to a hydrodynamic radius of 2.8 ± 0.1 nm, which corresponds to the expected hydrodynamic properties of a random-coil polymer of 8.8 kDa. These values were in agreement with those confirmed by fluorescence correlation spectroscopy experiments [39]. In 2003, for the first time, it was proposed that the *C. reinhardtii* CP12 belongs to the IDP family (formerly also called an "intrinsically unstructured protein") [25]. Indeed, CP12 possesses a range of properties that are landmarks of IDPs such as a bias in amino acid composition, is enriched in charged amino acid residues, is depleted in hydrophobic residues, and has a high proportion of disorder-promoting residues (Figure 3). Even if CP12 has a high proportion of disorder-promoting residues, the presence of cysteine residues was first surprising as cysteine residues were considered as "order-promoting residues" due to their ability to form inter- or intramolecular disulfide bridges.

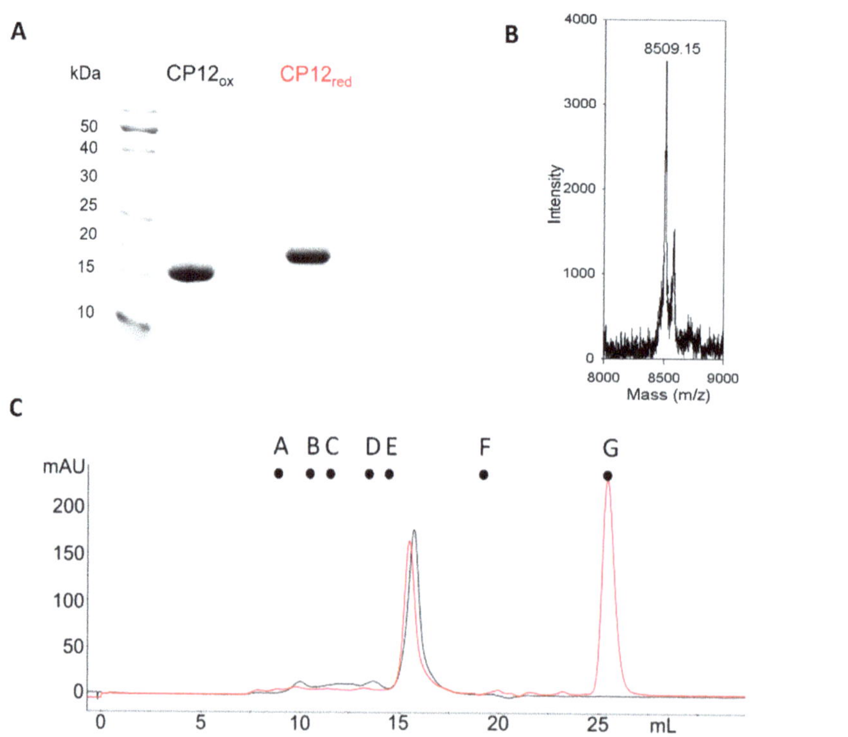


**Figure 2.** CP12 behaves as an IDP. (**A**) SDS-PAGE (12%) of 4 μg *C. reinhardtii* recombinant CP12 under its oxidized or reduced state. (**B**) MALDI-ToF mass spectrum of the native CP12 isolated from the PRK/GAPDH/CP12 complex of *C. reinhardtii*. (**C**) Size-exclusion chromatography profile

of *C. reinhardtii* recombinant CP12 under oxidized (black) or reduced (red) conditions (column: Superdex 200 10 × 300 mm). Above the chromatogram, the dots from A to G indicate the position of molecular-weight standard globular proteins. A: Ferritine (MW 440 kDa, $r_H$ 6.8 nm); B: Catalase (MW 240 kDa, $r_H$ 5.5 nm); C: dimer of Bovine Serum Albumin (BSA, MW 136 kDa, $r_H$ 4.5 nm); D: monomer of BSA (MW 68 kDa, $r_H$ 3.5 nm); E: Ovalbumin (MW 43 kDa, $r_H$ 3 nm); F: Cytochrome (MW 12.5 kDa, $r_H$ 2 nm) C; and G: oxidized form of DTT. MW and $r_H$ stand for molecular weight and hydrodynamic radius.

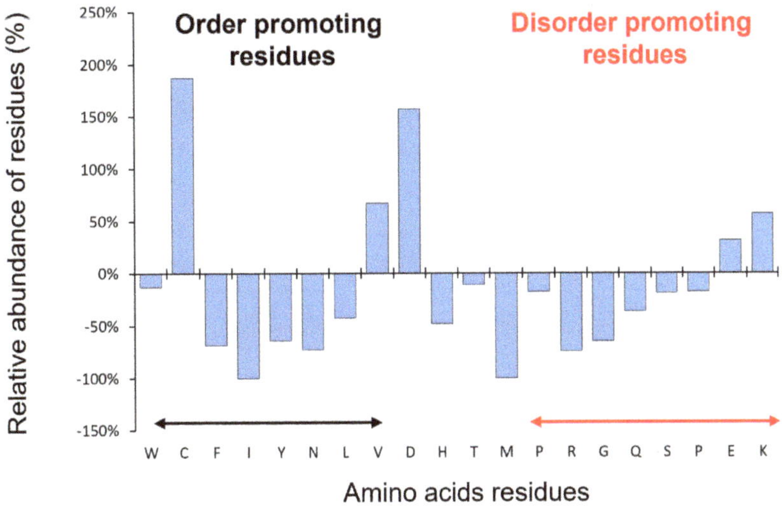


**Figure 3.** CP12 presents a bias in amino acid composition. Comparison of amino acid composition between globular proteins and *C. reinhardtii* CP12 using composition profiler (http://www.cprofiler.org (accessed on 25 July 2022)) [40]. The globular proteins dataset is from protein data bank (PDB) Select 25.

After the identification of CP12 as an IDP, a range of biophysical techniques confirmed that reduced CP12 completely lacks stable secondary and tertiary structural elements. The circular dichroism (CD) spectrum of reduced CP12 (or imitations of reduced CP12 using cysteine to serine mutants) showed a minimum ellipticity at 200 nm, as is characteristic for disordered proteins (Figure 4A) [25]. The Kratky representation of the small-angle X-ray scattering (SAXS) data of reduced CP12 exhibited a plateau at q.Rg > 2 typical of random polymers and characteristic of fully disordered proteins (Figure 4B) [41]. The $^1$H nuclear magnetic resonance (NMR) frequencies of all resonances from reduced CP12 showed minimal chemical shift dispersion (clustered from 7.5 to 8.5 ppm); their linewidths were sharp, and all the features of NMR data were typical of that disordered proteins (Figure 4C) [41]. In addition, NMR data confirmed that reduced CP12 exchanges between a myriad of possible conformations rapidly at a timescale of less than a nanosecond, as expected for an IDP. Other biophysical methods confirmed the IDP properties for reduced CP12, including Förster resonance energy transfer (FRET), fluorescence correlation spectroscopy (FCS), or mass spectrometry [42]. Moreover, it was shown that under oxidized conditions, CP12 was partially folded but still very flexible, and only a model structure obtained by sequence-based molecular modelling was available for many years [43]. CD analysis showed that it was much more helical than in its reduced form (Figure 4A), and an ion-mobility mass spectrometry study showed that the algal oxidized CP12 exists under two conformational states, a compact one and an extended one [34]. Later, experimental data obtained by SAXS also showed atypical features: the Kratky plot of oxidized CP12 was an intermediate between that of a well-folded protein (a bell-shaped curve with a maximum at q.Rg value of $\sqrt{3}$) and that of a fully disordered protein (such as that of reduced CP12), and these features are characteristic of protein with unstable structural properties (Figure 4B) [44,45]. The SAXS profile revealed the co-existence of two populations of conformers in solution, a compact one and a more disordered one, with all features being characteristic of protein with unstable structural properties. Similarly, the $^{1}H$-$^{15}N$ NMR spectrum of oxidized CP12 differed from that of reduced CP12 and showed a small number of dispersed resonances together with a large number of broad resonances clustered from 7.5 to 8.5 ppm (Figure 4C). All these experimental data could be reconciled with a two-state equilibrium for the algal oxidized CP12: (i) 60% of the oxidized CP12 molecules have two helices in the N-terminal half of the protein and a globular domain at the C-terminus; (ii) 40% of the oxidized CP12 molecules have only the globular fold at the C-terminus, while the N-terminal half remains disordered [44]. The multiple structural transitions and conformational flexibility of CP12 could provide a clue on how this protein can carry variable functions and bind multiple partners. When the stable C-terminal structural element of the *C. reinhardtii* oxidized CP12 binds to GAPDH, it induces a cryptic disorder, and its unstable N-terminal region is further destabilized to favor a disordered conformation [44]. This structural transition upon GAPDH binding contrasts to plant oxidized CP12, where the binding of GAPDH leads to a compaction of the N-terminal region [36,37]. These differences in the stability of the N-terminal region of oxidized CP12 correlate with the differences of relative affinity of CP12 for GAPDH between the algal and the plant species mentioned above with opposite entropic contribution to the binding.

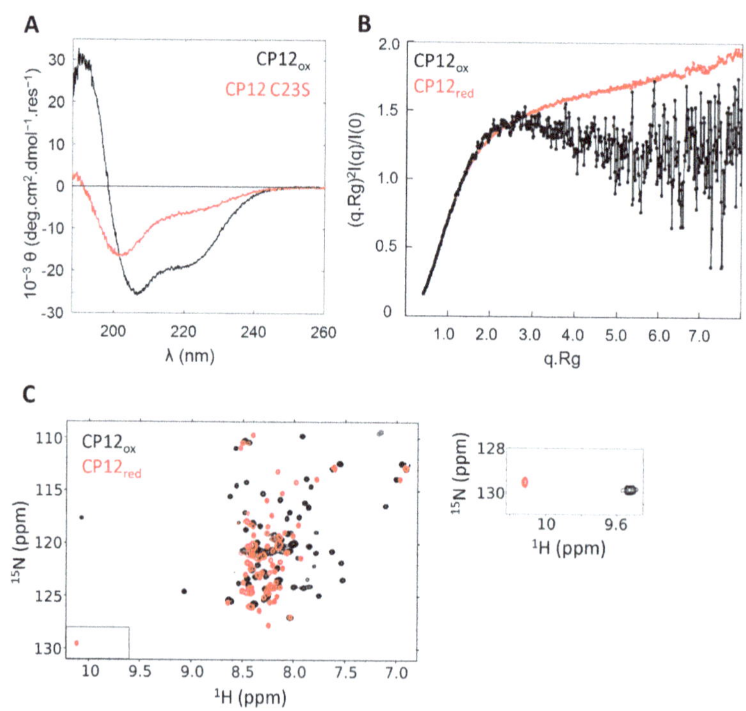


**Figure 4.** Biophysical analysis of CP12 confirmed that CP12 is an IDP. (**A**) Circular dichroism spectra of 10 μM recombinant *C. reinhardtii* oxidized CP12 (black), and of a CP12 mutant lacking the N-terminal disulfide bridge (mimicking reducing conditions, red). (**B**) Normalised Kratky representation of the SAXS data of the oxidized (black) and reduced (red) form of recombinant *C. reinhardtii* CP12. (**C**) NMR $^{1}H$-$^{15}N$-HSQC spectra of the oxidized (black) and reduced (red) form of recombinant *C. reinhardtii* CP12. The box between 9.5 and 10 ppm corresponds to the insert shown on the left.

Because the structural properties of CP12 vary significantly depending upon the redox conditions, the term conditionally disordered was coined for this protein. Structural properties of CDP such as CP12 are challenging to analyze [46]. Therefore, the only high-resolution structures available for CP12 are those of oxidized CP12 within the ternary complex and have been deciphered recently by crystallography and cryo-electron microscopy [47–50].

CP12 is not the unique protein that undergoes structural transitions upon oxidation/reduction, and it is predicted that redox-sensitive CDPs are widespread and have key roles in many eukaryotic processes [51]. Based on the computational platform, IUPred2A, it was predicted that cysteine-rich sequences display significant disorder in the reduced but not the oxidized form, increasing the potential for such sequences to function in a redox-sensitive manner [52]. In photosynthetic organisms where dark-light transitions are correlated to different oxido-reduction conditions, this concept is of paramount importance. The redox structural transitions that have been observed for CP12 might be highly relevant to CP12 being a redox switch of the CBB cycle [53].

## 4. CP12, a Widespread Protein with Sequence Variations on an Original Theme

After 2002, the number of manuscripts dealing with this protein started to increase, and CP12 has been found in many species such as higher plants, microalgae and cyanobacteria [54]. The canonical CP12 sequence contains one N-terminal cysteine residue pair separated by seven or eight residues, one C-terminal cysteine residue pair separated by eight residues encompassing a central proline residue (CxxxPxxxxC), and a core sequence AWD_VEEL (Figure 5). The two pairs of cysteine residues are capable of forming disulfide bridges required to form the ternary complex described above in green algae and higher plants [55]. However, in the glaucophyte *Cyanophora paradoxa*, CP12 lacks the two cysteine residues at the N-terminus [54]. The lack of the N-terminal pair of cysteine residues was also found in the red algal *Galdieria sulphuraria* CP12 and *Synechococcus elongatus* PCC7942, but it did not impair the formation of the ternary complex [56,57]. Though these two cysteine residues were claimed to be important to the PRK binding in higher plants, the presence of the disulfide bond at the N-terminus of CP12 might not be a requisite for PRK binding. It is, however, possible that the affinity between PRK and CP12 is much lower when this disulfide bond is absent and that its absence modulates the stability of the N-terminal helical hairpin described above. Indeed, the mutant of CP12 lacking this disulfide bond is less prone to interact with PRK, but a faint band is still present, indicating a degree of PRK and CP12 interaction [16].

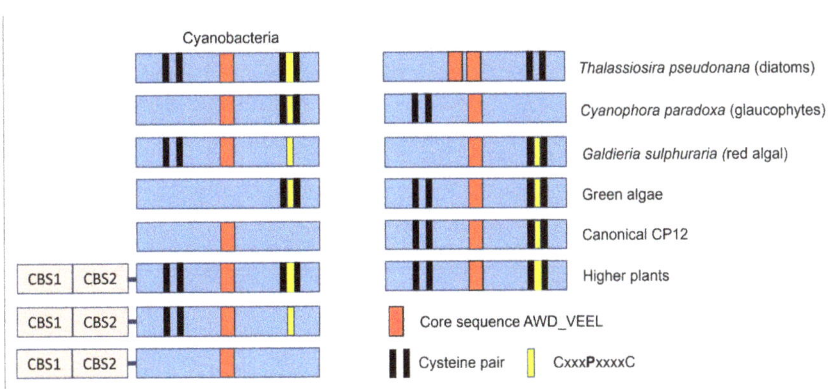


**Figure 5.** Schematic representation of structural variants of CP12. CBS (orange rectangles) stands for cystathionine β-synthase. The pairs of black lines represent cysteine residues pairs, the red rectangles represent the core sequences AWD_VEEL, and the yellow lines represent the central proline within the C-terminal residues pair. Adapted from D.N Stanley et al., 2013 [58].

In cyanobacteria, CP12 proteins fused to two cystathionine β-synthase (CBS) domains (CBS-CP12) were found beside the stand-alone CP12, and at present, CBS-fused CP12 has only been reported in these organisms. These CBS-proteins are widespread, and the analysis of 333 cyanobacterial genomes revealed the presence of many variants (Figure 5) [58].

A CP12-like protein was reported in the freshwater diatom, *Asterionella formosa*, that was associated with GAPDH and the ferredoxin NADP reductase, but the sequence of this protein is not available [59,60]. In contrast, in the marine diatom, *Thalassiosira pseudonana*, three CP12 proteins were identified, CP12-1 and CP12-2 were predicted to be localized in the chloroplast, and only CP12-2 was found in expressed sequence tags (ESTs) database and further characterized [61]. The gene coding for this protein in other diatoms was also found. In diatoms, nonetheless, PRK/GAPDH/CP12 has never been found [62], and it seems that the absence of two cysteine residues at positions 245 and 248 on diatom PRK could explain this [63]. Like the canonical CP12, the *T. pseudonana* CP12 possesses some intrinsically disordered regions, is highly dynamic but possesses a central coiled coil motif, and is dimeric, and these characteristics give *T. pseudonana* CP12-2 a form of an elongated cylinder with kinks [61].

The cyanophage-infecting marine picocyanobacteria of the genera Prochlorococcus and Synechococcus have been shown to express a protein that has a C-terminal extension similar to that of CP12. This protein shuts down the CBB cycle, as does the canonical CP12, and uses the NADPH produced by the host to fuel their own deoxynucleotide biosynthesis for replication [64]. Other proteins also possess a C-terminal extension similar to the C-terminus of CP12, such as the B subunit of the higher plant $A_2B_2$ GAPDH, the adenylate kinase 3 (ADK3) from *C. reinhardtii*, and the argininosuccinate lyase. This CxxxPxxxxC extremity interacts with GAPDH in the PRK/GAPDH/CP12, and this interaction is also conserved in the $A_2B_2$ GAPDH and the ADK3 [65,66]. In the prasinophycean green algae, *Ostreococcus tauri* and *Ostreococcus lucimarinus*, CP12 is not present, but they possess the redox-regulated B subunit of GAPDH, which is typical of *Streptophyta* [67].

Three isoforms of CP12 have been reported in higher plants [66]. In *A. thaliana*, the transcripts localization of the isoforms differs; CP12-1 and CP12-2 are mostly expressed in photosynthetic tissues, whereas CP12-3 is expressed in non-photosynthetic tissues such as in the roots. In contrast, in *C. reinhardtii*, one unique isoform has been reported to be localized in the chloroplast. In the C4 plant maize (*Zea mays*), a CP12 homolog was found in the bundle sheath and not in the mesophyll cells [68]. Recently, two CP12 proteins were found in sugarcane, another C4 plant [69]. Though yet never reported, it is very likely that plants with a crassulacean acid metabolism (CAM) also possess this protein. Therefore, it seems that this protein is ubiquitous in the plant kingdom.

## 5. One Gene, One Protein, Many Functions

### *5.1. CP12 Jack-of-All Trades but Master of the CBB Cycle*

As mentioned above, CP12 is known to be the master of the CBB cycle [9]. The involvement of oxidized CP12 in supramolecular complexes containing GAPDH and PRK has been demonstrated in several photosynthetic organisms though the strength of binding between these proteins differs among species. As mentioned above, the dissociation constant for GAPDH/CP12 is in the micromolar range in *A. thaliana* [33] but in the nanomolar range in *C. reinhardtii* [25]. The flexibility and the net negative charge of CP12 may increase its reactive area and 'stickiness' compared with rigid proteins, thus enhancing the ability of this protein to act as a 'scaffold protein' [70].

In *S. elongatus* PCC7942, CP12 forms the ternary complex in response to NADP(H)/NAD(H) ratio. Of interest, most CBB enzymes are not redox-regulated in cyanobacteria [57], whereas in higher plants and green algae, some CBB enzymes are redox-regulated via the thioredoxins (Trx). In Plantae, the Trx participate, in addition to the association-dissociation of the complex PRK/CP12/GAPDH, regulates PRK and GAPDH enzymes activities. PRK and CP12 are reduced by Trx f and m and could be oxidized by the newly characterized TrxLike2 [71,72]. The presence of CP12 has been shown to modify the PRK redox regulation, and, in particular, the formation of the ternary complex decreases the time required for PRK activation [73]. CP12 is also responsible for the redox regulation of the $A_4$ form of GAPDH in *C. reinhardtii* [35]. In contrast, the $A_2B_2$ form of GAPDH is autonomously redox-regulated, and CP12 therefore might not be required. Nevertheless, the $A_2B_2$ GAPDH is found in the ternary complex and is more easily activated in dimmer light than the $A_8B_8$ GAPDH oligomer mentioned above [29]. This suggests that CP12 can have other functions than the redox regulation of PRK and GAPDH.

The expression of the genes encoding GAPDH, PRK, and CP12-2 in *A. thaliana* was found to be coordinated, and this suggests that they are regulated at the transcriptional level [74,75]. This suggests that CP12 is involved in the post-translational regulation and at the transcriptional level. A recent study showed that reduced *C. reinhardtii* CP12 stabilizes PRK in vitro and in vivo, but the mechanism of this protection needs further investigation [76,77]. In the mutant strain of *C. reinhardtii*, where the CP12 protein is absent, while the abundance of numerous proteins increases (see below), the abundance of others, including PRK, involved in photosynthesis, decreases. This is in agreement with other studies on *N. tabacum*, *A. thaliana*, and *S. guianensis* that showed that photosynthetic efficiency is reduced in the CP12 deletion mutant [76–80].

### *5.2. CP12, Other Functions*

Like many IDPs, CP12 is a promiscuous protein, and in *C. reinhardtii*, in an oxidized state, it can bind other enzymes such as the malate dehydrogenase, the elongation factor 1α2, and 38 kDa ribosome-associated protein, but to a lesser extent than PRK, GAPDH, and the fructose-1,6-bisphosphate aldolase [81]. IDPs are well known to be a hub for the supramolecular complex, but it is surprising that, so far, no interacting partners have been identified for the disordered reduced CP12, and this has probably been overlooked.

Several studies have shown that the role of CP12 is beyond the CBB cycle. In *C. reinhardtii*, the deletion of the protein induced a re-routing of the metabolism under the light. In particular, metabolic pathways involving malate shuttles increased in the mutant such as the tricarboxylic acid cycle (TCA) and the glyoxylate pathway [77]. Malate shuttles, combined with other signaling factors, play a putative role in algal $CO_2$-concentrating mechanisms (CCM) [82,83]. In relation to this, it can be noticed that CP12 increases in low $CO_2$ conditions in *T. pseudonana*, conditions that trigger CCM [84]. In *N. tabacum* antisense plants, the activity of malate dehydrogenase and glucose-6-phosphate dehydrogenase decreased, and transcripts for polyamine metabolism and polyphenol oxidase were up-regulated [78,79]. A CP12-disrupted strain was engineered in *S. elongatus* PCC7942, and its growth was similar to that of wild-type cells under continuous light but was significantly reduced under the light/dark cycle (12 h/12 h) [57,85]. In the dark, the $O_2$ consumption by the mutant strain was lower, and the concentration of ribulose-1,5-bisphosphate, the product of the PRK reaction, was higher than for the wild-type. In cyanobacteria, the main metabolic pathway in the dark is the oxidative pentose phosphate (OPP) pathway that also encompasses the ribulose-1,5 -bisphosphate. By inhibiting the activity of PRK and GAPDH in the dark, CP12 thus regulates the carbon flow from the CBB cycle to the OPP cycle. The authors also found that the cyanobacterial CP12 can bind NADPH (not NADH), but this has not been reported and/or studied to our knowledge in any other CP12. All these results show that the role of CP12 is beyond the regulation of the CBB cycle. In the sugarcane, the expression of one of the isoform of the CP12—ShCP12-1—decreased immediately on the onset of sucrose accumulation that occurs under the yellow canopy syndrome, a specific pattern of leaf yellowing accompanied by abnormal and lethal accumulation of sucrose and starch in leaves [69]. This CP12 might therefore be the primary regulation point of sugar feedback regulation occurring in C4 plants, while the two carboxylating enzymes, ribulose-1,5-bisphosphate carboxylase oxygenase (RuBisCO) and phosphoenolpyruvate carboxylase, were only negatively regulated at a later stage and might be the secondary regulation points.

In cyanobacteria, CP12 has been found in a fusion protein with a CBS domain, as mentioned above. A study of CBS-CP12 from *Microcystis aeruginosa* revealed that the gene expression of this protein is clearly light-induced. In addition, CBS-CP12 oligomerizes and forms a hexamer but does not form the ternary complex with GAPDH and PRK. It can bind AMP and then inhibits the activation of PRK by thioredoxins [86]. The authors propose that this new architecture provides CP12 additional regulatory functions in cyanobacteria.

### 5.3. CP12, an Anti-Stress Protein

In both *A. thaliana* and *N. tabacum*, the antisense suppression of CP12 increased the expression of proteins related to oxidative stress [87]. Recently, it has been shown in *C. reinhardtii* that the suppression of CP12 leads also to an increase in the proteins involved in stress [77]. In addition, in cyanobacteria, CP12 might be involved in oxidative stress by controlling the electrons flux from Photosystem I. Indeed, while the growth at low light of the wild-type and the CP12 mutant strain were the same, at high light the mutant strain grew more slowly. The chlorophyll content also decreased in this strain, and the reactive oxygen species increased [85], while in A. thaliana and *C. reinhardtii*, it has been shown that CP12 provides the thiol groups PRK and GAPDH protection against oxidative damage [87]. In cyanobacteria, the defense mechanism could be different and independent of the thiol groups of these enzymes [57].

In *C. reinhardtii*, CP12 protected GAPDH against heat inactivation and aggregation and therefore plays the role of a specific chaperone [88]. As mentioned above, CP12 also protects PRK against irreversible inactivation in vitro [77]. Besides these roles as a specific chaperone, CP12 from other organisms is more abundant in stress conditions, and this is the case for the *T. pseudonana* CP12-2. The expression of this protein was higher under low $CO_2$ [84] but also under N, P, or Si limited conditions [89]. These results therefore indicate that CP12 is not specific to carbon metabolism.

In the tropical legume, *Stylosanthes guianensis*, the higher expression of CP12 increases growth and plant height. In addition to the expected functions, a potential role for CP12 in chilling tolerance has been suggested [80]. A recent transcriptomic analysis of maize also revealed the different regulations of cold-responsive genes and, among them, the CP12 gene is present [90].

All these results show that the role of CP12 is not restricted to the formation of the well-known supramolecular complex involving PRK and GAPDH but is probably more general and characterized not only by conformation heterogeneity but also by functional heterogeneity defining its moonlighting signature as many IDPs.

### 5.4. CP12 and Metal Ions

Metal binding is ubiquitous in biology, being important for folding, stability, transportation, and catalysis [91]. *C. reinhardtii* recombinant CP12 purified by affinity chromatography on nickel columns had a yellow color, even after dialysis with a buffer devoid of metal and imidazole. The absorption spectra from 280 to 600 nm showed the presence of a broad peak around 410 nm, and these spectra strongly resembled those of ferredoxin [92]. Using electrospray non-denaturing mass spectrometry, the authors showed that CP12 was specifically able to bind $Cu^{2+}$ and $Ni^{2+}$ with a low affinity (dissociation constants of 26 and 11 µM, respectively) [92], values close to those obtained for the binding of copper to prion proteins ($K_D$ of 14 µM) [93]. $Cu^{2+}$ catalyzed the oxidation of the reduced CP12, with the reformation of disulfide bonds leading to the formation of oxidized CP12, which was able to bind a $Cu^{2+}$ ion. In addition, a hydrophobic cluster analysis showed that CP12 had high similarity with copper chaperones from A. thaliana. Though many questions remain unanswered, one can hypothesize that CP12 may play a role in copper homeostasis like other copper chaperones [94]. Later, using top-down mass spectrometry, three regions were found to be involved in metal ion binding: Asp16-Asp23, Asp38-Lys50, and Asp70-Glu76 [88]. It has been suggested that the binding of copper led to a more rigid structure, but this requires further investigation. Later, using two-dimensional polyacrylamide gel electrophoresis separation of the stroma fraction of *A. thaliana* chloroplasts followed by calcium overlay assay, CP12 was identified as a calcium-binding protein [95]. Though this protein does not possess the canonical calcium-binding EF-hand motif, the authors suggested that negatively charged amino acid residues could be involved in this binding. The biological functions of the $Cu^{2+}$, $Ni^2$+, and $Ca^{2+}$ binding, however, remain unsolved and need to be further investigated.

## 6. Conclusions

Photosynthesis regulation depends on many signals, including pH, metabolite concentrations, and oxido-reduction conditions. For photosynthesis to be optimized, the signals received have to be transmitted in a rapid and specific manner and often involve protein-protein interactions; IDPs are well suited for such functions. The chloroplast protein, CP12, a redox dependent conditionally disordered protein, acting as a linker or scaffold between PRK and GAPDH, can integrate these multiple signals to regulate their activity. The redox state of CP12 conditions a severe structural transition of its structural properties from a completely disordered state under reducing conditions to a partially stable state under oxidizing conditions. This redox-dependent structural transition is also concomitant with the association-dissociation with PRK and GAPDH enzymes and thus the regulation of their activity under dark (inactive enzymes) or light (active enzymes). The two enzymes, PRK and GAPDH, do not catalyze consecutive reactions but are using ATP and NADPH, respectively, both products from the primary phase of photosynthesis. PRK produces the RuBisCO substrate, ribulose 1,5-bisphosphate, from the ribulose-5-phosphate, an intermediate of the OPP pathway. GAPDH uses NADPH to produce glyceraldehyde-3-phosphate, which can be exported and is also an intermediate of the OPP. Therefore, CP12 using as a regulatory protein of both PRK and GAPDH, thus "killing two birds with one stone", contributes to the fine tuning of metabolic pathways such as the CBB cycle, the glycolysis, and the OPP, avoiding futile cycling. It is also involved in the regulation of TCA and glyoxylate cycles involving the malate shuttle and possibly involved in CCM. Moreover, besides its role in controlling metabolic pathways, CP12 provides a cell-signaling pathway, triggers anti-stress responses and protects against oxidative damage. It is also able to bind metal ions, though hitherto the biological significance of this remains unknown (Figure 6). The pursuit of knowledge on these disordered proteins will probably produce new concepts in the sciences as the more we learn and the more questions we will find to ask. The discovery of disordered proteins and of CP12, 70 years after the discovery of the CBB cycle, offers new insights into the photosynthesis field, and this is probably not a dead end.

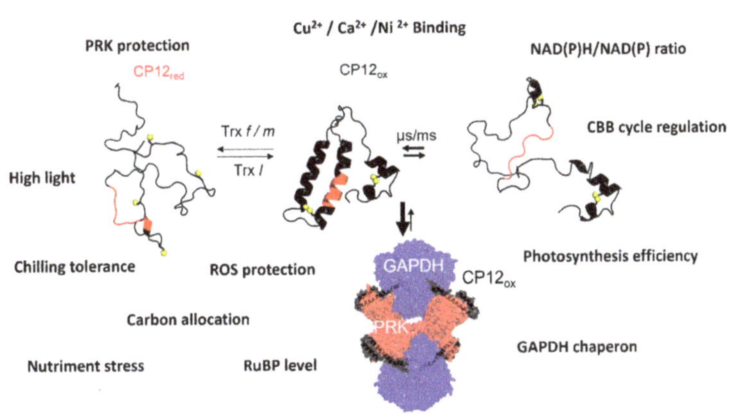


**Figure 6.** Atlas of CP12 functions. The core sequence of CP12, AWD_VEEL, is in red, and sulfur atoms are indicated in yellow. Reduced CP12 is fully disordered, and oxidized CP12 is partially ordered. Under oxidized state, CP12 forms a supramolecular complex with GAPDH and PRK. A non-exhaustive list of CP12 functions is indicated in the scheme.

**Author Contributions:** B.G., H.L. and C.G., writing—original draft preparation, B.G., H.L., C.G., V.R.-B. and F.C. writing—review and editing. All authors have read and agreed to the published version of the manuscript.

**Funding:** C.G.'s studentship (ED62) is supported by the Ministère de la Recherche, de l'Enseignement supérieur et de l'Innovation. Research from B.G., H.L., F.C. and V.R.-B. is supported by the Centre National de la Recherche Scientifique (CNRS), Aix-Marseille Université, Agence Nationale de la Recherche (ANR-21-CE20-0029) and the Federation of European Biochemical Societies (Launay FEBS Excellence Award, 2021).

**Institutional Review Board Statement:** Not applicable.

**Informed Consent Statement:** Not applicable.

**Data Availability Statement:** Not applicable.

**Acknowledgments:** This work received support from the French government under the France 2030 investment plan, as part of the Initiative d'Excellence d'Aix-Marseille Université—A*MIDEX and is part of the Institute of Microbiology, Bioenergies, and Biotechnology—IM2B (AMX-19-IET-006).

**Conflicts of Interest:** The authors declare no conflict of interest.

## References

1. Uversky, V.N. Protein Intrinsic Disorder and Structure-Function Continuum. *Prog. Mol. Biol. Transl. Sci.* **2019**, *166*, 1–17. [CrossRef] [PubMed]
2. Uversky, V.N. Natively Unfolded Proteins: A Point Where Biology Waits for Physics. *Protein Sci. Publ. Protein Soc.* **2002**, *11*, 739–756. [CrossRef] [PubMed]
3. Yruela, I.; Contreras-Moreira, B. Genetic Recombination Is Associated with Intrinsic Disorder in Plant Proteomes. *BMC Genom.* **2013**, *14*, 772. [CrossRef]
4. Rendón-Luna, D.F.; Romero-Pérez, P.S.; Cuevas-Velazquez, C.L.; Reyes, J.L.; Covarrubias, A.A. Determining the Protective Activity of IDPs Under Partial Dehydration and Freeze-Thaw Conditions. *Methods Mol. Biol.* **2020**, *2141*, 519–528. [CrossRef]
5. Zhang, Y.; Launay, H.; Schramm, A.; Lebrun, R.; Gontero, B. Exploring Intrinsically Disordered Proteins in *Chlamydomonas reinhardtii*. *Sci. Rep.* **2018**, *8*, 6805. [CrossRef] [PubMed]
6. Thieulin-Pardo, G.; Schramm, A.; Lignon, S.; Lebrun, R.; Kojadinovic, M.; Gontero, B. The Intriguing CP12-like Tail of Adenylate Kinase 3 from *Chlamydomonas reinhardtii*. *FEBS J.* **2016**, *283*, 3389–3407. [CrossRef] [PubMed]
7. Sena, L.; Uversky, V.N. Comparison of the Intrinsic Disorder Propensities of the RuBisCO Activase Enzyme from the Motile and Non-Motile Oceanic Green Microalgae. *Intrinsically Disord. Proteins* **2016**, *4*, e1253526. [CrossRef]
8. Launay, H.; Receveur-Bréchot, V.; Carrière, F.; Gontero, B. Orchestration of Algal Metabolism by Protein Disorder. *Arch. Biochem. Biophys.* **2019**, *672*, 108070. [CrossRef]
9. Gontero, B.; Maberly, S.C. An Intrinsically Disordered Protein, CP12: Jack of All Trades and Master of the Calvin Cycle. *Biochem. Soc. Trans.* **2012**, *40*, 995–999. [CrossRef]
10. Mackinder, L.C.M.; Meyer, M.T.; Mettler-Altmann, T.; Chen, V.K.; Mitchell, M.C.; Caspari, O.; Rosenzweig, E.S.F.; Pallesen, L.; Reeves, G.; Itakura, A.; et al. A Repeat Protein Links Rubisco to Form the Eukaryotic Carbon-Concentrating Organelle. *Proc. Natl. Acad. Sci. USA* **2016**, *113*, 5958–5963. [CrossRef]
11. Atkinson, N.; Mao, Y.; Chan, K.X.; McCormick, A.J. Condensation of Rubisco into a Proto-Pyrenoid in Higher Plant Chloroplasts. *Nat. Commun.* **2020**, *11*, 6303. [CrossRef]
12. He, S.; Chou, H.-T.; Matthies, D.; Wunder, T.; Meyer, M.T.; Atkinson, N.; Martinez-Sanchez, A.; Jeffrey, P.D.; Port, S.A.; Patena, W.; et al. The Structural Basis of Rubisco Phase Separation in the Pyrenoid. *Nat. Plants* **2020**, *6*, 1480–1490. [CrossRef] [PubMed]
13. Wunder, T.; Cheng, S.L.H.; Lai, S.-K.; Li, H.-Y.; Mueller-Cajar, O. The Phase Separation Underlying the Pyrenoid-Based Microalgal Rubisco Supercharger. *Nat. Commun.* **2018**, *9*, 5076. [CrossRef] [PubMed]
14. Mackinder, L.C.M.; Chen, C.; Leib, R.D.; Patena, W.; Blum, S.R.; Rodman, M.; Ramundo, S.; Adams, C.M.; Jonikas, M.C. A Spatial Interactome Reveals the Protein Organization of the Algal $CO_2$-Concentrating Mechanism. *Cell* **2017**, *171*, 133–147.e14. [CrossRef] [PubMed]
15. Pohlmeyer, K.; Paap, B.K.; Soll, J.; Wedel, N. CP12: A Small Nuclear-Encoded Chloroplast Protein Provides Novel Insights into Higher-Plant GAPDH Evolution. *Plant Mol. Biol.* **1996**, *32*, 969–978. [CrossRef]
16. Wedel, N.; Soll, J.; Paap, B.K. CP12 Provides a New Mode of Light Regulation of Calvin Cycle Activity in Higher Plants. *Proc. Natl. Acad. Sci. USA* **1997**, *94*, 10479–10484. [CrossRef] [PubMed]
17. Gontero, B.; Cardenas, M.L.; Ricard, J. A Functional Five-Enzyme Complex of Chloroplasts Involved in the Calvin Cycle. *Eur. J. Biochem.* **1988**, *173*, 437–443. [CrossRef]
18. Gontero, B.; Mulliert, G.; Rault, M.; Giudici-Orticoni, M.-T.; Ricard, J. Structural and Functional Properties of a Multi-Enzyme Complex from Spinach Chloroplasts. II: Modulation of the Kinetic Properties of Enzymes in the Aggregated State. *Eur. J. Biochem.* **1993**, *217*, 1075–1082. [CrossRef]

19. Rault, M.; Gontero, B.; Ricard, J. Thioredoxin Activation of Phosphoribulokinase in a Chloroplast Multi-Enzyme Complex. *Eur. J. Biochem.* **1991**, *197*, 791–797. [CrossRef]
20. Avilan, L.; Gontero, B.; Lebreton, S.; Ricard, J. Memory and Imprinting Effects in Multienzyme Complexes. *Eur. J. Biochem.* **1997**, *246*, 78–84. [CrossRef]
21. Avilan, L.; Gontero, B.; Lebreton, S.; Ricard, J. Information Transfer in Multienzyme Complexes.2. The Role of Arg64 of *Chlamydomonas reinhardtii* Phosphoribulokinase in the Information Transfer between Glyceraldehyde-3-Phosphate Dehydrogenase and Phosphoribulokinase. *Eur. J. Biochem.* **1997**, *250*, 296–302. [CrossRef]
22. Lebreton, S.; Gontero, B.; Avilan, L.; Ricard, J. Memory and Imprinting Effects in Multienzyme Complexes.2. Kinetics of the Bienzyme Complex from *Chlamydomonas reinhardtii* and Hysteretic Activation of Chloroplast Oxidized Phosphoribulokinase. *Eur. J. Biochem.* **1997**, *246*, 85–91. [CrossRef] [PubMed]
23. Lebreton, S.; Gontero, B.; Avilan, L.; Ricard, J. Information Transfer in Multienzyme Complexes.1. Thermodynamics of Conformational Constraints and Memory Effects in the Bienzyme Glyceraldehyde-3-Phosphate-Dehydrogenase-Phosphoribulokinase Complex of *Chlamydomonas reinhardtii* Chloroplasts. *Eur. J. Biochem.* **1997**, *250*, 286–295. [CrossRef] [PubMed]
24. Graciet, E.; Lebreton, S.; Camadro, J.-M.; Gontero, B. Characterization of Native and Recombinant A4 Glyceraldehyde 3-Phosphate Dehydrogenase. Kinetic Evidence for Confromation Changes upon Association with the Small Protein CP12. *Eur. J. Biochem.* **2003**, *270*, 129–136. [CrossRef] [PubMed]
25. Graciet, E.; Gans, P.; Wedel, N.; Lebreton, S.; Camadro, J.-M.; Gontero, B. The Small Protein CP12: A Protein Linker for Supramolecular Complex Assembly. *Biochemistry* **2003**, *42*, 8163–8170. [CrossRef] [PubMed]
26. Mouche, F.; Gontero, B.; Callebaut, I.; Mornon, J.P.; Boisset, N. Striking Conformational Change Suspected within the Phosphoribulokinase Dimer Induced by Interaction with GAPDH. *J. Biol. Chem.* **2002**, *277*, 6743–6749. [CrossRef] [PubMed]
27. Graciet, E.; Lebreton, S.; Camadro, J.-M.; Gontero, B. Thermodynamic Analysis of the Emergence of New Regulatory Properties in a Phosphoribulokinase-Glyceraldehyde 3-Phosphate Dehydrogenase Complex. *J. Biol. Chem.* **2002**, *277*, 12697–12702. [CrossRef] [PubMed]
28. Wedel, N.; Soll, J. Evolutionary Conserved Light Regulation of Calvin Cycle Activity by NADPH-Mediated Reversible Phosphoribulokinase/CP12/Glyceraldehyde-3-Phosphate Dehydrogenase Complex Dissociation. *Proc. Natl. Acad. Sci. USA* **1998**, *95*, 9699–9704. [CrossRef] [PubMed]
29. Scheibe, R.; Wedel, N.; Vetter, S.; Emmerlich, V.; Sauermann, S.M. Co-Existence of Two Regulatory NADP-Glyceraldehyde 3-P Dehydrogenase Complexes in Higher Plant Chloroplasts. *Eur. J. Biochem.* **2002**, *269*, 5617–5624. [CrossRef]
30. Howard, T.P.; Lloyd, J.C.; Raines, C.A. Inter-Species Variation in the Oligomeric States of the Higher Plant Calvin Cycle Enzymes Glyceraldehyde-3-Phosphate Dehydrogenase and Phosphoribulokinase. *J. Exp. Bot.* **2011**, *62*, 3799–3805. [CrossRef]
31. Clasper, S.; Easterby, J.S.; Powls, R. Properties of Two High-Molecular-Mass Forms of Glyceraldehyde-3-Phosphate Dehydrogenase from Spinach Leaf, One of Which Also Possesses Latent Phosphoribulokinase Activity. *Eur. J. Biochem.* **1991**, *202*, 1239–1246. [CrossRef] [PubMed]
32. Wara-Aswapati, O.; Kemble, R.J.; Bradbeer, J.W. Activation of Glyceraldehyde-Phosphate Dehydrogenase (NADP) and Phosphoribulokinase in *Phaseolus vulgaris* Leaf Extracts Involves the Dissociation of Oligomers. *Plant Physiol.* **1980**, *66*, 34–39. [CrossRef] [PubMed]
33. Marri, L.; Trost, P.; Trivelli, X.; Gonnelli, L.; Pupillo, P.; Sparla, F. Spontaneous Assembly of Photosynthetic Supramolecular Complexes as Mediated by the Intrinsically Unstructured Protein CP12. *J. Biol. Chem.* **2008**, *283*, 1831–1838. [CrossRef] [PubMed]
34. Kaaki, W.; Woudstra, M.; Gontero, B.; Halgand, F. Exploration of CP12 Conformational Changes and of Quaternary Structural Properties Using Electrospray Ionization Traveling Wave Ion Mobility Mass Spectrometry. *Rapid Commun. Mass Spectrom.* **2013**, *27*, 179–186. [CrossRef]
35. Erales, J.; Mekhalfi, M.; Woudstra, M.; Gontero, B. Molecular Mechanism of NADPH-Glyceraldehyde-3-Phosphate Dehydrogenase Regulation through the C-Terminus of CP12 in *Chlamydomonas reinhardtii*. *Biochemistry* **2011**, *50*, 2881–2888. [CrossRef]
36. Wright, P.E.; Dyson, H.J. Intrinsically Unstructured Proteins: Re-Assessing the Protein Structure-Function Paradigm. *J. Mol. Biol.* **1999**, *293*, 321–331. [CrossRef]
37. Dyson, H.J.; Wright, P.E. Coupling of Folding and Binding for Unstructured Proteins. *Curr. Opin. Struct. Biol.* **2002**, *12*, 54–60. [CrossRef]
38. Tompa, P. Intrinsically Unstructured Proteins. *Trends Biochem. Sci.* **2002**, *27*, 527–533. [CrossRef]
39. Moparthi, S.B.; Thieulin-Pardo, G.; Mansuelle, P.; Rigneault, H.; Gontero, B.; Wenger, J. Conformational Modulation and Hydrodynamic Radii of CP12 Protein and Its Complexes Probed by Fluorescence Correlation Spectroscopy. *FEBS J.* **2014**, *281*, 3206–3217. [CrossRef]
40. Vacic, V.; Uversky, V.N.; Dunker, A.K.; Lonardi, S. Composition Profiler: A Tool for Discovery and Visualization of Amino Acid Composition Differences. *BMC Bioinform.* **2007**, *8*, 211. [CrossRef]
41. Launay, H.; Barré, P.; Puppo, C.; Manneville, S.; Gontero, B.; Receveur-Bréchot, V. Absence of Residual Structure in the Intrinsically Disordered Regulatory Protein CP12 in Its Reduced State. *Biochem. Biophys. Res. Commun.* **2016**, *477*, 20–26. [CrossRef] [PubMed]
42. Moparthi, S.B.; Thieulin-Pardo, G.; de Torres, J.; Ghenuche, P.; Gontero, B.; Wenger, J. FRET Analysis of CP12 Structural Interplay by GAPDH and PRK. *Biochem. Biophys. Res. Commun.* **2015**, *458*, 488–493. [CrossRef] [PubMed]
43. Gardebien, F.; Thangudu, R.R.; Gontero, B.; Offmann, B. Construction of a 3D Model of CP12, a Protein Linker. *J. Mol. Graph. Model.* **2006**, *25*, 186–195. [CrossRef] [PubMed]

44. Launay, H.; Barré, P.; Puppo, C.; Zhang, Y.; Maneville, S.; Gontero, B.; Receveur-Bréchot, V. Cryptic Disorder Out of Disorder: Encounter between Conditionally Disordered CP12 and Glyceraldehyde-3-Phosphate Dehydrogenase. *J. Mol. Biol.* **2018**, *430*, 1218–1234. [CrossRef] [PubMed]
45. Receveur-Brechot, V.; Durand, D. How Random Are Intrinsically Disordered Proteins? A Small Angle Scattering Perspective. *Curr. Protein Pept. Sci.* **2012**, *13*, 55–75. [CrossRef] [PubMed]
46. Reichmann, D.; Jakob, U. The Roles of Conditional Disorder in Redox Proteins. *Curr. Opin. Struct. Biol.* **2013**, *23*, 436–442. [CrossRef] [PubMed]
47. McFarlane, C.R.; Shah, N.R.; Kabasakal, B.V.; Echeverria, B.; Cotton, C.A.R.; Bubeck, D.; Murray, J.W. Structural Basis of Light-Induced Redox Regulation in the Calvin-Benson Cycle in Cyanobacteria. *Proc. Natl. Acad. Sci. USA* **2019**, *116*, 20984–20990. [CrossRef] [PubMed]
48. Yu, A.; Xie, Y.; Pan, X.; Zhang, H.; Cao, P.; Su, X.; Chang, W.; Li, M. Photosynthetic Phosphoribulokinase Structures: Enzymatic Mechanisms and the Redox Regulation of the Calvin-Benson-Bassham Cycle. *Plant Cell* **2020**, *32*, 1556–1573. [CrossRef]
49. Matsumura, H.; Kai, A.; Maeda, T.; Tamoi, M.; Satoh, A.; Tamura, H.; Hirose, M.; Ogawa, T.; Kizu, N.; Wadano, A.; et al. Structure Basis for the Regulation of Glyceraldehyde-3-Phosphate Dehydrogenase Activity via the Intrinsically Disordered Protein CP12. *Structure* **2011**, *19*, 1846–1854. [CrossRef]
50. Trost, P.; Fermani, S.; Marri, L.; Zaffagnini, M.; Falini, G.; Scagliarini, S.; Pupillo, P.; Sparla, F. Thioredoxin-Dependent Regulation of Photosynthetic Glyceraldehyde-3-Phosphate Dehydrogenase: Autonomous vs. CP12-Dependent Mechanisms. *Photosynth. Res.* **2006**, *89*, 263–275. [CrossRef]
51. Erdős, G.; Mészáros, B.; Reichmann, D.; Dosztányi, Z. Large-Scale Analysis of Redox-Sensitive Conditionally Disordered Protein Regions Reveals Their Widespread Nature and Key Roles in High-Level Eukaryotic Processes. *Proteomics* **2019**, *19*, e1800070. [CrossRef] [PubMed]
52. Bhopatkar, A.A.; Uversky, V.N.; Rangachari, V. Disorder and Cysteines in Proteins: A Design for Orchestration of Conformational See-Saw and Modulatory Functions. *Prog. Mol. Biol. Transl. Sci.* **2020**, *174*, 331–373. [CrossRef] [PubMed]
53. Lopez-Calcagno, P.E.; Howard, T.P.; Raines, C.A. The CP12 Protein Family: A Thioredoxin-Mediated Metabolic Switch? *Front. Plant Sci.* **2014**, *5*, 9. [CrossRef] [PubMed]
54. Groben, R.; Kaloudas, D.; Raines, C.A.; Offmann, B.; Maberly, S.C.; Gontero, B. Comparative Sequence Analysis of CP12, a Small Protein Involved in the Formation of a Calvin Cycle Complex in Photosynthetic Organisms. *Photosynth. Res.* **2010**, *103*, 183–194. [CrossRef] [PubMed]
55. Avilan, L.; Puppo, C.; Erales, J.; Woudstra, M.; Lebrun, R.; Gontero, B. CP12 Residues Involved in the Formation and Regulation of the Glyceraldehyde-3-Phosphate Dehydrogenase–CP12–Phosphoribulokinase Complex in *Chlamydomonas reinhardtii*. *Mol. Biosyst.* **2012**, *8*, 2994. [CrossRef] [PubMed]
56. Oesterhelt, C.; Klocke, S.; Holtgrefe, S.; Linke, V.; Weber, A.P.M.; Scheibe, R. Redox Regulation of Chloroplast Enzymes in *Galdieria sulphuraria* in View of Eukaryotic Evolution. *Plant Cell Physiol.* **2007**, *48*, 1359–1373. [CrossRef] [PubMed]
57. Tamoi, M.; Miyazaki, T.; Fukamizo, T.; Shigeoka, S. The Calvin Cycle in Cyanobacteria Is Regulated by CP12 via the NAD(H)/NADP(H) Ratio under Light/Dark Conditions. *Plant J.* **2005**, *42*, 504–513. [CrossRef]
58. Stanley, D.N.; Raines, C.A.; Kerfeld, C.A. Comparative Analysis of 126 Cyanobacterial Genomes Reveals Evidence of Functional Diversity among Homologs of the Redox-Regulated CP12 Protein. *Plant Physiol.* **2013**, *161*, 824–835. [CrossRef] [PubMed]
59. Mekhalfi, M.; Puppo, C.; Avilan, L.; Lebrun, R.; Mansuelle, P.; Maberly, S.C.; Gontero, B. Glyceraldehyde-3-Phosphate Dehydrogenase Is Regulated by Ferredoxin-NADP Reductase in the Diatom *Asterionella formosa*. *New Phytol.* **2014**, *203*, 414–423. [CrossRef]
60. Boggetto, N.; Gontero, B.; Maberly, S.C. Regulation of Phosphoribulokinase and Glyceraldehyde 3-Phosphate Dehydrogenase in a Freshwater Diatom, *Asterionella formosa*. *J. Phycol.* **2007**, *43*, 1227–1235. [CrossRef]
61. Shao, H.; Huang, W.; Avilan, L.; Receveur-Bréchot, V.; Puppo, C.; Puppo, R.; Lebrun, R.; Gontero, B.; Launay, H. A New Type of Flexible CP12 Protein in the Marine Diatom *Thalassiosira pseudonana*. *Cell Commun. Signal. CCS* **2021**, *19*, 38. [CrossRef] [PubMed]
62. Wilhelm, C.; Büchel, C.; Fisahn, J.; Goss, R.; Jakob, T.; Laroche, J.; Lavaud, J.; Lohr, M.; Riebesell, U.; Stehfest, K.; et al. The Regulation of Carbon and Nutrient Assimilation in Diatoms Is Significantly Different from Green Algae. *Protist* **2006**, *157*, 91–124. [CrossRef] [PubMed]
63. Thieulin-Pardo, G.; Remy, T.; Lignon, S.; Lebrun, R.; Gontero, B. Phosphoribulokinase from *Chlamydomonas reinhardtii*: A Benson-Calvin Cycle Enzyme Enslaved to Its Cysteine Residues. *Mol. Biosyst.* **2015**, *11*, 1134–1145. [CrossRef] [PubMed]
64. Thompson, L.R.; Zeng, Q.; Kelly, L.; Huang, K.H.; Singer, A.U.; Stubbe, J.; Chisholm, S.W. Phage Auxiliary Metabolic Genes and the Redirection of Cyanobacterial Host Carbon Metabolism. *Proc. Natl. Acad. Sci. USA* **2011**, *108*, E757–E764. [CrossRef]
65. Zhang, Y.; Launay, H.; Liu, F.; Lebrun, R.; Gontero, B. Interaction between Adenylate Kinase 3 and Glyceraldehyde-3-phosphate Dehydrogenase from *Chlamydomonas reinhardtii*. *FEBS J.* **2018**, *285*, 2495–2503. [CrossRef]
66. Marri, L.; Pesaresi, A.; Valerio, C.; Lamba, D.; Pupillo, P.; Trost, P.; Sparla, F. In Vitro Characterization of Arabidopsis CP12 Isoforms Reveals Common Biochemical and Molecular Properties. *J. Plant Physiol.* **2010**, *167*, 939–950. [CrossRef]
67. Robbens, S.; Petersen, J.; Brinkmann, H.; Rouzé, P.; Van de Peer, Y. Unique Regulation of the Calvin Cycle in the Ultrasmall Green Alga Ostreococcus. *J. Mol. Evol.* **2007**, *64*, 601–604. [CrossRef]
68. Sun, Q.; Zybailov, B.; Majeran, W.; Friso, G.; Olinares, P.D.B.; van Wijk, K.J. PPDB, the Plant Proteomics Database at Cornell. *Nucleic Acids Res.* **2009**, *37*, D969–D974. [CrossRef]

69. Marquardt, A.; Henry, R.J.; Botha, F.C. Effect of Sugar Feedback Regulation on Major Genes and Proteins of Photosynthesis in Sugarcane Leaves. *Plant Physiol. Biochem. PPB* **2021**, *158*, 321–333. [CrossRef]
70. Cortese, M.S.; Uversky, V.N.; Keith Dunker, A. Intrinsic Disorder in Scaffold Proteins: Getting More from Less. *Prog. Biophys. Mol. Biol.* **2008**, *98*, 85–106. [CrossRef]
71. Marri, L.; Zaffagnini, M.; Collin, V.; Issakidis-Bourguet, E.; Lemaire, S.D.; Pupillo, P.; Sparla, F.; Miginiac-Maslow, M.; Trost, P. Prompt and Easy Activation by Specific Thioredoxins of Calvin Cycle Enzymes of *Arabidopsis thaliana* Associated in the GAPDH/CP12/PRK Supramolecular Complex. *Mol. Plant* **2009**, *2*, 259–269. [CrossRef] [PubMed]
72. Yoshida, K.; Hara, A.; Sugiura, K.; Fukaya, Y.; Hisabori, T. Thioredoxin-Like2/2-Cys Peroxiredoxin Redox Cascade Supports Oxidative Thiol Modulation in Chloroplasts. *Proc. Natl. Acad. Sci. USA* **2018**, *115*, E8296–E8304. [CrossRef] [PubMed]
73. Avilan, L.; Lebreton, S.; Gontero, B. Thioredoxin Activation of Phosphoribulokinase in a Bi-Enzyme Complex from *Chlamydomonas reinhardtii* Chloroplasts. *J. Biol. Chem.* **2000**, *275*, 9447–9451. [CrossRef] [PubMed]
74. Marri, L.; Trost, P.; Pupillo, P.; Sparla, F. Reconstitution and Properties of the Recombinant Glyceraldehyde-3-Phosphate Dehydrogenase/CP12/Phosphoribulokinase Supramolecular Complex of Arabidopsis. *Plant Physiol.* **2005**, *139*, 1433–1443. [CrossRef] [PubMed]
75. Terzaghi, W.B.; Cashmore, A.R. Light-Regulated Transcription. *Annu. Rev. Plant Physiol Plant Mol. Biol.* **1995**, *46*, 445–474. [CrossRef]
76. Elena López-Calcagno, P.; Omar Abuzaid, A.; Lawson, T.; Anne Raines, C. Arabidopsis CP12 Mutants Have Reduced Levels of Phosphoribulokinase and Impaired Function of the Calvin-Benson Cycle. *J. Exp. Bot.* **2017**, *68*, 2285–2298. [CrossRef] [PubMed]
77. Gérard, C.; Lebrun, R.; Lemesle, E.; Avilan, L.; Chang, K.S.; Jin, E.; Carrière, F.; Gontero, B.; Launay, H. Reduction in Phosphoribulokinase Amount and Re-Routing Metabolism in *Chlamydomonas reinhardtii* CP12 Mutants. *Int. J. Mol. Sci.* **2022**, *23*, 2710. [CrossRef]
78. Howard, T.P.; Fryer, M.J.; Singh, P.; Metodiev, M.; Lytovchenko, A.; Obata, T.; Fernie, A.R.; Kruger, N.J.; Quick, W.P.; Lloyd, J.C.; et al. Antisense Suppression of the Small Chloroplast Protein CP12 in Tobacco Alters Carbon Partitioning and Severely Restricts Growth. *Plant Physiol.* **2011**, *157*, 620–631. [CrossRef] [PubMed]
79. Howard, T.P.; Upton, G.J.G.; Lloyd, J.C.; Raines, C.A. Antisense Suppression of the Small Chloroplast Protein CP12 in Tobacco: A Transcriptional Viewpoint. *Plant Signal. Behav.* **2011**, *6*, 2026–2030. [CrossRef]
80. Li, K.; Qiu, H.; Zhou, M.; Lin, Y.; Guo, Z.; Lu, S. Chloroplast Protein 12 Expression Alters Growth and Chilling Tolerance in Tropical Forage *Stylosanthes guianensis* (Aublet) Sw. *Front. Plant Sci.* **2018**, *9*, 1319. [CrossRef]
81. Erales, J.; Avilan, L.; Lebreton, S.; Gontero, B. Exploring CP12 Binding Proteins Revealed Aldolase as a New Partner for the Phosphoribulokinase/Glyceraldehyde 3-Phosphate Dehydrogenase/CP12 Complex–Purification and Kinetic Characterization of This Enzyme from *Chlamydomonas reinhardtii*. *FEBS J.* **2008**, *275*, 1248–1259. [CrossRef] [PubMed]
82. Burlacot, A.; Dao, O.; Auroy, P.; Cuiné, S.; Li-Beisson, Y.; Peltier, G. Alternative Photosynthesis Pathways Drive the Algal $CO_2$-Concentrating Mechanism. *Nature* **2022**, *605*, 366–371. [CrossRef] [PubMed]
83. Dao, O.; Kuhnert, F.; Weber, A.P.M.; Peltier, G.; Li-Beisson, Y. Physiological Functions of Malate Shuttles in Plants and Algae. *Trends Plant Sci.* **2022**, *27*, 488–501. [CrossRef] [PubMed]
84. Clement, R.; Lignon, S.; Mansuelle, P.; Jensen, E.; Pophillat, M.; Lebrun, R.; Denis, Y.; Puppo, C.; Maberly, S.C.; Gontero, B. Responses of the Marine Diatom *Thalassiosira pseudonana* to Changes in $CO_2$ Concentration: A Proteomic Approach. *Sci. Rep.* **2017**, *7*, 42333. [CrossRef]
85. Tamoi, M.; Shigeoka, S. CP12 Is Involved in Protection against High Light Intensity by Suppressing the ROS Generation in *Synechococcus elongatus* PCC7942. *Plants* **2021**, *10*, 1275. [CrossRef] [PubMed]
86. Hackenberg, C.; Hakanpää, J.; Cai, F.; Antonyuk, S.; Eigner, C.; Meissner, S.; Laitaoja, M.; Jänis, J.; Kerfeld, C.A.; Dittmann, E.; et al. Structural and Functional Insights into the Unique CBS-CP12 Fusion Protein Family in Cyanobacteria. *Proc. Natl. Acad. Sci. USA* **2018**, *115*, 7141–7146. [CrossRef] [PubMed]
87. Marri, L.; Thieulin-Pardo, G.; Lebrun, R.; Puppo, R.; Zaffagnini, M.; Trost, P.; Gontero, B.; Sparla, F. CP12-Mediated Protection of Calvin-Benson Cycle Enzymes from Oxidative Stress. *Biochimie* **2014**, *97*, 228–237. [CrossRef] [PubMed]
88. Erales, J.; Lignon, S.; Gontero, B. CP12 from *Chlamydomonas reinhardtii*, a Permanent Specific "Chaperone-like" Protein of Glyceraldehyde-3-Phosphate Dehydrogenase. *J. Biol. Chem.* **2009**, *284*, 12735–12744. [CrossRef]
89. Chen, X.-H.; Li, Y.-Y.; Zhang, H.; Liu, J.-L.; Xie, Z.-X.; Lin, L.; Wang, D.-Z. Quantitative Proteomics Reveals Common and Specific Responses of a Marine Diatom *Thalassiosira pseudonana* to Different Macronutrient Deficiencies. *Front. Microbiol.* **2018**, *9*, 2761. [CrossRef] [PubMed]
90. Banovic Deri, B.; Bozic, M.; Dudic, D.; Vicic, I.; Milivojevic, M.; Ignjatovic-Micic, D.; Samardzic, J.; Vancetovic, J.; Delic, N.; Nikolic, A. Leaf Transcriptome Analysis of Lancaster versus Other Heterotic Groups' Maize Inbred Lines Revealed Different Regulation of Cold-Responsive Genes. *J. Agron. Crop Sci.* **2022**, *208*, 497–509. [CrossRef]
91. Bosco, G.L.; Baxa, M.; Sosnick, T.R. Metal Binding Kinetics of Bi-Histidine Sites Used in Psi Analysis: Evidence of High-Energy Protein Folding Intermediates. *Biochemistry* **2009**, *48*, 2950–2959. [CrossRef] [PubMed]
92. Delobel, A.; Graciet, E.; Andreescu, S.; Gontero, B.; Halgand, F.; Laprévote, O. Mass Spectrometric Analysis of the Interactions between CP12, a Chloroplast Protein, and Metal Ions: A Possible Regulatory Role within a PRK/GAPDH/CP12 Complex. *Rapid Commun. Mass Spectrom.* **2005**, *19*, 3379–3388. [CrossRef] [PubMed]

93. Stöckel, J.; Safar, J.; Wallace, A.C.; Cohen, F.E.; Prusiner, S.B. Prion Protein Selectively Binds Copper(II) Ions. *Biochemistry* **1998**, *37*, 7185–7193. [CrossRef] [PubMed]
94. Himelblau, E.; Mira, H.; Lin, S.J.; Culotta, V.C.; Peñarrubia, L.; Amasino, R.M. Identification of a Functional Homolog of the Yeast Copper Homeostasis Gene ATX1 from Arabidopsis. *Plant Physiol.* **1998**, *117*, 1227–1234. [CrossRef]
95. Rocha, A.G.; Vothknecht, U.C. Identification of CP12 as a Novel Calcium-Binding Protein in Chloroplasts. *Plants* **2013**, *2*, 530–540. [CrossRef]

MDPI
St. Alban-Anlage 66
4052 Basel
Switzerland
www.mdpi.com

*Biomolecules* Editorial Office
E-mail: biomolecules@mdpi.com
www.mdpi.com/journal/biomolecules

www.ingramcontent.com/pod-product-compliance
Lightning Source LLC
Chambersburg PA
CBHW061505220926
51850CB00004B/177
* 9 7 8 3 0 3 6 5 9 1 1 2 4 *